D1285811

OXFORD MEDICAL PUBLICATIONS

OXFORD TEXTBOOK OF PUBLIC HEALTH

EDITORS

WALTER W. HOLLAND, MD, FRCGP, FRCP Edin,
FRCP, PFPHM
Professor of Clinical Epidemiology and Social Medicine,
Department of Public Health Medicine, and
Honorary Director, Social Medicine and
Health Services Research Unit,
United Medical and Dental Schools,
St. Thomas's Campus,
London SE1 7EH, England.

ROGER DETELS, MD, MS
Professor of Epidemiology,
School of Public Health,
University of California, Los Angeles, CA 90024, USA.

GEORGE KNOX, MD, BS, FRCP, FFPHM, FFOM
Professor of Social Medicine, Department of Social Medicine,
Health Services Research Centre,
University of Birmingham Medical School,
Edgbaston, Birmingham, B15 2TJ, England.

BEVERLEY FITZSIMONS, BA
Research Assistant, Department of Public Health Medicine,
United Medical and Dental Schools,
St. Thomas's Campus,
London SE1 7EH, England.

LUCY GARDNER, BA
Research Assistant, Department of Public Health Medicine,
United Medical and Dental Schools,
St. Thomas's Campus,
London SE1 7EH, England.

OXFORD TEXTBOOK OF PUBLIC HEALTH

SECOND EDITION

VOLUME 2

Methods of Public Health

Edited by

WALTER W. HOLLAND,
ROGER DETELS, and GEORGE KNOX

with the assistance of
BEVERLEY FITZSIMONS and LUCY GARDNER

Oxford   New York   Toronto
OXFORD UNIVERSITY PRESS

Oxford University Press, Walton Street, Oxford OX2 6DP

Oxford New York Toronto
Delhi Bombay Calcutta Madras Karachi
Petaling Jaya Singapore Hong Kong Tokyo
Nairobi Dar es Salaam Cape Town
Melbourne Auckland
and associated companies in
Berlin Ibadan

Oxford is a trade mark of Oxford University Press

Published in the United States
by Oxford University Press, New York

© *The contributors listed on p. xvii, 1991*

First published 1991
Reprinted 1991

All rights reserved. No part of this publication may be reproduced,
stored in a retrieval system, or transmitted, in any form or by any means,
electronic, mechanical, photocopying, recording, or otherwise, without
the prior permission of Oxford University Press

British Library Cataloguing in Publication Data
Oxford textbook of public health.—2nd ed.
1. Public health
I. Holland, Walter W. (Walter Werner)   II. Detels, Roger
III. Knox, George 1926–

ISBN 0 19 261706 0   Vol. 1
ISBN 0 19 261707 9   Vol. 2
ISBN 0 19 261708 7   Vol. 3
ISBN 0 19 261926 8   (3 vol set)

Typeset by Promenade Graphics Ltd., Cheltenham
Printed in Great Britain by
Wm. Clowes Ltd., Beccles, Suffolk

Q 362,1
O98
1991
v. 2

Preface to the first edition

It is not an easy task to follow in the footsteps of such a renowned editor as Bill Hobson. We were however very honoured when, on the retirement of Professor Hobson, the Oxford University Press approached us about taking up the challenge of revising Hobson's *Theory and practice of public health*. Since this work first appeared in February 1961 Professor Hobson was responsible for taking it through no less than five editions. Many eminent public health academics and practitioners have contributed to this book and it has been recognized as a standard textbook on the subject. Sadly, Professor Hobson died after a long illness at the end of November 1982. After an early training in public health starting as a medical officer of health and then as a specialist in hygiene and epidemiology in the army, he went on to be a lecturer in social medicine at Sheffield University, becoming professor in 1949. From 1957 until his retirement, he served in a variety of posts at the WHO, where his major responsibilities were always concerned with education and training. His interest in this and in the international aspects of health were well exemplified by the first edition of *Theory and practice of public health*. One of the major strengths of the book has been its international nature and its link to the WHO.

On accepting the daunting task of revising this major work our first step was to look dispassionately at its role within public health, a field which has evolved and changed greatly over the last 25 years. We decided that although this book is held in great esteem in the western world it was appropriate now to introduce major revisions and thus increase its relevance to the problems facing us as we approach the twenty-first century. A particularly important advance has been the recognition in recent years that the problems in public health facing developing countries are quite different to those facing the developed world. The interests of WHO, quite correctly, have been focused on developing countries. We consider that this book should concentrate on presenting a comprehensive view of public health as it relates to developed countries. (Perhaps there is a place now for a comparable textbook concerned specifically with developing countries.) This is not to say however, that the content will not prove relevant and of interest to the student of public health from a developing country.

The *Oxford textbook of public health* attempts to portray the philosophy and underlying princples of the practice of public health. The methods used for the investigation and the solution of public health problems are described and examples given of how these methods are applied in practice. It is aimed primarily at postgraduate students and practitioners of public health but most clinicians and others concerned with public health issues will find some chapters relevant to their concerns. It is intended to be a comprehensive textbook present in the library of every institution concerned with the health sciences. The term 'public' is used quite deliberately to portray the field. Public health is concerned with defining the problems facing communities in the prevention of illness and thus studies of disease aetiology and promotion of health. It covers the investigation, promotion, and evaluation of optimal health services to communities and is concerned with the wider aspects of health within the general context of health and the environment. Other terms in common use, such as community medicine, preventive medicine, social medicine, and population medicine have acquired different meanings according to the country or setting. This gives rise to confusion and we have avoided their use since this book is directed to a worldwide audience. Public health, we believe, is more evocative of the basic philosophy which underlies this book.

The first volume aims to lead the reader through the historical determinants of health to the overall scope and strategies of public health. Through knowledge of historical aspects of the subject we may gain an understanding of what it is possible to achieve now and in the future. Only by grasping the underlying strategies of public health can we determine whether specific actions are feasible or not. In outlining the scope of public health we have emphasized that this covers not only the environment but also the social and genetic determinants of disease, which may ultimately enable us to identify those at greatest risk and thus to prevent the development and onset of disease and disability. The scope has now been broadened further by the growing concern with the provision and development of health services. This is clearly a function of public health. The approaches towards public health and the underlying political realities differ from country to country. However there are a few basic concepts common to every situation and these are outlined.

The major determinants of health and disease are dealt with in the second section of this volume, which paints, with a broad brush, the factors concerned with the development of disease, such as the physical environment, infectious agents, the social environment, war and social disorder. It is also important to have an understanding of the methods governments employ to control health hazards as well as their overall policies towards health. This we illustrate by describing the widely different approaches adopted on the two sides of the Atlantic. The strategies both for tackling modern hazards and for modifying behaviour through education are considered. To draw all these themes together into the same perspective, the final three chapters, prepared by three specialists with broad experience of public health in the context of the world as a whole, consider overall public health strategies relevant to the western as well as the developing world.

Volume 2 deals in depth with the processes of health promotion. Historically, public health has been primarily concerned with the identification and prevention of disease as the primary means to safeguard the health of a population. The need for this traditional approach still remains, but within the last decade public health has also come to incorporate a broader perspective as our understanding of the factors which determine the health of the public has increased.

This greater understanding of factors which enhance health has led to a reconceptualization of health as a condition which must be actively maintained. Thus public health has been able to adopt a proactive approach with health, rather than absence of disease, as a goal. This aim is reflected in the World Health Organization motto, 'Health for all by the Year 2000'. The agenda for public health in the 1980s will be to actively promote health and enhance the quality of life through a wide range of activities, both traditional and new.

Commensurate with this proactive approach is the assumption of new responsibilities. For example, public health is now also concerned with the quality of health care, adequate access to medical care by all segments of society, the health effects of the many chemicals and other agents which are released into the environment as a result of the many new technical advances, and promotion of healthy life-styles. These responsibilities call for both new strategies and processes to promote the health of the public and innovative use of traditional strategies.

In this volume we present a discussion of the processes currently being used to promote the health of the public. These range from the scientific and regulatory strategies used to control the physical environment and the spread of infectious diseases to the involvement of national governments and international organizations in promotion of public health.

In the first two chapters of the volume, the control of the physical environment and the control of infectious diseases are discussed. Although these have been important traditional concerns of public health, there have also been new strategies developed as new health problems emerge and former health problems are resolved.

The next several chapters address more recent concerns of public health: modifying the social environment, intervening in population dynamics in order to limit growth, and providing more effective health services to include provisions for specialty care and for care of the mentally and emotionally handicapped.

The third section discusses the organization of public health services, the staffing of these services, and the training of personnel for public health. The volume concludes with a discussion of the co-ordination and development of strategies and policy on the national level and to some extent, on the international level.

In conclusion, this volume should give the reader some appreciation of the dynamic, evolving nature of public health and the diversity of approaches which are being used to assure the highest levels of public health in developed countries.

Volume 3 deals with the investigative methods used in public health.

The final volume of this series attemps to pull together all the threads of the earlier volumes which have considered the theory, policies, strategies, and research methodologies that form the basis of public health endeavour. Specific diseases and the needs of specific groups within a community have been touched on in greater or lesser detail in earlier volumes. Here, the major disease groups, systems of the body, and special care groups are treated systematically to review the public health issues they raise, and the extent to which the methodologies described earlier have been applied to their control and prevention.

The volume is divided into three sections, the first of which is concerned with the application of public health methods to specific disease processes including acute infectious episodes, diseases related to nutrition, trauma, developmental defects, degenerative neurological diseases, psychiatric conditions, and neoplasms. Each chapter attempts to review the public health impact of the diseases in question, including their epidemiology and the contribution of public health measures. Of particular emphasis is the potential for prevention that enquiry into the aetiology has revealed, and the role of public health interventions in implementing programmes of prevention.

Leading on from specific disease states, the next section looks at the role of public health in relation to the various systems of the body. The intention is not to provide an exhaustive description of the conditions affecting each system, but to discuss the facets of the system where public health investigation has had (or indeed in some cases has failed to make) a major contribution. It will be evident to the reader that the extent to which the public-health approach has contributed to research and strategies to prevent and treat conditions affecting the various systems of the body varies tremendously. The cardiovascular system and the respiratory system provide classical examples of the application of the public health approach to a health problem. Extensive epidemiological research has clarified the causation of these conditions and public health measures form the basic framework for implementing control measures. Investigations of gastrointestinal disease, metabolic and endocrine disorders, and conditions affecting the genitourinary system have on the other hand, relied more on clinical than public health orientated research. The chapters dealing with these systems however describe the epidemiological work that has been carried out and its relevance to clinical practice. The chapter on the genito-urinary system, for example, illustrates the value of national and international comparative studies in informing policy decision makers of different management practices and outcomes. The problems presented by the investigation of musculoskeletal and dental conditions are clearly demonstrated in this section of Volume 4.

The final section of this volume treats the unique problems and health service needs of special client groups. Again, this is not intended to be comprehensive or exhaustive, the emphasis being placed on demonstrating the application of public health approaches to a variety of different problems. These include investigations to identify the needs of special client groups and the extent to which these needs are satisfied, and the application of public health measures to solving the problems of these groups.

The chapter on acute emergencies is an excellent example of the ways in which epidemiological investigation can be applied to surgical conditions, providing evidence from which conclusions can be drawn about the most appropriate form of treatment. The chapter on adolescence demonstrates the relevance of a multidisciplinary approach to health problems, and the difficulties of tackling some of the problems experienced by adolescents in the changing society around us. The discussion of handicap—both physical and mental—illustrates the application of epidemiological and public health methodology to both the prevention as well as the development of care policies for specific groups of individuals.

Health care policy decision-making is fraught with difficulties as is illustrated by the chapter on maternity care, theoretically the simplest of examples, but one which demonstrates the complexity of the issues involved. The chapter on the special needs of the elderly also shows how complicated the problems facing modern public health are. In outlining the difficulties faced by the unemployed and disadvantaged Illsley demonstrates the way the social scientist approaches a public health problem.

This volume should give the reader some idea of the vast scope of public health, the range of fields and problems for which public health has a major contribution to make, and the multidisciplinary nature of much of the work. It attempts to review how we can approach and develop a methodology for the investigation, prevention, and control of the major public health problems in our society at this time.

The development of public health policy is dependent upon a series of scientific methods, and we do not attempt in this book to cover all the methods and their applications. However it is to be hoped that those examples that have been chosen will illustrate to the reader the way in which particular problems can be approached. Each chapter includes a comprehensive list of further reading which should equip the reader with the means of obtaining a deeper knowledge should he or she wish to pursue any theme further.

This is the first of what we hope will be many editions. As each chapter was submitted to the editors we have attempted to identify gaps and areas of overlap. There is no doubt however that some remain. It is only through feedback from readers that we will be able to adapt, modify, and improve further editions. If the book is successful it will be entirely due to the effort of the contributors who undertook with great patience a tremendous amount of work. They were bombarded with instructions, advice, reminders, and modifications and we would like to express our thanks and extend our apologies to all of them. Our gratitude also goes to our secretaries and assistants who coped so admirably with the enormous task of compiling this work. We hope that it will be widely read by all those concerned with the formulation and execution of public health policy and that it will provide a suitable framework for devising approaches to some of the problems challenging public health today.

London
August, 1984

W.W.H.
R.D.
G.K.

Preface to the second edition

The first edition of the *Oxford textbook of public health* attempted to provide a sense of the history and philosophy of the subject, the underlying forces that condition the subject, as well as the basic methodologies and their application to specific problems or care groups.

It became clear to the editors from reviews and discussion with leaders in the field that the ideas they had about the format, content, and structure of the book did not achieve their original intentions as well as they had hoped. One of the problems was that the relationships between the philosophy of public health and its applications in the solution of individual problems were not clearly expressed in the book. Further, there were problems in effectively conveying an understanding of what the editors were trying to achieve.

In spite of these problems, most of the readers and reviewers were generally satisfied with the content of the first edition, partly because the chapters in themselves were extremely good, thanks to the quality of the contributing authors.

Furthermore, in the intervening time period, between the publication of the first edition and the plans for the second, there have been major changes in the way in which public health is regarded. The appearance of a major new infectious disease, AIDS, has influenced opinion on the importance of public health measures in the structure of services and the ways in which disease prevention is approached.

We recognize that the first edition did not clearly express our intention of considering primarily the problems of the developed world and had few contributors from non-English-speaking countries.

To try to overcome these problems, and to attempt to rectify the deficiences that were apparent to the editors in the first edition, we have undertaken a radical change in both the authorship and structure of the second edition. This in no way implies that the authors in the first edition were inadequate. It merely demonstrates the editors' wish to broaden the contributions and to make the second edition somewhat more coherent. We do not feel able to deal adequately with the problems of the developing world, but in an effort to broaden the scope of this edition, we have expanded Volume 1 to include more examples from non-English-speaking countries.

Two other changes have been introduced into this edition. The first has been to reduce the textbook from four volumes to three. The second, in response to comments from reviewers, has been to try to provide syntheses of different approaches to public health management and application of public health methodologies, thus we have added concluding chapters for the major sections on policy, management, and methodologies which both summarize and evaluate the different approaches currently in use to resolve public health problems.

Volume 1 examines various influences on public health. This volume attempts to set the scene of the subject of public health, describe its determinants and the methods used in its improvement. We have attempted to include most of the contributions from Volumes 1 and 2 of our previous edition within this volume. In addition we have broadened the description of a variety of health systems in the western world by providing more examples from countries from Europe and Asia, as well as the US, UK, and Australia. The volume gives a historical account of the subject's development. Similarly, we have tried to provide a description of some of the major forms of provision of public health services. Finally there is a series of chapters that illustrate how strategies and policies are co-ordinated and the legal and ethical public health issues.

The second volume is almost entirely concerned with the several methodologies basic to public health including information systems, epidemiologic approaches, social science approaches, and field investigations.

The third volume demonstrates the application of public health strategies and methods, presented in the first two volumes, both to broad areas of public health concern (the environment, social issues etc.) and to specific public health problems. These include both diseases of public health importance (for example mental and cerebrovascular illness) and groups requiring special attention from public health leaders. The volume concludes with two broad overviews on the performance of the public health function.

We hope that the discussion of the structure and philosophy of the book will assist the reader to place the individual contributions of the public health leaders contributing to the textbook into the broader context of public health.

London W.W.H.
September, 1990 R.D.
G.K.

Contents

Contents of volumes 1 and 3

Volume 1

E Provision of Public Health Services

F Co-ordination and Development of Strategies and Policy for Public Health

G The Ethics of Public Health

Volume 3

A Interventions in Public Health

Contributors

J. H. ABRAMSON

Professor, Department of Social Medicine, The Hebrew University-Hadassah School of Public Health and Community Medicine, PO Box 1172, Jerusalem, Israel.

R. M. ANDERSON BSc PhD DIC ARCS CBIOL FIBIOL FRS

Professor and Head, Department of Biology, Imperial College of Science, Technology, and Medicine, Prince Consort Road, London SW7 2BB, UK.

JOHN S. A. ASHLEY MB FFCM

Deputy Chief Medical Statistician, Officer of Population Censuses and Surveys, St Catherine's House, 10 Kingsway, London WC2B 6JP, UK.

RUTH L. BERKELMAN MD

Chief, Surveillance Branch, Center for Infectious Diseases, AIDS Program G29, Centers for Disease Control, 1600 Clifton Road, Atlanta, GA 30333, USA.

JAMES W. BUEHLER MD

Assistant Director for Science, Division of Surveillance and Epidemiologic Studies, Epidemiology Program Office, Centers for Disease Control, Department of Health and Human Services, Atlanta, GA 30333, USA.

NORMAN BRESLOW PhD

Professor and Chair, Department of Biostatistics SC–32, University of Washington, Seattle, WA 98195, USA.

ANDREW P. BROWN BSc MRCP FRCR

Consultant Radiotherapist and Oncologist, Department of Radiotherapy and Oncology, St Thomas's Hospital, Lambeth Palace Road, London SE1 7EH, UK.

JOHN P. BUNKER MD

Professor of Health, Research and Policy, Emeritus, Stanford University. Visiting Professor, Rayne Institute, King's College School of Medicine and Surgery, 123 Coldharbour Lane, London SE5 9NU, UK.

SUSAN K. COLE MD FRCOG FFCM

Consultant in Public Health Medicine, Information and Statistics Division, Scottish Health Service, Trinity Park House, South Trinity Road, Edinburgh EH5 3SQ, UK.

SHAN CRETIN SB PhD MPH

Associate Professor and Head, Division of Health Services, UCLA School of Public Health, 10833 LeConte, Los Angeles, CA 90024–1772, USA.

ROGER DETELS MD MS

Professor of Epidemiology, School of Public Health, University of California, Los Angeles, CA 90024–1772, USA.

M. F. DRUMMOND BSc MCom DPhil

Professor of Economics, University of York, Centre for Health Economics, York YO1 5DD, UK.

FRANK A. FAIRWEATHER QHP MBBS FRCPath FIBiol

Environmental Safety Officer, Research Division, Unilever PLC, PO Box 68, Unilever House, Blackfriars, London EC4 4BQ, UK.

JOHN W. FARQUHAR MD

Professor of Medicine and Professor of Health Research and Policy, Director, Stanford Center for Research in Disease Prevention, Stanford University, 1000 Welch Road, Palo Alto, CA 94304–1885, USA.

MANNING FEINLEIB MD DrPH

Director, National Center for Health Statistics, Department of Health and Human Services, Presidential Building, Room 1140. 6525 Belcrest Road, Hyattsville, MD 20782, USA.

STEPHEN P. FORTMANN MD FACP

Associate Professor of Medicine and Epidemiology, Head, Section on Clinical Epidemiology and Disease Prevention, Stanford University, 1000 Welch Road, Palo Alto, CA 94304–1885, USA.

JUNE A. FLORA PhD

Assistant Professor in the Communication Department, Associate Director, Stanford Center for Research in Disease Prevention, Department of Communication, McClatchy Hall, Stanford University, Stanford, CA 94305, USA.

JACOB J. FELDMAN PhD

Associate Director for Analysis and Epidemiology, National Center for Health Statistics, Presidential Building, Room 1000, 6525 Belcrest Road, Hyattsville, MD 20782, USA.

RALPH R. FRERICHS DVM DrPH

Professor and Chairman, Department of Epidemiology, School of Public Health, University of California at Los Angeles, Los Angeles, CA 90024–1772, USA.

PATRICIA M. GOLDEN MPH

Special Assistant to the Division of Epidemiology and Health Promotion, Office of Anaylsis and Epidemiology, National Center for Health Statistics, Presidential Building, Room 1000, 6525 Belcrest Road, Hyattsville, MD 20782, USA.

RAYMOND S. GREENBERG MD MPH PhD

Dean, Emory University School of Public Health, 1599 Clifton Road, NE, Atlanta, GA 30329, USA.
Chairman Designate, Department of Epidemiology and Biostatistics, Emory University School of Medicine, 246 Sycamore Street, Suite 100, Decatur, GA 30030, USA.

SANDER GREENLAND AB MA MS DrPh

Professor of Epidemiology, Department of Epidemiology, UCLA School of Public Health, Los Angeles, CA 90024–1772, USA.

MICHAEL B. GREGG MD

Deputy Director, Epidemiology Program Office, Centers for Disease Control, 1600 Clifton Road, NE, Room 5017, Atlanta, GA 30333, USA.

MICHEL A. IBRAHIM MD PhD

Dean, School of Public Health, The University of North Carolina at Chapel Hill, Rosenau Hall 201 H, Chapel Hill, NC 27514–6201, USA.

WILLIAM H. W. INMAN FRCP FFPHM FFPM

Director, Drug Safety Research Unit, Bursledon Hall, Southampton SO3 8BA, UK.

M. PAULA J. KILBANE MB MSc FFCM

Director of Public Health, Southern Health and Social Services Board, 20 Seagoe Industrial Estate, Portadown, Craigavon BT63 5QD, UK.

JOEL C. KLEINMAN PhD

Director, Division of Analysis, Office of Analysis and Epidemiology, National Center for Health Statistics, Presidential Building, Room 1080, 6525 Belcrest Road, Hyattsville, MD 20782, USA.

GEORGE KNOX

Professor of Social Medicine, Department of Social Medicine, University of Birmingham Medical School, Edgbaston, Birmingham B15 2TJ, UK.

NATHAN MACCOBY PhD

Janet M. Peck Professor of International Communication Emeritus and Associate Director of the Stanford Center for Research in Disease Prevention, Stanford University School of Medicine, Stanford, CA 94305, USA.

NICHOLAS MAYS

Lecturer in Medical Sociology, Department of Public Health Medicine, United Medical and Dental Schools of Guy's and St Thomas's Hospitals, St Thomas's Campus, London SE1 7EH, UK.

KLIM McPHERSON

Professor of Public Health Epidemiology, Department of Public Health and Policy, London School of Hygiene and Tropical Medicine, London WC1E 7HT, UK.

MYFANWY MORGAN

Senior Lecturer in Medical Sociology, Department of Public Health Medicine, United Medical and Dental Schools of Guy's and St Thomas's Hospitals, St Thomas's Campus, London SE1 7EH, UK.

MASAKI NAGAI MD DrMedSci Dip Public Health MSc

Associate Professor, Department of Public Health, Jichi Medical School, 3311–1 Yakushiji, Minamikawachi-machi, Tochigi-ken 329–04, Japan

NORMAN D. NOAH

Professor, Department of Public Health and Epidemiology, King's College School of Medicine and Dentistry, Bessemer Road, London SE5 9PJ, UK.

JAMES NOKES BSc PhD

Research Associate, Department of Biology, Imperial College of Science, Technology and Medicine, Prince Consort Road, London SW7 2BB, UK.

M. C. O'RIORDAN PhD

Head of Radiological Measurement, National Radiological Protection Board, Chilton, Didcot, Oxon OX11 0RQ, UK.

JULIE PARSONNET MD

Medical Epidemiologist, Division of Bacterial Diseases, Center for Infectious Diseases, Centers for Disease Control, Atlanta, GA 30333, USA.

NANCY D. PEARCE

Division of Data Policy, Office of the Assistant Secretary for Health, Department of Health and Human Services, Room 717–H, Humphrey Building, 200 Independence Avenue, SW, Washington, DC 20201, USA.

JAMES L. PEARSON MPH DrPH

Director, Consolidated Laboratory Branch, 2635 East Main Street, Bismarck, ND 58502–0937, USA.

PEKKA PUSKA MD MPolSc

Professor and Director, Department of Epidemiology, National Public Health Institute, Mannerheimintie 166, 00280 Helsinki, Finland

LEON S. ROBERTSON PhD

Lecturer, Department of Epidemiology and Public Health, Yale University School of Medicine, 2 Montgomery Parkway, Branford, CT 06405, USA.

BEATRICE J. SELWYN ScD

Associate Professor, Epidemiology Discipline, School of Public Health, University of Texas, Health Sciences Center at Houston, PO Box 20186, Houston, TX 77225, USA.

TONY SWAN BSc MSc PhD

Reader in Medical Statistics, Department of Public Health Medicine, United Medical and Dental Schools of Guy's and St Thomas's Hospitals, St Thomas's Campus, London SE1 7EH, UK.

PAUL R. TORRENS MD MPH

Professor of Health Services Administration, School of Public Health, University of California, 10833 Le Conte Avenue, Los Angeles, CA 90024–1772, USA.

ROBERT E. WALLER BSc
Toxicology and Environmental Health, Department of Health, Hannibal House, Elephant and Castle, London SE1 6TE, UK.

RONALD W. WILSON MA
Director, Division of Epidemiology and Health Promotion, Office of Analysis and Epidemiology, National Center for Health Statistics, Presidential Building, Room 1080, 6525 Belcrest Road, Hyattsville, MD 20782, USA.

HIROSHI YANAGAWA MD DrMedSci DipPublic Health
Professor, Department of Public Health, Jichi Medical School, 3311–1 Yakushiji, Minamikawachi-machi, Tochigi-ken 329–04, Japan.

Introduction to volume 2
The methods of public health

This volume is about method. It displays the technical and inductive basis for scientific studies, and other scientifically-informed enquiries, within the discipline of Public Health. These are the methods through which the public health practitioner—medical or non-medical—lays claim to the status of being an 'expert'. Investigations and analyses based upon the methods to be outlined here provided the first element of that characteristic sequence: enquiry → interpretation → advice → action: common to all areas of medical practice, and to many other forms of professional activity besides.

In reality, public health enquiries based upon these methods range very widely in terms of their scope, their duration, their costs and staffing, and the relative ambitiousness of their objectives; and therein lies a problem for the discipline. Public health practitioners are frequently asked to respond quickly to unforeseen problems on so short a time-scale that it would be impossible to assemble additional resources or comprehensive data, even if they were available. Enquiries of these kinds are associated typically with outbreaks of communicable disease, but appear in many other guises; for example, an abruptly-surfacing problem in coping with congenital heart surgery, an excess of asthma deaths, or a letter from the Minister complaining about recent perinatal mortality levels. At the other extreme lie enormous case–control studies pursued over many years seeking to illuminate the aetiologies of major disease groups; complex multi-centre preventive trials of chronic diseases; or post-hoc evaluations of national screening programmes. Each has its problems. Lcoal urgent enquiries must often compromise on their scientific standards; while larger studies may have difficulties surviving across a sequence of staff-retirement/recruitment/retraining cycles, political/ideological turnabouts, and repeated crises in the continuity of their financial support. Long-term routine monitoring programmes—the epidemiological surveillance of congenital malformations or of cancer incidence, or the continued investigation of the relationship between fetal X-ray and childhood cancer—have similar maintenance problems.

The wide range in terms of scale, scope, timing, and technique, also raises problems of communications within the discipline. Small ad hoc studies tend to be reported and distributed locally; they do not reach the scientific literature. Larger studies are published in scientific journals and monographs, at a pace which the busy practitioner may regard as over-leisurely, and they are concerned with 'grand issues'

which he sees as being without obvious relevance to the problem of this very day. Public health investigators tend then to divide into sub-groups communicating internally through their own pathways, but not with each other. This is a problem which requires assiduous attention if we are to avoid the dangers of mutual disrespect and alienation.

The reader will notice very quickly that this volume has adopted a multi-disciplinary approach. The various chapters deal with information system methods, statistical methods, epidemiological/analytical methods, and computer applications. They cover field methodologies as well as analytical methodologies. Clearly, epidemiology is one of the central issues. However, we have also included a series of chapters on Social Science techniques.

We use 'Social Science' as a generic term to cover a range of disciplines, including economics, sociology, communication and education, operational and systems studies, management-science and social administration, demography, law, and ethics. It is neither possible nor desirable to seek an absolute separation of method from subject matter, and some of these subjects seemed to sit more comfortably in Volumes 1 and 3. Therefore, although Volume 2 is *chiefly* about method, it cannot be read in isolation. There is, inevitably, a substantial degree of necessary cross reference between the volumes.

As we have said, epidemiology is the centre-piece of the constellation of disciplines presented here. We justify this priority by asserting that our discipline is concerned with the health and disease of populations, that these concerns are part of medicine, and that epidemiology is the basic support-science for these concerns. These assertions themselves have ancient precedent. Aesculapius had *two* daughters. One was called Panacea (cure-all); the other, our own 'patron saint' was called Hygeia. We accept, of course, that epidemiology as a discipline is not the sole prerogative of physicians; but the extensions of medicine and of other health professions must still to some extent be bound by the Aesculapian precedent.

Let us now admit that medicine, even preventive medicine, cannot lay claim to be the main determination of health in populations. Much more powerful influences are exerted through education, nutrition, housing, occupation, and limitations of exposures to a wide range of social, physical, and chemical hazards. Insofar as these can be influenced by a deliberately constructive approach, they are reached through social policy pathways, through politics, and through the media. We have felt it inappropriate to pursue a detailed dissertation on managing these pathways within these volumes, but we felt that we should at least display the nature and expose the credentials of the basic social sciences which proffer a degree of rationality in an environment not especially characterized by rational argument.

There is one other methodological issue—apart from politics and proselytization—which we have avoided in these volumes. It represents and imposes on of the current unresolved schisms within the discipline of public health. The traditional medical approach, as described above, begins with enquiry, proceeds through explanation and advice, and then offers an effective prescription, if such be available. It operates essentially through persuasion. There is however a separate tradition. It too begins with enquiry, possibly based upon a complaint, but it proceeds then to a warning and then to enforcement through legal prosecution. There is a whole zone of public health practice which is based upon the enforcement of layed-down standards. The investigative methods employed here—especially field methods—are sometimes similar, but they are directed towards quite different objectives, operating through different institutional mechanisms. In the UK, for example, the 'medical analogue' approach is pursued through various institutional organs and personnel located within the national health service; while environmental control and enforcement is pursued chiefly through local government authorities. In other cases, the enforcement mechanisms are pursued through national institutions, for example through the Health and Safety Executive, in relation to safety at work and related matters.

The two classes of institutional pathways are separate. They are staffed separately and have different training pathways for their personnel. Their communications with each other are generally limited to cross-institutional advisory activities. Their philosophies differ. One is based firmly upon the notion of scientific enquiry and professional advice and the other on policing, and upon enforcing conformity with regulations.

The problems arising from this schism have not in the past been adequately addressed. It might be beneficial if the whole of the public health effort of a nation could be unified within a common system, or at least managed within an operational federation. However, the fundamental difference of approach and the different legal and adminstrative powers of the parent institutions are bound to raise difficulties. Indeed, they *have* raised difficulties. Closer to home, they have influenced the choice of subject matters in the following chapters. We have ourselves not tackled enforcement methodologies; by implication, our account is based upon the alternative medical advisory model.

One further comment is needed in introducing this volume. Although the subject matters are essentially technical and scientific, we cannot escape the fact that this is applied science; that application brings the scientific practitioner into contact with those who exercise value-judgements, and allocate resources; that these responsibilities for enquiry and advice extend some professional roles at the expense of others, and that these interactions invoke considerations which are resolved in a political rather than in a scientific milieu.

Thus, scientific results restrict the scope for political judgement, on which politicians pride themselves, and so may sometimes be less than welcome. They also confront existing beliefs and prejudices, and certainly offend those who indulge in wishful thinking. They indicate beneficial reallocations of resources and of authority, and thus threaten existing deployments and responsibilities.

If an investigator recommends that rubella vaccination in infancy which fails to eradicate rubella will actually make things worse, and that vaccination should therefore be compulsory, he sets up an awkward problem for a politician who might not appreciate the enhanced clarity of his dilemma between a duty to protect individuals rights and a duty to maximize benefit. If an investigator recommends that lead in petrol should be reduced but need not be totally abolished, then he/she reduces the prospect of catalytic converters eventually being required by law, and this affects prospects for increases in the price of platinum. Someone, somewhere, will try to rubbish the work, and perhaps the investigator as well. If a public health investigator demonstrates that following the introduction of screening, the incidence of late diagnosed congenital dislocation of the hip has not declined, then he/she inevitably comments upon the hopes, aspirations, self-esteem, and perceived responsibilities of paediatricians. The results will be received with less than acclamation. If epidemiologists wish to investigate the spread of HIV, their plans will conflict with those of STD physicians who may regard this as their own job, and who regard themselves as the unique custodians of absoluted confidentiality in this field. A proposal to investigate the effectiveness of screening for neural tube defects, and to answer the question whether the declining incidence followed or preceded the institution of the screening service, threatens the self esteem of the biochemists who have made themselves responsible for running the programme. Their budgets are also threatened! And so on. Applied research is *always* potentially threatening and there is no avoiding this. Anything which threatens the status quo will elicit a reaction.

For these reasons, scientists in this field can *never* forget the professional and political interactions of their work, or the budgetary or institutional or even personal repercussions of publishing their results. In one of our chapters we indulge in a short polemic, in extension of this theme.

A

Information Systems and Sources of Intelligence

1

Information needs in public health

E.G. KNOX

I must Create a System, or be enslav'd by another Man's;
I will not Reason and Compare: my business is to Create.
William Blake (1757–1827) *Jerusalem*

Introduction

The purpose of this chapter is to review *needs* for public health information systems, rather than to describe what exists. It will be necessary to refer to services now provided, but chiefly in order to identify mismatches between them and those which are required. The chapter constructs a short taxonomy of public health information systems, and relates the different classes to the purposes for which they are designed. It reviews their quality and their fitness for their purposes. Some are relatively successful and others chaotic. The picture varies in different places but some common patterns of success and failure emerge. We try to identify the reasons and to suggest remedies.

Purpose and quality

Information systems supporting the public health functions offer a commentary of extremes. Many western countries have long established and efficient processes for collecting health-relevant civil data and presenting periodic statistical digests of the demographic structure and the overall health-status of the population. These processes are based upon censuses, and upon birth, marriage, and death certificates, and they are supplemented by occupational and communicable disease notifications, by nutritional and social surveys, by periodic health surveys, and by activity statistics derived from health insurance and from health care systems. In general, these data-acquisitive systems are entirely 'passive', offering no direct feedback to the respondents. They serve a wide range of social planning purposes, of which health planning is only one aspect. They are maintained by the force of law and regulation.

In contrast with these simple monolithic systems, directed towards global reviews of health, we also recognize information systems of a more intimate nature, directed towards the management and surveillance of the day-to-day problems of delivering preventive, curative, and caring services. They frequently involve the registration of patients by name, the scheduling of appointments, and other quite complex transfers of information between professional workers. They are used to control vaccination services, infant health-surveillance, screening, communicable disease surveillance, mental health care, the care of the aged, hospital waiting lists, hospital discharge procedures, follow-up schedules, and an extended range of similar subjects and processes. They are driven not so much by regulation and legislation as through motivating the practitioners involved, trying to meet their expressed needs, exploiting their desire to produce a better service for patients and populations: and through feeding back information of interest. Many of these systems provide diagnostic and other indexing, furnishing the means of tracing original medical records, selecting samples and carrying out research. They are also sometimes linked—an additional incentive—to managing payments for particular items of service.

In contrast with the routine global public health information systems referred to earlier, these purpose-directed operational systems are frequently chaotic, badly designed, inadequately resourced, badly managed, unreliable in their information content, and ineffective in meeting their stated or their apparent objectives. They fall into disuse because they are unreliable. They are unreliable because they are little used. We shall therefore have to ask why it is that systems which flout all the rules of good practice—with no feedback, no motivation, no precisely defined objectives, and little professional involvement—should be relatively efficient and successful: while those which are most directly related to the end-purposes of a health care system—which directly involve the professionals providing the care, which assist them in scheduling their professional activities, and which provide them with feedback information and the means of carrying out scientific studies—should so often fail to meet the objectives for which they were set up.

Levels of information

This picture of extremes is possibly contentious. We therefore present these contrasts in an alternative way: as the

Table 1.1. Health information systems and their applications: five levels

Data sources	Application
1. Demographic/civil	Global: Overall health policy
2. Environmental/occupational monitoring	Design: Education: Enforcement of regulations
3. Health care activities	Resource deployment: Plant: Staffing: Financial accountability
4. Population health needs and demands: Assessments of service effectiveness	Identification of unmet needs and inadequate provisions. Priority allocations
5. Registers, population lists, admission and attendance indexes	Scheduling of care, follow-up and recall, preventive surveillance, screening, vaccination, *ad hoc* investigations

limits of a scale of information activities, as set out in Table 1.1. Most health service professionals will probably recognize that this table embodies several different forms of graduation between the top and the bottom level. They run in parallel and are clearly correlated with each other.

From top to bottom, the table displays an increasing involvement with the day-to-day management and scheduling of professional activities, and direct concern with the problems of individual patients. There is a parallel increase of requirements for professional and scientific judgements, and a progressive decrease in the need for administrative, legal, and management interventions. The information systems themselves become progressively more complex and detailed, more involved in day-to-day problems, and they require a quicker turn-around of information. It is also a general fact that descent through the levels represented in the table is accompanied by less investment, less equipment, less staff, a less clear-cut managerial responsibility and, in general, a less satisfactory level of performance. Standards of implementation are sometimes appalling. From general practitioner notes through hospital case files to population records of who is vaccinated and who is not, these systems frequently fail to meet the most rudimentary requirements.

Disasters

These problems are far from being trivial matters. Information system failures have cost many thousands of lives. We can illustrate this with three examples, all taken from the United Kingdom. We use the traditional medical approach of the post-mortem examination!

Control of cervical cancer in the UK

Technical methods for detecting early cancerous processes in the cervix uteri were developed by Papanicolaou in the 1940s. Successful applications in the USA, Canada, Sweden, Denmark, Finland, Iceland, north-east Scotland, and elsewhere, have been followed by substantial falls in mortality. In England and Wales, by contrast, the rate of decline has

been little greater than that obtaining before the screening service was introduced. Computer-based studies have shown that, for the investment deployed, the mortality should by now have been reduced to about 30 per cent of its original value. The shortfall has cost 10 000 to 20 000 deaths spread over about 20 years.

This was not a failure to invest or to act. Current levels of provision amount to over three million smears per year and cervical intra-epithelial neoplasia (CIN) lesions have been excised at the rate of about 10 000 annually over the last 20 years, in a population which was originally generating only 4 500 cervical cancers per year. Other countries with similar per capita investments have achieved major reductions in incidence and mortality. Why then has the UK system failed?

The primary failure, probably, was a failure to recognize that this was a public health programme with a public health objective—to reduce mortality. Consequently there was no need for a 'head office' or for a manager accountable in terms of attainments in mortality terms, and the question of resource levels or information systems did not then arise. Responsibilities became hopelessly divided between health authorities, general practitioners, public health physicians, and cytological laboratories. Success came to be measured in terms of activity levels (numbers of tests) rather than in terms of outcomes.

Central control was limited to the issue of 'guide-lines' formulated by unresourced committees working without adequate statistics or any real monitoring or research capability. The central declaration of 'policy'—that tests should be performed preferentially in age groups suffering the greatest and most immediate risks of invasive cancer (over 35)—was widely ignored. General practitioners were paid an item-of-service payment for tests performed not more frequently than five-yearly in women over 35, but this led in practice to a running battle between the Department of Health and the clinical and laboratory specialties who saw themselves responsible for developments in the field and who thought they knew better.

The parallel information system failures were both contributory and symptomatic. There could be no effective programme without an adequate supporting information and scheduling network; and there could be no such network without a proper resolution of the managerial ambiguities. In the event, the main internal NHS data-acquisition network was based upon the laboratories and was limited essentially to an account of individual tests, with no adequate information on the numbers of women screened at different ages, after different intervals, and for different purposes.

No systematic linkage was possible between different test-results belonging to the same woman. There was no register of women who had not been screened. A scheduling system was set up to assist in the re-screening of women who had already been screened and who were initially negative (the lowest-risk group it would be possible to find) but there was no provision for the primary invitation and scheduling of women who had never attended. The re-screening system was contracted out to the Office of Population Censuses and Surveys (OPCS) who had no access to service data, and no provision was made for the development of native scheduling expertise within the NHS itself. The OPCS operation was entirely manual with no provision for the development of a computer system. The statistics of activity were based upon measuring the heights of piles of requisition forms with a ruler.

The information flow for re-screening was tortuous beyond belief, designed more to satisfy the perceived professional interests of general practitioners, laboratories, and health authorities, than the prevention of cervical cancer. Duplicates of the requisition form were actually designed, and explicitly so, to *prevent* public health physicians from seeing clinical data. No recall register was maintained, so that it was never possible to see what proportion of recall notices was honoured. The system was based upon sending back, to Family Practitioner Committees, the carbon duplicates of laboratory-test requisitions of five years ago.

The only available population register which might have been used for primary scheduling was the 'Family Practitioner Committee' list, established to maintain general practitioner patient registration lists, and to facilitate the payment of general practitioners. In the event it was in practice barred to many District Health Authorities who were (eventually) made responsible for organizing primary scheduling. These lists were in any case managed manually (until about 1987), and even when they were transferred to computers there was no provision for the incorporation of clinical data; and the legal and professional restrictions upon the uses permitted for these lists, including screening and other public health scheduling and monitoring, are even to this day unresolved.

Throughout the whole of this period, from about 1966 to 1989, the basic uncertainty in the optimal deployment of this resource hinged upon a lack of exact knowledge regarding the natural history of cervical intraepithelial neoplasia; the solution of this problem would have supplied the single most valuable contribution to the control of this cancer. This has been known for twenty-five years. Despite the enormous amount of information held within the screening system, and published demonstrations of the manner in which these data might have answered these questions, the screening service made not a single contribution to this problem.

Congenital dislocation of the hip

Screening for congenital dislocation of the hip was increasingly used in the UK from about 1968 (Department of Health and Social Security 1969). Maternity units all over the country began to apply the Barlow–Ortellani test to newborn infants with increasing, although variable, enthusiasm. A number of paediatric centres led the way in developing the technique, educating their peers and, to their own satisfaction, identifying large numbers of 'early cases'. From 1964, the newly established national congenital malformations

notification system had begun to provide a basis for measuring progress. Unfortunately, this system was limited to notifications made within 7 days of birth. In effect, it recorded dislocations whose persistence had not yet been verified, together with a large number of 'clicking' hips whose future course was not then charted, and there was no provision for recording dislocations detected for the first time at a later age. OPCS was careful to chart clicking hips separately from reported dislocations but, inevitably, the success of the system became identified with its activities (i.e. detections) rather than its benefits. Any measures of success became progressively disassociated from the true preventive target, namely the late diagnosis. It was noted that notification levels varied tenfold in different parts of the country and, after a time, that the high rates associated with the initial enthusiasms began to decline.

The first real attempts to measure true outcomes appeared from about 1976 (Klinberg *et al.* 1976; Leck 1986; Knox *et al.* 1987). In several such studies, conducted mainly by University departments rather than by Health Services themselves, it was discovered that the incidence of late diagnosed congenital dislocation had scarcely altered. In some studies, it appeared to have increased.

The causes of the systems failure were, in essence, the same as those for cervical cancer. First, no public health objective had been identified and accepted; no one was put in charge or was accountable; no information system had been set up; no responsibility for overall intelligence was established; not even the existing hospital information systems associated with orthopaedic admissions had been brought to bear upon the problem.

Control of HIV

The human immunodeficiency virus and AIDS will visit the race with a disaster comparable with the worst epidemics in the whole of human history. In contrast with the last two examples, where the information services had to cope, using data-handling techniques and equipment which were frankly not up to the task, the HIV epidemic supplies the present-day public health service with an opportunity of being wise *before* the event. There are, however, other hindrances. So far, the developing catastrophe has been hidden beneath a long latent interval and a slow pattern of epidemic growth. The scale and the inevitability of the process have thus been difficult to grasp, and reactions to the emergency have been almost anything but rational. This has been remarkably documented by Shilts in his book *And the band played on* (Shilts 1987). However, our chief purpose, here, is to examine the health care information systems designed to monitor the progress of the disease and to support control measures.

We find, characteristically, that the large-scale data-acquisitive processes have been managed admirably. Thus, the information system for monitoring the world epidemic of the terminal phases of HIV infection (the syndrome AIDS itself) is in good shape. Most countries in the world have by now an operational system of accumulating numbers of AIDS cases so that they can keep a running statement of their own experience, and almost all of them report these results to the WHO centre in Paris where they are very efficiently assembled and quickly reported. A great deal of additional information is assembled there, such that it can now be estimated that there are 5 to 10 million people infected with HIV. Patterns of increase in different countries and in different risk groups can be monitored and projected. We can already be sure that many millions will die of the disease before the end of the century.

By contrast, systems for co-ordinating different preventive activities and different professional and administrative functions *within* individual countries are not in such good shape. We must accept, of course, that the currently available control procedures are very limited, and there is no single overriding control activity which would serve to bring these functions into sharp focus. In the absence of a vaccine or of a treatment which reduces infectivity, they consist of (a) the control of blood products, (b) the control of needle sharing by intravenous drug users, (c) education aimed at limiting promiscuous heterosexual and homosexual behaviour, (d) education promoting the use of condoms and discouraging the more dangerous forms of homosexual behaviour, (e) identification and effective sexual isolation of infected persons, and (f) the control of prostitution.

However, there has been little in the way of strong imperatives to assemble the intelligence which might more accurately guide policies in these individual fields and little towards developing optimal policies when more powerful tools, such as vaccines, eventually become available. Instead, we see a growth of role-competition between different professional groups, fragmentation of preventive responsibilities, failures to face up to imminent resource dilemmas, and the development of information processes which do more to restrict access to information than to enable it.

In the UK, for example, the Minister decided very early on that this disease should *not* be notifiable. Since the diagnosis is always made at one of a limited number of laboratories, this may not have mattered. But many of the specimens, especially those from STD clinics, are sent bearing only code numbers with no indication of address, risk group, or other details necessary for proper epidemiological monitoring, and with no means of telling whether they belong to people who have been tested before or with any means of checking whether their HIV status is already known through another route, or whether they are members of known risk groups. The manner in which the proper use of records for epidemiological purposes might be reconciled with requirements for confidentiality has simply not been resolved.

Furthermore, large-scale epidemiological monitoring using surplus material from maternity examinations (e.g. rubella antibody testing) may not be used unless the identities attached to the specimens are first removed. (Even that is an improvement on earlier restrictions!) No follow-up or counselling or checking against known-positives from other sources, or risk group membership, is therefore possible. Public health physicians are denied access to individual records. Even the general practitioners and obstetricians may not be informed. Epidemiologists seeking to understand and predict the course of the epidemic, and to supply guidance on its control with existing or future techniques are denied access to this essential material.

Laboratory work on the biology of the virus and on therapeutic drugs and vaccine developments are well supported, but studies of sexual behaviour (on which we have almost no contemporary information) have not been supported well. One major study, having successfully completed its pilot phases was unceremoniously axed by none less than the Prime Minister. The reasons were political/social rather than scientific. The true scientific and control necessities, and the scale of the human disaster which must follow upon their failure, were not addressed. There was at least a precedent for this action. A similar axe had been wielded a few months earlier, in similar circumstances, in the USA.

It is not difficult now to read into this picture all the characteristic conflicts and deficiencies which from previous experience are known to result in disaster. We can be sure by now that disaster will follow.

A mirage of progress

Dissatisfaction with health service information processes within the UK is widespread. Prevalent attitudes are those of despair rather than complacency. When the government decided to set up a fundamental enquiry into these problems there was widespread acclaim that something at last was being done. 'The Steering Group on Health Services Information' was set up by the Department of Health in the UK in 1980 and produced the so-called 'Körner' report (NHS/DHSS Steering Group 1984). Its purpose was to declare a set of minimum standards, including a minimum data set, for use by the 200 District Health Authorities of the NHS in England and Wales. This seemed reasonable. It was also reasonable that the enquiry should be instituted by a government department with the duties and powers to impose the standards recommended.

There were reservations in that it was generally believed that the primary motivation probably sprang from problems of financial control, rather than an open recognition of the kinds of disaster described above. Faith in the exercise was reduced during the seven years the committee took to make its recommendations and to begin implementation. Finally, no sooner had the implementation begun than the planning environment on which it was based, was fundamentally changed. The primary basis of management and control is now to be moved from one based upon central and delegated planning, to one based upon internal markets. We do not yet know which side of the bargaining process will own the information! It would appear that this exercise, conducted in this particular way, was hopelessly outrun by the pace of events and overwhelmed by the scale of the problem.

The size of the problem had earlier forced the committee to limits its own terms of reference very severely. It focused explicitly upon the functions of the District General Managers of the District Health Authorities and upon their (then) information needs; as a matter of deliberate choice it turned away from the information needs of public health practitioners (as represented by the Faculty of Community Medicine of the Royal Colleges of Physicians). As a result, the recommendations of the Körner Group, when they appeared, were not universally accepted.

This resulted in an extended subsequent discussion such that it became necessary to set up a secondary committee jointly representing the original Körner Group and the Public Health Practitioners, which constructed a supplementary set of recommendations. The Körner exercise was then dissolved—without communicating the fact to the Joint Committee—and so far as can be judged, no heed was ever taken of the supplementary recommendations. The supplementary report was not acknowledged; no thanks were proffered; the Joint Committee had to make private arrangements for the publication of its report (Knox 1987a).

Despite these difficulties, there is no doubt that some progress was made. The NHS had established for the first time a degree of uniformity; and adapted itself to the idea that there *should be* a minimum standard for its information services; and that management decisions—even those restricted to the institutional mechanics of the service—needed to be informed. However, there was no real resolution or even recognition of the true nature of the basic underlying problems of maintaining a health intelligence system directed towards achieving fundamental health purposes.

A diagnosis of the problem

These short histories of disaster, and of ineffective attempts to repair disaster, show that there is something desperately wrong. But what? These disasters were selected from a particular health care system, but commentators in other countries have little difficulty in identifying parallel absurdities elsewhere. They say something, possibly, about the futility of setting up committees and working parties as substitutes for activities which demand continuing professional commitment, and quite substantial resources. But, then, why do we persist in denying public health responsibilities to public health specialists and opting instead for committee mechanisms for whom such processes are inappropriate and which can be guaranteed in advance to fail?

The answer must almost certainly be expressed in terms of social and professional rivalries, and in terms of a primary concern for power and status among individuals and groups. This is what creates the need for a bargaining mechanism, a committee rather than a team of professionals trained in the art; and this is the cause of failure. Blake (*supra*) was concerned with political systems rather than information systems, but his couplet is incisive. We are mistaken if we believe that health care systems and public health information systems are directed uniquely towards improved health; they serve also as implements of administrative and managerial control, and as expressions and extensions of the positions and the powers of individuals and of professional groups. They are used to restrict the roles and activities of some would-be actors, as well as to enable the roles and activities of others.

We see, now, that the reasons for disasters of the kinds described are never simply technical, and indeed never simple. In very few cases can these losses of life be attributed to defects of the information systems alone. Information system failures are undoubtedly crucial, in the sense that these services could never have worked without a more effective formulation; but they are also *symptomatic* in that they arise from equally fatal failures in the design, management and objective-setting of the health care service, of which they are part.

Principles

At this point we turn aside from the role-assertion/role-denial and other power functions which have so largely determined the nature and content of our current public health information systems. We consider instead those functions which might quite genuinely serve the health of the population, and the standards of care given. They were set out in the report of the Joint Committee of the Körner committee and of the Faculty of Public Health Medicine (Knox 1987a). This has been declared in similar terms by Donabedian (1989). Three main principles can be identified.

The axiomatic principle

The Joint Committee (see above) was explicit in identifying the primary justification of its approach. It was essentially axiomatic. Its argument was that it is not possible to manage and develop a health service without clearly identifying its health and health care objectives and then assembling information on the manner and degree in which these objectives are achieved. The principle, it was argued, is universal, and applies to any major enterprise whatever, whether public or private.

Information must subsequently be analysed and presented, and collated with non-information-dependent components including value judgments and political financial and professional considerations; and then brought to bear upon policy and management issues. If the objectives of a health service are to go beyond its social and political aims (such as providing medical care free at the point of demand, and achieving a reasonable social and geographical equity of resource distribution) and if the service is to exhibit genuine concern for health, then its appraisals and its information must go beyond the enumeration of its activities. The Joint Committee regarded these requirements as absolute. There was no way of avoiding them.

The utilitarian principle

The objectives of public health services and of public health information systems—including those designed to service health care systems which are publicly financed and organized—are plainly utilitarian. Their overall aim is to attain the maximum good for the maximum number within current resource constraints. The design and use of such systems not only permits, but in effect demands, adjudications between one resource use and another: and between one client group and another. Adjudications have sometimes been based upon the number of lives to be saved, or the number of life years to be saved, or upon Quality Adjusted Life Years (QALYs) or (less acceptably as a rule) upon the economic benefit to society as a whole. This approach departs fundamentally from the ethic of clinical practice, where the doctor is expected to act on behalf of the patient rather than the health care system or the national economy, or indeed *other* patients. It also conflicts with the basic principles of social justice where, in stark and very explicit contrast, one person's life and one person's rights must *not* be differentiated from another's.

A dis-engagement of this moral conflict can be maintained so long as the allocation processes are restricted to functions and services whose individual uptakes are not yet determined, such that questions of social justice and clinical advocacy do not yet arise. However, there are some situations where conflict can and does occur, and many of the difficulties of seeking agreement on public health information systems have in the past stemmed from failures to recognize and resolve these issues (Knox 1987*b*).

The health-objective principle

The third main premise, springing from the original axiomatic declaration, is that each information system must be an integral part of the care-system which it is designed to serve, and that the objectives must be declared in the true 'outcome' terms of the care-system; that is, relating to the maintenance of health or the care of the sick. The successes of the information system and of the care system are to be seen as joint successes, and their failures as combined failures. If a vaccination programme can produce no record of its uptake rates or the level of disease control which it has achieved, or if a notification and vaccination record system can identify no control programme whose needs must be serviced, then the whole thing must, from a public health point of view, be deemed a combined failure. Any attempts to allocate primary responsibility for this failure is fruitless; failure is as unitary as success.

Surmounting the block

Multiple causes: holistic failures

Public Health information system failures do not generally result from a single critical circumstance. They are not, as a rule, the simple consequence of a failure to invest, or of a technical failure in a service whose organization is otherwise adequate. They are seldom such as might be put right through firing and hiring. Inadequate reporting, inadequate communication, inadequate research, and inadequate investment all play their parts; but in none of the cited failures was there one single critical cause.

The failures seldom arise from lack of effort on the parts of public health or other professional staff. Hundreds of committees and working groups have pursued the topics described under our 'post-mortem' examinations. They have consumed tens of thousands of travel miles and working hours. This is a universal experience.

The origins of such failures are always complex, and inevitably holistic. This, it would appear, is the first lesson to be learned.

Conceptual inadequacies

Part of the problem arises from a widespread conceptual inadequacy. Public health 'information systems' have in the past been associated with a simple image of a centripetal flow of information and a subsequent presentation of statistical digests. Many such large-scale systems are described in other chapters in this book. This monolithic image has been reinforced by success. It was strengthened further in the 1960s through adaptations to mainframe computer systems: these depending in turn upon large-scale centralized data-processing operations. It is nevertheless a serious error to suppose that this represents their necessary structure, or that their operation should continue to depend upon the technical fashions and capabilities of earlier decades.

On the contrary, public health information systems should be conceived and presented in 'systems' terms, comprising control and scheduling and feedback functions, as well as the ascertainment functions. This can be illustrated by imagining a communicable disease control programme which first assembles a register through the afferent flow of birth notifications, maintains it through accessing information on in-and-out migration, schedules the vaccination dates and sends out appointments, records completion, pays the doctor, maintains vaccine batch number deliveries, registers side-effects, accepts and links disease notifications, maintains continuing records of population immune status, identifies imminent epidemics, issues warnings, and modifies the age schedules for vaccination in response to the situation. This image—for nowhere is it a reality—provides a more appropriate general basis for our aspirations.

Moral and ethical problems

Unfortunately, it is precisely in the interactive systems of modern public health practice that rivalries for responsibilities between public health and clinical practitioners are most acute; where the traditional independent status of the clinicians may seem most threatened by the new technologies. The defensive reactions which follow are expressed in several different ways, but most often in moral and ethical terms. For example, they may be expressed as a concern for the confidentiality of medical records, accompanied by a reluctance to define the specific purposes for which medical records may legitimately be used—apart from their direct primary purpose for assisting the care of the patient. Clinical research is usually accepted as legitimate; but public health research and other public health usages often invokes reluctance (Knox 1984).

The basic fact of the matter is that the public health 'principles' outlined in the last section, and in the report of the Joint Committee, have not been universally and enthusiastically accepted. They incite a deep-seated ambivalence. The ethical framework of clinical practice is delineated within the terms of the clinical contract between the doctor and the individual patient, and this does not encompass any necessary concern for public health. If the duality of medical responsibility is recognized at all, then the utilitarian principle and the population health objective principle—and thereafter the monitoring and measuring principle—are forced very firmly into a position subsidiary to that of the clinical encounter. Although positive formulations of the ethics of public health are now being constructed (Doxiadis 1987) this is a fairly recent development. Conflicts, when they occur, are as a rule settled only in one way.

It would seem, therefore, that the really serious failures of public health information systems arise for two main reasons: conceptual inadequacies and ethical ambivalences. They are most obtrusive in those situations depending upon complex information networks involving the 'co-ordination' of activities of different professional groups, appearing to pursue a common objective, but with as many cross-purposes as good intentions. There is frequently a serious conflict (not necessarily well articulated) between operational philosophies stemming from local objectives; and health care and health information systems strung together from a multiplicity of self-perceived professional and administrative roles. Achieving a reasonable accommodation between the two approaches stands as the most important single problem awaiting solution in the whole of modern public health practice. The solutions must depend partly upon an acceptance of a more rounded concept of the role of medicine, on the parts of its practitioners, and partly upon an informed and explicit public/political insistence that an effective accommodation be reached. Unfortunately, there is little evidence at present, from either quarter, that the problem has been explicitly recognized; and no effective solutions are yet in sight.

References

Department of Health and Social Security (1969). Screening for the detection of congenital dislocation of the hip in infants. Prepared by the Standing Medical Advisory Committee for the Central Health Services Council and the Secretary of State for Social Services. Department of Health and Social security, London.

Donabedian, A. (1989). Institutional and professional responsibilities in quality assurance. *Quality Assurance in Health Care* **1** (1), 3–11.

Doxiadis, S. (1987). *Ethical dilemmas in health promotion* (ed. S. Doxiadis). Wiley, New York.

Klinberg, M.A., Chen, R., Chemke, J., and Levin, S. (1976). Rising rates of congenital dislocation of the hip [Letter]. *Lancet* **i**, 298.

Knox, E.G. (1984). *The confidentiality of medical records: the principles and practice of protection in a research dependent environment.* Commission of the European Communities, Luxembourg.

Knox, E.G. (1987*a*). *Health Care Information.* Report of a joint working group of the Körner committee on health services information and the Faculty of Community Medicine. Nuffield Provincial Hospitals Trust, London.

Knox, E.G. (1987*b*). Personal and public health care: Conflict, congruence or accommodation? In *Ethical dilemmas in health promotion* (ed. S. Doxiadis), Chapter 6, pp. 59–68. Wiley New York.

Knox, E.G., Armstrong, E.H., and Lancashire, R.J. (1987). Effectiveness of screening for congenital dislocation of the hip. *Journal of Epidemiology and Community Health* **41** (4), 283.

Leck, I. (1986). An epidemiological assessment of neonatal screening for dislocation of the hip. *Journal of the Royal College of Physicians* **20**, 56.

NHS/DHSS Steering Group on Health Services Information (1984). *A series of six reports, an interim report, and two Supplements.* HMSO, London.

Shilts, R. (1987) *And the band played on: Politics, people and the Aids epidemic.* First published by St. Martin's Press, New York. Also Penguin Books, 1988, London.

2

Health information resources: United States—health and social factors

N. PEARCE

Introduction

This chapter reviews health information systems that provide data for routine monitoring of the health of the population of the US. Since it is possible to discuss only a few of the existing systems, the material in this chapter represents a sampling of the total universe of data systems. Several criteria were applied in the selection of the systems to be included. Each system must be: (i) national in coverage, representative of the situation in the country as a whole; (ii) currently operational, although it may be periodic in its collection of data (e.g. conducted biennially rather than continuously); (iii) operated by either the Federal Government or the private sector; and (iv) produce primary data rather than a secondary compilation of data from other sources. Systems excluded are those that provide data for specific programmes, with the exception of the Medicare programme's statistical system which has been included because it covers virtually the entire population aged 65 years and older. Also excluded are those programmes that operate at the State level, providing data only for the State and some or all of its subdivisions, and important and highly useful compendia that obtain highlights from a number of primary data sources. Examples of secondary data sources in the health area are the *Statistical Abstract of the United States*, published annually by the Bureau of the Census and *Health, United States*, the annual report to Congress from the Secretary of the Department of Health and Human Services, prepared annually by the National Center for Health Statistics.

To provide a framework for the organization of these diverse data sources, they are presented in three broad groupings:

(1) health status, including morbidity and mortality;

(2) health care resources and their utilization;

(3) health economics.

Placement of individual data sources in the categories is somewhat arbitrary, but every effort has been made to ensure that the placement is as logical as possible. For example, while the National Health Interview Survey produces information on both the utilization and financing of health care services, its primary purpose is provision of information on morbidity and so it is located in the section on health status.

The relative paucity of information on health care professionals is noteworthy. Numerous *ad hoc* studies have been conducted for several health occupations but they have typically been limited to the membership of a particular national professional organization and conducted on an irregular basis. Anyone interested in the most recent data for any particular health occupation should contact the relevant national professional association.

There are limited sources that meet the criteria for inclusion for health economic data because most health economics studies use secondary data, involve secondary analyses of data from various sources, or are based on single-time studies.

For each data system the material is typically divided into two sections. The first section describes the purpose and scope of the system; the second section provides an overview of the data collection procedures and data items. At the end of each description a reference is given to one or two sources for additional information about each system. The references also frequently identify other data collection programmes of the respective organizations.

Health status

This section considers national morbidity and mortality reporting systems; the Drug Abuse Warning Network; the National Health Interview Survey; the National Health and Nutrition Examination Survey; the National Electronic Injury Surveillance System; the Annual Survey of Occupational Injuries and Illness; and basic vital statistics and vital statistics follow-back surveys.

National morbidity and mortality reporting systems

Purpose and scope

The Centers for Disease Control (CDC) maintain national surveillance programmes for selected diseases, with the co-operation of State and local health departments. Over the years the surveillance systems maintained by these centres have expanded, and emphasis has shifted as certain diseases have lower incidence rates and other diseases have taken on new aspects. The data are used to identify outbreaks of communicable diseases and to monitor trends in those diseases.

In 1878 an Act of Congress authorized the collection of morbidity reports by the Public Health Service for use in connection with quarantine measures against pestilential diseases such as cholera, smallpox, plague, and yellow fever. The following year a specific appropriation was made for the collection and publication of reports of notifiable diseases, principally from foreign ports; in 1893, an Act provided for the collection each week from State and municipal authorities throughout the US. To secure uniformity for the registration of these morbidity statistics Congress enacted a law in 1902 directing the Surgeon General of the Public Health Service to provide forms for the collection, compilation, and publication of such data.

Data collection procedures and data items

Reports on notifiable diseases were received from very few States and cities before 1900, but gradually more States submitted monthly and annual summaries. In 1913 the State and territorial health authorities recommended weekly telegraphic reporting by States for a few diseases, but it was not until after 1925 that all States reported regularly. In 1950 the Association of State and Territorial Health Officers authorized a conference of state and territorial epidemiologists for the purpose of determining the diseases that should be reported by the States to the Public Health Service. Following approval of the list of diseases by the Association of State and Territorial Health Officers the first manual on reporting procedures was issued; since then recommendations and revisions made by this conference have been incorporated into the *Manual of Procedures for National Morbidity Reporting and Surveillance of Communicable Diseases.*

The following notifiable diseases are those agreed upon by the State and territorial epidemiologists and approved by State and territorial health departments: acquired immune deficiency syndrome (AIDS), anthrax, aseptic meningitis, botulism, brucellosis, chickenpox, diphtheria, encephalitis, hepatitis, leprosy, leptospirosis, malaria, measles (rubeola), meningococcal infections, mumps, pertussis, poliomyelitis, psittacosis—ornithosis, rabies in animals and humans, rubella (German measles), congenital rubella syndrome, tetanus, trichinosis, tuberculosis, tularaemia, typhoid fever, typhus, and venereal diseases. Physicians report cases of these diseases to their local city, county, or State health departments. Although notification of a case of one of the quarantinable diseases may have been made by telephone to CDC, each case of cholera, plague, smallpox, and yellow fever is reported by the State epidemiologist to CDC in the weekly morbidity report.

Completeness of reporting varies greatly, since not all cases receive medical care and not all treated conditions are reported by physicians. Thus, the data should be interpreted with caution. Some diseases, such as plague and rabies, that cause severe clinical illness and are associated with serious consequences are probably reported quite accurately. However, diseases such as salmonellosis and mumps that are clinically mild and infrequently associated with serious consequences are less likely to be reported. Estimates of under-reporting were made several years ago by CDC for two diseases—measles and viral hepatitis. At that time, before the institution of the Measles Elimination Program in 1978, it was generally accepted that about 10–15 per cent of cases of measles and about 15–20 per cent of all cases of viral hepatitis occurring in the US were reported to CDC. The degree of completeness of reporting is also influenced by the diagnostic facilities available, the control measures in effect, and the interests and priorities of State and local officials responsible for disease control and surveillance. Finally, factors such as the introduction of new diagnostic tests (e.g. for hepatitis B) and the discovery of new disease entities (e.g. infant botulism and legionellosis) may cause changes in disease-reporting independent of the true incidence of disease.

In addition to the weekly report of notifiable diseases, a weekly mortality report is made to CDC by City Health Officers or Vital Statistics Registrars from 121 major cities. This report is in the form of a table in which total deaths by age are cross-classified by number of deaths assigned to pneumonia and influenza. An annual summary of the reported incidence of each listed nationally notifiable disease is also submitted by each State and Territorial Health Department to CDC for the previous calendar year (see Centers for Disease Control (weekly)).

Drug Abuse Warning Network

Purpose and scope

The Drug Abuse Warning Network (DAWN) is operated by the National Institute of Drug Abuse (NIDA) to monitor drug abuse trends and patterns, to identify licit and illicit drugs and new substances associated with drug abuse morbidity and mortality, and to provide data for national, State and local drug abuse policy and programme planning. DAWN was initiated in 1972 by the Drug Enforcement Administration (DEA) under authorization of the Comprehensive Drug Abuse Prevention and Controlled Substance Act 1970. Responsibility for the system was transferred to NIDA in 1981.

Data collection procedures and data items

DAWN collects drug abuse information from panels of hospital emergency departments and medical examiners located primarily in metropolitan statistical areas throughout the continental US. The areas selected are those for which NIDA, DEA, and the Food and Drug Administration felt the most valuable information could be obtained to meet the objectives of the system. Geographical location was also taken into consideration to maintain representative regional coverage across the country. Abuse information is also collected from a randomly selected sample of emergency departments drawn from the remaining emergency department population segment outside these areas, but data are not collected from medical examiners outside the selected areas.

Separate forms are completed by emergency departments and by medical examiners. The data items collected concern specific drugs being abused, the magnitude of the abuse, abuse problems unique to certain specific geographical areas, the source of the abused substance, the form (e.g. tablet, capsule, powder) of the abused substance, and the age, race, sex, and employment status of abusers. No patient identifiers or names are ever transcribed on DAWN data collection forms. Data collection forms are completed for all drug-related events.

Eligible emergency departments are defined as those that are open 24 hours a day and 7 days a week, located in non-federal, short-stay general hospitals; eligibility is determined primarily from information in the American Hospital Association's *Guide to the Health Care Field* and is updated with each new annual edition. Upon implementation in 1990, a new sample of emergency departments will include approximately 600 facilities located primarily in 21 metropolitan statistical areas, a subset of the 27 primary metropolitan statistical areas in the former sample. All medical examiners in the 27 metropolitan statistical areas in the old sample are considered eligible for continued participation; information on medical examiners is compiled from medical professional society organization membership lists and other available sources (National Institute on Drug Abuse (annual, semi-annual)).

National Health Interview Survey

Purpose and scope

The National Health Interview Survey (NHIS) conducted by the National Center for Health Statistics (NCHS) is a principal source of information on the health of the population of the US. The survey was initiated in July 1957 in response to the National Health Survey Act 1956 which provided for a continuing survey and special studies to secure on a voluntary basis accurate and current statistical information on the amount, distribution, and effects of illness and disability in the US and the services rendered for or because of such conditions.

The purpose of the survey is to provide national data on the incidence of acute illness and accidental injuries, the prevalence of chronic conditions and impairments, the extent of disability, the utilization of health care services, and other health-related topics. Data collected over the period of a year

form the basis for the development of annual estimates of the health characteristics of the population and for the analysis of trends in those characteristics.

The survey covers the non-institutionalized, civilian population of the US alive at the time of the interview. Persons excluded are: patients in long-term care facilities (data are obtained on patients in some of these facilities through the National Nursing Home Survey conducted by the NCHS); persons on active duty with the Armed Forces (though their dependants are included); US nationals living in foreign countries; and persons who have died during the calendar year preceding interview.

Data procedures and data items

NHIS is a cross-sectional household interview survey which consists of continuous sampling and interviewing of the population. The sampling plan follows a multi-stage probability design which yields national estimates, although some estimates are obtained for the four geographical regions. Currently, the first stage consists of a sample of 198 primary sampling units drawn from approximately 1900 geographically defined primary sampling units that cover the 50 States and the District of Columbia. A primary sampling unit consists of a county, a small group of contiguous counties, or a metropolitan statistical area; within primary sampling units, smaller units called segments are defined in such a manner that each segment contains an expected eight households.

Households selected for interview each week are a probability sample representative of the target population. Each calendar year, data are collected from approximately 48 500 households including about 122 000 persons. The annual response rate is usually at least 95 per cent of the eligible households in the sample; the 5 per cent who do not respond is divided equally between refusals and households where no eligible respondent could be found at home after repeated calls.

Data are collected through a personal household interview conducted by interviewers employed and trained by the Bureau of the Census according to procedures specified by NCHS. All adult members of the household 17 years of age and older who are at home at the time of the interview are invited to participate and to respond for themselves; the mother is usually the respondent for children. For individuals not at home during the interview, information is provided by a responsible adult family member (aged 19 years or older) residing in the household. On some occasions, a random sub-sample of adult household members is selected to respond to questions on selected topics. Follow-up supplements are sometimes completed for either the entire household or for individuals identified as having particular health problems. As required, these supplements are either left for the appropriate person to complete and return by mail, or the interviewer calls again in person or by telephone to secure the information directly.

On average, the interviews require about 80 minutes in the household. The questionnaire consists of two basic parts: (i) a 'core' set of health, socio-economic, and demographic items; and (ii) one or more sets of 'supplemental' health items. The core items are repeated each year. The arrangement of core items complemented by rotating as well as single-time supplements allows the survey to respond to changing needs for data and to cover a wider variety of topics, while providing continuous information on fundamental topics.

The questionnaire includes the following type of 'core' questions: the basic demographic characteristics of household members, including age, sex, race, education, and family income; disability days, including restricted activity and bed days, and work-loss and school-loss days occurring during the two-week period prior to the week of interview; physician visits occurring during the same two-week period; the acute and chronic conditions responsible for these days and visits; long-term limitation of activity resulting from chronic disease or impairment and the chronic conditions associated with the disability; short-stay hospitalization data, including the number of persons with hospital episodes during the past year and the number of discharges from short-stay hospitals; and the interval since the last visit to a doctor. The questionnaire also includes six lists of chronic conditions; each concentrates on a group of chronic conditions involving a specific system of the body (e.g. digestive, circulatory, respiratory).

Supplements to the questionnaire change in response to current interest in special health topics. For example, throughout 1987 a cancer risk factor supplement was included, and in 1988 there were supplements on occupational health, medical device implants, child health, and AIDS knowledge and attitudes.

Suggestions and requests for special supplements are solicited and received from many sources, including university-based researchers, administrators of national organizations and programmes in the private and public health sectors, and other parts of the Department of Health and Human Services. A lead time of at least one year is required to develop and pre-test questions for new topics to include as special supplements (National Center for Health Statistics 1981, 1988, 1989).

National Health and Nutrition Examination Survey

Purpose and scope

The first National Health and Nutrition Examination Survey, referred to as the NHANES I, was initiated by the NCHS in 1970, with data collection beginning in April 1971. NHANES I was a modification and expansion of the earlier Health Examination Survey (HES) which had been initiated a decade earlier and had carried out three separate programmes. The restructuring and modification of the HES reflected the assignment to the NCHS of an additional specific responsibility—the measurement of the nutritional status of the population and the subsequent monitoring of changes

in that status over time. A second National Health and Nutrition Examination Survey (NHANES II) began in February 1976 and ended in February 1980. NHANES III began data collection in 1988 and will end in 1994.

NHANES is designed to collect data that can be obtained best, or only, by direct physical examination, clinical and laboratory tests, and related measurement procedures. This information is of two kinds: (i) prevalence data for specifically defined diseases or conditions of ill-health; and (ii) normative health-related measurement data which show distributions of the total population with respect to particular parameters such as blood pressure, visual acuity, or serum cholesterol level.

Successive surveys in the HES and NHANES programmes have been directed to different segments of the population and have had different sets of target conditions. The first HES 'cycle' involved examining a sample of adults with the focus primarily on selected chronic diseases. The second and third cycles of the HES were directed respectively to children between the ages of 6 and 11 years and to youths between the ages of 12 and 17 years; both of these surveys emphasized growth and development data and sensory defects. The nutrition component of the first NHANES programme was directed to a probability sample of the broad age range of 1-74 years, while the detailed health examination component focused on the population between ages 25 and 74 years. NHANES II was again directed to a broad population aged from 6 months to 74 years, and the nutritional data collected can be used in conjunction with the earlier NHANES I data to monitor changes in nutritional status over time. A special Hispanic Health and Nutrition Examination Survey (HHANES) directed to persons in families with one or more members of Hispanic origin or descent who lived in areas with a high concentration of Hispanics was conducted in 1983 and 1984. The study population for NHANES III is of persons aged 2 months and older; blacks and Mexican Americans are being over-sampled, as are children and older persons.

Data collection procedures and data items

The samples for all of the HES and NHANES programmes have been multi-stage, highly clustered probability samples, stratified by broad geographical region and by population density grouping. Within strata the sampling stages employed have been the primary sampling unit, the census enumeration district, the segment, the household, and lastly the individual person. Until the household stage is reached, all sampling is carried out centrally.

The final stage of the sampling is conducted in the field in the particular chosen area. It involves interviewer visits and questionnaire completion at each one of the selected households, with the final selection of individuals included in the sample being dependent upon information elicited by the household interview questionnaire. The size of the sample in the survey programme has varied. In each of the three HES programmes the sample size was approximately 7500 persons.

In NHANES I the sample selected for the major nutrition components of the examination contained approximately 28 000 persons, of whom 21 000 were examined. A comparably sized sample for NHANES II again yielded approximately 21 000 examined persons. The sample for NHANES III is projected to be approximately 40 000 persons, of whom about 30 000 will be examined.

Data collection teams consist of specially trained interviewers and examiners including physicians, nurses, dentists, dietitians, and medical and laboratory technicians. The examinations take place in the survey's specially constructed mobile examination centres, each consisting of three truck-drawn trailers which are interconnected and which provide a standardized environment and equipment for the performance of specific parts of the examination. This standardized environment is necessary, for example, for such components of the examination as audiometry, which requires hearing chambers within which the ambient noise level conforms to the American Speech Association's standards for acoustical measurements. In NHANES III, the inclusion of young infants down to 2 months of age and of significant numbers of elderly persons over age 75 years (both subgroups with relatively lower response rates) has resulted in use of a home examination comprised of a core set of questions and examinations.

The general pattern of data collection has meant that each survey has been conducted over a period of 3–4 years. This is due to the constraints that limit the number of persons examined in a given time-span (e.g. the number of field teams and the number of sample areas). This imposes a limitation on the kinds of data to be collected by this mechanism, since conditions that might show marked year-to-year variation or seasonal patterns cannot be included. However, many important chronic diseases and health-related measurements are not subject to such changes in prevalence within short-run periods.

In the HES and NHANES programmes there has been and continues to be much attention devoted to the question of the response rate, which is the proportion of sampled persons who are actually examined. Both previous NHANES programmes succeeded in obtaining household interview data on about 99 per cent of the sample population. More detailed health data appear in the medical history questionnaires; these were completed for 90 per cent of the selected sample persons in NHANES I, NHANES II, and HHANES. Overall, NHANES I, NHANES II, and HHANES averaged about a 74 per cent examination response rate. There is considerable ancillary information on most of the non-examined persons in the sample population, and it is possible to make use of those data in the process of imputation and analysis of non-response bias. There is, moreover, some evidence that data obtained through examinations, tests, and measurements such as used in these surveys are less susceptible to

potential bias from a given rate of non-response than data provided by the individuals themselves.

The kinds of information collected in the NHANES and other examination surveys are so varied and extensive that they are only illustrated here. With respect to nutrition, four types of data are included: (i) information concerning dietary intake—the mechanisms used have included 24-hour recall interviews and food frequency questionnaires, both administered by an interviewer who is a trained dietitian; (ii) haematological and biochemical tests—a sizeable battery of such tests has been performed, with processing at the mobile examination centres where necessary, but for the most part at a central nutrition laboratory established at CDC; (iii) body measurements—the battery used is especially important in connection with infants, children, and youths where growth may be affected by nutritional deficiencies; and (iv) various signs of high risk of nutritional deficiency, based on clinical examinations.

The health component of the NHANES programme includes detailed examinations, tests, and questionnaires, which have been developed to obtain a measure of prevalence levels of specific diseases and conditions. These vary with the particular programme and have included such conditions as chronic rheumatoid arthritis and hypertensive heart disease. Important normative health-related measurements, such as height, weight, and blood pressure are also obtained (National Center for Health Statistics 1981, 1988, 1989).

National Electronic Injury Surveillance System
Purpose and scope

The Consumer Product Safety Commission, an independent regulatory agency established under the Consumer Product Safety Act 1972 and activated in May 1973, is responsible to protect the public against unreasonable risk of injury or illness associated with consumer products, to assist consumers in evaluating the comparative safety of consumer products, to develop uniform safety standards for consumer products and minimize conflicting State and local regulations, and to promote research and investigation into causes and prevention of product-related deaths, illnesses, and injuries. While the Commission relies on a number of sources to meet its data needs, its primary source of data is the National Electronic Injury Surveillance System (NEISS) which became operational in 1972 under the auspices of the Bureau of Product Safety of the Food and Drug Administration, which became the nucleus of the Consumer Product Safety Commission.

Data collection procedures and data items

NEISS operates at two levels and is designed and maintained as an intelligence-gathering system which provides decision-making data. The first level of the system, surveillance, is comprised of 64 hospital emergency departments, a statistical sample representative of all 50 States and US Territories. The sample design is composed of five strata, four based on hospital size, as measured by the annual number of emergency department visits, and a fifth stratum including those hospitals with some type of specialized burn care unit. In order to ensure reasonable geographical distribution in the selection process, the hospitals were ordered by zip code within each stratum of the hospital frame.

Each hospital participating in the system is expected to report all injuries treated within its emergency room that involved a consumer product in any way. It is the usual procedure in most emergency rooms to provide a brief description of the accident within the medical record, e.g. 'amputated finger on lawn mower'. However, the emergency department staff of each participating hospital is urged to provide an adequate description of any product involved. Daily, all emergency department records are reviewed and a coded record is generated for those cases involving a consumer product. This record includes the following data: age of victim; sex of victim; up to two products involved in the accident; whether a third product was involved in the accident; type of injury—a simple 31-element coding scheme in which the injury is described in broad lay terms; body part injured; disposition of the patient after emergency department treatment; accident locale; fire/motor vehicle involvement; comments—space to provide a brief narrative description of the accident or other pertinent information such as the product's brand name; and other supplementary data—a section of the record is set aside to allow for coding of specific information for defined accident types.

At the end of each day's coding, the coded data are entered into a computer terminal installed for this purpose. Simultaneously, the computer system edits the entered data and identifies errors for correction.

This leads up to the second level of the system, investigation. When possible, effort is made to select cases representative of the NEISS surveillance data. Most cases assigned for investigation are selected to provide data for directing regulatory action, for monitoring existing standards, or for providing a basis for response to petitions submitted to the Consumer Product Safety Commission. Accident investigations are based on personal contact (on-site or telephone interviews) and provide information on the accident sequence, ways in which the product is being used, environmental circumstances related to the accident, and behaviour of the person or persons involved. Whenever possible, investigators document the product's brand name, indicate the involvement of the product or its component parts in the accident, and include photographs and diagrams. Police, fire, and coroners' reports may be included as supplementary data. Abbreviated cases, generally conducted by telephone, contain most but not all of these items. None of the investigations includes identifying information on the victims or other respondents (Consumer Product Safety Commission 1986).

Annual Survey of Occupational Injuries and Illnesses

Purpose and scope

The Annual Survey of Occupational Injuries and Illnesses is conducted by the Bureau of Labor Statistics in response to the requirements of the Occupational Safety and Health Act of 1970 for the collection, compilation, and analysis of occupational safety and health statistics. The survey builds upon a record-keeping system, also mandated under the Act, which requires virtually all employers to keep records on all work-related deaths and illnesses and those injuries that result in one or more of the following: loss of consciousness, restriction of work or motion, transfer to another job, or medical treatment beyond first aid.

The present annual survey produces data to reflect the job-related injury and illness experience of the work-force as a whole. Survey data are solicited from a sample of employers with 11 or more employees in agriculture, forestry, and fishing, and from all employers in oil and gas extraction, construction, manufacturing, transportation and public utilities, and wholesale trade. Since December 1982, low-risk industries in retail trade; finance, insurance, and real estate; and services industries have been exempt from routine record-keeping of the log and supplementary record. The sample is selected to represent private industries in the States and Territories. Data for employees covered by other federal safety and health legislation are provided by the Mine and Health Safety Administration of the Department of Labor and the Federal Railroad Administration of the Department of Transportation. The Occupational Safety and Health Administration is responsible for the collection and compilation of comparable data for federal agencies; State and local government agencies are not surveyed for national estimates. Self-employed persons are excluded because they are not considered to be 'employees' under the Act.

Data collection procedures and data items

The universe frame of employers is first stratified into industries and then into employment-size groups. Because the survey is a federal–State co-operative programme, and the data must also meet the needs of participating State agencies, the universe is further stratified by State before sample selection. The Bureau of Labor Statistics designs and identifies the survey sample for each State, and through its regional offices validates the survey results and provides technical assistance to the State agencies on a continuing basis. In each State participating in the survey on an operational basis, an agency collects and processes the data, prepares State estimates, and provides the data from which the Bureau produces national results.

The national sample for the survey consists of approximately 280 000 sample units in private industry. Report forms are mailed to selected employers in February of each year to cover the experience of the previous calendar year. Each employer completes a single report form which is then used for national and State estimates. Information for the illness and injury portion of the report form is copied directly from the Log and Summary of Occupational Injuries and Illnesses, which is required record-keeping by each employer under the Act. Recorded on the log is information for each recordable case on the number of lost work days following the day of injury or onset of illness. The days are recorded either as days away from work or as days of restricted work activity (days when the employee is assigned to another job on a temporary basis, works at a permanent job less than full time, or works at a permanently assigned job but cannot perform all duties normally connected with it). The form also contains questions about the number of employee-hours worked (needed to calculate incidence rates), the reporting unit's principal products or activity, and average employment, to ensure that the establishment is classified in the correct industry and employment-size class.

Using a weighting procedure, sample units are made to represent all units in their size class for a particular industry. Data are further adjusted to reflect the actual employment in an industry during the survey year. Since the universe file which provides the sample frame is not current to the reference year of the survey, it is necessary to 'benchmark' the data to reflect current employment levels.

Although the reported data are carefully edited and reviewed it is recognized that there are undoubtedly errors of interpretations of record-keeping definitions by employers that are not uncovered. Therefore a quality assurance programme is conducted to evaluate the extent of this type of error in the records. A sample of establishments that have participated in the annual survey is visited, and federal and State survey personnel compare entries on the log with supplementary records and other available information to evaluate the reliability of the log entries which provide the basic data for the reporting system.

Despite progress in identification, recording, and reporting of illnesses there are some known limitations inherent in the system. These include the fact that cases are recorded only in the year in which they are diagnosed and recognized as work related. Many occupational illnesses may develop years after an employee has left the firm where he or she contracted the illness, and many illnesses that may be of occupational origin are not as yet commonly recognized as such (Bureau of Labor Statistics 1988).

Basic vital statistics

Purpose and scope

Basic vital statistics provided through the registration system come from records of live births, deaths, fetal deaths, and induced terminations of pregnancy. Registration of these events is a local and State function, but uniform registration practices and use of the records for national statistics have been established over the years through co-operative agreements between the States and the NCHS and its predecessor agencies.

The purpose of the basic vital statistics programme is to formulate and maintain a co-operative and co-ordinated vital records and vital statistics system, promoting high standards of performance. The programme is nationwide in scope, covering the entire population of the US.

Both provisional and final vital statistics are derived from the registration system. The provisional data are obtained from counts of vital records registered without reference to the date the event occurred and the final data are obtained from the record and its contents, processed by date of occurrence of the event.

The civil laws of every State provide for a continuous and permanent birth, death, and fetal death registration system. In general, the local registrar of a town, city, county, or other geographical place collects the records of births and deaths occurring in the area, inspects, queries, and corrects if necessary, maintains a local copy, register, or index, and transmits them to the State health department. There, the vital statistics office inspects the records for promptness of filing and for completeness and consistency of information; queries if necessary; numbers, indexes, and processes the statistical information for State and local use; and binds the records for permanent reference and safe keeping. Microfilm copies of the individual records or machine-readable data are transmitted to NCHS for use in compiling the final annual national vital statistics volumes.

Provisional vital statistics are collected and published monthly and summarized annually. They are derived from monthly reports from the States to the NCHS giving the number of certificates accepted by the State for filing between two dates a month apart, without regard to actual date of occurrence. These reports to NCHS are to be mailed on or before the 25th day of the month following the data month. They are the source of the provisional vital statistics published in the *Monthly Vital Statistics Report* and the annual summary of the monthly reports. Provisional data also include a 10 per cent sample of death certificates, the current mortality sample, which provides provisional cause-of-death data on a monthly basis. The sample is selected by NCHS from the regular data-file of deaths for those States submitting their entire month's file by the end of the following month. Otherwise, the State is asked to provide a sample of records on a current basis. The sample is selected by including each record with a given last digit in the certificate number.

To promote uniformity in the statistical information collected from States and local areas for national statistical purposes, NCHS recommends standard certificates or reports for birth, death, fetal death, and induced termination of pregnancy. The standard certificates and reports are developed co-operatively with the States and local areas, taking into account the needs and problems expressed by the major providers and users of the data. They are reviewed about every 10 years to assure that they meet, to the fullest extent feasible, current needs as legal records and as sources of vital and health statistics.

Although the use of standard certificates and reports by States is voluntary and their form and content may vary according to the laws and practices of each State, the certificates and reports in most States closely follow the standard. The birth certificate includes information on the child's birth date and place of birth; the parents' State of birth, age, race, and educational attainment; previous pregnancy history of the mother; information about the mother's prenatal care, any complications of pregnancy and labour, and any congenital malformations or anomalies noticed at birth. Items on the death certificate include the deceased's date and place of birth and of death, place of residence, usual occupation, and immediate and underlying cause of death (National Center for Health Statistics 1981, 1988, 1989).

Vital statistics follow-back surveys

Purpose and scope
National natality and mortality surveys are periodic data collections based on samples of registered deaths and births occurring during a calendar year. Mortality surveys were conducted by NCHS annually from 1961 to 1968, and natality surveys from 1963 to 1969, in 1972 and in 1985. A National Infant Mortality Survey was also conducted from 1964 to 1966. For 1980, a National Natality Survey was conducted, which was expanded to include fetal mortality; for 1988 a National Maternal and Infant Health Survey is being conducted for which a 1990 Longitudinal Follow-up Survey is planned.

The national follow-back surveys extend for statistical purposes the range of items that are normally included in the vital records. They provide national estimates of births and deaths by characteristics not available from the vital registration system. They also serve as a basis for evaluating the quality of information reported on the vital records.

Data collection procedures and data items
The birth or death record serves as the sampling unit, and samples of these units are selected from a frame of records representing births or deaths registered during a given period, usually a calendar year. The sampling frame for the National Mortality Survey is the current mortality sample, the 10 per cent systematic sample of death certificates received each month by NCHS from the registration areas in the US. The sample for the National Mortality Survey is subselected monthly from the current mortality sample. The sampling frame for the National Natality Survey is the file of birth certificates from each of the registration areas of the US. Each registration area assigns a file number to each birth certificate; these file numbers run consecutively from the first to the last birth occurring in that area during that year. The survey samples are based on a probability design that makes use of these certificate numbers.

Data for all the follow-back surveys are collected primarily by mail. In the natality surveys, from addresses given on the birth certificates, questionnaires are sent to the mother, to the physician who delivered the baby, and to the medical facility where the baby was born. In addition, the 1980 and 1988 surveys obtained data from the medical sources the woman had named as having given her radiation treatment or examination in the year before the delivery.

In the mortality surveys, a questionnaire is sent to the person who provided the funeral director with the decedent's personal information for recording on the certificate. This questionnaire requests socio-economic information about the decedent as well as the names and addresses of hospitals and institutions that might have provided care for the decedent at any time during the last year of life. If the death occurred in a hospital or institution, a hospital questionnaire is sent directly to the hospital or institution asking for information about the care provided and for the names and addresses of other medical facilities providing care.

The questionnaire for national mortality surveys have contained questions concerning the patient's last year of life. The surveys have included questions on hospital utilization, diagnoses, operations performed, institutions in which hospitalized, income, whether working or retired during most of the last year of life, household composition, education, health insurance, and the smoking habits of the deceased.

The questionnaire for the 1964–6 National Infant Mortality Survey included questions on hospitalization of the infant who died, information about other children of the mother, household composition, income, employment of mother, education of mother and father, and health insurance.

The national natality surveys collect information from mothers who had live births during a given year. They have gathered information on the medical and dental care and radiological treatment of the mother, employment and education of mother and father, family income, pregnancy history of the mother, expectations of having more children, household composition, income, whether this was a first or later marriage, whether mother was employed and when during her pregnancy she stopped working, and health insurance coverage.

The 1980 and 1988 national natality surveys were broadened to include fetal mortality, and the 1988 survey included infant mortality; these surveys included many of the same or similar questions as previous surveys to allow trend studies in the areas of smoking habits, marriage and pregnancy history, education, income, health status of mother and infant, sterilization, radiological treatment, employment, childbearing expectations, and breast-feeding. Many new areas of study were added such as alcohol consumption, electronic fetal monitoring, amniocentesis, additional maternal and infant health indicators, occupation of mother and father, and ethnicity. Furthermore, there was an oversampling of low birth-weight infants in the natality survey which will enable more in-depth of this high risk group (National Center for Health Statistics 1981, 1988, 1989).

Health care resources and their utilization

This section considers the National Master Facility Inventory; the Annual Survey of Hospitals; the National Hospital Discharge Survey; the National Nursing Home Survey; the National Ambulatory Medical Care Survey; the statistical system of the Medicare programme; the Physician Masterfile; and health occupation data from the decennial census of the population.

National master facility inventory

Purpose and scope

The National Master Facility Inventory (NMFI), compiled and updated regularly by NCHS, is a comprehensive file of information on virtually all health facilities in the US. Health facilities are those that provide medical, nursing, personal, or custodial care to groups of unrelated persons on an in-patient (at least overnight) basis. Facilities in the NMFI are categorized into three broad classes: hospitals, both short and long stay; nursing homes, including all types of nursing, personal, and domiciliary care facilities, not just those certified by Medicare or Medicaid; and other facilities of a remedial or custodial nature, such as homes or resident schools for the deaf, blind, mentally retarded, and emotionally disturbed, resident treatment centres for alcohol and drug abusers, and homes for unmarried mothers.

NMFI is the only comprehensive source of information on the nation's in-patient health facilities. Its major purpose is to provide data for the analysis of the supply, distribution, and utilization of health resources.

Data collection procedures and data items

NMFI was initiated in 1962–3 by combining the lists of health facilities maintained by four federal agencies, directories of health facilities prepared by various national associations and organizations, and files from State licensure agencies. At present, NMFI produces hospital data annually and data for nursing homes and other health facilities on a less frequent basis. Generally, the data are collected by mail. The administrator of the facility is sent a questionnaire and asked to complete and return it by mail.

Two mechanisms are used to keep the data in NMFI as current as possible. For all facilities except hospitals, NCHS conducts a series of mail surveys to: (i) insure that the data on file on the basic characteristics of the facilities are accurate; and (ii) identify and then delete those facilities that have gone out of business or are no longer eligible for inclusion. In addition, at regular intervals State licensure agencies, national voluntary associations, and other appropriate sources send to the NCHS their most recent directories or lists of new facilities. These lists are then clerically matched with the most current NMFI file and facilities not already included are added.

For hospitals, data were gathered annually in a joint survey of the American Hospital Association and NCHS. The con-

tractual arrangement which in effect merged the hospital portion of NMFI with the American Hospital Association's annual survey of hospitals began in 1968. Through 1975, the Association performed the data collection (from October to January each year) for its member hospitals, while NCHS performed the data collection for the approximately 400 hospitals not registered with the Association. Both portions of the survey were edited and processed to the same specifications by the Association, which delivered to NCHS two edited tapes, one for Association hospitals and one for non-member hospitals. In 1976 and 1977, the Association surveyed all hospitals as part of its annual survey. The contractual relationship with the Association ceased in 1978; since that time NCHS has purchased hospital data-tapes annually from the American Hospital Association, which are solely for NCHS's internal statistical use.

NMFI contains the following types of data for the three categories of facilities (National Center for Health Statistics 1981, 1988, 1989):

1. *Hospitals*: ownership; major type of service offered; whether various facilities and services are offered; number of beds, admissions, in-patient days of care, and discharges; patient census; number of bassinets, live births, and newborn days of care; out-patient utilization; number of surgical operations; revenue, expenses, and assets; and staffing.

2. *Nursing homes*: ownership; major type of service; licensed and staffed beds; beds certified for Medicare and Medicaid; admission policy with regard to age, sex, and various conditions; patient census by age and sex; in-patient days of care; number of admissions, discharges, and deaths; staffing; who is in charge of nursing care; number of patients receiving nursing care; services routinely provided; basic monthly charge; and operating expenses.

3. *Other facilities*: ownership; major type of service; licensed and staffed beds; beds certified as intermediate care beds; admission policy regarding age and sex; patient census by age and sex; in-patient days of care; number of admissions, discharges, and deaths; staffing; basic monthly charge; and operating expenses.

Annual Survey of Hospitals

Purpose and scope

The Annual Survey of Hospitals is conducted by the American Hospital Association. The primary purpose of the survey, which has been conducted since 1946, is to provide a cross-sectional view of the hospital industry and to make it possible to monitor hospital performance over time; it also provides a sampling frame for other Association surveys and special studies. Information is gathered from a universe of approximately 7000 hospitals in the country and includes information on the availability of services, utilization, personnel, finances, and governance.

Data collection procedures and data items

The questionnaire for the annual survey is mailed to both American Hospital Association registered and non-registered hospitals in the US and its associated areas (i.e. American Samoa, Guam, the Marshall Islands, Puerto Rico, and the Virgin Islands); US Government hospitals located outside the US are not included.

Survey questionnaires are mailed in late October of each year. After follow-up, a response rate of about 90 per cent is achieved. Response rates vary between groups of hospitals categorized by size, ownership, service, geographical location, and membership status. For example, the response rate for community hospitals, defined as all non-federal, short-stay general, and other special hospitals, is generally higher than that of non-community hospitals, and the response rate for registered hospitals averages about 90 per cent, while the rate for non-registered hospitals averages less than 60 per cent (registered hospitals comprise approximately 98 per cent of the mailing universe).

When a questionnaire is incomplete or when the information provided does not pass specified edits, the individual hospital is contacted for clarification and confirmation. If it is not possible to obtain information from a hospital, estimates for most missing data items are generated on the basis of their values in the previous year, whether those values were actual or estimated, and on the basis of information reported by hospitals similar to the non-respondents in size, type of control, principal medical service provided, and length of stay (long or short term).

Since its beginning the survey questionnaire has been kept in the same format in order to provide continuity and to permit important time-series and trend analysis. This means that similar definitions have been used, similar arrangement of questions has been followed, and a relatively consistent data-set has been collected. In 1980, the survey questionnaire was considerably expanded, however, in response to the recognition that additional detail was needed on some topics in order to provide a more adequate profile of the hospital industry. To facilitate response to the items the organization of the 1980 questionnaire was identical to that of previous years. In addition, all data items requested on the 1979 questionnaire were also requested on the new questionnaire and similar definitions were used.

The questionnaires for 1979 and 1980 were both divided into seven sections: reporting period; classification by governance and principal medical service provided; facilities and services available; beds set up and staffed within distinct in-patient service areas of the hospital and utilization of those units in terms of discharges and patient days; total number of beds set up and staffed, admissions, discharges, patient days, discharge days, out-patient utilization, and surgical operations for the entire reporting period; financial data on total patient and non-patient revenue, payroll and non-payroll expenses, restricted and unrestricted assets, and liabilities; hospital personnel. Significant changes in 1980 were: expansion from 44 to 110 specific services, information on the manner in which the services are provided, and a special subsection devoted to ambulatory care and the manner in which ambulatory services are provided; expansion of

general utilization items to include a subsection relating to Medicare and Medicaid admissions and patient days; expansion of the financial section to request information about the sources of patient revenues by type of payer and information on capital expenditures and disposals and retirements of capital assets; expansion of the personnel section to obtain information about full-time and part-time employees for 35 categories (as opposed to the six categories included in the past) and about the number of budgeted staff vacancies for each category. Subsequent annual questionnaires have remained essentially the same.

Information gathered from the survey, while it includes many specific services, does not necessarily include all of each hospital's services, and thus the data do not reflect an exhaustive list of all services offered by all hospitals. Similarly, because respondents are asked to provide data for a 12-month period beginning on 1 October and ending on 30 September of the following year, most of the survey data are not for the calendar year; about one-quarter of the responding hospitals use a reporting period of July through June and about one-half use a reporting period of October through September (American Hospital Association (annual); Kralovec and Mullner 1981).

National Hospital Discharge Survey

Purpose and scope

The National Hospital Discharge Survey (NHDS) of the NCHS is the principal source of information on in-patient utilization of short-stay hospitals. Data collection began in 1965 and has been continuous since then.

The purpose of NHDS is to produce statistics that are representative of the experience of the US civilian non-institutionalized population discharged from short-term hospitals. Specifically, the survey provides information on the characteristics of patients, the lengths of stay, diagnoses, and surgical operations, and patterns of use of care in hospitals of different sizes and type of ownership in the four geographical regions of the country. The scope of NHDS encompasses discharges from non-federal hospitals in the 50 States and the District of Columbia; only hospitals with six or more beds and an average length of stay for all patients of less than 30 days are included in the sample.

Data collection procedures and data items

The unit of enumeration in the survey is a hospital discharge. The sampling frame is each hospital's daily listing of discharges. In 1987, the sample consisted of 558 hospitals from a universe of approximately 8000 short-stay hospitals. Of the 492 in-scope hospitals, information was collected from 400 participating hospitals (an 81 per cent response rate) on about 181 000 discharges.

Beginning in 1985 two data collection procedures have been used for the survey. The first is the traditional manual system of sample selection by hospital staff or representatives of NCHS; the daily listing sheet of discharges is used as the sampling frame. Sample discharges are selected by a random technique, usually using the terminal digit or digits of the medical record number for the patient. Data are then abstracted from the face sheet and discharge summary in the medical record. The second method involves purchase of data-tapes from commercial abstracting services; in 1987 this method was used in approximately 17 per cent of sample hospitals. For these hospitals, tapes containing machine-readable medical record data are purchased and subjected to NCHS sampling, editing, and weighting procedures.

The medical abstract form contains items relating to the personal characteristics of the patient, including birth date, sex, race, and marital status, but not name and address; administrative information, including admission and discharge dates, discharge status, and medical record number; and medical information, including diagnoses and surgical operations or procedures. It takes an average of five minutes for medical records personnel to sample and complete each form. The contents of the medical abstract form did not change from the inception of the survey until 1977 when modifications were made so that it more nearly paralleled the Uniform Hospital Discharge Data-Set. The items added to the abstract at that time were residence of patient (Zip code), expected source of payment, disposition of patient, and date of procedures (National Center for Health Statistics 1981, 1988, 1989).

National Nursing Home Survey

Purpose and scope

Between 1963 and 1969, NCHS conducted surveys of nursing homes and their residents on an *ad hoc* basis. With the implementation of the Medicaid and Medicare programmes, the increased utilization of nursing homes, and the projected increases in the aged population, those who set standards for, plan, provide, and assess long-term care services needed comprehensive national data on a continuing basis. To meet their needs, NNHS was developed in 1972, with the initial survey conducted in 1973–4, the second in 1977, and a third in 1985.

This periodic data collection system is a series of nationwide sample surveys of nursing homes, their residents, and staff. The purposes of the surveys are: to collect national baseline data on characteristics of the nursing home, its services, residents, and staff for all nursing homes in the nation, regardless of whether or not they are participating in federal programmes such as Medicare or Medicaid; to collect data on the costs incurred by the facility for providing care by major components such as labour, fixed, operating, and miscellaneous costs; to collect data on Medicare and Medicaid certification (such as utilization of certified beds and the health of residents receiving programme benefits) so that all data can be analysed by certification status; and to provide comparable data for valid trend analyses on a variety of topics.

For the initial survey conducted in 1973–4, the universe included only those nursing homes that provided some level of nursing care, regardless of whether or not they were participating in the Medicare or Medicaid programmes. Thus, homes providing only personal or domiciliary care were excluded. Beginning with the 1977 survey, the universe was expanded to include all nursing, personal care, and domiciliary care homes, regardless of their participation in Medicare or Medicaid. Homes that provide only room and board are excluded. In all three surveys, homes in the universe included those that were operated under proprietary, non-profit, and governmental auspices. The universe included homes that were units of a larger institution (usually a hospital or retirement centre).

Data collection procedures and data items

The National Master Facility Inventory (NMFI) is the universe from which the sample homes are selected. The NMFI listing, maintained by NCHS, contains basic information about the home (such as name, address, size, ownership, number of residents, and number of staff) that is needed to design efficient sampling plans.

Resident data are collected by reviewing medical records and questioning the nurse who usually provides care for the resident. Residents are not interviewed directly. Response rates for the surveys differ according to the type of questionnaire, but range between 98 per cent for the resident questionnaires and 81 per cent for staff questionnaires. The initial survey, conducted from August 1973 to April 1974, had a nationally representative sample of 2100 nursing homes, with a subsample of 25 100 staff and 19 400 residents. The second survey, conducted from May to December 1977, had a total sample of 1700 nursing homes, with a subsample of 16 800 staff, 7100 residents, and 5300 discharged residents. The 1985 NNHS was conducted from August 1985 to January 1986 and had a total sample of about 1150 nursing and related care homes, 3400 registered nurses, 4600 current residents, and 6000 discharged residents.

NNHS uses several questionnaires. The facility questionnaire includes questions on number of beds and residents, services provided, certification status, and various utilization measures. The expense questionnaire includes questions on the facility's expenses by major components, such as labour, fixed, operating, and miscellaneous expenses. The staff questionnaire includes questions about the employee's demographic characteristics, work experience, education, and salary. The residents questionnaire includes questions about the resident's demographic characteristics, health status, functional status, participation in social activities, monthly charge, and source of payment. Included in the 1977 and 1985 surveys was the discharged resident questionnaire which included some of the same questions as the current-resident questionnaire, selected on the basis of their availability in the medical record.

The survey has a number of respondents in a home and is a combination of personal interview and self-administered questionnaires. Facility information is secured through a

20-minute personal interview with the administrator. Expense data are collected on a self-administered questionnaire, requiring about 30 minutes to answer, completed by the facility's accountant under authorization from the administrator or by the administrator. Sampled staff members (registered nurses in the 1985 survey) complete a brief form that requires about 5 minutes to complete. Information on sample current residents is secured by the interviewer in a personal interview with the nurse who provides care to the resident and who refers to information in the medical record. About 15 minutes is required for each sample resident. Beginning in the 1977 survey, information on the sample discharged residents was secured by the interviewer in a personal interview with the nurse who was most familiar with the medical records and who referred to them for replying to all questions. In the 1985 survey, a telephone interview with next-of-kin of current and living discharged residents was conducted to obtain further information not available from the nursing home; this next-of-kin interview was repeated twice at approximately 6-monthly intervals to obtain information on vital status, current living arrangements, and nursing home and hospital utilization (National Center for Health Statistics 1981, 1988, 1989).

National Ambulatory Medical Care Survey

Purpose and scope

In May 1973, the National Center for Health Statistics inaugurated the National Ambulatory Medical Care Survey (NAMCS) on a continuing basis to gather and disseminate statistical data about ambulatory medical care provided by office-based physicians to the population of the US.

The purpose of NAMCS is to meet the needs and demands for statistical information about the provision of ambulatory medical care services in the US. Ambulatory services are rendered in a wide variety of settings, including physicians' offices, neighbourhood health centres, and hospital outpatient facilities. It is expected that NAMCS will encompass all of these settings in the future, and appropriate survey instruments and methodologies will be developed as resources permit.

The NAMCS target population consists of all office visits within the conterminous US made by ambulatory patients to non-federal physicians who are in office-based practice and engaged in direct patient care. Excluded are visits to hospital-based physicians, visits to specialists in anaesthesiology, pathology, and radiology, and visits to physicians who are principally engaged in teaching, research, or administration. Telephone contacts and other non-office visits are also excluded. Since about 70 per cent of all direct ambulatory medical care visits occur in physicians' offices, this office-based NAMCS design provides data on the majority of ambulatory care services. NAMCS was conducted annually through 1981, in 1985, and annually beginning in 1989. It is hoped that within the next few years it will be possible to include non-office-based ambulatory visits (e.g. hospital outpatient departments and emergency rooms) in the survey.

Data collection procedures and data items

The NAMCS sampling frame is a list of licensed physicians in office-based, patient care practice compiled from files that are classified and maintained by the American Medical Association and the American Osteopathic Association. These files are continuously updated by both Associations, making them as current and correct as possible at the time of sample selection.

Through 1981, the sample consisted of approximately 3000 office-based physicians per year, with sample physicians randomly distributed across the 52 weeks of the year so that the resulting data reflect seasonal variations. In 1985 the sample was about 5000 physicians; beginning in 1989 the annual sample is 2500 office-based physicians. Since the assignment of the reporting week is an integral part of the sample design, each physician is required to report during a predetermined period, and no substitute reporting periods are permitted. Approximately 70 per cent of the eligible physicians in the sample participate in the survey.

The final stage involves sampling patient visits within a physician's practice. The sampling rate, determined at the time of the interviewer's appointment, is dependent on the number of days during the reporting week that the physician is in practice and the number of patients he or she expects to see. In actual practice, the sampling procedure is handled through the use of a patient log.

Actual data collection for NAMCS is carried out by the participating physician, aided by office assistants when possible. The physician completes a patient record for a sample of patients seen during the assigned reporting week. Sampling procedures are designed so that patient records are completed each day of practice for at most ten patient visits. Physicians expecting ten or fewer visits per day record data for all of them, while those expecting more than 10 visits per day record data after every second, or third, or fifth visit, observing the same predetermined sampling interval continuously. Each form requires only one or two minutes to complete.

Two data collection forms are employed by the participating physician—the patient log and the patient record. The patient log is a sequential listing of patients that serves as a sampling frame to indicate for which visits data should be recorded. The patient record contains the following items of information about the visit: date and duration of the visit; patient's date of birth; sex, race, and problem; expected sources of payment; whether the patient has been seen for the particular problem before and whether the patient was referred by another physician; length of time since onset of the problem; diagnoses; diagnostic and therapeutic services; seriousness of the condition; and disposition. Periodically, supplemental items are added to the basic patient record to investigate specific health conditions or other aspects of ambulatory care. For example, questions about medications were added for 1980–1 and questions about accident or product-related illnesses were added for 1979 (National Center for Health Statistics 1981, 1988, 1989).

Statistical system of the Medicare programme

Purpose and scope

The Medicare programme, enacted on 30 July 1965 as Title XVIII of the Social Security Act, became effective on 1 July 1966. The programme, which is administered by the Health Care Financing Administration, makes available two separate but complementary health insurance programmes: (i) hospital insurance, covering nearly all persons aged 65 years and over and disabled beneficiaries aged under 65 years entitled to benefits for at least 24 consecutive months, and covered workers and their dependents with end-stage renal disease who require renal dialysis or a kidney transplant; and (ii) supplementary medical insurance, covering those persons who voluntarily pay the premiums.

The primary objective of the Medicare statistical system is to provide data to measure and evaluate programme operation and effectiveness. Benefit payment operations furnish information about the amount and kind of hospital and medical care services used by disabled persons and those aged 65 years and over, as well as the expenditures for such services. Applications by hospitals, skilled nursing facilities, home health agencies, independent laboratories, and suppliers of portable X-ray and out-patient physical therapy services to participate in the programme provide data on the characteristics of such providers of services. The claim number assigned to each individual serves as the link between the services used under Medicare and the demographic characteristics of individual beneficiaries.

System components and operation

The statistical system is based on four related computer records: the health insurance master file, the provider record, the hospital insurance (Part A) utilization record, and the medical insurance (Part B) payment record.

The health insurance masterfile identifies each aged and disabled person eligible for health insurance benefits and indicates whether he or she is entitled to hospital insurance benefits, to supplemental medical insurance benefits, or to both. The entitlement record provides the population data for each part of the programme and serves as the base for the computation of a variety of utilization rates. The enrolment file contains demographic data including age, sex, race, State, county and Zip code of residence, and eligibility information for enrollees. The file also contains information on date of death so that an important health outcome measure can be linked to records on use of services.

Every hospital, skilled nursing facility, home health agency, independent clinical laboratory, and supplier of portable X-ray or out-patient physical therapy services must apply for participation in the Medicare programme; data on the application forms are stored in the central provider record and are updated as facilities are re-certified periodically, as new ones apply for participation, and as some leave the programme. When the information in the provider file is combined with utilization data, it relates the characteristics of facilities and agencies that provide care to the kinds and amounts of services used by the persons insured under Medicare. Information in the provider file includes the institution's size, location, and type of control. Just as each enrollee has a unique identification number, each institutional provider also has a unique identification number.

Administration of the hospital insurance programme requires that two things be known about each person at the time of admission to a hospital—the individual's entitlement under the programme and the extent to which he or she has used the benefits available. When a patient is admitted to a hospital, the admission section of the in-patient hospital admission and billing form is completed by the hospital and forwarded through its intermediary to the Health Care Financing Administration's (HCFA) central record. As soon as the record is checked, normally in less than 24 hours, the intermediary is informed of the patient's benefit status and of the number of days of in-patient care to which he or she is entitled during the current benefit period.

This information is then forwarded to the hospital. At discharge the hospital completes the billing section of the form and sends it to the intermediary for payment. When payment is approved, the intermediary forwards the claim to the HCFA for inclusion in the central record.

As part of this process, information on diagnoses and surgical procedures is coded for a 20 per cent sample of beneficiaries based on specified combinations of digits in the health insurance claim number. Admission and billing forms are handled in a comparable manner by home health agencies and skilled nursing facilities. The out-patient billing form is also transmitted to HCFA for entry in the central record after the bill is approved for payment by the intermediary.

All information on utilization experience in hospitals and skilled nursing facilities needed to administer the 'benefit period' provision is recorded centrally. This information includes stays in certain non-participating institutions that meet the definition of a hospital or skilled nursing facility under the law and days of care not covered or reimbursable under the programme.

Each admission and billing form contains both the beneficiary's claim number and the provider's identification number. The resulting record can be readily matched to the beneficiary and provider files. By this process, a statistical tape record is created for the sample of insured persons; it contains the information needed for tabulation from the three files related to hospital insurance use.

Payment or reimbursement under the supplementary medical insurance programme is made only after receipt of the carrier's (intermediaries involved in Part B of the Medicare programme) bills with allowed charges in excess of a set amount during a calendar year. For the enrolled population, carriers need to know from a central source the amount of the deductible that has been met; thereafter, during the remainder of the calendar year, the only additional information required from the HCFA for reimbursement or payment purposes is whether the person is still enrolled under the supplementary medical insurance programme.

Administration and operation of the programme requires accurate and complete information on the amounts paid by the carriers for physician services and for other services and supplies under this part of the programme. To meet these needs, carriers furnish a payment record consisting of tape, punched card, or other machine-readable form of each bill paid. A bill is defined as a request for payment from, or on behalf of, a beneficiary as the result of services provided by a single physician or supplier.

As with hospital bills, the bills for Part B services are linked to the information in the enrolment file through the enrollee's Medicare number. At this time no system operates nationally to identify the physician supplying the service. Thus, physician service data cannot be aggregated on physician characteristics as hospital data can be on hospital characteristics. This is why patient-origin data for Medicare physician services cannot yet be developed at the national level (Health Care Financing Administration 1981).

Physician Masterfile

Purpose and scope

A masterfile of physicians has been maintained by the American Medical Association (AMA) since 1906. In the early days of its existence, the masterfile was primarily a listing of physicians maintained as a record-keeping device for membership and mailing purposes.

The AMA Physician Masterfile is considered the most comprehensive and complete source of physician data in the US. It includes information on every physician in the country and on graduates of American medical schools who are temporarily practising overseas. The file includes members and non-members of the Association, and graduates of foreign medical schools who are in the US and meet US education standards for primary recognition as physicians. Thus, all physicians comprising the total physician personnel pool are included on the AMA Physician Masterfile.

Data collection procedures and data items

A file is started on each individual upon entry into medical school or, in the case of foreign and Canadian graduates, upon entry into the US. As a physician's training and career develop, additional information is added to the file, e.g. internship and residency training, licensure, board certification, professional affiliations, and other characteristics. These characteristics, while they may change over time, are not subject to constant change and are included in the historical portion of the masterfile.

There is also a current professional activities portion of each physician's record that identifies current address, professional activity, specialties, and employment status. By definition, this current portion of the file is subject to constant change and is updated regularly. Between 1969 and 1985, a mail survey of all those in the masterfile was conducted approximately every four years by the AMA using a questionnaire entitled Record of Physicians' Professional Activities (PPA). Effective in 1988, approximately one-quarter of the masterfile receive a PPA each year; by the end of a four-year period, all those in the masterfile will have received a PPA for update.

Between periodic updates, a computerized weekly updating system keeps the masterfile current. Each physician's record is updated to reflect the most recent change, which may be signalled by AMA mailings or publications, by physician correspondence, or by hospitals, governmental agencies, medical schools, medical agencies, medical societies, specialty boards, and licensing agencies. Any indication of a change in professional status or address triggers a questionnaire similar to the one used in the most recent census.

While the data collected from the PPA questionnaire represent a major input to the masterfile, data from other sources are also incorporated. These other data sources include: information on year of graduation, name, address, place of birth, and date of birth, from medical schools; information on interns and residents, place of birth, and foreign medical graduates in training from hospitals; information on physicians in Government provided annually by the Surgeons General of the US Government; data on board certification of physicians from American specialty boards; information on membership in specialty, State, and county societies provided by the societies; and data on foreign medical graduates provided by the Educational Commission for Foreign Medical Graduates (American Medical Association (annual)).

Health occupation data from the decennial census of the population

Purpose and scope

The Constitution of the US provides for an enumeration of the population every 10 years. It was quickly recognized that enumeration offered an opportunity to obtain important information on characteristics of the population. The first census, taken in 1790, therefore included questions on the age and sex of each enumerated individual, and over the years many additional items have been added to the census. Information on occupation was added in 1850 and has been included since. Beginning in 1940 many questions, including occupation, have been asked of only a sample of the households in the total population in order to reduce the burden on the public. The censuses of 1960 and 1970 obtained information on occupation for persons 14 years of age and over in the experienced labour force or in the labour reserve; the census of 1980 obtained information on occupation for each person aged 15 years and over in households that fell into the sample. For the 1990 census, essentially the same questions and procedures will be used as were used in 1980.

Data collection procedures and data items

Like all information on the questionnaire, occupation is self-reported by household members in sample households where a long form of the questionnaire is received. Upon completion of the questionnaires in the households and their return to census field offices, the questionnaires are edited by hand for completeness and consistency. The questionnaires are then transferred to the Bureau of the Census facilities in Jeffersonville, Indiana, where information provided in narrative form, such as information on occupation, is coded by clerks.

Persons queried about their occupation in the 1980 census were asked whether they worked at all the week before 1 April, and if so how many hours they worked that week at all jobs (if they worked at more than one job); they were also asked at what location they worked that week, and a number of questions about the journey to work. Those who did not work at all in the prior week were asked other questions about their availability for employment in that week.

Specific occupational questions were asked of those who had worked in the prior week and of those who did not work the past week but had worked for at least a few days in the period 1975–80. If the person had had more than one job or business the prior week, the one at which the most hours were worked was to be described; if the person had no job or business that week, information give was to pertain to the last job or business since 1975. These questions included the name of the company or employer for whom the person worked, the kind of business or industry, the kind of work they were doing, their most important duties, and whether they were employees of a private company for wages, salary or commission, of government (federal, State, or local), self-employed in their own business (either incorporated or not incorporated), or working without pay in a family business or farm.

The occupation classification system used in the 1980 census is similar to that used in each decennial census since 1940. However, the changes made for each of the censuses affect comparability of data from one census to another, and care is required in making such comparisons. There was a major change in the classification system used for coding occupations between the 1970 (441 specific categories) and 1980 (503 specific categories) censuses.

The 1980 census was the first in which the Standard Occupational Classification system was used. Before 1980, the Census Bureau used its own occupational classification system but the Standard Occupational Classification was developed in 1977 for use by all federal agencies to provide greater comparability of occupation data among governmental agencies.

Analysis of occupational data from the decennial census requires some caution, since the data are self-reported by individuals in response to open-ended questions. Since census data on occupations are based on a sample of the population, there is also sampling error associated with the estimates (Bureau of the Census 1984; National Center for Health Statistics 1985).

Health economics

This section considers the National Medical Expenditure Survey and the Consumer Price Index.

National medical expenditure survey

Purpose and scope

The 1987 National Medical Expenditure Survey (NMES), conducted by the National Center for Health Services Research, is a unique source of detailed national estimates on the utilization and expenditures for various types of medical care. NMES builds on the experience of the former Current Medicare Survey, the National Health Interview Survey, the 1977 National Medical Care Expenditure Survey, and the 1980 National Medical Care Utilization and Expenditure Survey.

NMES was designed to be directly responsive to the continuing need for statistical information on the health care expenditures associated with health services utilization for the entire US population. The survey was designed and conducted in collaboration with the Health Care Financing Administration to provide additional utilization and expenditure data for persons in the Medicare and Medicaid populations. NMES will produce estimates for the evaluation of the impact of legislation and programmes on health status, costs, utilization, and illness-related behaviour in the medical care delivery system. A separate household survey of American Indians and Alaska Natives (that is, Aleuts and Eskimos) (SAIAN) was conducted under sponsorship of the Indian Health Service; SAIAN was designed to provide previously unavailable data on health services utilization and health care financing for these population subgroups. SAIAN used a probability sample of approximately 2000 households in a sample of the 482 counties served by the Indian Health Service. To be eligible for inclusion in the survey, a household had to include one or more Americans or Alaska Natives eligible for services from the Indian Health Service.

Data collection procedures and data items

NMES was composed of several related surveys. The household portion of the survey consisted of a national survey of approximately 14 000 households in the civilian non-institutionalized population and over-sampled several population subgroups of particular policy interest—blacks, Hispanics, the poor and near poor, the elderly, and persons with functional limitations. The sample for this survey was a multi-stage area probability sample.

The household survey had a screener interview plus a series of five interviews over an approximately 18-month period in order to obtain all relevant data for calendar year 1987. The screener and interviews for rounds 1, 2, and 4 were conducted by face-to-face interview and rounds 3 and 5 were conducted by telephone when possible.

Collection of data from the households was facilitated by the use of a calendar/diary and a summary. At the time of the first interview, the household respondent was given a calendar on which to record information about health problems and health services utilizations between interviews. Following each household interview, information about health provider contacts and the payment of charges associated with them was used to generate a computer summary of information provided. This summary was used by the interviewer to obtain information not available during a previous interview. In addition to a core questionnaire, each interview included supplements on topics such as: functional limitations; health status, health habits and health opinions; caregiving (for those living with or caring for functionally impaired persons within or outside the household); income during 1987, details of home ownership, and income tax filing status; and use of long-term care services.

A medical provider survey, a patient-identified physician survey, and a health insurance plans survey were also conducted as follow-back surveys to augment and enrich the person-related and family-related data collected in the household surveys. Signed permission forms were obtained from individuals in the household surveys before conducting these other surveys to obtain additional information about their use of health care services or their health insurance coverage.

The medical provider survey had the purposes of adding data on charges and sources and amounts of payment where such information could not be provided by a household respondent or where experience has shown household responses are not reliable and of providing separate estimates of use and expenditures for validation of estimates derived from enriched household data. Information on the demographic and practice characteristics of physicians seen by participants in the household surveys was the focus of the patient-identified physician survey; analysis will focus, for example, on mix of treatment by specialty and practice characteristics for different population subgroups. Through the health insurance plans survey, the private health insurance coverage of household survey participants will be verified; information will also be obtained on the total premium costs of private health insurance, the division of these costs between employers and employees, the kinds of coverage and benefits held by privately insured persons, the availability of health insurance and other fringe benefits at the workplace, the extent of retiree health insurance benefits, and the degree to which employer-provided health insurance is self-insured. A Medicare records component was included to check the eligibility status and claims data for Medicare recipients in the household survey, in SAIAN, and in the institutional population component.

A separate institutional population component was developed to obtain for the first time information on the population in nursing and personal care homes and in facilities for the mentally handicapped that would parallel and complement the information on the non-institutionalized population from the household survey. This institutional component surveyed about 13 000 sampled persons in a sample of eligible facilities. The universe for this survey was all persons in nursing homes and facilities for the mentally handicapped for any part of 1987.

At the first visit to a sampled facility the interviewer developed a sample of 'current' residents as of 1 January 1987. For each of these individuals, information on their residential history in the facility, health status, and demographic information was obtained. There were three subsequent quarterly visits to the facilities to update information on the 1 January sample of residents and to select samples of 'new' admissions during the preceding quarter. For each sampled resident, information on facility charges and use of and expenditures for other medical care services was obtained. In addition, for each sampled resident, the interviewer identified a person outside the facility who was best able to answer questions about the sampled patient's life outside the institution—their residential history outside the facility, their last household residence, health status before admission to the sampled facility, health insurance coverage, further demographic information, and 1986 income (National Center for Health Services Research 1987).

Consumer Price Index

Purpose and scope

The Consumer Price Index (CPI) is a measure of the average change in prices over time for a fixed 'market basket' of goods and services. The market basket is revised periodically to reflect changes in what Americans buy and in the way they live. The latest CPI revision, effective with the release of January 1987 data, continues a CPI for All Urban Consumers (CPI-U), which was introduced in 1978 and is representative of about 80 per cent of the total non-institutional population, and a CPI for Urban Wage Earners and Clerical Workers (CPI-W), which covers about 42 per cent of the non-institutional population. The CPI-U includes, in addition to wage earners and clerical workers, groups such as professional, managerial, and technical workers, the self-employed, short-term workers, the unemployed, and retirees and others not in the labour force. CPI includes an identifiable medical care component.

Effective with the 1978 revision, 'medical care' represents a major expenditure group and is no longer a sub-listing under 'health and recreation' as it had been. The general definition of 'medical care' continues, however, to include both medical care services and other commodities. The medical care expenditure group is one of seven major groups; the major groups are further divided into 69 expenditure classes, which in turn are divided into 184 item strata.

Data collection procedures and data items

CPI is based on prices of food, clothing, shelter, and fuels, transportation fares, charges for doctors' and dentists' services, drugs, and other goods and services that people buy for day-to-day living. Prices are collected in 85 urban areas across the country from persons in about 57 000 housing units and from about 19 000 retail and service establishments—grocery and department stores, hospitals, filling stations, an other types of stores and service establishments. Foods, fuels, and a few other items are priced monthly in all 85 locations. Prices of most other goods and services are obtained monthly in the five largest geographical areas, and every other month in other areas by Census Bureau representatives.

In calculating the index, price changes for the various items in each location are averaged together with weights that represent their importance in the spending of the appropriate population group. Local data are then combined to obtain a US city average. Price indexes are published by size of city, by region of the country, for cross calculations of regions and population-size classes, and for 27 local areas.

The base period for the general-purpose federal index series is revised about every 10 years. Base periods are changed in order to facilitate the visual comprehension of rates of change from a base period that is not too distant in time. Beginning with release of data for January 1988, the standard reference base period for both CPI-U and CPI-W is 1982–4. An increase of 22 per cent, for example, is shown as 122.0. This change can also be expressed in dollars: the price of a base period 'market basket' of goods and services in the CPI has risen from $10 in 1982–4 to $12.20.

The weights for items in the 'market basket' were developed from the Consumer Expenditure Survey conducted for the Bureau of Labor Statistics by the Bureau of the Census in 1982–4. The 1982–4 Consumer Expenditure Survey shows that as a proportion of total consumption, the medical care services component was smaller than that of the most recent previous survey, conducted in 1972–3. The decrease reflects changes in the ways by which consumers pay for medical care, with major medical expenses frequently paid for partially (and sometimes completely paid for) by health insurance, and many insurance premiums fully or partially paid by employers or by Government. Although medical care prices rose at a rapid rate over the decade, average consumer unit expenditures for medical care rose less rapidly because of these employer-and Government-provided benefits.

The 1987 restructuring of the medical care component created three new indexes by separating previously combined items: eye care was separated from other professional services, and in-patient and out-patient treatment were separated from other hospital and medical care services. Eye care has been combined with physician's services, dental services, and other professional services to form the 'professional medical services' index. The previous distinction between the purchase price of eyeglasses and contact lenses (commodities) and the charge for fitting eyeglasses and contact lenses (services) is increasingly difficult to make, so both eye care commodities and services are included in a single index in the medical care service component. Also, fees for laboratory tests and X-rays have now been removed from professional services and, along with emergency room charges, make up the new out-patient services category.

Since the 1987 revision to CPI, the expenditure weight for health insurance in CPI reflects only payments that employees or consumer units make toward health insurance premiums. For computing CPI, contributions of employers are treated as income to consumers, not as a consumer expenditure. Changes in the price that consumers pay for health insurance are calculated through an indirect method, instead of through direct pricing (Bureau of Labor Statistics (monthly); Ford and Sturm 1988; Mason and Butler 1987).

References

American Hospital Association (annual). *Guide to the health care field*. Chicago, Illinois.

American Hospital Association (annual). *Hospital statistics*. Chicago, Illinois.

American Medical Association (annual). *Physician characteristics and distribution in the US*. AMA, Chicago, Illinois.

Bureau of the Census (1984). 1980 Census of population Vol. 2, Subject reports: *Earnings by occupation and education* (PC 80–2–8B). US Department of Labor, Washington, DC.

Bureau of Labor Statistics (monthly). *CPI detailed report*. US Department of Labor and US Government Printing Office, Washington, DC.

Bureau of Labor Statistics (1988). *Occupational safety and health statistics: Annual survey of occupational injuries and illnesses*. Bulletin 2285. US Department of Labor, Washington, DC.

Centers for Disease Control (weekly). *Morbidity and mortality weekly report*. US Department of Health and Human Services, Public Health Service, Atlanta, Georgia.

Consumer Product Safety Commission (1986). *The National Electronic Injury Surveillance System: A description of its role in the US Consumer Product Safety Commission*. US Consumer Product Safety Commission, Washington, DC.

Ford, I.K. and Sturm, K. (1988). CPS revision improves pricing of medical care services. *Monthly Labor Review* **111**, 4.

Health Care Financing Administration (1981). *Medicare data system*. HCFA Pub. No. 03111, Baltimore, Maryland.

Kralovec, P.D. and Mullner, R. (1981). The American Hospital Association's Annual Survey of Hospitals: continuity and change. *Health Services Research*. **16**, 351.

Mason, C. and Butler, C. (1987). New market basket for the Consumer Price Index. *Monthly Labor Review* **100**, 1.

National Center for Health Services Research (1987). *The 1987 National Medical Expenditure Survey: Its design and analytic goals*. Public Health Service, Rockville, Maryland.

National Center for Health Statistics (monthly). Monthly vital statistics report. DHHS Pub No. (PHS) 90–1120. Public Health Service, Washington, DC.

National Center for Health Statistics (1989). *Data systems of the National Center for Health Statistics*. Vital and Health Statistics. Series 1, No. 23, DHHS Pub No. (PHS) 89–1325. Public Health Service, Washington, DC.

National Center for Health Statistics (1988). *Catalog of publications of the National Center for Health Statistics*. DHHS Pub. No. (PHS) 88–1301. Public Health Service, Washington, DC.

National Center for Health Statistics (1981). *Data systems of the National Center for Health Statistics*. Vital and Health Statistics. Series 1, No. 16, DHHS Pub No. (PHS) 82–1318. Public Health Service, Washington, DC.

National Center for Health Statistics (1985). *Decennial census data for selected health occupations, United States, 1980*. DHEW Pub No. (HRA) 86–1826. Public Health Service, Washington, DC.

National Institute on Drug Abuse (annual). *Data from the Drug Abuse Warning Network*. Series I. Public Health Service, Rockville, Maryland.

National Institute on Drug Abuse (semi-annual). *Trend data from the Drug Abuse Warning Network*. Series G. Public Health Service, Rockville, Maryland.

3

Health information resources: United Kingdom—health and social factors

JOHN S.A. ASHLEY, SUSAN K. COLE, and M. PAULA J. KILBANE

Introduction

This chapter describes the various sources of statistical information on health and social factors in the UK, but excludes those that primarily contribute information about the health care, rather than the health, of the population. Thus it does not cover sources of data on health service activity or facilities, health personnel, or expenditure. The text concentrates on the legislative and other background, and reviews matters relevant to the validity of the data systems that are described. In many cases the sources concerned have different characteristics within the different countries of the UK and particular attention is drawn to such variations.

For many of the topics considered there may be access to additional analyses beyond those normally presented in the standard, and usually annual, publications. It may be possible to consult unpublished supplementary tabular material or to request *ad hoc* analysis of the data. Many of the reports indicate whether these facilities are available, usually with the provision of a contact address. In some cases hard-copy publications are enhanced or supplemented by microfiche tables in greater detail.

The chapter is divided into four sections, the first of which provides a brief description of population statistics. These not only provide a general overview of the demographic and social status of the population, but also provide the basic denominator material required for the calculation of appropriate rates from the event data covered in the subsequent sections. The following section covers mortality statistics, including specific contributions relating to occupational and area mortality, and the availability of information from linked files.

In England and Wales, the Office of Population Censuses and Surveys (OPCS) is the central governmental department responsible for the collection and compilation of the Census; most other population statistics; and mortality and other vital statistics. This was formerly the General Register Office, and its director is the Registrar General. Similarly there are General Register Offices in Scotland (GRO(S)) and Northern Ireland (GRO(NI)), directed by the respective Registrars General.

The third section covers a range of morbidity statistics set out in approximate order of inauguration. It includes the notification of infectious disease, cancer registration, hospital discharge statistics, morbidity statistics from general practice, and the notification of congenital malformations and abortions. This order is used as it is not easy to classify the various sources by alternative means: they do not follow any particular pattern in terms of their contribution to morbidity information, nor do they relate to any logical sequence in respect of, say the age group to which they primarily refer.

OPCS, and to a lesser extent GRO(S) AND GRO(NI), also collect and compile some of the available morbidity statistics; but in other cases responsibilities lie with the appropriate Health Departments:

1. England—the Department of Health, formerly the Department of Health and Social Security (DHSS), previously the Ministry of Health;
2. Wales—the Welsh Office;
3. Scotland—the Scottish Home and Health Department; a major role is also played by the Information and Statistics Division of the Common Services Agency of the Scottish Health Service;
4. Northern Ireland—the Department of Health and Social Services (DHSS(NI)).

The final section describes the General Household Survey and its Northern Ireland equivalent, which are designed as sources of a wide range of annual social information, and which usually include data specifically related to health. In conclusion, brief mention is also made of other relevant sources.

Population statistics

The main UK sources of medical data, although important in their own right, are enhanced when combined with appropriate population statistics to give rates. Of special value are rates that take into account variation in the distribution of certain demographic characteristics in the base population from which the event data are derived.

Population data for the UK pre-date those both for mortality and morbidity; thus it is opportune to commence with a description of the national population censuses and the annual estimates of the population which emanate from them.

National population censuses

Background

In the middle of the eighteenth century, acknowledgement of the need for a count of the population led to the introduction of a Bill in Parliament, but this was allowed to lapse. However, the Population Act 1800 led to the first of four censuses (1801–31) conducted under the guidance of the Clerk to the House of Commons. From 1841, when responsibility passed to the newly appointed Registrar General, the method of enumeration was changed from a count of houses and persons by local Overseers of the Poor, to self-enumeration by household directed by the registration service.

By 1901 a standard legislative and administrative pattern had evolved. Authorization was by an Act of Parliament about a year before each census. The Registrars General of England and Wales, Scotland, and Ireland were responsible for its conduct, with registrars organizing local arrangements. The head of each household was responsible for making the full return for the house, with the information being designated confidential, and with prescribed penalties for disclosure. Subsequently, population figures were published for local government areas down to civil parishes.

The Census Act 1920 now governs the taking of censuses in Great Britain (England, Wales, and Scotland). Additionally, an Order in Council has to be laid before both Houses of Parliament and approved prior to submission to the Sovereign for each census. Apart from questions named in the 1920 Act, any others must have the purpose of ascertaining the 'social or civil conditions of the population'. In determining the frequency of censuses, the need for timely information about changes in the size, distribution, and structure of the population has to be balanced against the resources available, taking into account public attitudes to compliance. This has resulted in a decennial pattern, although an additional 'mid-term' census was carried out on a 10 per cent sample of dwellings in 1966. General descriptions of the development of censuses in England and Wales can be found in Redfern (1981) and Mills (1987). Plans for the 1991 census were published as a White Paper in 1988 (CM430) (Great Britain 1988a), and were described in outline by Whitehead (1988).

The situation in Northern Ireland is somewhat different. Until 1911, censuses in Ireland were organized from Dublin, but the Census Act (Northern Ireland) 1925 transferred responsibility to the Registrar General for Northern Ireland. In consequence, the first census under the Act was carried out in 1926 and subsequently in 1937. From 1951, censuses were contemporaneous with those for Great Britain but, until more permanent legislation in the form of the Census (Northern Ireland) Act 1969, separate Acts preceded each census. The 1969 Act is still in force, but has to be accompanied by an Order in Council for each subsequent census.

Method

A similar procedure is used both in Great Britain and in Northern Ireland whereby the census is usually carried out on a Sunday in April, and the vast majority of the population is counted in their (or someone else's) home. Each householder is required to complete a census schedule, delivered by an enumerator before census day, and subsequently collected. In 1981 the enumerators listed addresses during an advance round, before delivering the forms.

Thus the household is the unit to which the census form normally relates, being defined in 1981 as one or more persons living at the same address with common housekeeping (i.e. sharing at least one meal a day or a common living room), implying that a dwelling may contain one or more households. Individuals may choose to make personal returns, rather than have their data included on the household schedule. There are also arrangements for enumerating people present in institutions such as hospitals or prisons.

In Great Britain the range of topics on which information is collected was extended from the beginning of the century (Table 3.1). The number of questions asked of respondents reached a peak of 30 in 1971, and was reduced to 21 in 1981. From the point of providing a denominator for calculating rates, the main demographic items collected about individuals are sex, age, and marital status, but also included are household composition; the educational level of all those who have left school; the occupation of those who are presently employed, temporarily out of work, or retired; migration within the UK; birth outside the UK; and other minor aspects such as travel to work. The household schedule also seeks information about accommodation and amenities.

The enumerator can, if required, help the householder to complete the schedule, but in any case should identify the majority of missing answers and obtain the information. Some editing at punching occurs, by checking that the values lie in an acceptable range. Further correcting of completion errors occurs with an auto-edit system—by replacing missing or unacceptable values of the simpler items by an acceptable one using the last processed complete record with similar demographic or other characteristics. 'Hard-to-code' items, for example occupation and higher education, are clerically corrected where necessary.

Sampling was first used in the UK in 1951 when advanced tables of results were based on 1 per cent of households. In 1961, to reduce coding and processing costs, only 10 per cent of households were asked complete a full-length questionnaire, involving 11 additional questions covering topics such as occupation. The mid-decade census of 1966 involved a different approach, with lists of all 1961 dwellings and local

Table 3.1. Various items collected in the population censuses in Great Britain, 1801–1981

Item	1801	1811	1821	1831	1841	1851	1861	1871	1881	1891	1901	1911	1921	1931	1951	1961	1971	1981
Name	—	—	—	—	GB	GB	GB	GB	GB	GB	GB	GB	GB	GB	GB	GB	GB	GB
Age (any)	GB	GB	GB	GB	GB	GB	GB	GB	GB	GB	GB	GB	GB	GB	GB	GB	GB	GB
Date of birth	—	—	—	—	—	—	—	—	—	—	—	—	—	—	—	—	GB	GB
Marital status	—	—	—	—	—	GB	GB	GB	GB	GB	GB	GB	GB	GB	GB	GB	GB	GB
Relationship to head of household	—	—	—	—	—	GB	GB	GB	GB	GB	GB	GB	GB	GB	GB	GB	GB	GB
Usual address	—	—	—	—	—	—	—	—	—	—	—	—	—	—	GB	GB	GB	GB
Migration 1 year ago	—	—	—	—	—	—	—	—	—	—	—	—	—	—	—	GB	GB	GB
Country of birth/birthplace	—	—	—	—	GB	GB	GB	GB	GB	GB	GB	GB	GB	GB	GB	GB	GB	GB
Nationality*	—	—	—	—	GB	GB	GB	GB	GB	GB	GB	GB	GB	GB	GB	GB	—	—
Education: scholar/student	—	—	—	—	—	GB	GB	GB	GB	S	S	GB†	GB†	—	GB†	—	—	—
Whether in job/unemployed/retired	—	—	—	—	—	GB	GB	GB	GB	GB	GB	GB	GB	GB	GB	GB	GB	GB
Employment status	—	—	—	—	—	GB‡	GB‡	GB‡	GB‡	GB	GB	GB	GB	GB	GB	GB	GB	GB
Occupation	GB§	—	—	GB§	GB	GB	GB	GB	GB	GB	GB	GB	GB	GB	GB	GB	GB	GB
Marriage and fertility (any)	—	—	—	—	—	—	—	—	—	—	—	GB	—	—	GB	GB	GB	—
Infirmity: deaf/dumb/blind etc.	—	—	—	—	—	GB	GB	GB	GB	GB	GB	GB	—	—	—	—	—	—
Number of rooms¶	—	—	—	—	—	S	S	S	GB	GB	GB	GB	GB	GB	GB	GB	GB	GB
Household amenities (any)	—	—	—	—	—	—	—	—	—	—	—	—	—	—	GB	GB	GB	GB
Families per house	GB	GB	GB	GB	—	—	—	—	—	—	—	GB	GB	GB	GB	GB	—	—

* 1841: only for persons born in Scotland or Ireland: 1851–91: whether British subject or not.
† Also whether full-time or part-time.
‡ Asked of farmers and tradesmen only.
§ Only distinguishing (a) agriculture, (b) trade, manufacture, or handicraft, (c) others.
¶ 1891–1901 only required if under five rooms.
GB = Great Britain.
S = Scotland.

valuation records of new buildings erected in the interim used to draw a 10 per cent sample. In 1971 and 1981 coding and processing of hard-to-code items was predominantly on a 10 per cent sample of census schedules, all of which were fully completed.

The Welsh Office conducted a sample Intercensal Survey of Wales in 1986. Whilst some of its characteristics are those of the General Household Survey (p. 48) in that it was conducted by interview, its content and aims were very similar to those of the national census. A 5 per cent sample of addresses was included, and a response rate of nearly 85 per cent was achieved. Tabulations covering population, housing amenities, and economic activity were presented in the report (Welsh Office 1987).

Validity of population censuses

In a census some individuals may be missed, and others counted twice. Characteristics such as age or occupation can be wrongly recorded or incorrectly coded. In England and Wales since 1961 it has become standard to carry out a census evaluation programme. The aims of this have been summarized as: to allow users to appreciate the broad limits to which the data can be put; and to give the Census Office some guidance on the topics and questions which may need attention in a future census (OPCS, Census Division 1985). The two main areas checked are the coverage and the quality of the replies entered on the schedule. The method used is an attempt to repeat the enumeration for a sample of households, shortly after census day by using a skilled team of field staff.

In 1981, a most extensive programme was carried out and reported (Britton and Birch 1985). With regard to coverage, it was estimated that 0.62 per cent of people had been missed, but about 0.17 per cent had been counted twice, indicating that net underenumeration was less than 0.5 per cent. As to validity, the questions subject to the most response error were on rooms, various aspects of economic activity, and the main means of travel to work. This compares with findings after the 1966 census reported by Gray and Gee (1972) that the misclassification rate varied from 16.7 per cent for the topic 'rooms' (using the 1966 definition) to 0.5 per cent for *de facto* household size.

In Northern Ireland, the 1981 census was taken under difficult circumstances since it coincided with the hunger strike in the Maze prison. A campaign of non-co-operation culminating in the murder of one enumerator prompted the Registrar General to ask for the return of uncollected forms by post. By June 1982 it was estimated by the Census Office that approximately 6000 forms (representing 20 000 population) remained unaccounted, implying a coverage of about 98 per cent. However, comparison of the number of children enumerated with the numbers recorded in comparable school census and child benefit data made it clear that nearly 30 000 under 15-year-olds were not included (Morris and Compton 1985). On the assumption that the non-enumerated population was a typical sample of their area of residence, the authors found it difficult to justify an estimate of total non-enumeration of less than 70 000. On an alternative assumption, that the non-enumerated were drawn from restricted groups within the population (areas with high non-enumeration), lower estimates of the shortfall (26 000–61 000) may be appropriate.

Annual population estimates

Background

The census publications provide the main source of detailed statistics suitable for use as denominators in calculating rates. Annual population estimates are then produced, which take account of: (i) the increment in age of the population; (ii) the occurrence of births and deaths; and (iii) estimates of migration into and out of the country or locality. These have been prepared for the country and large towns since the nineteenth century, and since 1911 for local authority areas.

The method used in England and Wales for making subnational estimates is set out briefly in an occasional paper (OPCS, Population Division 1980) and summarized by Whitehead (1985). It involves tuning the adjustment for aging and the incorporation of births and deaths together with available data on migration. International migration is estimated by: (i) the International Passenger Survey for countries outside the British Isles; (ii) the National Health Service (NHS) Central Register for movement to and from Scotland and Northern Ireland; and (iii) comparison of censuses for the Republic of Ireland and Great Britain. The International Passenger Survey has the difficulty of a sample survey, the NHS Central Register relies on initiation of action from individuals re-registering with doctors, whilst the census comparisons form a very approximate guide to Irish migration.

None of these sources of information is adequate to provide local estimates of population movement. These have to be derived from changes in the size of the electoral role; estimates of movement between Family Practitioner Committee areas (as recorded at the NHS Central Register); and cohort comparison of certain education statistics.

Although of less direct relevance to the calculation of current rates, mention should also be made of the fact that the same parameters affect the projection of the future population, with the added complication of the need to make assumptions as to future trends in mortality, fertility, and migration. For a description of the methods used for the projection of the UK population as a whole see Daykin (1986), and for local authority areas see Armitage (1986). Some of the difficulties of projecting mortality are described by Alderson and Ashwood (1985), and those relating to fertility by Werner (1983).

Validity of population estimates

The validity of the annual estimates will depend on the quality of the census enumeration, which is the base for the estimates, and the precision with which changes in the population can be gauged. There is no cause for concern over births or deaths, but migration can be very difficult to quantify—both for movement across national boundaries and of persons moving from one locality to another. The difference between the population estimates which have been carried forward from the 1971 census and the estimates derived from 1981 census data was calculated for England and Wales for each local government area. The differences, for seven age groups as well as the total population, have been published in an OPCS Occasional Paper (OPCS 1982a).

For England and Wales as a whole the estimates were 0.2 per cent lower than the census, but there were larger errors for specific age groups. Those aged below 25 years were over-estimated and those above were under-estimated, but, apart from those aged from 0 to 4 years and those aged from 25 to 44 years, the errors were all less than about 1 per cent.

At the local level there were considerably larger differences because of the inadequate information about people moving between local authority areas. Overall the differences were 5 per cent or more for about 1 in 20 of the 403 districts; 2.5 per cent or more for 1 in 4; and 1 per cent or more for just over one-half of the districts. The mean of the differences was approximately 1.8 per cent—ten times as large as for England and Wales. The larger ones tended to be an over-estimate and the smaller ones an underestimate.

For individual age groups the differences were again much larger than overall, especially in two age groups that are particularly difficult to deal with. For the very mobile 15–24-years group the average error was 8 per cent. An even greater average error (13 per cent) affected those aged 80 years or over—a very small group, many of whom live in institutions.

Publications relating to population

Many of the census reports published by OPCS, GRO(S), and GRO(NI) are relevant for use as denominators for calculating mortality rates, or for preparing national or local population estimates. Also of particular interest may be an OPCS monitor issued in 1982 (OPCS 1982b), which reviewed national and area data on ethnic groups, although the principal source of these data remains the Labour Force Survey (Shaw 1988). National population estimates for England and Wales, and those for local and health authorities, are published annually by OPCS (Series PP1), and similarly for projections (Series PP2). Similar statistics are available for Scotland and Northern Ireland.

Mortality statistics

The most long-standing of health-related information is that relating to mortality. This has been available in England and Wales for over a century and a half, and for well over a century in the rest of the UK. Based as it is on a statutory process, without which there cannot be legal disposal of a body, there are few anxieties regarding its completeness, although there are some qualms as to certain aspects of its validity.

Legislative background and procedures

In England and Wales the registration of deaths dates from the Births and Deaths Registration Act 1836. This included provision for the recording of cause of death and came into operation on 1 July 1837. In consequence, the Registrar General has collected and published mortality statistics since this date. Failure to register a death was initially neither subject to penalty nor was it a requirement that the cause should be supplied by a doctor, although a medical opinion as to this was sought wherever possible. With the enactment of the Births and Deaths Registration Act 1874, death registration was made compulsory, and a specific duty was placed on the medical practitioner who attended the deceased during the last illness to provide the cause of death, unless there was an inquest.

Various subsequent modifying enactments were eventually consolidated into the Births and Deaths Registration Act 1953 and its companion legislation relating to the organization of the registration service—the Registration Service Act 1953. Current regulations are embodied in a Statutory Instrument (1987), although proposals for further changes to the Registration Services were made in a discussion document (CM531) issued in 1988 (Great Britain 1988*b*); and a subsequent White Paper in 1990 (CM939) (Great Britain 1990). One further important enactment affecting the provision and promulgation of mortality data is the Population Statistics Act 1938, and its revision in 1960, the history and provisions of which have been described by Whitehead (1987). These Acts follow the confidential collection at registration of certain additional items of information regarding the marital status of the deceased, which are not entered in the public record (the register) and may be used only by the Registrar General for statistical purposes.

Registration of deaths in Scotland has occurred since the Registration of Births, Deaths and Marriages (Scotland) Act 1854. A new Act in 1965 relaxed some of the previous regulations, giving, for example, the Registrar General for Scotland power to correct errors, which previously had been allowed only to a Sheriff. The current effective legislation in Northern Ireland is the Births and Deaths Registration (Northern Ireland) Order 1976, although births and deaths have had to be registered since 1864 following the passing of the Registration of Births and Deaths (Ireland) Act 1863. Throughout the UK, mortality data are derived from a certificate of cause of death usually issued by a medical practitioner in combination with information given to the registrar by an informant. In England and Wales the regulation medical certificate currently required in respect of deaths after the neonatal period is similar to that presently recommended by the World Health Organization (WHO) (1977). Furthermore, the format for the cause of death information, distinguishing diseases or conditions directly leading to death from other significant conditions contributing to death, differs in only minor detail from that initiated by the then General Register Office in 1927 (Swerdlow 1987). Also in line with WHO recommendations, provision is made for the supply of information as to the duration of each condition mentioned, but, unlike the basic cause information, this is not transcribed on to the public record.

After an initial pilot study (Gedalla and Alderson 1984), a new neonatal death certificate was introduced in 1986 for deaths occurring within the first 28 days of life. This also conforms with WHO recommendations and requires the certifier to distinguish between fetal and maternal contributions to the cause of death. In many cases this precludes the assignment of a single underlying cause of death normally associated with the traditional certificate of cause.

Both types of certificate include other items, again not for the public record, such as whether information from a post-mortem has been taken into account, or whether the death has been reported to the coroner. Neither Scotland nor Northern Ireland has introduced the alternative form of a neonatal death certificate, so that in both these countries the medical certificate of cause issued by a practitioner is similar to that issued in England and Wales at older ages. In Scotland the medical practitioner can indicate whether a post-mortem has been carried out, or if proposed or not.

For deaths certified in England and Wales the certifier may indicate that further information regarding the cause of death is likely to be available at a later date. If such an offer is made, a standard enquiry form is then sent by the registrar at the time of registration of the death. On receipt of a reply (there is an 85 per cent response rate), the cause of death details are amended in the statistical system where appropriate, but not in the public record. Annually about 3 per cent of total deaths are affected in this way. In addition, there is a system whereby queries are sent by OPCS to certifiers, seeking clarification when there are difficulties in coding the cause of death. These account for a further 2.5 per cent of all deaths. Taken together, however, some causes are differentially affected in this way, particularly in the case of cancers (Swerdlow 1989). In Scotland further enquiries are made by the GRO(S) if the doctor completing a death certificate indicates that further information may be available from performing a post-mortem, resulting in about a 65 per cent response rate. Another 6 per cent of deaths are subject to other enquiries seeking clarification of the medical diagnoses, to which there is a 90 per cent response rate.

In England and Wales the death certificate has to be delivered to the registrar of the sub-district in which the death occurred. Registration districts and sub-districts are geographical subdivisions of local authority areas—a suitable arrangement since registrars are appointed and paid by local authorities. In Scotland registration of death is allowed in a district of choice or in the district of usual residence as a matter of convenience, although statistical analyses of death are attributed to the district of residence. The Registrar General for Scotland also has power to instruct local registrars, but this is a more informal process than the comprehensive regulations governing local registrars in England and Wales.

In Northern Ireland deaths are registered by the registrar for the district in which the person died, or in which he or she was ordinarily resident immediately before death. Since the 1973 reorganization of local government in Northern Ireland,

Table 3.2. Content of public record of death

Item	England and Wales	Scotland	Northern Ireland
Name of deceased	+	+	+
Maiden name if female	+	−	+
Usual address	+	+	+
Sex	+	+	+
Date of birth (and/or age)	+	+	+
Place of birth	+	−	+
Marital status	−	+	+
Details of spouse	−	+	−
Details of father	−	+	−
Details of mother	−	+	−
Occupation	+	+	+
Place of death	+	+	+
Date of death	+	+	+
Cause(s) of death	+	+	+
Details of informant	+	+	+
Details of certifier	+	+	−
Details of registrar	+	+	+
Date of registration	+	+	+

+, Present.
−, Absent.

each of the 26 districts has been administered by one registrar. This is in sharp contrast to the situation in 1864 when, as in England and Wales, registrars were assigned to areas based on the districts of the dispensary doctors of the Poor Law Unions, with approximately 120 registrars covering the area now known as Northern Ireland.

Those qualified as an informant of the death to the registrar are similarly defined throughout the UK. For example, in England and Wales he or she may be a relative of the deceased, a person present at death, the occupier of the institution in which death occurred or an inmate of the institution, or a person disposing of the body. Where death occurs in a public place, the informant may be the relative, any person present at death, any person who found the body, any person in charge of the body, or the person disposing of the body (i.e. the chain of responsibility stretches out until a responsible person is identified).

The informant is expected to provide information about the deceased for the public record, and, if required, also confidentially for statistical purposes. The public record has a slightly different content in each of the different countries of the UK (Table 3.2).

Some categories of death must be notified to the appropriate legal authority who was the power to enquire into them. Thus, in England and Wales, a registrar will report a death in one of these categories to the coroner; i.e. if (i) the death is in respect of a deceased person not attended during his or her last illness by a medical practitioner; (ii) the registrar has been unable to obtain a certificate of cause of death; (iii) the particulars of the certificate of cause of death indicate that

the deceased was seen neither after death nor within 14 days before death by the certifying practitioner; (iv) the cause of death is unknown; (v) the registrar has reason to believe that death was due to an unnatural cause or violence, neglect, abortion, or from suspicious circumstances; (vi) death appears to have occurred from an operation or before recovery from an anaesthetic; or (vii) from the contents of the certificate, it appears that death was due to industrial disease or poisoning. The history of the development of the office of coroner is set out in the Brodrick report (1971), whilst the coroner's particular role with regard to suicides has been described by Jennings and Barraclough (1980). The coroner may provide a notification of the cause of death without an inquest, having in some circumstances had a post-mortem carried out. This will depend on the results of preliminary enquiries and decision as to whether death is free from natural causes.

In circumstances similar to those in England and Wales, the local registrar in Scotland must report certain deaths to the Procurator Fiscal whose duty it is to enquire into cases that might be criminal and, also in the public interest, to eradicate dangers to health and life, to allay public anxiety, and to ensure that full and accurate statistics are compiled. The Procurator Fiscal does not always decide to order an autopsy. If he or she is satisfied that death is due to natural causes and that there is no element of criminal negligence, the general practitioner or hospital doctor can be invited to complete a death certificate, or if they are unwilling to do so, the police surgeon can be asked to view the body and issue the certificate or advise whether an autopsy should be carried out.

In Northern Ireland the provisions of the Coroners' Act (Northern Ireland) 1959 are somewhat different. The coroner must be notified if a person has died either directly or indirectly:

(1) as a result of violence or misadventure, or by unfair means;

(2) as a result of negligence, misconduct, or malpractice on the part of others;

(3) from any cause other than natural illness or disease for which a person was seen and treated by a registered medical practitioner within 28 days prior to death;

(4) in such circumstances as may require investigation (including death as a result of the administration of an anaesthetic);

(5) if it is believed that the deceased person has died as a result of an industrial disease of the lungs.

In England and Wales the proportion of deaths certified by coroners (24 per cent in 1980) is somewhat higher than for their counterparts elsewhere in the UK. In Scotland (in 1985) notifications to the Procurator Fiscal comprised 19 per cent of deaths, whilst in Northern Ireland (in 1986) coroners considered 16 per cent of all deaths.

Despite a suggestion to a House of Commons Select Committee in 1893 (House of Commons 1893), still births were first registerable in England and Wales in 1927, under the provisions of the Births and Deaths Registration Act 1926. Different documentation is used whereby a certifier, who may be the medical attendant or the midwife, completes a certificate where a child has been 'completely expelled or extracted from its mother after the 28th week of pregnancy and which did not at any time after such expulsion or extraction breathe or show any other evidence of life'. The current regulations are as for deaths, dating from 1968, but the format of the certificate was revised in 1986 to be in line with the requirements for cause of death information for neonatal deaths. However, in addition to the cause particulars, the certifier is also asked to give, not for the public record, the weight of the fetus and the estimated duration of pregnancy. It should also be indicated whether information has or will be obtained from post-mortem.

In Scotland still births have been registered since the beginning of 1939, under the provision of the Registration of Stillbirths (Scotland) Act 1938, which was superceded in 1965 by the Registration of Births, Deaths and Marriages (Scotland) Act 1965. Registration of still births in Northern Ireland began in 1961 following the passage of the Registration of Stillbirths Act (Northern Ireland) 1960, and is currently administered under the Births and Deaths Registration (Northern Ireland) Order 1976.

In both these countries the qualifications of certifiers and the requirements for certification are similar to those for England and Wales, but the cause of death information is not collected in the new format. However, in Scotland an annual national perinatal death enquiry (1983–4) and still birth neonatal death enquiry (1985 to date) has been carried out by the Information and Statistics Division. This enquiry also led to the decision that there was no need to adopt the WHO recommended perinatal death certificate, either for still births or for neonatal deaths in Scotland.

Validity of mortality statistics

Any routine data collection system is liable to incur inaccuracies, and in respect of death certification Alderson *et al.* (1983) identified many of the practical issues involved. This was in response to a report on the medical aspects of certification by a joint working party of the Royal College of Physicians and the Royal College of Pathologists (1982). Inaccuracies may occur at any of the several separate steps in the chain of events leading to the production of mortality statistics. These range from the allocation of a clinical diagnosis and the completion of a death certificate by a clinician, through the transcription of this information on to the death notification together with its classification and coding, to the processing, analysis, and interpretation of the statistics produced.

It is usual to examine the accuracy of the diagnostic information at the time of death by comparison of data derived from autopsy, as in a major study sponsored by the then General Register Office (Heasman and Lipworth 1966). One

cannot, however, assume the infallibility of the autopsy diagnosis or that differences found for hospital deaths can be extrapolated to the certification of persons dying at home. Alderson (1981), reviewing the literature on this topic, pointed out that some studies indicate a worrying degree of variation between autopsy and clinical diagnoses. Other studies using this approach have been by Busuttil *et al.* (1981), and a number of interrelated ones in Scotland by Cameron and McGoogan (1981).

It is usual to investigate variation in certification practice by circulating 'dummy' case histories to clinicians and requiring them to complete 'mock' death certificates. Reid and Rose (1964) used this technique with the collaboration of physicians from Norway, the UK, and the US. Similarly, in two related studies (Diehl and Gau 1982, Gau and Diehl 1982) a total of 25 fictitious case histories was circulated to samples of general practitioners and junior hospital doctors in England and Wales for 'certification'. The number of different causes of death assigned varied widely, as did the proportion designated 'refer to coroner', but there was no marked regional variation in certification habits. McGoogan and Cameron (1978) obtained information from clinicians via questionnaires on their attitudes to the value of an autopsy in different diagnoses.

Others have compared death certificates with a review of detailed case histories (see, for example, Alderson 1965; Alderson and Meade 1967; Clarke and Whitfield 1978; Moriyama *et al.* 1958; Pole *et al.* 1977; Puffer and Griffith 1967). Edouard (1982) compared the clinical case notes with the certified cause of death for 200 still births in 1973–7. Using 14 categories of cause of death, there was concordance for 69 per cent of the still births; the discrepancies were most frequently omission of particulars known at the time of death.

More recently, Cole (1989) compared the causes recorded on certificates of neonatal deaths with those assigned in the Scottish neonatal survey, and found general agreement in the functional, if not the precise, cause of death. A more extensive validation study of deaths due to ischaemic heart disease was carried out in Belfast by McIlwaine *et al.* (1985). Relevant death certificates over a 1-year period were verified not only from hospital sources and post-mortem, but also the coroner's office, general practitioners, and ambulance service records. The study found that the number of deaths recorded as being due to this cause was numerically accurate, but that most of the inaccurate certification occurred in hospital. It was thus suggested that accuracy might improve if consultants issued death certificates in hospital, or countersigned those completed by junior doctors. Another source, a confidential reporting system to the Communicable Disease Surveillance Centre, was used by McCormick (1988) in the assessment of the validity of mortality statistics for acquired immune deficiency syndrome (AIDS). That AIDS and human immunodeficiency virus (HIV) infection were noted to be considerably understated as a cause of death may, however, be due to a variety of reasons—not least the particular sensitivity of this disease.

The influence of coding was studied by Wingrave *et al.* (1981) using death data from the Royal College of General Practitioners (RCGP) study of oral contraception. The authors compared the coding by the RCGP with that of the OPCS/GRO(S) of 205 death certificates using 50 aggregated categories, the B list (WHO 1967). There was considerable agreement for non-violent deaths, but some discrepancies occurred from using different sources of information and there were ten codes not in accordance with International Classification of Disease (ICD) rules.

Because of the importance of coding and classification, the WHO Regional Office for Europe (1966) investigated coding consistency by circulating standard certificates amongst coders in different European vital statistics offices. It was suggested that there was considerable disagreement in the selection and coding of the underlying cause of death, both between coders in different countries, and between these countries' coders and the appropriate WHO Centre for Classification. There was relatively less variation between coders working within the same office in each of the participating countries.

Nearly all published national statistics are based on tabulations of single causes for individual deaths—the underlying cause of death—although from time to time in England and Wales multiple cause coding of all diseases mentioned has been carried out (e.g. in 1985 and 1986). It has been pointed out a number of times that converting the information on the death certificate to single cause of death may fail to identify the combination of different diseases that are or are not related, but are thought by the certifier to contribute to death (see, for example, Abramson *et al.* 1971; Cohen and Steinitz 1969; Farr 1854; Markush 1968; Moriyama 1952).

It is generally accepted that mortality statistics, because of these problems, must be interpreted with caution. Obviously there are some conditions where the medical knowledge and facilities for diagnosis have altered markedly over time, or vary from place to place. This will have a major impact on the interpretation of the data, as will major alterations in the ICD with the splitting or amalgamation of various cause groups.

Publications relating to mortality

From 1840 until 1974 the Registrar General published annual statistics for England and Wales. In recent times this consisted of three parts, with Part I containing medical tables, Part II population tables, and Part III being a commentary volume. A range of detailed tables was provided, giving particulars of deaths and death-rates by cause, sex, age, locality, and a number of other variables. This series was replaced in 1974 by separate volumes incorporating subsets of the information. This mortality (DH) series comes out in five parts: DH1, general; DH2, cause; DH3, childhood and maternity; DH4, accidents and violence; and DH5, area.

In the cause volume (DH2) the main tables present, at ICD 3 and 4 digit level, the numbers of deaths by age and sex, with more limited analyses providing rates appearing in the general (DH1) volume. The majority of published statistics by cause of death are based on underlying cause coding, although for years in which multiple coding of causes is carried out, it is usual to include in the DH2 volume a table of the number of mentions of each cause on the certificates.

Prior to 1974 the Registrar General also published a Weekly Return and a Quarterly Return, after which the Weekly Return was issued as a monitor particularly devoted to statistics on infectious disease (see p. 40). The Quarterly Return is now replaced by a quarterly journal, *Population trends*, which includes articles on specific topics and regular limited analyses on mortality. In Scotland and Northern Ireland the traditional series of annual reports continue to be produced and published by the appropriate Registrar General.

Geographical mortality

Place of usual residence is one of the items recorded at death registration, and regular analyses of mortality by area appear in publications of mortality statistics. Also included are basic rates using the appropriate estimates of the local populations as denominators. In addition, local mortality statistics are routinely made available to health and local authorities.

Around the period of each decennial census it is possible to examine the geographical pattern more closely than the annual data usually allow (Britton 1990). The census provides more detailed, and more accurate, figures of the size and distribution of the population at risk, and aggregation of deaths for several surrounding years can more often yield sufficient numbers of deaths to permit analysis of individual causes of death in relation to small local areas. Internal migration of ill (or healthy) people may affect the interpretation of such statistics, but Fox and Goldblatt (1982) suggested that differential migration of those with varying health is not a major source of bias in these area analyses.

Place of usual residence is not the only geographical variable associated with records of death, as both the location of death and the place of birth, which although not routinely coded, are potentially available for study. The former can be used to cast some light on the geographical distribution of treatment by residents of a particular locality, whilst the latter can contribute to studies of 'migrants'. However, it should be remembered that when a patient or resident dies in a long-stay institution, after a stay of 6 months or over, his or her usual area of residence is deemed to be that in which the institution is situated, even if he or she was originally admitted from elsewhere.

Decennial area mortality analyses have been produced for England and Wales since 1851. In recent times, the publication of tabulations has been separated from the subsequent publication of a commentary. Area is also one of the variables used in tables for the decennial supplement on occupational mortality (see next section). No decennial area mortality analyses have been published for Scotland or Northern Ireland.

Occupational mortality

The first publication in the Registrar General's *Decennial supplement on occupational mortality for England and Wales* (1855) incorporated information from the 1851 census and the mortality returns for that year. In introducing the analyses the Registrar General suggested that 'the professions and occupations of men open a new field of enquiry, on which we are now prepared to enter, not unconscious, however, of peculiar difficulties that beset all enquiries in the mortality of limited, fluctuating and sometimes ill-defined sections of the population'. Since that time, at approximately 10-year intervals, tables have been produced showing the various mortality rates by occupation; the one gap in the series is due to the Second World War, as no census was carried out in 1941. Comparable reports in the Scottish series have appeared intermittently since 1895.

For many years it has been recognized that there are a number of difficulties in the calculation and presentation of occupational mortality rates, and each of the supplements on occupational mortality provides a careful discussion about the problems of collecting and interpreting this material.

The data for numerator and denominator in the rates are derived from two sources. The denominator is obtained from the record of current occupation inserted on the census schedule; the information from this is used to derive counts of males and females by age and occupation. The numerator is a count by occupation and cause of death for persons dying around the time of census. In 1951, the period on which the rates were based was extended to deaths occurring in a 5-year period around the time of the census. Problems are created as the occupation recorded at census and at death registration are provided under very different circumstances and are collected in rather different ways, and are not necessarily statistically comparable. For example, Heasman *et al.* (1958), in a survey on the accuracy of occupational descriptions on records used for the calculation of indices of mortality and morbidity in the mining industry, commented on errors in the descriptions of occupations used in the numerator and denominator of the rates, including errors introduced by the coding system.

A more generally based study, discussed in the General Report of the 1961 Census, concerned the matching of the information recorded at the death registration with the census schedule for a sample of deaths occurring shortly after the census. Of 2196 males, 63 per cent were assigned to the same occupation unit at death registration and at census, 10 per cent were assigned to different units within the same occupational order, whilst 27 per cent were assigned to different orders.

To overcome some of these problems emphasis has also been given to the routine use of proportional mortality. Such analyses of the proportion of different causes of death within an occupational data-set do not involve the potential biases introduced by using two different sources. Furthermore, the

1 per cent Longitudinal Study (see p. 38) is proving to be a reliable source of secular occupational mortality material.

Another difficulty in the interpretation of the data concerns the situation where an individual has developed a fatal chronic disease (whether or not occupationally induced) and has had to change his or her occupation because of impaired ability. In such an instance mortality will be shown against the final occupation. Should the change of job have been due to onset of occupationally induced disease, this will not be reflected in the mortality rate of the principal occupation; there will also be an erroneously high mortality rate for the final occupation.

A further particular problem relates to the classification of women at census. Single, widowed, or divorced women are classified by their own occupation; married women, recorded with their husband, are classified by his occupation. Similarly, occupation is not systematically recorded for all married women at death registration. Thus the only complete analyses have to make use of the husbands' occupation and obviously an indicator of 'way of life', rather than the environment of those women who go out to work.

Mention should also be made of analyses by social class, a traditional component of decennial occupational analyses since 1911. The history of this composite variable derived from occupational data has been critically reviewed by Jones and Cameron (1984) who advocated its abandonment. However, Alderson (1984) pointed to its ability to identify variation in exposure to hazard, use of health services, and risk of disease.

After the 1981 census, Scotland participated in the Great Britain occupational mortality study, and tables are also available showing Scottish results on their own. The Registrar General for Northern Ireland arranged for an analysis of occupational mortality for the years 1960–2, using the 1961 census as the denominator; no official report has been published. Age-standardized mortality ratios were calculated for males aged 15–64 years for 27 occupation orders and social classes for all causes and four broad causes of death. Crude mortality rates were also shown for men and married women for 27 occupation orders. Park (1966) published these limited results, but emphasized that detailed tabulations were available upon request.

Linked mortality records

It has been recognized for many years that some of the limitations of analysis of event data can be overcome by linking records of events occurring to individuals. This was foreshadowed by some of the writings of Farr, particularly in relation to estimation of the outcome of care in different hospitals (Farr 1864) and in recommendations for health statistics to be compiled for the Army (Farr 1861). Heady and Heasman (1959) examined the social and biological factors affecting still birth and infant mortality by linking data from the birth registrations to appropriate death details for all infants who died having been born in 1949–50. A similar approach was used by Spicer and Lipworth (1966), who investigated regional and social factors influencing infant mortality for infant deaths during the period from 1 April 1964 to 31 March 1965, matched to appropriate birth registrations.

In England and Wales in 1975 the national routine linkage of infant deaths to births commenced. This linkage extends the range of items by making available those recorded at birth and those recorded at death. The system uses the NHS number, sex, and date of birth as matching factors, and now only a very small proportion of infant deaths cannot be matched with birth details. The cause of death can be tabulated against social class; place of residence; country of birth; mother's age and parity (for legitimate children); legitimacy; whether singleton or multiple birth; birth-weight; and certification. Annual statistical reports are now being made available from this infant linked file. In Scotland birth and infant death registrations are linked within the GRO(S), but are not used for statistical analysis. However, since 1980 routine record linkage also occurs between infant deaths and hospital maternity records, covering 97 per cent of births in Scotland. These are used for statistical and epidemiological purposes only.

A major new venture began in England and Wales with the selection of a 1 per cent sample of respondents from the 1971 national census. Arrangements were made to assemble event data for the sample into a cohort file including cancer registration, psychiatric hospitalization of more than 2 years' duration, emigration, or death. The sample was enhanced by 1 per cent of identified immigrants as well as 1 per cent of all births. The data-set was later enhanced by the addition of data from the 1981 census. The outlines of the system, now known as the Longitudinal Study, were first described in 1973 (OPCS 1973a). A major category of analysis that the system permits is tabulation of mortality in relation to the variables recorded at the census. A comprehensive report (Fox and Goldblatt 1982) presented a wide range of results from mortality of the cohort over the period 1971–5. Comparisons of mortality for subgroups of the population were presented, each being derived from broad questions; this permitted examination of mortality in relation to: economic activity (economic position the week before the census, alternative methods of classifying women, economic activity, and transport to work); household structure (with emphasis on the elderly and those in non-private households); housing characteristics (tenure, density, amenities, etc.); marriage and fertility; education; area of residence; internal migration; and immigrants. Particular probes were made of the influence of various forms of health-related selection (e.g. for occupation, housing, and migration), of the influence of those 'permanently sick' upon subsequent mortality patterns, and of alternative ways of classifying individuals by non-occupationally derived socio-economic status.

Morbidity statistics

From the bedrock of mortality statistics in 1895 the then General Register Office extended its role to become the national repository of a range of related morbidity information, which has in general been paralleled by similar arrangements in Scotland and Northern Ireland. For data collection systems to be included in this section certain criteria have to be met. Each record has to relate to a single individual or client (with the obvious exception of maternity records), and should explicitly or implicitly include diagnosis (or other medical cause) as a parameter (Ashley and McLachlan 1985). There is no requirement that all national events of a particular character have to be covered, and various sample or selective schemes are included. Finally, the schemes are not easily classified, and are considered roughly in chronological order of implementation.

Notification of infectious diseases

Compulsory notification of infectious disease was first introduced in England and Wales in Huddersfield in 1876; many other towns soon obtained local powers. The Infectious Diseases Notification Act 1889 gave a list of diseases that were to be notified and laid down how, when, and by whom they were to be notified. The original object of the Act was to provide powers to combat 'dangerous infectious disorders'. It was to apply immediately to London and to be extended to other areas thereafter. In 1895, the Registrar General first included in his Weekly Return the numbers of cases of five infectious diseases which had been admitted to certain London hospitals.

Before 1922, the Ministry of Health compiled an unpublished document based on a weekly summary of returns from sanitary authorities. The Registrar General's Weekly Return continued to include information about infectious diseases notified by the Metropolitan Asylum Board and the London Fever Hospital. In about 1920 proposals were made for publication, by the then General Register Office of the material handled by the Ministry of Health, and the first issue of this was in January 1922. This was based on provisional weekly returns supplied by each Medical Officer of Health; these statistics were subject to amendment and did not show numbers of patients with infectious diseases by sex or age. However, from 1944 Medical Officers of Health were asked to submit quarterly returns containing corrected figures grouped by sex and age (Registrar General 1949), although this was not a statutory requirement.

The legislation for Scotland is somewhat different, being based on the Infectious Disease (Notification) Act 1889 and the Public Health (Infectious Diseases) (Scotland) Act 1897; the latter Act made notification compulsory, 2 years before similar legislation was introduced in England and Wales. Notification of infectious disease in Northern Ireland also dates from the Infectious Disease Notification Act 1889, the current legislation being the Public Health Act (Northern Ireland) 1967.

The list of notifiable diseases has enlarged with time, and there are some differences between the countries of the UK (Table 3.3). New or amended regulations were introduced

Table 3.3. Infectious diseases currently statutorily notifiable, with date when first notifiable nationally

Disease	England and Wales	Scotland	Northern Ireland
Acute encephalitis	1915	—	1949
Acute poliomyelitis	1912	1932	1913
Anthrax	1960	1960	1949
Chicken-pox	—	1988	—
Cholera	1889	1889	1889
Continued fever	±	1889	±
Diphtheria	1889	1889	1889
Dysentry	1919	1919[a]	1919
Erysipelas	±	1889	±
Food poisoning	1938	1956	1949
Gastro-enteritis	—	—	1949[b]
Legionellosis	—	1988	—
Leprosy	1951	±	±
Leptospirosis	1961	1975	1949[c]
Malaria	1919	1919	±
Measles	1915	1919	1949
Membranous croup	±	1889	±
Meningitis*	1912[d]	1932[e]	1949[f]
Mumps	1988	1988	1988
Ophthalmia neonatorum	1914	±	±
Parathyphoid fever	1938[g]	1889[h]	1949[g]
Plague	1900	1932	1949
Puerperal fever	±	1889	±
Rabies	1956	1976	1976
Relapsing fever	1889	1889	1889
Rubella	1988	1988	1988
Scarlet fever	1889	1889	1889
Smallpox	1889	1889	1889
Tetanus	1968	1988	—
Tuberculosis	1911	1919	1909
Typhoid fever	1938[g]	1889	1889
Typhus	1901	1889	1889
Viral haemorrhagic fever†	1976[j]	1977[j]	1976[k]
Viral hepatitis	1968[l]	1965[l]	1949
Whooping cough	1939	1949	1949
Yellow fever	1968	1975[m]	1949

—, Never notifiable.

±, Previously but not currently notifiable.

* Variously named as cerebrospinal fever, meningococcal infection, acute meningitis, or meningococcal septicaemia.

† Viral haemorrhagic fevers means Argentine haemorrhagic fever (Junin), Bolivian haemorrhagic fever (Machupo), Chikungunya haemorrhagic fever, Congo/Crimean haemorrhagic fever, Dengue fever, Ebola virus disease, haemorrhagic fever with renal syndrome (Hantaan), Kyasanur Forest disease, Lassa fever, Marburg disease, Omsk haemorrhagic fever, and Rift Valley disease.

[a] From 1988 as bacillary dysentery.

[b] Aged under 2 years only.

[c] As leptospiral jaundice until 1967.

[d] Currently as meningitis and meningococcal septicaemia without meningitis.

[e] Currently as meningococcal infection.

[f] Currently as acute meningitis.

[g] From 1889 as enteric fever.

[h] As enteric fever—now includes paratyphoid fever.

[j] Until 1988 only Lassa fever and Marburg disease.

[k] Only Lassa fever and Marburg disease.

[l] Until 1988 as infective jaundice.

[m] From 1988 yellow fever is included within viral haemorrhagic fevers.

throughout the UK in 1988 (Statutory Instruments 1988*a*, 1988*b*; Statutory Rules 1988). Changes included the inclusion of mumps and rubella coincident with the national introduction of an immunization programme using measles, mumps, and rubella vaccine.

Notification by medical practitioners in England and Wales are to the proper officers (usually the Medical Officer for Environmental Health) of the local authority in which the disease is identified or suspected. In Scotland and Northern Ireland, notification is to the Chief Administrative Medical Officer for the area in which the case was diagnosed, and for the area of residence, respectively.

In England and Wales, prior to 1982, the Proper Officers submitted a weekly statistical return to the OPCS, giving counts of the notifications that had been received during the week. This was followed at quarterly intervals by a statistical return giving extra details of the sex and age group of the cases and incorporating any corrections to the original weekly figures. From the beginning of 1982 a new reporting system has operated, whereby the Proper Officers continue to send in a weekly statistical summary and now also provide information about individual notifications. Sex and exact age, rather than age group, are included, as is extra detail for tuberculosis and food poisoning. The system retains the provision of quarterly corrections and hence the data should be comparable with those for earlier years. Although individual records are now collected centrally, neither names nor addresses are included in the details submitted by Proper Officers.

In Scotland primary responsibility for the national collection and processing of the notifications lies with The Information and Statistics Division. In Northern Ireland the DHSS(NI) collates and publishes the information.

Other infectious diseases, although not notifiable under statute, have generated a need for specific reporting arrangements in relation to their surveillance, in particular the systems set up for reporting AIDS and HIV infection. The Communicable Disease Surveillance Centre collates data for England and Wales, Northern Ireland, the Channel Isles, and the Isle of Man. In Scotland the Communicable Diseases (Scotland) Unit operates a similar scheme. Three sources are used: the voluntary registration of AIDS cases by clinicians; laboratory reports of HIV-positive results; and death certificate data. Details of registered cases include age, sex, places of residence, diagnosis, mode of infection (if known), and certain clinical details (McCormick *et al.* 1987). Regular statistics from these systems are available through the Department of Health in England, and are published both by the Scottish Home and Health Department and in a supplement of the Communicable Disease Report in Scotland by the Communicable Diseases (Scotland) Unit.

In England and Wales the Royal College of General Practitioners (RCGP) collects data on patients with various infectious diseases (RCGP 1968). The College's Research Unit runs a surveillance system which consolidates data on these and other conditions on a weekly basis of returns from a number of selected 'spotter' practices with a population of about 200 000 patients. It is acknowledged that these practices are not evenly spread throughout the country (they are predominantly concentrated in or near large conurbations), and the data are based on clinical diagnosis for most of the patients.

In Scotland, a similar but somewhat larger surveillance

system was established in 1971, principally to collect data on influenza incidence; information from general practices in 11 health boards is supplemented by information from laboratories, schools, and workplaces. In 1986, 124 general practices were involved, providing information from 657 000 patients (16 per cent of all Scottish patients within the participating health boards). The returns are collated by the Communicable Diseases (Scotland) Unit. A further development envisaged for Scotland is that a list of 'reportable' infections be established, primarily aimed at diagnostic laboratories.

Validity of notifications of infectious disease

The validity of statistics from notification has periodically been questioned; it is acknowledged that reporting is frequently far from complete, especially for the more common conditions. So, although notifications may not give precise estimates of the incidence of individual infectious diseases, they do provide crude indicators of changes in prevalence in the community.

Stocks (1949) compared notification statistics with data from the Survey of Sickness and various research reports on incidence or cumulative attack rates. He suggested that completeness varied with the illness, from fairly complete notification of acute poliomyelitis to only fractional notification of dysentery. A discussion by Taylor (1965) indicated that a number of western countries have variable completion of notification of infectious disease, and that for some milder infections notification is poor. This issue has also been discussed by Benjamin (1968b), and a study from general practice (Haward 1973) suggested that there was substantial under-reporting of certain infectious diseases.

Lambert (1973) used Hospital In-patient Enquiry (HIPE) data to cross-check the statistics from notification of meningococcal infection. Assuming that the hospital data provide a fairly reliable estimate of the total number of cases, he suggested that notifications identify about half of the total. Clarkson and Fine (1985) also used this source to assess the efficacy of measles and pertussis reporting in England and Wales and concluded that, whereas for measles it was 40–60 per cent, for pertussis only between 5 and 25 per cent of cases seemed to be notified. Goldacre and Miller (1976) used a variety of sources of information to study notification of acute bacterial meningitis in children in one hospital region in 1969–73, concluding that only half the cases of meningococcal and less than one-quarter of other types of bacterial meningitis had been notified.

Stewart (1980) suggested that there was strong circumstantial evidence to question the validity of whooping cough notification, partly due to confusion with infection from various respiratory viruses. He pointed out (Stewart 1981) that 18 per cent of the practitioners in Glasgow notified all the whooping cough cases reported in 1977–9, with over one-third of the cases being reported by 2 per cent of the practitioners. In addition, many of the patients admitted to hospital were not notified. Davies *et al.* (1981) identified a number of problems with notification of tuberculosis, including duplicate notifi-

cation, changes in diagnosis and definition of respiratory disease, and posthumous registration. They concluded that the OPCS statistics had an impressive level of accuracy, which they estimated as being within 10 per cent. Possible solutions were discussed for some of the problems.

Although influenza is not a notifiable disease, there is considerable interest in identifying epidemics of the infection. Tillett and Spencer (1982) discussed the surveillance of routine statistics of the disease in England and Wales. They indicated that, as an epidemic occurred, the use of the term influenza in death certification changed. There was initial undercertification until it became common knowledge that an epidemic was in progress, but as the epidemic waned there was possible overcertification by way of 'compensation'. Paradoxically, therefore, the absolute numbers of deaths ascribed to influenza might be correct.

It is plausible that the same type of change in practice may occur with other acute infectious diseases, and thus the rise and fall in reported numbers may be incorrectly timed, though whether the overall reporting approximates to the correct figure is not clear. When using notification to monitor progress of an epidemic, it must also be remembered that bank holidays result in a fall in the count of weekly notifications (of 15–30 per cent), which may not be completely made up in succeeding weeks. It is impossible to validate notification information using laboratory reports, because of the very biased referral for investigation of appropriate specimens. Where the infection is only rarely fatal, there is also no chance of cross-checking the statistics with those of mortality.

Publications relating to infectious diseases

In England and Wales there are three series of OPCS publications at present—weekly, quarterly, and annual. The weekly monitor (which is still entitled *The Registrar General's weekly return for England and Wales* and is identified by the serial WR) includes: notifications of the major infectious diseases and deaths from certain of these for the preceding 6 weeks in England and Wales; notifications in the current week in England and Wales, standard regions, administrative areas, and health districts; and newly diagnosed episodes of communicable and respiratory disease, giving weekly numbers and rates, from returns from the 'spotter' general practices. Corrected data are provided in the quarterly report, which also includes data from the Communicable Disease Surveillance Centre on the laboratory identification of various diseases for England and Wales and from Regional Health Authorities.

The OPCS annual publication appears in the series 'Communicable Disease Statistics' (series MB2), which provides an appreciable number of tables on notifications and deaths from infectious disease, with some data on trends for the preceding years. It includes a commentary from the Communicable Disease Surveillance Centre, which also publishes a weekly and a quarterly report on communicable disease, with data obtained from hospital laboratories and the Public Health Laboratory Service. These data refer to specific episodes of infection with bacterial confirmation, often providing background discussion of some detail of particular cases of interest.

In Scotland the weekly report *Communicable diseases in Scotland* is produced by the Communicable Diseases (Scotland) Unit, and contains counts of notifications of infectious diseases and food poisoning by health board area, and 'current notes' on particular topics of interest. Data also appear in the weekly return of the Registrar General, Scotland. Annual figures appear in the Scottish Health Statistics. In Northern Ireland statistics for 1935–76 were published in the Annual Report of the Registrar General; subsequently they have been published by the DHSS(NI).

Cancer registration

Background

During the nineteenth century a number of authorities pointed out the drawbacks of mortality data in studying malignant disease, and at the beginning of the present century attempts were made to collect morbidity data (e.g. Dollinger 1907). In 1923 the then Ministry of Health in England set up a system, through the national Radium Commission, for the follow-up of patients treated with radium based on the premise that statistical information about such patients was essential for planning and operating the cancer care services. The introduction of the NHS in 1948 was accompanied by further reaffirmation of this view, and the move towards a national cancer registration scheme was stimulated by the General Register Office taking over responsibility for cancer records in 1947. Subsequently (OPCS 1970), it was suggested that studies of epidemiology and the cause of cancer should be considered as additional objectives of cancer registration.

Despite the desire for national registration at the introduction of the NHS, it was not until 1962 that all regions in England and Wales become incorporated in the national voluntary scheme, which relies initially upon co-operation at the clinic and hospital level. The data are then organized through regional cancer registries in England and the Welsh Office in Wales, which forward the material to the OPCS. Since 1971 patients registered with cancer have also been traced and flagged on the NHS Central Register. This permits automatic identification of the fact of death and permits calculation of survival statistics. Prior to this the regional registries had to carry out *ad hoc* follow-up at hospital level, or write to general practitioners. For a full description of the system in England and Wales, see Swerdlow (1986).

Cancer Registration in Scotland evolved from a similar scheme to that for England and Wales initiated by the Radium Commission in 1938. Complete registration began in 1947 when reports were obtained from cancer treatment centres, but the scheme was reorganized and computerized in 1958. All hospitals, using local data obtained from medical records and pathology and radiology departments, now register cases with the five regional cancer registries. At the registries additional material from case listings of cancer patients from the Scottish Hospital In-patient System and cancer deaths supplied by the GRO(S) is incorporated, and all the data are checked for accuracy. The registries then forward their data to the Information and Statistics Division for the production and publication of Scottish national statistics.

A limited scheme was set up in Northern Ireland in 1959. The primary sources of registrations are hospital doctors, the system being linked neither locally with pathology laboratories or hospital record systems, nor centrally to the master patient index for Northern Ireland maintained by the Central Services Agency. Information from the GRO(NI) serves to notify new cases and update the records with date of death, but only where cancer has been mentioned on the death certificate. However, the diagnosis is never checked and there is no follow-up of registered cases.

Validity of cancer registration data

Consideration has to be given to three different aspects of validity of cancer registration—the completeness of registration, the accuracy of the particulars registered, and the effectiveness of follow-up—from which statistics of survival are computed. Since 1948 the number of registrations in England and Wales has gradually risen, and it is difficult to determine what proportion of this rise is due to extension of the scheme, alteration in the efficiency of registration, or variation in the incidence of malignant disease. Benjamin (1968a) suggested that, over the country as a whole, it was doubtful whether more than two-thirds of all malignant cases were registered. Some studies have utilized mortality data to check the validity of registration for persons dying from malignant disease; Faulkner *et al.* (1967) suggested that 3.1 per cent of individuals dying from a malignant disease were not known to the registry in Bristol.

Using various sources of records, Alderson (1973) concluded that about 10 per cent of patients might not be registered, a result comparable with that obtained by Gillis (1971) in an examination of the data for the western region of Scotland. Leck *et al.* (1976) reported a 7 per cent deficiency in the registration of childhood cancers in Manchester. Whilst many factors can be associated with 'failure to register', West (1973) and Alderson *et al.* (1976) found that the methods for organizing the system for cancer registration can also have an impact on the overall registration rate.

Turning to the validity of the items recorded, diagnosis is relatively accurate, but errors and omissions of importance occur with items such as date of registration, place of birth, histology, and occupation. It appears that the major problem is variation in the completeness of registration, rather than abstraction of particulars for patients that have been identified; thus interpretation of the data is still feasible. (Alderson 1974; West 1976).

As far as survival statistics are concerned, Silman and Evans (1981) commented on the considerable variation in completeness of registration in different regions and suggested that this might be associated with systemic differences in the prognosis of the registered patients. They questioned whether differences in handling notification of cancer deaths could be a factor, but did not specifically indicate all the possible variations in practice. They suggested that, when analysing survival, the rates based on post-death registrations should be distinguished from those of other patients, and advocated that the national system should record stage of treatment as a classifying variate (this had been deleted from the national scheme in 1971 as a result of the Advisory Committee's Report; OPCS 1970). This potential error in survival statistics has been discussed further in a publication from the Cancer Research Campaign (1982).

Publications on cancer registration data

In England and Wales there are annual OPCS national published statistics from cancer registration (series MB1) for registrations and survival. The Longitudinal Study (p. 38) has also recently been used for this latter purpose (Kogevinas 1990). Cancer registration for the countries as a whole and individual registries also appear in the main international source book of cancer registration *Cancer incidence in five continents* (Muir *et al.* 1987). Scotland first achieved entry in the 1963–6 edition, lost it because of incomplete registration in 1967–70, but thereafter has appeared in successive editions. The cancer registry in Northern Ireland produces only an annual bulletin for internal circulation within the province.

Hospital discharge statistics

Background

A leading article in the *Lancet* (1841) pointed out that data collection on hospital in-patients had been suggested in 1732. In the nineteenth century there were some developments initiated by the Statistical Society (1842, 1844, 1862, 1866), and another notable advocate of collecting and utilizing such data was Florence Nightingale (1863). Limited progress was made in the UK until the period between the two World Wars, when there was a reawakening of interest for large-scale analysis of hospital statistics (Spear and Gould 1937).

With the advent of the NHS in 1948, the opportunity arose to collect more extensive statistics relating to the morbidity aspects of patient care in hospital. After initial trials the Ministry of Health and the General Register Office jointly set up two parallel schemes, one for mental hospitals and the other for the rest of the NHS hospitals, since it had been concluded that differences in length of stay dictated a separate mechanism for collecting data relating to mentally ill or mentally handicapped patients. The need for an enhanced data-set for deliveries eventually led to a partially separate scheme for maternity patients. From 1955 to 1985 the Hospital In-Patient Enquiry (HIPE) collected and processed an anonymous 10 per cent sample of non-psychiatric discharges for England, with Wales also included until 1981. Initially, hospitals completed forms for individual patients and sent them directly to the General Register Office, but latterly the appropriate samples were drawn from Hospital Activity

Analysis (HAA) tapes held by Regional Health Authorities (for original descriptions of the HAA scheme see Benjamin 1965; Rowe and Brewer 1972). Private patients in NHS hospitals were excluded until 1979, and day cases were accepted from 1974, but analysed and published separately from 1975.

The data collected centrally included demographic, administrative, and clinical items. Examples of demographic data were sex, age, marital status, and area of residence. The administrative particulars included the dates when entered on the waiting list, admitted, and discharged (or died)—and hence waiting time and length of stay; source of admission; category of patient (NHS or private); specialty; and disposal. The brief clinical particulars were the main condition treated, together with other relevant conditions, and the provision to record up to four surgical operations, with the principal one designated. Since 1982 similar data have continued to be collected for Wales.

From 1949 to 1960, an individual patient return was also sent to the General Register Office for each admission to, departure from, and death in the main designated hospitals for patients with mental disorder. Following the Mental Health Act 1959, the Ministry of Health undertook the collection of psychiatric statistics for general planning and administrative purposes on an individual patient basis, with the General Register Office enquiry continuing in a modified form. From 1964, the two schemes were combined into the Mental Health Enquiry under the responsibility of the Ministry of Health and all psychiatric hospitals and units were then included. This enquiry was in the form of a two-part submission, the first part of which was returned on admission and which contained the demographic information, reference to previous psychiatric admissions, whether the admission was under the terms of the Mental Health Act, and some medical details. A second submission was completed on discharge and recorded further administrative details and the final diagnosis. The provision for some of the data-set to be submitted on admission permitted analyses relating to hospital 'residents' to be carried out, an essential requirement for information on long-term care.

Although initially combined with the 'general' HIPE, an essentially separate enquiry, again on a 10 per cent basis, covered NHS hospitals and units exclusively treating maternity patients, but not admissions for abortion or other conditions in early pregnancy which were normally included in the non-maternity enquiry. As maternity patients were not routinely included in most regional HAA schemes, latterly the data were assimilated from three different sources. A limited national maternity HAA scheme in selected hospitals provided about 20 per cent; a further 20 per cent came from other local detailed schemes; and the remainder were submitted directly to the OPCS on forms.

The particulars available included the standard demographic, administrative, and clinical items in the non-maternity system, augmented by maternal, delivery, and infant parameters. The maternal and delivery items included the date of the last menstrual period (and hence of gestation); previous pregnancies; diagnosis of general, ante-natal, delivery, or puerperal complications; onset of labour; presentation; and mode and date of delivery. Items related to the infant included sex; birth-weight; outcome; and diseases or abnormalities present. Provision was also made to record such items separately for each of multiple births. For a more detailed description of the scheme, see Ashley (1980).

In England the first report of the Steering Group on Health Services Information (SGHISI) (1982a) recommended unification of the three separate elements and a change in the 'unit of account'—from stays in hospital to consultant episodes. Many data items were reclassified and all were precisely defined. With the emphasis being placed on the needs of local management, arrangements were suggested whereby there would be timely assimilation and availability of items after the 'transaction' to which they related. It was also recommended that a data-set somewhat similar to HIPE should be submitted centrally to the OPCS as the data-keeper for all consultant episodes. In a supplement to the first and fourth reports (SGHSI 1985), an appropriate data-set for maternity episodes was also defined, and again it generally resembled maternity HIPE. General implementation of the provisions of the first report was in April 1987 and the maternity provisions came into force in April 1988.

In Scotland general hospital and mental illness and mental handicap episode data were first collected in 1959. Anonymous summary data on all hospital admissions were collected by the Research and Intelligence Unit of the Scottish Home and Health Department, the precursor of the Information and Statistics Division. Maternity hospital episodes started to be collected in 1969, and a neonatal discharge summary was started in 1975 by a paediatrician who ran it until it came under the governmental management in 1980.

A record (SMR1) form is generated for each discharge or transfer to a different specialty or consultant within a hospital. Identifying data were added in 1968, and from 1975 post-code was used to identify area of residence. In 1978 day cases were specifically collected for the first time. Prior to that year information on day cases had been submitted unsystematically. There have been periodic revisions of the data items collected, but, currently, in addition to personal information, the demographic, administrative, and clinical details are similar to those for England and Wales. The mental illness and mental handicap return (SMR4) is a two-part form with similar context and method of transmission to that used with Mental Health Enquiry in England. Again the file can be used to produce statistics on episodes and on current residents in hospital. The residents file is also periodically checked by a census.

The maternity discharge record (SMR2) is an episode record which contains personal administrative and demographic data, and a maternity oriented section that is more comprehensive than either maternity HIPE or the implemented recommendations of the SGHSI. The section on the past obstetric history also includes the number of previous perinatal deaths and the number of previous Caesarean sec-tions. Data on current pregnancy include the date of booking and the type of antenatal care, the number of previous admissions during pregnancy, and maternal height and blood group. There is also a section for recording details of an abortive outcome to the pregnancy. Although the record of labour is similar to that for England, the details for the baby are enhanced to include apgar scores, and the case reference number of the baby to facilitate linkage to the neonatal discharge record. If the baby dies, the underlying cause of death is also recorded. As well as being capable of analysis on an episode basis, records which include delivery details can be segregated to give pregnancy-based statistics. The system was more fully described by Cole (1980).

The Scottish neonatal discharge record (SMR11) is an individual record, started at birth and continued until the infant is discharged from the neonatal service of a maternity hospital. It has personal and administrative data, including birth-weight, and the maternal case reference number to facilitate linkage to the SMR2. There are data on the condition at birth and if resuscitation was required. Head circumference is recorded, and transfer to special care and length of stay are noted, as is the need for transfer to another neonatal unit. Other clinical observations additional to diagnostic data include if jaundice was present and its severity, and the presence or absence of significant hypotonia, convulsions, recurrent apnoea, assisted ventilation lasting longer than 30 minutes, or feeding difficulties.

Northern Ireland has an HAA system, but collection of diagnostic data is incomplete. There is also a Mental Health Record Scheme, begun in 1960, which covers psychiatric patients. The hospitals concerned complete forms and forward them direct to the DHSS(NI). The forms outline in-patient and out-patient admissions and discharges, including diagnoses for both categories of patient. Mentally handicapped patients are due to come under this scheme in 1990. At present, mental handicap information is returned to the DHSS on a quarterly and annual basis.

Validity of hospital discharge data

Comprehensive studies of hospital discharge data (Lockwood 1971; Martini et al. 1976) have demonstrated a high level of accuracy for demographic and administrative items, although Gruer (1970) and Ashley (1972) questioned the quality of geographical data before the use of post-codes for this purpose. Thus, most exercises that have checked the accuracy of hospital discharge data have concentrated on the diagnostic information. Alderson and Meade (1967) compared hospital discharge data with mortality statistics for patients who had died in hospital, and suggested there were appreciable errors in the principal conditions treated for about 13 per cent of completed forms. Dunnigan et al. (1970) examined 1093 Scottish hospitals records where the discharge diagnosis had been coded to 'heart disease specified as involving coronary arteries', and judged 92 per cent to be appropriately assigned. Other studies on the accuracy of such data have produced a range of opinions: McNeilly and Moore (1975) observed appreciable errors in codes for a small sample of Welsh patients; Patel *et al.* (1976) suggested there was a considerable proportion of errors in Scottish morbidity data in Glasgow, but Parkin *et al.* (1976) seriously questioned some of their conclusions. Martini *et al.* (1976) quantified the errors for patients treated in the Nottingham area and pointed out that the statistics were almost as good as the clinical notes from which they were derived. Cameron *et al.* (1977) contrasted autopsy findings with data from the medical records, and indicated that fresh findings might arise at autopsy that were not incorporated in the Scottish statistical system.

Turning to other parameters, a series of small studies (Crawshaw and Moss 1982; Hole 1982; Rees 1982; Whates *et al.* 1982) quoted appreciable error rates for specific operative data in HAA, but Butts and Williams (1982) questioned some of the rather general conclusions reached. Forster and Mahadevan (1981) compared the data on the Mental Health Enquiry forms of 824 mental health patients in the north of England with that in their records. There were serious discrepancies for source of referral (20.5 per cent), outcome (10.8 per cent), and previous psychiatric care (6.3 per cent). In checking the validity of Scottish maternity data, Cole (1980) identified some clerical misconceptions from internal inconsistency in the computer-held data. In a comparison between the statistical return and research data abstracted from 1000 records, there was close agreement (97–98 per cent) on factual items such as past obstetric history, but about 70 per cent of discrepancies on items such as 'certainty of gestation'.

Perhaps as a result of the equivocal nature of the evidence about the accuracy of hospital morbidity data, much more emphasis is now being placed on the audit of data quality as envisaged by the SGHSI (1982*b*).

Linkage of repeat events

Reference has already been made to the value of record linkage event data (see p. 37), and the general concept applies equally to linkage of morbidity records. Acheson (1967) has described the history of medical record linkage and the pioneering efforts in assembling a cumulative record of hospital admissions in the Oxford Region. He described the system used, and the medical care and other epidemiological studies that could be facilitated by such work. The only element of linkage involving morbidity records that occurs nationally in England and Wales is the tracing of individuals with cancer registered in the NHS Central Register, so as to identify deaths that occur and permit calculation of survival rates (see p. 41).

The medical record linkage facility developed in Scotland has been described by Heasman and Clarke (1979). The objects are: (i) to provide economical collection of events occurring to an individual to permit statistical analysis; (ii) to enable individuals to be followed up; and (iii) to produce person-based statistics. The Scottish system includes: (i) general hospital discharge records from 1968; (ii) obstetric discharge records from 1970; (iii) mental and mental deficiency hospital admissions and discharges from 1963; (iv) cancer registrations from 1968; (v) school entrant and leaver medical examinations from 1967; (vi) handicapped children's register from 1973; and (vii) death and still birth registrations from 1968. An important point is that the system contains linkable records but only assembles a linked file for a particular (restricted) purpose, such as for epidemiological and health services research, and for follow-up of specific groups of individuals. Although the intention is to facilitate production of person-based statistics, these have yet to be published as a routine; however, it is thought that the system would enable examination, for example, of length of stay for either consecutive spells in different hospitals or multiple admissions into the same hospital. The authors stressed the importance of considering the issues of privacy and steps taken to ensure confidentiality of the data.

Publication of hospital discharge data

The division between 'acute', maternity, and psychiatric events was reflected in the publication policy for England and Wales. Latterly, the data relating to the first two of these types of event were published as HIPE jointly by the DHSS and OPCS (and by the Welsh Office until 1981) in annual OPCS volumes in the MB4 series. Psychiatric events were published by the DHSS as the Mental Health Enquiry within its Statistics and Research Report Series. No central publications are yet available from the Hospital Episode Statistics system which replaced HIPE and The Mental Health Enquiry from 1987. Corresponding publications for Scotland are produced by the Information and Statistics Division, and data for Wales since 1982 can be obtained from the Welsh Office, but no directly equivalent national data are published for Northern Ireland.

Morbidity statistics from general practice
Background

In 1951 a pilot study was launched to test methods for collecting and analysing medical records kept by a small number of general practitioners. From this work, Logan (1953) concluded that it was possible, without exceptional difficulty, to keep records over a 12-month period that were suitable for analysis. Various alternative methods of recording were advocated, but when this study was extended to 1954 a specially modified NHS continuation card was used.

The RCGP with the General Register Office initiated what was to become the first National Morbidity Study from 1955–6, involving 171 doctors and 106 practices in England and Wales. A standard method of recording limited particulars was used for each patient contact, including identification particulars of the individual, date of birth, diagnosis, date of consultation, and date of admission to hospital if this occurred.

A second National Morbidity Study was carried out in 1970–1, with the aim of collecting data compatible with the 1955–6 study (RCGP *et al.* 1974). In addition to morbidity information, particulars about the use of community and hospital facilities were recorded, and the patients' consultation patterns were identified. For each episode of care, the name and date of birth were recorded together with the date of first consultation and the episode type. Up to six face-to-face consultations could be coded on the same line of the record sheet, with continuation if required. Referral to other agencies was also identified. Fifty-four practices were initially involved, although one withdrew during the survey. Unlike the previous study, diagnostic and other information was coded at practice level, rather than centrally, using the RCGP classification for coding morbidity. This classification was modified to conform to the 8th revision of the ICD, having been derived initially from the 7th revision.

A linkage exercise with the census was also carried out to relate census variables to morbidity rates. Using census schedules for the enumeration districts that represented the catchment area of the practices, a computer matching was carried out without access to full names of the subjects in the study (RCGP *et al.* 1982). The non-matches from the age-sex registers were then examined clerically and possible matches were added to the computer file. This resulted in a 65–75 per cent initial match for most practices with up to a further 20 per cent added by the clerical exercise. The number of enumeration districts was extended for the few practices with a low match rate. The final proportion of the practice register population that could be matched was 78 per cent. Tabulations were presented on morbidity in relation to population characteristics; marital status; country of birth; urban/rural residence; social class; occupation order; housing tenure; and household amenities.

The third National Morbidity Study (RCGP *et al.* 1986) took place in 1981–2 and involved 48 practices providing care for over 300 000 patients. Information similar to the previous studies was collected, but the diagnostic classification was modified to conform with the ICD. Linkage of patient records to census data was again carried out for a proportion of the practices.

A number of other studies from general practice have collected data on contact with patients; these vary from pioneer efforts where individual doctors or groups of doctors have invested effort to codify material on the patients they see, up to major collaborative studies, such as that carried out by 68 doctors in South Wales in 1965–6 (Williams 1970). In the various surveys the practices were taken mainly from England and Wales. There has not been Scottish input throughout, but in the second national study three practices from Northern Ireland (representing a population of 15 000) took part.

Validity of data

A vital element in such studies is the count of the number of individuals on the doctor's NHS practice list, and the age distribution of these subjects; this is required as the denominator in calculating rates. Lees and Cooper (1963) identified problems of definition, and net change in the practice list.

The first National Morbidity Study initially showed practice list inflation of 1.1 per cent, although this was reduced to 0.7 per cent on investigation. Other studies have identified variation of 20 per cent (Backett *et al.* 1954) and 14 per cent (Morrell *et al.* 1970). Less extreme variation was found by Fraser (1978) and by Fraser and Clayton (1981). Another issue is the extent to which contact with a general practitioner indicates the prevalence of morbidity in the population. A number of studies have indicated that there is a varying degree of self-care and undetected disease (Cartwright 1967; Horder and Horder 1954; Kessel and Shepherd 1965; Last 1963; Wadsworth *et al.* 1971).

Equally important is the validity of data recording, as the accuracy of the analyses may be constrained by errors in the capture, coding, or processing of the material. This has been discussed by Morrell and his colleagues in a series of papers (Morrell 1972; Morrell *et al.* 1970, 1971; Morrell and Kasap 1972; Morrell and Nicholson 1974), and other aspects have been investigated by Clarke and Bennett (1971), Dawes (1972), Farmer *et al.* (1974), Hannay (1972), Kay (1968), and Munroe and Ratoff (1973). The effect of different systems of classification has been examined by Martini *et al.* (1976), who highlighted the contrast between presenting problems and underlying morbidity.

In the second National Morbidity Study the validity of the practice registers was checked, indicating a true inflation of about 1.1 per cent (RCGP *et al.* 1974). An attempt was made to check the morbidity data, which indicated some diagnostic problems and variation between the practices; coding errors resulted in only marginal variation in the morbidity that was recorded. It appeared that about 3.5 per cent of consultations and 2.4 per cent of episodes recorded on practice records were not on the computer records, whilst 0.35 per cent of the events on the computer file could not be identified in the practice notes. There was no independent check of the diagnostic labels used as specific contacts in any of the practices. A similar post-study evaluation exercise was carried out on the third National Morbidity Study (RCGP *et al.* 1986), with broadly similar results. The under-reporting of consultations was 2.2 per cent, and of episodes was 0.8 per cent. There were greater discrepancies for home visits and referrals.

It is not known how typical are studies which involve volunteer general practitioners. For example, among the general practitioners who volunteered for the second National Morbidity Study there was a slight excess from practices with more than four principals, from those with large list sizes, and from younger doctors, and of those practising from health centres or with ancillary help. Crombie (1973) has suggested that the diagnostic patterns and differences in the case-mix that a family doctor draws towards him or her can have some influence, but provided the number of doctors is sufficiently large, the overall population that is served approximates closely to the general population. A report (RCGP *et al.* 1979) examined some of the characteristics of the practices participating in the second survey and concluded that morbidity rates were likely to vary in relation to the location of the practice, size of partnership, doctor's age, and availability of ancillary staff.

The exercise whereby data from the second National Morbidity Study were linked to the census also facilitated a check of the representativeness of the study sample. The matched sample showed some deviation from the characteristics of the total study sample: the older age groups were slightly under-represented, as were the widowed and divorced. In contrast, social classes I, II, III non-manual, and those in owner-occupied housing were over-represented. It was concluded that these differences were insufficient to introduce bias such that the consultation rates of the census-matched sample were appreciably distorted.

Crombie and Fleming (1986) compared consultation rates for the second National Morbidity Study with those derived from the General Household Survey of 1971. They found a broad level of agreement, despite the different approaches of the two surveys.

Publications of general practice morbidity

The results from the first National Morbidity Study appeared in three reports (Logan and Cushion 1958; Logan 1960; Research Committee of the Council of the College of General Practitioners 1962). Later publications were issued jointly by the RCGP, OPCS, and DHSS.

Congenital malformations

Background

Two pioneering local schemes for collecting information about congenital malformations began in Birmingham in 1949 (Charles 1951) and in Liverpool in 1960 (Smithells 1962). Together, these covered only about 20 000 births in each year, but were sufficient to study the more common serious abnormalities.

In part stimulated by the thalidomide epidemic in 1960, a national scheme was launched in England and Wales in 1964 in which the doctor or midwife notifying a birth to the local Medical Officer of Health was asked to include particulars about any identifiable congenital abnormality. A standard form was completed by the Medical Officer of Health in respect of every child with an observable malformation identified at birth or within 7 days thereof. Each Medical Officer of Health forwarded to the GRO forms relating to mothers resident in their area, with appropriate arrangements being made where the mother was resident elsewhere.

The process was described in detail by Weatherall (1978) and, subject to changes in the channels of communication resulting from various reorganizations of the NHS, the method of data acquisition is essentially unchanged. As well as descriptions of all abnormalities present, details required include, for example, maternal date of birth, whether a multiple birth, whether a live or still birth, length of gestation, birth-weight, and place of birth.

A similar scheme has been operating in Northern Ireland since 1964. Area Boards complete a separate form for each case and transmit it to the DHSS(NI), which completes statistics for the province. In Scotland there is no separate reporting scheme, but a congenital malformation register, started in 1988 by the Information and Statistics Division, uses data from SMR11, supplemented by information derived from records of hospital episodes for infants aged less than 1 year.

As the main aim of the schemes is to monitor trends in malformation in England and Wales, there are two separate aspects to the analyses: (i) the routine tabulation of the data for the publication of statistics; and (ii) the statistical examination of the material to identify significant variations in reporting. Tabulations relating to reported malformations are sent monthly to each District Health Authority, which is also warned if the actual numbers of a particular malformation are greater than the expected number derived from the rate for that malformation in that month in the country as a whole. A comparison is also made with the number of the same malformation reported from the locality in the previous time period. Once a given malformation has been identified as having increased, the level in the locality is reviewed monthly for the following 11 months; if the reporting remains significantly higher, a further warning is automatically generated within a further 3 months.

Validity of congenital malformation data

As one objective is to detect 'epidemics' by finding changes in reporting, some degree of incompleteness does not materially affect the value of this work. However, it is of obvious importance to know the degree to which malformations are identified, and also whether diagnoses are appropriately classified. By comparing the notification of anencephaly with the number of deaths recorded for this condition, Weatherall (1969) suggested that notification was fairly complete for this (serious and obvious) condition.

In Exeter and Devon a more thorough investigation of the child population has indicated that the proportion of malformations found and reported in the national scheme varies with the type of malformation. The levels detected for all but a few internal malformations were thought to be sufficient to expose any increase in incidence (Vowles et al. 1975). In a case–control study Greenberg et al. (1975) found, from a sample of 2867 notifications, that further enquiry by the local health authority indicated that 77 were normal babies who should not have been included. Ericson et al. (1977) compared the information provided by: (i) specific notifications of congenital malformations, identified at birth; and (ii) computer records of routine birth diagnosis for all infants. Both systems appeared to 'lose' about 20 per cent of the malformed infants, but the quality of diagnostic information was better for the specific notification system, and the speed of data transmission was greater.

Nevin et al. (1978) compared the system for notifying congenital malformation in Northern Ireland with data recorded in the Child Health System and other sources (e.g. autopsy records, the regional spina bifida clinic, and the local association counselling clinics for spina bifida and hydrocephalus). Of a total of 686 infants with neural tube defects identified in 1974–6, 83.6 per cent were found in the notification system, whilst only 63.3 per cent were found in the Child Health System.

Initially, annual volumes were not published by the OPCS for this topic for England and Wales, quarterly and annual material being issued in the monitor series (MB3). There were, however, occasional analyses covering a longer period, the latest volume covering the quinquennium 1981–5 (OPCS 1988), with tabular material and maps provided for each malformation giving notification rates by a variety of parameters. It is now intended that this will be an annual volume commencing with the information for 1987.

For Northern Ireland some material appeared in the annual reports of the Registrar General until 1975, and latterly has been published by the DHSS(NI) in its annual Health and Personal Social Service Statistics. An informal International Clearing House for Birth Defects Monitoring System (1981) also exists, to which England and Wales and Northern Ireland contribute data and from which an annual report is produced, e.g. for 1985.

Abortion
Background
The Abortion Act 1967 came into force in Great Britain in April 1968, but there is no corresponding Act in Northern Ireland. This Act requires notifications of termination of pregnancy to be made within 7 days on a standard form as presented in subsequent regulations (Statutory Instruments 1968) which is sent to the Chief Medical Officer of England, Wales, or Scotland. Amendment regulations were made in 1969 (SI 1969) and 1976 (SI 1976). Further amendments in 1980 (SI 1980a, 1980b) made certain adjustments to the content of the notification forms for both England and Wales and for Scotland.

The aim of the information notified to the Chief Medical Officers is primarily to ensure that the requirements of the Act are being properly observed, but provision is also made for subsequent processing by the OPCS (for England and Wales) and the Information and Statistics Division (for Scotland), acting as agents. This enables basic epidemiological and demographic data about women undergoing terminations to be collected and regularly published. It is interesting to note that in this system the particulars are recorded in the unit where the termination takes place and the forms are transmitted direct to central Government, with no provision for collating or checking the material at local or regional level.

The notification forms for England and Wales and for Scotland are similar. They include such items as the age of the woman, her place of residence, marital status, and the number of her previous children. Also included are the length of gestation, the statutory grounds for termination, and the method used. These items are utilized in a range of cross-tabulations.

Publications on abortion
Preliminary information for England and Wales is published quarterly by the OPCS as an Abortion Monitor (AB series), which provides relatively quick issue with limited tables. More detailed statistics are presented in an annual publication in the AB series. Material for Scotland, which includes terminations carried out in England and Wales on Scottish residents, is published in the Scottish Health Statistics and in the Annual Report of the Registrar General for Scotland.

Household surveys

Routine information about morbidity derived from the kind of events described in the previous section has the inbuilt disadvantage that it is mostly associated with a contact between a patient and a health professional, usually a doctor. It thus fails to cover minor or other illnesses not associated with consultation, information about which can be obtained only by direct enquiry of population sample. Thus, traditionally, household interview surveys conducted within the UK have, in recent years, included such enquiries. Furthermore, the opportunity can be taken to use this mechanism to provide corroborative evidence as the accuracy of alternative sources of information about the use of health services, such as outpatient consultations.

Background
From 1944 to 1953 a monthly Survey of Sickness was carried out in England and Wales, questioning a sample of people about their health in the preceding 2–3 months (see Stocks 1949). The sample size was initially 2500 persons aged from 16 to 64 years, but increased to 4000 with the addition of those aged over 64 years.

The final report of the survey (Logan and Brooke 1957) had a considerable section devoted to some of the problems of the survey, with a detailed examination of the bias introduced by the memory factor. This was followed by a general critical appraisal which considered the validity and general usefulness of the data that were obtained. One interesting issue is the comment that the tables provided became heavily weighted with data on minor and trivial illness and, in particular, that it was unrewarding to examine for time trends. Apart from the number of comments about interpretation of the data, it was emphasized that the survey's positive contribution was to identify the large amount of ill-health which people suffer for which no medical advice is sought. It was concluded that the main contribution of such a survey should not be to provide a permanent operation identifying the total load of ill-health; it is more appropriate, as the need arises, to quantify particular items of information on illness and its effects that cannot be readily obtained in a routine way.

A major alteration occurred in the collection of data in Great Britain with the initiation of a General Household Survey in 1971 (OPCS 1973b). The aim of this survey was to provide a substantially improved flow of social statistics to complement the wide range of routine material currently processed by central Government, and to develop an instrument to examine the interaction between different policy areas. The intention was to create a survey that was developed in close collaboration with a number of Government departments in order to link their needs for information on population, housing, employment, education, health, and social services. Consideration of the predominant objective of the survey indicated that a very long interview would be required with all respondents; at the same time there were constraints on resources available. These two factors markedly influenced the study design. It was essential that respondents co-operated willingly in order to obtain the detailed answers called for in the interview.

A nationally representative sample is required, and the current design is a two-stage rotating stratified one, with electoral wards forming the primary sampling units. The sampling is similar for the whole of Great Britain, but the number of households contacted is doubled for Scotland to provide the minimum of precision thought necessary for separate analysis. The sampling procedure identifies addresses from the electoral register; the interviewer converts the list of addresses to the identification of households. For addresses containing more than one household, set procedures are used to try to ensure that there is the correct probability of selection of multi-households (see OPCS 1982c). The initial report provided considerable detail on the sampling errors of the survey; further detailed consideration of this appears in the report for 1976 (see OPCS 1978).

This social survey uses a detailed, structured interview which was tested in pilot trials during the development phase; throughout, the interviewers have been carefully trained and supervised. The minimum response rate (i.e. completely co-operating households) in 1971 was 71 per cent; including all partial responses, a maximum rate of 85 per cent was achieved. The corresponding figures for 1985 were 69 per cent and 84 per cent respectively, based on a total effective sample of nearly 12 000 households (excluding the supplementary Scottish sample). There appears to be marked stability in the response rate from month to month, but modest variation occurs across the country. The reports regularly provide comment on the quality of the information provided. The simplest checks that can be carried out on the data are by comparison with other material or by examination of internal consistency of response. For example, comparison with the 1971 census data suggested that the General Household Survey provides good representation of the population in private households. However, an appreciable proportion of the morbidity in the total population is amongst residents in institutions.

An analogous Continuous Household Survey, based on a sample of the general population resident in private households in Northern Ireland, has been running since 1983. The nature and aims of the Continuous Household Survey are similar to those of the General Household Survey, covering the subject areas of population, housing, employment, education, health, and social services. It provides information primarily for Government purposes, but there are plans in collaboration with local universities to make data-sets available to academics and other researchers, perhaps within the next 5 years. Sampling and other methods used are similar to those of the General Household Survey.

The intention was always to change the items included in the General Household Survey in response to users' requirements. Over time, the health section has collected data on activity limitation caused by acute or chronic sickness and contacts with health and personal social services. These have regularly included questions on consultations with doctors, and patient attendance at hospitals and details of in-patient care. In some years, questions on the use of other health and personal social services by the elderly have also been included, as have enquiries regarding sight and hearing. Also, of direct relevance to health, has been the establishment of patterns of smoking and alcohol consumption. The main changes in content are detailed cumulatively in the published volumes and those for the past decade are summarized in Table 3.4, which also shows the general compatibility of regular and occasional topics between the General and the Continuous Household Surveys. Each annual volume of the General Household Survey includes the questionnaire used and from which the exact nature of definitions can be determined.

An important point to emphasize is that the nature of the General Household Survey provides a powerful analytical tool, by virtue of its ability to cross-tabulate the health data against the wide range of terms collected on 'social' topics. Some of these topics overlap with material collected at census so as to provide continuous information, and include demographic particulars about the population and information on housing, employment, education, income, and family structure.

Validity of General Household Survey data

All the material is collected by direct questioning and no independent validation of the data occurs (such as detailed study of a subsample of the respondents, including medical investigations). The influence of memory bias in probing for events in the past has been indicated above. Obviously some of the general points about validity of surveys that record morbidity statistics apply (see Alderson and Dowie 1979 for a review of this topic).

A Scandinavian study of the validity of health contact data (Brorsson and Smedby 1982) showed false-positive reporting of 12.8 per cent and false-negative reporting of 13.4 per cent. The misclassification of reported visits varied with age and self-reported state of health, but it is not known whether this applies equally to the UK.

Table 3.4. Major items covered in the health section of the General Household Survey, Great Britain, 1977–86

Item	1977	1978	1979	1980	1981	1982	1983	1984	1985	1986	1987
Chronic sickness	–	–	+	+	+	+	++	++	++	++	++
Acute sickness in the last 2 weeks	–	–	+	+	+	+	++	++	++	++	++
Health in general in the last 12 months	+	+	+	+	+	+	++	++	++	++	++
GP consultation in past 2 weeks	+	+	+	+	+	+	++	++	++	++	++
Out-patient attendance in past 3 months	+	+	+	+	+	+	++	++	++	++	++
In-patient stays in past year	–	–	–	–	–	+	++	++	++	++	++
Use of various health and welfare services in past month	–	–	(+)	(+)	(+)	(+)	(+)	(++)	(++)	–	–
Difficulty with sight	+	+	+	(+)	+	+	–	–	(++)	–	–
Difficulty with hearing	+	+	+	(+)	+*	–	–	–	(++)	–	–
Dental health	–	–	–	–	–	–	+	–	++	–	++
Accidents at home	–	–	–	–	+	–	–	++	–	–	++
Smoking	–	+	–	+	–	+	–	++	–	++	–
Alcohol consumption	–	+	–	+	–	+	–	++	–	++	–

+, Included.
++, Also included in Continuous Household Survey in Northern Ireland.
–, Not included.

( ), Included only for the elderly.
* Additional information collected on tinnitus.

Publications from the General Household Survey

The General Household Survey is published annually by the OPCS, with sections relating to each of the topics covered. Of more recent innovation are annual monitors (series SS) with preliminary results. Annual monitors are also produced from the Continuous Household Survey by the Policy and Planning Research Unit of the Department of Finance and Personnel in Northern Ireland.

Other information sources

Some aspects have not been covered in this chapter. For example, the definition of the UK has been taken to include only England, Wales, Scotland, and Northern Ireland. Some of the sources mentioned have counterparts in the Channel Isles and the Isle of Man. For example, regular censuses are carried out on each island, the most recent being in 1981 on the Isle of Man and in 1986 on both Jersey and Guernsey. Similarly, regular mortality information can be found in the annual reports of the Medical Officers of Health of the Channel Isles and the Chief Registrar of the Isle of Man.

Over time some other national, but limited, sources of morbidity information have been available, such as occupational disease and injury data, issued by the Health and Safety Executive, and on sickness absence statistics, hitherto published by DHSS.

Similarly, there are a wide range of sources of social information with some relevance to health, for example the OPCS National Food Survey and the Family Expenditure Survey give food and nutrition data. Also, information is available on occupation from the OPCS Labour Force Survey, which may contribute information either in its own right or provide appropriate denominators. Many of the more topical aspects of these kinds of information are presented annually in the Central Statistical Office's publication *Social trends*.

Acknowledgements

The authors gratefully acknowledge the valuable contribution of the late Dr Michael Alderson to this chapter.

References

Abramson, J.H., Sacks, M.I., and Cahana, E. (1971). Death certification data as an indication of the presence of certain common diseases at death. *Journal of Chronic Diseases* **24**, 417.

Acheson, E.D. (1967). *Medical record linkage*. Oxford University Press, London.

Alderson, M.R. (1965). *The accuracy of the certification of death, and the classification of underlying cause of death from death certificate*. MD thesis, University of London, London.

Alderson, M.R. (1973). Cancer registration. In *Cancer priorities* (ed. G. Bennette) p. 101. British Cancer Council, London.

Alderson, M.R. (1974). Central Government routine health statistics. In *Reviews of United Kingdom statistical sources* (ed. W.F. Maunder) Vol. II. p. 1. Heineman, London.

Alderson, M.R. (1981). *International mortality statistics*. Macmillan, London.

Alderson, M.R. (1984). A comment on social class analysis. *Community Medicine* **6**, 1.

Alderson, M. and Ashwood, F. (1985). Projection of mortality rates for the elderly. *Population Trends* **42**, 22.

Alderson, M.R. and Dowie, R. (1979). *Health surveys and related studies*. Pergamon, Oxford.

Alderson, M.R. and Meade, T.W. (1967). Accuracy of diagnosis on death certificates compared with that in hospital records. *British Journal of Preventive and Social Medicine* **21**, 22.

Alderson, M.R., Bayliss, R.I.S., Clarke, C.A., and Whitfield, A.G.W. (1983). Death certification. *British Medical Journal* **287**, 444.

Alderson, M.R., Bradley, K., Rushton, L., and Thaker, P. (1976).

Cancer registration as a by-product of hospital activity analysis. *Hospital and Health Service Review* **72**, 118.

Armitage, R.I. (1986). Population projections for English local authority areas. *Population Trends* **43**, 31.

Ashley, J.S.A. (1972). Present state of statistics from hospital in-patient data and their uses. *British Journal of Preventive and Social Medicine* **26**, 135.

Ashley, J.S.A. (1980). The maternity hospital in-patient enquiry. In *Perinatal audit and surveillance* (ed. I. Chalmers and G. McIlwaine) p. 61. Royal College of Obstetricians and Gynaecologists, London.

Ashley, J. and McLachlan, G. (ed.) (1985). *Mortal or morbid? A diagnosis of the morbidity factor*. Nuffield Provincial Hospitals Trust, London.

Backett, E.M., Heady, J.A., and Evans, J.C.G. (1954). Studies of a general practice. The doctor's job in an urban area. *British Medical Journal* **i**, 109.

Benjamin, B. (1965). Hospital activity analysis. *The Hospital* **61**, 221.

Benjamin, B. (1968a). *Demographic analysis*. Allen and Unwin, London.

Benjamin, B. (1968b). *Health and vital statistics*. Allen and Unwin, London.

Britton, M. (ed.) (1990). *Mortality and geography. A review in the mid-1980s, England and Wales*. HMSO, London.

Britton, M. and Birch, F. (1985). *1981 Census: Post Enumeration Survey—An inquiry into the coverage and quality of the 1981 Census in England and Wales*. HMSO, London.

Brodrick, N. (1971). *Report of the committee on death certification and coroners* (Cmd 4810). HMSO, London.

Brorsson, B. and Smedby, B. (1982). Validity of health survey interview data concerning visits to doctors at Swedish health centre. *Stat Tidsksrift* **1**, 31.

Busuttil, A., Kemp, I.W., and Heasman, M.A. (1981). The accuracy of medical certificates of cause of death. *Health Bulletin* **39**, 146.

Butts, M.S. and Williams, D.R.R. (1982). Accuracy of hospital activity analysis data. *British Medical Journal* **285**, 506.

Cameron, H.M. and McGoogan, E. (1981). A prospective study of 1152 hospital autopsies. *Journal of Pathology* **133**, 273.

Cameron, H.M., Clarke, J., and Melville, A. (977). Autopsies and medical records. *Health Bulletin* **35**, 113.

Cancer Research Campaign, Cancer Statistics Group (1982). *Trends in cancer survival in Great Britain*. Cancer Research Campaign, London.

Cartwright, A. (1967). *Patients and their doctors—a study of general practice*. Routledge Kegan Paul, London.

Charles, E. (1951). Statistical utilisation of maternity and child welfare records. *British Journal of Social Medicine* **5**, 41.

Clarke, C. and Bennett, A.E. (1971). Problems in the measurement of hospital utilisation. *Proceedings of the Royal Society of Medicine* **64**, 795.

Clarke, C. and Whitfield, A.G.W. (1978). Death certification and epidemiological research. *British Medical Journal* **ii**, 1063.

Clarkson, J.A. and Fine, P.F.M. (1985). The efficiency of measles and pertussis morbidity reporting in England and Wales. *International Journal of Epidemiology* **14**, 153.

Cohen, J. and Steinitz, R. (1969). Underlying and contributory causes of death of adult males in two districts. *Journal of Chronic Diseases* **22**, 17.

Cole, S. (1980). Scottish maternity and neonatal records. In *Perinatal audit and surveillance* (ed. I. Chalmers and G. McIlwaine) p. 39. Royal College of Obstetricians and Gynaecologists, London.

Cole, S.K. (1989). Evaluation of a neonatal discharge record as a monitor of congenital malformations. *Community Medicine* **11**, 1.

Crawshaw, C. and Moss, J.G. (1982). Accuracy of HAA operation codes. *British Medical Journal* **285**, 210.

Crombie, D.L. (1973). Research and confidentiality in general practice. *Journal of the Royal College of General Practitioners* **23**, 863.

Crombie, D.L. and Fleming, D.M. (1986). Comparison of Second National Morbidity Study and General Household Survey 1970–71. *Health Trends* **18**, 15.

Davies, P.D.O., Darbyshire, J., Nunn, A.J., *et al.* (1981). Ambiguities and inaccuracies in the notification system for tuberculosis in England and Wales. *Community Medicine* **3**, 108.

Dawes, K.S. (1972). Survey of general practice records. *British Medical Journal* **iii**, 219.

Daykin, C. (1986). Projecting the population of the United Kingdom. *Population Trends* **44**, 28.

Diehl, A.K. and Gau, D.W. (1982). Death certification by British doctors: A demographic analysis. *Journal of Epidemiology and Community Health* **36**, 146.

Dollinger, J. (1907). Statistique des personnes attientes de cancer. *Publication Statistique Hongerie, Nouvelle Serie 19*, Budapest.

Dunnigan, D.G., Harland, W.A., and Fyfe, T. (1970). Seasonal incidence and mortality of ischaemic heart disease. *Lancet* **ii**, 793.

Edouard, L. (1982). Validation of the registered underlying cause of stillbirth. *Journal of Epidemiology and Community Health* **36**, 231.

Ericson, A., Kallen, B., and Winberg, J. (1977). Surveillance of malformations at birth: A comparison of two record systems run in parallel. *International Journal of Epidemiology* **6**, 35.

Farmer, R.D.T., Knox, E.G., Cross, K.W., and Crombie, D.L. (1974). Executive council lists and general practitioner files. *British Journal of Social and Preventive Medicine* **28**, 49.

Farr, W. (1854). Letter to the Registrar General. In *13th annual report of the Registrar General of Births, Deaths and Marriages in England*, p. 129. HMSO, London.

Farr, W. (1861). *Report of the committee on the preparation of army medical statistics, and on the duties to be performed by the statistical branch of the army medical department*. British Parliamentary Papers XXXVII.

Farr, W. (1864). Hospital mortality. *Medical Times Gazette* **i**, 242.

Faulkner, K., Leyland, L., and Wofinden, R.C. (1967). Cancer registration. *Medical Officer* **118**, 147.

Forster, D.F. and Mahadevan, S. (1981). Information sources for planning and evaluating adult psychiatric services. *Community Medicine* **3**, 160.

Fox, A.J. and Goldblatt, P.O. (1982). *Longitudinal study: Socioeconomic mortality differentials, 1971–75 England and Wales*. OPCS, London.

Fraser, R.C. (1978). The reliability and validity of the age–sex register as a population denominator in general practice. *Journal of the Royal College of General Practitioners* **28**, 283.

Fraser, R.C. and Clayton, D.G. (1981). The accuracy of age–sex registers, practice medical records and family practitioner committee registers. *Journal of the Royal College of General Practitioners* **31**, 410.

Gau, D.W. and Diehl, A.K. (1982). Disagreement among general practitioners regarding cause of death. *British Medical Journal* **284**, 239.

Gedalla, B. and Alderson, M.R. (1984). Pilot study of revised stillbirth and neonatal death certificates. *Archives of Diseases in Childhood* **59**, 976.

Gillis, R.C. (1971). *9th annual report of the Regional Cancer Committee for 1968*. Western Regional Hospital Cancer Registration Bureau, Glasgow.

Goldacre, M.J. and Miller, D.L. (1976). Completeness of statutory notification for acute bacterial meningitis. *British Medical Journal* **ii**, 501.

Gray, P. and Gee. F.A. (1972). *A quality check on the 1966 ten per cent sample census of England and Wales*. HMSO, London.

Great Britain (1988*a*). *1991 Census of Population* (CM430). HMSO, London.

Great Britain (1988*b*). *Registration: A modern service* (CM531). HMSO, London.

Great Britain (1990). *Registration: proposals for change* (CM939). HMSO, London.

Greenberg, G., Inman, W.H.W., Weatherall, J.A.C., and Adelstein, A.W. (1975). Hormonal pregnancy tests and congenital malformations. *British Medical Journal* **ii**, 191.

Gruer, R. (1970). Hospital discharges in relation to area of residence. *British Journal of Preventive and Social Medicine* **24**, 124.

Guralnick, L. (1966). Some problems in the use of multiple causes of death. *Journal of Chronic Diseases* **19**, 979.

Hannay, D.R. (1972). Accuracy of health-centre records. *Lancet* **ii**, 371.

Haward, R.A. (1973). Scale of under notification of infectious diseases by general practitioners. *Lancet* **i**, 873.

Heady, J.A. and Heasman, M.A. (1959). *Social and biological factors in infant mortality*. General Register Offices Studies on Medical and Population subjects No 15. HMSO, London.

Heasman, M.A. and Clarke, J.A. (1979). Medical record linkage in Scotland. *Health Bulletin* **37**, 97.

Heasman, M.A. and Lipworth, L. (1966). Accuracy of certification of cause of death. *General Register Office studies of medical and population subjects*, No. 20. HMSO, London.

Heasman, M.A., Liddell, F.O.K., and Reid, D.D. (1958). The accuracy of occupational vital statistics. *British Journal of Industrial Medicine* **15**, 141.

Hole, R. (1982). Accuracy of HAA operation codes. *British Medical Journal* **285**, 210.

Horder, J. and Horder, E. (1954). Illness in general practice. *Practitioner* **173**, 177.

House of Commons (1893). *First and second reports from the Select Committee on death certification*. HMSO, London.

International Clearing House for Birth Defects (1981). A communication from the International Clearing House for Birth Defects monitoring systems. *International Journal of Epidemiology* **10**, 245.

International Clearing House for Birth Defects (1985). *Annual Report 1985*. International Clearing House for Birth Defects, San Francisco, California.

Jennings, C. and Barraclough, B. (1980). Legal and administrative influences on the English suicide rate since 1900. *Psychological Medicine* **10**, 407.

Jones, I.G. and Cameron, D. (1984). Social class analysis—an embarrassment to epidemiology. *Community Medicine* **6**, 37.

Kay, C.R. (1968). A comparison of two methods of determining social status. *Journal of the Royal College of General Practitioners* **16**, 162.

Kessel, W.I.N. and Shepherd, M. (1965). The health and attitudes of people who seldom consult a doctor. *Medical Care* **3**, 6.

Kogevinas, E. (1990). *Longitudinal study 1971–83. Socio-demographic differences in career survival*, Series LS No. 5. HMSO, London.

Lambert, P.M. (1973). Recent trends in meningococcal infection. *Community Medicine* **129**, 279.

Lancet (1841). Hospital physicians and surgeons. *Lancet* **i**, 649.

Last, J.M. (1963). The iceberg: 'Completing the clinical picture' in general practice. *Lancet* **ii**, 28.

Leck. I., Birch, J.M., Marsden, H.B., and Steward, J.K. (1976). Methods of classifying and ascertaining children's tumours. *British Journal of Cancer* **34**, 69.

Lees, D.S. and Cooper, M.H. (1963). The work of the general practitioner. *Journal of the College of General Practitioners* **6**, 408.

Lockwood, E. (1971). Accuracy of Scottish Hospital Morbidity data. *British Journal of Preventive and Social Medicine* **25**, 76.

Logan, W.P.D. (1953). General practitioners' records. *General Register Office studies on medical and population subjects*, No. 7. HMSO, London.

Logan, W.P.D. (1960). Morbidity statistics from general practice—Volume II (occupation). *General Register Office studies on medical and population subjects*, No. 14. HMSO, London.

Logan, W.P.D. and Brooke, E.M. (1957). The survey of sickness—1943–1952. *General Register Office studies on medical and population subjects*, No. 12. HMSO, London.

Logan, W.P.D. and Cushion, A.A. (1958). Morbidity statistics from general practices—Volume I (general). *General Register Office studies on medical and population subjects*, No. 14. HMSO, London.

McCormick, A. (1988). Trends in mortality statistics in England and Wales with particular reference to AIDS from 1984 to April 1987. *British Medical Journal* **296**, 1289.

McCormick, A., Tillett, H., Bannister, B., and Emslie, J. (1987). Surveillance of AIDS in the United Kingdom. *British Medical Journal* **295**, 1466.

McGoogan, E. and Cameron, H.M. (1978). Clinical attitudes to the autopsy. *British Medical Journal* **23**, 19.

McIlwaine, W.J., Donnelly, M.D.I., Chivers, A.T., Evans, A.E., and Elwood, J.H. (1985). Certification of death from ischaemic heart disease in Belfast. *International Journal of Epidemiology* **14**, 560.

McNeilly, R.H. and Moore, F. (1975). The accuracy of some hospital activity analysis data. *Hospital and Health Services Research* **71**, 93.

Markush, R.E. (1968). National chronic respiratory disease mortality study. *Journal of Chronic Diseases* **21**, 129.

Martini, C.J.M., Hughes, A.O., and Patton, V.J. (1976). A study of the validity of the Hospital Activity Analysis. *British Journal of Preventive and Social Medicine* **30**, 180.

Mills, I. (1987). Developments in census-taking since 1841. *Population Trends* **48**, 37.

Moriyama, I.M. (1952). Needed improvements in mortality data. *Public Health Reports* **67**, 851.

Moriyama, I.M., Baum, W.S., Haenszel, W.M., and Mattison, B.F. (1958). Inquiry into diagnostic evidence supporting medical certifications of death. *American Journal of Public Health* **48**, 1376.

Morrell, D.C. (1972). Symptoms interpretation in general practice. *Journal of the Royal College of General Practitioners* **22**, 297.

Morrell, D.C. and Kasap, H.G. (1972). The effect of an appointment system on demand for medical care. *International Journal of Epidemiology* **2**, 148.

Morrell, D.C. and Nicholson, S. (1974). Measuring the results of changes in the method of delivering primary medical care—a cautionary tale. *Journal of the Royal College of General Practitioners* **24**, 111.

Morrell, D.C., Gage, H.G., and Robinson, N.A. (1970). Patterns of demand in general practice. *Journal of the Royal College of General Practitioners* **19**, 331.

Morrell, D.C., Gage, H.G., and Robinson, N.A. (1971). Referral to hospital by general practitioners. *Journal of the Royal College of General Practitioners* **103**, 77.

Morris, C. and Compton, P. (1985). 1981 census of population in Northern Ireland. *Population Trends* **40**, 16.

Munroe, J.E. and Ratoff, L. (1973). The accuracy of general practice records. *Journal of the Royal College of General Practitioners* **23**, 821.

Muir, C., Waterhouse, J., Mack, T., Powell, J., and Whelen, S. (ed.) (1987). *Cancer incidence in five continents, Volume V*. International Agency for Research on Cancer, Lyon.

Nevin, N.C., McDonald, J.R., and Walby, A.L. (1978). A comparison of neural tube defects identified by two routine recording systems for congenital malformations in Northern Ireland. *International Journal of Epidemiology* **7**, 319.

Nightingale, F. (1863). *Notes on hospitals* (3rd edn). Longman Green, Longman, Roberts and Green, London.

Office of Population Censuses and Surveys (1970). *Report of the Advisory Committee on Cancer Registration*. OPCS, London.

Office of Population Censuses and Surveys (1973a). Cohort studies: New developments. *Studies on medical and population subjects*, No. 25. HMSO, London.

Office of Population Censuses and Surveys (1973b). *General Household Survey: Introductory report*. HMSO, London.

Office of Population Censuses and Surveys (1978). *General Household Survey: 1976*. HMSO, London.

Office of Population Censuses and Surveys (1982a). *A comparison of the Registrar General's annual population estimates for England and Wales compared with the results of the 1981 census*. Occasional Paper No. 29. OPCS, London.

Office of Population Censuses and Surveys (1982b). *Sources of statistics on ethnic minorities*. OPCS Monitor PP1 82/1. OPCS, London.

Office of Population Censuses and Surveys (1982c). *General Household Survey, 1980*. HMSO, London.

Office of Population Censuses and Surveys (1988). *Congenital malformation statistics—notification, England and Wales 1981–1985*. Series MB No. 2 HMSO, London.

Office of Population Censuses and Surveys, Census Division (1985). Census evaluation programme. *Population Trends* **40**, 21.

Office of Population Censuses and Surveys, Population Division (1980). *Local authority population estimates methodology*. Occasional paper No. 18. OPCS, London.

Park, A.T. (1966). Occupational mortality in Northern Ireland (1960–62). *Journal of the Statistical Society and Inquiry Society of Ireland*, **XXI**(4), 24.

Parkin, D.M., Clarke, J.A., and Heasman, M.A. (1976). Routine statistical data for the clinician. Review and prospect. *Health Bulletin* **34**, 279.

Patel, A.R., Gray, G., Lang, G.D., Baillie, G.G.H., Fleming, L., and Wilson, G.M. (1976). Scottish hospital morbidity data. 1: Errors in diagnostic returns. *Health Bulletin* **34**, 215.

Pole, D.J., McCall, M.G., Reader, R., and Woodings, T. (1977). Incidence and mortality of acute myocardial infarctions in Perth, Western Australia. *Journal of Chronic Disease* **30**, 19.

Puffer, R.R. and Griffith, W.G. (1967). *Patterns of urban mortality*. Pan American Health Organization, Washington.

Redfern, P. (1981). Census 1981—and historical and international perspective. *Population Trends* **25**, 3.

Rees, J.L. (1982). Accuracy of hospital activity analysis data in estimating the incidence of proximal femoral fracture. *British Medical Journal* **284**, 1856.

Registrar General (1949). *The Registrar General's statistical review of England and Wales for the six years 1940–45, Text, Volume I, Medical, statistics of infectious diseases*, p. 85. HMSO, London.

Reid, D.D. and Rose, G.A. (1964). Assessing the comparability of mortality statistics. *British Medical Journal* **ii**, 1437.

Research Committee of the Council of the College of General Practitioners (1962). *Morbidity statistics from general practice* Volume III, *(Diseases in general practice)*. HMSO, London.

Rowe, R.G. and Brewer, W. (1972). *Hospital activity analysis*. Butterworth, London.

Royal College of General Practitioners (1968). Returns from general practice. *British Medical Journal* **iv**, 63.

Royal College of General Practitioners, Office of Population Censuses and Surveys, and Department of Health and Social Security (1974). Morbidity statistics from general practice, 2nd national study, 1970–71. *Studies on medical and population subjects*, No. 26. HMSO, London.

Royal College of General Practitioners, Office of Population Censuses and Surveys, and Department of Health and Social Security (1979). Morbidity statistics from general practice 1971–72. Second national study. *Studies on medical and population subjects*, No. 36. HMSO, London.

Royal College of General Practitioners, Office of Population Censuses and Surveys, and Department of Health and Social Security (1982). Morbidity statistics from general practice 1970–71: Socio-economic analyses. *Studies of medical and population subjects*, No. 46. HMSO, London.

Royal College of General Practitioners, Office of Population Censuses and Surveys, and Department of Health and Social Security (1986). *Morbidity statistics from general practice 1981–1982. Third national study*, Series MB5 No. 1. HMSO, London.

Royal College of Physicians and Royal College of Pathologists (1982). Medical aspects of death certification. *Journal of Royal College of Physicians (London)* **16**, 205.

Shaw, C. (1988). Latest estimates of ethnic minority populations. *Population Trends* **51**, 5.

Silman, A.J. and Evans, S.J.W. (1981). Regional differences in survival from cancer. *Community Medicine* **3**, 291.

Smithells, R.W. (1962). The Liverpool Congenital Abnormalities Register. *Developmental Medicine and Child Neurology* **4**, 320.

Spear, B.E. and Gould, C.A. (1937). Mechanical tabulation of hospital records. *Proceedings of the Royal Society of Medicine* **XXX**(1), 633.

Spicer, C.C. and Lipworth, L. (1966). Regional and social factors in infant mortality. *General Register Office studies on medical and population subjects*, No. 19. HMSO, London.

Statistical Society (1842). Report of the committee on hospital statistics. *Journal of the Statistical Society* **5**, 168.

Statistical Society (1844). Second report of the committee on hospital statistics. *Journal of Statistical Society* **7**, 214.

Statistical Society (1862). Statistics of general hospitals of London 1861. *Journal of the Statistical Society* **25**, 348.

Statistical Society (1866). Statistics of metropolitan and provincial general hospitals for 1865. *Journal of the Statistical Society* **29**, 596.

Statutory Instruments (1968). *Medical Profession. The Abortion Regulations 1968*. SI No. 390. HMSO, London.

Statutory Instruments (1969). *Medical Profession. The Abortion (Amendment) Regulations 1969*. SI No. 636. HMSO, London.

Statutory Instruments (1976). *Medical Profession. The Abortion (Amendment) Regulations 1976*. SI No. 15. HMSO, London.

Statutory Instruments (1980a). *Medical Profession. The Abortion (Amendment) Regulations 1980*. SI No. 1724. HMSO, London.

Statutory Instruments (1980b). *Medical Profession. The Abortion*

(Scotland) (Amendment) Regulations 1980. SI No. 1964 (SI69). HMSO, London.

Statutory Instruments (1987). *Registration of births, deaths, marriages, etc. England and Wales. The Registration of Births and Deaths Regulations 1987*. SI No. 2088. HMSO, London.

Statutory Instruments (1988*a*). *Public Health, England and Wales. The Public Health (Infectious Diseases) Regulations 1988*. SI No. 1546. HMSO, London.

Statutory Instruments (1988*b*). *Public Health, Scotland. The Public Health (Notification of Infectious Diseases) (Scotland) Regulations 1988*. SI No. 1550 (SI55). HMSO, London.

Statutory Rules (1988). Public Health. *The Public Health Notifiable Diseases Order* (Northern Ireland) 1988. SR No. 319. HMSO, Belfast.

Steering Group on Health Services Information (1982*a*). *First report to the Secretary of State* (Chairman E. Körner). HMSO, London.

Steering Group on Health Services Information (1982*b*). *Converting data into information*. King's Fund, London.

Steering Group on Health Services Information (1985). *Supplement to the first and fourth reports to the Secretary of State* (Chairman M.J. Fairey). HMSO, London.

Stewart, G.T. (1980). Whooping cough in the United Kingdom, 1977–78. *British Medical Journal* ii, 451.

Stewart, G.T. (1981). Whooping cough in relation to other childhood infections in 1977–79 in the United Kingdom. *Journal of Epidemiology and Community Health* 35, 139.

Stocks, P. (1949). *Sickness in the population of England and Wales 1944–47*. HMSO, London.

Swerdlow, A.J. (1986). Cancer registration in England and Wales: Some aspects relevant to interpretation of the data. *Journal of the Royal Statistical Society* A 149(2), 146.

Swerdlow, A.J. (1987). 150 years of Registrar Generals' medical statistics. *Population Trends* 48, 20.

Swerdlow, A.J. (1989). Interpretation of England and Wales cancer mortality data: The effect of enquiries to certifiers for further information. *British Journal of Cancer* 59, 787.

Taylor, I. (1965). The notification of infectious disease in various countries. In *Trends in the study of mortality and morbidity*. WHO Public Health Papers 27, p. 17. World Health Organization, Geneva.

Tillett, E. and Spencer, I.L. (1982). Influenza surveillance in England and Wales using routine statistics. *Journal of Hygiene* 88, 33.

Vowles, M., Pethybridge, R.J., and Brimblecombe, F.S.W. (1975) Congenital malformations in Devon, their incidence, age and primary source of detection. In *Bridging in health—Reports on studies for health services in children*, p. 201. Nuffield Provincial Hospitals Trust, Oxford.

Wadsworth, M.E.J., Butterfield, W.J.H., and Blaney, R. (1971). *Health and sickness: The choice of treatment*. Tavistock, London.

Weatherall, J.A.C. (1969). An assessment of the efficiency of notification of congenital malformations. *Medical Officer* 121, 65.

Weatherall, J.A.C. (1978). Congenital malformations: Surveillance and reporting. *Population Trends* 11, 27.

Welsh Office (1987). *Welsh intercensal survey, 1986*. Welsh Office, Cardiff.

Werner, B. (1983). Family size and age at childbirth: Trends and projections, *Population Trends*, 33, 4.

West, R.R. (1973). Cancer registration by means of Hospital Activity Analysis. *Hospital and Health Services Review* 69, 372.

West, R.R. (1976). Accuracy of cancer registration. *British Journal of Preventive and Social Medicine* 30, 187.

Whates, P.D., Birzgalis, A.R., and Irving, M. (1982). Accuracy of hospital activity analysis operation codes. *British Medical Journal* 284, 1857.

Whitehead, F. (1985). Population statistics in the United Kingdom. *Population Trends* 41, 26.

Whitehead, F. (1987). The use of registration data for population statistics. *Population Trends* 49, 12.

Whitehead, F. (1988). How the 1991 census should improve Government statistics. *Population Trends* 52, 18.

Williams, W.O. (1970). A study of general practitioners' workload in South Wales, 1965–66. *Reports from General Practice*, No. 12. Royal College of General Practitioners, London.

Wingrave, S.J., Beral, V., Adelstein, A.M., and Kay, C.R. (1981). Comparison of cause of death coding on death certificates with coding in the Royal College of General Practitioners Oral Contraceptive Study. *Journal of Epidemiology and Community Health* 35, 51.

World Health Organization (1967). *Manual of the international statistical classification of diseases, injuries and causes of death*, (8th revision). World Health Organization, Geneva.

World Health Organization (1977). *Manual of the international statistical classification of diseases, injuries and causes of death*, (9th revision). World Health Organization, Geneva.

World Health Organization Regional Office for Europe (1966). *Studies on the accuracy and comparability of statistics on causes of death*. Unpublished WHO Document EURO–215. 1/16. World Health Organization, Copenhagen.

4

Health information resources: Japan—health and social factors

HIROSHI YANAGAWA and MASAKI NAGAI

Introduction

This chapter reviews health information systems that periodically provide population data on health and social factors in Japan. The criteria for selection of information systems in this review are that they are: (1) national in coverage; (2) currently operational; (3) periodic in data collection; (4) operated by governmental organizations; and (5) primary data rather than compiled data from other sources. For each system, we describe the background, the method of data collection, and the resulting official publications.

Population census

Population censuses of Japan have been conducted every 5 years since 1920, except for the suspension of 1945 census due to the confusion of the Second World War. In its place, an extraordinary census was carried out in 1947.

The censuses conducted every 10 years, starting in 1920, are comprehensive whereas those conducted between each decennial census are simplified censuses. The items investigated in the simplified censuses were limited to basic characteristics of population. However, the items included in the simplified censuses after the Second World War, such as in 1955, 1965, and 1975, were expanded in response to the demands of the users from various fields. The census covers the whole territory of Japan excluding the islands ruled by foreign countries.

Method of data collection

The whole area of Japan was divided into about 780 000 enumeration districts each comprising 50 households, on average. The census was carried out by about 800 000 census enumerators and census supervisors who were temporarily appointed by the Director-General of the Management and Co-ordination Agency of the Japanese Government. The enumerators distributed the census questionnaires and the head of household was required to fill in the form as of 1

Table 4.1. Items investigated in 1985 Japanese census

For household members
  Name
  Sex
  Year and month of birth
  Relationship to the head of the household
  Marital status
  Nationality
  Type of occupational activity
  Name of establishment and classification of business
  Type of occupation
  Employment status
  Place of work or location of school
For households
  Type of household
  Number of household members
  Type and tenure of dwelling
  Number of dwelling rooms
  Area of floor space of dwelling rooms
  Type of building and number of stories

October of that year. The Statistics Bureau of the Management and Co-ordination Agency of the Government takes charge of the tabulation of the results.

The items investigated in the 1985 census are shown in Table 4.1.

Official publications

The official publications of the census data are as follows:

1. The preliminary counts of the population by sex, by prefecture (47 prefectures), and by municipalities (about 3300 cities, towns, and villages) are published 2 months after each survey. These counts are based on the entries on the summary sheets of the census that are prepared by prefectures and municipalities. The title of this report is *Preliminary counts of the population on the basis of summary sheets*.

2. A prompt tabulation is made of the principal statistics based on a 1 per cent sample of households. This is

published as the *Population census of Japan, prompt report of the basic findings (results of one per cent sample tabulation)*.

3. The final counts of population and households for the whole country, prefectures, and municipalities are published as: *Population and households (final counts)*.

4. The first complete tabulation is of basic characteristics of population and households. These are published in: *Population census of Japan, Volume 1 Total population* and *Volume 2 Results of the first basic complete tabulation: Part 1 Japan* and *Part 2 Prefectures and municipalities*.

The second complete tabulation includes basic statistics on economic structure within the municipalities. The publications are: *Population census of Japan, Volume 3 Results of the second basic complete tabulation: Part 1 Japan and Part 2 Prefectures and municipalities*.

The third tabulation includes basic statistics on the occupational structure of population at the level of municipalities as well as statistics on specific households such as single-elderly-person households and mother–child households. The publications are: *Population census of Japan, Volume 4 Results of the third basic complete tabulation: Part 1 Japan* and *Part 2 Prefectures and municipalities*.

5. Detailed tabulation of statistics covering the whole country is carried out for a 20 per cent sample of households. Statistics on detailed classification of industry and occupation, and on socio-economic groups, are included. These are published in *Population census of Japan, Volume 5 Results of detailed sample tabulation: Part 1 Whole of Japan* and *Part 2 Prefectures*.

6. Statistics on place of work or schooling are published by sex, age, occupation, and place of work. This tabulation identifies the daily movements of workers and students between their homes and places of work or study, and the daytime population in each locality. Data are published in *Population census of Japan, Volume 6 Commutation: Part 1 Place of work or schooling of population by sex, age, and industry; Part 2 Place of work or schooling by occupation;* and *Part 3 Place of work or schooling of population by industry and occupation*.

Vital statistics

Vital Statistics in Japan date from 1872. The statistics are compiled from the notification of birth, death, still birth, marriage, and divorce. They are nation-wide, cover the entire population of Japan, and are collected for promoting health services.

Method of data collection

The local health centres (municipal facilities for personal and environmental health services established under the jurisdiction of Health Centre Act) collect vital statistics records from

Table 4.2. Items included in birth records

Sex
Marital status of parents
Date of birth
Address
Age of parents
Date of marriage
Nationality
Occupation of head of household
Weight at birth
Plurality
Place of delivery
Length of gestation
Number of previous births
Attendant at birth

Table 4.3. Items included on death records

Name
Sex
Date of birth
Address
Date of death
Nationality
Marital status
Occupation of head of household
Place of death
Causes of death

Table 4.4. Items included on fetal death records

Name of parents
Age of parents
Marital status of parents
Date of delivery
Address of mother
Occupation of head of household
Number of previous live births and fetal deaths
Length of gestation
Weight of fetus
Place of delivery
Plurality
Type of delivery (artificial or natural)
Attendant at birth
Cause of fetal death

legal documents: birth, death, still birth, marriage, and divorce certificates. These records are sent to the Prefectural Health Department every month and then to Ministry of Health and Welfare of the Japanese Government. Statistical analyses are made by the Statistics and Information Department of this Ministry. The five record forms compiled in Japan and their principal items are listed in Table 4.2–4.6. Statistics for people in Japan (Japanese and other nationalities) and for Japanese people abroad are analysed.

Official publications

Annual *Vital statistics reports* (three volumes) are published for each year not more than 18 months after the end of the

Table 4.5. Items included on marriage records

Name of bride and groom
Dates of birth
Address of groom
Nationality
Date of marriage
Marriage order
Occupation of household heads before marriage

Table 4.6. Items included on divorce records

Name of husband and wife
Date of birth of husband and wife
Address of husband and wife
Legal type of divorce (agreement, mediation, decision, adjudgement)
Number of children
Date of marriage
Date of divorce
Address (before divorce)
Main sources of income in the family before divorce

Table 4.7. Contents of Annual Vital Statistics Reports

Volume 1 (summary)
Volume 2 (detailed statistics)
  Live births
  Deaths
  Infant deaths
  Fetal deaths
  Perinatal deaths
  Marriages
  Divorces
  Appendix (vital statistics for foreigners in Japan and for Japanese people in foreign countries)
Volume 3 (detailed statistics)
  Cause of death
  Cause of infant death
  Cause of fetal death
  Cause of perinatal death
  Appendix (cause of death of foreigners in Japan and of Japanese people in foreign countries)

year of data collection. The contents of the reports are summarized in Table 4.7.

Vital statistics monthly reports, which include basic monthly statistics and cumulative data for the year, are published not later than 6 months after data collection. A special series of vital statistics reports tabulated for socio-economic factors is published every year, and those for occupation and industry are published every 5 years. These series are based on sampling surveys.

Statistics on age-adjusted mortality rate, by sex and prefecture, for leading causes of death have been published every 5 years since 1960.

Communicable disease statistics

Weekly statistics of communicable diseases were first compiled in 1880. The diseases included in this series are those

Table 4.8. Diseases included in statistics of communicable diseases

Cholera	Dysentery (amoebic or bacillary)
Typhoid fever	Paratyphoid fever
Smallpox	Typhus
Scarlet fever	Diphtheria
Epidemic encephalomyelitis	Plague
Japanese encephalitis	Poliomyelitis
Lassa fever	Influenza
Rabies	Anthrax
Infectious diarrhoea	Whooping cough
Measles	Tetanus
Malaria	Tsutsugamushi disease
Filariasis	Yellow fever
Relapsing fever	Schistosomiasis
Syphilis	Gonorrhoea
Chancroid	Lymphogranuloma inguinale
Tuberculosis	Leprosy

listed by the Infectious Disease Prevention Act, the Parasite Prevention Act, the Venereal Disease Prevention Act, the Tuberculosis Prevention Act, and the Leprosy Prevention Act.

The infectious diseases currently included are shown in Table 4.8. The purpose of notification is to examine trends in communicable diseases and to facilitate the planning and evaluation of control programmes.

Method of data collection

Based on the Acts relating to infectious disease prevention, physicians who diagnose the diseases mentioned above are required to notify their local health centre. The director of each local health centre reports to the Prefectural Health Department. The records are then sent to the governmental Statistics and Information Department, Ministry of Health and Welfare, by the end of following month.

Doctors complete one of three different notification forms which are used according to which disease is diagnosed. One is for the majority of infectious diseases, but excludes tuberculosis, leprosy, and venereal disease. The items on this form are shown in Table 4.9. The items on the form for tuberculosis and leprosy are shown in Table 4.10. Items on the venereal disease notification form are shown in Table 4.11.

Official publications

A report is published every year by the Statistics and Information Department, Ministry of Health and Welfare. Numbers and case rates of each disease are tabulated by year, month, sex, age, and prefecture. Detailed tabulations are made for certain diseases. For example, a table by group of bacillus is made for the disease shigellosis; tabulation by stage of disease is made for cases of syphilis.

Food poisoning statistics

Statistics on cases of food poisoning in Japan have been available since the reporting system for food poisoning and

Table 4.9. Items included in the communicable disease notification form

Name of disease
Date of onset
Date of doctor visit
Date of diagnosis
Method of diagnosis
Date of death
Name
Sex
Date of birth
Address
Present location of patient
Occupation
Name and occupation of household head
Doctor's name and address

Table 4.10. Items included in the tuberculosis and leprosy notification form

Name of disease
Date of onset
Date of doctor visit
Date of diagnosis
Name
Sex
Date of birth
Address
Present location of patient
Occupation of household head
Doctor's name and address

Table 4.11. Items included in the venereal diseases notification form

Name of disease
Type and stage (syphilis and gonorrhoea only)
Sex
Age
Occupation
Occupation of the infection source person
Name of facility
Doctor's name and address

Table 4.12. Items included in the food poisoning notification form

Type of food poisoning
Date of onset
Date of doctor's visit
Date of diagnosis
Method of diagnosis (bacteriological, serological, clinical examination)
Date of death
Name
Sex
Date of birth
Address
Present location of patient
Occupation
Name and occupation of household head
Doctor's name and address

Table 4.13. Items included in the food poisoning investigation form

Name of causative food
Place where the causative cooking was undertaken
Type of cooking of causative food
Microbiological examination
Prognosis
Principal symptoms
Description of place of occurrence
Type of food poisoning

Table 4.14. Pathogenic substances classified in the report

Bacteria
   Salmonella species,   *Staphylococcus aureus*,   *Clostridium botulinum*,
   Vibrio   *Escherichia coli*   *Clostridium*
      parahaemolyticus,   (pathogenic),   *perfringens*,
   Bacillus cereus,   *Yersinia enterocolitica*,   *Campylobacter jejuni/*
   Non-O$_1$ *Vibrio*   *coli*
      cholerae,
   others
Chemical substances
   Methanol, others
Natural poisons
   Poisonous plants,
   Poisonous animals
Unknown

misapplication of drugs was established in Japan in 1878. The Food Sanitation Act which was enacted in 1947 made it compulsory for doctors to report cases of food poisoning.

Method of data collection

Doctors who diagnose food poisoning notify local health centres. On notification, the health centre investigates the causative food and pathogenic substance, and completes a Food Poisoning Investigation Form for the individual patient and summary report of the food poisoning episode. The report is submitted to the Statistics and Information Department, Ministry of Health and Welfare.

Items included on the notification form are shown in the Table 4.12, and those additional items on the investigation Form are listed in Table 4.13.

Official publications

A report is published every year by the Statistics and Information Department, Ministry of Health and Welfare. The pathogenic agents in the report are classified as shown in Table 4.14.

Health administration survey

From 1953, the Health Administration Survey was conducted every year until 1985. The purpose of the survey was to collect basic information on the structure of households and socio-economic background of household members.

Table 4.15. Items included in the Health Administration Survey Form

Family members
  Relation to the family head
  Date of birth
  Sex
  Marital status
  Type of social insurance and pension
  Occupation
  Type of employment
Households
  Type of household
  Monthly household expenditure
  Size of area under cultivation (farmers only)

Table 4.16. Items included in the National Health Survey Form

Items for household summary sheet (all family members)
  Sex
  Date of birth
  Type of health insurance
  Occupation
  Days of bed rest in a year
  Days under treatment at medical institutions in a year
Items for individual sheet
  Name of illness
  Duration of illness
  Type of treatment
  Bed rest and sickness absence

The basic sampling frame was the same as that used for the national health and disease surveys conducted by the Ministry of Health and Welfare.

Method of data collection

About 88 000 households (280 000 household members) in 1800 areas randomly sampled from census records with stratification were investigated. The questionnaire was completed by an interviewer who visited the households. The forms were collected by health centres and sent to the Statistics and Information Bureau, Ministry of Health and Welfare, via the prefectural government.

The principal items collected for each family member and household are shown in Table 4.15.

Official publications

The report of the health administration survey was published every year from 1953 to 85. The results are available for all households and specified households such as households with elderly people, those with children, and those with a fatherless or motherless family.

The survey was incorporated into the Basic Survey on Living Conditions which was newly established in 1986.

National health survey

The National Health Survey is the principal source of information on levels and distribution of illness and impairment, and on treatment. The purpose of the survey is to provide national data on the prevalence of illness and impairment, and on the utilization of medical facilities. The survey was initiated in 1948 and continued until 1985. It was incorporated into the Basic Survey on Living Conditions in 1986.

Method of data collection

Sampling has followed two stages. The first is the sampling frame for the Health Administration Survey (1800 areas). The second stage uses 700 of these areas, in which 16 000 households are located.

Before the interview survey, a calender on which to record information about illness and impairment over a 3-day period

of investigation was distributed to the households. All household members in the survey areas were interviewed by interviewers who were trained and supervised by local health centre officials in the district.

The survey forms consisted of a Household Summary Sheet and an Individual Sheet for family members with disease during the specified 3-day observation period. The items included on these two forms are listed in Table 4.16.

Official publications

A report had been published every year until 1985. Since 1986 the report has been included in the Basic Survey on Living Conditions.

Daily life survey

The Daily Life Survey is intended to collect information on family income, housing conditions, and various other aspects of family life. The questions on income are included every year, but other questions vary between years. The information obtained by this survey is useful for the planning of social security and social welfare. The survey began in 1962.

Method of data collection

About 9000 households in 360 areas taken from the sample frame for the Health Administration Survey (approximately 1800 areas) are investigated by trained interviewers.

The items studied in the 1985 survey, for example, included source of family income, taxation, participation in community affairs, and balance of the household economy.

Official publications

The report was published every year from 1962 until 1985. Publications include a detailed report of income levels.

The basic survey on living conditions

The Health Administration Survey, the National Health Survey, and the Daily Life Survey were incorporated into Basic Survey on Living Conditions in 1986. The results of this survey provide the fundamental data on levels of health, utilization of medical care, social welfare, pensions, and life-styles.

Method of data collection

In the 1987 survey, 45 000 households (150 000 household members) in 940 areas sampled from the 1985 census records were studied. Each household was visited by an interviewer who completed the form. The items studied are a combination of the three studies mentioned above.

Official publications

The results of the Basic Survey on Living Conditions are published yearly.

Patient survey

The Patient Survey is the principal source of information on utilization of hospitals and clinic services on one specified day and on hospital discharges over a one-month period. The survey provides information on the numbers and distribution of these patients, their diagnosis, and the type of health insurance plan.

Since 1953, the Patient Survey has been carried out annually on a sampling basis. About 1 in 100 of clinics and dental clinics, and one in ten hospitals, are randomly selected and stratified by prefecture. The purpose of the study is to obtain fundamental information on patients treated in the medical institutions. In 1984, the sampling frame of the survey was enlarged three times so that estimates could be made for 47 prefectures, and the frequency of the survey was reduced to once every 3 years. The total numbers of medical facilities sampled for the 1984 survey was 2377 hospitals, 2905 clinics, and 771 dental clinics.

Method of data collection

In-patients and out-patients on one specified day (One-Day Patient Survey) and in-patients who were discharged during one specified month (One-Month Discharge Survey) in randomly sampled medical facilities stratified by prefecture throughout Japan are the subjects for this survey. The items included in these surveys are listed in Table 4.17.

Official publications

The report of this survey was published every year until 1983. Since 1984, when the method of data collection changed, the report has been published once every 3 years. The report in 1984 consisted of two volumes: data for all Japan and for prefectures. Estimated numbers and rates of admissions and discharged patients according to The International Classification of Diseases were tabulated for the items observed. Average length of hospital stay for discharged patients is also assessed.

Tuberculosis statistics

Tuberculosis statistics were first collected in 1961 when the registration of tuberculosis was introduced by the revision of

Table 4.17. Items included in Patient Survey

One-day patient survey
  Sex
  Date of birth
  Address
  Type of treatment (in-patient or out-patient)
  Name of illness or injury
  Specialty of medical facility
  Type of health insurance
  Time of last visit
  Route of consultation (referred from hospital or clinic, by ambulance, others)
One-month discharge survey
  Sex
  Date of birth
  Address
  Name of illness or injury
  Specialty of medical facility
  Type of health insurance
  Date of hospitalization
  Date of discharge
  With or without surgical operation

Table 4.18. Items included in the statistical tables for tuberculosis registration

Diagnostic category
Sex
Age
Address
Bacteriological finding
Hospitalized or not
Social health insurance

the Tuberculosis Control Act. Physicians who diagnose tuberculosis are required to notify the case to local health centre. Confirmed tuberculous cases are registered and followed up by the health centre. Medical expenses for the treatment of tuberculosis are publicly subsidized by the Tuberculosis Control Act.

The registration of tuberculosis cases ensures that basic epidemiological data on tuberculosis such as incidence and prevalence rates according to the type of disease by sex, age, and prefecture, and type of treatment are available. The data are useful for the evaluation of countermeasures for tuberculosis provided by the Tuberculosis Control Act.

Method of data collection

Health centres are responsible for the registration of patients. Statistics on newly registered cases in a year and total registered cases at the end of year are reported by health centres to the Statistics and Information Department, Ministry of Health and Welfare, through the prefectural government. The items included in the statistical tables are shown in Table 4.18.

Official publications

The Ministry of Health and Welfare publishes tuberculosis data every year. The tables included in the report are tabulated for prefectures and selected large cities.

Table 4.19. Items included on the form for medical facility statistics

Name of the institution
Address
Ownership
Type of medical service offered
Medical specialty
Number of licensed beds
Number of staff (medical and paramedical)
Type of social health insurance accepted

Medical facility statistics

The purpose of these statistics is to provide fundamental data for the analysis of the supply and geographical distribution of hospitals, clinics, and dental clinics. The survey first started in 1948.

Method of data collection

All the hospitals, clinics, and dental clinics that have been newly established, closed, or altered are required to submit a notification form (Table 4.19) to the governor of the prefecture.

Official publications

The Ministry of Health and Welfare publishes a yearly report on medical facilities. The number of hospitals, clinics, and dental clinics are tabulated by prefecture, ownership, medical speciality, and size of facility.

Hospital patient statistics

These data were first collected in 1945 and published as a weekly report of hospital patients; since 1948 the report has been monthly. At present, monthly statistics of hospital patients and the number of working staff at 1 November every year are collected.

Method of data collection

Information is gathered from all hospitals in the country. Hospitals are required to submit a monthly report of hospital patients and an annual report of hospital employees. Table 4.20 shows the data collected on the monthly and yearly report forms.

Official publications

The Ministry of Health and Welfare publishes a yearly report of medical facility statistics (see above). The average daily and monthly numbers of patients are calculated, based on monthly reports, as are numbers per unit population. These numbers are tabulated by type of hospital and prefecture. The occupancy rate and average length of patient stay in hospital is tabulated by type of bed and prefecture.

The number of hospital employees based on annual reports is tabulated as number per 100 beds. The tabulation for each

Table 4.20. Items included in the report forms for hospital patient statistics

Monthly report form
    Cumulative number of monthly out-patients
    Cumulative number of monthly in-patients
    Number of newly hospitalized patients
    Number of newly discharged patients
    Number of hospital beds
    Type of hospital
    Ownership
    Number of new-born babies
Yearly report form
    Number of physicians
    Number of dentists
    Number of pharmacologists
    Number of nurses
    Number of assistant nurses
    Number of midwives
    Number of medical social workers
    Number of other medical staffs
    Number of clerical workers

Table 4.21. Items included in the physician, dentist, and pharmacologist report form

Address
Name
Address of family
Registration number
Type of licence
Type of professional activities
Specialties
Name of institution
Address of institution

specialty is made on type of hospital, size of hospital, prefecture, and ownership.

Survey of physicians, dentists, and pharmacologists

The purpose of this survey is to obtain basic information on the number and distribution of physicians, dentists, and pharmacologists in Japan and thus aid medical administration. The reporting of physicians and dentists began in 1948 when the Medical Act and the Dental Act became effective. The reporting of pharmacologists started 6 years later when the Drugs, Cosmetics, and Medical Instruments Act became effective. The reporting was primarily on an annual basis, but it was revised to alternate years in 1982.

Method of data collection

Physicians, dentists, and pharmacologists are required to report their professional activities for the year to local health centres. The items included in the forms are shown in Table 4.21.

Official publications

A report of this survey was published every year by the Statistics and Information Department, Ministry of Health and Welfare, until 1982, and is now produced biannually. Data on employment status, sex, age, speciality, and prefecture are also available.

National nutrition survey

The National Nutrition Survey has been conducted every year since 1952 under the Nutrition Improvement Act. The purpose of the survey is to investigate the intake of nutrients and health status in order to obtain fundamental data to improve nutrition.

At the beginning of the survey, the main object was the assessment of malnutrition. The dietary habits of the Japanese have changed greatly with the increase of national income levels. The intake of fat and animal protein has increased sharply. The current nutritional aims in Japan are to improve eating habits, notably to prevent hypertension and heart diseases.

Method of data collection

All households and household members in about 300 areas taken from the Japanese census are studied. The number of households and household members studied in 1985 were 7000 and 20 000 respectively. The survey consists of three parts: physical examinations, a nutrient intake survey, and a dietary habit survey.

The physical examination surveys focus on chronic diseases of adults such as hypertension and heart disease. The measurements taken in the 1985 survey were height, weight, skin-fold thickness, and blood pressure.

Nutrient intake was estimated by recording the total food intake of each household for three consecutive days. A nutritionist specially trained for the survey visited each household at least once a day during the survey period and checked the records.

The dietary habit surveys vary between years. The 1985 survey included items on daily frequency and amounts of consumption of selected foods such as confectionery, coffee, and tea, the time of taking the evening meal, the frequency of meal-taking after regular supper or dinner, and so on.

Official publications

The yearly reports are published by the Ministry of Health and Welfare. The average amount of energy, food, and nutrient intake per caput are calculated from the food intake records. These data are tabulated by geographical area, income level, whether the area is urban or rural, and size and type of household. The weight, height, blood pressure levels, and skin-fold thickness were tabulated by age and occupation.

Table 4.22. Functions of health centres

Health education
Matters concerning vital statistics
Improvement in nutrition and sanitation of food
Environmental health
Community activity of public health nurse
Medical social services
Maternal, child, and adult health
Dental health
Laboratory examination for personal and environmental health
Mental health
Control of communicable diseases
Other functions regarding community health

Table 4.23. Statistical tables included in the statistics of health centre activities

Number of health examinations (mental disorders, stomach and uterine cancers, hypertension, pregnant women, new-born babies, 3-year-old children and other examinations)
Number of pregnancies notified
Amount of health guidance for mother and children
Amount of guidance for disabled people
Amount of consultation and home visits by public health nurses (mental disorders, medical social work, health education)
Number of vaccinations
Activities on tuberculosis control
Number of bacteriological examinations
Number of parasitological examinations
Number of food sanitation inspections
Number of environmental sanitation inspections
Number of drinking water examinations

Data on health centre activity

Health centres that are run by prefectural or city government, are core institutions in disease prevention and health promotion for the people in the community. The functions of health centres are shown in Table 4.22.

When the Health Centre Act was first enforced in 1948, health centres were asked to submit a monthly report on their activities. In 1954, the format of the report was revised and has continued, with minor revisions, to the present. The purposes of the statistics are to obtain fundamental information on activities of health centres and on the status of public health in the population of each administrative area.

Methods of data collection

Statistical tables covering the whole spectrum of health centre activities are prepared by the Ministry of Health and Welfare from data supplied by the director of the Health Centre every year (with some of the data being supplied quarterly). The tables are collected and tabulated at the Statistics and Information Department, Ministry of Health and Welfare. The statistical table forms are shown in Table 4.23.

Official publications

The report describes all the aspects of the activities of more than 800 health centres throughout Japan. The report is pub-

Table 4.24. Information required from local government for the report of health services for the elderly

Number of health record books issued
Number of health consultations
Amount of health education
Number of fundamental health examinations (basic and detailed examinations for hypertension, liver disorders, diabetes, and anaemia)
Number of cancer examinations (early detection for stomach, uterine, lung, and breast cancer)
Amount of functional training
Number of visits for guidance (mainly by public health nurses)

Table 4.25. Items for eugenic and induced abortion forms

Eugenic operation form
   Name
   Sex
   Address
   Reason for operation
   Date of operation
   Type of operation
Induced abortion form
   Age
   Address
   Week of pregnancy
   Date of abortion
   Reason for abortion
   Public assistance of medical expense

lished every year. The statistical tables are given by prefecture and selected big cities.

Health services for the elderly

With the enactment of the Health and Medical Services Act for the Elderly in 1983, a variety of health services for the elderly were provided by local governments throughout Japan (about 3300 cities, towns, and villages). The Report on Health Services for the Elderly is intended to provide fundamental statistics regarding health activities directed at the aging population.

Method of data collection

Local governments are required to submit details of health activities for the elderly for each fiscal year, using a form prepared by the Ministry of Health and Welfare. The information required is shown in Table 4.24.

Official publications

The report on health services for the elderly is published every year and information is given by prefecture. The number of staff by specialty engaged in each activity is also tabulated for each prefecture.

Eugenic protection statistics

These statistics provide information on eugenic operations (tubal sterilization, vasoligation, vasectomy) and induced abortions, and survey maternal health. The statistics were first collected by monthly reporting of eugenic operation and induced abortion in 1948 and the present system of yearly collection of statistics started in 1969.

Method of data collection

Physicians who perform eugenic operations are required to submit a reporting form to the local health centre under the Eugenic Protection Act. Each prefecture collects all the reports from local health centres and prepares a yearly report, and the Statistics and Information Department, Ministry of Health and Welfare, summarizes the reports from all prefectures.

The items required for the Eugenic Operation Form and the Induced Abortion Form are shown in Table 4.25.

Official publications

The Ministry of Health and Welfare publishes a yearly report summarizing these data. The number and rate per unit population of eugenic operations and induced abortions by reason, sex, age, prefecture, and week of pregnancy (induced abortion only) are tabulated.

National medical care expenditure estimates

National medical care expenditure is the total amount of money paid to hospitals, clinics, dental clinics, midwives, pharmacies, and other medical institutions. Medical care expenditure in Japan has continued to expand at faster pace than the increase in national income. The factors that have effected this increase include the increase in chronic diseases due to population aging, popularization of highly developed medical techniques, and higher percentage of pharmaceutical expenses in the total medical expense.

Method of estimation of medical care expenditure

Annual national medical care expenditure is estimated from the annual amount of medical care carried out at public expense under the Daily Life Security Act, the Tuberculosis Control Act, the Mental Health Act, etc.; social insurance (medical care insurance, national health insurance); medical services for the elderly; and personal expenses. Statistics from the Patient Survey are also used.

Official publications

The report on medical care expenditure is published every year and includes tables for total amount, percentage increase, expenditure per caput, proportion of gross national product and national income, amount according to public liability of cost, amount by type of medical care, and amount by classification of disease.

Age-adjusted mortality rate—a special report of vital statistics

Since 1960, the Statistics and Information Bureau, Ministry of Health and Welfare, has published age-adjusted mortality rates as a Special Report of Vital Statistics every 5 years. The aging of the population varies widely among prefectures in Japan. Therefore, this report is useful for the comparison of mortality rates between prefectures and between years.

Official publications

The report includes age-specific death-rates and age-adjusted death-rates by prefecture for selected causes. The standard population of this series is the 1960 Japanese population by sex for prefectural observation and the 1935 population for yearly observation.

Life-tables

The first Japanese complete life-tables were calculated based on deaths for the period 1891–8 and were published in 1902. Since then, the life-tables have been calculated 16 times, approximately every 5 years. Since 1948, an abridged life-table has been calculated every year.

The latest complete life-table in Japan was constructed on the basis of mortality data from the vital statistics of 1985, and on the census population on 1 October 1985. The latest abridged life-table was calculated on the basis of mortality data and estimated population at 1 October 1987.

Official publications

The complete life-tables are published every 5 years, and the abridged life-tables are published every year.

National survey of public assistance recipients

The survey provides a unique source of data on the number and distribution of households that receive public assistance under the Daily Life Security Act. The results were first reported in 1956.

Method of data collection

The report is based on two forms—the inclusion and exclusion forms. The inclusion form is for the households that start to receive public assistance, and the exclusion form is for those that have stopped receiving assistance. Local social welfare organizations complete the survey forms. The items in the inclusion and exclusion forms are shown in Table 4.26.

Official publications

The report is published every year by the Statistics and Information Department, Ministry of Health and Welfare. In the summary of the report, yearly changes, type of household,

Table 4.26. Items on inclusion and exclusion forms of public assistance recipients

Inclusion form
    Type of household
    Past recipient
    Reason
    Duration of unemployment
    Source of income
    Type of assistance
    Housing conditions
    Medical assistance
    Family members
Exclusion form
    Date of starting assistance
    Reason for assistance and for stopping
    Type of assistance
    Medical assistance
    Family members

reasons for giving assistance and its withdrawal, period of assistance, type of assistance, and distribution by prefecture are described.

Summary table

All the information sources described in this chapter are summarized in Table 4.27.

Table 4.27. Official publications of health and social information

All publications are available from the Statistics and Information Department, Ministry of Health and Welfare of the Japanese Government unless otherwise stated.

Population census of Japan (every 5 years, 1920–)*
  Preliminary counts of the population on the basis of summary sheets
  Prompt report of the basic findings (results of one per cent sample tabulation)
  Population and households (final counts)
  Basic complete tabulations
    Volume 1 Total population
    Volume 2 Results of the first basic complete tabulation
    Volume 3 Results of the second basic complete tabulation
    Volume 4 Results of the third basic complete tabulation
  Detailed sample tabulations
  Tabulation on place of work or schooling

Vital statistics reports (yearly, 1872–)
  Volume 1 Summary of the results
  Volume 2 Birth, total death, fetal death, perinatal death, marriage, and divorce
  Volume 3 Death, infant death, fetal death, and perinatal death by cause
  Special series of vital statistics
    Socio-economic characteristics (every year)
    Age-adjusted mortality rate (every 5 years)
    Statistics according to occupation and industry (every 5 years)
Communicable disease statistics (yearly, 1880–)
Food poisoning statistics (yearly, 1880–)
Health Administration Survey (yearly, 1953–)
  Included in Basic Survey on Living Conditions since 1986
National Health Survey (yearly, 1948–)
  Included in Basic Survey on Living Conditions since 1986
Daily Life Survey (yearly, 1962–)
  Included in Basic Survey on Living Conditions since 1986
Basic Survey on Living Conditions (yearly, 1986–)
Patient Survey (yearly, 1953–1983; every 3 years, 1984–)
Tuberculosis statistics (yearly, 1961–)
Medical facility statistics (yearly, 1948–)
Hospital patient statistics (yearly, 1948–)
Survey of physicians, dentists, and pharmacologists (yearly, 1948–)
National Nutrition Survey (yearly 1952–)
Data on health centre activity (yearly, 1948–)
Report on health services for the elderly (yearly, 1983–)
Eugenic protection statistics (yearly, 1948–)
National medical care expenditure estimates (yearly, 1954–)
Life-tables (complete life-table about every 5 years, 1891–; abridged life-table yearly, 1948–)
National survey of public assistance recipients (yearly, 1956–)

* Available at Statistics Bureau, Management and Co-ordination Agency of the Japanese Government.

5

Health information as a guide to the organization and delivery of services

KLIM McPHERSON and JOHN BUNKER

Introduction

The quality of medical decision-making can be no better than the quality of the information on which such decisions are based. Florence Nightingale (1859), Codman (1914) and Hey Groves (1908) in a previous era recognized the need for information on the outcome of medical and surgical intervention, but only recently has their plea for the evaluation of treatments been heeded. A major impetus to the current interest in the evaluation of health services comes from the rapid increases in the cost of medical care and the large differences in services provided among countries (Office of Health Economics 1987) and among regions (Department of Health and Social Security 1976; see also American College of Surgeons 1975) and cities (Wennberg *et al.* 1987*a*). Some of these differences are a manifestation of differences in wealth (as between the developed and developing world), but close examination among similar regions within a given country reveals large systematic differences that cannot be explained by differences in demand or need alone. Such differences, instead, may reflect differences in the organization of health care or in resources, such as available beds and trained staff, together with professional uncertainty (Eddy 1984; Wennberg and Gittelsohn 1982), an uncertainty that underlies widespread disagreement as to the choice of optimal diagnostic and therapeutic procedures (Bunker 1985).

The quality of health services is, in part, dependent on the organization and financing of that care, i.e. on health policy. Health policy, in its turn, is also highly dependent on the quality of health information. This chapter considers how the wide variety of information described in previous chapters can be used to improve the health of individuals and of the population as a whole, including discussion of where those data fall short and some methodological problems with their interpretation. The major emphasis will be on practical uses of information in the UK and in the US.

Health information as a guide to clinical policy

This century has seen an enormous increase in the size and strength of the database on which the scientific practice of medicine depends. Beeson (1980) has estimated that no more than 6 per cent of recommended treatments in the first edition of Cecil's *Textbook of medicine* (1927) can, in retrospect, be considered to have been effective, whereas 50 per cent of the treatments recommended in the fourteenth edition (edited by Beeson and McDermott 1975) he judged to be effective in suppressing or controlling disease, or highly effective therapy or prevention. While acknowledging these advances, Thomas (1987) reminds us that it is important to bear in mind how much there remains to be learned. (Indeed, medicine's record of achievement in 1975, as assessed by Beeson, may appear considerably smaller in another 50 years).

Advances in medical knowledge have come, to a large extent, from undirected basic research (Comroe and Dripps 1977). Such basic knowledge, when translated into potential diagnostic or therapeutic advances, is introduced into clinical practice by means of an inconsistent process of diffusion, sometimes following evaluation by carefully controlled clinical trials, sometimes by poorly controlled studies, and sometimes without evaluation. The clinical practices that emerge may be relatively quickly rejected, but, if accepted, form the basis for clinical guide-lines or standards; these, in turn, may be explicitly formulated as decision rules or algorithms, or merely adopted as rough rules-of-thumb. This process, from research and development of basic knowledge and its formulation into a clinical usable form, through evaluation with or without clinical trials, to release and diffusion into clinical practice, is one which begins with a strong science base and is often successively weakened as it progresses through its several stages to clinical practice (Bunker and Fowles 1982; Fineberg 1985).

We know more about what ought to work from the

laboratory than we know about what does work (its efficacy) from reliable clinical trials; and we know a great deal more about what does work in the ideal world of the clinical trial than is actually achieved in clinical practice (its effectiveness) (Bunker 1988). What we do not yet know from the laboratory, coupled with serious shortcomings in the process of clinical evaluation, allows a large measure of residual uncertainty for the practising physican faced with responsibility for patient care and clinical decisions. This uncertainty underlies the large reported disagreement exhibited by physicians in their practices (Bunker 1985; Wennberg and Gittelsohn 1973, 1982).

Randomized clinical trials

To reduce uncertainty and resultant professional disagreement, more reliable information is needed on which to judge the efficacy of diagnostic and therapeutic technology. Randomized controlled clinical trials provide by far the most reliable evidence of efficacy (Chalmers 1974; Hill 1962), although there are limitations to their application and interpretation, as we discuss below. Properly controlled trials, with random assignment of the treatments or diagnostic procedures to be compared, are required by the Food and Drug Administration as part of the pre-marketing approval process of new drugs in the US. In the UK, the Committee on Safety of Medicines, which licenses all drugs and devices, does not necessarily require randomized clinical trials, but relies heavily on controlled trials in its judgements. It will license products for which the evidence is less secure so long as it is clear that adequate phase I and phase II testing has been undertaken and that fuller randomized comparison between available alternatives will be made. While the introduction of new surgical and non-pharmacological medical procedures is, by contrast, unregulated in either country, the evaluation of surgical and medical innovation by randomized clinical trial has become well established Gray et al. 1989; McPeek et al. 1989).

The process by which drugs are introduced is commonly held up as a model for how other medical technologies should be judged. Despite the rigour with which drugs are evaluated, however, there remain substantial limitations to what can be achieved. Pre-marketing clinical trials of drugs necessarily include too few patients to detect even serious rare complications or adverse reactions. An example of this in practice was the failure to detect the risk of Guillain–Barré syndrome in the field trials of the swine flue vaccine (Dutton 1988). Although more that 5000 subjects were included in the trial of the vaccine, the incidence of Guillain–Barré syndrome after the inoculation of 45 million persons turned out to be less than one in 100 000. Thus, the clinical trial, large though it was, could not detect even serious complications, if rare.

Similarly, clinical trials cannot detect complications that occur in the distant future. The adverse effects of diethylstilboestrol (DES), notably clear-cell adenocarcinoma of the vagina in the daughters of women who had been prescribed

DES during pregnancy, and breast cancer among the women themselves (Greenberg et al. 1984; Herbst et al. 1971), did not manifest themselves until many years later. The clinical trials that had been carried out to test the efficacy of DES in preventing miscarriage, and which failed to demonstrate a beneficial therapeutic effect, completely missed—and could not have found—the serious harm associated with its use.

The harmful effects of DES are now widely known by the public, as well as by physicians. It is also well known that the clinical trials failed to show a beneficial effect. What is not generally appreciated is that the small size of the clinical trials of DES could have missed a small, but possibly important, therapeutic effect. In this particular case, it is highly unlikely that any benefit could justify the subsequent risk, but had such an effect been demonstrated, many more pregnant women would have been prescribed the drug, and long before the risks became known. In other trials, the number of patients or subjects might be large enough to detect a substantial positive effect of a treatment, but the trial may, if relatively small, miss subtle side-effects as a consequence of weak power. This is, in practice, a real deficiency, because practical and ethical considerations tend to restrict clinical trial size, and treatment effects on which power is calculated tend, if anything, to be optimistic (McPherson 1982).

For these and other reasons, clinical trials tend to be too small. One obvious solution is to mount larger trials, as is beginning to happen (for example Collins et al. 1988). Another option is the attempt to gain statistical power by combining in a meta-analysis the results of similar trials in a statistically efficient way, pooling all evidence from randomized groups bearing on the comparison in question (Laird and DerSimonian 1982). Such efforts can lead to greatly increased power (Yusuf et al. 1985) for the primary assessment and can often, in addition, test the efficacy within subgroups (for example, tamoxifen among oestrogen receptor positive and negative patients, Early Breast Cancer Trialists' Collaborative Group 1988).

The potential of meta-analysis is realized only to the extent that all relevant data are available. In the real world of medical science, the problem of publication bias (Begg and Berlin 1989) presents a serious obstacle. Publication bias may take many forms. The work submitted by famous physicians or scientists, or from prestigious institutions, is more apt to be accepted. Another form of publication bias is the greater tendency for positive and interesting results to be published and for negative or boring results to be ignored, giving a biased impression or assessment (Pocock and Hughes 1990). Increasingly comprehensive data banks are being created in particular subject areas of randomized trials, the aim of which is to include all trials, whether completed, unpublished or not. In this way, the unbiased assessment of efficacy can be accomplished with greater rigour (Chalmers et al. 1986). In order to ensure completeness it has been suggested, for example, that ethical committees be obliged to keep registers of all applications for clinical trials. The results of clinical trials would, thus, become more legitimately a part of available health information, whether or not they appear in the scientific medical literature. (It should be noted, however, that the resultant registry will consist of the results, and partial results, of trials that are unrefereed, as well as those that have been reviewed for publication.)

In the actual practice of clinical trials, an important component of their execution in the UK is the existence of the National Health Service central registry. This facility enables all randomized patients to be routinely flagged. In this way, all cancer registrations and all deaths among patients who have not emigrated will subsequently be automatically reported to the investigators. Thus, the identification of a randomized series can be cheaply and conveniently followed indefinitely. Increasingly, this is becoming routine in clinical trials in the UK. A roughly comparable service can be provided in the US by the National Death Registry, to which an investigator can check for the status of subjects who have been lost to follow-up.

Despite the very real advantages of controlled clinical trials, clinical opinion is often too entrenched, or patient expectation is such that randomized trials may seem impractical, and possibly unethical. A randomized clinical trial may, indeed, be considered unethical in situations where the well-informed physician believes that one treatment is superior to a second to which it is to be compared, or it may be considered unethical to complete a randomized trial when it begins to appear that one treatment is superior. On the other hand, when there is genuine uncertainty as to which of two treatments is the better, or whether a proposed treatment is better or worse than a placebo, to pursue entrenched opinion without evaluation may be less 'ethical' than randomization might appear to be (Cochrane 1972).

It is ironic that much unevaluated treatment can be, and is, offered without any requirement for informed consent by the patient, while randomized studies always carry such requirements in the US, and usually in the UK. In the US, new drugs must always be introduced with controlled clinical trials, and such clinical trials must be carried out with the informed consent of the patients or subjects. Once approved for clinical use for the condition evaluated by controlled clinical trial, the drug may then be used by physicians for other conditions as they see fit, whether or not the efficacy of the drug for that condition has been tested, and with or without informed consent. The introduction of new surgical procedures is not regulated by the Food and Drug Administration, or other governmental agency, unless a medical device is involved; nevertheless, when physicians choose to introduce new medical technologies by way of a controlled clinical trial, they must submit to a rigorous informed consent procedure as required by the institutional review committee, in compliance with National Institutes of Health regulations. For the same surgical innovation, introduced without a clinical trial, 'informed consent' is also required. In this case, however, it is the much less rigorous consent procedure used for routine hospital procedures. Thus, there is a clear double standard in the requirements and restrictions placed on medi-

cal and surgical innovations that are introduced with careful scientific evaluation, compared with the freedoms associated with innovations introduced without evaluation.

In the UK, informed consent is not a legal requirement in clinical trials or in clinical practice. Guide-lines for clinical trials are set by the local ethical committee and vary from institution to institution. Despite such inconsistency and some resultant confusion (for a discussion see Faulder 1985), failure to obtain the informed consent of patients in clinical trials can be a source of strong public criticism. Informed consent is neither expected nor often obtained when a medical or surgical innovation is introduced without a clinical trial: here, again, we see the double standard.

Observational or non-experimental data

Randomized clinical trials can tell us whether a treatment or procedure is efficacious under the circumstances of that particular trial and for the particular population studied. They may tell us little or nothing of the 'effectiveness' of the treatment under the less ideal conditions of everyday clinical practice, when the average physician may be less skilled, the facilities and resources may be suboptimal, and the patients less compliant. To determine the effectiveness of health services, i.e. the effect of medical care on the population as a whole, we must, if large pragmatic trials are impractical, turn to observational or non-experimental data. The distinction between efficacy and effectiveness is crucial to judging the quality of individual medical care decisions, effectiveness representing the more appropriate standard of comparison (Bunker 1988). It is on effectiveness, as reflected in observational, non-experimental data, that physicians must also rely in making those many medical decisions for which there are no controlled clinical trials to guide them.

If the results of observational studies are less reliable than those of randomized clinical trials—and there is a large history of erroneous conclusions drawn from observational studies that were corrected by subsequent randomized clinical trials—observational studies can offer a number of real advantages. The advantages can be summarized as: (a) sample sizes can be large and, therefore, high power may be achieved; (b) 'natural experiments' comparing alternative treatments prescribed by physicians holding strong, but differing, clinical opinions (however diverse) do not evoke the ethical objections that are raised by randomization (since these decisions are made in good faith the notion of uncertainty, and therefore experimentation, does not appear to arise); (c) these differences can, under some circumstances, be investigated historically, allowing the possibility of comparing long-term consequences; (d) selection of participants is less apt to be a problem because data sources may include all relevant patients; (e) for the same reason, selection of institution or doctor may sometimes not present problems (see below); (f) in circumstances where the nature of appropriate medical care is unambiguous such data can provide reliable information on quality; and (g) possibly most importantly, such investigations allow observation of the relationship between structure, process, and outcome as they actually occur (Donabedian 1966; Lohr 1988).

On the other hand, routine data sources, unlike randomized clinical trials, generally provide little reliable evidence on outcome as the basis for comparing competing therapies. There are two reasons for this. First, such sources usually concentrate on recording aspects of process or structure, since they derive from health care facilities, and providing effective patient linkage to mortality statistics or to subsequent medical care performed elsewhere is expensive and difficult. Second, information on patients not receiving care is almost always lacking; as a result, the natural history of the condition is not known and it may not be possible to determine whether observed effects are the result of treatment or are random fluctuations of the disease process. Thus, unless the relationship of process and structure to outcome is extremely well established, such information may be relatively useless for evaluating quality or relative effectiveness. It is often tempting to believe that these relationships are self-evident, but to be persuasive they require empirical validation. Only if we can be sure that the patients being treated are comparable and that a procedure is optimally provided will it be possible to show that one procedure is superior or inferior to another.

Use of administrative and ad hoc medical data sources

Within the theoretical and practical constraints discussed in the previous section, questions of effectiveness can sometimes be reliably, and often cheaply, investigated using routine or *ad hoc* health information systems. Routinely available sources differ among countries, and are reviewed in previous chapters. They typically consist of hospital use data, reimbursement claims, and vital statistics of various kinds. *Ad hoc* surveys consist, for example, of registries, special data banks, post-marketing surveillance, special record linkage studies, or registries of medical visits. In some countries, linked data banks are routinely available and are often the most useful source for a full assessment of outcome.

The case-mix problem

Observational comparisons are often referred to as natural experiments. If, in such a natural experiment, it is observed that 10 per cent of patients in one category die before discharge in hospital A while only 5 per cent do in hospital B, it is tempting to assume that hospital B is safer than A. The recent release of hospital-specific mortality data by the American Health Care Financing Administration presented exactly this situation. However, such a difference could merely be a manifestation of the selective assignment of patients to one or the other hospital (as a consequence of different referral processes, for example) rather than of the quality of care provided. The possibility of systematic and often unrecognized differences in the populations to be compared must be considered for all observational comparisons of outcome. One would want to adjust for such differences, the so-called 'case-mix', to achieve an approximation to random allocation, and then compare outcome. Indeed, increasingly sophisticated methods have been developed (Charlson *et al.* 1987; Horn 1983) to do just this. It is important, however, in making such adjustments, to bear in mind their intrinsic limitations. The limitations of the attempt to adjust for case-mix are strikingly illustrated by the large clinical trial of the cholesterol-lowering drug, clofibrate, in the Coronary Drug Project.

The Coronary Drug Project (1980) was a large prospective, double-blind, randomized, controlled clinical trial undertaken in the 1960s in the US and Canada. The objectives were to evaluate the effectiveness of drug treatment in the secondary prevention of myocardial infarctions. Of men aged from 30 to 64 years with a recent myocardial infarction, 2789 were randomly allocated to placebo and 1103 to clofibrate. During the course of the trial careful documentation was made of adherence to treatment, which was blind of actual drug and also blind to outcome. Thus, each patient was continuously assessed by pill counts and other means for their adherence to prescribed therapy.

The 5-year mortality rate of patients randomized to clofibrate was 20.0 per cent, and of patients randomized to placebo was 20.9 per cent, a difference that was not statistically significant. However, on inspection, it was discovered that in the clofibrate group, among those men who had a better than 80 per cent adherence to treatment, the 5-year mortality rate was 15 per cent, compared with 24.6 per cent among those with an adherence record of less than 80 per cent. This difference was highly significant. But similar results were also found in the placebo group: a 15.1 per cent mortality rate for adherers compared with 28.3 per cent for poor adherers. As an example of an observed difference in relevant outcomes, such an effect is of special interest precisely because we know that this difference must be due to case-mix; placebo prescribed in a double-blind trial cannot have a differential therapeutic effect on mortality, depending on whether people take it or not.

Adjustment for case-mix had an interesting effect on the comparison that is worth bearing in mind each time such adjustment is attempted. There were some 40 characteristics measured at the time of randomization which, it was thought, might effect the outcome. These included the number of involved coronary arteries, previous history, electrocardiographic abnormalities, smoking, drinking, obesity, and others. Some of these were indeed associated with adherence, but adjustment has little effect on the results: the mortality rates for placebo adherers and non-adherers, after adjustment for all of these characteristics, were 16.4 and 25.8 per cent, respectively. Thus, while we know that these differences in outcome must be attributable to case-mix, they could not be attributed to any of the 40 measured characteristics and, therefore, must be attributable to characteristics of patients which affected outcome, but were not measured and remain unknown.

The conclusions to be drawn from the Coronary Drug Project are important and sobering. Large differences in mortality rate occurred between recognized and identifiable groups that appear to be biologically plausible: adherers to a potentially active drug (clofibrate) were observed to have a markedly better survival than similar patients who did not adhere. In an observational epidemiological study, such evidence would be taken as a strong indication that the drug had the predicted effect of preventing death, but in this possibly unique case such an inference is precluded by knowing the outcome among placebo takers. If, in this carefully conducted study, in which there were 40 known risk factors identified as possible confounders, and with the advantages of hindsight, it was not possible to explain such large residual differences in mortality, we conclude that all attempts at case mix adjustment, and therefore many observational studies and uncontrolled comparisons, have intrinsic limitations that cannot be overcome with certainty.

While appropriate scepticism of observational comparisons is always necessary, it is reasonable to assume that unknown indices or indices that measure some aspects of patients which could not be expected to affect outcome are less important than those known or designed to affect outcome. But it is also important to consider what other unmeasured and unknown factors and characteristics might affect outcome and might differ systematically. In the Coronary Drug Project, adherence to placebo could be a correlate of outcome, regardless of correlation with other risk factors, simply because it might measure how the patient felt and how much the patient wanted to get better. On the other hand, clinical decisions might, because of external influences (to do, perhaps, with health policy, education, or culture), favour intervention in one place but not in another; if such decisions are truly independent of the prognosis of the patient, place of care, as a determinant of therapy, need not be a confounding variable, and meaningful comparisons of outcome may be possible.

However, if place of care effectively determines who is to receive care, and observation includes only those who do, outcome measures could be strongly affected by confounding with case-mix. In particular, the comparison between hospitals or individual clinicians presents problems when referral to place of care is itself a choice made (by patient or physician) with some knowledge of the likely clinical outcome. Then, the possible confounding of all of these choices with prognosis could make interpretation of outcome data, among treated patients, extremely difficult. The problem of small numbers associated with individual institutions or personnel adds to the problem of interpretation. Hence, while comparison of outcomes between institutions and individual professionals will continue, such data will remain difficult to interpret until the choice of hospital or doctor is made randomly, and all randomized patients are carefully followed.

Population-based information systems

To mitigate the problems of the confounding effects of choice and selection in the evaluation of health care, comparison is most sensibly made between populations or communities who receive all, or almost all, of their care in a health care system associated with a defined catchment area (Ellwood 1988; Wennberg 1987a). This has two quite distinct advantages: the first, as discussed in the previous section, is that selective referral processes and choice of treatments that result in a biased distribution of patients need not be confounding influences; the second is that such populations can be treated as an epidemiological entity subject to known morbidity patterns and comparable illness rates. Thus, without knowing the detailed incidences of particular diseases, patterns and rates of morbidity can reasonably be considered to be consistent with known epidemiological principles. Unfortunately, in considering the effect of a proposed intervention on patient care outcome, we find that baseline outcomes of health care are rarely routinely observed and recorded. We may know the baseline status of patients on discharge from hospital, for example, but what happens after discharge is mostly unrecorded, or if recorded is not linked to previous care and, hence, not available as an outcome measure. Thus, even mortality six months after a surgical operation is generally not known without special study or *ad hoc* databases, or unless linked files are available.

Mortality statistics

For the evaluation of health care, death rates by age and sex or by cause are usually available but rarely sufficient. As a measure of outcome they are the most complete, but possibly the least specific. The determinants of a population's mortality include the environment, life-style, and occupation, as well as medical care, the latter's effect being almost too small to identify. Thus, McKeown (1976) attributed the doubling of life expectancy in the years from the mid-nineteenth century to 1971 primarily to improvements in sanitation, nutrition, and housing, and attributed to medical care an unmeasured, but negligible, contribution to the markedly improved health status.* Similarly, Sir Douglas Black and the working group on inequalities in health could not demonstrate any lessening of the discrepancy in mortality across social classes in the UK following introduction of the National Health Service in 1948 (Townsend and Davidson 1982; Whitehead 1987).

In reports of cross-national comparisons of health resources and personnel, mortality may even move in the opposite direction from that predicted. Thus, Bunker (1970) and Vayda (1973) reported that, while rates of surgery in the

---

* In the second edition (1979), McKeown gave greater credit to medicine's potential role, writing that ' . . . it is not possible to estimate with any precision the contribution which therapeutic and other advances have made to the decline of the multiple noninfective causes of death which together were associated with about a quarter of the reduction of mortality in this century'.

US and Canada were approximately double those of the UK, mortality rates in the US and Canada were greater than in the UK. Similarly, Cochrane and colleagues (1978) reported strong positive correlations between all-cause age- and sex-specific mortality and the number of doctors per thousand population between developed countries. These and similar observations may be explained, at least in part, by artefact of secondary association, and, at best, can be a useful basis for hypothesis generation.

The role of mortality statistics in generating hypotheses may be enhanced at times of rapid change in mortality, particularly when such change occurs in conjunction with dramatic social events. An example is provided by the five-week strike of physicians in Los Angeles County in 1976 (Roemer, 1981; Roemer and Schwartz, 1979) in which almost all elective surgery was suspended. The strike was accompanied by an approximately 30 per cent reduction in mortality rates for Los Angeles County that began two weeks after the start of the strike. There was a sharp increase in mortality rates when the strike ended and elective surgery resumed, with a return to the pre-strike mortality level two weeks after the end of the strike. Such data generate, but do not in any way prove, the hypothesis that elective surgery increases a population's mortality, since they do not address the possibility of a subsequent rebound increase in mortality rates.

The possibility of iatrogenic mortality using death registration data was addressed by Bunker and Wennberg (1973) in an investigation of risks associated with varying surgical rates. Death-rates in the US before age 65 years are somewhat higher than in the UK. At the same time, rates of common surgical procedures are often two to three times as high in the US. The question explored by these authors was whether the excess mortality could be explained by anaesthetic or operative mortality among populations subjected to higher rates of surgery. Given known mortality risks, they estimated that such an explanation could be responsible for one-third to one-half of the difference in age-specific mortality once known differences in cause-specific mortality had been taken into account. Even if this was the explanation for part of the mortality difference, it remains unclear to what extent such a cost can be compensated by gain in quality of life among the large majority who survive discretionary surgery. The possibility of iatrogenic mortality in surgery has received some additional support by the report of a positive correlation of population mortality with operation rates among geographical areas (Roos 1984), but is questioned by others (Vayda *et al.* 1982).

Variations in mortality rates are observed within, as well as between, countries. An analysis of hospital deaths in England and Wales has recently been published by Kind (1988) in which he analysed the Hospital In-Patient Enquiry (HIPE) for 1985. HIPE is a one-in-ten sample of all in-patient records, and documents reason for admission, together with age and sex and area of residence, and status and destination of discharge, the only measure of outcome. In this analysis the overall death rate was 5.5 per cent and varied between

4.5 and 5.9 per cent according to the Regional Health Authority. At a more local level, age-standardized mortality varied according to district by as much as twofold, but some of this variation must be considered to be random. In terms of understanding health care, such analyses offer little insight into quality until excess mortality rates can unambiguously be attributed to care, as opposed to chance or intrinsic prognosis. To accomplish this, more data than are recorded in such routine systems as HIPE (Ashley 1972) will be required. Until such time, the use of HIPE and similar data is largely limited to assist in formulating hypotheses for further investigation.

One such hypothesis, that 'avoidable surgical or anaesthetic factors' contribute to deaths during or after surgery, has been investigated by a special study of perioperative deaths co-ordinated by the Nuffield Hospital Trust and the King's Fund, and published as the report 'A Confidential Enquiry into Perioperative Deaths' (Buck *et al.* 1988). The major purpose was to monitor all deaths associated with surgery or anaesthesia in a specified period and area so that their cause could be established. Each death, on notification, was individually investigated by anonymous postal questionnaire to the responsible surgeon or anaesthetist. Such an exercise is fraught with problems associated with response bias, however. Of 4034 perioperative deaths among 555 258 (0.7 per cent) surgical operations, the surgical data were recovered in only 2784 cases (69.0 per cent). Such a response is too low to allow strong inferences, for non-response is most unlikely to be independent of the circumstances of death. Despite this serious shortcoming of their study, the authors were able to conclude that 'there were important differences in clinical practice between the three regions studied', and that there were a number of deaths attributable to inadequate supervision of junior surgeons or anaesthetists.

Large and unexplained variations in post-operative death rates among hospitals in the US have been known for some time. First recognized during the National Halothane Study (Bunker *et al.* 1969; Moses and Mosteller 1968), their causes were investigated by the Institutional Differences Study, from which it was concluded that there is considerable intra-hospital variation in the quality of care (Flood and Scott 1987). 'Hospitals providing more than average specific services to patients exhibited better outcomes than hospitals supplying fewer services'; better outcomes were found in hospitals 'in which the surgical staff was in a better position to regulate the behaviour of its own members'; and 'hospitals that exhibited a strong medical staff structure, had more experience in treating similar categories of patients, were affiliated with a medical school or had a lower rate of house staff to patients, or reported higher expenditures per patient episode were more likely to provide better care'.

In conducting the Institutional Differences Study, the investigators compared data obtained from direct, on-site observation in 17 hospitals with routinely collected hospital discharge data for the same hospitals. That they obtained roughly similar results from the routine data as for prospectively collected 'intensive' observation offers considerably increased confidence in the reliability of routinely collected data. The use of such data in the evaluation of the quality of care, as the US government's Health Care Financing Administration is attempting to do, is currently being hotly debated in the US (Blumberg 1987b; Dubois et al. 1987). While the reliability of such data remains to be determined, their release to the general public represents a large step forward in the quest for professional accountability.

From the point of view of understanding the consequences of individual clinical decisions, or of determining the quality of care provided by individual hospitals, mortality rates as a measure of outcome will usually lack precision and possibly accuracy (Luft and Hunt 1986). On the other hand, comparison of deaths between large aggregations of population for particular causes may provide insight into overall effectiveness. For example, Vayda (1973), in a study comparing surgical rates between the UK and Canada, noted a threefold difference in the rates at which cholecystectomy was performed. In such an example, as long as death certification is comparable, differences in mortality rates (an excess was observed in Canada) must reflect morbidity differences and treatment effectiveness, at least as measured by mortality from a related condition. In fact, autopsy studies have suggested that the morbidity rates were similar between countries (Brett et al. 1976) and, hence, such analyses support the notion that an increased propensity to operate ought to provide aggregate benefit to compensate for iatrogenic mortality. However, in the UK variations in cholecystectomy rates between widely separated districts do appear to correlate with illness rates (McPherson et al. 1985). Such analyses take their model from the original work of Lichtner and Pflanz (1971) who studied appendicectomy rates and mortality from acute appendicitis in the Federal Republic of Germany.

Rates of use of medical and surgical services

Hospital use and procedure rates have, in recent years, become the subject of intensive investigation in many countries with a view to describing and understanding the nature of clinical decision-making (Aaron and Schwartz 1984; Bunker et al. 1977). As long ago as the 1930s, differences in the propensity to undergo tonsillectomy were documented between school districts of southern England (Glover 1938). Glover's work was essentially prescriptive rather than descriptive; the assumption that the higher rates observed for some children were suboptimal was accepted, but not demonstrated. The work of Bunker (1970), Vayda (1973), and McPherson et al. (1981, 1982) has documented the extent of cross-national variation in population-based hospital use rates and drawn attention to the generally higher rates in North America compared to with the UK or other European countries for which data exist. These differences are often of such magnitude that intrinsic differences in illness rates

(Plant et al. 1973) cannot plausibly provide the entire explanation.

As an example of the magnitude of differences in cross-national rates for a specific procedure, McPherson (1988) reported an age-standarized rate for hysterectomy of 700 per 100 000 in the US, approximately 600 in Canada, 450 in Australia, 250 in England and Wales, and 110 in Norway. More recently, Coulter et al. (1988) reported a rate of hysterectomies of 130 per 100 000 in Sweden, in contrast with 360 in neighbouring Denmark. Several possible explanations for these differences have been discussed at length by McPherson et al. (1981). Method of payment, supply of resources, staffing, reimbursement, and referral patterns may all play a part. As the real causes of these differences remain, in most cases, unknown, the important question about the outcome differences associated with these variations and whether the benefit is commensurate with the cost also remains unresolved without further study.

The possible role that variations in the demand for medical and surgical services might play in the observed variations in hospitalization rates has received considerable attention. Bunker and Brown (1974) demonstrated that the wives of men in different professions did have significantly different rates for hysterectomy and for several other discretionary surgical procedures. Of special interest was the observation that the wives of physicians reported operation rates as high, or higher, than those for the other professional groups. Whether this was demand lead or a manifestation of more available (and less expensive) supply is difficult to tell. Bloor et al. (1978), in an extensive study of childhood tonsillectomy, also failed to discern a demand component in the decision-making process, and a study by Coulter and McPherson (1985) found little social class difference in the probability of discretionary surgery in the Oxford region, or support for the notion of differential demand.

Wennberg and Gittelsohn (1982) have proposed that the observed variations in procedure rates are consistent with the economic notion of supplier-induced demand. In support of this idea, they have called attention to a 'surgical signature' associating individual, identifiable clinicians with a pattern of procedure-specific, population-based rates in hospital service areas. These rates tend to differ in a way that cannot be explained by the characteristics of the populations served, but are sustained over several years, often until there is a change of clinical personnel. Such evidence is enhanced by studies of physician feedback of information on rates, to which a common physician response is a reduction of procedure rates in geographical areas in which such rates were found to be high (American Medical Association 1986; Dyck et al. 1977; Lembcke 1956; Wennberg and Gittelsohn 1973).

The notion that the observed variations in population-based rates are largely the result of differences in physicians' clinical judgement, rather than simply a reflection of differences in the prevalence of disease, or in demographic differences, is supported by studies comparing rates between relatively homogeneous communities. For variations between neighbouring hospital service areas, many of the possible exogenous influences are held constant. This has become known as small area variation, developed, in this context, mainly by Wennberg and Gittelsohn (1973). The hypothesis on which their studies are based is that if there are variations in hospital use rates between communities that are ostensibly similar with respect to major determinants of health need or use, then variations that are larger than could be explained by chance are likely to be a manifestation of differences in clinical opinion; these differences in opinion and judgement, it is posited, reflect the uncertainty underlying much of clinical practice (Wennberg 1990). This argument leads to comparison of the amount of variation between neighbouring small areas, both among procedures or reasons for admission and among countries or health care systems. Such comparisons require a metric for measuring variation which is robust and excludes the random component of variation (McPherson *et al.* 1982, cf. appendix; Roos *et al.* 1988).

Some procedures are found to be much more variable than others (Wennberg *et al.* 1984) and, in particular, many medical admissions exhibit the greatest variation. It was found in Wennberg's study that 85 per cent of medical admissions are more variable between neighbouring communities than are rates of hysterectomy. Such a phenomenon can arise as a consequence of nomenclature ambiguities, as well as choices between out-patient (ambulatory) and in-patient care, but otherwise could be a manifestation of the amount of clinical uncertainty. In the absence of plausible alternative explanations, clinicians can, from such observations, gain insight into varying practice styles and hence come to question their own therapeutic decisions. When large variations are demonstrated, such doubts can lead to acceptance of prospective clinical trials to 'settle an issue' that had previously been unquestioned.

Hysterectomy is, for example, much more variable than appendicectomy. These procedures exhibit as much relative variation in the centralized system of the UK as they do in the US, despite higher aggregate rates in North America (McPherson *et al.* 1982). From this, one may conclude that the clincial uncertainty concerning the indications for these procedures and for hospitalization for medical diagnoses is a function of the procedure or diagnosis itself, rather than the system through which it is provided.

Such analyses can, when the appropriate data exist, be cheap and reasonably quick, and they may indicate the need for further study. While studies of variation provide strong clues to the amount of uncertainty associated with hospital admissions for a particular procedure or medical diagnosis, they do not inform us which rates are appropriate or whether it is better to be a patient in an area where the rates are low or where the rates are high. It is generally assumed, nevertheless, that in under-resourced systems such as the British National Health Service, low rates reflect inadequate 'under' supply, and that high rates in North America represent overtreatment. The assumed relationship may appear to be self-evident, but remains to be established.

Performance indicators

In response to a growing awareness of variations in rates of hospital use, and the attendant uncertainty, the British National Health Service has implemented a programme of 'performance indicators' by districts. Each year, a whole series of such indicators is required. These are in an early stage of development and, at the moment, suffer from the shortcoming that their implications for outcome are unclear. Goldacre and Griffin (1983) have written a useful review of their development. The indicators consist of readily available data reduced to catchment population rates. They consist of items such as length of stay, discharges and deaths per 1000 population, and total out-patients per 1000 population; unfortunately, they measure little that has to do directly with quality or efficacy. As a result, it is possible, as suggested in a recent review in the *British Medical Journal* (Lowry 1988) that 'Doctors . . . if they are clever will turn the figures to their own advantage.' The evolution of routine health statistics will have to progress considerably before these data can provide much information about the quality of health care (McPherson, 1985).

Studies of outcome

The American and British Governments' investment in outcome evaluation and assessment is incommensurate with their investment into the provision of health services, falling far short of need (Bunker 1980; Sanders *et al.* 1989). This may reflect the tendency to assume the efficacy of treatment unless there is strong anecdotal reason to suspect otherwise. Cochrane (1972) and others have argued that treatment is not known to be useful until randomized comparison has demonstrated it to be so, and that expenditure on new and untested treatments should await such validation. While randomization of all new procedures may be unrealistic, the Health Care Financing Administration and private insurers in the US are increasingly demanding evidence of efficacy as a condition of reimbursement. In the UK, a less formal restraint makes expenditure on unproven new technology much more difficult than expenditure on proven technology. Thus, the building and equipping of additional facilities for coronary bypass grafting continues to await further evidence of efficacy (Smith 1984).

'Established' diagnostic and therapeutic procedures that have long been in use, but never evaluated, continue to be paid for in both countries until demonstrated to be useless or harmful. Some such procedures may be without benefit under any circumstances, and many may benefit some patients but not others. Priorities as to which of such procedures to evaluate should be based, at least as a first step, on the degree of uncertainty, as reflected in variations between small areas in rates at which the procedure is carried out, combined with its economic and potential therapeutic importance. Thus, large variations in tonsillectomy may not achieve a particularly high priority if the costs of the operation are small and the risks low, particularly if the benefit bestowed can only be small. Hysterectomy may, on the other hand, achieve a higher priority.

If uncertainty is a major criterion in selecting procedures for evaluation, it is worth noting that uncertainty is usually associated with relatively small net benefits. This is simply because those treatments that are very effective tend to be associated with relatively unambiguous evidence, and therefore of greater certainty as to their value. It is only when the evidence from small case series is ambiguous that more rigorous evaluation is required and then, usually, only if the potential gains for patients are important. The relative advantage of observational comparison or of a randomized experiment for evaluation, if judged necessary, is discussed in earlier sections.

If a comparison of outcomes based on routinely available administrative data is the evaluation strategy chosen, it is important to include longitudinal observation of sufficient duration to identify the outcome in question. The Mayo Clinic database covering the medical histories of the residents of Olmsted County for 81 years (Mayo Clinic 1987), the Oxford record linkage study covering two million people for more than 20 years, and the claims files of the Manitoba Health Services Commission (Roos et al. 1987) are examples of successful longitudinal studies of outcome. Hospital utilization statistics, morbidity surveys, and registries and claims data, if they can be linked, may also serve this purpose. For the Medicare population in the US, this can be done using data from three computer files maintained by the Health Care Financing Administration:

(1)   the Medicare Provider Analysis and Review (MEDPAR) file, which contains computerized discharge abstracts for each hospitalization for Medicare enrollees and includes patient and hospital identifiers and data on diagnoses, procedures, and resource utilization;
(2)   the Health Insurance Skeletonized Eligibility Writeoff (HISKEW) file, which records whether a patient died and the date of death, regardless of whether that death occurred within a hospital;
(3)   the Provider of Services file, which describes the relevant characteristics of all Medicare institutional providers (Roos et al. 1988; Wennberg et al. 1987b).

Mosteller (1989) and Cook and Ware (1983) have summarized the advantages and pitfalls of using such non-experimental data together with comprehensive reviews of methodology.

To illustrate how one might identify a procedure for evaluation, and an appropriate evaluation strategy, we summarize recent observational studies of prostatectomy for benign prostatic hypertrophy, a procedure that is performed at variable rates and that is costly. Prostatectomy in the UK was, in the 1970s, about one-third as common as in the US. For the National Health Service to perform prostatectomy at rates comparable to those in the US would cost, at 1986 prices, nearly £50 million (US$85 million) each year, without taking into consideration the extra costs of facilities and staffing

beyond those of the existing establishment. To do this would, of course, beg questions of priorities and efficiency in health care provision, as well as questions about the effectiveness of the treatment. To perform three times as many prostatectomies within current budgets would mean forgoing other medical needs. Similar questions are raised in the US about the foregone opportunities associated with the observed higher procedure rates.

Notwithstanding these differences in overall rates, the relative small area variation in a region of the UK appears to be greater than in a region of the US. This implies that the uncertainty associated with the decision is at least similar in the two countries, while the external constraints and medical consensus are quite different. Questions of effectiveness and outcome are therefore raised. To address these questions, Wennberg and his colleagues in New England, Canada, and the UK have evaluated the cost-effectiveness of prostatectomy, including patient preferences (Barry et al. 1988; Fowler et al. 1988; Roos et al. 1989; Wennberg et al. 1987b, 1988). Their analyses included the following steps.

First, the claims data were linked between all hospitalizations and death registration so that mortality and readmission could be studied. Then, using routine statistical methods such as Cox' proportional hazard survival analysis (Cox, 1972), mortality and readmission rates were determined. When death-rates 90 days after discharge were found to be considerably greater than anticipated, the need for a full-scale evaluation became apparent. A cross-sectional study was started in which patients referred for prostatectomy were interviewed and their physical status was determined before and after the operation. Many patients were found not to experience the expected symptomatic improvement following surgery (Fowler et al. 1988). A second important observation was that mortality and readmission rates differed markedly between patients undergoing prostatectomy by the relatively newer transurethral method (TURP) and the older open procedure.

The mortality rate among men treated for benign disease at 3 months of follow-up was approximately 30 per cent higher in the TURP group (Roos et al. 1987). The re-operation rate was more than twice as high up to 8 years of follow-up. Such information is difficult to interpret without adequate adjustment for case-mix. In this study adjustment could be made for age and previous illness, as recorded in the linked file. Finer adjustment could not be made, and it seemed possible that selection of patients for one or the other operation explained some or all the difference in mortality. Subsequent analyses in the Province of Manitoba, in the Oxford region using the Oxford Record Linkage Study, and in linked files from Norway have confirmed the higher late mortality following TURP (Roos et al. 1989). The Manitoba data-set included pre-operative assessment of risk and previous illness, providing additional strength to the observation.

We cannot conclude that open prostatectomy is a safer procedure than the closed, transurethral approach. The possibility that the closed procedure is selected preferentially and consistently for sicker patients must still be entertained: only a randomized clinical trial of adequate size would allow stronger conclusions. It is doubtful whether it will be considered appropriate to carry out such a clinical trial, however, and this series of observational studies must be considered the only answer to an important clinical question, and a model of how such studies can be carried out using routinely available administrative data.

Conclusions and policy implications

Individual medical care decisions must, finally, be a matter for the individual patient and his or her doctor (Hoffenberg 1987). Access to relevant and accurate health information will greatly assist both in making a decision, but the decision will often not be clear-cut. Medicine remains an inexact science, and decisions are a matter of balancing probabilities from existing information. Recently developed decision analytical methods now allow physicians to invoke the summation of existing knowledge to inform their patients of the odds associated with particular outcomes. Patients will then have the opportunity to balance their own utilities and preferences (Barry *et al.* 1988) with what is possible and what is likely; and most importantly, purchasers of medical care (such as Regional Health Authorities in the UK, the Health Care Financing Administration in the US, and private insurance companies) will have the opportunity to consider what is affordable in the context of other legitimate uses of health care resources. For that to happen reliably, the information sources required will need to be far greater than those available today.

Formulation of health policy at local, regional, or national levels is severely constrained by the limitations of existing data sources. At the national level, existing data may help to identify broad trends and problems requiring further study. The most urgent national and regional need is for reliable outcome data for specific diagnostic and therapeutic interventions as the basis for reimbursement by fee-for-service. In managed medical systems, and where reimbursement is made by prepayment or by capitation, the need for procedure-specific outcome data as the basis for resource allocation is no less urgent. At the local level, the audit of individual physician or individual hospital performance and the feedback of outcome data is of greatest use as an instrument of continuing professional education. The use of data obtained by medical audit as the basis for disciplinary action has proved unreliable to be in the US and is probably counter-productive (Berwick 1989).

Almost all western governments are watching increases in expenditures over which many have limited control and for which the aggregate commensurate benefit is unclear. Their concerns are heightened by a growing awareness that the actual per caput expenditure or proportion of gross domestic product spent, and the amount of clinical services provided, are so variable between and within countries that they cannot be exclusively in response to clinical need.

Compounding the problems of the large variations in clinical services provided, as demonstrated by small area analyses, are the large inter-institutional and geographical differences in mortality reported in the US, and, to a lesser extent, in the UK. Areas of high utilization and expenditure may or may not be the targets of over-treatment; areas of low utilization may or may not have unmet medical needs. The inter-institutional mortality differences are almost universally confounded by the difficulty of case mix adjustment. Diagnosis- and procedure-specific outcome research is needed in order to resolve these questions. Short of procedure-specific outcome research, a potentially important advance in the interpretation of these data may be achieved by linkage of institution mortality rates to the utilization rates for the small areas in which they are located. Thus, if a hospital with a relatively high mortality rate is determined to be located in an area of low utilization, the high mortality may reflect the fact that admissions are limited to sicker patients.

The problem of hospital case-mix and the variability in criteria for hospital admission can, in theory, be bypassed by the study of condition-specific causes of death among particular populations. Thus, it should be possible to determine the quality of health care by comparing death rates for particular causes between countries or systems of provision or time. Such studies have been reported by Charlton *et al.* (1983) which examine differences between districts in England and Wales. Their study consisted of an examination of mortality from 14 conditions selected by Rutstein *et al.* (1980) as amenable to effective medical intervention, but excluding conditions for which causes are preventable. They found large differences in mortality rates between districts, after adjusting for social factors, which could be due to quality of health provision.

Similarly, Poikolainen and Eskola (1986) reported a fall in mortality from 'amenable' causes under the age of 65 years of around 65 per cent between 1969 and 1981, compared with a fall in non-amenable causes of around 25 per cent. These authors concluded that health services account for 50 per cent of the decline in mortality from amenable causes among both sexes in Finland. The validity of their conclusion requires the assumption that the contribution of other factors to the observed change in mortality is of the same magnitude. Given the importance of these other factors—social, political, economic, or genetic—first recognized by McKeown (1976), some reservation in accepting these conclusions must be entertained.

In the absence of reliable evidence of the efficacy of medical technology, new and old, society will continue to assume its effectiveness. Continuing advances in medical technology of potential benefit to patients will place further fiscal demands on the medical budget. Rationing of medical care that is effective, or possibly effective, is already widespread in the UK and is considered to be imminent in the US. To minimize resultant harm to the public's health, it is essential that society increase its investment in the acquisition of health information, and that existing information be more fully exploited to develop optimal health policies.

In practice, this means monitoring the quantity and quality of those aspects of health care that are believed to be efficacious and effective. It requires provision of the resources necessary to support assessment of the relative utility of multiple clinical options. It means providing information by which the extent of uncertainty can be measured. It means comparing the outcome associated with alternative treatments.

References

Aaron, H.J. and Schwartz, W.B. (1984). *The painful prescription: Rationing hospital care.* Brookings Institute, Washington DC.

American College of Surgeons and American Surgical Association (1975). *Surgery in the United States: A summary report of the study on surgical services in the United States (SOSSUS).* American College of Surgeons and American Surgical Association, Chicago, Illinois.

American Medical Association (1986). *Confronting regional variations: The Maine approach.* Publication No. OP–007. American Medical Association, Chicago, Illinois.

Ashley, J.S.A. (1972). Present state of statistics from hospital in-patient data and their uses. *British Journal of Preventive Social Medicine* **26**, 135.

Barry, M.J., Mulley, A.G. Jr, Fowler, F.J., and Wennberg, J.W. (1988). Watchful waiting vs. immediate transurethral resection for symptomatic prostatism: The importance of patients' preferences. *Journal of the American Medical Association* **259**, 3010

Beeson, P.B. (1980). Changes in medical therapy during the past half century. *Medicine* **59**, 79.

Beeson, P.B. and McDermott, W. (ed.) (1975). *Textbook of medicine* (14th edn). W.B. Saunders, Philadelphia, Pennsylvania.

Begg, C.B. and Berlin, J.A. (1988). Publication bias: A problem in interpreting medical data. *Journal of the Royal Statistical Society, A*, **151** (3), 419.

Berwick, D.M. (1989). Continuous improvement as an ideal in health care. *New England Journal of Medicine* **320**, 53.

Bloor, M.J., Venters, G.A., amd Samphier, M.L. (1978). Geographical variation in the incidence of operations on the tonsils and adenoids: An epidemiological and sociological investigation (Parts 1 and 2). *Journal of Laryngology and Otology* **92**, and 791 and 883.

Blumberg, M.S. (1987*a*). Inter-area variations in age-adjusted health status. *Medical Care* **25**, 340.

Blumberg, M.S. (1987*b*). Comments on HCFA hospital death rate statistical outliers. *Health Services Research* **21**, 715.

Brett, M. and Barker, D.J.P. (1976). The world distribution of gallstones. *International Journal of Epidemiology* **5**, 335.

Brook, R.H., Lohr, K.N., Chassin, M., Kosecoff, J., Fink, A., and Solomon, D. (1984). Geographic variations in the use of services: Do they have any clinical significance? *Health Affairs* **3**, 63.

Buck, N., Devlin, H.B., and Lunn, J.N. (1987) *The report of a confidential enquiry into perioperative deaths.* The Nuffield Provincial Hospitals Trust and The King's Fund, London,.

Bunker, J.P. (1970). Surgical manpower: A comparison of operations and surgeons in the United States and in England and Wales. *New England Journal of Medicine* **282**, 135.

Bunker, J.P. (1980). Hard Times for the National Centres. *New England Journal of Medicine* **303**, 580.

Bunker, J.P. (1985). When doctors disagree. *New York Review of Books* **32**, 77.

Bunker, J.P. (1988). Is efficacy the gold standard for quality assessment? *Inquiry* **25**, 51.

Bunker, J.P. and Brown, B.W. (1974). The physician-patient as an informed consumer of surgical services. *New England Journal of Medicine* **290**, 1051.

Bunker, J.P. and Fowles, J. (1982). Between the laboratory and the patient. *Nature* **298**, 405.

Bunker, J.P. and Wennberg, J.E. (1973). Operation rates, mortality statistics and the quality of life. *New England Journal of Medicine* **289**, 1249.

Bunker, J.P., Barnes, B.A., and Mosteller, F. (1977). *Costs, risks, and benefits of surgery.* Oxford University Press.

Bunker, J.P., Fowles, J., and Schaffarzick, R. (1982). Evaluation of medical technology strategies. *New England Journal of Medicine* **306**, 620 and 687.

Bunker, J.P., Forrest, W.H. Jr., Mosteller, F., and Vandam, L.D. (ed.) (1969). *The National Halothane Study: A study of the possible association between halothane anesthesia and postoperative hepatic necrosis.* United States Government Printing Office, Washington, DC.

Cecil, R.L. (ed.) (1927). *A textbook of medicine* (1st edn). W.B. Saunders, Philadelphia, Pennsylvania.

Chalmers, T.C. (1974). The impact of controlled trials on the practice of medicine. *The Mount Sinai Journal of Medicine* **41**, 753.

Chalmers, I., Hetherington, J., Newdick, M., *et al.* (1986). The Oxford database of perinatal trials: Developing a register of published reports of controlled trials. *Controlled Clinical Trials. Design, Methods, and Analysis* **7**, 306.

Charlson, M.E., Pompei, P., Ales, K.L., and MacKenzie, C.R. (1987). A new method of classifying prognostic comorbidity in longitudinal studies: Development and validation. *Journal of Chronic Disease* **40**, 373.

Charlton, J.R.H., Hartley, R.M., Silver, R., and Holland, W.W. (1983). Geographical variation in mortality from conditions amenable to medical intervention in England and Wales. *Lancet* **i**, 691.

Cochrane, A.L. (1972). *Effectiveness and efficiency: Random reflections on health services.* The Nuffield Provincial Hospitals Trust, London.

Cochrane, A.L., St. Leger, A.S., Moore, F. (1978). Health service 'input' and mortality 'output' in developed countries. *Journal of Epidemiology and Community Health* **32**, 200.

Codman, E.A. (1914). The product of a hospital. *Surgery, Gynecology and Obstetrics* **18**, 491.

Collins, R., Scrimgeour, A., Yusuf, S., and Peto, R. (1988). Reduction in fatal pulmonary embolism and venous thrombosis by perioperative administration of subcutaneous heparin. Overview of results of randomised trials in general, orthopedic and urological surgery. *New England Journal of Medicine* **318**, 1162.

Comroe, J.H. Jr and Dripps, R.D. (1977). *The top ten clinical advances in Cardiovascular Pulmonary medicine and surgery, 1945–1975.* Volumes I and II United States Department of Health, Education and Welfare (DHEW Publication No (NIH) 78–1521 and 78–1522). United States Government Printing Office, Washington, DC.

Cook, N.R. and Ware, J.H. (1983) Design and analysis methods for longitudinal research. *Annual Review of Public Health* **4**, 1.

Coronary Drug Project Research Group (1980). Influence of adherence to treatment and response of cholesterol on mortality in the Coronary Drug Project. *New England Journal of Medicine* **303**, 1038.

Coulter, A. and McPherson, K. (1985). Socioeconomic variations in the use of common surgical operations. *British Medical Journal* **7**, 186.

Coulter, A., McPherson, K., and Vessey, M. (1988). Do British women undergo too many or too few hysterectomies? *Social Science and Medicine* **27**, 987.

Cox, D.R. (1972). Regression models and life tables. *Journal of the Royal Society. Series B* **34**, 187.

Department of Health and Social Security (1976). *Report of resource allocation working party*. HMSO, London.

Donabedian, A. (1966). Evaluating the quality of medical care. *Millbank Memorial Fund Quarterly* **44(2)**, 166.

Dubois, R.W., Rogers, W.H., Moxley, J.H., Draper, D., and Brook, R.H. (1987). Hospital inpatient mortality: Is it a predictor of quality? *New England Journal of Medicine* **317**, 1674.

Dutton, D.B. (1988). *Worse than the disease: Pitfalls of medical progress*. Cambridge University Press.

Dyck, F.J., Murphy, F.A., Murphy, J.K., *et al.* (1977). Effect of surveillance on the number of hysterectomies in the province of Saskatchewan. *New England Journal of Medicine* **296**, 1326.

Early Breast Cancer Trialists' Collaborative Group (1988). Effects of adjuvant Tamoxifen and cytotoxic chemotherapy on mortality in early breast cancer: An overview of 61 randomised trials among 28 896 women. *New England Journal of Medicine* **319**, 1681.

Eddy, D.M. (1984). Variations in physician practice: The role of uncertainty. *Health Affairs* **3, 74**.

Ellwood, P.M. (1988). Shattuck Lecture—Outcome management: A technology of patient experience. *New England Journal of Medicine* **318**, 1549.

Faulder, C. (1985). *Whose body is it?* Virago Press, London.

Fineberg. H.V. (1985). Effects of clinical evaluation on the diffusion of medical technology. In *Assessing medical technologies* (ed. Committee for Evaluating Medical Technologies in Clinical Use. Institute of Medicine) p. 176. National Academy Press, Washington, DC.

Flood, A.B. and Scott, W.R. (1987). *Hospital structure and performance*. The Johns Hopkins University Press, Baltimore, Maryland.

Fowler, F.J., Wennberg, J.E., Timothy, R.P., Barry, M.J., Mulley, A.G., and Hanley, D. (1988). Symptom status and quality of life following prostatectomy. *Journal of the American Medical Association* **259**, 3018.

Glover, J.A. (1938). The incidence of tonsillectomy in school children. *Proceedings of the Royal Society of Medicine* **31**, 1219.

Goldacre, M. and Griffin, K. (1983). *Performance indicators: A commentary on the literature*. Unit of Clinical Epidemiology, University of Oxford.

Gray, D.T., Hewitt, P., and Chalmers, T.C. (1989). The evaluation of surgical therapies. In *Socioeconomics of surgery* (ed. I.M. Rutkow), pp. 228–56. C.V. Mosby, St Louis, Missouri.

Greenberg, E.R., Barnes, A.B., Resseguie, L., *et al.* (1984). Breast cancer in mothers given diethylstiebestrol in pregnancy. *New England Journal of Medicine* **311**, 1393.

Herbst, A.L., Ulfelder, H., and Poskanzer, D.C. (1971). Adenocarcinoma of the vagina: Association of maternal stilbestrol therapy with tumour appearance in young women. *New England Journal of Medicine* **284**, 878.

Hey Groves, E.W. (1908). A plea for a uniform registration of operation results. *British Medical Journal* **ii**, 1008.

Hill, A.B. (1962). *Statistical methods in clinical and preventive medicine*. E. & S. Livingstone, Edinburgh and London.

Hoffenberg, R. (1987). *Clinical freedom*. The Nuffield Provincial Hospital Trust, London.

Horn, S.D. (1983). Measuring severity of illness: Comparisons across institutions. *American Journal of Public Health* **73**, 25.

Kind, P. (1988). *Hospital deaths—the missing link: Measuring outcome in hospital activity data* (Discussion paper 44). Centre for Health Economics, Health Economics Consortium, University of York, England.

Laird, N.M. DerSimonian, R. (1982). Issues in combining evidence from several comparative trials of clinical therapy. Proceedings of the 11th Institute of Biometrics Conference. *Biometrics* **38**, 91.

Lembcke, P.A. (1956). Medical auditing by scientific methods: Illustrated by major female pelvic surgery. *Journal of the American Medical Association* **162**, 646.

Lichtner, S. and Pflanz, M. (1971). Appendectomy in the Federal Republic of Germany: Epidemiology and medical care patterns. *Medical Care* **9**, 311.

Lohr, K.N. (1988). Outcome measurement: Concepts and questions. *Inquiry* **25**, 37.

Lowry, S. (1988). Focus on performance indicators. *British Medical Journal* **296**, 992.

Luft, H.S. and Hunt, S.S. (1986). Evaluating individual hospital quality through outcome statistics. *Journal of the American Medical Association* **255**, 2780.

McKeown, T. (1976). *The role of medicine: Dream, mirage or nemesis?* The Nuffield Provincial Hospitals Trust; 2nd edn (1979), p. 39. Blackwells, Oxford, and Princeton University Press.

McPeek, B., Mosteller, F., and McKneally, M. (1989). Randomized clinical trials in surgery. *International Journal of Technology Assessment in Health Care* **5**, 317.

McPherson, K. (1982). Choosing the number of interim analyses in clinical trials. *Statistics in Medicine* **1**, 25

McPherson, K. (1985). The political argument on health costs. *British Medical Journal* **290**, 1679.

McPherson, K. (1988). Variations in hospitalisation rates: Why and how to study them. In *Health care variations: Assessing the evidence* (ed. C. Ham), pp. 15–20. King's Fund Institute, London.

McPherson, K., Strong, P.M., Epstein, A., and Jones, L. (1981). Regional variations in the use of common surgical procedures: Within and between England and Wales, Canada and the United States of America. *Social Science and Medicine* **15a**, 273.

McPherson, K., Strong, P.M., Jones, L., and Britton, B.J. (1985). Do cholecystectomy rates correlate with geographic variations in the prevalence of gallstones? *Journal of Epidemiology and Community Health* **39**, 179.

McPherson, K., Wennberg, J.E., Hovind, O.B., and Clifford, P. (1982). Small area variation in the use of common surgical procedures: An international comparison of New England, England and Norway. *New England Journal of Medicine* **307**, 1310.

Maynard, A. (1988). *The political dynamics of physician manpower policy: The case of Britain*. King's Fund Centre, London.

Mayo Clinic (1987). *Fact Sheet*. Mayo Clinic, Rochester, Minnesota.

Moore, F.D. (1985). Small area variations studies: Illuminating or misleading? *Health Affairs* **4**, 96.

Moses, L.E. and Mosteller, F. (1968). Institutional differences in postoperative death rates: Commentary on some of the findings of the National Halothane Study. *Journal of the American Medical Association* **203**, 492.

Mosteller, F. (1990). Improving research methodology: An overview. In *Research Methodology: Strengthening causal interpretations of non-experimental data* (ed. L. Sechrest, E. Perrin, and J. Bunker). US Department of Health and Human Services,

Public Heath Service, Agency for Health Care Policy and Research, Rockville, Maryland.

Nightingale, F. (1859). *Notes on hospitals*. John W. Parker, London.

Office of Health Economics (1987). *Compendium of health statistics*, 6th edn. Office of Health Economics, 12 Whitehall, London, SW1A 2DY.

Plant, J.C.D., Percy, I., Bates, T., Gastard, J., and Hita de Nercy, Y. (1973). Incidence of gallbladder disease in Canada, England and France. *Lancet* ii, 249.

Pocock, S. and Hughes, M. (1990). Estimation issues in clinical trials and overviews. *Statistics in Medicine* 9, 657.

Poikolainen, K. and Eskola, J. (1986). The effect of health services on mortality: Decline in death rates from amenable and non-amenable causes in Finland, 1969–81. *Lancet* i, 199.

Roemer, M.I. (1981). More data on post-surgical deaths related to the 1976 Los Angles doctor slowdown. *Social Science and Medicine* 15C, 161.

Roemer, M.I. and Schwartz, J.L. (1979). Doctor slowdown: Effects on the population of Los Angeles County. *Social Science and Medicine* 13C, 213.

Roos, L.L. Jr. (1984). Surgical rates and mortality: A correlational analysis. *Medical Care* 22(6), 586.

Roos, L.L. (1989). Non-experimental data systems in surgery. *International Journal of Technology Assessment in Health Care* 5, 341.

Roos, L.L. Jr., Nicol, J.P., and Cageorge, S.M. (1987). Using administrative data for longitudinal research: Comparisons with primary data collection. *Journal of Chronic Diseases*, 40, 41.

Roos, N.P., Wennberg, J.E., and McPherson, K. (1988). Using diagnosis-related groups for studying variations in hospital admissions. *Health Care Financing Review* 9, 53.

Roos, N.P., Wennberg, J.E., Dalenka, D.J., *et al.* (1989). Mortality and reoperation after open and transurethral resection of the prostate for benign prostatic hyperplasia. *New England Journal of Medicine* (320), 1120.

Rutstein, D.D., Berenberg, W., Chalmers, T.C., *et al.* (1980). Measuring the quality of medical care: Second revision of tables of indexes. *New England Journal of Medicine* 302, 1146.

Sanders, D., Coulter, A., and McPherson, K. (1989). *Variations in hospital admission rates: A review of the literature*. King's Fund, London.

Smith, T. (1984). Consensus on cabbage. *British Medical Journal* 289, 1477.

Thomas, L. (1987). What doctors don't know. *New York Review of Books* 34, 6.

Townsend, P. and Davidson, N. (1982). *Inequalities in health. The Black Report*. Penguin Books, Harmondsworth, Middlesex, England.

Vayda, E. (1973). A comparison of surgical rates in Canada and in England and Wales. *New England Journal of Medicine* 289, 1224.

Vayda, E., Mindell, W.R., and Rutkow, I.M. (1982). A decade of surgery in Canada, England and Wales and the United States. *Archives of Surgery* 117, 846.

Wennberg, J.E. (1990). Small area analysis and the medical care outcome problem. In *Research Methodology: Strengthening causal interpretations of non-experimental data* (ed. L Sechrest, J.J. Bunker, and E. Perrin). US Department of Health and Human Services, Public Health Service, Agency for Health Care Policy and Research, Rockville, Maryland.

Wennberg, J. and Gittelsohn, A. (1973). Small area variations in health care delivery. A population based health information system can guide planning and regulatory decision-making. *Science* 182, 1102.

Wennberg, J. and Gittelsohn, A. (1982). Variations in medical care among small areas. *Scientific American* 246, 120.

Wennberg, J.E., McPherson, K., and Caper, P. (1984). Will payment based on diagnosis-related groups control hospital costs? *New England Journal of Medicine* 311, 295.

Wennberg, J.E., Freeman, J.L., and Culp, W.J. (1987a). Are hospital services rationed in New Haven or over-utilised in Boston? *Lancet* i, 1185.

Wennberg, J.E., Roos, N., Sola, L., Schori, A., and Jaffe, R. (1987b). Use of claims data systems to evaluate health care outcomes: Mortality and reoperation following prostatectomy. *Journal of the American Medical Association* 257, 933.

Wennberg, J.E., Mulley, A.G., Hanley, D., *et al.* (1988). As assessment of prostatectomy for benign urinary tract obstruction. Geographic variations and the evaluation of medical care outcomes. *Journal of the American Medical Association* 259, 3027.

Whitehead, M. (1987). *The Health divide: Inequalities in health in the 1980s*. The Health Education Council, London.

Yusuf, S., Peto, R., Lewis, J., Collins, R., and Sleight, P. (1985). Beta blockade during and after myocardial infarction: An overview of the randomized trials. *Progress in Cardiovascular Diseases*, 27 335.

6

Health statistics for health promotion: the United States perspective

JOEL C. KLEINMAN, MANNING FEINLEIB, JACOB J. FELDMAN, PATRICIA GOLDEN, and RONALD W. WILSON

Introduction

In 1974, the US Congress approved legislation (Public Law 93–353) which required that the Secretary of the Department of Health, Education, and Welfare prepare an annual report to Congress on the nation's health and health system. This report was to be prepared under the auspices of the National Center for Health Statistics. The legislation specified that the report cover four topics: health status, health services utilization, health care resources, and health expenditures. The first volume of this report, *Health, United States*, appeared in 1976 (NCHS National Center for Health Statistics 1976). This report has served to focus attention on health progress and problems in the US.

During the latter part of the 1970s the interest generated by these reports on the nation's health was accompanied by increasing emphasis on the promotion of health and prevention of disease. As interest in improving the health of Americans grew, so too did recognition of the need to measure the dimensions of the health of the population and to document the effects over time of efforts to alleviate or avoid identified problems. In 1978, legislation was enacted (Public Law 95–626) that called for the triennial preparation of a national disease prevention data profile. The first three editions of the *Prevention profile* were published together with *Health, United States* in 1980, 1983, and 1986.

The heightened appreciation for the role of health promotion and disease prevention provided the theme of *Healthy people—The Surgeon General's report on health promotion and disease prevention*, published in 1979 (Office of the Assistant Secretary for Health and Surgeon General 1979). This report not only stressed how important health promotion and disease prevention could be in reducing unnecessary death and disability in the US but also described a number of important measures that could contribute to further improvements in health status. *Healthy people* delineated a set of broad national goals for improving the health of the American people during the decade of the 1980s. These goals, one for each of five major life stages, are:

(1) to continue to improve infant health, and, by 1990, to reduce infant mortality by at least 35 per cent, to fewer than nine deaths per 1000 live births;

(2) to improve child health, to foster optimal childhood development, and, by 1990, to reduce deaths among children aged 1–14 years by at least 20 per cent, to fewer than 34 per 100 000;

(3) to improve the health and health habits of adolescents and young adults, and, by 1990, to reduce deaths among people aged 15–24 years by at least 20 per cent, to fewer than 93 per 100 000;

(4) to improve the health of adults, and, by 1990, to reduce deaths among people aged 25–64 by at least 25 per cent, to fewer than 400 per 100 000;

(5) to improve the health and quality of life for older adults, and, by 1990, to reduce the average annual number of days of restricted activity due to acute and chronic conditions by 20 per cent, to fewer than 30 days per year for people aged 65 years and older.

Subsequent to the publication of *Healthy people*, a comprehensive strategy was developed for attaining the goals enunciated in the Surgeon General's report. This strategy took into account both the actions that individuals could carry out for themselves to improve their health and the factors over which individuals had little control. Thus, the strategy addressed actions that encouraged individuals to lead healthier lifestyles as well as actions that encouraged decision-makers in the public and private sectors to promote and maintain a safer and healthier environment.

This comprehensive strategy took the form of 226 objectives that were published in *Promoting health/preventing disease: Objectives for the nation* (US Department of Health and Human Services 1980). These 226 objectives set forth specific and quantifiable targets, which, when taken together, would permit the realization of the overall national goals set down in the Surgeon General's earlier report. The objectives were established for each of 15 areas: high blood pressure control; family planning; pregnancy and infant health; immunization; sexually transmitted diseases; toxic agent and radiation control; occupational safety and health; injury prevention; fluoridation and dental health; surveillance and control of infectious diseases; smoking and health; alcohol and drug misuse prevention; improved nutrition; physical fitness and exercise; and control of stress and violent behaviour. The 15 areas are grouped under three broad headings—preventive health services, health protection, and health promotion—and specific, often quantitative, targets were established for the objectives.

The *Prevention profile* was designed to direct attention to the five major goals published in *Healthy people* and to report on progress toward a substantial number of the measurable objectives formulated in *Promoting health/preventing disease: Objectives for the nation*. The *Prevention profile* plays an important part in the oversight of Public Health Service initiatives in disease prevention and health promotion. The kind of measurement provided offers a useful form of accountability and a basis for considering the need for programme modifications. The availability of baseline data and data reported for subsequent intervals makes possible a more specific and coherent statement of the problems that must be faced in moving toward the goals and objectives in health promotion and disease prevention.

However, despite the broad array of data presented in *Health, United States* and the *Prevention profile*, there are several problems involved in using health statistics effectively to monitor progress in health promotion and disease prevention. In this paper we consider five such problems:

1. What aspects of health promotion can be measured with validity and precision?
2. How can these measures be assessed by characteristics of place and person?
3. How can past trends be used to guide the setting of goals and objectives?
4. How can we estimate the resources needed to achieve goals?
5. In monitoring progress towards the goals, at what points should policy-makers be alerted to the need for further intervention to achieve goals?

Measurement

The content of *Health, United States* and the *Prevention profile* has evolved over time to maximize its utility while keeping the resources required to produce it within reasonable bounds. Initially, there was an attempt to make *Health, United States* exhaustive, i.e. recent interesting data across a wide variety of data systems were incorporated. This was an extremely time-consuming task. It also resulted in very different content from one year to the next which prevented

effective tracking of changes in health indicators. As a result, it was decided in 1978 to redesign *Health, United States* so that it could focus more effectively on surveillance. Thus, a standard set of tables which would appear in each volume of *Health, United States* was developed, to reflect the most recent data available.

This decision had two advantages. First, it focused attention on change so that progress or lack of progress in key indicators could be assessed. Second, from a management point of view, it made the production of the report much less resource-intensive. After the initial choice of data, a great deal of staff time could be saved by not scouring all data systems to look for tables to include. Of course, it is still necessary to be aware of changes in data systems in order to be sure that comparability is maintained and to add data reflecting new public health issues (e.g. acquired immune deficiency syndrome).

The major criterion for inclusion of data in *Health, United States* is the availability of comparable data collected regularly over several points in time. The actual content of the tables included was driven in large measure by this requirement. A list of the detailed tables in *Health, United States, 1988* is presented in the Appendix.

Similarly, the *Prevention profile* has evolved into a document that emphasizes change. The initial volume was primarily devoted to historical background and data related to health promotion and disease prevention. The next two issues focused on graphic displays to show progress toward the 1990 goals and objectives.

Despite the wide variety of data presented in the reports, there are important unresolved measurement issues in choosing appropriate indicators for monitoring progress. The ultimate goal of health promotion activities is to improve health and quality of life. The difficulty in monitory progress in health promotion is that these characteristics are very difficult to define and measure. As a result, indicators that are easily available but not entirely relevant are often chosen. For example, although it could be argued that the infant mortality rate is highly relevant to infant health, the death-rate for children, adolescents, and young adults affects such a small proportion of the target population that the Surgeon General's goals could easily be met without substantial improvement in the important health problems of these groups (e.g. injuries, developmental disabilities, substance abuse). The prominence given to the death-rate in the statement of these goals is primarily due to its easy availability and quantification.

The 226 objectives specified in *Objectives for the nation* represented an attempt to move beyond such artificial considerations. Although an attempt was made to define most of the objectives in quantifiable terms, neither the availability nor absence of data took precedence over considerations of the magnitude of the problems and opportunities for intervention. In fact, a number of objectives for which no baseline data were initially available were included. For example, data necessary to measure public awareness of the objectives were not available. However, new data collection efforts were initiated and existing data sources were expanded to include new questions measure these objectives. The most notable effort along these lines is the battery of questions included in the 1985 National Interview Survey (NHIS).

The information obtained from these questions provided the first baseline data for many of the 1990 public awareness objectives, and results were presented in the *1986 Prevention profile*. For example, one of the objectives (US Department of Health and Human Services 1980) stated: 'By 1990, at least 85 per cent of women should be aware of the special health risks for women who smoke, including the effect on outcomes of pregnancy and the excess risk of cardiovascular disease with oral contraceptive use'. The data from the 1985 NHIS Health Promotion Supplement showed that we had come close to attaining part of this objective: 'In 1985 the proportion who knew that smoking during pregnancy increased the chance of miscarriage was 74 percent; of low birth weight, 85 percent; of stillbirth, 67 percent; of premature birth, 76 percent'. (National Center for Health Statistics 1986.)

A supplement to the 1990 NHIS will measure progress in achieving these objectives.

Assessment by place and person

Even for those objectives for which data are available, assessment by place and person is often difficult. Vital statistics provide the most extensive geographical coverage. For example, existing files now allow the relatively easy tabulation of death-rates by age, sex, race, county of residence, and underlying cause of death for 1968–85 (National Center for Health Statistics 1988). However, development of meaningful analyses of these data for health promotion presents rather daunting problems: there are more than 3000 countries in the US, each with 18 years of deaths, four race–sex groups and 15 age groups. The statistical problems in analysing a database of this size in order to detect unusual patterns for targeting interventions are formidable (see Feinleib 1988; Kleinman 1986).

Apart from vital statistics, availability of data by place is much more problematic. For example, the National Health Interview Survey (NHIS), a major source of health data in the US, is based on a sample design of 200 primary sampling units across the US. The design allows for estimates by region and type of residence (e.g. size of place) but further geographical specificity is generally impossible.

On the other hand, the survey data are richer than vital statistics with respect to person. Survey data generally include the respondent's education, income, occupation, and ethnic background in addition to age, race, and sex. Analyses of these data often point to the need for more targeted interventions.

The monitoring of trends in cigarette smoking provides an interesting example. Data on smoking have been collected in the NHIS periodically since 1965 and are presented in

Health, United States each year. These data illustrate substantial declines in smoking since the 1964 Surgeon General's Report on Smoking and Health (US Department of Health, Education and Welfare 1964). In detailed analysis of the NHIS data from 1974 to 1985 Fiore *et al.* (1989) showed that the prevalence of smoking has declined among both blacks and whites, males and females. However, the decline has been greater among men than women. As a result, smoking prevalence was only 6 percentage points higher among men than women in 1985, down from 12 percentage points in 1974. The decline in prevalence was somewhat greater for blacks than whites but the 1985 prevalence remained higher for blacks (35 versus 29 per cent). Black men showed the largest decline in smoking prevalence.

The NHIS collects information on whether the respondent ever smoked, in addition to current smoking status. Thus, it is possible to disaggregate trends in smoking prevalence to examine the effects of smoking cessation (defined as the proportion of ever-smokers who are former smokers) from smoking initiation (measured by the prevalence of current smoking among 20–24-year-olds). Fiore *et al.* (1989) reported that smoking cessation increased among all race–sex groups but was highest among black men and white women. Smoking initiation declined more among black men than white men, so that in 1985 the prevalence of smoking among 20–24-year-olds was lower among blacks (28 per cent) than whites (32 per cent). On the other hand, there was essentially no change in smoking prevalence among females aged 20–24 years between 1974 and 1985. The authors concluded that differences in initiation rather than cessation are primarily responsible for the convergence of smoking prevalence among men and women.

Further analysis of these NHIS data provides additional information which can be used to target anti smoking initiatives. Although smoking prevalence has declined among individuals at all educational levels, the decline has been much more rapid among those with greater education. This is the result of both higher cessation and lower initiation of smoking among the more educated group. Education has therefore become the major socio-demographic predictor of smoking status (Pierce *et al.* 1989). Although educational differentials in smoking have not in the past been presented regularly in *Health, United States*, these results are now presented, beginning in 1989.

These results provide important information for planning health promotion activities regarding smoking. It is clear that more attention must be focused on women, especially young women, and on those with less education. Antismoking programmes need to be targeted with these groups in mind.

Although primary prevention forms the major focus of health promotion activities, the importance of secondary prevention in minimizing the impact of chronic illness on health and functional status should not be ignored. The data systems of the National Center for Health Statistics provide several ways of monitoring progress in these activities. Using data from NHIS, Makuc *et al.* (1989) investigated changes

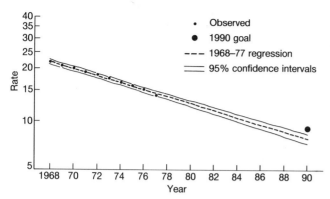

Fig. 6.1.

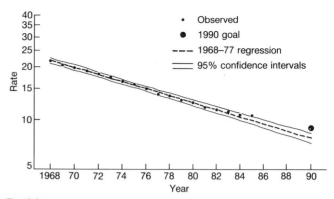

Fig. 6.2.

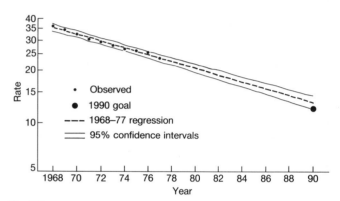

Fig. 6.3.

between 1973–4 and 1985 in women's use of three preventive health services: breast examination, Pap tests, and blood pressure tests. Questions on these specific services were included as supplements to the NHIS in these years. Although progress has been made over this time period in increasing access to these services, the poor remained less likely than the non-poor to have had recent preventive care. Furthermore, despite the increasing risk of cancer and hypertension with age, older women were least likely to receive preventive care. The analysis also showed that most women without recent cancer screening tests had a recent physician contact, highlighting the need for greater emphasis on cancer prevention by health care providers. In a separate analysis of NHIS data, Woolhandler and Himmelstein (1988) found that 'inadequate health insurance coverage leads to "reverse targeting" of preventive care—that is, populations at highest risk are least likely to be screened'.

Setting goals and objectives

Health statistics have not been used to their fullest potential to set goals and objectives. The initial expert deliberations in 1978 which led to *Objectives for the nation* were not always tied to empirical evidence. Many of the problems which arose were identified and revised in the *Midcourse review* (Office of Disease Prevention and Health Promotion 1986), but certain problems remain.

Examples are provided by considering (a) the national goal on infant mortality: by 1990, the national infant mortality rate should be reduced to no more than 9 per 1000; and (b) one of the specific objectives in the priority area 'pregnancy and infant health': by 1990, no county and no racial group should have an infant mortality rate in excess of 12 per 1000.

The goal and the specific objective were set without any formal assessment of past trends in the infant mortality rates. More careful analysis of these trends would have revealed certain inconsistencies. After a plateau during the 1950s and 1960s, infant mortality rates began to decline at a rapid pace. If a formal regression analysis of the annual rates from 1968 to 1977 (using a log–linear model) was projected to 1990, Figure 6. 1 shows what would have been obtained. The 1990 goal of 9 deaths per 1000 live births was conservative in being

somewhat above what the recent trends would have predicted.

As infant mortality rates continue to be monitored through 1985, the observed rates fell within the predicted confidence intervals but tended to drift toward the upper limit until it was surpassed in 1985 (Fig. 6.2). However, since the 1990 goal was well above the upper confidence limit, this goal might still be met and, in fact, the provisional data for 1986 and 1987 indicate that the observed points will be within the confidence limits for those years.

The 1990 race-specific objective presents a very different situation with respect to infant mortality rates among black babies. In 1978, the 1990 objective for black infant mortality was set at 12 deaths per 1000 live births. If a formal regression analysis had been done, Figure 6.3 shows that the 1990 objective would have been seen to be an optimistic projection on the basis of historical trends. It was clear that, in order to achieve the 1990 objective, special efforts (illustrated by the 'intervention' line in Figure 6.4) would be needed to improve the historical trend of a 4.5 per cent per year decline, to a decline of 5.1 per cent per year. New programmes, technologies, or other methods would have to be put in place to reach the objective.

By using the desired trajectory to the objective as the trend to be monitored, appropriate confidence limits could be fitted and the progress toward achieving this objective moni-

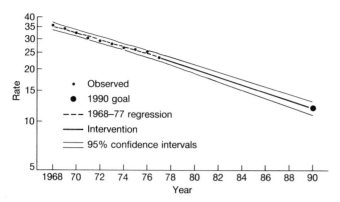

Fig. 6.4.

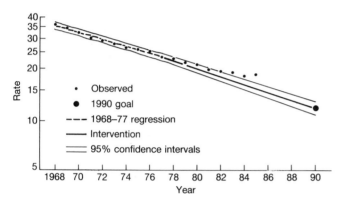

Fig. 6.5.

tored. Unfortunately, the observed mortality rates for black infants began to depart from the targeted trajectory in 1980 and have exceeded the upper confidence limits since 1982 (Fig. 6.5). It seems doubtful whether the 1990 objective will now be met for black infants.

Formal analysis would also have been helpful in pointing out inconsistencies between the overall infant mortality rate goal of 9 deaths per 1000 and the race-specific objective of 12 deaths per 1000 live births. The latter objective was specifically set to accelerate progress in reducing the infant mortality rate among blacks. However, the overall goal of 9 together with the objective of 12 among blacks implies a 1990 rate of 8.5 deaths per 1000 among whites. This would represent only a 2.8 per cent per year decrease from the 1977 infant mortality rate. The observed decrease from 1968 to 1977 was actually 4.5 per cent per year. Clearly, the overall goal should have reflected both an acceleration of the black decline to 5.1 per cent and at least a maintenance of the white decline of 4.5 per cent, which would result in a 1990 rate of about 6.8 per 1000 for whites and a reduction in the black: white ratio from 1.92 in 1977 to 1.76 in 1990. The overall goal under this scenario should have been set at about 7.5 rather than 9. Analysis of past trends and interrelationships among objectives can therefore be helpful in understanding the magnitude of interventions necessary to reach objectives.

It is important to note, however, that analysis of past trends cannot be used mechanically to set goals or objectives. We need to have some idea of the interventions that are available and feasible to meet the objectives. A sophisticated approach to setting objectives using alternative intervention models was recently developed with respect to the objectives for US cancer mortality (Levin *et al.* 1986). In the infant mortality example there was some speculation even at the time the objectives were formulated that the 4.5 per cent per year decline observed during the 1970s was not likely to continue without a major breakthrough in reducing the incidence of prematurity. The implied acceleration of the decline among black infants when setting the objective was laudable but perhaps not based on a realistic assessment of the intervention opportunities.

In setting national health objectives for the year 2000, greater efforts are being made to base them on careful analysis of trends and differentials in health statistics. Consider once again the objective for black infant mortality. The average annual decline in the black infant mortality rate slowed from 4.1 per cent per year between 1970 and 1981 to 2.2 per cent between 1981 and 1986. Extrapolation of recent trends to the year 2000 results in a rate of 13.0 (above the 1990 objective). In order to use this information to help set objectives, it is important to understand why the infant mortality rate decline slowed and to disaggregate the data to examine differences among subpopulations.

The reasons for the slow-down are not entirely clear. The rapid decline in infant mortality rates during the 1970s was fuelled in large part by reduction in birth-weight-specific mortality (due to improvements in neonatal intensive care and the dissemination of that technology throughout the US). Changes in the birth-weight distribution were small during the early part of the 1970s and the incidence of very low birth-weight infants those under 1500 g who are at highest risk of death) actually increased in the latter half of the decade. Changes in the distribution of maternal risk, especially an increase in the proportion of births to unmarried mothers, contributed to the increase in very low birth-weight infants.

A new data system now being implemented by the National Center for Health Statistics will help to disentangle the effects of such maternal and infant factors. Beginning with the 1983 birth cohort, a national file of linked birth and infant death records will be produced annually. The linkage of these records will allow more in-depth analysis of maternal and infant factors that affect infant mortality. For example, the 1983 births were stratified according to race, maternal risk, and infant's birth-weight. Maternal risk was classified into three levels based on age, parity, education, marital status, and prenatal care (Kleinman and Kessel 1987). Birth-weight was divided into three groups (under 1500 g, 1500–2499 g, and 2500 g or more). Within each race the infant mortality rate increased steadily with increasing maternal risk. In fact, if the entire population experienced the infant mortality rate of the lowest risk group, 30 per cent of the infant deaths in each racial group would be averted. By disaggregating infant deaths into its components, we find that (a) the incidence of very low and moderately low birth-weight increases steadily with increasing maternal risk; (b) birth-weight-specific neonatal mortality rates do not vary much by maternal risk except for the births above 2500 g; and (c) birth-weight-specific post-neonatal mortality rates increase with increasing maternal risk for every birth-weight group.

Thus, five areas have been identified in which the performance of the low risk group is better than the performance of the moderate and high risk groups: (1) birth-weight distribution; (2) neonatal mortality among births weighing 2500 g or more; (3)–(5) post-neonatal mortality in each birth-weight group (less than 1500 g, 1500–2499 g, and 2500 g or more).

It is then possible to calculate how each of these factors contributes to the 30 per cent of infant deaths that would be averted if the performance of the lowest risk group could be achieved in the overall population. The birth-weight distribution accounts for almost half the excess white deaths and one-third the excess black deaths. Unfortunately, improving the birth-weight distribution, especially in terms of reducing very low birth-weight births, has proven to be difficult. However, half of the excess deaths in both white and black populations is accounted for by mortality among infants weighing 2500 g or more. Preventive strategies to deal with this problem are much more feasible. For example, increased access to primary paediatric care could have a substantial impact on reducing post-neonatal mortality among normal birth-weight infants (World Health Organization Collaborating Centre 1987). Analyses of this type are now being carried out by the work developing national health objectives for the year 2000 in maternal and infant health in order to develop objectives that are realistic yet challenging.

Estimating resources

Detailed implementation plans were developed to achieve the 1990 objectives (US Department of Health and Human Services 1983). Unfortunately the current state of our knowledge on how to achieve many of the objectives is rudimentary. For example, although the objective for black infant mortality requires an acceleration in the decline, we have little understanding of the factors responsible for the current rate of decline, let alone what types of interventions would be required to accelerate it (US Congress Office of Technology Assessment 1988). The implementation plan for this objective includes programmes that conventional wisdom suggests are appropriate, yet the intensity and duration of such programmes that would be required to affect black infant mortality is unknown. As a result, estimating the resources required to achieve an objective is extremely difficult. It follows that the net cost of such programmes (i.e. the cost minus the benefits) is impossible to estimate.

Monitoring progress

There are 67 objectives for which data are available on a regular periodic basis in the US. Monitoring these data to detect departures from trend (either historical or modified due to interventions) presents statistical and practical problems. The statistical problems are familiar ones faced by any type of surveillance or quality control system. A balance is required between sensitivity (the power to detect departures) and specificity (the probability of false alarms).

For example, the objective discussed earlier states that no county and no racial or ethnic group should have an infant mortality rate in excess of 12 deaths per 1000 live births. It was recognized in the midcourse review of the objectives (Office of Disease Prevention and Health Promotion 1986) that the reference to counties should be deleted because of the problem of interpreting rates based on small numbers. Let us suppose instead that we monitor infant mortality rates by State. For monitoring the black rate, we limit attention to the 26 States with 5000 or more black births per year. If we begin by comparing the 1970–81 decline (based on log–linear regression) with the 1981–5 decline, 19 of the 26 States had slower declines during the more recent period. However, only nine of these States had a statistically significant slow-down (corresponding to a one-sided p-value of 0.05). As noted in a previous article, the power to detect departures from trend is very limited in all but the largest States (Kleinman 1986). Yet the dilemma remains. Even in Texas, a relatively large State, the average annual decline in 1981–5 was 3.2 per cent (with 95 per cent confidence interval (CI) 0.8–5.5 per cent) compared with a 1970–81 decline of 4.9 per cent (with 95 per cent CI 4.3–5.4 per cent). The difference between these slopes was substantial, but not statistically significant. Interpretation of such data is extremely difficult, especially in terms of determining appropriate action.

The practical problems involved are even greater. Even if it is clear that the trajectory of the trend will not meet the objective it is often unclear what the appropriate response should be. Should the intervention plan be modified or do we merely need more of the interventions that were specified?

Conclusion

We have outlined many problems in using health statistics for health promotion and disease prevention. At this stage of development, the most effective and feasible use of health statistics involves the setting of goals and priorities. We have illustrated how past trends can be used in formal statistical analysis to help define reasonable goals and objectives and to clarify the magnitude of interventions required. Analysis can also help us avoid inconsistencies and clarify the interrelationships among goals.

The problems we discussed should not, however, detract from the benefits of using health statistics to plan, target, and monitor health promotion activities. It is only with more extensive use of data-based planning that solutions to many of these problems will be forthcoming.

The data presented annually in *Health, United States* and triennially in the *Prevention profile* have served to alert policy-makers to progress and problems in the health system. These reports have achieved a great deal of visibility, even beyond health professionals. The media have always covered the report prominently. Such reports should be an integral part of every nation's health promotion effort.

Appendix: the list of detailed tables published in *Health, United States*, 1988

Health status and determinants

Population

1. Resident population, according to age, sex, and race: United States, selected years 1950–86.

Fertility and natality

2. Live births, crude birth-rates, and birth-rates by age of mother according to race of child: United States, selected years 1950–86.

3. Birth-rates for women 15–44 years of age, according to live-birth order and race of child: United States, selected years 1950–86.

4. Completed fertility rates and parity distribution for women 50–54 years of age at the beginning of selected years 1930–87, according to race of child and birth cohort; United States, selected birth cohorts 1876–1937.

5. Life-time births expected by currently married women and percentage of expected births already born, according to age and race: United States, selected years 1967–87.

6. Characteristics of live births, according to hispanic origin of mother and race of child: Selected States, 1980–86.

7. Live births, according to race of child and selected characteristics: United States, selected years 1970–86.

8. Infants weighing less than 2500 g at birth, according to race of child, geographical division, and State: United States, average annual 1974–76, 1979–81, and 1984–6.

9. Legal abortion ratios, according to selected patient characteristics: United States, selected years 1973–85.

10. Legal abortions, according to selected characteristics: United States, selected years 1973–85.

11. Legal abortions, abortion-related deaths, and death-rates, and relative risk of death, according to period of gestation: United States, 1974–6, 1977–9, 1980–2, and 1983–5.

12. Methods of contraception for ever-married women 15–44 years of age, according to race and age: United States, 1973, 1976, and 1982.

Mortality

13. Life expectancy at birth and at 65 years of age, according to race and sex: United States, selected years 1900–87.

14. Infant, mortality rates, fetal death-rates, and perinatal mortality rates, according to race: United States selected years 1950–87.

15. Infant mortality rates, according to race, geographical division, and State: United States, average annual 1974–6, 1979–81, and 1984–6.

16. Neonatal mortality rates, according to race, geographical division, and State: United States, average annual 1974–6, 1979–81, 1984–6.

17. Post-neonatal mortality rates, according to race, geographical division, and State: United States, average annual 1974–6, 1979–81, and 1984–86.

18. Fetal death-rates, according to race, geographical division, and State: United States average annual 1974–6, 1979–81, 1984–6.

19. Infant mortality rates, perinatal mortality ratios, and average annual percentage change: Selected countries, 1980 and 1985.

20. Life expectancy at birth, according to sex: Selected countries, selected periods.

21. Death rates for all causes, according to sex, race, and age: United States, selected years 1950–86.

22. Age-adjusted death rates for selected causes of death, according to sex and race: United States, selected years 1950–86.

23. Years of potential life lost before age 65 years for selected causes of death, according to sex and race: United States, 1980, 1985, and 1986.

24. Death-rates for diseases of heart, according to sex, race and age: United States, selected years 1950–96.

25. Death-rates for cerebrovascular diseases, according to sex, race, and age: United States, selected years 1950–86.

26. Death-rates for malignant neoplasms, according to sex, race, and age: United States, selected years 1950–86.

27. Death-rates for malignant neoplasms of respiratory system, according to sex, race, and age: United States, selected years 1950–86.

28. Death-rates for malignant neoplasm of breast for females, according to race and age: United States, selected years 1950–86.

29. Maternal mortality rates for complications of pregnancy, childbirth, and the puerperium, according to race and age: United States, selected years 1950–86.

30. Death-rates for motor vehicle accidents, according to sex, race, and age: United States, selected years 1950–86.

31. Death-rates for homicide and legal intervention, according to sex, race, and age: United States, selected years 1950–86.

32. Death-rates for suicide, according to sex, race, and age: United States, selected years 1950–86.

33. Deaths for selected occupational diseases for males, according to age: United States, selected years 1970–86.

34. Provisonal death-rates for all causes, according to race, sex, and age: United States, 1985–7.

35. Provisional age-adjusted death-rates for selected causes of death: United States, 1985–7.

36. Provisional death-rates for the three leading causes of death, according to age: United States, 1985–7.

Determinants and measures of health

37. Progress toward 1990 health promotion goals: 1977–86.

38. Vaccinations of children 1–4 years of age for selected diseases, according to race and residence in metropolitan statistical area: United States, 1970, 1976, and 1983–5.

39. Selected notifiable disease rates, according to disease: United States, selected years 1950–87.

40. Acquired immune deficiency syndrome cases, according to age, sex, and race/ethnicity: United States, 1982–8.

41. Acquired immune deficiency syndrome deaths, according to age, sex, and race/ethnicity: United States, 1982–8.

42. Acquired immune deficiency syndrome cases, according to race/ethnicity, sex, and transmission category for persons 13 years of age and over: United States, 1982–8.

43. Acquired immune deficiency syndrome deaths, according to race/ethnicity, sex, and transmission category for persons 13 years of age and over: United States, 1982–8.

44. Acquired immune deficiency syndrome cases, according to geographical division and State: United States, 1982–8.

45. Acquired immune deficiency syndrome deaths, according to geographical division and State: United States, 1982–8.

46. Age-adjusted cancer incidence rates for selected cancer sites, according to sex and race: Selected years 1973–86.

47. Five-year relative cancer survival rates for selected sites, according to race: 1974–6, 1977–9, and 1980–5.

48. Limitation of activity caused by chronic conditions, according to selected characteristics: United States, 1983 and 1987.

49. Disability days associated with acute conditions and incidence of acute conditions according to age: United States, 1982–7.

50. Self-assessment of health, according to selected characteristics: United States, 1983 and 1987.

51. Cigarette smoking by persons 20 years of age and over, according to sex, race, and age: United States, 1965, 1976, 1983, and 1987.

52. Use of selected substances in the past month by youths 12–17 years of age and young adults 18–25 years of age, according to age and sex: United States, selected years 1972–85.

53. Alcohol consumption status of persons 18 years of age and over, according to sex: United States, selected years 1971–85.

54. Borderline or definite elevated blood pressure for persons 25–74 years of age, according to race, sex, and age: United States, 1960–2, 1971–4, 1976–80.

55. Definite elevated blood pressure for persons 25–74

years of age, according to race, sex, and age: United States, 1960–2, 1971–4, and 1976–80.

56. High-risk serum cholesterol levels for persons 25–74 years of age, according to race, sex, and age: United States, 1960–2, 1971–4, and 1976–80.

57. Overweight persons 25–74 years of age, according to race, sex and age: United States, 1960–2, 1971–4, and 1976–80.

58. Air pollution, according to source and type of pollutant: United States, selected years 1970–86.

59. Employees with potential exposure to continuous noise without controls in selected industries, according to size of facility: United States, 1972–4 and 1981–3.

60. Health and safety services in manufacturing industries, according to size of facility: United States, 1972–4 and 1981–3.

Utilization of health resources
Ambulatory care

61. Physician contacts, according to place of contact and selected patient characteristics: United States, 1983 and 1987.

62. Interval since last physician contact, according to selected patient characteristics: United States, 1964, 1982, and 1987.

63. Office visits to physicians, according to physician specialty and selected patient characteristics: United States, 1980 and 1985.

64. Office visits to physicians, according to selected patient characteristics: United States, 1980 and 1985.

65. Dental visits and interval since last visit, according to selected patient characteristics: United States, 1964, 1981, and 1986.

In-patient care

66. Discharges, days of care, and average length of stay in short-stay hospitals, according to selected characteristics: United States, 1964, 1981, and 1987.

67. Discharges, days of care, and average length of stay in non-Federal short-stay hospitals, according to selected characteristics: United States, 1980–7.

68. Discharges, days of care, and average length of stay in non-Federal short-stay hospitals for patients discharged with the diagnosis of acquired immune deficiency syndrome and for all patients: United States, 1984–7.

69. Rates of discharges and days of care in non-Federal short-stay hospitals, according to sex, age, and selected first-listed diagnosis: United States, 1980 1985, and 1987.

70. Discharges and average length of stay in non-Federal short-stay hospitals, according to sex, age, and selected first-listed diagnosis: United States, 1980, 1985, and 1987.

71. Operations for in-patients discharged from non-Federal short-stay hospitals, according to sex, age, and surgical category: United States, 1980, 1985, and 1987.

72. Diagnostic and other non-surgical procedures for in-patients discharged from non-Federal short-stay hospitals, according to sex, age, and procedure category: United States, 1980, 1985, and 1987.

73. Admissions, average length of stay, and out-patient visits in short-stay hospitals, according to type of ownership: United States, selected years 1960–86.

74. Nursing home and personal care home residents 65 years of age and over and rate per 1000 population, according to sex and race: United States, 1963, 1973–4, 1977, and 1985.

75. Nursing home residents, according to selected functional status and age: United States, 1977 and 1985.

76. Admissions to mental health organizations and rate per 100 000 civilian population, according to type of service and organization: United States, selected years 1969–86.

77. In-patient and residential treatment episodes in mental health organizations, rate per 100 000 civilian population, and in-patient days, according to type of organization: United States, selected years 1969–86.

78. Admissions to selected in-patient psychiatric organizations and rate per 100 000 civilian population, according to sex, age, and race: United States, selected years 1970–80.

79. Admissions to selected in-patient psychiatric organizations, according to selected primary diagnoses and age: United States, 1975 and 1980.

Health care resources

Personnel

80. Persons employed in health service sites: United States, selected years 1970–87.

81. Active non-Federal physicians per 10 000 civilian population, according to geographical division, State, and primary specialty: United States, 1975, 1985, and 1986.

82. Active physicians, according to type of physician, and number per 10 000 population: United States and outlying US areas, selected 1950–86 estimates and 1990 and 2000 projections.

83. Physicians, according to activity and place of medical education: United States and outlying US areas, selected years 1970–86.

84. Active health personnel and number per 100 000 population, according to occupation and geographical region: United States, 1970, 1980, and 1986.

85. Full-time equivalent employment in selected occupations for community hospitals: United States, 1981 and 1984–6.

86. First-year enrolment and graduates of health professions schools and number of schools, according to profession: United States, selected 1950–87 estimates and 1990 and 2000 projections.

87. Total and first-year enrolment of minorities and women in schools for selected health occupations: United States, academic years 1976–7 and 1986–7.

88. Total and first-year enrolment and percentage of women in schools of medicine, according to race and ethni-

city: United States, academic years 1971–2, 1977–8, and 1986–7.

Facilities

89. Short-stay hospitals, beds, and occupancy rates, according to type of ownership: United States, selected years 1960–86.

90. Long-term hospitals, beds, and occupancy rates, according to type of hospital and ownership: United States, selected years 1970–86.

91. In-patient and residential treatment beds in mental health organizations and rate per 100 000 civilian population, according to type of organization: United States, selected years 1970–86.

92. Community hospital beds per 1000 population and average annual percentage change, according to geographical division and State: United States, selected years 1940–86.

93. Occupancy rate in community hospitals and average annual percentage change, according to geographical division and State: United States, selected years 1940–86.

94. Full-time equivalent employees per 100 average daily patients in community hospitals and average annual percentage change, according to geographical division and State: United States, selected years 1960–86.

95. Nursing homes with 25 or more beds, beds, and bed rates, according to geographical division and State: United States, selected years 1940–86.

Health care expenditures

National health expenditures

96. Gross national product and national health expenditures: United States, selected years 1929–96.

97. Total health expenditures as a percentage of gross domestic product: Selected countries, selected years 1960–86.

98. National health expenditures and percentage distribution, according to type of expenditure: United States, selected years 1950–86.

99. National health expenditures average annual percentage change, according to type of expenditure: United States, selected years 1950–86.

100. Personal health care expenditures average annual percentage change and percentage distribution of factors affecting growth: United States, 1965–86.

101. Consumer Price Index and average annual percentage change for all items and selected items: United States, selected years 1950–87.

102. Consumer Price Index for all items and medical care components: United States, selected years 1950–87.

103. Consumer Price Index average annual percentage change for all items and medical care components: United States, selected years 1950–87.

Sources and types of payment

104. Hospital expenses and personnel and average annual percentage change: United States, 1971–86.

105. National health expenditures and average annual percentage change, according to source of funds: United States, selected years 1929–86.

106. Personal health care expenditures and percentage distribution, according to source of funds: United States, selected years 1929–86.

107. Expenditures on hospital care, nursing home care, and physician services and percentage distribution, according to source of funds: United States, selected years, 1965–86.

108. Nursing home average monthly charges per resident and percentage of residents according to primary source of payments and selected facility characteristics: United States, 1977 and 1985.

109. Nursing home average monthly charges per resident and percentage of residents, according to selected facility and resident characteristics: United States, 1964, 1973–4, 1977, and 1985.

110. National funding for health research and development and average annual percentage change, according to source of funds: United States, selected years 1960–87.

111. Federal obligations for health research and development and percentage distribution, according to agency: United States, selected fiscal years 1970–86.

112. Obligations for human immunodeficiency virus related activities by National Institutes of Health and other Public Health Service agencies: United States, fiscal years 1982–7.

113. Public health expenditures by State and territorial health agencies, according to source of funds and programme area: United States, selected fiscal years 1976–86.

114. Personal health care per caput expenditures and average annual percentage change, according to geographical division and State: United States, selected years 1966–82.

115. Hospital care per caput expenditures and average annual percentage change, according to geographical division and State: United States, selected years 1966–82.

116. Nursing home care per caput expenditures and average annual percentage change, according to geographical division and State: United States, selected years 1966–82.

Health care coverage and major federal programmes

117. Health care coverage for persons under 65 years of age, according to type of coverage and selected characteristics: United States, 1980, 1982, and 1986.

118. Health care coverage for persons 65 years of age and over, according to type of coverage and selected characteristics: United States, 1980, 1982, and 1986.

119. Health maintenance organizations and enrolment, according to model type, geographical region, and Federal programme: United States, selected years 1976–87.

120. Medicare enrollees and Medicaid recipients and expenditures and percentage distribution, according to type of service: United States, selected years 1967–86.

121. Medicare enrolment, persons served, and reimbursements for Medicare enrollees 65 years of age and over,

according to selected characteristics: United States, selected years 1967–86.

122. Selected rates of non-Federal short-stay hospital utilization and benefit payments for aged and disabled Medicare enrollees, according to geographical division: United States, 1980, 1984, and 1986.

123. Recipients and Medicaid medical vendor payments, according to basis of eligibility: United States, selected years 1972–87.

124. Veterans medical care expenditures and percentage distribution, according to type of service: United States, selected fiscal years 1965–87.

125. State mental health agency per caput expenditures for mental health services, by State: United States, fiscal years 1981, 1983, and 1985.

Source: National Center for Health Statistics (1989).

References

Feinleib, M. (1988). *Mortality surveillance systems (3rd annual Harry A. Feldman lecture)*. The 61st Annual Meeting of the American Epidemiological Society. March 14–15, 1988. San Diego, California.

Fiore, M.C., Novotny, T.E., Pierce, J.P., Hatziandreu, E.J., Patel, K.M., and Davis, R.M. (1989). Trends in cigarette smoking in the United States. *Journal of the American Medical Association* **261**, 49.

Kleinman, J.C. (1986). State trends in infant mortality, 1968–83. *American Journal of Public Health* **76**, 681.

Kleinman, J.C. and Kessel, S.S. (1987). Racial differences in low birth weight: Trends and risk factors. *New England Journal of Medicine* **317**, 749.

Levin, D.L., Gail, M.H., Kessler, L.G., and Eddy, D.M. (1986). A model for projecting cancer incidence and mortality in the presence of prevention, screening, and treatment programs,. In *NCH monograph No.2: Cancer control objectives for the nation, 1985–2000*. National Cancer Institute, Bethesda, Maryland.

Makuc, D.M., Freid, V.M., and Kleinman J.C. (1989). National trends in the use of preventive health care by women. *American Journal of Public Health* **79**, 21.

National Center for Health Statistics (1976). *Health, United States, 1975*. DHEW publication No. (HRA) 76–1232. Health resources Administration, Rockville, Maryland.

National Center for Health Statistics (1986). *Health, United States, 1986 and prevention profile*. DHHS publication No. (PHS) 87–1232. Public Health Service, Hyattsville, Maryland.

National Center for Health Statistics (1988). Compressed mortality file, 1968–85. *Public use data tape documentation*. Public Health Service, Hyattsville, Maryland.

National Center for Health Statistics (1989). *Health, United States, 1988*. DHHS publication No. (PHS) 88–1232. Public Health Service, Hyattsville, Maryland.

Office of the Assistant Secretary for Health and Surgeon General (1979). *Healthy people—Surgeon General's report on health promotion and disease prevention, 1979*. DHEW Publication No. (PHS) 79–55071. Public Health Service. US Government Printing Office, Washington, DC.

Office of Disease Prevention and Health Promotion (1986). *The 1990 health objectives for the nation: A midcourse review*. Public

Health Service. US Government Printing Office, Washington, DC.

Pierce, J.P., Fiore, M.C., Novotny, T.E., Hatziandreu, E.J., and Davis, R.M. (1989). *Trends in cigarette smoking in the United States. Journal of the American Medical Association* **261**, 56.

US Congress, Office of Technology Assessment (1988). *Healthy children: Investing in the future*. OTA-H-345. US Government Printing Office, Washington, DC.

US Department of Health and Human Services (1980). *Promoting health/preventing disease: Objectives for the nation*. US Government Printing Office, Washington, DC.

US Department of Health and Human Services (1983). *Promoting Health/preventing disease. Public Health implementation plans for attaining the objectives for the nation. Public health reports.* Supplement to the September/October 1983 issue. US Government Printing Office, Washington, DC.

US. Department of Health, Education and Welfare (1964). *Smoking and Health*. Report of the Advisory Committee to the Surgeon General of the Public Health Service. US Government Printing Office, Washington, DC.

Woolhandler, S. and Himmelstein D.U. (1988). Reverse targeting of preventive care due to lack of health insurance. *Journal of the American Medical Association* **259**, 2872.

World Health Organization Collaborating Center in Perinatal Care and Health Service Research in Maternal and Child Care (1987). Unintended pregnancy and infant mortality morbidity. In *Closing the gap: The burden of unnecessary illness* (ed. R.W. Amler and H.B. Dull). Oxford University Press, New York.

B

Epidemiological Approaches

7

Spatial and temporal studies in epidemiology

E.G. KNOX

Fundamentals

Classical epidemiology was concerned almost entirely with communicable disease; and the epidemiology of the last thirty years very largely with chronic disease. These at least were the dominant if not the exclusive themes of the two eras. The scope of epidemiology is recognized as universal (or so most epidemiologists would claim) but these two basic themes have left its methodology divided. Epidemiologists tend to specialize in one or the other; and these original preoccupations are still reflected in two separate technical repertoires.

Indeed, the division of the subject is *better* expressed in methodological terms than in terms of subject matters. The technical division is between the epidemiology of 'events' and the epidemiology of 'states'. The first is associated with temporal graphs and plots on maps; the second with contingency tables. States and events are dimensionally different and the clue to understanding the different technical requirements of different investigations lies in understanding this distinction. It is this which dictates the observational and the analytical techniques to be employed. Events are necessarily dimensioned in time while states, whether categorical or quantitative, are not. Records of either, at the investigator's option and according to the objectives of the enquiry, may or may not be located in space.

We elaborate on these points as follows.

Dimensionalities

States (whether measured or categorical) exist over a *period* of time. The appropriate adjective is 'static'! The observer can intercept a 'state' at any single point in time between its onset and its termination. He needs to visit only *once*: and there is no point in visiting twice. Consequently, the frequency of a state is expressed as a 'point prevalence' (= 'point prevalence rate') which in technical terms is a simple proportion, undimensioned in time. This is as befits a record based upon a single observation, without the benefit of a clock and without the measurement of passing time.

Events (deaths, traffic accidents, measles) occur at a *point* in time. Unlike states, they cannot be intercepted by an observer who himself attends only at a single point in time. To intercept and record an event, the investigator must observe over a *period* of time. He must observe continuously, or else return repeatedly. Frequencies of events are expressed as an event-rate (or 'attack-rate' or 'incidence' or 'incidence-rate'), which is expressed per unit time as well as per 1000 population. This befits a *period* of observation between *two* declared points in time, whose duration must be measured.

The inversion of the relationship between *points* and *periods* is important. A time-extended phenomenon demands a point observation; a point phenomenon demands time-extended observation. We should note that it is the dimensionality of the observing mode which imposes the dimensionality of the frequency measure: not the duration of the phenomenon itself.

Events and states are related to each other through processes analogous with mathematical integration and differentiation. If a constant incidence rate is multiplied by the time during which it operates, the result is a 'prevalence', or, more properly, a 'prevalence increment'. Thus, the multiplication by 'time' (t) of a unit initially measured 'per unit time' (t^{-1}), results in a cancelling-out of the time element; the resulting 'prevalence' is dimensionless. Where the incidence varies over time, the accumulation process corresponds with integration rather than simple multiplication, but the dimensionality consequences are the same. Conversely, where the 'state' of an individual is ascertained at two different times, and where the state has changed, we can claim to have detected an 'event'. The analogy here is with algebraic differentiation.

An event is then formally definable as a 'change of state'. When an 'event' is observed indirectly, thus, it is usual to locate it arbitrarily at a point in time mid-way between the two observations. It is possible through the use of record linkage operations to create a file of observations from which the occurrence of many events can be inferred without any having been directly witnessed. For a series of such paired observations, and through a differencing process (to ascertain the change of state and to measure the period of time) we generate a time-dimensioned 'incidence'.

Our concern in this chapter is more with the pragmatic solution of practical problems than with the syntax of medical and scientific observations, but we begin with the formal differentiation of different kinds of observation because this provides an orderly framework within which to locate a varied constellation of questions. Epidemiologists are constantly bombarded with questions relating to incidence and others relating to prevalence; to disease events and to disease states; to questions relating to temporal and to geographical patterns, and other questions relating to neither. The patterns may take the forms of trends or cycles or steps or irregular clusters. They may be presented as 'pure' phenomena, without reference to anything else on the map or on the calendar; or they may be presented in conjunction with putative hazards, raising questions of co-variations between them and the putative consequences. They may also relate to space and time jointly . . . a conceptual distribution of events and hazards within a space–time block.

In all these circumstances it is necessary to distinguish legitimate analytical operations from illegitimate and to distinguish what it is about a particular question which demands one technical approach as opposed to another. These distinctions have also guided the construction of this chapter.

Real and random

A second necessary preoccupation is to distinguish between 'real' and 'random' phenomena in space, or in time, or in space and time jointly. As in other branches of epidemiology, this is subsumed within the subject-matter of 'sampling theory'. We ask whether the phenomena displayed in our sample are the products of the sampling process alone, or whether they reflect real world phenomena in the universe from which the sample was drawn. This is the second major theme which has guided the construction of this chapter.

Styles of analysis

Investigators analysing chronic disease states and exploring aetiologies through the medium of contingency tables, seldom pay much attention to the timing or the geographical spacing of their observations. These analyses concentrate upon measuring the frequencies of coexistence of different *pairs* of states (smokes, has cancer) within the same subject. 'Events', as recorded in this kind of study, are generally represented as static attributes: has *had* 'x' (e.g. has had previous tonsillectomy, has had recent poliomyelitis). Contingency analyses such as these have formed a very large proportion of the corporate epidemiological effort reported in the scientific literature over the last 30 years. This style of working, both the collection of data and the subsequent analysis, is quite different from the methods discussed in this chapter. Why should this be? What is the technical basis of the distinction?

The clue to understanding is in the formal file- and record-operations used in each style of working. Briefly, one style of analysis is concerned with *intra*-record relationships and the other with *inter*-record relationships.

Thus, contingency analysis depends upon the simultaneous coexistence of different attributes within the same record; and where these attributes are noted in separate records, a major element of such investigations lies in bringing them together within a new record, through record-linkage procedures. Thus, the original technical basis of contingency studies was 'the consolidated medical record', supported by such devices as the Hollerith card and (later) by efficient record handling computer systems. These processes have been well supported by the development of statistical tabulation processes and multivariate analysis methods (e.g. in SPSS and GLIM) imported from other disciplines. By contrast, spatial/temporal investigations study the relationships *between* records, distances apart and times apart, and the dispersion of different recorded observations to different locations and different dates on the calendar. Often, the *different* records note the *same* class of event in different people; as opposed to the same record noting different classes of events in the same person.

Inter-record operations invoke complexities beyond those encountered in intra-record analyses. In record-linkage operations, for example, n records offer n^2 relationships between different pairs of records, far greater than for the types of record-linkage (strictly file-merging) operations referred to earlier. If we are dealing with larger groupings, for example triplets and quadruplets, then the analytical task is very onerous indeed.

There is also a serious problem when we try to relate records through matching personal identifiers. Problems of these kinds arise in kindred-studies, including the assembly of sibships, and in the investigation of contacts in the study of communicable diseases. For simple file-merging operations we can cope with a proportion of near-miss matches and with varied spelling and clerical errors; but for intra-file inter-record linkage operations a very large number of false positive near-miss matches can outnumber and bury the true links.

Finally, the investigator is faced with model-fitting problems relating to natural-histories, disease-transmission, latent intervals and contact/diffusion mechanisms of daunting complexity. They are far more difficult to formalize within generalized software systems than those commonly encountered within the framework of multivariate analysis.

Geographical display and analysis

The majority of geographical patterns are interpreted intuitively, at least in the first instance. This is commonly referred to as 'eyeball' epidemiology. The most pervasive restriction upon the display and interpretation of geographically located morbidity is the uneven distribution of the populations at risk. This nearly always forces us to use rates (prevalences, incidences, etc.) so that we take simultaneous account of the distribution of denominators as well as numerators.

The most characteristic forms of presentation are the choropleth map, the isopleth (or isoline) map, and the spot map. We also frequently need to use a technical procedure described as 'map-on-map' in which (for example) a disease-distribution map is overlaid upon a hazard distribution map. (See McGlashan (1972) for a discussion of geographical techniques.)

Choropleth mapping

This most familiar of all geographical formats uses geo-political or other boundaries related to known populations-at-risk. Attack rates of events, or prevalences of states, are calculated for each sub-area. The different rates are represented by differential shading or hatching or colouring (see Gardner *et al.* (1983) for a good example).

There is a hazard of interpretation. A large sparsely populated rural area will tend to catch the eye of the investigator, while a small densely populated city may almost escape notice. A useful modification which corrects for this is the 'demographic' map in which the different zones are shrunk or expanded so that their areas are proportional to the populations at risk.

For sparse observations, individual events can be marked as spots on the map. On a demographic map the exact locations can be difficult to identify, but once achieved, there are some special advantages. Appearances of spot-clusters on a non-demographic map chiefly represent the uneven distribution of the population at risk; but on a demographic map they represent genuine concentrations of risk.

Distances on a demographic map do not represent physical distances (e.g. kilometres), so biological interpretation is not straightforward. Distance-deformation will be a disadvantage in examining the data for the presence of a 'toxic cloud' diffusion-model, but an advantage in the study of a case-to-case infection hypothesis. People (and animals such as foxes) tend to travel further for their social contacts when they live in sparsely populated areas, than they do when they live at high densities. The rate of geographical spread of rabies by foxes is almost independent of the density of foxes.

Isopleth mapping

An isopleth (or isoline) map uses contours of equal disease density. Its form is not then dependent upon geo-political outlines. It looks like an ordinary elevation contour or an isobar or isotherm map. The mapping technique draws its own lines, and does not rely upon arbitrary boundaries.

Isopleth mapping avoids dependence upon a direct knowledge of local populations by collecting a series of 'control observations', as well as the disease observations. For example, it compares the distribution of malformed births against a sample of normal births. It can use one or several normal controls for each abnormal.

This technique can be used where adequate civil data do not exist at all, as in underdeveloped regions, but it is also useful where the epidemiological hypothesis under investigation is not easily expressed in terms of the available geo-political divisions. For example, if we wished to see whether anophthalmos occurred in a localized industrial zone of a city, or whether hydrocephalus tended to occur in small clusters irrespective of any map-feature, we might find that available census or electoral divisions were entirely unsuited to our purposes.

Choropleth presentations cannot show fine detail, whereas an isopleth map can in principle display small clusters as concentric nests of closely spaced isolines. In practice, however, isopleth maps have a serious *dis*advantage. They are often based upon small numbers of control observations compared with the large denominators available for choropleth rates. This leads to a loss of resolution. Suspected disease concentrations visible on a choropleth map can disappear altogether within the noise of sampling variations, on an isopleth map.

Total population denominators, set out in small areas to permit flexible reassembly, offer the best of both worlds. Automatic map referencing of postal codes will then provide a basis for coarse-resolution isopleth mapping, for example (in a city) in 0.5-km squares. Commercial pressures for marketing-information are likely in the very near future to supply population-wide address-data at a high level of resolution in computer-processable form.

Map-on-map techniques

We have not yet considered what else may be found on the map apart from disease observations and the populations at risk. Yet we frequently need to bring disease-observations into visual apposition with atmospheric pollution levels, income distributions, other indices of affluence and poverty, power stations, high tension transmission lines, roads, railways and canals, or recreation areas. For this, at the visual level, we have to combine *two* maps. The *method* depends upon the formats of the original maps.

If death-rates from chronic bronchitis, and a range of socio-economic indices such as car ownership statistics, are both available on a congruent choropleth basis, then we can find a means of expressing them side by side; or we might even re-draw the mortality-map as an observed/expected mortality ratio. However, in relating bronchitis deaths to particulate smoke deposition, we might find that the pollution data had been set out in isopleth form, having been derived from a large number of measuring stations which were geographically unrelated to electoral population boundaries. We might then simply superimpose the pollution contours upon the choropleth map, and interpret them visually as best we can.

The problems of relating different map formats in any strictly formal manner are extremely difficult. It is usually necessary to fall back upon parallel correlation analyses with no intrinsic geographical dimension. The problems are heightened because disease phenomena are not always represented best as single points; the biological model under investigation which will itself dictate the necessary mode of presentation.

In childhood leukaemia we may be interested in several alternative points, namely the address at birth, the address at onset and the address at death. For tuberculosis we will be interested in a complete trace of addresses at different ages. For suspected Legionnaires' disease we would trace in great detail all the journeys undertaken and all the places visited, within a restricted period of time.

The choice of format for representing map-features also presents many options, but the choice is not *entirely* arbitrary. In general, map-features representing hazards come in three main forms, namely (a) points (for example, nuclear processing plants); (b) linear features (canals, roads, power lines); and (c) areal features (zones of affluence, high/low pollution zones). For points and for linear features we will be chiefly interested in 'distance from'; for areal features we will be chiefly interested in 'within-or-without'.

Because of these difficulties, many investigators have had to coerce reconciliations between different types of map-on-map plots. For example, background radiation levels collected at scattered points and subsequently converted into an isopleth map, have been related to choropleth data for childhood cancer death-rates whose distribution is set out in quite another manner (for example, civil registration areas). Both distributions were forced into yet another mode, namely the 10-km squares of the National Cartographic Grid (Knox *et al.* 1988). In other examples, civil-area data on leukaemias and populations at risk have been forced into a correspondence with circles of different radii centred upon nuclear processing plants (Cooke-Mozaffari *et al.* 1987).

Computer generated displays

Geographers use a wide variety of map-on-map techniques at a purely visual level. Epidemiologists, wistful still about the possibilities of formal inference, have been less prolific in devising or using them. The availability of computers, and especially micro-computers with coloured screens and relatively inexpensive graph plotters with coloured pens, have engendered a growth of new methods in which the form of presentation, if not the mode of inference, follows formal rules. We are likely to see a major development of these techniques over the next decade. Our purpose in this section is to describe a few examples.

For example, there are now formal algorithms for converting an ordinary choropleth into a demographic map. There is no longer any need to construct demographic maps on a trial and error basis using large quantities of graph paper (Selvin *et al.* 1987).

There is also an engaging format of display for congruent choropleth maps of two different variables, one set over the other. One variable is represented in a shaded colour system transversing the columns of a matrix; from left to right it uses shades graduating from yellow (low incidence) through inter-

mediate shades (orange) to red. The other variable is represented in the rows of the matrix, moving upwards from yellow (low social index) through shades of green to blue. The left-to-right and the bottom-to-top scales, superimposed, produce a chequer-board matrix, with any combination of levels on each of the two scales marked as a unique colour-mix. The colour-mixes in the cells of the matrix are then transferred to the choropleth divisions of the map. The map thus represents a superposition of two coloured maps. If the two variables are positively correlated, then the predominant colours will stretch along the yellow-to-purple diagonal of the matrix; if they are negatively correlated, then they will stretch along the red to blue diagonal.

With a little practice, this permits immediate recognition of the type of correlation (positive or negative), without any necessary reference to numerical data, and at the same time locates the zones of greatest positive or negative association. It also instantly shows the atypical outliers. A large number of combinations of different social, physical and morbidity variables may be inspected seriatim.

Isopleth presentations can be performed on simple monochrome devices such as dot matrix printers, but the ideal device is the colour-pen graph-plotter. The value which is plotted, and represented by the isolines, is the 'relative density' of cases and controls. There are several alternative algorithms for estimating relative density. One uses the sums of the inverse squared distances from the map point in question, to the locations of all the events, and to the locations of all the controls. The granularity of the resulting map derived from all these points can then be smoothed and the contours traced in order to separate the different density zones. For isopleth-on-isopleth plots, the two sets of lines may simply be traced in different colours, for inspection. Several commercial packages for calculating isopleths exist, but none has yet won a routine place in epidemiological analysis.

A computer-generated cluster-display method described by Openshaw *et al.* (1988) and used to study childhood leukaemias, is based upon a hybrid presentation. The denominators (populations) are derived from choropleth data while the numerators consist of exactly located events. Every point on the map is scanned. Circles of different radii around each point are used to estimate alternative 'populations-at-risk' and significant departures of the associated event-counts from a random expectation are marked. The authors use the drawn perimeter of the circle itself as the map 'marker', and clusters are displayed as sets of concentric/overlapping circles. The algorithm displays points of 'statistical significance' rather than points of high incidence, *per se*.

Spot maps, clustering, and proximity analyses

Where large numbers of events are available for mapping, such that pre-defined sub-areas can have 'prevalences' attached to them, then we can apply correlation techniques to pairs of areas, to see whether similarities of rates are correlated with adjacency or non-adjacency (Kemp *et al.* 1985).

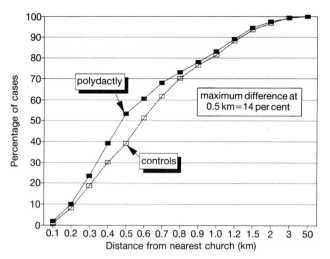

Fig. 7.1. Polydactyly and churches in Birmingham; cumulative proximity distribution

Such an approach will detect geographical heterogeneities which have a granularity of larger diameter than the individual areas themselves.

However, geographic plots of disease events or disease states may be so sparse that we can scarcely think of them in terms of continuous variables such as prevalences or incidences. They arrive in distinct quanta. We must then base our analyses upon the measured distances between pairs of events, pairs of controls and case–control pairs. For example, we might plot a distribution of the distances between all possible pairs of cases, and another between all possible pairs of controls. If the proportional distributions differ, such that there is a relative excess of short inter-pair distances among the cases, we can conclude that there is a relative degree of geographical clustering.

If we wish to study distance-relationships between cases, controls, and putative hazards, we compare the distributions of control-to-hazard distances with case-to-hazard distances. We refer to this as a 'proximity analysis'. In effect, we follow the perimeter of an expanding circle centred on the position of the hazard, and plot the progressive accumulation of 'cases' as the radius enlarges. We do the same for controls and plot the two incremental curves on a single graph. If there is more than one hazard point then we expand the radii simultaneously around each of these points, accumulating cases and controls as we do so. If the hazard is a linear feature such as a high tension powerline, we drop perpendiculars from the positions of the events and the controls to the nearest point of the nearest element of the linear feature in question. We then compare the cumulative acquisition curves for the cases and the controls as we move away from the suspected hazard. We should use several different sets of controls. If the tested hypothesis is correct, then the 'case-curve' will differ from the 'control-curves' (see Fig. 7.1).

Two main tests of significance for such differences are available. One seeks and tests the significance of the *maximum ratio* between the cumulative observed and the cumu-

lative expected values (Stone 1988); the other is a version of the Kolmogorov–Smirnov statistic which seeks and tests the significance of the *maximum difference* between two separate curves (see Siegel (1956) for details of this and related tests). The Stone method is probably best suited to the testing of precise postulated diffusion-hazards, for example occurrences of childhood leukaemias at different distances from nuclear processing plants. The second method is probably best suited to demonstrating less specific proximity hypotheses; for example, the increased birth prevalence of polydactyly in Birmingham close to churches, as shown in Fig. 7.1, probably reflects the high prevalence among Afro/Caribbean immigrants, and their geographical concentration in dense mid-nineteenth century housing areas with plenty of churches.

For areal map features we may be interested either in distance from the boundary (e.g. lakes, uninhabited wooded areas) or alternatively (e.g. polluted zones) with the question whether the events and controls occur inside or outside the boundary. The distinction between 'inside' and 'outside' turns out to be an intractable philosophical question. The simplest route to its pragmatic solution in programming terms seems to be through the use of Jordan's theorem. An arbitrary point is declared as 'outside'—for example the grid-origin of the map—and a line is drawn from it to each event and to each control. We count the number of points where this line crosses a boundary segment of any of the areas in question. If the number is even then the test-point is also 'outside' all of them; if it is odd then it is 'inside' one of them.

Temporal patterns

The temporal scale is uni-dimensional and it is used to seek uni-dimensional phenomena. It is usual first to seek (a) linear trends (upwards or downwards), and (b) cycles (hour of day, month of year, etc.). Trends, sometimes referred to as 'secular' trends, may extend across the full range of the time-period: or they may be quite localized, when they are usually referred to as 'steps'. Curvilinear trends, as well as linear trends, may be sought. Finally, the investigator looks for non-linear, non-cyclical, non-random residual phenomena. This, he often refers to as 'clustering'. He tends to regard it as something to be sought *after* the more systematic forms of variation have been detected and excluded (i.e. de-trended: de-cycled).

Basic formats

There are two basic formats for representing a time-series of events. First, we can set a series of rates or of integer values on different dates, or in different weeks or in other short periods. The durations of the periods are arbitrary. The representation is analogous with a choropleth map. Second, we can simply record a series of exact dates and times, to whatever resolution we wish, and we can measure the time intervals between successive pairs, or between all possible pairs. The point-format can be converted to the first, with some loss of information, but the first may not be convertible to the second.

Trends

Trends are usually sought using the first of the two formats; each time-period records a number of events (e.g. cases of meningitis in each week). A simple linear regression coefficient can be calculated using the numbers or rates as the y-variable, and the sequence number of the time periods (e.g. weeks) as the x-variable. The line of best fit is calculated and drawn and a standard significance test is applied. Residual variations of frequency ('de-trended' values) can be calculated as the differences between the observed values and those predicted from the regression line.

It is often sufficient with calendar data to operate on absolute counts. However, if the population at risk varies over the period of investigation, or is suspected of varying, then we must convert the absolute frequencies to attack rates through reference to the appropriate denominators.

Cycles

Periodicities are also generally sought using the first basic format. Sometimes we can use numbers, or de-trended numbers, but it is usually safer to use rates based upon appropriate denominators. For example, mortality from cerebrovascular accident in the UK has in recent years exhibited both a steady year-by-year decline and a well-marked seasonal variation. The curve can be 'de-trended' using the first technique, and periodicities displayed on the basis of the residual variations. The raw numbers should be sufficient here, since equal numbers of persons are at risk each month. However, if we were examining the birth prevalence of a malformation, and if we knew or suspected that the numbers of all births varied seasonally, then we would need to calculate the rates for each month. Edwards (1961) devised a method for testing departure from an even non-cycling distribution of uncorrected numerators. A modification devised by Walters and Elwood (1975) incorporated a consideration of varying denominators.

Steps

Steps, which in truth are very short trends, are frequently of interest in relation to environmental hazards. One particular detection procedure, with a long tradition of industrial usage, has recently been adapted towards the monitoring of the frequency of congenital malformations; this is the so-called Cusum procedure (Weatherall and Haskey 1976). An incremental curve of acquisitions of events is simultaneously decremented at each 'tick of the clock' by the *expected* increase. The cumulative curve of net acquisition, the balance between increment and decrement, is then represented as a horizontal line. Any stepwise change in the level of risk occurring during the period of surveillance causes the horizontal curve to take off at an upward or downward angle from its otherwise course. Formal tests have been developed

for optimizing the sensitivity and specificity of decision criteria and to adjust the ratio of true alarms and false alarms to particular needs; but the main virtue of this approach is probably the sensitive and immediate visual trigger which it supplies when plotted on a wall chart. It is not, however, particularly useful for the detection of short-term alternations of risk of the switch-up/switch-back variety.

General analyses

As we saw earlier, an investigator seeking cyclical phenomena can nominate a particular periodicity in advance (day of week, hour of day, month of year) and test for its presence. There are, however, more general approaches which do not require such prior nomination, for which efficient analytical programs are now available for microcomputers. The first general method is called Fourier analysis and the second is called polynomial analysis.

Fourier analysis programs decompose a curve of any complexity into a series of sine-waves of successive 'integer-frequencies' and they calculate a coefficient (i.e. a weighting) and a significance test appropriate to each of these cycles. By an integer-frequency $(1, 2, \ldots, n)$ we mean a wave-form with $1, 2, \ldots, n$ cycles within the total period of observation. 'Frequency' is the reciprocal of the length of the cycle, expressed as a fraction of the total period of observation. We can extend the integer frequency analysis to include all frequencies up to the point where n is half of the total number of data-containing cells within the total observation period. It is possible to reconstruct the original curve exactly, from these coefficients; or, alternatively, to construct an approximation to it from a limited number of statistically significant coefficients.

The second method, polynomial analysis, is generally simpler to use. The curve is expressed in the form $y = a + bx + cx^2 + dx^3 + ex^4 + \cdots + tx^n$.

This is an extension of linear and quadratic regression: but it provides a more flexible curve which can be fitted to more complex serial data-sets. The fitting is done through minimizing the least-squares deviation of the observations from the regression line. This method is capable of providing a concise summary of the basic wave-forms behind a curve with a large amount of random variation.

These techniques are for special cases and such generality of analysis is not often justified. We usually have sufficient prior indications of cyclical patterns or other forms of variation to know in advance the form which we have to test.

Clusters in time

A series of integers representing events in successive time periods (first format, above) can be treated, under the terms of the null hypothesis, as elements of a Poisson distribution. The observed frequency distribution of time periods with $0, 1, 2, \ldots, n$ events can be compared with a calculated Poisson distribution, to see whether the observed pattern differs from it. For a reasonably large average value per interval (say $n > 8$) we can use the theorem that the variance of a Poisson distribution is equal to its mean, and the standard deviation of the sampling distribution is the square root of the variance. Therefore, we do not expect individual values to depart very often from n by more than $2\sqrt{n}$ in either direction.

We can alternatively calculate a chi-square value for departures from the mean monthly expectation, attaching $n - 1$ degrees of freedom for n time-periods. For example, a four-weekly count of infants diagnosed with congenital dislocation of the hip, over a ten-year period, gave a chi-square of 603.7, with 521 degrees of freedom. We use a normalizing transformation to estimate significance. First, we calculate $\sqrt{(2X^2)} \ldots 34.75$; and we calculate the random expectation for this value from $\sqrt{(2df-1)} \ldots 32.2$. The latter value has unit variance (and unit standard error), from which we see that the observed transformation (34.75) is 2.54 standard errors outside the null expectation.

These methods are commonly used, but this is an intrinsically weak form of examination. It takes no account of any 'order', and loses all information relating to sequence. It will frequently fail to demonstrate genuine non-randomness. Negative findings cannot be trusted.

In this kind of work, however, positive findings are almost as difficult to interpret as negative ones. Suppose that in a series of 80 serial values we encountered two cells with 'significant' excesses ($p < 0.05$) of cases. This is what we would expect by chance. Suppose now that one of them was quite exceptional, with $p = 0.001$. If this particular cell had been nominated for testing in advance, we would regard the excess as significant. However, the question we really have to ask is the probability of such an event occurring by chance among *any* of the 80 values.

First, we note that the probability of so many events *not* occurring in a single nominated cell is $(1 - p)$. Therefore, the probability of it *not* occurring in *any* of 80 cells is $(1 - p)^{80}$. Finally, the probability of it occurring in *any* cell is $1 - (1 - p)^{80}$. Then, $p = 1 - 0.999^{80} = 0.077$. We now see that while this still a *rather* unusual observation on a purely random basis, it not *so* unusual that we would feel compelled to accept it as a 'cluster'.

It is not difficult in such situations to 'manufacture' significant findings. We might double up the weeks to fortnights, or measure the weeks from Thursday to Wednesday instead of Monday to Sunday, or some such procedure. This is, of course, illegitimate. It is better, as a rule, to recognize from the beginning the difficulties which will be encountered, both with respect to false-negatives and false-positives, and to use instead a form of investigation which makes use of 'sequence' information right from the beginning.

Sequence-based tests

The 'sliding window' technique was devised in order to evade the problem of locating the boundaries of successive time intervals in a particular way. Suppose we have records of a sparse set of events, located to individual days, and suppose we wish to search for the 'maximum seven-day period', without committing ourselves to the day of the week on which this might begin. Probability calculations based upon an arbitrary (e.g. Monday-to-Sunday) week would not then be legitimate. The 'sliding window' method is an extension of the probability conversion technique described above. A mathematical solution for this manoeuvre had indeed been found, but the time required for its computation is impractical on existing computers (Naus 1965; Knox and Lancashire 1982).

Approximations have been worked out to relieve the problem, but yet another problem appears beyond it. Although the window may be located at any position, it is a window of pre-defined size. Unless we have good prior information as to what that size might be, the real problem is to look for clusters of an *un*defined length at *any* position. This is not the question which the sliding window test answers. Perhaps for obvious reasons, these methods are not widely used.

The example has been discussed at disproportionate length because it illustrates a very general point; namely, that *first* questions in such situations are almost always of such generality, and so lacking in prior specificity, as to defeat the significance test approach, altogether. The best to be expected is a sufficient clarification of the question, so that it may be asked afresh and more precisely on another occasion.

There are other kinds of sequencing tests from which to choose, but they all run into similar problems. For example, the 'runs test' classifies successive intervals as containing more than or less than the median value (e.g. coded +, or −). There is a very simple test for measuring the probability that n consecutive runs of '+' or '−' might be encountered. This test will detect periods of increased or decreased incidence spanning (e.g.) 8 or more intervals; a greater *length* of the runs reduces the total number of runs, and it is this reduction that we test. However, the runs test would be of little use if an epidemic oscillation of higher frequency produced something approaching an alternation of short and long intervals. A number of other tests, for example the Kolmogorov–Smirnov test, also interact with cyclical or clustering frequencies; this particular test will detect slow long-term changes but is unsuitable for detecting other forms of heterogeneity, including tight clusters. The technique of serial correlation, in which the number of events (or the rate) in one time period is correlated with the number in the next, is another sometimes-useful test; but one which is also susceptible to interactions between the periodicities of the measuring scale and of the underlying process.

An analogous method is to calculate 'parallel' correlations between two separate series of events. For example, if we wished to examine the hypothesis that outbreaks of meningococcal meningitis are triggered by inhibitory cross-immunity reactions involving a particular form of *Escherichia coli*, then we would set week-by-week isolations of the two organisms alongside each other and look for cross-correlations between the two. We might also wish to experiment with a series of offsets between the two date-sets.

Exact timings and the detection of clusters

In the previous section, time was generally presented as the *x*-coordinate and the *number of events* as the enumerated value. In this section we consider the alternative class of presentation (the second format, above), where the lengths of the *intervals* between the events are treated as a quantitative independent variable (*x*), and the *number of intervals* of different lengths, as the enumerated value (*y*). This is the natural mode of presentation for calendar plots with sparse distributions of events, such that most of the days contain no events at all, and the remainder contain only one each.

For a first approach, the observed distribution of interval-lengths between successive pairs can be set against a calculated negative-exponential distribution. Clustering—and indeed *any* heterogeneity of frequency—would be represented here as an excess of observed over expected at both the left-hand (short) and the right-hand (long) ends of the scale. This approach may sometimes be fruitful but is a priori inefficient. It fails to use any of the information contained in the adjacency or non-adjacency of different short/long intervals; that is, it wastes information about sequence.

The runs test can again be used in this context and repairs some of this deficiency (Siegel 1956). The intervals are classified according to whether they are greater than (+) or less than (−) the median interval and the numbers of alternating 'runs' of pluses and minuses are counted. However, as mentioned earlier, it is *not* a good test for the detection of isolated or sparsely repeated tight clusters of (say) 2 to 5 events.

An alternative and more fruitful method of taking cognizance of sequential relationships between shorter and longer intervals is to construct a distribution of intervals between *all* possible pairs of events rather than (as earlier) between successive events. The form of the distribution turns out to be quite simple. It is triangular in shape, high to the left and tapering downwards towards the right. For a series of indefinite duration it approaches the form of a rectangular flat-topped distribution. A plot of the observed distribution of interval lengths between all possible pairs, set against an expected set of values, clearly contains much more information relevant to infective and toxic models than does a distribution of intervals between successive pairs, alone. This kind of display is capable in principle of displaying not only an excess of the very shortest intervals—such as we might expect from a point source infective epidemic or a 'toxic cloud' escape—but also of intervals of other 'preferred' lengths such as might occur in the onward transmission of an infective process through secondary and tertiary waves, each separated by one latent interval.

If we have a period of *D* days, with *n* events, and $n(n-1)/2$ possible pairs of events, then the number of pairs separated by less than *t* days is

$$\frac{n(n-1)}{2}(1 - v^2) \qquad \text{where } v = \frac{D - t}{D - 0.5}$$

Time on time

Exactly as for our earlier discussions on geographical distributions we have so far considered temporal distributions mainly *in vacuo*, without reference to anything else which might have been marked on the calendar. Exactly as for geographical distributions, there may be considerable advantages in setting one temporal plot upon another. As with geographically distributed hazards, the representation of temporal hazards may take different forms ranging from point events (such as earthquakes), through point events with extended aftermaths (such as Hiroshima, Nagasaki, or Chernobyl), to prolonged periods of exposure with well-defined or with ill-defined onsets and terminations (the marketing of thalidomide, or the practice of X-raying pregnant women). For short exposures we must introduce a concept which has no clear geographical analogue, namely the concept of the latent period. If we have prior evidence as to what that latent interval might be, the power of the examination is greatly increased. Otherwise, we must investigate parallel correlations between putative hazards and effects by using a series of different offsets. This, unfortunately, introduces additional degrees of freedom to any prior hypotheses which we wish to test. The tests themselves are thus degraded. In practice, we find ourselves returning to a now familiar problem; we must use a first data set only to formulate hypotheses, and a second independent set of data to test them.

Interactions

In addition to detecting and interpreting heterogeneities of disease frequency according to place or time or person, epidemiologists may wish to seek interactions *between* these different forms of heterogeneity. Such interactions are among the most fruitful of analytical tools. We can ask three general questions:

1. Is a geographical heterogeneity constant in time: or does it change? Conversely (but equivalently) formulated: is the temporal heterogeneity constant everywhere, or does it differ in extent or in its configuration in different places?

2. Is the temporal distribution constant for different kinds of affected person, or does it differ? Conversely put, do different kinds of affected person display the same form and the same degree of temporal heterogeneity?

3. Is the geographical heterogeneity constant for different kinds of affected person, or does it differ? Conversely put, do different kinds of person display the same or a different geographical pattern?

We refer here to person–time interaction (2 above) and person–place interaction (3 above) only for the purpose of completeness and for illustrating the generality of the interaction concept. We pursue the examination of space–time interactions (1) in greater detail.

Space–time interaction

The main application of space–time analysis is to detect sparse mobile clusters. A cluster may be conceived as being located within the spatial and temporal dimensions jointly. Within a three-dimensional space–time block, a cluster is conceived as having both geographical and temporal limits. Where there are several clusters, each is located within different time and geographical limits.

If a narrow time-scale is taken from such a time–space block, then it will appear as a simple geographical-cluster pattern. Alternatively, if a sufficiently small geographical sub-area is examined, then it will appear as a temporal cluster pattern. A space–time cluster pattern may be regarded either as a time-dispersed set of different geographical patterns—a series of maps; or as a spatially-dispersed set of time cluster patterns—a group of time-plots. It is possible to simplify a space–time distribution by collapsing the time-dimension, producing a simple map. Alternatively, the geographical dimensions can be collapsed and the result is then projected on to the time-scale alone. Thus, so far as *events* are concerned, each of the simple cluster formats already presented—temporal and geographical—can be seen as condensations of this more general one.

The 'sub-slice' and 'collapsing' techniques do however raise a problem. If a sufficiently small time-period is taken, then there may be insufficient cases for the observer to be reasonably sure that the geographical clustering is real. If a wider time-slice is taken, incorporating more months and years, then the inconstant position of the clusters may effectively 'even out' the geographical heterogeneities. The phenomenon, even though it be real, will disappear. Conversely (but equivalently) stated, a sufficiently small geographical area may contain insufficient cases for the observer to be sure that the temporal clustering is real. If the geographical area is enlarged, then the out-of-phase nature of the different temporal-cluster components may 'even out' the total, so that the phenomenon dissipates.

One solution consists of an expansion of the 'all possible pairs' treatment of time sequences, which was described earlier. The 'all possible pairs' are categorized simultaneously in terms of time apart and distance apart. These differencing operations create new variables with new spatial and temporal dimensionalities. Distance, being non-directional, effectively gets rid of both of the geographical co-ordinate scales. A two-dimensional table of observations relating to *pairs* of events is constructed in order to display the relative numbers of pairs with different combinations of long and short time-separations, and long and short distance-separations.

The overall distribution of temporal intervals may be expected to be triangular (provided that the population at risk and the incidence remain fairly constant) but the distance distribution is idiosyncratic and depends upon the geographical situation of the particular population at risk. The *expected* values for the cells of the table can be obtained from the mar-

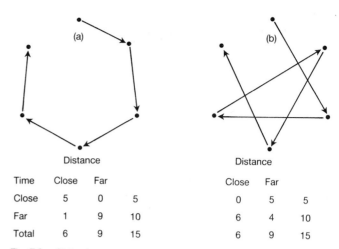

Time	Distance		
	Close	Far	
Close	5	0	5
Far	1	9	10
Total	6	9	15

	Distance		
	Close	Far	
Close	0	5	5
Far	6	4	10
Total	6	9	15

Fig. 7.2. Stylized space–time interactions for six events

ginal totals (time intervals and distance intervals) and compared with the observed values. Short-distance short-time clustering is displayed as an excess of observed pairs over expected pairs in the short-distance short-time cells. A positive finding means that, following an index case, any subsequent cases occurring within a short period of time are more likely than otherwise to occur within a short distance-range.

The interaction format is illustrated in Fig. 7.2(a) and (b). We imagine six events located at the vertices of a regular hexagon. Six events generate 15 possible pairs. Six of the pairs are adjacent on the perimeter and can be classified as 'close', while the remaining nine pairs are 'distant'. These considerations determine the grand total and the column totals of the tabular representations. In Fig. 7.2(a) we imagine a contagious process in which events at times 1,2, . . . , 6 follow a 'contagious' pattern, each event following on at the position closest to the previous event. In Fig. 7.2(b) we envisage an 'anti-contagious' process in which successive events consistently skip the nearest position and strike at the next. Inspection of the figures and the derived tables will clearly illustrate the manner in which these two hypothetical sequences produce different numerical interactions within a table whose marginal totals are fixed by the circumstances.

This form of analysis offers an unexpected benefit to the analyst. Because it is designed to display only interactions, it can operate upon absolute numerator frequencies rather than rates, and the observed/expected ratios can be obtained from marginal totals without reference to the distributions in the population at risk. The technique is effectively 'denominator-free'. This freedom also allows the analyst to pursue higher levels of interaction. For example, in the investigation of childhood cancers we may first of all seek evidence of interaction of all possible pairs irrespective of the type of cancer. We then repeat the analysis for leukaemia/leukaemia pairs; and again for solid-tumour/solid-tumour pairs; finally by subtracting the last two tabulations from the first, we display an interaction pattern for leukaemia/solid-tumour pairs.

The author has also experimented with the reproduction of epidemics in compressed real-time on the screen of a micro-computer. The objective is to see whether 'eyeball real-time' techniques are capable of picking up space–time cluster patterns intuitively.

Different applications of the space–time principle have used different modes of display and of analysis. The simplest (the Knox method, as described above) uses simple contingency tables (Knox 1963, 1964). The Mantel method (Mantel 1967; Klauber 1971) is based upon parametric statistics and is essentially a correlative representation of the associations between the time and space intervals. However, because of the non-normal distribution of the time and distance intervals, Mantel first converts times and distances to their reciprocals (which values can be referred to as 'closeness'), converting the original asymmetric distributions to forms more appropriate to testing through standard correlative/regression methods.

There is, however, a special problem in testing significance for any method based on all possible pairs. The pairs are not independent of each other; each of the n individual events has been incorporated into $(n - 1)$ of the $n(n - 1)/2$ pairs. This can invalidate the use of Poisson-distribution-based methods such as the chi-square statistic, which are based upon the premise of *in*dependence. These tests are valid only when the dichotomies between short and long times and between short and long distances are sufficiently asymmetric to render the short-time/short-distance pairs effectively independent of each other. The parametric method partly overcomes this problem, and circumvents the problems of what dichotomies to choose, but can be less efficient at distinguishing highly localized (very-short-time/very-short-distance) clustering from within a large number of more widely spaced pairs. As with other choices between alternative modes of presentation and testing, some techniques are more effective in one situation, and others more effective elsewhere.

Space–time interactions do not *have* to be based upon pairs, alone. They can be based upon an examination of all possible triplets, all possible quadruplets . . . *n*-tuplets . . . , using some composite value of spatial and temporal distance for purposes of tabulation. For example, the mean distance and the mean time interval between all possible pairs within the *n*-tuplet, can be used for an overall space–time tabulation. These methods are expensive in computer time and the tabulation must usually be restricted to a search for interactions within one part of the total tabulation, setting upper limits to the *n*-tuplet-times which will be incorporated within the analysis.

Another 'refinement' for space–time studies is to combine them with the principles of demographic mapping and to 'adjust' the distance measures according to the local geographical density of events; thus converting map distances, in effect, to inter-personal contact distances.

Spatial and temporal models

There is a constant pressure upon investigators of spatial/temporal data to move beyond the scope of traditional statistical analysis and into the domain of model construction, of model validation, and the estimation of model parameters. This arises partly from the limited relevance and validity, in this field, of sampling theory: and the limited value which can be placed upon significance tests and the calculation of confidence limits. It is reflected in the wide usage of 'eyeball interpretation', for example in the inspection of maps. Such interpretations usually depend upon the investigator bringing external knowledge and externally derived theoretical constructs to bear upon his data. Indeed, this is the main point of drawing maps and plotting graphs. If analysis could be handled entirely through formal techniques, then plotting for visual inspection would not be necessary at all.

Another area where model fitting, whether intuitive or formal, seems obligatory, is in the study of natural histories. For example, cervical cancer screening policies still suffer from the inadequacy of the available natural history formulations. It was recognized many years ago that much of this problem sprang from the manner in which screening services had become focused upon the individual test-result; and from the technical and operational difficulties of assembling these observations into sequences. Clearly, a 'natural history' was dimensioned in time, and it was not possible to infer the natural history of this disease until time-dimensioned records of the process had been assembled through linking successive examinations. But, even then, how was a natural history to be calculated from temporally-linked data? Where was the computational algorithm for doing so?

Except for the simplest components of a natural history, the only practical way of handling this problem is to 'propose' a natural history model, to set it out in the form of a matrix of transmission probabilities between different stages of the disease, and to adjust the values until the observations are successfully 'predicted'. Several quite different models may be capable of achieving this! That is to say, the data might be *in*capable of distinguishing between these alternative models. This can be something of a shock to planners of preventive services; but to scientists it may indicate what kinds of data must next be collected and analysed in order to solve an unsolved problem.

A modelling approach also permits a more satisfactory definition of a 'cluster' than that afforded by effectively defining the phenomenon as a residual statistical heterogeneity after the exclusion of trends and cycles. Negative definitions are seldom of much use, yet this is a position into which many investigators have been forced. In 'model' terms, we can define a cluster in an alternative and positive manner, as a temporally (and/or spatially) bounded set of events which are related to each other through some biological or social mechanism, or which have a common relationship with some other event or circumstance. Declared in this way, the definition owes nothing to probability theory or to random distributions or to departures from them. It does not mention statistical criteria at all. This has the advantage that we are now entitled to use statistical descriptors as inductive criteria to decide when such a cluster might be present, without getting into the circularities which occur when we use statistical concepts both for our definition *and* for our decision process. It also serves as a reminder that, having aroused our suspicions, we must still seek to uncover the biological or social mechanisms; and that the 'reality' or otherwise of our clusters depends ultimately upon our success in doing so.

References

Cooke-Mozaffari, P.J., Ashwood, F.L., Vincent, T., Forman D., and Alderson, M. (1987). *Cancer incidence and mortality in the vicinity of nuclear installations. England and Wales, 1950–1980*, Studies on Medical and Population Subject No. 51. HMSO, London.

Edwards, J.H. (1961). The recognition and estimation of cyclic trends. *Annals of Human Genetics* **25**, 83.

Gardner, M.J., Winter, P.D., Taylor, C.P., and Acheson, E.D. (1983). *Atlas of cancer mortality in England and Wales, 1968–78*. Wiley, Chichester.

Kemp, I., Boyle, P., Smans, M., and Muir, C. (ed.) (1985). *Atlas of cancer in Scotland 1975–1980: Incidence and epidemiological perspective*. IARC Scientific Publications, Vol. 72, Lyons.

Klauber, M.R. (1971). Two-sample randomization tests for space-time clustering. *Biometrics* **27**, 129.

Knox, E.G. (1963). Detection of low intensity epidemicity: Application to cleft lip and palate. *British Journal of Preventive and Social Medicine* **17**, 121.

Knox, E.G. (1964). Epidemiology of childhood leukaemia in Northumberland and Durham. *British Journal of Preventive and Social Medicine* **18**, 17.

Knox, E.G. and Lancashire, R.J. (1982). Detection of minimal epidemics. *Statistics in Medicine* **1**, 83.

Knox, E.G., Stewart, A.M., Gilman, E.A., and Kneale, G.W. (1988). Background radiation and childhood cancers. *Journal of Radiological Protection* **8**(1), 9.

Mantel, N. (1967). The detection of disease clustering and a generalised regression approach. *Cancer Research* **27**, 209.

McGlashan, N.D. (ed.) (1972). *Medical geography*. Methuen, London.

Naus, J.L. (1965). The distribution of the size of the maximum cluster of points on a line. *Journal of the American Statistical Association* **60**, 532.

Openshaw, S., Craft, A.W., Charlton, M., and Birch, J.M. (1988). Investigation of leukaemia clusters by use of a geographical analysis machine. *Lancet* **i**, 272.

Selvin, S., Shaw, G., Schulman, J., and Merrill, D.W. (1987). Spatial distribution of disease: Three case studies. *Journal of the National Cancer Institute* **79**, (3), 417.

Siegel, S. (1956). *Nonparametric statistics for the behavioural sciences*. McGraw-Hill, New York.

Stone, R.A. (1988). Investigations of excess environmental risks around putative sources: Statistical problems and a proposed test. *Statistics in Medicine* **7**, 649.

Walters, S.D. and Elwood, J.M. (1975). A test for seasonality of events with a variable population at risk. *British Journal of Preventive and Social Medicine* **29**, 18.

Weatherall, J.A.C. and Haskey, J.C. (1976). Surveillance of malformations. *British Medical Bulletin* **32**, 39.

8

Cross-sectional studies

J.H. ABRAMSON

Introduction

This chapter deals with prevalence and other cross-sectional studies, i.e. with surveys of the situation existing at a given time in a group or population, or set of groups or populations. These surveys may be concerned with:

(1) the presence of disorders, such as diseases, disabilities, and symptoms of ill-health;

(2) dimensions of positive health, such as physical fitness;

(3) other attributes relevant to health, such as blood pressure and body measurements;

(4) factors associated with health and disease, such as exposure to specific environmental factors, defined social and behavioural attributes, and demographic characteristics; the correlates may be determinants, predictors, or effects of health and disease states.

Such a study may be descriptive, analytical, or both. At a descriptive level it yields information about a single variable (diabetes, haemoglobin concentration, capacity to work, cigarette smoking, etc.) or about each of a number of separate variables, in a total study population or in specific population groups. At an analytical level, it provides information about the presence and strength of associations between variables, permitting the testing of hypotheses about such associations.

Most cross-sectional surveys are individual-based, i.e. they use information about the individuals in the group or sample studied. There are also group-based surveys, which use information about groups or populations. As an example, Poikolanen and Eskola (1988) found that mortality rates from various causes in 25 developed countries had strong negative associations with the per capita gross domestic product (a measure of economic development) but not with the numbers of doctors, nurses, or hospital beds per 10 000 population, nor with health expenditure per head. Analytical group-based surveys are sometimes referred to as ecological or correlational studies.

Cross-sectional surveys may be contrasted with incidence and other 'time-span' studies (Lilienfeld and Lilienfeld 1980)

that require information relating to the state of the individual at two or more points in time. The latter studies, which are discussed in later chapters, measure changes in status (e.g. disease onset, growth, changes in blood pressure) or they examine associations between variables with a defined temporal relationship, e.g. between childhood experiences and health in adulthood. The difference between cross-sectional surveys and these surveys is often likened to the difference between snapshots and motion pictures.

The distinction between these two kinds of survey is not always strict. Although the essential feature of cross-sectional studies is that they collect information relating to a single specified time, they are often extended to include historical information that can easily be collected at the same time. This may lead to the demonstration of statistical associations with past experiences, e.g. a relationship between varicose veins and the number of pregnancies (Maffei et al. 1986), a negative association (in women only) between weight and the frequency of drinking alcohol during the previous year, and a tendency for weight to be highest in people who have quit smoking (Williamson et al. 1987).

The uses of cross-sectional studies can be categorized as follows:

1. The findings may be used to promote the health of the specific group or population studied, i.e. the study can be used as a tool in community health care.

2. The findings may contribute to the clinical care of individuals.

3. The study may provide 'new knowledge'—generalizable inferences that can be applied beyond the specific group or population studied. This knowledge may relate, for example, to the aetiology of a disease or the value of a type of health care.

These uses are, of course, not mutually exclusive; a single study may fulfil more than one purpose.

This chapter briefly considers the terms prevalence and incidence, and then reviews the methods used in cross-sectional studies, paying special attention to the statistical

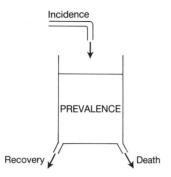

Fig. 8.1. Relationship between prevalence and incidence.

measures commonly used, including prevalence rates of various kinds. The main body of the chapter has three sections, which give separate consideration to the uses listed above. The first of these sections, on uses in community health care, considers community diagnosis, surveillance, community education and community involvement, and the evaluation of a community's health care. The subsection on community diagnosis deals with studies of health status, determinants of health and disease, associations between variables (including the measurement of impact, risk markers, and community syndromes), and the identification of groups requiring special care. The section on uses in clinical practice briefly describes applications in individual and family care and in community-oriented primary care. The section on studies yielding new knowledge reviews studies of growth and development, studies of aetiology, and programme trials.

Prevalence and incidence

Prevalence refers to the number of individuals who have a given disease or other defined attribute at a specific time, as opposed to incidence, which refers to the number of events that occur in a given period. The event may be the onset of a new disease, death, etc.

The prevalence of a disease in a population at any point in time depends on the prior incidence of new cases and on the average duration of the disease from onset to recovery or death. This relationship is shown diagrammatically in Fig. 8.1, in which the content of the container represents prevalence, and the time spent in the container is the duration of the disease.

If incidence and the average duration have remained constant over a long period (a condition seldom encountered in real life), the point prevalence rate (defined below) is the product of the incidence rate of new cases per time unit t and the average duration of the disease (mean t per case). For a disease that runs an episodic course, the point prevalence rate of active disease is (under certain assumptions) the product of the incidence rate, the average duration of an episode, and the average number of episodes per case (Von Korff and Parker 1980).

Methods

Like any other kind of study, a cross-sectional study can yield useful findings only if sound methods are used. At all stages —in the planning phase, during the collection of data, and when the data are processed and interpreted—there is a need for detailed attention to the minutiae of survey technique, so as to minimize bias and ensure that the results will be as accurate as practical constraints permit (Abramson 1984).

A cross-sectional study may be performed in a total target group or population, or in a representative sample. Simple random sampling or systematic, stratified, or cluster sampling may be used (see Volume 2, Chapter 12).

Methods of collecting information may be broadly classified as follows:

1. Clinical examinations, special tests, and other observations.

2. Interviews and questionnaires. The subjects themselves may be questioned, or proxy respondents, e.g. household informants, may be used.

3. Clinical records and other documentary sources. Sources of information on the prevalence of diseases include hospital and other medical records, disease registers, records of routine examinations (in schools, army induction centres, health insurance schemes, etc.), and published statistics based on these and other records.

Disease prevalence may be studied in two stages: by using a screening test to identify people who are likely to have a given disease, and then by subjecting them to more elaborate and specific tests.

Each method of data collection has its own advantages and limitations and carries its own possible biases. If information on the prevalence of a disease is obtained from hospitals, for example, people with mild disease are likely to be under-represented. The degree of bias may vary for different categories of the study population as a result of variation in the accessibility or use of health services, or of differences between clinical services in their diagnostic and recording procedures. In a rural region in the US, Anderson et al. (1988) found that 42 per cent of the cases of Parkinson's disease found in a survey based on screening questions and subsequent neurological examinations had not been diagnosed previously, and would have been missed by a survey based on medical records. Had people in institutions been omitted from the survey, a quarter of the cases would have been missed.

Associations observed in a study population may differ from those in the general population if admission rates to the study population are connected with the variables whose association is studied. This is a common form of bias (Berksonian bias) in studies of hospital or clinic patients and autopsy samples.

In addition to selection bias (where the individuals for whom data are available are not representative of the target population), there may be information bias, caused by shortcomings in the gathering or handling of information. A catalogue of biases, most of them applicable to cross-sectional studies, has been published by Sackett (1979).

In cross-sectional surveys that set out to examine causal associations, bias is commonly caused by the respondents' or investigators' knowledge that there has been exposure to the putative cause (a source of bias in cohort studies also) or that the putative effect is present (a source of bias in case-control studies also). In a study of the association between coffee-drinking and digestive symptoms, for example, a respondent who drinks much coffee and believes that this beverage causes digestive upsets may be more likely to recall and report symptoms, or an interviewer may tend to be more persistent when asking such a respondent about symptoms. Also, subjects who have symptoms and believe these are caused by coffee may tend to provide a fuller (or exaggerated) account of their consumption of the beverage, and interviewers may tend to be especially persistent when questioning subjects whom they know to have symptoms. The subjects' awareness of their symptoms may also have led them to avoid coffee. These and other kinds of bias can be minimized by suitable survey procedures.

Some biases can be avoided or measured; others cannot, but must still be taken into account when inferences are drawn from the findings. Sometimes bias can be corrected. Smith and Hill (1974), for example, have suggested a technique for estimating the number of patients with a chronic disorder—the example they use is Buerger's disease—who are missed because their disease is early or mild; their procedure requires information on the prior course of the disease.

When analysing the findings of a cross-sectional study, the exploration of cause–effect relationships may be based on the retrospective approach that characterizes case-control studies (e.g. comparing the coffee consumption of people who do and do not have symptoms) or on the prospective approach used in cohort studies (comparing the prevalence of symptoms in people who do and do not drink coffee). As in other epidemiological studies, causal effects can be inferred only after careful consideration has been given to the possibility of fortuitous and artefactual associations and confounding effects (Abramson 1988; Susser 1973). The results of group-based (correlational) studies may be especially difficult to interpret, since they are usually based on official statistics, which may not provide satisfactory measures, particularly for the control of confounding. Moreover, associations found at a group level may not exist at the individual level; there is more malaria in poor countries than in rich countries, but it does not follow that poor people are at higher risk than rich people in the same region (the so-called 'ecological fallacy').

Statistical measures

The statistical measures used to summarize the findings of descriptive cross-sectional studies include means and standard deviations; medians, percentiles, and other quantiles;

Table 8.1. Measures of association and impact (cross-sectional study of a population)

	Disease present	Disease absent	Total
Factor present	a	b	$a + b$
Factor absent	c	d	$c + d$
Total	$a + c$	$b + d$	N

Odds ratio = ad/bc or bc/ad
Rate ratio = $[a/(a + b)]/[c/(c + d)]$ or $[a/(a + c)]/[b/(b + d)]$
Rate difference = $a/(a + b) - c/(c + d)$ or $a/(a + c) - b/(b + d)$

Measures of impact
If the factor is a risk factor:
　Excess risk among exposed = $a/(a + b) - c/(c + d)$
　Population excess risk = $(a + c)/N - c/(c + d)$
　Attributable fraction among exposed = $[a/(a + b) - c/(c + d)]/[a/(a + b)]$
　Attributable fraction in population = $[(a + c)/N - c/(c + d)]/[(a + c)/N]$
If the factor is a protective factor:
　Excess risk among unexposed = $c/(c + d) - a/(a + b)$
　Population excess risk = $(a + c)/N - a/(a + b)$
　Prevented fraction among exposed = $[c/(c + d) - a/(a + b)]/[c/(c + d)]$
　(This is also the preventable fraction among the unexposed.)
　Prevented fraction in population = $[c/(c + d) - (a + c)/N]/[c/(c + d)]$
　Preventable fraction in population = $[(a + c)/N - a/(a + b)]/[(a + c)/N]$

and proportions (including the rates described below). Ratios other than proportions are occasionally used, e.g. the sex ratio (usually the male: female ratio) of people with a specific disease. Separate statistical measures may be provided for specific sex and age categories, ethnic groups, social classes, regions, etc.

In analytical cross-sectional studies, associations between variables may be measured by correlation and regression coefficients, differences between means, and other statistics. The most commonly used measures of the strength of an association, applicable if both variables are dichotomies, are the odds ratio, rate ratio, and rate difference (Table 8.1). If the variables are exposure to some factor and the presence of a disease, an odds ratio of three means that the odds in favour of the disease are three times as high among people exposed to the factor as they are among those not exposed. It also means that the odds in favour of exposure are three times as high among people with the disease as they are among people free of it. A rate ratio of three, based on the prevalence of the disease, means that the rate of the disease in exposed people is three times as high in exposed as in unexposed people. A similar rate ratio based on the prevalence of exposure means that exposure to the factor is three times as prevalent among people with the disease as it is among those free of it. If the frequency of the disease in the population is low, the odds ratio is a good estimator of the rate ratio for the disease (risk ratio).

Measures of the impact of the factor on the prevalence of the disease (assuming that the association is a causal one) are also shown in Table 8.1. These will be discussed under 'Measurement of impact', p. 112. For a fuller explanation of various measures and their uses, see Fleiss (1981).

In analytical cross-sectional studies that aim to explain as well as to describe associations, a variety of measures and techniques may be used to control confounding factors and determine whether other variables modify the association. These procedures range in complexity from stratification and standardization to sophisticated multivariate techniques that permit the simultaneous consideration of a large number of variables and their relationships (see Volume 2, Chapter 12).

If the data were obtained from a random sample, confidence intervals may be calculated in order to obtain an interval that has a high probability of containing the true value of the measure in the total target population. Confidence intervals are also often calculated, even in the absence of a 'chance' process such as random sampling, to permit generalization to a broad 'reference' population (e.g. 'White men in the US'), but in this instance the procedure is open to criticism.

Prevalence rates

A prevalence rate is used to measure the relative frequency of a disease or other qualitative attribute in a group or population; it is a proportion. It should be noted that this use of the term prevalence rate, which conforms with common usage by epidemiologists, is disparaged by some experts who prefer to confine the term rate to measures of the rapidity of change, and therefore use 'prevalence' rather than 'prevalence rate', claiming that prevalence rate is an impossible concept (Elandt-Johnson 1975).

There are different kinds of prevalence rate. When the term is used without qualification, it usually refers to a *point prevalence rate*, i.e. the prevalence at a specified point of time. The point prevalence rate of a disease per 1000 population, for example, is calculated by the formula:

$$\frac{\text{Number of individuals with}}{\text{the disease at a specified point of time}} \times 1000$$
$$\frac{}{\text{Population at that time}}$$

The numerator of this rate is the total number of people who have the disease at the stated time, whenever the disease commenced. The denominator is the total population (actual or estimated) at that time, including affected and unaffected people. The multiplier used in this and other rates may be 100, 1000, or any other convenient or conventional multiple of ten.

The point of time to which prevalence refers need not be a fixed calendar time. The reference may be to a fixed point in the experience of each individual, e.g. birth, entry to a job or army service, immigration, death, or (in a prevalence survey where interviews or examinations are staggered over a period) the date of examination. In such instances the formula is:

$$\frac{\text{Number of individuals with}}{\text{the disease at the time the individual is studied}} \times 1000$$
$$\frac{}{\text{Number of individuals studied}}$$

A rate expressing the frequency of a finding in an autopsy study is a point prevalence rate that refers to the time of death. Such a rate is especially useful to indicate the prevalence of a disorder whose presence does not affect the risk of

dying or the probability of an autopsy. Autopsy studies have shown, for example, that 45 per cent of young US soldiers killed in battle had coronary atherosclerotic lesions (Mac-Namara *et al.* 1971) and that most women have cystic disease of the breast (macroscopic or microscopic) when they die (Frantz *et al.* 1951).

Like other rates, the point prevalence rate may express the findings in a specific subgroup of the population; when so used, the numerator and denominator must both refer to the same population category. As an example of a sex- and age-specific rate, the point prevalence rate of a disease per 1000 men aged 45–54 years is calculated by the formula:

$$\frac{\text{Number of men aged 45–54 years}}{\text{with the disease at a specified point of time}} \times 1000$$
$$\frac{}{\text{Total number of men aged 45–54 years in the}}$$
$$\text{population at that time}$$

Paradoxically, there are some point prevalence rates that can be accurately measured only by a longitudinal study. An example is the rate of congenital anomalies per 100 births. This may be regarded as a point prevalence rate referring to the moment of birth. Many anomalies become manifest only weeks, months, or years after birth, so that reasonably full case-finding requires long-term follow-up.

A *period prevalence rate* measures prevalence not at a single point of time but during a defined period (usually a specific year). The *period prevalence rate (persons)* represents the proportion of the population manifesting the disease at any time during the period. The formula is:

$$\frac{\text{Number of individuals}}{\text{manifesting the disease in the stated time period}} \times 1000$$
$$\frac{}{\text{Population at risk}}$$

The numerator is the number of people with the illness during the specific period, including those whose illness started before this period. The denominator is the average size of the total population during the specified period. It is often estimated by using the population at the middle of the period, or by averaging the size of the population at the beginning and end of the period. Other methods may be needed if the change in population size during the period was large and did not occur at an approximately even pace. It is usually more helpful to know the point prevalence rate at the beginning of the period and the incidence rate of new cases during the period, than the period prevalence rate.

There is also a type of period prevalence rate that refers not to a defined calendar period but to a defined period of the individual's life, e.g. the period of pregnancy. A study of a sample of healthy pregnant women in a US city, for example, revealed that the prevalence of reported physical battering during the current pregnancy was 8 per cent (Helton *et al.* 1987).

The numerator of the little-used *period prevalence rate (spells)* is the number of spells of an illness observed during the specified period (including spells that commenced before the start of the period); the same person may be ill more than once. The denominator is the population at risk. For a short-term disease, this rate is usually similar to the incidence rate (spells).

A *lifetime prevalence rate* is a period prevalence rate referring to the whole of the subject's prior life. It differs from the point prevalence rate only if the disorder is one that may not persist. It refers to the presence of the disorder or of a scar, antibodies, or other evidence that the disorder was present in the past. The formula is:

$$\frac{\text{Number of individuals with evidence of the disorder (past or present)}}{\text{Number of individuals studied}} \times 1000$$

This rate is usually useful only if it refers to a specific age, and if valid information on prior occurrence is available. As an example, a cross-sectional study in Jerusalem revealed that the point prevalence rate of inguinal hernia among men aged 65–74 years was 30 per cent. An additional 10 per cent had scars of hernia repair operations, with no current evidence of hernia. The lifetime prevalence rate for this age-group was thus 40 per cent (Abramson *et al.* 1978). It could be inferred that in this cohort the risk of developing a hernia, among men surviving to the age of 65–74 years, was 40 per cent. Such information would be of little value if the disorder were one with an important association with survival, such as cancer.

As a further example, the lifetime prevalence rate of definite psychiatric disorder among men aged 60–69 years was 11 per cent, according to the Stirling County Study in Canada, and the lifetime prevalence rate of probable psychiatric disorder was 41 per cent (Leighton *et al.* 1963). As the historical information was probably incomplete, these rates were probably underestimates of the risk.

The lifetime prevalence rate of a disorder among the blood relatives of an index case may be used as a measure of familial risk, especially in genetic studies.

Uses in community health care

Cross-sectional studies can fulfil important functions in the health care of a community. They can contribute to the planning of services, to the effective implementation of care, and to decision-making on the continuance and modification of services. In this discussion, 'community' may be taken to refer to any aggregation of people for whose care a physician, health care team, agency or authority is responsible; it may be a nation or region, a local neighbourhood, a list of registered patients, a defined group of school children or workers, inmates of an institution, etc.

We will give separate attention to the use of cross-sectional studies in community diagnosis, in ongoing surveillance, in community health education and the promotion of community involvement, and in the evaluation of the community's health care.

Community diagnosis

Cross-sectional studies can provide a major part of the epidemiological foundation for community diagnosis (Kark 1981;

McGavran 1956; Morris 1975), i.e. for determining the health status of a community and the factors responsible for producing it. They can supply information on the nature, extent, and impact of health problems, as a basis for the identification of priorities and the planning of intervention. Such studies may relate to a broad spectrum of health states and their correlates, or may be limited in their scope. Even the simplest of information, e.g. about the population's size and its age and sex distribution, may be of help in planning the allocation of resources.

Health status

Cross-sectional studies may yield useful information on a variety of dimensions of health and disease, including self-appraised health, mental health status, growth and development, physical fitness, the distribution of blood pressures, etc. The following remarks refer only to the prevalence of disorders; the cross-sectional method for the study of growth and development is discussed on p. 117.

Information on the prevalence of diseases and disabilities in the community, supplemented by incidence and mortality data, provides a basis for inferences about needs for curative and rehabilitative care, i.e. for secondary and tertiary prevention. It must be remembered that the picture provided by prevalence studies alone, especially by studies of point prevalence, may be incomplete because of the under-representation of conditions with a short duration; these include not only the acute non-fatal diseases that usually provide a considerable load for the health services, and acute episodes of long-term or recurrent diseases, but also severe and rapidly fatal conditions, such as fatal strokes and sudden deaths from coronary heart disease.

The most direct evidence of a need for improved secondary and tertiary prevention, at least for long-term diseases that are not rapidly fatal, is an unduly high prevalence of remediable disease that is undiagnosed, or that is diagnosed but untreated or inadequately treated. Prevalence surveys providing such evidence may be based on examinations, interviews, clinical files or other documentary sources, or a combination of these. An examination survey, for example, may reveal a high prevalence of untreated dental caries, or hypertension that (according to the subjects' statements or their medical files) is not under treatment. An interview survey in which many respondents report that they have been told they have high blood pressure, but say they are not currently receiving care for it, may provide similar if less valid evidence of a need for improved care. Or a review of clinical files may reveal that there are many known hypertensives who are currently not under control or are not receiving treatment.

Needs for primary prevention can be inferred from the presence of disorders known to be preventable, i.e. those whose incidence can be reduced by known preventive measures. These include ischaemic heart disease and lung cancer, which are presently of epidemic proportions in many parts of the world. For this purpose too, the prevalence data should be supplemented by data on incidence and mortality, both because diseases with a high fatality rate will otherwise be under-represented and because prevalent cases may be long-standing ones and may not reflect present preventive needs. A high prevalence of crippling caused by poliomyelitis, for example, does not necessarily mean that current preventive procedures are ineffective. Information on the recent incidence of new cases is to be preferred for this purpose. If data on prevalence are to be used, it is advisable to seek information (if this can be obtained) on the duration of the disorder, so that the prevalence of disease of recent onset can be measured. For some diseases, information on their stage of activity may fulfil the same purpose. In institutional and other settings where people who develop a disorder are especially likely to remain in the study population, prevalence data may overestimate the need for primary prevention. In a hospital, for example, patients who develop nosocomial infections are for this reason likely to have a longer hospital stay. Such patients may thus be over-represented in a prevalence study of such infections in a hospital, giving an exaggerated idea of the need for primary prevention.

The use of highly valid measures of the presence of a disease often presents practical difficulties, and reliance may be placed on a proxy measure that is of lower validity, but is simple, inexpensive, and acceptable. The result of a screening test may be used for this purpose. If the prevalence of a proxy attribute in the population or a sample is known, methods are available for estimating the confidence limits of the prevalence of the disease itself in the population (Peritz 1971; Rogan and Gladen 1978).

Determinants of health and disease

Information on modifiable factors that are known to affect health is of obvious relevance to the planning of health care. These may be factors that affect the community's health in a general way, e.g. dietary, infant rearing and family planning practices, and (presumably) the use of health services, and they may be factors that affect the risk of developing specific disorders. They may be *risk factors*, e.g. cigarette smoking or obesity, which increase the risk of ill health, or *protective factors*, e.g. physical activity or specific immunity (natural or acquired) to a pathogenic agent.

To collect information on the current prevalence and distribution of such determinants, there is no substitute for descriptive cross-sectional studies.

Associations between variables

When associations are investigated in a cross-sectional study in the context of community health care, the dependent variable is usually the presence of a disease or disability or a measure of health. The aim of such analyses is usually to throw light on determinants or predictors. The dependent variable may also be a supposed risk factor or protective factor, as in studies of the determinants of cigarette smoking by schoolchildren, or the use of a health service, or compliance with medical advice. Attention may also be paid to associations among diseases or other dimensions of health, or among determinants of health.

In this context, analyses of associations with diseases are usually undertaken in order to determine what causal factors (of those known to be potentially important) are active in the community studied, and to measure their impact. The primary aim is to obtain information that will be useful in practice, not to generate new knowledge about aetiology, although this may be a secondary gain.

Measurement of impact

In a situation where it is believed that an association of a risk factor with the prevalence (or incidence) of a disease expresses a causal relationship, the factor's impact may be measured by the attributable fraction (or aetiological fraction) in the population. This is the proportion of the disease in the population that can be attributed to exposure to the factor. Among workers aged 20–64 years in a population in Jerusalem, for example, the fraction of the prevalence of varicose veins that could be attributed to work involving much standing was 16 per cent in each sex, after controlling for effects connected with age, region of birth, weight, and height (Abramson *et al.* 1981). Such values must be interpreted with caution, as part or all of the apparent causal effect may be due to other (uncontrolled) factors associated with the apparent causal factor. The attributable fraction among the exposed may also be of interest; 31 per cent of the prevalence of varicose veins among men whose work involved much standing could be attributed to their work posture; for women, this fraction was 32 per cent. Formulae are shown in Table 8.1.

For a protective factor, the corresponding measures are the prevented fraction, which is the proportion of the hypothetical total prevalence that has been prevented by exposure to the factor, and the preventable fraction, which is the proportion of the observed prevalence that would be prevented if everyone was exposed to the factor (Table 8.1).

Attributable, prevented, and preventable fractions in the population are influenced by the prevalence of the risk or protective factor.

Risk markers

Interest may not be confined to cause–effect relationships. *Any* attribute or exposure that is strongly associated with a disease or other disorder, even non-causally, has potential value as a predictor, provided there is reason to believe that it precedes the appearance of the disorder. Such predictors may be used as risk markers (Grundy 1973) to identify vulnerable individuals or groups.

A risk marker may be a factor that itself influences the risk, or a precursor or early manifestation of the disorder, or it may be secondarily associated with the disorder because it is associated with a cause or precursor of the disorder. Risk markers for stroke, for example, may include hypertension (a modifiable risk factor), age and ethnic group (non-modifiable determinants), and episodes of transient cerebral ischaemia (a precursor condition). Among the elderly, risk markers for early mortality include impaired cognitive functioning and the use of hypoglycaemic agents (Jagger and Clarke 1988). These are markers, but not necessarily causes, of an increased mortality risk.

Risk markers are best identified by longitudinal studies, but can also be detected by cross-sectional ones. A survey of the development of two-year-olds in a Jerusalem neighbourhood, for example, revealed that poor development was more prevalent if the mother had had little education or there were many other children in the family, and especially if these factors were present together (Palti *et al.* 1977). These two attributes and their combination could be used as risk markers to identify children requiring special care in their early infancy. In drug addicts in Italy, a prevalence study showed that the risk of being infected with the AIDS virus was more than eight-fold in one of the provinces studied; this difference, which was not lessened by controlling for possible confounding factors, had no obvious explanation (Franceschi *et al.* 1988).

The value of a risk marker or combination of risk markers depends on the following considerations.

1. Is its use practical? Questions of cost, resources, acceptability, safety, and convenience must be considered. The most useful risk markers are those whose measurement can be built into ordinary clinical care.

2. Is detection of vulnerability likely to be beneficial? Are resources and techniques for reducing the risk available? Is the anticipated benefit outweighed by the harm that intervention may cause, such as toxic side-effects, anxiety, and iatrogenic invalidism? What is the predictive value of the presence of the risk marker, i.e. what proportion of the individuals with the risk marker are likely to develop the disorder?

3. How prevalent is the risk marker? If more than half the children in a community fall into a high-risk group needing special attention, might it not be more efficient and possibly more effective to modify the routine care programme so as to give extra attention to *all* children? A risk marker for a disorder may have a high prevalence because the disorder itself has a high prevalence, or because the marker has a low specificity (occurs in a high proportion of people who will not develop the disorder).

4. What proportion of the individuals with the disorder will be identified? That is, what is the sensitivity of the marker as a predictor? If the marker can identify only a small minority of the prospective cases, a different approach to the problem may be preferable.

The answers to these questions may vary in different contexts, as may the associations between specific factors and diseases. A given risk marker may be useful in one setting but not in another.

Community syndromes

The term 'community health syndrome' may be used to refer to diseases or other health characteristics found to occur together in a community. Examples described by Kark (1974, 1981), who introduced the community syndrome concept and has emphasized its potential importance for the development of community medicine programmes, are a syndrome of malnutrition, communicable diseases, and mental ill-health in a poor rural community undergoing rapid change, and the syndrome of hypertension, coronary heart disease, and diabetes frequently found in affluent communities characterized by nutritional imbalance and excesses, limited physical activity, and a drive for achievement.

The components of a syndrome may occur together because they posses shared or related causes, or because they are themselves causally interrelated. The syndrome points to a nexus of causal processes in the community. Even if the 'web of causation' (McMahon *et al.* 1960) is not completely understood, a health programme directed at the syndrome as a whole may be more effective and efficient than an endeavour to deal separately with the individual components.

Associations between disease or other health states may be detected by observing that they occur together in the same population or population groups. They may also, and more convincingly, be detected by a tendency to affect the same individuals. As an example of the latter approach, an analysis of co-prevalence in a Jerusalem neighbourhood revealed an unexpected cluster of mutually associated disorders—migraine, chronic bronchitis, congestive heart failure, gallbladder disease, and chronic arthritis (Abramson *et al.* 1982). The clustering was especially strong when people with clear objective evidence of these diseases were removed from consideration; that is, the clustering was essentially between complaint-based disorders. People with one or more of these common conditions made especially heavy use of the primary care service. These disorders were frequently associated with emotional symptoms and with family disharmony or other stressful situations. The analysis drew attention to the occurrence of a community syndrome that represented a considerable burden of discomfort for the many affected individuals and their families, and for which there was no organized programme in the health service.

Identification of groups requiring special care

Community diagnosis may focus not only on the community as a whole but on its component groups. Comparisons of different population categories may serve to identify groups that require special care. This identification may be based on the presence of disorders, on screening tests that point to a high probability of having a disorder, on the presence of modifiable risk factors, and/or on the presence of known risk markers indicative of vulnerability and a need for preventive care.

A differential approach in community diagnosis is of basic importance for the identification of priorities and the allocation of resources. Sometimes simple descriptive findings suffice for these purposes, and in other circumstances the planning of effective care requires an understanding of the reasons for the differences found, requiring the use of analytical epidemiological techniques.

Surveillance

Ongoing surveillance permits the identification of changes in health status and its determinants in the community, and updating of the community diagnosis. Both cross-sectional and incidence studies have a role in surveillance. For some purposes, e.g. to detect changes in a community's health habits or blood pressure distributions, the only practicable method of surveillance is the performance of repeated cross-sectional studies. Repeated population surveys of smoking status, blood pressure, serum cholesterol, medication for hypertension, and other variables are an essential element of the World Health Organization's MONICA project (multinational monitoring of trends and determinants in cardiovascular disease), which aims to examine cardiovascular trends and their relationships to known risk factors; this project is now being implemented in over 100 sub-populations (WHO MONICA Project Principal Investigators 1988).

Surveillance of the prevalence of chronic disorders may be based on repeated prevalence surveys, or on the use of a case register that is updated as new cases are found or old ones recover, die, or leave. Changes in the prevalence of chronic disorders may be of importance as an indication of changing needs for curative and rehabilitative care facilities.

Changes in the prevalence of a chronic disease cannot, however, be glibly taken to indicate changes in the risk of developing the disorder; for this purpose, incidence data should be used. There are a number of possible reasons for the changes in prevalence. As demonstrated in Fig. 8.1, these changes reflect the interplay of incidence, recovery, and fatality rates. Changes in prevalence may be caused by changes in the demographic characteristics of the population, as a result of aging or inward or outward migration. Especially in studies of small local communities, prevalence may be influenced by a tendency of affected persons to leave or enter the neighbourhood, e.g. because of changes in admission or discharge policies of hospitals or other institutions, or in the availability of these institutions. Often, apparent changes in prevalence are artefacts caused by changes in methods of case identification (e.g. the introduction of a case-finding programme), in the use of medical services, in diagnostic procedures or definitions, or in recording, notification, or registration practices. They may also be caused by incomplete updating of a case register.

Community education and community involvement

Community surveys can be used as tools for community health education. This may be done not only by communicating the findings and their implications to the community and its leaders, but also by using the educational potential of the survey situation itself, e.g. by explaining to participants why the collection of specific information is important. If accurate results are required, such explanations should preferably be given after the information has been collected, to minimize bias in the responses.

An example of the use of the survey as an educational tool is provided by the 'Know Your Body' programme, a school health education programme aimed at motivating children to adopt a healthier lifestyle (Williams *et al.* 1977). An essential component of this programme is a simple set of measurements of chronic disease risk factors. Each child receives a feedback of results on a 'health passport', together with explanations of desirable ranges for each test, in the hope that this may enhance the effect of the curriculum. Quasi-experimental evaluation indicates that this approach effectively modifies the children's knowledge and beliefs, and (less strikingly) their health behaviour (Marcus *et al.* 1987). Besides their educational use, the survey provides a picture of risk factor prevalence.

Involvement of key community members in the planning and conduct of a health survey may be a useful way to motivate them to a more active participation in the promotion of their community's health. A community's interest and involvement in its own health care may find expression in the performance of community self-surveys, even without the participation of professional health workers. Such studies are usually simple descriptive surveys, and may not collect very accurate or sophisticated information.

Evaluation of a community's health care

In the context of community health care, the purpose of an evaluative study is to yield a factual basis for decisions concerning the provision of care to a specific community. This kind of evaluative study, the programme review, can be contrasted with the programme trial (see p. 118), which aims to provide generalizable inferences about the value of a given type of health programme. In programme reviews, considerable attention is given to evaluation of the process of care (the performance of activities by providers and recipients of care), as well as to measurements of desirable and undesirable effects, especially more immediate outcomes.

Certain findings that are used in the process of community diagnosis as indicators of needs for health care, such as a high prevalence of preventable or remediable disorders, may by the same token be seen as indicators of the value of prior health care. In some instances a prevalence survey may reveal more direct evidence of the quality of previous care; the quality of the dental work found in subjects' mouths may be appraised, for example, or the presence of inguinal hernia recurrences may be recorded (in one study, one in five operated hernias showed evidence of recurrence; Abramson *et al.* 1978).

Evaluative judgments that relate to the subjects' prior health care as a whole, however, may be less helpful than those relating to recent or current health care, and especially to a specific health programme or service. Cross-sectional and other studies designed for the latter purpose focus on the performance of specific planned activities and the achievement of specific outcomes. They might deal, for example, with compliance with medical advice (what proportion of hypertensives are taking the medicines prescribed for them?), with satisfaction with medical care, or with the immunization status of one-year-olds or two-year-olds in the community.

In these as in other cross-sectional studies, separate attention is usually paid to various population sub-groups. The impact of a health programme often varies with age, sex, social class, and other characteristics.

The demonstration of outcomes usually requires evidence of change. This may be provided not only by incidence and other longitudinal studies but also, for many purposes, by repeated cross-sectional studies. Surveys of the contraceptive practices of post-natal women in a Jerusalem neighbourhood, for example, revealed that during a six-year period the proportion using the contraceptive pill, an intra-uterine device, or condom increased from 33 to 62 per cent (Kark 1981). Such information would be of obvious interest to a health service operating a family planning programme in the neighbourhood, even in the absence of rigorous proof that the change was actually an outcome of the programme and not an effect of other influences. It is usually assumed that such changes (or their absence) are, at least to some extent, reflections of programme effectiveness. They are hence often used as a basis for decisions on the continuation or modification of programmes. At the very least, they may indicate whether there is a need for more detailed evaluative study.

More rigorous proof that it was the programme that produced the change requires a comparison with controls. For example, surveys of infant feeding practices (based on clinic records and telephone interviews) showed that in a community where a structured programme for the promotion of breast-feeding was integrated into a maternal and child health service, the prevalence of breast-feeding among infants aged 26 weeks rose from 10 per cent in 1978 (before the programme was started) to 29 per cent in 1985. Confidence that this was attributable to the programme and not to a secular change was enhanced by the finding that in a neighbouring community with no such programme the corresponding rate in 1985 was 12 per cent (Palti et al. 1988).

Uses in clinical practice

Epidemoliogical studies serve important functions in clinical care. We will consider the role of cross-sectional studies in (1) individual and family care, and (2) community-oriented primary care.

Individual and family care

Textbooks of clinical epidemiology emphasize that epidemiology is a basic science for clinicians (Fletcher et al. 1982; Sackett et al. 1985). 'A great many routine clinical decisions

about individual patient care relating to matters such as the choice of diagnosis and treatment options, and advice on prognosis can only be based upon information from properly designed and executed studies in groups or populations' (Roberts 1977).

For the clinician who bases decisions on epidemiological facts rather than on dogma or impressions, the information required may be derived from any epidemiological studies whose results can validly be generalized to the population in which the clinician works. If information about this specific population is available, it is of course of especial value. Such information (usually largely based on cross-sectional studies) may deal with the prevalence of various diseases and their causes, the frequency distributions of biochemical and other measurements in the population and in patients with specific disorders, patterns of child growth, customary practices of significance to health, and so on. This information is, of course, seldom the fruit of the clinician's own labours. Sometimes, however, even a clinician whose sole concern is with the care of individual patients may conduct a small-scale epidemiological survey (usually cross-sectional). The clinician may, for example, need to survey a patient's contacts for evidence of a specific communicable disease (such as gonorrhoea), if only to reduce the patient's risk of reinfection.

A family physician, whose responsibility extends to the care of the whole family, may need to perform such investigations more often. Whenever a patient is found to have a disease with a known tendency to 'run in families' (e.g. rheumatic heart disease, amoebiasis, diabetes mellitus, or acquired immune deficiency syndrome), this may be seen as a signal that the family as a whole should be surveyed for the disease.

Family diagnosis

The process of appraising a family's health status and the factors that affect it (Kark and Kark 1962; Medalie 1978) is an exercise in small-group epidemiology. Its aim is to determine a family's health needs as a basis for the planning of a family care programme. This involves the elucidation of the health of the family's members and appraisal of relevant features of the family life situation, e.g. the family's structure and composition, the role performance and health-relevant behaviour of its members, relationships, material resources and their use, and the family's social and physical environment. At an analytical level, it involves assessment of the ways in which individuals' health may be affected by other family members and by the family life situation.

Community-oriented primary care

Community-oriented primary care refers to the combination, in a single integrated practice, of the health care of the community and of its individual members—the clinician or practice team providing primary clinical care for the individuals in a defined community also initiates specific programmes to deal in a systematic manner with the community's main health problems (Abramson and Kark 1983; Kark 1974,

1981). There is a growing awareness of the potential of this form of practice for improving health in both developing and developed countries (Connor and Mullan 1983; Kark and Abramson 1981; Nutting 1987).

Epidemiology provides an indispensable basis for the planning, development, and evaluation of the community health programmes that characterize community-oriented primary care. The uses listed above of cross-sectional studies in community health care, and their uses in individual and family care, are both relevant to this form of practice.

The integration of individual-oriented and community-oriented functions in the same primary care setting carries important implications for the conduct of epidemiological studies. An important feature is that a good deal of the information needed for community diagnosis, ongoing surveillance, and programme evaluation can be obtained in the course of routine clinical care. The collection of data can thus serve a double function. When a child is weighed or a question is asked about smoking or a diagnosis is determined, the results may be used both in the management of the patient and as data for subsequent analysis at a group level. Such information may be derived either from routine data collected in the ordinary diagnostic investigation and surveillance of patients, or from questions or tests specially added to clinical procedures for epidemiological purposes. In a practice where periodic health examinations are conducted, these provide an especially useful opportunity for the collection of such information.

This use of clinical data demands careful attention to methods of obtaining, recording, and retrieving data, to ensure that the information will be as accurate and complete as possible. Standardized procedures are required, and definitions and diagnostic criteria should be standardized as rigorously as possible.

The performance of a cross-sectional study in this way may, of course, extend over a considerable span of time. However long the time allotted, there will often be a need for supplementary survey procedures to obtain information about members of the community who have not attended for clinical care. These people may be invited to attend for health examinations, visited at home, or asked to supply information by mail or telephone. In population groups with very high attendance rates (say infants and their mothers, pregnant women, and the elderly), there may be little bias if non-attenders are excluded.

For chronic diseases, a common technique is the maintenance of a register of cases (this may be a list, a card index, or a computerized database). Such a register, like other practice records, has a dual purpose. At a group level, it permits the calculation of point prevalence rates or other epidemiological indices, and may provide a means of monitoring the performance of planned tests and other activities; and at an individual level, it can be used as a tool for ensuring that specific patients receive the care they need.

Information may also be obtained by special surveys. These may be conducted primarily for purposes of community diagnosis or surveillance, or to evaluate health programmes. For some purposes (say to measure smoking habits) it may be satisfactory to investigate a representative sample of the population. More usually, surveys in the context of community-oriented primary care aim not only to provide information at a community level, but also to identify individuals who need care. In such surveys the use of a sample is clearly unsatisfactory.

A survey that has obvious relevance to health may be a useful means of stimulating the community's interest and involvement in its own health care. The purpose of the survey should be explained to the community, and a 'feedback' of results supplied to community leaders and the community at large, using local newsletters and whatever other media may be available.

Sackett and Holland (1975) have contrasted the roles of epidemiological surveys, screening, and 'case-finding' in the detection of disease. They define epidemiological surveys as studies of carefully selected population samples, aimed at generating new knowledge and implying no health benefits to the participants. Screening involves the testing of apparently healthy people who respond to an invitation to be examined, and it carries an implied promise of benefit. 'Case-finding' is performed among patients who have sought health care, and is concerned with tests to reveal disorders that may be unrelated to their complaints.

In the context of community-oriented primary care, such distinctions become blurred or may vanish. An epidemiological survey may aim to provide information that will benefit individual participants, as well as providing a basis for decisions about care at a community level. The survey may be conducted in a clinical setting. In this form of practice, people are seldom invited to attend solely for screening purposes. Screening tests may be added to the routine procedures for patients who attend for treatment, or incorporated in health examinations that provide an opportunity for appraisal and surveillance of the individual's health status and life situation and for counselling, and that are not concerned solely with screening for disease.

The collection of information by the providers of care in a care setting carries advantages and disadvantages. On the one hand, if relationships with patients are sound it may be relatively easy to obtain answers to awkward questions, and to achieve a high response rate. On the other hand, possibilities of bias must be recognized. The respondents may be aware of what replies are acceptable to care-providers, and their desire to please may lead to biased responses. Also, care-providers conducting an evaluation of the care they themselves provide may tend to make biased observations and inferences. Precautions are needed to reduce bias by using objective measures when possible, and by obtaining the assistance of independent observers or investigators.

It is often difficult to obtain satisfactory denominator data, i.e. information about the size and demographic characteristics of the population that is the target of health care and the denominator for calculating rates and other health indices. If information on age, sex, and other attributes of this population is not available from census and administrative sources, the health service may have to devise its own data-gathering mechanism. An incremental approach may be used, by starting data collection in an 'initial defined area' and gradually expanding the defined segment of the practice population (Kark 1966; Nutting 1987). One approach is to incorporate the collection of demographic data in a broader household health survey. In a practice where voluntary or other community health workers are active, demographic surveillance often becomes one of their functions.

Studies yielding 'new knowledge'

Many cross-sectional studies are performed to expand the horizons of knowledge, rather than solely to promote the health of the specific group or population studied. They aim to yield generalizable inferences that are of broad application and not relevant only to a specific local context.

We will briefly consider studies of growth and development and of aetiology, and programme trials. Other topics include the natural history of health and disease (usually better investigated by longitudinal studies) and methodological issues. Cross-sectional studies are commonly used to examine the effects of differences in operational definitions or methods of study and to appraise the validity of screening tests and proxy measures. These methodological studies may aim to develop and evaluate procedures for use in both community and individual diagnosis.

Studies of growth and development

Growth and development, and age trends in the prevalence of disorders, can be studied cross-sectionally as well as longitudinally. The cross-sectional method compares different age-groups observed at one point of time, whereas the longitudinal method makes repeated observations of a single cohort through various age levels. The cross-sectional method is simpler, but has limitations. It can provide information about average changes, but not about intra-individual changes or inter-individual differences in change.

The main limitation of the cross-sectional method is that the age-groups that are compared may differ in respects other than age, so that the effects of age and other influences may be confounded. There is always potential confounding of age changes with differences between birth cohorts, as the age-groups that are compared must belong to different cohorts. Cohort differences in growth may be negligible, but they may be significant if cohorts were exposed to very different circumstances, e.g. different infant-feeding or child-rearing fashions, changes in economic prosperity, or war. As a result the cross-sectional method may yield a misleading picture. If there has been a secular increase in height, a cross-sectional study may show a decrease in average height throughout adult life; but young adults will be taller because they belong to a more recently born generation, not because of their youth. Similar confounding has been demonstrated in studies

of the changes in intelligence during adult life. Cross-sectional studies have indicated a decline beginning at about 30 years of age, whereas longitudinal studies have shown increases or no change in intellectual performance until the age of 50 or 60 years (Baltes *et al.* 1977). If a series of cross-sectional studies has been done, suitable rearrangement of the data may permit examination and comparison of the longitudinal changes in different cohorts.

In studies (cross-sectional or longitudinal) that include the middle-aged and elderly, selective survival may be important. The mean blood pressure may be lower in the very old, not because blood pressure tends to drop with age, but because hypertensive people are more likely to have died and thus left the study population. The possibility that the validity of measures may vary with age also requires consideration. The results of a memory test in the elderly may be expressions of hearing ability, attentiveness, or depression, rather than of memory capacity.

Studies of aetiology

Cross-sectional studies often provide useful clues to aetiological processes, especially with respect to influences on long-term disorders and relatively stable measurements and health habits. They have two features, however, that often restrict their value for the testing of causal hypotheses.

First, any associations they reveal are with the *presence*, not the *appearance*, of the disorder or other variable studied. Transient or rapidly fatal cases of the disorder are inevitably under-represented. The causes that determine the appearance of the disorder are confounded with those that influence its duration, and it may be difficult to draw clear inferences about either set of causes. Such confounding is relatively unimportant if the disorder studied is seldom fatal and if it has a high chronicity or data on lifetime prevalence are used. In such instances the main difficulty is that the causal factors may no longer be apparent because of the time-lag since initiation.

Second, in a strictly cross-sectional study the absence of information on time relationships may render it difficult to separate effects *on* a dependent variable from effects *of* the dependent variable. The influence of blood pressure, serum cholesterol, and cigarette smoking on the occurrence of myocardial infarction, for example, may be confounded with changes ensuing from the disease episode. The demonstration in a cross-sectional study that fat abdomens (based on a comparison of waist with hips size) are associated with hypertension, hypertensive heart disease, and diabetes (controlling for sex, age, and ponderal index) is difficult to interpret without knowing which came first, the fat abdomen or the disease (Gillum 1987). The discovery of an inverse association in San Francisco's bus drivers between hypertension and reported job-related problems—a relationship that was not explained by confounding factors and was specific to hypertension (gastro-intestinal, respiratory, and musculo-skeletal problems were positively associated with the

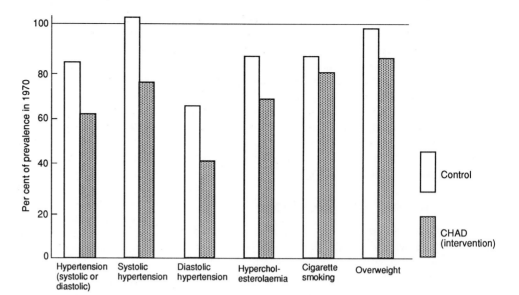

Fig. 8.2. Prevalence of risk factors in CHAD (intervention) and in control populations in 1975, expressed as a percentage of the prevalence in the same population in 1970. Based on age- and sex-standardized rates.

self-reported stressors)—was given two competing explanations. On the one hand, 'emotional states or coping mechanisms of a person may play a role in the pathogenesis of hypertension through the repression of anger and hostility . . . hypertension can result from repressing conflicts . . . '. On the other, 'the hemodynamic consequences of elevated blood pressure may lead to a physiologic alteration of perception . . . elevation in blood pressure reduces reactivity to noxious stimuli' (Winkleby *et al.* 1988).

The value of a cross-sectional study in the search for causes and precursors is limited whenever there is a possibility that the disease may change the subject's life style, bodily functions and characteristics, or circumstances. To throw light on time sequences, cross-sectional studies are often extended to include relevant historical information on times of disease onset or of other occurrences. Repeated cross-sectional studies of the same population can sometimes provide information on the order of events.

A cross-sectional study may form the first stage of a longitudinal study, for which it provides baseline measurements of dependent and independent variables. If the study is concerned with the incidence of a long-term disease, the baseline prevalence study identifies affected people, who must be excluded from the population at risk of subsequently developing the disease. The affected individuals may be followed up in order to study the natural history of the disease.

Programme trials

To obtain convincing evidence of the effectiveness of a health programme that aims to modify the distribution of a characteristic in a population or to reduce the prevalence of a disease, it is necessary to measure the change in the population and to demonstrate that this can be attributed to the programme rather than to other causes. Changes in the population can be appraised by repeated cross-sectional studies. In a

population where the prevalence of anaemia in pregnant women was originally 12.0 per cent, the introduction of an intervention programme was followed by a progressive drop in prevalence, in successive periods, to 8.8, then to 3.3, then to 1.6 per cent (Kark 1981).

The cause-and-effect relationship between a programme and its apparent outcome may be difficult to substantiate without observations of a control population not exposed to the programme. For this purpose, a simple comparison of the changes seen in people who voluntarily participate or do not participate in a programme is seldom justifiable, as the comparison may be confounded by differences between the groups.

In trials of programmes directed at populations, it is unfortunately seldom possible to randomize. There is often little or no choice as to *what* population will be exposed to the programme under trial, and a restricted choice concerning a control population. Most programme trials are therefore quasi-experiments, in which the control group or groups are selected so as to be as similar as possible to the intervention group.

As an illustration, Figure 8.2 summarizes some results of a controlled evaluation of the CHAD programme for the control of cardiovascular risk factors in a Jerusalem neighbourhood (Abramson and Kark 1983). This is a community health programme integrated into primary care, implemented by family physicians and nurses; its name is an acronym representing 'Community syndrome of Hypertension, Atherosclerosis, and Diabetes'. The figure is based on age- and sex-standardized prevalence rates derived from cross-sectional surveys of all adults aged at least 35 years in 1970 (before the introduction of the programme) and in 1975, in the intervention (CHAD) population and in a neighbouring control population. It shows the rates in 1975, expressed as percentages of the baseline rates in 1970. For each of the risk factors shown, the decrease in prevalence (the 'prevented fraction' defined in Table 8.1) was greater in the CHAD population. The fact that the prevalence of some risk factors also declined in the control population—apparently as a result of changes with time in awareness and care—underlines the need for control groups in such studies.

The use of population samples in surveys designed to evaluate health programmes is discussed by Salonen *et al.* (1986), who compare the pros and cons of surveying separate samples on each occasion, compared with repeated surveys of the same samples.

References

Abramson, J.H. (1984). *Survey methods in community medicine* (3rd edn). Churchill Livingstone, Edinburgh.

Abramson, J.H. (1988). *Making sense of data: a self-instruction manual on the interpretation of epidemiological data.* Oxford University Press, New York.

Abramson, J.H. and Kark, S.L. (1983). Community oriented primary care: meaning and scope. In *Community-oriented primary care: new directions for health services delivery* (ed. E. Connor and F. Mullan) p. 21. National Academy Press, Washington, DC.

Abramson, J.H., Hopp, C., and Epstein, L.M. (1981). The epidemiology of varicose veins: a survey in western Jerusalem. *Journal of Epidemiology and Community Health* **35**, 213.

Abramson, J.H., Gofin, J., Hopp, C., Makler, A., and Epstein, L.M. (1978). The epidemiology of inguinal hernia. *Journal of Epidemiology and Community Health* **32**, 59.

Abramson, J.H., Gofin, J., Peritz, E., Hopp C., and Epstein, L.M. (1982). Clustering of chronic disorders—a community study of coprevalence in Jerusalem. *Journal of Chronic Diseases* **35**, 221.

Anderson, D.W., Schoenberg, B.S., and Haerer, A.F. (1988) Prevalence surveys of neurologic disorders: methodological implications of the Copiah County study. *Journal of Clinical Epidemiology* **41**, 339.

Baltes, B.P., Reese, H.W., and Nesselroade, J.R. (1977). *Lifespan developmental psychology: introduction to research methods.* Brooks/Cole, Monterey, California.

Connor, E. and Mullan, F. (ed.)(1983). *Community oriented primary care; new directions for health service delivery.* National Academy Press, Washington, DC.

Elandt-Johnson, R.C. (1975). Definitions of rates: some remarks on their use and misuse. *American Journal of Epidemiology* **102**, 261.

Fleiss, J.L. (1981). *Statistical methods for rates and proportions* (2nd edn.). Wiley, New York.

Fletcher, R.H., Fletcher, S.W., and Wagner, E.H. (1982). *Clinical epidemiology—the essentials.* Williams and Wilkins, Baltimore, Maryland.

Franceschi, S., Tirelli, U., Vaccher, E., *et al.* (1988). Risk factors for HIV infection in drug addicts from the Northeast of Italy. *International Journal of Epidemiology* **17**, 162.

Frantz, V.K., Pickren, J.W., Melcher, G.W., and Auchincloss, H., Jr. (1951). Incidence of chronic cystic disease in so-called 'normal breasts': a study based on 225 postmortem examinations. *Cancer* **4**, 762.

Gillum, R.F. (1987). The association of body fat distribution with hypertension, hypertensive heart disease, diabetes and cardiovascular risk factors in men and women aged 18–79 years. *Journal of Chronic Diseases* **40**, 421.

Grundy, P.F. (1973). A rational approach to the 'at risk' concept. *Lancet* **ii**, 1489.

Helton, A.S., McFarlane, J., and Anderson, E.T. (1987). Battered and pregnant: a prevalence study. *American Journal of Public Health* **77**, 1337.

Jagger, C. and Clarke, M. (1988). Mortality risks in the elderly: five-year follow-up of a total population. *International Journal of Epidemiology* **17**, 111.

Kark, S.L. (1966). An approach to public health. In *Medical care in developing countries* p. 5:1 (ed. M. King). Oxford University Press, Nairobi.

Kark, S.L. (1974). *Epidemiology and community medicine.* Appleton-Century-Crofts, New York.

Kark, S.L. (1981). *The practice of community-oriented primary health care.* Appleton-Century-Crofts, New York.

Kark, S.L. and Abramson, J.H. (ed.)(1981). Community-focused health care. *Israel Journal of Medical Sciences* **17**, 65.

Kark, S.L. and Kark, E. (1962). A practice of social medicine. In *A practice of social medicine: a South African team's experience in different African communities* (ed. S.L. Kark and G.W. Steuart), p. 3. Livingstone, Edinburgh.

Leighton, D.C., Harding, J.S., Macklin, D.B., MacMillan, A.M., and Leighton, A.H. (1963). *The character of danger.* Basic Books, New York.

Lilienfeld, A.M. and Lilienfeld, D.E. (1980). *Foundations of epidemiology* (2nd edn). Oxford University Press, New York.

McGavran, E.G. (1956). Scientific diagnosis and treatment of the community as a patient. *Journal of the American Medical Association* **162**, 723.

MacNamara, J.J., Molot, M.A., Stremple, J.F., and Cutting, R.T. (1971). Coronary artery disease in combat casualties in Vietnam. *Journal of the American Medical Association* **216**, 1185.

McMahon, B., Pugh, T.F., and Ipsen, J. (1960). *Epidemiologic methods.* Little Brown, Boston, Massachusetts.

Maffei, F.H.A., Magaldi, C., Pinho, S.Z., *et al.* (1986). Varicose veins and chronic venous insufficiency in Brazil: prevalence among 1755 inhabitants of a country town. *International Journal of Epidemiology* **15**, 210.

Marcus, A.C., Wheeler, R.C., Cullen, J.W., and Crane, L.A. (1987). Quasi-experimental evaluation of the Los Angeles Know Your Body program: knowledge, beliefs, and self-reported behaviours. *Preventive Medicine* **16**, 803.

Medalie, J.H. (ed.)(1978). *Family medicine: principles and applications.* Williams and Wilkins, Baltimore, Maryland.

Morris, J.N. (1975). *Uses of epidemiology* (3rd edn.). Churchill Livingstone, Edinburgh.

Nutting, P.A. (ed.)(1987). *Community oriented primary care: from principle to practice.* Health Resources and Services Administration, Public Health Services, Washington, DC.

Palti, H., Adler, B., Flug, D., *et al.* (1977). Community diagnosis of psychomotor development in infancy. *Israel Annals of Psychiatry* **15**, 223.

Palti, H., Valderama, C., Pugrund, R., Jarkoni, J., and Kurtzman, C. (1988). Evaluation of the effectiveness of a structured breast-feeding promotion programme integrated into a maternal and child health service in Jerusalem. *Israel Journal of Medical Sciences* **24**, 342.

Peritz, E. (1971). Estimating the ratio of two marginal probabilities in a contingency table. *Biometrics* **27**, 223; correction note in *Biometrics* **27**, 1104.

Poikolanen, K. and Eskola, J. (1988). Health services resources and their relation to mortality from causes amenable to health care intervention: a cross-national study. *International Journal of Epidemiology* **17**, 86.

Roberts, C.J. (1977). *Epidemiology for clinicians*. Pitman Medical, Tunbridge Wells.

Rogan, W.J. and Gladen, B. (1978). Estimating prevalence from the results of a screening test. *American Journal of Epidemiology* **107**, 71.

Sackett, D.L. (1979). Bias in analytic research. *Journal of Chronic Diseases* **33**, 51.

Sackett, D.L. and Holland, W.W. (1975). Controversy in the detection of disease. *Lancet* **ii**, 357.

Sackett, D.L., Haynes, K.B., and Tugwell, P. (1985). *Clinical epidemiology: a basic science for clinical medicine*. Little Brown, Boston, Massachusetts.

Salonen, J.T., Kottke, T.E., Jacobs, D.R., Jr., and Hannan, P.J. (1986). Analysis of community-based cardiovascular disease prevention studies—evalution issues in the North Karelia Project and the Minnesota Heart Health Program. *International Journal of Epidemiology* **15**, 176.

Smith, S.H. and Hill, G.L. (1974). Corrections for sampling bias in the epidemiological survey of a chronic disease. *International Journal of Epidemiology* **3**, 63.

Susser, M. (1973). *Causal thinking in the health sciences: concepts and strategies in epidemiology*. Oxford University Press, New York.

Von Korff, M. and Parker, R.D. (1980). The dynamics of the prevalence of chronic episodic disease. *Journal of Chronic Diseases* **33**, 79.

WHO MONICA Project Principal Investigators (1988). The World Health Organization MONICA project (monitoring trends and determinants in cardiovascular disease): a major international collaboration. *Journal of Clinical Epidemiology* **41**, 105.

Williams, C L., Arnold, C.B., and Wynder, E.L. (1977). Primary prevention of chronic disease beginning in childhood, the 'Know Your Body' program: design of study. *Preventive Medicine* **6**, 344.

Williamson, D.F., Forman, M.R., Binkin, N.J., Gentry, E.M., Remington, P.L., and Trowbridge, F.L. (1987). Alcohol and body weight in United States adults. *American Journal of Public Health* **77**, 1324.

Winkleby, M.A., Ragland, D.R., and Syme, S.L. (1988). Self-reported stressors and hypertension: evidence of an inverse association. *American Journal of Epidemiology* **127**, 124.

9

The case–control study

RAYMOND S. GREENBERG and MICHEL A. IBRAHIM

Introduction

The case–control study is used primarily to assess risks and to study causes of disease in general. This chapter is devoted to the latter or aetiological use. There are several terms that are sometimes used in lieu of the case–control study. Terms such as case–referent, case–compeer, and trohoc have been preferred by some investigators. However, the term case–control seems to be preferred by most epidemiologists and, therefore, is the one used throughout this chapter. The nature, setting, design, biases, and analysis of case–control studies will be covered. Finally, some scientific standards and obligations of scientists are proposed. Examples from the literature are cited to illustrate the various concepts and methodological issues.

The nature of case–control research

As with most epidemiological research, case–control studies involve observations of naturally occurring exposures and disease. In these non-experimental or observational methods, the investigator observes and studies, but does not intervene with, the natural history of disease processes. This contrasts with experimental studies, in which the investigator intervenes to influence the exposure status of subjects.

Within the broad category of observational research, a further subdivision can be made. Certain types of observational studies are 'exploratory' since they are used for the generation of research hypotheses. This type is particularly useful at an early stage of inquiry, when there is relatively little known about the condition under investigation. In these studies, a variety of potential aetiological factors are assessed to identify promising areas for further research. The exploratory approach is typified by the case series (a summary of individuals with a particular disease), the ecological study (a correlation of exposure and disease distributions), or the proportional mortality study (a comparison of cause-specific contributions to total mortality in exposed and unexposed groups). Occasionally, case–control studies are used for exploratory purposes, by evaluating possible associations between a single disease entity and a variety of different

exposures. Ideally, exposure–disease associations detected in an exploratory study should be considered tentative until more definitive evidence is obtained.

A research hypothesis concerning a specific cause-and-effect relationship may be tested in an 'explanatory study'. Although such studies focus on a single exposure–disease association, it is usually necessary to collect information on other factors which might distort or modify that relationship. There are three different approaches to observational explanatory research: cross-sectional, cohort (also prospective, follow-up), and case–control studies. These three approaches differ with regard to the sequence of observations on exposure and disease status. In cross-sectional research, both variables are measured simultaneously. A cohort study begins with a measurement of exposure status and then follows the subjects forward in time for subsequent disease outcome. In contrast, a case–control study begins with a determination of disease status and then traces the subjects backward in time for prior exposure history.

Regardless of the sequence of observations in an explanatory study, an observed exposure–disease association does not necessarily imply a causal relationship. Even when chance and known systematic errors are reasonably excluded, it is possible that unrecognized factors influenced the relationship of interest. This problem is somewhat alleviated in experimental studies, where randomization tends to balance the distribution of both known and unrecognized factors between study groups. Therefore, experimental studies are often considered the best evidence of a causal relationship. Unfortunately, many exposure–disease associations cannot be studied with human experimentation, since it would be unethical to expose people intentionally to detrimental agents. In these situations, explanatory observational studies provide the best available assessment of causation.

The anatomy of case–control research

In a case–control study, two types of subjects are sampled: persons with (cases) and without (controls) the disease of interest. The source population from which cases and controls are drawn has been referred to variously as the *study*

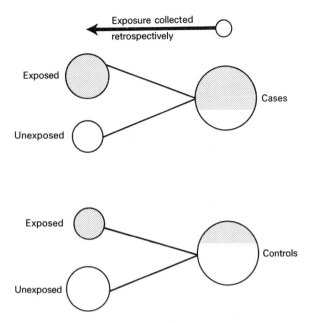

Fig. 9.1. The case–control study. Shaded areas represent exposed persons and unshaded areas represent unexposed persons.

base (Miettinen 1985), the *candidate population* (Kleinbaum *et al.* 1982), and the *target population* (Schlesselman 1982). In some circumstances, the source population is defined by residency in a particular community. In other situations, the source population is characterized by affiliation with a specific group (e.g. an occupational cohort). In still other settings, the source population consists of patients who attend certain medical facilities. Regardless of the source of subjects in a case–control study, samples are obtained contingent upon the presence or absence of the disease of interest. After the cases and controls have been selected, information is collected on each individual's prior exposure history (Fig. 9.1). Then the relative frequency of exposure is compared between cases and controls.

The structure of a case–control study affords certain advantages as well as certain limitations (Table 9.1). The most obvious advantage is that the investigator can choose the ratio of cases to controls. Thus, a disease that is rare in the general population can be heavily sampled in the study population. In this manner, uncommon diseases can be studied efficiently, with relatively few subjects and a modest expenditure of resources.

Example 1—The efficiency of case–control studies for rare diseases
As a simple illustration, consider two different approaches to the study of leukaemia in children. The crude annual incidence rate of this disease is about 3.4 cases for every 100 000 children under 15 years of age (Silverberg 1982). A cohort study of leukaemia in children would require a year of observations on a million children to identify 34 cases of this disease. In a study that evaluates the association between leukaemia and a proposed aetiological agent, these 34 chil-

dren with leukaemia would be subdivided into two or more exposure categories. Consider how much easier it would be to identify the 34 leukaemia cases as they are diagnosed and then select an appropriate comparison group of children without leukaemia to evaluate possible aetiological exposures.

The case–control approach is also advantageous when there is a long induction period between the exposure and clinical onset of disease. Rather than waiting years for the prospective accrual of cases, the investigator may 'compress time' by using historical documents to evaluate earlier exposures. Indeed, the case–control approach was first widely utilized for the study of chronic diseases such as cancer and cardiovascular disease.

The limitations of case–control studies are related principally to the sampling scheme and the retrospective sequence of observations. These two features make case–control studies susceptible to errors in the evaluation of exposure–disease relationships. These systematic errors, also termed 'biases', may arise in the selection of subjects, collection of information, or the mixing of multiple effects. While biases may occur in any study design, case–control studies are especially vulnerable to these errors. Later in this chapter, we will present an overview of bias and methods for its containment or elimination.

The setting of case–control research

Aetiological research is traditionally performed in one of three different contexts: clinical, community, or occupational settings. These research environments differ with regard to the goals and logistics of investigation. Clinically oriented research is usually directed at the mechanisms of disease causation and modes of management. Typically, the subjects of clinical investigations are patients who seek care at a particular hospital or medical centre. Since the conditions of interest may be relatively uncommon, clinical studies are often necessarily based upon small samples. The 'exposure' under study might be a specific physiological state, a past medical event, a pharmacological agent, or a therapeutic procedure.

Table 9.1. Advantages and limitations of case–control studies

Advantages
Efficient sampling of rare diseases
Rapid evaluation of chronic diseases
Economy of expense and personnel
May serve either exploratory or explanatory purposes

Limitations
Not practical for rare exposures
Subject sampling prone to systematic errors
Historical information often cannot be validated
Relevant co-factors may be difficult to control
Temporal sequence of exposure and disease may be obscured

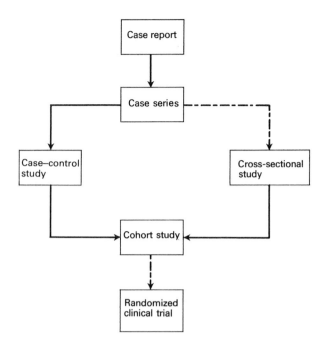

Fig. 9.2. Typical progression of study designs in clinical research.

In contrast, community-based aetiological research is usually directed towards public health concerns. The emphasis in community research is on the population health impact of various exposures. This might be a particular lifestyle, diet, socio-economic status, or place of residence. To evaluate public health consequences of these exposures, it is usually necessary to study relatively large populations. Since public health impact is directly related to exposure prevalence, community studies are often directed at common exposures and diseases.

The major aim of occupational research is to identify associations between certain conditions and chemical, physical, or radiological exposures in the workplace. This information may be used to remove potential hazards and develop epidemiological and environmental surveillance systems. Although common diseases have been considered in occupational studies (e.g. lung cancer), emphasis has generally been on rare conditions (e.g. angiosarcoma of the liver).

A typical progression of study designs employed in clinical research is depicted in Figure 9.2. In the clinical setting, aetiological research often begins with isolated case reports, followed by a larger case series. These investigations are intended to describe the characteristics of a group of patients with a specific disease. After the cases are described, it is necessary to determine how the cases differ from other persons. These differences are defined through comparative studies using case–control or cross-sectional methods. If the hypothesized aetiological association is supported, then that relationship may be confirmed by a subsequent case–control (or possibly a cohort) study. Ultimately, logistics and ethics permitting, an intervention study may be undertaken so that the most convincing evidence of causation may be realized,

or *in vitro* and/or animal studies are conducted for confirmatory evidence.

Example 2—The sequence of study designs in clinical research
The role of oxygen in retrolental fibroplasia (RLF) among premature infants was shown by a progression of studies beginning in 1941. The first in a series of RLF cases was noted on 14 February of that year by a Boston paediatrician Dr Stewart H. Clifford (Silverman 1980). That case was a premature infant girl suffering from nystagmus and scar tissue behind the lens of the eye. A case–control study was subsequently conducted on 53 RLF children and 298 normal children. In spite of the association between longer hours of oxygen used and RLF, it was postulated that poor health of infants necessitated longer hours of oxygen. That is, poor health and not oxygen use *per se* 'caused' RLF (Silverman 1980).

The frequency of RLF in a cohort of infants exposed to high oxygen was compared to that in a cohort of infants exposed to moderate oxygen in several international studies. Some studies confirmed the oxygen–RLF association and some did not. These contradictory findings stimulated the initiation of randomized clinical trials. The first was at the Gallinger Municipal Hospital in Washington, DC, and the second was a collaborative multi-centre trial. Although the results of these trials confirmed the role of oxygen in the aetiology of RLF, questions concerning a safe exposure level remain unresolved (Silverman 1980).

Figure 9.3 illustrates a typical progression of study designs utilized in community research. Since data collection in the community setting can be an expensive proposition, one often begins with descriptive data which were collected originally for some other purpose. For example, one may undertake an ecological study to correlate mortality data with

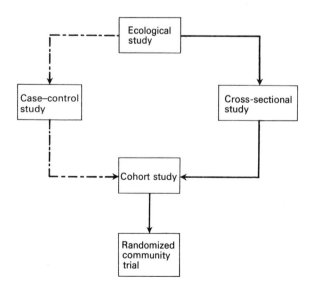

Fig. 9.3. Typical progression of study designs in community research.

aggregate exposure data, such as per caput consumption of dietary ingredients. If a promising exposure–disease correlation is observed, the next step is to conduct a cross-sectional survey of a defined community population. Often, results from a cross-sectional study are used as the baseline observations for a cohort investigation. Alternatively, from the ecological study, one may proceed to a case–control study. If the hypothesized relationship is confirmed, the investigator would have some justification to consider undertaking a cohort study. The definitive study design in community aetiological research is an intervention study in the form of a randomized community trial.

Example 3—The sequence of study designs in community research

An analysis of death rates from coronary artery disease according to per caput fat consumption in 20 countries represented an important beginning for the generation of the lipid–atherosclerosis hypothesis (Joliffe and Archer 1959). Cross-sectional studies such as the initial cross-sectional survey of the Framingham and Evans County heart studies (Cassel 1971; Dawber *et al.* 1957) provided evidence of associations between serum cholesterol levels and coronary artery disease. These findings were confirmed by a number of case–control studies.

Cohort studies, such as the long-term prospective investigations in Framingham, Massachusetts, and in Evans County, Georgia, were the logical next step (Truett *et al.* 1967; Tyroler *et al.* 1971). Coronary artery disease incidence was clearly shown to be related to high levels of cholesterol in these and other studies. Community-based controlled trials of lipid reduction were undertaken and in one such trial, the treatment of elevated serum cholesterol produced a 19 per cent. reduction in death from coronary artery disease and/or non-fatal myocardial infarction (Lipid Research Clinics Program 1984).

Figure 9.4 depicts a typical progression of study designs utilized in occupational research. At the exploratory level, the cause-specific distribution of deaths within an industry may be considered in a proportional mortality study. This type of descriptive study can be performed rapidly, without enumerating the entire workforce or evaluating individuals' exposures. If a particular cause of death appears elevated, then mortality rates within the industry may be calculated and compared with another industrial group or the general population (viz. standardized mortality ratio). If the mortality excess is confirmed, specific aetiological relationships may be investigated by collection of exposure information on a historical cohort of workers. The relative risk for the hypothesized aetiological factor may be estimated directly in a cohort, or estimated indirectly with a case–control analysis.

The latter approach involves selecting cases that arise within the cohort as well as controls from the same occupational group. Mantel (1973) described this type of investigation as a 'synthetic retrospective study', but it has become more commonly referred to as a 'case–control study nested

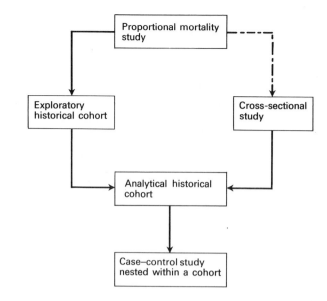

Fig. 9.4. Typical progression of study designs in occupational research.

within a cohort'. Although widely used in occupational research, this type of case–control study has been recently applied to community research as well. A case–control analysis was conducted in a cohort of homosexual men who were seropositive for the human immunodeficiency virus (Polk *et al.* 1987). Five seropositive controls were matched to each of the 59 cases of the acquired immunodeficiency syndrome (AIDS) that developed in the cohort over a 15-month follow-up period. A multi-variate analysis revealed several factors to be associated with the development of AIDS.

The recommended approach for selecting subjects in such a study depends upon the desired measure of effect. Some studies are designed to make inferences about the ratio of *rates* of disease development in exposed and non-exposed persons. For this purpose incident cases should be sampled at the time of diagnosis and matched to one or more persons who are free from disease (controls) at the time (Prentice and Breslow 1978). In other studies, inferences are intended on the ratio of the overall probabilities (*risks*) of disease occurrence in exposed and non-exposed persons over a specified time period. For that purpose all cases that arise during the interval are selected, and persons who remain free-from-disease at the end of the observation period are sampled as controls (Lubin and Gail 1984).

The primary motivation for conducting a case–control study nested within a cohort is to achieve greater efficiency in data collection and analysis than would be entailed in a traditional cohort analysis. In order to perform a cohort analysis, the exposure and disease status must be determined for each person in the source population. With the case–control approach, however, it is necessary to collect information only on those individuals who develop the disease and on a sample of non-diseased persons.

When the disease of interest is rare, the nested case–

control design can save considerable time and expense in data collection, and also simplify the computational effort.

Example 4—The sequence of study designs in occupational research

The sequence of study designs in occupational research may be illustrated by studies of cancer in rubber workers. Mancuso (1949) reported a proportional elevation of respiratory, genito-urinary, and central nervous systems cancer mortality in the rubber industry. A subsequent standardized mortality study revealed excess rates of colon, prostate, and haematopoietic cancers in rubber workers (McMichael *et al.* 1974). Further examination of the haematopoietic cancer deaths indicated that the greatest elevation in risk corresponded to lymphatic leukaemia in middle-aged workers (McMichael *et al.* 1975). This observation led the investigators to conduct a matched case–control study within the industrial cohort. This analysis revealed a consistent association between solvent exposure and lymphatic leukaemia (McMichael *et al.* 1975).

While clinical, community, and occupational research often follow the respective sequences outlined above, many exceptions to these rules could be cited. In practice, the course of scientific inquiry is less predictable than is suggested by Figures 9.2–9.4. For some associations, such as tampon use and toxic shock syndrome, case–control studies may be viewed as exploratory research (Davis *et al.* 1980). For other associations, such as prenatal diethylstilboestrol exposure and adenocarcinoma of the vagina, case–control studies may be considered as more definitive research (Herbst *et al.* 1971). However, most case–control studies fulfil an intermediate role between these two extremes and should be viewed as complementary to the other observational research strategies available in epidemiology.

The conduct of case–control research

The decision to conduct a case–control study is determined by characteristics of the exposure and disease under investigation, the current state of knowledge about the hypothesized aetiological relationship, the immediate goals of the study, the research setting, and the resources available. Case-control studies are especially useful at a preliminary stage of investigation, for the study of rare and chronic diseases, and when resources are scarce. After the decision to conduct a case–control study is made, the investigator must next consider the methods of subject enrolment, data collection, and analysis.

Case identification

The first task for the case–control investigator is to define the disease state. In some situations, the designation of a case may be relatively simple and straightforward. For example, the identification of children with cleft palates could be based entirely upon a simple physical examination. In other situations, it may be difficult to decide what constitutes a disease state. For example, should hypertension be defined by systo-

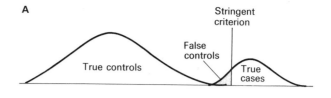

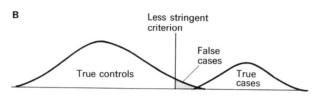

Fig. 9.5. Misclassification errors in the dichotomization of a continuous variable.

Table 9.2. Methods of disease identification with illustrative examples related to coronary artery disease

Method of disease identification	Cardiovascular example
Report of symptoms	Angina pectoris
Physical examination	Hypertension
Passive observation of abnormality	Cardiac arrhythmia
Induced observation of abnormality	Exercise stress testing
Indirect marker of abnormality	Abnormal myocardial enzymes
Response to pharmacological agent	Nitroglycerine trial
Direct observation of abnormality	Coronary angiography

lic, diastolic, or mean arterial blood pressure? Even if one decides on the appropriate diagnostic marker, what level of blood pressure should one consider abnormal? In case–control studies, continuous variables, such as blood pressure, must be categorized and this process is apt to be somewhat arbitrary. The process of disease categorization is schematically represented in Figure 9.5. The most stringent criterion for a disease is likely to misclassify some mild cases as normals. On the other hand, a less stringent criterion may misclassify some normals as diseased. There is no universal rule for balancing these risks of disease misclassification. Either form of misclassification is likely to lead to a biased study result. Cole (1979) has argued for the use of homogeneous case groups, which would support the use of stringent case definitions. However, as Figure 9.5A illustrates, this may diminish the number of eligible cases of an already rare disease. While homogeneous case groups may be desirable, the investigator may be forced to use more liberal definitions of disease in order to obtain a sufficient number of cases.

A variety of methods may be employed alone or in combination to establish the presence of a disease, as depicted in Table 9.2. It is often possible to obtain a direct confirmation of the disease process, as in certain forms of cancer, which can be objectively diagnosed by histopathology. Other diseases, such as mental illness, cannot be defined in terms of structural alterations. For conditions in which the diagnosis cannot be made with reasonable precision, it may be useful to require several diagnostic procedures obtained by different means before accepting a case into the study.

In addition, the diagnostic criteria established by convention of previous investigators and/or consensus committees may provide a standard for disease classification. The designation of standard diagnostic schemes would be useful in comparing the results of independent studies.

It is often recommended that case–control studies should be limited to 'incident' or newly diagnosed cases (Sackett 1979). There are several advantages to the study of incident cases. First, the aetiological exposures in these persons are presumably more reliably recalled. In addition, it is likely that the aetiological milieu of incident cases is relatively homogeneous. Finally, the exclusion of older, surviving cases may prevent the selection bias which is introduced by factors which affect clinical prognosis.

The procedure for locating cases is largely dictated by the research setting. In the medical care setting, it is often convenient to identify potential cases from clinical records, hospital discharge rosters, or institutional registries. In the community, case identification can be a more tedious and time-consuming process. Potential cases may be located from surveillance programmes, employment records, or death certificates. Often, it is necessary to use several sources to identify a sufficient number of cases. Similarly, several sources may be needed to locate cases in occupational studies. In almost all instances where a search for the deceased is required, the investigator must supplement employment records with other data sources, such as state health departments and the social security administration.

Controls

Perhaps the greatest challenge of case–control research is to identify an appropriate comparison population. In a strict sense, comparison subjects in a retrospective study are not true controls. The term 'control' derives from experimental research, where the treatment and comparison populations originate from the same population, and thus only differ with regard to the experimental manipulation. The notion of 'control' is that extraneous influences on the outcome have been eliminated by studying two groups that differ only in a randomly allocated treatment.

Since cases and controls are sampled from two different populations, extraneous influences on the outcome cannot be entirely eliminated. It is possible to account for differences in known relevant co-factors, but there is always the uncertainty that other unknown influences were not considered. With the recognition that a comparison group in a retrospective study is not necessarily a true 'control' group, a valid comparison requires that extraneous variables associated with the disease are not differentially distributed between exposure groups (Feinstein 1979). There are two basic approaches to the control of such extraneous variables:

1. restrict the sampling of subjects to certain levels of relevant co-factors;
2. sample the comparison population without restriction and perform *post hoc* adjustment.

Restricted sampling of subjects can involve either absolute or partial limitations on subject enrolment. With absolute restriction, all participants are limited to narrowly defined co-factor levels (e.g. white males in the fifth decade of life). Absolute restriction may sacrifice important information and limit generalizability of study results. Partial restriction, also known as matching, involves selecting a comparison group which parallels the relevant co-factor distribution of the case group.

The usual intent of matching in case–control studies is to adjust for the effects of extraneous variables. Perhaps the strongest argument for matching is that it is comparatively simple to execute and understand. In addition, matching may allow for control of extraneous variables which are difficult to accommodate with other adjustment methods. For example, matching on place of residence may provide adequate control for socio-economic status.

Matching may be accomplished in either of two different mechanisms:

1. *Pair-wise matching.* For each case, a specific comparison subject (or subjects), with similar values of the matching factors is (are) selected.
2. *Frequency matching.* The group of comparison subjects is chosen so that the overall distribution of matching factors parallels the distribution in the cases.

Despite apparent differences in subject selection, pair-wise and frequency matching usually achieve the same result, since pairs are typically not unique and could be pooled into larger generic strata (Kupper *et al.* 1981). However, in situations where there are natural unique pairs, such as monozygotic twins, pair matching may be advantageous.

The usually perceived intent of matching is to adjust for the effects of relevant co-factors. In prospective studies, where matching occurs prior to disease onset, known relevant factors may be adequately controlled. However, in case–control studies, where subject selection transpires after disease onset, appropriate adjustment of extraneous variables is not always obtained by matching (Kupper *et al.* 1981). When matching is used in a case–control study, it should be accompanied by an analysis which accounts for the method of subject selection. In particular, a pair-wise matched study should account for the pairings when the data are analysed.

Perhaps the most common misunderstanding about matching is the erroneous impression that the goal is to make the case and control groups similar in all respects, except for disease status (Cole 1979). An optimal matching scheme involves only those variables that improve statistical efficiency or eliminate bias from the effect of interest. Often it is difficult to predict in advance which variables are appropriate for matching. When prior work is available, independent risk factors for the disease may be identified for matching. In some situations, matching by interviewer or hospital may be used to balance out the effects of interviewer and observer errors. For exploratory studies, usually it is best to limit matching to basic descriptors, such as age, race, and sex, or not match at all.

Over-zealous matching may have two adverse effects: first, matching on a strong correlate of the exposure, which is not an independent risk factor for the outcome (overmatching) may lead to an underestimate of the study effect (Miettinen 1970); second, matching may lead to a false sense of security that a particular variable is adequately controlled. The following example illustrates how over-matching can lead to an underestimate of the study effect.

Example 5—Overmatching in case–control research
Over-matching may be illustrated by a hypothetical study of prenatal exposure to diethylstilboestrol and the occurrence of clear cell adenocarcinoma of the vagina in young women, whose mothers were matched for history of threatened spontaneous abortion. Since the matching factor was the indication for diethylstilboestrol adminstration, it is a strong correlate of exposure, but is not known to be an independent risk factor for clear cell adenocarcinoma of the vagina. By matching on history of threatened spontaneous abortion, one over-represents diethylstilboestrol exposure in the control group and thereby diminishes any observed exposure–disease association.

Inadequate control of a variable by matching may be illustrated by the treatment of a continuous variable, such as age. To match on age it is usually necessary to categorize the range of possible values into discrete intervals. A common strategy is to employ five-year age-spans for matching, In essence, this assumes that the persons within a particular age interval have the same age effect. For certain situations, such as the study of childhood diseases, there may be residual age effects within five-year age intervals. As a result, such matching intervals will not adequately control the effect of age for these diseases.

Ultimately, a decision on whether or not to use matching in a particular study depends upon the circumstances of that investigation. Table 9.3 indicates some of the advantages and disadvantages of matching. As a general rule, matching is most beneficial when there are only a few matching factors that otherwise would be difficult to control. Furthermore, the logistics of matching are favoured when there is a limited number of cases and a large pool of eligible controls. In other situations, it may be preferable to obtain unmatched samples and adjust for extraneous effects in the analysis (McKinley 1977).

The investigator must also decide upon the source of control subjects. Two different types of comparison populations may be utilized: hospital and community controls. A hospital control group often is employed in clinical studies, since it is convenient to sample controls from the institutions that con-

Table 9.3. Advantages and disadvantages of a matched subject selection

Advantages

In some situations, there are natural partners for comparison (e.g. identical twins)

Matching may account for extraneous variables which are otherwise difficult to control (e.g. neighbourhood matching to adjust for socio-economic status)

Matching may improve the statistical precision of study results, especially for studies with small sample sizes and when the matching factors are strong predictors of the outcome

Matching may enhance the validity of study results

Matching results are often easier to interpret than those obtained with alternative procedures

Disadvantages

When there are a large number of matching variables, it may be difficult to find suitable matches

Unmatched cases and controls cannot be analysed, thereby entailing a loss of potential information

Over-matching may lead to an underestimate of the study effect

Matching may increase the costs of a study

tribute cases. Table 9.4 provides a summary of advantages and disadvantages of a hospital control group.

The term 'community controls' is used as a generic label for comparison groups obtained from non-clinical sources. Within this broad category, several specific types of controls may be recognized. For instance, the cases may be matched to acquaintances, such as co-workers, classmates, friends, or neighbours. These acquaintances may be identified by the cases, by the investigator, or by a third party, such as an employer. In other situations, it may be preferable to sample the general population, either through a simple random sample, or with a probability sample to adjust for extraneous variables.

One popular technique for sampling subjects is referred to as random digit telephone dialling (Hartge *et al.* 1984a). With this procedure, telephone numbers appropriate to a particular geographical area are generated randomly and then dialled to determine whether they correspond to residences and if so whether any occupants meet the eligibility criteria for the study. Of course, selection through this process is limited to persons who reside in households with telephones. In the US, the overall completeness of telephone coverage of households was estimated recently to be 93 per cent, although variation in coverage was observed by socio-economic status, race, and geographical region (Hartge *et al.* 1984a). Since controls sampled through random digit telephone dialling are limited to persons who have a household telephone, in this type of study it may be appropriate to restrict eligibility of cases to those who have a household telephone.

The advantages and limitations of community controls are indicated in Table 9.5. One frequently cited shortcoming of community controls is a low participation rate. It is possible, however, to achieve reasonably high response rates when interviewers are trained in methods to avert refusals and repeated attempts are made to solicit participation. For

Table 9.4. Advantages and disadvantages of a hospital control group

Advantages

Subjects are easily accessible

Patients usually have time to participate in a study

Patients are often motivated to co-operate with investigators

Controls and cases may be drawn from similar social and geographical environments

Differential recall of prior exposure is likely to be minimized

Disadvantages

Differential hospitalization patterns may introduce selection bias

Difficult to blind disease status of cases and controls

An underestimate of the study effect may be obtained if control diseases are aetiologically similar to case disease

example, in a recent large population-based case–control study of bladder cancer, 88 per cent of persons who resided in 25 826 households contacted through random digit telephone dialling agreed to provide a household census, and 84 per cent of the sampled eligible controls agreed to participate in the study (Hartge *et al.* 1984*b*).

From the preceding discussion, it should be apparent that the choice of a hospital or community control group depends upon the research question and study setting, as well as logistical considerations. An increasingly popular approach is to collect two separate control populations, one from hospitals and one from the community. For example, Schoenbaum *et al.* (1975) performed a case–control study to evaluate the association between maternal parity and congenital rubella syndrome. These authors utilized a community control group matched from birth certificate registration, as well as a hospital control group matched from discharge records. When the cases were compared separately with each control group, the results indicated that primiparae are at increased risk of congenital rubella syndrome. Although the consistency of these results does not guarantee their validity, it does increase confidence in the reported association. Whenever possible, the use of more than one control population is encouraged to provide an opportunity for replication of study results.

Information on exposure

Exposure data may be self-reported (in response to an interview or self-administered questionnaire), obtained either from medical or occupational records, or by the use of a biological marker (e.g. serological determination). Information elicited should pertain to the presence of exposure, its intensity, and duration. With self-reported information, problems of inaccuracy and incompleteness may arise. The validity and reliability of self-reported data continue to be important issues in case–control studies. When information about exposures is collected through subject recall, it may be helpful to utilize memory aids for the respondents. For example, in a large population-based case–control study of cancers of the breast, endometrium, and ovary, cases (women with the

cancers of interest) and controls were asked to recall prior use of oral contraceptives. Two memory aids were employed: (1) a major life events calendar including dates of marriages, births, divorces, menarche, menopause, and other events that might have affected contraceptive practices; and (2) a catalogue of colour photographs of all oral contraceptives marketed in the US (Wingo *et al.* 1988). The calendar helped women to focus on the time periods of possible exposure, and the catalogue helped women to identify the specific brands of oral contraceptives that were used. In that study, it also was possible to validate self-reported information on key gynaecological data from information in medical records for 75 per cent of participants (Wingo *et al.* 1988).

Medical records can be more valid and reliable than self-reports, but only when events related to medical care are always recorded in hospital charts. Unfortunately, many relevant events are not recorded, and the researcher is often at the mercy of the quality of the medical record. Medical records, even when accurate, may not contain the necessary data in a form required to answer the specific research questions.

The method of obtaining exposure data is an important source of information bias. For example, the numerous studies of the reserpine–breast cancer association employed a variety of methods to determine drug exposure (Labarthe 1979). Some studies of this association relied upon the mere mention of reserpine use, while others required at least six months of reserpine use to be classified as 'exposed'. Similarly, some studies based the drug exposure determination on self-reporting, while other studies used medical records.

In conducting case–control studies the investigator must decide on the best means of data collection. Occasionally, self-reported data may be verified against information in medical records to assess the validity of subject reports. It is incumbent upon the investigator to consider possible deficiencies that may exist in the information. This requires the assurance of proper data collection, standardized definitions of exposure and disease, as well as criteria that minimize misclassification, and conclusions that are tempered by the possible presence of information bias.

Table 9.5. Advantages and disadvantages of a community control group

Advantages

May reduce the opportunity for selection biases

Study results may be generalizable to a wider target population

For some research questions, a 'natural' community control group will exist

May provide convenient control of extraneous variables

Disadvantages

Usually time-consuming and expensive to obtain

May suffer from low participation rates

Cases and controls may exhibit differential recall of prior exposures

Sample size determination

In general, the statistical confidence in study results is strengthened as the sample size increases. This relationship argues for the use of large study populations. At the same time, there are several feasibility considerations which may restrict the number of subjects examined. First, case–control studies are typically undertaken for the investigation of rare diseases. For extremely uncommon conditions, there may be a limited number of cases available for study. Furthermore, there are usually constraints upon the time, personnel, and financial resources that may be devoted to a particular study. Thus, a balance must be achieved between the interest of statistical precision and matters of practicality.

The traditional approach to sample size (n) estimation is to perform a calculation based upon:

(1) the anticipated prevalence of exposure in the control population, p_0;
(2) the magnitude of association between exposure and disease, as measured by the relative risk, RR; (<1, negative association; $=1$, no association; >1, positive association);
(3) the probability of erroneously finding an exposure–disease association when none exists in reality (Type I error), α;
(4) the probability of erroneously not finding an exposure–disease association, when one exists in reality (Type II error), β.

For unmatched studies, with equal numbers of cases and controls, the approximate number of subjects required in a group is given by (Schlesselman 1982):

$$n = 2\bar{p}\,\bar{q}(z_\alpha + z_\beta)^2/(p_1 - p_0)^2 \qquad (9.1)$$

where

$$p_1 = p_0 RR/[1 + p_0(RR - 1)]$$
$$\bar{p} = \tfrac{1}{2}(p_1 + p_0) \qquad \bar{q} = 1 - \bar{p}$$

and z_α and z_β are the values of the standard normal distribution that are exceeded by α (one-sided test) and β, respectively. Applying equation 9.1, we can generate sample size estimates for specified values of p_0, RR, α and β. In Table 9.6, sample size calculations are summarized for various levels of p_0 and RR using conventional values for the probabilities of Type I and II errors. This table may be used to estimate sample sizes, as indicated in the following illustrative example.

Example 6—Sample size estimation in case–control research
Suppose we wish to perform a case–control study of the relationship between cigarette smoking and lung cancer. From previously published work, we anticipate that smokers have a tenfold excess risk (RR) of lung cancer (Doll and Hill 1950) and the prevalence of cigarette smoking (p_0) is about 0.4 (Schuman 1977). From these values, with an $\alpha=0.05$ (two-sided) and a $\beta=0.10$, it would be necessary to study at least

Table 9.6. Sample size for each group in an unmatched case–control study with equal numbers of cases and controls, $\alpha = 0.05$ (two-sided), $\beta = 0.10$[*]

RR	p_0 0.01	0.1	0.2	0.4	0.6	0.8	0.9
0.1	1420	137	66	31	20	18	23
0.5	6323	658	347	203	176	229	378
2.0	3206	378	229	176	203	347	658
3.0	1074	133	85	71	89	163	319
4.0	599	77	51	46	61	117	232
5.0	406	54	37	35	48	96	194
10.0	150	23	18	20	31	66	137
20.0	66	12	11	14	24	54	115

[*] Adapted from Schlesselman (1982).

20 cases and 20 controls. Note that this relatively small sample size is possible because of the strength of the cigarette smoking–lung cancer association and the substantial prevalence of smoking.

Several general relationships are apparent in sample size calculations:

(1) for a given level of α, β, and p_0, the sample size requirement decreases as the strength of the exposure–disease association increases;
(2) for a given level of α, β, and RR, the sample size requirement is usually minimized at intermediate levels of control exposure prevalence;
(3) for a given level of β, RR, and p_0, the sample size requirement decreases as the probability of a Type I error increases;
(4) for a given level of α, RR, and p_0, the sample size requirement decreases as the probability of the Type II error increases.

When there are a limited number of cases available for study, it may be advantageous to include a larger number of controls. Although the discriminatory power of a study will progressively improve with the addition of more controls, the marginal gain in power is small when the study group sizes are greatly unbalanced. As a rule of thumb, it is often recommended that the case:control ratio should not exceed 1:4 (Ury 1975). A simple modification of equation 9.1 allows the calculation of sample size for studies with different numbers of cases and controls (Schlesselman 1982).

$$n = (1 + 1/c)\bar{p}'\bar{q}'(z_\alpha + z_\beta)^2/(p_1 - p_0)^2, \qquad (9.2)$$

where c = number of controls per case

$$\bar{p}' = (p_1 + cp_0)/(1 + c),$$

and $$\bar{q}' = 1 - \bar{p}'$$

For sample size calculations in pair-matched case–control studies, the end-point is not the total number of cases, but rather the number of case–control pairs that are discordant with regard to exposure status. The necessary number of discordant pairs (m) can be calculated as (Schlesselman 1982):

$$m = [z_\alpha/2 + z_\beta\sqrt{P(1-P)}]^2/(P - \tfrac{1}{2})^2 \qquad (9.3)$$

where

$$P \simeq RR/(1 + RR)$$

From equation 9.3, one can predict the total number of case–control pairs required if the probability of a discordant pair is known or can be estimated. However, this information is not usually known prior to data collection.

All of the sample size calculations discussed thus far are based upon the assumption of a fixed sample size. That is, the size of the study population is set and maintained prior to the collection of any data. An alternative approach, referred to as group sequential sampling, involves accumulating subjects until there is a sufficient amount of information to accept or reject the null hypothesis at some specified level of α and β. The group sequential method may be especially useful for the surveillance studies in which there is no prior estimate of the magnitude of an exposure–disease association and data collection proceeds in a continuous fashion.

In practice, the group sequential method involves enrolling a subgroup of K cases and an equal number of controls and then testing for statistical significance of an exposure–disease association. If the result is significant at the predetermined α level, data collection is terminated. However, if the observed effect is consistent with no association, another group of K cases and K controls is enrolled. This process is repeated until either the null hypothesis is rejected, or a specified number of subgroups are entered without rejection of the null hypothesis. In light of the repeated statistical significance tests, it is necessary to test for the exposure–disease association at a nominal α level which is smaller than the overall significance level. Pocock (1977) has published a reference table for rapid determination of appropriate nominal significance levels in group sequential studies.

The sample size for a group sequential study may be calculated for specified values of the maximum number of subgroups, α, β, p_0, and RR. Pasternack and Shore (1981) have published tables of maximum group sequential sample size calculations and the interested reader is referred to their work. It should be remembered that these values represent upper limits of required subjects, since early detection of significant associations may permit a smaller sample.

Estimation of relative risk in unmatched case–control research

The traditional measure of an exposure–disease association is the relative risk (risk ratio). Relative risk may be defined as the probability of disease development in exposed persons, divided by the probability of disease development in unexposed subjects. As the ratio of two probabilities, the relative risk has a range of possible values from zero to positive infinity. The value of relative risk which corresponds to 'no associa-

Table 9.7. Guide-lines for the interpretation of a relative risk value, in terms of exposure–disease association

Relative risk range	Interpretation
0–0.3	Strong benefit
0.4–0.5	Moderate benefit
0.6–0.8	Weak benefit
0.9–1.1	No effect
1.2–1.6	Weak hazard
1.7–2.5	Moderate hazard
> 2.6	Strong hazard

Table 9.8. Simple fourfold table for an unmatched case–control study

	Exposure status		
	Exposed	**Unexposed**	**Total**
Disease status			
Case	a	b	m_1
Control	c	d	m_0
Total	n_1	n_0	$t = m_1 + m_0$

tion' is unity. A relative risk significantly greater than unity implies an excess disease risk in exposed persons, whereas a relative risk significantly less than unity corresponds to a diminished disease risk in exposed persons. While there are no universal rules for the interpretation of relative risks, Table 9.7 may serve as a rough guide-line. The probability of disease development conditional upon exposure status can be estimated directly in prospective cohort studies. However, in case–control studies, the subjects are sampled conditional on disease status, and thus one cannot obtain a direct estimate of relative risk. In the simplest unmatched situation, consider a fourfold table with disease status on one axis and a dichotomous classification of exposure history on the other axis (Table 9.8). Each subject in the study fits into one and only one of the cells labelled a through d. For example, a case with a positive exposure history would be placed into a cell a, whereas another case with a negative exposure history would be placed in cell b. When all of the subjects have been classified in this manner, the table marginal totals ($m_1 + m_0$ or $n_1 + n_0$) may be summed to obtain the value t, the total number of study subjects. In the case–control sampling method, subjects enter the study based upon their disease status. Thus, the investigator chooses the ratio of cases-to-controls, and therefore the ratio of m_1 to m_0 is fixed. Notice the contrast with a prospective cohort study in which the investigator determines the ratio of exposed to unexposed persons (ratio of n_1 to n_0 fixed).

Although one cannot obtain a direct estimate of relative risk in case–control studies, Cornfield (1951) demonstrated that an indirect estimate may be obtained. This indirect estimate, subsequently termed the (exposure) odds ratio (OR), requires two assumptions:

1. The controls are representative of the candidate population from which the cases arose (Miettinen 1976).

2. The disease under investigation is rare (Cornfield 1951).

The odds ratio is calculated as the simple cross-product of the fourfold table:

$$\hat{OR} = \frac{ad}{bc} \qquad (9.4)$$

Where $\hat{OR}$ indicates that the odds ratio is an estimate from the study population. An advantageous property of the odds ratio is that its value does not depend on the underlying research design. This advantage allows direct comparison of odds ratio estimates from cohort, cross-sectional, and case–control studies (Fleiss 1981).

Example 7—Calculation of odds ratio in unmatched case–control research
Suppose a case–control study is performed to test a postulated association between prenatal irradiation and childhood leukaemia. One hundred cases of leukaemia are identified from a tumour registry and then 200 neighbourhood controls are obtained. From interviews with the parents of all subjects it is determined that 30 cases and 45 controls were exposed to intra-uterine diagnostic irradiation. These data may be represented in the format shown in Table 9.9. Thus, if the assumptions of the odds ratio calculation hold, we estimate a relative risk of 1.48. This means that children irradiated *in utero* have almost a 50 per cent excess risk of developing leukaemia.

Measures of population health impact in case–control research

In aetiological research, the emphasis of analysis is usually on evaluating the strength of an exposure–disease association, as indicated by relative risk. However, for public health concerns, it is often useful to consider measures that address the population impact of an exposure, or the social benefit of curtailing an exposure.

The attributable risk proportion (ARP) may be defined as the proportion of total disease risk in exposed persons which may be attributed to their exposure. Cole and MacMahon (1971) have shown that the ARP may be estimated in case–control studies by:

$$\hat{ARP} = \frac{\hat{OR} - 1}{\hat{OR}} \qquad (9.5)$$

When the $\hat{OR}$ is >1, the ARP has a range of possible values from zero to unity. In the limiting case when $\hat{OR}$ is equal to unity (no exposure–disease association), none of the disease risk in exposed persons will be attributed to that exposure ($\hat{ARP}=0$). At the other extreme, when the $\hat{OR}$ is very large, much of the total disease risk in exposed persons may result from that exposure.

Another population health impact measure of special note is the population attributable risk proportion (PARP). This

Table 9.9. Data for unmatched case–control study of prenatal irradiation and childhood leukaemia

	Prenatal irradiation		
	Exposed	Unexposed	Total
Childhood leukaemia			
Case	30	70	100
Control	45	155	200
Total	75	225	300

measure, also known as the aetiological fraction, corresponds to the proportion of disease risk in all persons which may be attributed to the exposure under investigation. Under the assumption that the exposure histories of controls are typical of the target population, Cole and MacMahon (1971) have demonstrated that the PARP may be estimated in case–control studies by:

$$\hat{PARP} = \frac{\hat{p}_0(\hat{OR} - 1)}{1 + \hat{p}_0(\hat{OR} - 1)} \qquad (9.6)$$

where $\hat{p}_0$ represents the estimated prevalence of exposure in controls (and by inference the exposure prevalence in the target population). Again, when the $\hat{OR}$ is >1, the limits of this expression are zero to unity. Notice, however, that the PARP is determined by both the frequency of exposure and the strength of association between exposure and disease. Taylor (1977) has shown that the PARP may be simply estimated from case–control studies by:

$$\hat{PARP} = 1 - \left[\frac{b(c + d)}{d(a + b)} \right] \qquad (9.7)$$

Example 8—Calculation of attributable risk proportion and population attributable risk per cent in case–control research
Using the data presented in Example 7 ($\hat{OR}=1.48$), we can calculate measures of childhood leukaemia population impact for intra-uterine irradiation. First, the ARP is estimated by:

$$\hat{ARP} = \frac{1.48 - 1}{1.48} = 0.32$$

Thus, to the extent that the $\hat{OR}$ provides a valid estimate of the strength of the exposure–disease association, we conclude that about one-third of childhood leukaemia in the irradiated children may be attributed in part to prenatal irradiation. Similarly, we may estimate the PARP with Taylor's formula:

$$\hat{PARP} = 1 - \left[\frac{70(45 + 155)}{155(30 + 70)} \right] = 1 - 0.90 = 0.10$$

Thus, we conclude that about 10 per cent of all childhood leukaemia may be attributed in part to intra-uterine irradiation.

Stratified analysis of case—control research

Up to this point, the simplest evaluation of an exposure–disease association has been considered. In most research situations, it is also necessary to take into account the influence of other factors, such as age, race, and sex. One approach to the accommodation of these co-variables is to stratify the data into subsets by co-factor level. For example, if we are interested in assessing the relationship between dietary fibre intake and colon cancer risk, we might examine the fibre–colon cancer association separately in four race–sex groups: white males, white females, non-white males, and non-white females. If the association does not vary across the race–sex specific categories, then it may be desirable to obtain a summary measure of this association. This condition of uniformity of an exposure–disease association is termed 'homogeneity' and means in this instance that the effect of dietary fibre on colon cancer risk is not modified by race or sex. For homogeneous data, all of the four race–sex stratum-specific odds ratio estimates would be the same. A statistical test for homogeneity of the natural logarithm of OR is available (Woolf 1955), and may be used to determine whether the strata are sufficiently homogeneous to allow summarization.

The most popular approach to calculating a summary $\hat{\text{OR}}$ was first introduced by Mantel and Haenszel (1959). This measure may be obtained in the following manner:

$$\hat{\text{OR}}_{\text{MH}} = \sum_{i=1}^{I} \left(\frac{a_i d_i}{t_i} \right) \bigg/ \sum_{i=1}^{I} \left(\frac{b_i c_i}{t_i} \right) \qquad (9.8)$$

where the subscript MH designates a Mantel–Haenszel estimate, i is an index for co-variate-specific strata, I is the total number of summarized strata, and the remaining symbols are the stratum-specific equivalents of those specified in Table 9.8. It can be shown that the $\hat{\text{OR}}_{\text{MH}}$ is a weighted average of the stratum-specific values, where the weights are $b_i c_i / t_i$. In this approach, the strata with the most information and hence the greatest statistical precision receive the most weight (Mantel and Haenszel 1959). Furthermore, this summary measure has the advantage that it can be used without modification when there are 'zero' entries in some of the stratum-specific tables. When there are multiple co-variates to accommodate it is likely that the study subjects will become thinly spread over the relevant strata. Thus, the Mantel–Haenszel summary odds ratio is often preferred over alternative approaches on the basis of precision (Kleinbaum *et al.* 1982).

Hypothesis testing in unmatched case—control research

Once the odds ratio is calculated in a case–control study, the next step is to perform a test of statistical significance. The purpose of a significance test is to assess whether the observed odds ratio is sufficiently different from unity to exclude chance as a likely explanation for the observed exposure–disease association. The hypothesis test is formulated in either a one-or two-sided direction:

Two-sided alternative hypothesis:
H_0: There is no association between the exposure and disease (OR = 1).
H_a: There is an association, either positive or negative, between the exposure and diseases, (OR < 1 or OR > 1).

One-sided alternative hypothesis:
H_0: There is either a negative or no association between the exposure and disease (OR ≤ 1).
H_a: There is a positive association between the exposure and disease (OR > 1).

If the direction of the tested association can be predicted in advance, the use of a one-sided test may be justified. In other circumstances, it is preferable to use the two-sided version.

Mantel and Haenszel (1959) proposed the use of a large sample test statistic based upon the hypergeometric model. This test statistic (without continuity correction) may be calculated with the following formula:

$$\chi^2_{\text{MH}} = \left(\sum_{i=1}^{I} \frac{a_i d_i - b_i c_i}{t_i} \right)^2 \bigg/ \left[\sum_{i=1}^{I} \frac{m_{1i} m_{0i} n_{1i} n_{0i}}{(t_i - 1) t_i^2} \right] \qquad (9.9)$$

The Mantel–Haenszel procedure was shown to have optimal statistical features when the stratum-specific odds ratios are homogeneous (Radhakrishna 1965). Under the large sample assumptions, the Mantel–Haenszel test statistic follows a chi-squared distribution with one degree of freedom. It should be noted that the large sample assumption pertains to the total amount of data aggregated over all strata. Mantel and Fleiss (1980) have proposed the following criteria for appropriate use of tabled chi-squared distributions with one degree of freedom:

$$\left[\sum_{i=1}^{I} \frac{n_{1i} m_{1i}}{t_i} \right] - \left[\sum_{i=1}^{I} \max\left(0, m_{1i} - n_{0i}\right) \right] \geqslant 5$$

and

$$\left[\sum_{i=1}^{I} \min\left(n_{1i}, m_{1i}\right) \right] - \left[\sum_{i=1}^{I} \frac{n_{1i} m_{1i}}{t_i} \right] \geqslant 5$$

When both of these conditions are satisfied, the sample size is considered sufficient to use the large sample calculation.

Confidence interval estimation in unmatched case—control research

In addition to the point estimate of an odds ratio, it is often useful to calculate a measure of variability of the estimate. The standard approach is to calculate a confidence interval for the odds ratio estimate. A 95 per cent confidence interval is a range of values such that in 95 out of 100 repeated samples, the odds ratio estimate would lie between the upper and lower bounds.

Several methods for calculating odds ratio confidence intervals have been proposed. For small sample sizes, it is often recommended that 'exact' type confidence intervals be used (Kleinbaum *et al.* 1982). However, the exact method requires complex calculations, which may become prohibitive for even modest sample sizes. Thomas (1971) has provided a computer algorithm to assist in the determination of exact confidence intervals.

For larger samples sizes, it may be more practical to calculate 'approximate' confidence intervals. Several different approximation methods have been proposed, which may differ with regard to the estimation of the odds ratio variance. A Taylor series approximation yields a $(1 - \alpha)$ 100 per cent confidence interval of the following form (Woolf 1955):

$$\hat{OR} \exp\left(\pm z_{1-\alpha/2} \sqrt{1/a + 1/b + 1/c + 1/d}\right) \quad (9.10)$$

Miettinen (1976) proposed an alternative, test-based confidence interval, in which the odds ratio variance may be estimated from the square of the odds ratio divided by the Mantel–Haenszel test statistic. From this approximation it is possible to construct a $(1 - \alpha)$ 100 per cent confidence interval of the following form (Miettinen 1976):

$$\hat{OR}^{(1 \pm z_{1-\alpha/2} \sqrt{\chi_{MH}})} \quad (9.11)$$

In general, the test-based confidence intervals tend to be slightly narrower than the Taylor series intervals. Either of these estimation procedures may provide unsatisfactory approximations for small samples (Kleinbaum *et al.* 1982). The test-based procedure has also been shown to introduce a systematic bias as the magnitude of the observed odds ratio increases (Brown 1981). Therefore, this procedure should be used with caution for strong exposure–disease associations.

Example 9—Calculation of unmatched summary odds ratio, confidence interval, and test statistics
Returning to the prenatal irradiation–childhood leukaemia data presented in Table 9.9, let us consider a stratified analysis which accounts for the effect of maternal gravidity. In the simplest situation, we might stratify the crude data into two levels: primigravidae and multigravidae. Table 9.10 demonstrates the gravidity-specific fourfold tables for the association between prenatal irradiation and childhood leukaemia.

As a preliminary assessment of homogeneity of effect between levels of gravidity, stratum-specific odds ratios may be calculated and compared. Using equation 9.4 we obtain the following measures of effect:

$$\hat{OR}_{primigravidae} = \frac{(23)(38)}{(37)(22)} = 1.07$$

$$\hat{OR}_{multigravidae} = \frac{(7)(117)}{(33)(23)} = 1.08$$

Since these stratum-specific effect measures are homogenous, it is appropriate to determine a summary odds ratio,

Table 9.10. Data for unmatched case–control study of prenatal irradiation and childhood leukaemia, stratified by maternal gravidity

	Prenatal irradiation		
	Exposed	**Unexposed**	**Total**
Primigravidae			
Childhood leukaemia			
Case	23	37	60
Control	22	38	60
Total	45	75	120
Multigravidae			
Childhood leukaemia			
Case	7	33	40
Control	23	117	140
Total	30	150	180

with corresponding confidence intervals and an associated hypothesis test.

The point estimate of the Mantel–Haenszel summary odds ratio is calculated with equation 9.8:

$$\hat{OR}_{MH} = \frac{\left[\dfrac{(23)(38)}{(120)} + \dfrac{(7)(117)}{(180)}\right]}{\left[\dfrac{(37)(22)}{(120)} + \dfrac{(33)(23)}{(180)}\right]} = 1.08$$

This measure may be interpreted to suggest that prenatal irradiation results in an 8 per cent increased risk of childhood leukaemia, after the effect of maternal gravidity is taken into account. Notice that this summary odds ratio is less than the crude odds ratio previously obtained in Example 7. The difference in these measures may be attributed to the confounding effect of maternal gravidity, which exaggerates the apparent effect of prenatal irradiation in the crude data.

The statistical significance of the summary measure may be tested with equation 9.9:

$$\chi^2_{MH} = \frac{\left[\dfrac{(23)(38) - (37)(22)}{(120)} + \dfrac{(7)(117) - (33)(23)}{(180)}\right]^2}{\left[\dfrac{(60)(60)(45)(75)}{(119)(120)^2} + \dfrac{(40)(140)(30)(150)}{(179)(180)^2}\right]}$$

$$= 0.06, p_{(two\text{-}tailed)} > 0.80$$

This test statistic may be interpreted as indicating that the remaining weak association between prenatal irradiation and childhood leukaemia could easily have arisen by chance in these data.

Test-based 95 per cent confidence limits for the summary odds ratio may be calculated with equation 9.11:

$$95 \text{ per cent CI} = 1.08^{(1 \pm 1.96/\sqrt{0.06})} = (0.58, 2.00)$$

Notice that this confidence interval includes the null value of

Table 9.11. Simple fourfold table for a pair-matched case–control study*

	Control		
	Exposed	**Unexposed**	**Total**
Case			
Exposed	A	B	$A + B$
Unexposed	C	D	$C + D$
Total	$A + C$	$B + D$	$N = A + B + C + D$

* Each cell entry in this table represents a pair of subjects, comprised of a case and his/her control.

unity, as would be expected because of the lack of statistical significance of the odds ratio point estimate.

Estimation of relative risk in pair-matched case–control research

When matching is used to choose comparison subjects, the matching process should be considered in the analysis. The actual form of analysis depends upon the case-to-control ratio. In this discussion, we present the simplest situation in which one control is individually matched to one case. For the analysis of more complex situations, such as multiple controls per case, the reader is referred to the excellent discussion by Schlesselman (1982).

The analysis of a pair-matched case–control study may be considered as a logical extension of stratified analysis. In the pair-matched situation, each stratum is composed of one case and one control. For any case–control pair there are four possible exposure histories:

A. case exposed, control exposed;
B. case exposed, control unexposed;
C. case unexposed, control exposed;
D. case unexposed, control unexposed;

After each case–control pair is classified into one of these exposure categories, a fourfold table may be constructed to summarize all of the pairs (Table 9.11).

Notice that Table 9.11 is different from the previous layout in Table 9.8. For the pair-matched study, each entry into the fourfold table represents a pair of subjects. For example, the cell labelled A includes all pairs in which both the case and control were exposed. The cell labelled B includes all exposed case/unexposed control pairs and so forth. Thus, the total number of entries (N) in the table represents the number of case–control pairs, not individual subjects.

When the exposure histories of a case and control are similar, the pair is termed 'concordant'. When the exposure histories of a case and control differ, the pair is termed 'discordant'. Since we are interested in differential exposure histories within each case–control pair, the primary focus is upon the cells in Table 9.11 labelled B and C. Kraus (1960)

demonstrated that the maximum likelihood estimate of the odds ratio is given by:

$$\hat{OR} = B/C \qquad (9.12)$$

An identical expression for the Mantel–Haenszel odds ratio would be obtained from a stratified analysis, where each stratum consists of a case–control pair (Mantel and Haenszel 1959).

Hypothesis testing in pair-matched case–control research

A simple statistical significance test for matched case–control studies can be constructed with a normal approximation to the binomial distribution. This test statistic (without continuity correction) may be calculated with the following formula (Schlesselman 1982):

$$\chi^2_{(1)} = (B - C)^2/(B + C) \qquad (9.13)$$

This statistic is distributed approximately as a chi-square with one degree of freedom for matched studies with large numbers of discordant pairs. If the number of discordant pairs is small, an exact hypothesis test must be performed (Schlesselman 1982).

Confidence interval estimation in pair-matched case–control research

For matched case–control studies with large numbers of discordant pairs, an approximate $(1 - \alpha)$ 100 per cent confidence interval for the odds ratio may be estimated by (Schlesselman 1982):

$$(B/C) \exp\left[\pm z_{1 - \alpha/2} \sqrt{1/B + 1/C}\right] \qquad (9.14)$$

If the number of discordant pairs is small, an exact $(1 - \alpha)$ 100 per cent confidence interval may be calculated (Schlesselman 1982).

Example 10—Calculation of pair-matched summary odds ratio, confidence interval, and test statistic
Suppose we wish to conduct a matched case–control study of the association between infectious mononucleosis (IM) and the subsequent development of lymphoma. It is decided that controls (persons without lymphoma) will be pairwise matched to lymphoma patients by age (±five years), race, and sex. After enrolment of cases and controls, histories of prior IM infection are obtained. The resulting data are displayed in Table 9.12. The corresponding maximum likelihood estimate of the odds ratio is calculated with equation 9.12:

$$\hat{OR} = 60/35 = 1.71$$

To the extent that this odds ratio is a valid estimate of relative risk, this measure suggests that persons exposed to IM have

Table 9.12. Data for pair-matched case–control study of infectious mononucleosis (IM) and lymphoma[*]

	Control		
	IM infection	No IM infection	Total
Lymphoma patient			
IM infection	15	60	75
No IM infection	35	40	75
Total	50	100	150

[*] Each cell entry in this table represents a pair of subjects, comprised of a case and his/her control.

about a 70 per cent excess risk of lymphoma when compared with persons not exposed to IM.

The statistical significance of this result may be assessed with the test statistic in equation 9.13:

$$\chi^2_{(1)} = (60 - 35)^2/(60 + 35) = 6.58$$

This statistic yields a two-sided p-value of about 0.01. Thus, we conclude that the association between IM and lymphoma is unlikely to have occurred by chance alone.

A 95 per cent confidence interval for the estimated odds ratio may be calculated with equation 9.14:

$$95 \text{ per cent CI} = 1.71 \exp [\pm 1.96 \sqrt{(1/60) + (1/35)}]$$
$$= (1.13, 2.59)$$

Notice that the entire range of this confidence interval is greater than unity, as anticipated by the statistical significance of the odds ratio point estimate.

Logistic regression analysis of case–control research

In the study of complex aetiological processes, there may be many variables that influence the disease risk. Even relatively direct exposure–disease associations may be affected by variables such as age, race, sex, and socio-economic status, as well as other exposures. As already indicated, these relevant co-factors may be accounted for with a stratified analysis. However, as the number of co-factor-specific strata increases, the study sample may be sparsely distributed across each level. For example, suppose we have conducted a case–control study of the relationship between hepatitis infection and subsequent risk of liver cancer. It is determined that the following co-variables require control: age (20–39, 40–59, ≥60 years), race (white, non-white), sex (male, female), socio-economic status (high, middle, low), and alcohol consumption (heavy, moderate, occasional). A stratified analysis accounting for these co-factors would require $3 \times 2 \times 2 \times 3 \times 3 = 108$ strata. Even with a relatively large sample size, it is likely that there will be insufficient data in many of these strata. Moreover, if the stratum-specific effects are heterogeneous, then it may be a challenge to summarize and interpret these study results.

For these multi-variable situations, it may be desirable to estimate independent and joint effects with a mathematical model. A popular model for analysis of case–control studies is the logistic regression approach (Schlesselman 1982). The theory and appropriate use of logistic analyses require some familiarity with regression techniques. The principles of logistic analysis are briefly summarized in this chapter and the reader who is interested in a more detailed discussion is referred to three advanced textbooks (Breslow and Day 1980; Kleinbaum *et al.* 1982; Schlesselman 1982).

In the logistic model, the conditional probability of disease (D) occurrence given exposure (E) is represented by a linear function (Kleinbaum *et al.* 1982):

$$\text{logit } P(D|E) = \ln \left[\frac{P(D|E)}{1 - P(D|E)} \right] = \alpha + \beta E \quad (9.15)$$

where logit is an abbreviation of logarithmic unit, ln is the natural logarithm, $P(D|E)$= probability of disease given exposure; α, β=regression coefficients; and E may be dichotomized: (0=no exposure, 1=exposure), or expressed as a continuous variable.

From equation 9.15 it is possible to derive an expression for the odds ratio which reduces to the following form (Kleinbaum *et al.* 1982):

$$\hat{\text{OR}} = e^\beta \quad (9.16)$$

The simple model presented in equation 9.15 can be extended to account for potential confounders (V_i) and effect modifiers (W_j) (Kleinbaum *et al.* 1982):

$$\text{logit } (P|E, V_i, W_j) = \ln \left[\frac{P(D|E, V_i, W_j)}{1 - P(D|E, V_i, W_j)} \right] \quad (9.17)$$
$$= \alpha + \beta E + \sum_{i=1}^{I} \gamma_i V_i + \sum_{j=1}^{J} \delta_j W_j$$

where α and β have their former definitions; γ_i and δ_i are the regression coefficients for the potential confounders and effect modifiers respectively.

From equation 9.17 the following expression for the odds ratio may be obtained (Prentice 1976):

$$\hat{\text{OR}} = \exp [\beta + \sum_{j=1}^{J} \delta_j W_j] \quad (9.18)$$

Thus, the odds ratio can be expressed as the exponential of the sum of the main effect and the effect modifiers. Although the confounders are not explicitly represented in this odds ratio estimate, their inclusion in equation 9.17 will influence the estimated values of α, β, and δ_i, thereby indirectly influencing the odds ratio estimate. Moreover, a matched study design may be accommodated in logistic analysis.

The coefficients in a logistic analysis may be estimated by either of two approaches: discriminant analysis (Cornfield 1962), or maximum likelihood estimation (Prentice and Pyke 1979). For the usual situation in which the data are not multivariate normally distributed, discriminant analysis can produce misleading results (Press and Wilson 1978). Thus, maximum likelihood estimation is usually the preferred approach to parameter estimation (Schlesselman 1982). The details of maximum likelihood estimation and statistical inference may be found in three advanced textbooks (Breslow and Day 1980; Kleinbaum *et al.* 1982; Schlesselman 1982).

While logistic regression is a powerful analytical method, it should be recognized that there are several potential drawbacks to such modelling procedures. First, the linear model is most useful when disease risk can be expressed as a monotonic function of the independent variables (Gordon 1974). Occasionally, curvilinear exposure–disease relationships will be encountered and alternate models should be employed, such as the use of quadratic independent variables or logarithmic transformation of the independent variables. A second limitation of logistic regression is that it assumes a multiplicative statistical relationship between independent variables (Greenland 1979). In certain situations, the data may fit better with an additive statistical relationship between independent variables. For any particular research application, the appropriateness of a logistic model will depend upon the nature of the exposure–disease and exposure–co-variate relationships.

Consistency of results

Most of the criticism of case–control research has centred around the possibility of systematic errors in these studies (Hayden *et al.* 1982). It should be emphasized that all types of observational research and even randomized experiments (May *et al.* 1981) are subject to potential systematic errors (viz. bias). Nevertheless, some authors have suggested that case–control studies have a special vulnerability to bias because of the retrospective timing of observations. For instance, Horwitz and Feinstein (1979) compiled a list of 17 substantive research topics for which case–control studies yielded discordant results.

While such contradictory findings are of concern, they do not constitute an incontrovertible indictment of case–control studies. Many instances of conflicting study results can be cited from other observational and experimental work. As a single illustration, consider two well-designed and executed randomized, placebo-controlled clinical trials which reached opposite conclusions about the efficacy of sodium nitroprusside in the reduction of mortality from acute myocardial infarction (Cohn *et al.* 1982; Durrer *et al.* 1982). The apparent disagreement between these investigations may be explained in part by differences in the study populations, such as the prevalence of left ventricular failure and the time to initiation of treatment. Thus, such disparate findings do not imply that randomized clinical trials are inherently subject to bias. By the same token, occasional discordant results in case–control studies cannot be construed as evidence that case–control studies are inherently prone to errors.

Scientists have attempted to resolve the issue of contradictory findings by carefully reviewing the literature or holding consensus conferences on specific topics. They would reconcile, whenever possible, differences in methodology and findings and arrive at the best conclusion based on the best evidence available. An emerging approach, meta-analysis, for combining results of studies to derive more definitive and generalizable conclusions is gaining wider acceptance (Thacker 1988). The method requires a systematic, explicit, and quantitative approach to the selection and analysis of available evidence. Although the approach has been largely used in integrating results of randomized controlled trials, it shows promise for application to observational studies including case–control studies.

Biases in case–control studies

Sackett (1979) provided a catalogue of biases that may occur in analytical research. This serves as a handy reference for the design and conduct of a proposed investigation. In the following sections we present an abbreviated overview of some systematic errors that may occur in case–control studies. We divide these errors into three general categories:

1. **Selection bias.** A distortion in the study effect which results from the manner in which subjects are sampled for investigation.

2. **Misclassification (information) bias.** A distortion in the study effect which results from inaccurate determination of exposure or disease status.

3. **Confounding.** A distortion in the study effect which results from the mixing of the exposure–disease association with the effect(s) of extraneous variable(s).

Selection bias

Kleinbaum *et al.* (1981) proposed a conceptual framework for the consideration of selection bias. From this model it is possible to define the conditions which give rise to selection bias in the estimation of the odds ratio. In the present context, we emphasize two situations which may result in a biased odds ratio estimate:

Situation 1. The odds of selection into the sample conditional on exposure status is different for cases and controls. In case–control studies the outcome variable (exposure) occurs prior to subject selection and consequently exposure status may differentially affect the selection probabilities of cases and controls.

Situation 2. The odds of selection into the sample conditional on disease status is different for exposed and unexposed persons.

Potential selection biases may be controlled either in the design or analysis stage of a case–control study. With either approach to bias control, the first step is to recognize the manner in which such errors arise. Toward this end, we present a brief review of some classical selection biases, along with suggestions for avoiding these potential errors.

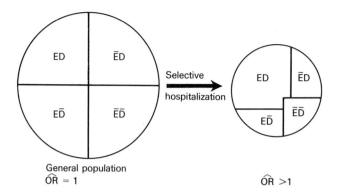

Fig. 9.6. A schematic representation of the Berkson paradox creating a spurious association; E = exposed; Ē = unexposed; D = case; D̃ = control.

Berkson's paradox

In 1946, Berkson constructed a theoretical argument that hospital samples may systematically differ from general populations because of factors that influence the likelihood of hospitalization. As a result, hospital samples may exhibit spurious associations between two variables, even though these variables are independently distributed in the general population. Figure 9.6 provides a schematic representation of the Berkson paradox. This type of selection bias was demonstrated empirically by Roberts and co-workers (1978) who studied the associations between several conditions in a hospital sample and in the general population. For example, respiratory and bone diseases were not associated in the general population ($\widehat{OR}$=1.06), but a strong positive association ($\widehat{OR}$=4.06) was observed in a hospital sample. The authors also demonstrated that distorted medication–disease associations could arise in hospitalized populations. For instance, laxative use and arthritic disease were only weakly associated ($\widehat{OR}$=1.48) in the general population, but a strong relationship ($\widehat{OR}$=5.00) was observed in the hospital sample.

Walter (1980) has specified two particular circumstances for which the Berkson paradox will be negligible:

(1) the exposure under investigation is not a direct cause of hospitalization; and/or

(2) the case and control populations are mutually exclusive.

These two conditions can be used as guide-lines for evaluating the potential of a Berkson paradox in any particular hospital-based case–control study.

Neyman fallacy

In 1955, Neyman proposed that distorted exposure–disease associations could be obtained from the study of prevalent disease cases, if exposure is related to disease prognosis. As an illustration, consider the relationship between sex and risk of colorectal cancer. The incidence rate of colorectal cancer is slightly higher in males than females (Devesa and Silverman 1978). However, the survival from colorectal cancer is significantly longer in females than males (Koch *et al.* 1982).

Since the female colorectal cancer patients live longer than males, a sample of prevalent cases will include a higher proportion of women than a corresponding sample of incident cases (Fig. 9.7). To remove the influence of selective survival (or demise), it is often recommended that the cases in a case–control study should be limited to newly diagnosed patients (Schlesselman 1982).

Selective referral

For investigations of rare diseases, it is often necessary to obtain the study cases from tertiary care centres, or population surveillance programmes. Either of these referral networks can introduce a selection bias if cases within the population are differentially reported. It is likely that patients at tertiary care centres tend to have complicated or severe forms of their diseases, which may differ aetiologically from other cases. For example, it is estimated that only one in seven poisoning incidents is officially reported (Illinois Department of Public Health 1975). As a result, studies based upon reported poisonings may under-represent benign and illicit substances (Greenberg and Osterhout 1982).

Even when a concerted effort is made to identify most of the disease cases within a population, selection factors may arise. For instance, in a case–control study of the toxic-shock syndrome, the Wisconsin Division of Health mailed questionnaires to 3500 licensed physicians in that state (Davis *et al.* 1980). The mailing included a description of toxic-shock syndrome and indicated a possible association with menstruation. Of the 38 cases thus identified, 35 occurred during menses. Since studies in other geographical areas have reported a lower proportion of menstruation-related cases (Clayton 1982), it is possible that the information included in the Wisconsin Survey caused preferential reporting of menstrual cases. As a general rule, case surveillance will be most reliable when there are clear definitions of disease, adequate access to medical care, appropriate diagnostic facilities, and uniform reporting practices.

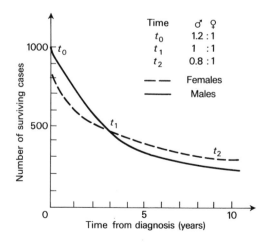

Fig. 9.7. Sex-specific survival from colorectal cancer as an illustration of selective survival.

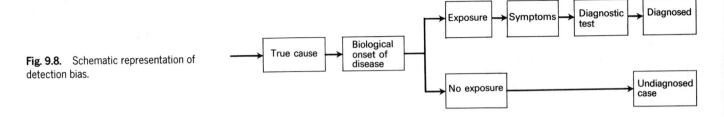

Fig. 9.8. Schematic representation of detection bias.

Detection bias

A non-causal exposure–disease association may occur in observational studies if exposure status influences the likelihood of clinical recognition of the disease. For example, Horwitz and Feinstein (1978) proposed that the association between use of exogenous oestrogens and endometrial cancer may be partially attributed to the preferential detection of this cancer in exposed women. These authors speculated that oestrogen use may lead to dysfunctional bleeding, thus prompting an intra-endometrial diagnostic examination and diagnosis of an otherwise asymptomatic endometrial cancer. In contrast, women who do not take oestrogens might be less likely to manifest bleeding per vagina, and thus a greater proportion of endometrial cancer cases may be undetected in unexposed women. Note that a premise for this bias is that some cases will never be diagnosed (Fig. 9.8). Hutchison and Rothman (1978) argued that a progressive disease, such as endometrial cancer, would ultimately produce symptoms independent of exposure status. As a result, they suggest that the proportion of detected cases will be comparable in oestrogen-exposed and unexposed women. Indeed, detection bias is more likely to occur in the study of diseases with documented asymptomatic cases, such as 'silent' gallstones (Feinstein 1979). Case–control studies of these diseases should attempt to evaluate the extent to which exposure brings an otherwise asymptomatic case to clinical detection.

Non-response

Both experimental and observational studies of human populations may suffer from poor subject co-operation. Selection bias may be introduced if enrolled subjects systematically differ from non-participants. In the conduct of a case–control study, there are several stages at which subject enrolment may be impaired. First, some cases may be omitted because of limited access to medical care. Second, some diagnosed persons may be overlooked because of incomplete reporting practices. Third, some diagnosed persons may be excluded because of restricted access to their records. Finally, some patients, or their attending physicians, may decline to participate in a research project.

Many of these selection factors were illustrated by a case–control study of colon cancer (Herrmann *et al.* 1981). In that investigation, over 20 per cent of eligible hospitals denied permission to contact their patients. In addition, 25 per cent of attending physicians placed restrictions on patient contact. The authors noted that the most commonly cited reasons for non-participation were concerns about confidentiality and lack of personal advantage for the patient involved.

Criqui and co-workers (1979) have demonstrated that subjects who participated in a cardiovascular disease survey tended to be a biased subset of the target population. Respondents were characterized as 'worried well' individuals, that is to say, persons with risk factors but without disease. This pattern of non-response produced minor to moderate errors in estimated odds ratios between specific risk factors and disease occurrence. However, the authors concluded that larger errors could occur with a similarly biased sample and a lower level of response. The non-response effects can be minimized by repeated attempts at subject enrolment and the collection of auxiliary data to evaluate the comparability of participants and non-participants. In addition, subject response may be increased in studies with protections of confidentiality and demonstrable personal advantage to participation.

Misclassification bias

The second major source of potential bias in analytical research is referred to as misclassification (information) bias. This type of error may result from inaccurate assignment of either exposure or disease status. Copeland and colleagues (1977) developed a conceptual framework for the consideration of misclassification in outcome variables (exposure status in case–control studies). Other authors (Gladen and Rogan 1979; Shy *et al.* 1978) have considered the effects of exposure misclassification on risk estimates in environmental studies. In the present context, we note that two types of information error may occur: non-differential and differential misclassification.

Non-differential misclassification

The errors in classification of one variable (e.g. exposure status) do not depend on the level of the other variable (e.g. disease status). As an example, consider a sphygmomanometer which always reads 10 mm Hg higher than the true blood pressure value. With these erroneous measurements, some normotensive persons will be misclassified as hypertensive, but these errors will not depend on disease status (e.g. myocardial infarction).

Differential misclassification

The errors in classification of one variable (e.g. exposure status) depend on the level of the other variable (e.g. disease status). As an example of differential misclassification, consider an interviewer who systematically over-reports prior history of hypertension in myocardial infarction cases compared with control subjects.

The distinction between non-differential and differential misclassification may provide information on the anticipated direction of bias. It has been demonstrated that non-differential misclassification errors will tend to decrease the observed odds ratio (Gullen et al. 1968). In this situation, the study will provide an underestimate of the true exposure–disease association. Diffential misclassification can produce either an overestimate or an underestimate of the true odds ratio (Copeland et al. 1977).

A variety of errors may result in non-differential misclassification bias. The following two examples illustrate the general considerations of non-differential misclassification of exposure in case–control studies.

Exposure specification

One source of potential non-differential misclassification of exposure may arise if the exposure under investigation is not measured directly. For example, in occupational research it is often impossible to identify or quantify specific exposures to individual workers. As a surrogate of exposure status, a common practice is to classify subjects by job title. Without accurate information on individual exposures, it is possible that these surrogate measures may misrepresent the true odds of exposure in cases and controls. This type of error in case–control studies may be minimized when detailed information on specific exposures is available.

Unacceptability bias

A second potential source of non-differential misclassification may occur when the exposure under investigation is a behaviour or characteristic which subjects are inclined to under-report. For example, subjects often answer questions about sexual practices, alcohol consumption, or use of illicit drugs with the perceived 'socially acceptable' response. More reliable answers to such sensitive questions may be obtained with self-administered questionnaires and strict guarantees of confidentiality.

A large number of systematic errors may lead to differential misclassification of exposure status in case–control studies. The following examples provide an overview of some common differential misclassification biases.

Recall (anamnestic) bias

When historical exposure information is collected in case–control studies, it is possible that the subjects' memory of earlier events will be influenced by disease status. Patients are often highly motivated to remember any event that may have contributed to their disease aetiology. In their eagerness to identify possible casual factors, they may even inadvertently over-report certain exposures. For example, in a case–control study of gestational drug use and adverse pregnancy outcome, the mothers of affected babies reported more unsubstantiated drug use than mothers of normal neonates (Klemetti and Saxen 1967).

There are several strategies that may be used to minimize the effects of differential recall of exposures. First, subjects and interviewers should not be aware of the specific exposure–disease association under investigation. Second, whenever possible, subject responses should be compared with other data sources, such as employment or medical records. Third the use of a hospital control group may select comparison subjects who are similarly motivated to report possible aetiological exposures.

Protopathic bias

Even when accurate historical exposure data are obtained. it is possible that the disease onset may actually precede the exposure. This error, referred to as 'protopathic bias' (Horwitz and Feinstein 1980), may result when the early manifestations of a disease cause a change in the pattern of exposures in cases. In case–control studies, the retrospective sequence of observation makes it difficult to determine whether a particular exposure occurred prior to the disease under investigation. This temporal confusion can produce erroneous conclusions about either suspected protective or deleterious exposure. For instance, physicians may refuse oral contraceptive prescriptions to women with breast 'lumps'. If these women are subsequently studied as breast cancer cases, then the relative lack of oral contraceptive use in these women may be mistakenly interpreted as a protective effect of the medication (Feinstein 1979). On the other hand, a particular exposure may be spuriously labelled as a causal factor if early disease signs or symptoms increase the likelihood of exposure. For example, Feinstein and colleagues (1981) speculated that an erroneous association between coffee consumption and pancreatic cancer could arise if pancreatic cancer cases increased their coffee drinking as a result of anxiety about vague abdominal discomfort.

Interview bias

A third common source of differential misclassification may arise in case–control studies with exposure status determined by subject interview. In particular, the circumstances under which cases and controls are interviewed should be comparable. This requirement for comparability of data collection procedures includes such features as:

(a) the time from suspected exposure to interview;

(b) the setting of interview (e.g. hospital versus subject's residence);

(c) the format of the interview;

(d) the manner in which questions are phrased;

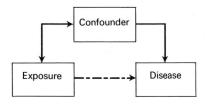

Fig. 9.9. Schematic representation of confounding.

(e) the amount of interviewer prompting for specific answers;

(f) subject and interviewer knowledge about the research hypothesis.

Examples of potential interview bias are common in the case–control literature. For instance, in a study of tampon use and toxic-shock syndrome, Davis *et al.* (1980) conducted personal interviews with cases, but presented controls with self-administered questionnaires. In a study of salicylate use and Reye's syndrome (Halpin *et al.* 1982), the time to interview of parents of controls was systematically longer than time to interview of parents of cases, which might have given rise to differential abilities to recall exposures.

One must also recognize that intentional or inadvertent differential misclassification errors may occur when the interviewers are aware of the research hypothesis and the disease status of individual subjects. To avoid this type of differential misclassification, it is advisable to 'blind' the interviewers to the research hypothesis and, if possible, to the disease status of individual subjects. Similarly, cases and controls should be 'blinded' to the specific exposure–disease hypothesis under investigation.

Confounding bias

The third major source of potential bias in analytical research is referred to as confounding bias. This relates to the distortion that may occur when the study effect is mixed with another effect. This mixing process may introduce bias if the resulting exposure–disease association is meaningfully altered from its 'uncontaminated' value.

Under the prevailing notion of this bias, a confounder (contaminating variable) in a case–control study must be:

(a) extraneous to the exposure–disease association under investigation; and

(b) predictive of the disease; and

(c) unequally distributed between exposure groups.

These conditions for a confounder are schematically represented in Figure 9.9.

Confounding may be illustrated by the data presented in Example 7. In that example, the study exposure was gestational irradiation and the disease was childhood leukaemia. A crude odds ratio of 1.48 indicated a modest positive association between these variables. However, when the effect of maternal gravidity was taken into account (Example 9), the observed irradiation–leukaemia odds ratio was 1.08. The

latter value, which may be described as an 'adjusted' measure, indicated little or no association between gestational irradiation and leukaemia. In this situation, we conclude that the apparent irradiation–leukaemia association may be attributed to the confounding effect of maternal gravidity. It can be demonstrated easily with these data that maternal gravidity was associated conditionally both with the exposure and the disease under investigation.

The problem of confounding may become particularly acute in the study of diseases with multiple risk factors. For example, in planning a case–control study of the relationship between oral contraceptive (OC) use and acute myocardial infarction (MI), a list of potential confounders might include: age, race, hypertension, cigarette smoking, alcohol use, sedentary lifestyle, and serum triglycerides. Since all of these variables are known MI risk factors, they should be considered as potential confounders of the association between OC use and MI. However, it should be recognized that those risk factors that are not associated with OC use in the data will not confound the study results. Thus, it may be possible to arrive at a convenient subset of those risk factors that actually confound the association between OC use and MI. In other words, a valid study result will be obtained by controlling for only the confounding risk factors. Controlling for a minimal number of co-variables has the additional appeal of a possible gain in precision (Kleinbaum *et al.* 1982).

The effects of potential confounders may be controlled in the design or analysis of a case–control study. At the design stage, one may select subjects so that the potential confounder is not associated with disease status. For example, if age is an anticipated risk factor, then the cases and controls may be restricted to a narrow age-band. As indicated earlier in this chapter, partial restriction (matching) is another mechanism for eliminating co-variable–disease associations. Indeed, matching is frequently employed with the intent of adjusting for potential confounders. However, the risk factor–disease association criterion for confounding relates to unexposed persons. Thus, matching on a co-variate does not necessarily guarantee the elimination of confounding by that factor (Kupper *et al.* 1981).

At the analysis stage of investigation, confounding may be controlled by stratification or mathematical modelling. The data in Examples 7 and 9 demonstrate the use of stratification to obtain an adjusted odds ratio. When there are multiple co-variates, an analogous adjusted odds ratio may be obtained from summarization over multiple strata. Alternatively, when there are many potential confounders, a modelling technique such as logistic regression may be preferred. Equation 9.17 illustrates the manner by which potential confounders may be entered into a logistic model. The reader interested in a more detailed description of mathematical modelling strategies is referred to two excellent textbooks (Breslow and Day 1980; Schlesselman 1982).

In summary, there are a number of approaches that may be employed to adjust for potential confounders. With each method, one must first anticipate which factors might con-

Table 9.13. Characteristics recommended for the evaluation of case–control studies

A statement of the research question
Identification of the source of controls
Identification of the source of cases
A statement of the conclusions reached from the data
Exclusion criteria for cases
A statement of the non-response rate
Exclusion criteria for controls
Information on the matching procedure, if used
Information on the method of data collection (interviewer, self-administered questionnaire, record review)
Information on the presence of possible confounding variables
Information on the investigation of possible sources of bias
Information on whether interviewers or record abstractors, if used, were blinded to the status of cases and controls
Information on the methods used for dealing with confounding variables
A description of the analytical methods
A description of the sampling technique
Information on whether the cases are incident or prevalent
Diagnostic procedure used to identify cases
Presentation of confidence limits
Information on duration and intensity of exposure
A statement on whether the controls experienced the same diagnostic procedures as the cases

found the exposure–disease association under study. Then, sufficient data must be collected to evaluate each of these co-variates. Finally, the effects of co-variates that actually confound the study results should be controlled, either by subject selection or analytical adjustment. Control of non-confounders is discouraged since this may entail a loss of precision.

Concluding remarks

In 1978, the editors and the publisher of the *Journal of Chronic Diseases* invited thirty scientists to attend a symposium in Bermuda on case–control studies. One of us (MAI) was responsible for moderating the symposium and editing the proceedings (Ibrahim and Spitzer 1979). The discussions at the symposium clearly highlighted the controversial issues surrounding the strategies used in case–control studies, the interpretation of these studies, and the policy action to be taken on the basis of such findings. The symposium took place at a time when several health issues, such as those of reserpine use and breast cancer or oestrogen use and endometrial cancer, were vigorously debated in scientific journals and national meetings. Since the publication of that special issue of the *Journal*, numerous case–control studies on important and controversial issues have been conducted and a book entirely devoted to case–control studies was published (Schlesselman 1982). The question today is not whether case–control studies should be used, but how to conduct one properly.

Toward that end, Lichtenstein and colleagues (1987) surveyed 37 experts to determine the features that were considered crucial for the evaluation of case–control studies. The 20 most commonly chosen characteristics are listed in Table 9.13 in descending frequency of citation. Although the collective judgements of these experts are neither comprehensive nor entirely objective, they do represent guide-lines that may be useful in the design and publication of case–control research. When Lichtenstein and colleagues (1987) then applied these criteria to evaluate 48 case–control studies published in 1984, they found that 88 per cent of the studies did not include information on one or more of these criteria. The items that were most frequently missing included information on the blinding of interviewers (58 per cent), investigation of possible sources of bias (40 per cent), a description of the sampling technique (29 per cent), a statement of the non-response rate (27 per cent), and exclusion criteria for controls (25 per cent).

Originated in the nineteenth century (Lilienfeld and Lilienfeld 1979), the case–control study has matured to a prominent place in epidemiological research. Additional advances in the design and analysis will continue to occur and thus make the case–control study an even more attractive investigative tool. It is fitting to conclude this chapter by pointing out that however rigorous and complex the case–control design and analysis methodology might become, the case–control study remains *only* a tool for addressing critical questions, which are the hardest and most essential aspect of epidemiological research.

References

Berkson, J. (1946). Limitations of the applications of fourfold table analysis to hospital date. *Biometrics Bulletin* **2**, 47.

Breslow, N.E. and Day, N.E. (1980). *Statistical methods in cancer research, Vol. 1, The analysis of case–control studies.* IARC Scientific Publications No. 32. International Agency for Research on Cancer, Lyon.

Brown, C.C. (1981). The validity of approximation methods for interval estimation of the odds ratio. *American Journal of Epidemiology* **113**, 474.

Cassel, J.C. (1971). Review of 1960 through 1962 cardiovascular disease prevalence study. *Archives of Internal Medicine* **128**, 890.

Clayton, A.J. (1982). Toxic shock syndrome in Canada. *Annals of Internal Medicine* **96** (Part 2), 881.

Cohn, J.N., Franciosa, J.A., Francis, G.S., *et al.* (1982). Effect of short-term infusion of sodium nitroprusside on mortality rate in acute myocardial infarction complicated by left ventricular failure: Results of a Veterans Administration Cooperative Study. *New England Journal of Medicine* **306**, 1129.

Cole, P. (1979). The evolving case–control study. *Journal of Chronic Diseases* **32**, 15.

Cole, P. and MacMahon, B. (1971). Attributable risk percent in case–control studies. *British Journal of Preventive and Social Medicine* **25**, 242.

Copeland, K.T., Checkoway, H., Holbrook, R.H., *et al.* (1977). Bias due to misclassification in the estimate of relative risk. *American Journal of Epidemiology* **105**, 488.

Cornfield, J. (1951). A method of estimating comparative rates from clinical data: applications to cancer of the lung, breast, and cervix. *Journal of the National Cancer Institute* **11**, 1269.

Cornfield, J. (1962). Joint dependence of risk of coronary heart disease on serum cholesterol and systolic blood pressure: a discriminant function analysis. *Federation Proceedings* **21**, 58.

Criqui, M.H., Austin, M., and Barrett-Connor, E. (1979). The effect of non-response on risk ratios in a cardiovascular disease study. *Journal of Chronic Diseases* **32**, 633.

Davis, J.P., Chesney, P.J., Wand, P.J., *et al.* (1980). Toxic-shock syndrome: epidemiologic features, recurrence, risk factors and prevention. *New England Journal of Medicine* **303**, 1429.

Dawber, T.R., Moore, F.E., and Mann, G.V. (1957). Coronary heart disease in the Framingham Study. *American Journal of Public Health* **47**, 4.

Devesa, S.S. and Silverman, D.T. (1978). Cancer incidence and mortality trends in the United States: 1935–1974. *Journal of the National Cancer Institute* **60**, 545.

Doll, R. and Hill, A.B. (1950). Smoking and carcinoma of the lung: preliminary report. *British Medical Journal* **ii**, 739.

Durrer, J.D., Lie, K.I., Van Capell, F.J.L., *et al.* (1982). Effect of sodium nitroprusside on mortality in acute myocardial infarction. *New England Journal of Medicine* **306**, 1121.

Feinstein, A.R. (1979). Methodologic problems and standards in case–control research. *Journal of Chronic Diseases* **32**, 35.

Feinstein, A.R., Horwitz, R.I., Spitzer, W.O., *et al.* (1981). Coffee and pancreatic cancer: the problems of etiologic science and epidemiologic case–control research. *Journal of the American Medical Association* **246**, 957.

Fleiss, J.L. (1981). *Statistical methods for rates and proportions* (2nd edn.) Wiley, New York.

Gladen, B. and Rogan, W.J. (1979). Misclassification and the design of environmental studies. *American Journal of Epidemiology* **109**, 607.

Gordon, T. (1974). Hazards in the use of the logistic function with special reference to data from prospective cardiovascular studies. *Journal of Chronic Diseases* **27**, 97.

Greenberg, R.S. and Osterhout, S.K. (1982). Seasonal trends in reported poisonings. *American Journal of Public Health* **72**, 394.

Greenland, S. (1979). Limitations of the logistic analysis of epidemiologic data. *American Journal of Epidemiology* **110**, 693.

Gullen, W.H., Bearman, J.E., and Johnson, E.A. (1968). Effects of misclassification in epidemiologic studies. *Public Health Reports* **53**, 1956.

Halpin, T.J., Holtzhauer, F.J., Campbell, R.J., *et al.* (1982). Reye's syndrome and medication use. *Journal of the American Medical Association* **248**, 687.

Hartge, P., Brinton, L.A., Rosenthal, J.F., *et al.* (1984*a*). Random digit dialing in selecting a population-based control group. *American Journal of Epidemiology* **120**, 825.

Hartge, P., Cahill, J.I., West, D., *et al.* (1984*b*). Design and methods in a multicenter case–control interview study. *American Journal of Public Health* **74**, 52.

Hayden, G.F., Kramer, M.S., and Horwitz, R.I. (1982). The case–control study: a practical review for the clinician. *Journal of the American Medical Association* **247**, 326.

Herbst, A.L., Ulfelder, H., and Poskanzer, D.C. (1971). Adenocarcinoma of the vagina: association of maternal stilbestrol therapy with tumor appearances in young women. *New England Journal of Medicine* **284**, 878.

Herrmann, N., Amsel, J., and Lynch, E. (1981). Obtaining hospital and physician participation in a case–control study of colon cancer. *American Journal of Public Health* **71**, 1314.

Hortwitz, R.I. and Feinstein, A.R. (1978). Alternative analytic methods for case–control studies of estrogens and endometrial cancer. *New England Journal of Medicine* **299**, 1089.

Horwitz, R.I. and Feinstein, A.R. (1979). Methodologic standards and contradictory results in case–control research. *American Journal of Medicine* **66**, 556.

Horwitz, R.I. and Feinstein, A.R. (1980). The problem of 'protopathic bias' in case–control studies. *American Journal of Medicine* **111**, 389.

Hutchinson, G.B. and Rothman, K.J. (1978). Correcting a bias? *New England Journal of Medicine* **299**, 1129.

Ibrahim, M.A. and Spitzer, W.O. (1979). *The case–control study: consensus and controversy.* Pergamon Press, London.

Illinois Department of Public Health (1975). *Poison Control Program report: 1974.* Released by the State of Illinois, July.

Jolliffe, N. and Archer, M. (1959). Statistical associations between international coronary heart disease death rates and certain environmental factors. *Journal of Chronic Diseases* **9**, 636.

Kleinbaum, D.G., Kupper, L.L., and Morgenstern, H. (1982). *Epidemiologic research: principles and quantitative methods.* Lifetime Learning Publications, Belmont, California.

Kleinbaum, D.G., Morgenstern, H., and Kupper, L.L. (1981). Selection bias in epidemiologic studies. *American Journal of Epidemiology* **113**, 452.

Klemetti, A. and Saxen, L. (1967). Prospective versus retrospective approach in the search for environmental causes for malformations. *American Journal of Public Health* **57**, 2071.

Koch. M., McPherson, T.A., and Edgedahl, R.D. (1982). Effects of sex and reproductive history on the survival of patients with colorectal cancer. *Journal of Chronic Diseases* **35**, 69.

Kraus, A.S. (1960). Comparison of a group with a disease and a control group from the same families, in search for possible etiologic factors. *American Journal of Public Health* **50**, 303.

Kupper, L.L., Karon, J.M., Kleinbaum, D.G., *et al.* (1981). Matching in epidemiologic studies: validity and efficiency considerations. *Biometrics* **37**, 271.

Labarthe, D.R. (1979). Methodologic variation in case–control studies of reserpine and breast cancer. *Journal of Chronic Diseases* **32**, 95.

Lichtenstein, M.J., Mulrow, C.D., and Elwood, P.C. (1987). Guidelines for reading case–control studies. *Journal of Chronic Diseases* **40**, 893.

Lilienfeld, A.M. and Lilienfeld, D.E. (1979). A century of case–control studies: progress? *Journal of Chronic Diseases* **32**, 5.

Lipids Research Clinics Program (1984). The lipid research clinics coronary primary prevention trial results. 1. Reduction in incidence of coronary heart disease. *Journal of the American Medical Association* **251**, 351.

Lubin, J.H. and Gail, M.H. (1984). Biased selection of controls for case–control analyses of cohort studies. *Biometrics* **40**, 63.

McKinley, S.M. (1977). Pair-matching—a reappraisal of a popular technique. *Biometrics* **33**, 725.

McMichael, A.J., Spirtas, R., and Kupper, L.L. (1974). An epidemiologic study of mortality within a cohort of rubber workers, 1964–72. *Journal of Occupational Medicine* **16**, 458.

McMichael, A.J., Spirtas, R., Kupper, L.L., *et al.* (1975). Solvent exposure and leukemia among rubber workers: an epidemiologic study. *Journal of Occupational Medicine* **17**, 234.

Mancuso, T.F. (1949). Occupational cancer survey in Ohio. *Proceedings of the Public Health and Cancer Association of America* **56**.

Mantel, N. (1973). Synthetic retrospective studies and related topics. *Biometrics* **29**, 479.

Mantel, N. and Fleiss, J.L. (1980). Minimum expected cell size requirements for the Mantel–Haenszel one-degree-of-freedom chi-square test and a related rapid procedure. *American Journal of Epidemiology* **112**, 129.

Mantel, N. and Haenszel, W. (1959). Statistical aspects of the analysis of data from retrospective studies of disease. *Journal of the National Cancer Institute* **22**, 719.

May, G.S., DeMets, D.L., Friedman, L.M., *et al.* (1981). The randomized clinical trial: bias in analysis. *Circulation* **64**, 669.

Miettinen, O.S. (1970). Matching and design efficiency in retrospective studies. *American Journal of Epidemiology* **91**, 111.

Miettinen, O.S. (1976). Estimability and estimation in case–referent studies. *American Journal of Epidemiology* **103**, 226.

Miettinen, O.S. (1985). The 'case–control' study: valid selection of subjects. *Journal of Chronic Diseases* **38**, 543.

Neyman, J. (1955). Statistics—servant of all sciences. *Science* **122**, 401.

Pasternack, B.S. and Shore, R.E. (1981). Sample sizes for group sequential cohort and case–control study designs. *American Journal of Epidemiology* **113**, 182.

Pocock, S.J. (1977). Group sequential methods in the design and analysis of clinical trials. *Biometrika* **64**, 191.

Polk, B.F., Fox, R., Brookmeyer, R., *et al.* (1987). Predictors of the acquired immunodeficiency syndrome developing in a cohort of seropositive homosexual men. *New England Journal of Medicine* **316**, 61.

Prentice, R. (1976). Use of the logistic model in retrospective studies. *Biometrics* **32**, 599.

Prentice, R.L. and Breslow, N.E. (1978). Retrospective studies and failure time models. *Biometrika* **65**, 153.

Prentice, R.L. and Pyke, R. (1979). Logistic disease incidence models and case–control studies. *Biometrika* **66**, 403.

Press, S.J. and Wilson, J. (1978). Choosing between logistic regression and discriminant analysis. *Journal of the American Statistical Association* **73**, 699.

Radhakrishna, S. (1965). Combination of results from several 2 × 2 contingency tables. *Biometrics* **21**, 86.

Roberts, R.S., Spitzer, W.O., Delmore, T., *et al.* (1978). An empirical demonstration of Berkson's bias. *Journal of Chronic Diseases* **31**, 119.

Sackett, D.L. (1979). Bias in analytic research. *Journal of Chronic Diseases* **32**, 51.

Schlesselman, J.J. (1982). *Case–control studies: design, conduct, analysis*. Oxford University Press, New York.

Schoenbaum, S.C., Biano, S., and Mack, T. (1975). Epidemiology of congenital rubella syndrome. *Journal of the American Medical Association* **233**, 151.

Schuman, L.M.(1977). Patterns of smoking behavior. In *Research on smoking behavior* (ed. M. E. Jarvik, J. W. Cullen, E. R. Gritz, *et al.*) p. 36. Department of Health, Education, and Welfare. US Government Printing Office, Washington, DC.

Shy, C.M., Kleinbaum, D.G. and Morgenstern, H. (1978). The effect of misclassification of exposure status in epidemiological studies of air pollution health effects. *Bulletin of the New York Academy of Medicine* **54**, 1155.

Silverberg, E. (1982). Cancer statistics, 1982. *Ca–A Cancer Journal Clinic* **32**, 15.

Silverman, W.A. (1980). *Retrolental fibroplasia: a modern parable*. Grune and Stratton, New York.

Taylor, J.W. (1977). Simple estimation of population attributable risk from case–control studies. *American Journal of Epidemiology* **106**, 260.

Thacker, S.B. (1988). Meta-analysis: a quantitative approach to research integration. *Journal of the American Medical Association* **259**, 1685.

Thomas, D.G. (1971). Exact confidence limits for an odds ratio in a 2 × 2 table. *Applied Statistics* **20**, 105.

Truett, J., Cornfield, J., and Kannel, W. (1967). A multivariate analysis of the risk of coronary heart disease in Framingham. *Journal of Chronic Diseases* **20**, 511.

Tyroler, H.A., Heyden, S., Bartel, A., *et al.* (1971). Blood pressure and cholesterol as coronary heart disease risk factors. *Archives of Internal Medicine* **128**, 907.

Ury, H.K. (1975). Efficiency of case–control studies with multiple controls per case: continuous or dichotomous data. *Biometrics* **31**, 643.

Walter, S.D. (1980). Berkson's bias and its control in epidemiologic studies. *Journal of Chronic Diseases* **33**, 721.

Wingo, P., Ory, H.W., Layde, P.M., *et al.* (1988). The evaluation of the data collection process for a multicenter population-based, case–control design. *American Journal of Epidemiology* **128**, 206.

Woolf, B. (1955). On estimating the relation between blood group and disease. *Annals of Human Genetics* **19**, 251.

10

Cohort studies

MANNING FEINLEIB, NORMAN E. BRESLOW, and ROGER DETELS

Introduction

The cohort study is an observational epidemiological study which, after the manner of an experiment, attempts to study the relationship between a purported cause (exposure) and the subsequent risk of developing disease. As in other observational epidemiological studies, and unlike experimental studies, the suspected causal factor or exposure is not randomly assigned to the study population. However, the cohort study follows the same time direction of an experiment in that the suspected exposure is identified as having or not having occurred in the study population before the occurrence of disease is investigated. Thus, certain biases that may occur in other forms of epidemiological studies can be avoided, specifically those concerned with ascertaining the exposure status of the population. Furthermore, because disease occurrence is identified subsequent to enumeration of exposure groups, this type of study allows direct estimation of the risk of developing disease and how risk varies with time since exposure.

Cohort studies have been called by a variety of names including incidence studies, prospective studies, follow-up studies, longitudinal studies, and panel studies, although the latter two terms have more generally been applied to studies involving repeated measurements of the same variables over time. They are similar to the usual scientific experiment in that they proceed from the suspected cause or aetiological agent to the disease outcome with controls or comparison groups selected on the basis of absence of exposure to the putative cause. As a type of observational study there is no randomization to exposure classes nor is there any attempt to manipulate the exposure. Case–control studies, on the other hand (also known as case–referent studies and, formerly, as retrospective studies), have no counterpart in experimental science since they work from the outcome event back towards the supposed aetiological factor. Indeed, case–control studies are often best viewed conceptually in terms of sampling data from an ongoing (and possibly fictitious) cohort study. In this way the cohort studies offer the possibility of studying the full range of effects of the suspected aetiological factor. Frequently the suspected aetiological factor is related not only to the occurrence of the disease of

primary interest but may influence the natural history of the disease, and may be related to a variety of other health conditions which may not have been suspected at first. A particularly important aspect of cohort studies is that they provide direct estimates of the risk of disease for each exposure group separately. These separate estimates of risk can then be used to estimate a variety of measures of interest to epidemiologists such as the attributable risk, relative risk, and aetiological fraction. (These measures of risk are discussed in the analysis section below and in Table 10.3.) Although these risks can often be estimated from other types of studies when certain assumptions are made or ancillary information is available, cohort studies permit direct estimates of these measures from the data obtained in the study itself.

The disadvantages of cohort studies are primarily logistical and administrative. Often, relatively large populations have to be followed for long periods of time, thus entailing considerable expense in terms of funding and professional resources. If the disease outcome of interest is rare the sample sizes required for prospective studies may be prohibitive. If the follow-up period is long, which is often the case for chronic diseases, the problem of attrition of the study group due to loss from follow-up, migration, competing causes of death, or gradual deterioration of interest in participation may present serious analytical problems which might negate the value of the overall study. Longitudinal follow-up requires careful attention to maintaining standardized diagnostic methods and criteria. And, of course, the longer the study is continued the more difficult it is to maintain a committed investigative team and stable funding for the project.

In the first part of this chapter we discuss the major methodological aspects of cohort studies: forms of cohort studies, selection of study cohorts, gathering of baseline information, follow-up, and analysis. To illustrate these points we use examples from three studies: a historical cohort study of artificial menopause and breast cancer using available hospital and death certificate information (Feinleib 1968); a prospective cohort study of cigarette smoking and mortality among British doctors using mail questionnaires (Doll and Peto 1976); and a prospective cohort study of heart disease in

Framingham, Massachusetts, using periodic medical examinations (Kannel *et al*. 1961; Dawber *et al*. 1963). The second part of the chapter presents the various types of bias that can confound interpretation of cohort studies and suggests ways to identify, reduce and/or resolve these biases. In this section examples are drawn from a wider range of studies.

Forms of cohort studies

Cohort studies may take a variety of forms. The key distinction that has been established in the past is based primarily on the availability of data. In *prospective cohort studies* data on exposure status and disease outcome are not available at the outset of the study; they must be ascertained through the direct efforts of the investigator in the future. In *ambispective cohort studies* data on exposure status have been collected in the past and are available from existing records while disease outcome is unknown or incompletely known; the investigator is obliged to follow the cohort for subsequent occurrence of the disease. In *historical cohort studies* data on exposure status and disease outcome have been collected in the past and are available from existing records; the investigator's efforts are devoted primarily to linking the relevant data-files. Each type of study involves certain basic steps which will be described briefly below. These include the selection of the study and comparison groups; obtaining baseline information with regard to exposure and health status initially; follow-up of the members of the cohort and surveillance for disease outcome; and analysis of the results.

Selection of the study cohorts

Objective

Selection of representative samples of exposed and non-exposed groups.

There are two approaches to the selection of the groups to be followed in a cohort study:

1. The identification of a *special exposure group* defined because of (a) unusual exposure to a suspected causative (aetiological) factor, or (b) unusual life-style or work experience.

2. Using a *general population sample* in which there is heterogeneity of exposure to the suspected aetiological factor.

Where the starting point is with a special exposure group it is necessary to find appropriate comparison groups or the means to make comparisons with the general population. When the general population sample is used as a starting point the various levels of exposure within the study group provide the basis for internal comparisons. Each approach also takes into consideration various logistical constraints, for example accessibility and co-operativeness of the study groups, availability of medical and other records, and anticipated completeness and cost of end-point surveillance.

Example 1—A historical cohort study of the relation between artificial menopause and breast cancer

Seven case–control studies done between 1926 and 1962 all reported that the occurrence of artificial menopause (surgical removal of the uterus and/or ovaries) occurred significantly less frequently among breast cancer patients than among a variety of controls. Because the case–control studies did not present information about the extent of surgery (the effect of removal of only the uterus versus removal of the ovaries) nor about the effects of the age at which the artificial menopause occurred, it was decided to investigate these issues by means of a cohort study. The disadvantage of using a prospective cohort method in elucidating the relation between artificial menopause and breast cancer is that there is a long period between the gynaecological procedure and the appearance of the disease in appreciable frequency. To abridge this delay it was decided to use the historical cohort approach. The cohorts were selected from the records of two technical hospitals in the Boston area. The study cohort included all eligible patients seen at these hospitals from 1920 through 1940. Women aged 55 years or younger were eligible for inclusion in the study if they had undergone any of the following procedures as determined from the surgical and pathological records: (i) hysterectomy; (ii) unilateral oophorectomy; (iii) bilateral oophorectomy; (iv) radium or X-ray treatment of the ovaries or uterus; or (v) cholecystectomy. The last group served as a control cohort.

Certain patients were excluded from the study; (i) women who had a prior mastectomy or a prior breast malignancy or who had undergone castration as part of the treatment for an existing breast tumour; (ii) women treated for pelvic malignancies; (iii) women who had previous removal of their ovaries or a history of natural menopause before the age of 40 years; (iv) women who did not survive their index admission; and (v) all who were not residents of Massachusetts at the time of their index procedure. At the final editing of the study abstract forms and the elimination of duplicate records there were 8387 patients in the study populations. They were subdivided into four 'exposure' categories:

1. Natural menopause—1479 women (including 953 women who underwent cholecystectomy and 526 women who were post-menopausal at the time of the gynaecological procedures for benign conditions).

2. Hysterectomy and bilateral oophorectomy—3241 women (this constitutes the surgically castrated group who were believed to have no residual ovarian activity).

3. Those undergoing hysterectomy and/or unilateral oophorectomy and who, as far as could be ascertained from the surgical and pathological records, retained at least one intact ovary—2149 women (referred to as the 'partial surgery' group).

4. Radiation-induced artificial menopause—1518 women.

The partial surgery group constituted a second control cohort and 'sham operations' with which to contrast the women subjected to hysterectomy and bilateral oophorectomy.

It should be noted that it is not possible to relate the actual cohort studied to a clearly definable population. Although in this case adequate records were available for virtually every woman admitted to these hospitals who was eligible for the study, it is not known from what source population the women using this hospital came. However, it is assumed that the reasons for coming to these particular hospitals were not correlated with both the type of procedure and the subsequent risk of developing breast cancer, that is, they were not confounding factors (see section on 'biases').

Example 2—A prospective cohort study of the relation between cigarette smoking and mortality (see Breslow and Day 1987, Appendix IA)

By 1950, several case–control studies had been published and were in agreement in showing that a larger proportion of lung cancer patients had been heavy cigarette smokers and a smaller proportion had been non-smokers than were patients with other diseases. Because of the possibility of a variety of biases in these case–control studies, a prospective study was launched in 1951 among the members of the medical profession in the UK. This group was chosen because it was felt that physicians would respond to mailed questionnaires, would report their smoking histories accurately, and could be followed economically through the death records of the Registrars-General and through the registries of the General Medical Council and the British Medical Association. It was felt that the relation of smoking to health among physicians would be similar to that in the general population. A simple questionnaire was mailed out on 31 October 1951 to 59 600 men and women on the Medical Register.

The replies received from 40 637 doctors were sufficiently complete to be used—34 445 from men and 6192 from women. From a one-in-ten random sample of the register, it was estimated that this represented answers from 69 per cent of the men and 60 per cent of the women alive at the time of the inquiry. The degree of self-selection in those who replied was assessed in terms of the overall mortality using this one-in-ten sample. The standardized death-rate of those who replied was only 63 per cent of the death-rate for all doctors in the second year of the inquiry, and 85 per cent in the third year. In the fourth to tenth years the proportion varied about an average of 93 per cent and there was no evidence of any regular change with the further passage of years. Evidently the effect of selection did not entirely wear off, but after the third year it had become slight.

Example 3—Prospective cohort study of risk factors for heart disease: the Framingham Heart Study

The Framingham Heart Study is a long-term follow-up study of a sample of adults who lived in the town of Framingham, Massachusetts, in 1950. The first participants were actually examined in 1948 as part of an effort to conduct a demon-stration programme in the detection and natural history of cardiovascular diseases. In 1950, however, the study was reconstituted as a long-term epidemiological investigation of coronary heart disease, hypertension, and other cardiovascular diseases, and the original voluntary participants were incorporated into a random sample drawn from all adults aged 30–60 years living in the town. Of the eligible random sample of 6507 persons, 4469 (68.7 per cent) participated in the examinations, and when this number was supplemented with the volunteers a total cohort of 5209 was obtained. The possible effects of supplementing the cohort to replace the originally selected participants who refused to participate is discussed in Example 9.

Although it was recognized from the outset that the town of Framingham could be considered neither a random nor completely representative sample of the US, the town did have certain characteristics that made it extremely suitable for a long-term epidemiological study. The town was of adequate size (28 000) to provide enough individuals in the desired age range. It was compact enough that the study population could be observed conveniently by means of an examination at a single examining facility, and most of the residents received their hospital care at a single central hospital in the town. Due in part to a relatively stable economy supported by a diversity of employment opportunities, the population was relatively stable so as to enable adequate follow-up for a long period of time. Both the general community and the medical profession of the town were felt to be co-operative. The town was not believed to be 'grossly atypical in any respect that appeared relevant'.

Since only 68.7 per cent of the eligible random sample participated in the examination, it is possible that they might not be representative of the total population. This is a serious concern in all epidemiological studies where participation is voluntary and may be subject to self-selection. In this study it was felt that reasons for not participating were not appreciably related simultaneously to both the characteristics to be studied in the investigation and the risk of developing heart disease (see section on 'biases').

Gathering of baseline information

Objectives

1. Valid assessment of the exposure status of the members of the cohort groups.

2. Define the individuals 'at risk'; exclude those individuals with known disease at baseline.

3. Establish a basis for follow-up, obtaining identifying data, informed consent, commitment to co-operate in the follow-up (for example, permission to contact family members, physicians, and to obtain hospital and employment records).

4. Obtain data on important co-variables (that is, other exposures that may be associated with the risk of acquiring the disease) so that adjustments can be made for their

contribution to the incidence of disease in analysis (see section on confounding variables).

Sources of baseline information

Existing records

Baseline information about the cohorts can be obtained from a variety of sources such as available records from hospitals or employment records; interviews of the cohort members or other informants; direct medical and other special examinations; and indirect measures of exposure estimated from investigations of the environment. The availability of written records such as medical or employment records may provide useful information to select and define the cohort. If high quality records are available, they may permit the study to begin from the point of the recording of the information thereby adding a considerable period of follow-up time before the actual initiation of the investigation. Studies based on such records with follow-up of patients from such prior point in time to the present have been called by special names such as retrospective cohort studies, non-concurrent cohort studies, and historical prospective studies. There are several other advantages for using previously recorded information. The data are apt to be free from certain biases since they are recorded before any knowledge of the particular study for which they are used. Written records may provide details that are not fully known to the subject such as details on medical conditions or actual levels of exposure. However, such records may also have certain drawbacks. Records may not be uniformly available for all cohort members. Even when available, the detail and quality of the data in the records are not controllable by the investigator and it is difficult to verify the accuracy of questionable items.

Interviews

One of the more common methods of obtaining information is to interview the cohort members or other informants. A variety of techniques may be used: direct personal interviews, mailed questionnaires, telephone interviews, and, recently, having the subject complete a questionnaire administered by computer. When approaches are made to individual cohort members there are varying rates of response to requests to participate in the study. A wide variety of cohort studies has reported response rates of approximately 65–75 per cent for direct interviews. Mail questionnaires, depending on the length of the questionnaire and motivations of the group, often have appreciably lower rates of response. The advantages of interviewing the cohort members include the ability to obtain information on a wide variety of topics. Interviews can provide data on attitudes and permit the asking of quite complex questions with the possibility of probing to ensure accurate recording of responses (such as eliciting histories about diet, exercise, or measures of stress). On the other hand, interview data may not always be reliable because the subject may fail to recall information or may not be aware of his or her own habits or history. There is also the possibility that the information may be biased by the subject's knowledge of the aims of the investigation.

Examinations

Medical and other special examinations are necessary to obtain information of which the subject cannot be expected to be aware. Direct examination is often necessitated by the nature of the aetiological factor to be investigated and may be the only way to obtain biologically meaningful information. Subjects often appreciate the availability of an examination and this may enhance the response rate to certain types of investigations. On the other hand, special examinations are usually expensive and require attention to standardization of procedures, training of appropriate observers or laboratory personnel, and quality control across observers and over time. It has also been reported that response rates to medical examinations tend to be biased towards subjects who are relatively free from disease. Direct examination can also be used to validate information obtained from interviews. For example, testing for urinary thiocyanate has been a useful adjunct to smoking studies.

Measure of environment

The fourth type of baseline information is that obtained for each of the groups as a whole, especially when one is dealing with special exposure groups. Thus, it might be appropriate to measure air pollution, exposure to radiation or other toxicological substances, or exposures on the job for an entire group of workers and to apply this measure to each of the individuals in the group. Although this type of information is usually quite useful, especially when individual measures of exposure cannot be obtained directly, one should be aware that it essentially constitutes 'ecological data', that is, the measurement of a mean or modal value for a group which may conceal individual variability within the group.

Example 4—Artificial menopause and breast cancer

All of the baseline information for this investigation was obtained from the available surgical and pathological records already filed in the record rooms of the two hospitals between 1920 and 1940. The data were felt to be adequate for providing a valid assessment of the exposure status of the members of the cohorts in terms of whether or not they had received the indicated operation. Furthermore, as indicated above, those individuals with known disease could be identified from the available records. In part, the high quality of the records is due to the fact that the hospitals chosen were teaching hospitals for a major medical school and were generally filled out by medical students and interns who provided careful and detailed histories. However, if there was no mention of existing or pre-existing breast cancer, there was no way of confirming this independently of the available records. Likewise, the existence of breast cancer was based solely on the report of the patient to the interviewing physician. Co-variables that were available from the hospital records were the age at the time of the index procedure and the parity of women. Other co-variables of possible interest, since they could have been related both to the risk of cancer and the risk of gynaecological procedures, were not available from the records. These included body-weight, history of breast-feeding, and exposure to diagnostic X-rays.

Example 5—The British Doctors Study

The initial 'mail' questionnaire was intentionally kept short and simple to encourage a high proportion of replies. The doctors were asked to classify themselves into one of three groups: (1) whether they were, at the time, smoking; (2) whether they had smoked but had given up; or (3) whether they had never smoked regularly (that is, had never smoked as much as one cigarette a day, or its equivalent in pipe tobacco, for as long as one year). Present smokers and ex-smokers were asked additional questions. The former were asked the age at which they had started smoking, the amount of tobacco that they were currently smoking, and the method by which it was consumed. The ex-smokers were asked similar questions but relating to the time just before they had given up smoking.

'In a covering letter, the doctors were invited to give any information on their smoking habits or history that might be of interest, but, apart from that, no information was sought on previous changes in habit (other than the amount smoked prior to last giving up, if smoking had been abandoned). The decision to restrict question on amount smoked to current smoking habits was based mainly on the results of [an] earlier case–control study . . . [which showed] that the classification of smokers according to the amount that they had most recently smoked gave almost as sharp a differentiation between the groups of patients with and without lung cancer as the use of smoking histories over many years—theoretically more relevant statistics, but clearly based on less accurate data' (quoted from Breslow and Day 1987, Appendix IA).

Example 6—The Framingham Heart Study

On the basis of an initial examination and detailed interview the sample was characterized according to a variety of 'risk factors', i.e. blood cholesterol, blood pressure, cigarette smoking status, body mass index, and the presence of a variety of other diseases and conditions. Careful attention was given to standardization of the examination procedures and the structure of the interview.

On the basis of a medical history and examination, electrocardiogram, and other medical tests, it was found that 82 individuals in the base cohort of 5209 had a cardiovascular event before the baseline examination. Thus, the cohort of individuals 'at risk' for the key cardiovascular end-point of coronary heart disease numbered 5127.

To establish the basis for follow-up 'each of the subjects was advised at the initial interview that it was intended to re-examine him at two-year intervals, and that he would be approached directly at the appropriate time. The names of a relative, a friend, and the family physician were all recorded so that the subject could be traced in case he moved during the interval. An abstract of the initial examination was sent to the family physician and the subject was advised by letter as to whether the physician should be consulted or not. The objective of this procedure was to provide some tangible benefit to the subject other than the knowledge of his contribution to medical science. At the same time, care was taken not to become involved in the medical management of the subjects and to avoid interfering in any way with the relationship between the subject and his physician. This helped to maintain rapport, not only with the subjects themselves, but with the medical community as well' (Dawber *et al.* 1963).

Follow-up

Objectives

1. Uniform and complete follow-up of all cohort groups.
2. Complete ascertainment of outcome events.
3. Standardized diagnosis of outcome events.

One of the key criteria by which the quality of a longitudinal incidence study can be judged is the extent to which the investigator achieves complete ascertainment of outcome events in all exposure classes. A variety of methods is available for follow-up but to ensure uniform ascertainment across all subgroups it is desirable that the follow-up methods be independent of the method used to classify the exposure category. Methods of follow-up include correspondence with the subject and other informants, periodical re-examination of the subjects, and indirect surveillance of hospital records and death certificates. (Some countries such as the UK, the US, and some Scandinavian countries maintain central death registers that facilitate efficient and complete mortality follow-up.) The duration of follow-up will be governed primarily by the natural history of the disease and the length of the incubation period between exposure and the onset of illness. It is important that the criteria for diagnosis of end-points be standardized early in the follow-up period. Although criteria for the end-points may change in the clinical community during the study, it is important that some criteria remain stable over time so that the incidence of cases occurring early in the period of follow-up can be compared with similar cases occurring later on in the observational period. Attention should be paid to criteria to verify the absence as well as the presence of the study end-points (that is, to minimize both false-positive and false-negative diagnoses).

Unequal loss of follow-up across different exposure categories presents serious problems in the analysis and every method possible should be used to assure uniform surveillance of each group. Because of the possibility of ascertainment bias resulting from knowledge of the exposure class it is often desirable to have objective end-point criteria which can be measured by 'blinded' observers. Information used in these criteria should be sought with equal diligence in all exposure classes. This is particularly important when the exposure class is defined by a variable that may lead to different degrees of medical observation, especially medical examinations that are not under the direct control of the study investigators. For example, if in a study of cardiovascular diseases there is a tendency for participants with high cholesterol levels to receive more frequent electrocardiograms or other examinations by cardiologists, there may be a tendency to diagnose more cardiovascular events, especially milder events, in this group than in the group with low cholesterol levels. Repeat examination of the subjects, besides providing standardized information on the illnesses under investigation, can often yield additional information about co-variables which may be of importance and also allows studies of longitudinal changes in the exposure status.

Example 7—Artificial menopause and breast cancer

All patients in the study were followed from their index admission to 1 December 1961 so that the potential period of observation ranged from 21 to 42 years. The follow-up information was obtained from three sources of which the first was the hospital records. All information relating to a given patient from any and all admissions to either hospital in the study was located and the data for each patient were then combined into a single record. The second source was the death certificates registered at the Massachusetts Division of Vital Statistics from 1 January 1920 to 31 December 1961. Alphabetical listings of the study patients' names were compared with those in the index of vital records. Whenever a possible match was obtained, the death certificate was located and the information on the certificate and the identifying data obtained from the hospital chart were compared according to a prescribed set of criteria designed to minimize false matches. Therefore, there may have been increased risk of discarding acceptable matches due to some discrepancies in the available identifying information. All conditions mentioned on the death certificates were coded according to a uniform system. In addition, the underlying cause of death was coded according to the revision of the International Classification of Diseases in use at the State Division of Vital Statistics at the time of the patient's death. Thus, direct comparison could be made with published mortality statistics. The third source of follow-up information was the Massachusetts Tumour Registry, a unit of the Bureau of Chronic Disease Control of the Massachusetts Department of Public Health. Since 1927 this registry had recorded all patients diagnosed with, or treated for, malignancies at State or State-aided cancer clinics. Possible matches were obtained according to rules similar to the criteria for death certificate matching. With regard to mortality follow-up, the assumption was made that all patients dying during the study period should be registered at the Division of Vital Statistics. If no death certificate was located then one of three situations may have occurred: (i) the patient was still alive; (ii) before death she had emigrated from the State and was not a resident of Massachusetts at the time of death; or (iii) she had died, but no record could be located because of reporting or matching errors (misspellings, changes of name, failure to file a death certificate, etc.). From the three sources of information the status of 19 per cent of the women was known as of January 1962. It was noted that those receiving pelvic radiation had slightly more complete follow-up to death than those surgically treated—20 per cent versus 18.9 per cent. This difference was statistically significant but there was no significant difference in completeness of follow-up among the surgically treated groups. The relative success of the follow-up procedure was estimated by comparing the percentages of those in the cohorts known to have died before 1962 with those expected on the basis of published mortality rates and estimated migration rates. It was estimated that the deaths observed comprise 72.8 per cent of the expected number after allowance for migration.

With the advent of automated data-files in hospitals and the creation of national automated databases, including central death registries, follow-up of a cohort such as this should become easier and more complete. Although, as in this study, only a small proportion of the original cohort may be known to have died, if one is confident that those known are nearly all of those who did die and that there is not bias for better ascertainment of deaths in one group as compared to the other, the results should be valid for mortality endpoints. The British Doctors Study (Example 8) is an illustration of the use of multiple questionnaires, linkage to other files (physician registries), and other forms of contact to assure that complete follow-up has been attained. It should be noted that the Artificial Menopause Study, using historical records, took less than three years to complete while the next two examples of prospective studies took several decades to achieve similar follow-up.

Example 8—The British Doctors Study

'During the study, further questionnaires were sent out on three separate occasions to men and on two occasions to women. The purpose was partly to obtain detailed information on smoking habits, in particular giving up smoking, and also to ask additional questions, the relevance of which had emerged during the period of follow-up. Degree of inhalation was asked in these questionnaires, and the use of filter-tipped or plain cigarettes asked in the last questionnaire'.

'Information about the death of doctors was obtained at first directly from the Registrars-General of the United Kingdom, who provided particulars of every death as referring to a medical practitioner. Later, lists of deaths were obtained from the General Medical Council, and these were complemented by reference to the records of the British Medical Association and other sources at home and abroad. Some deaths came to light in response to the questionnaires. Others were discovered in the course of following up doctors who had not replied to or who had not been sent subsequent questionnaires. Of the 34 440 men studied, 10 072 were known to have died before 1 November 1971, 24 265 were known to have been alive at that date, and 103 (0.3 per cent) were not yet traced.'

'Many of the 103 untraced doctors were not British, and 67 (65 per cent) were known to have gone abroad. It was felt unlikely that more than about a dozen deaths relevant to the study could have been missed.'

'Information on the underlying cause of death in the 10 072 doctors known to have died before 1 November 1971 was obtained for the vast majority from the official death certificates. Except for deaths for which lung cancer was mentioned, the certified cause was accepted and (unless otherwise stated) the deaths classified according to the underlying cause. (In only four cases was no evidence of the cause obtainable.) The underlying causes were classified according to the seventh revision of the International Classification of Diseases . . . except that a separate category of "pulmonary heart disease" was created.'

'Cancer of the lung, including trachea or pleura, was given as the underlying cause of 467 deaths and as a contributory cause in a further 20. For each of the 487 deaths, confirmation of the diagnosis was sought from the doctor who had certified the death and, when necessary, from the consultant to whom the patient had been referred. Information about the nature of the evidence was thus obtained in all but two cases. Doubtful reports were interpreted by an outside consultant, with no knowledge of the patient's smoking history. As a result, carcinoma of the lung was accepted as the underlying cause of 441 deaths and as a contributory cause of 17' (quoted from Breslow and Day 1987, Appendix IA).

Example 9—The Framingham Heart Study

The key method of follow-up in the Framingham Heart Study was through repeated medical examinations on a two-year cycle. The greatest loss due to drop-outs occurred between the first and second examinations and those who came in most reluctantly for the initial examination (that is, toward the end of the recruitment period) seemed to have the highest drop-out rate during the next 30 years. During the first 14 years of follow-up, more than 85 per cent of the participants who were still alive at any examination cycle came in for their examinations. During the subsequent 12 years the examination rates fell to about 80 per cent of the surviving cohort. The chief reasons for non-examination were believed to be the increasing numbers of people who were physically incapacitated or had migrated from the Framingham area.

Indirect follow-up through secondary sources of information was also pursued. The Framingham Union Hospital, the major source of hospital care for the Framingham community, identified each of the Framingham Study participants and notified the study staff of admissions of participants to the hospital. This is particularly important for allowing standardized examination of stroke cases while symptoms of the disease are still present. Mortality follow-up was maintained through regular perusal of vital records at the Town Registrar and following up of obituary notices in newspapers. Mortality follow-up after 30 years was virtually complete with the vital status of less than 2 per cent of the cohort being unknown.

Fig. 10.1. Division of the study period time j time intervals.

The criteria for diagnosis of cardiovascular and other endpoints investigated in the Framingham Study have been precisely defined and the utility of the various sources of information in providing diagnostic information according to the study criteria has been investigated. Throughout the follow-up period the core criteria for the major cardiovascular endpoints have remained fixed and all potential cases are reviewed by a panel of trained medical reviewers.

It should be noted that the rate of disease occurrence in this cohort might have been altered by the subjects' continued participation in the biennial series of examinations. Although no direct advice or treatment was offered the participants, they were informed through their physicians of abnormal findings such as high blood pressure. If effective preventive measures were instituted in such subjects, then rates of overt cardiovascular diseases would be lowered and would interfere with estimating the 'true' effects of the risk factors. It was felt that during the early period of the study such treatment was not widely offered in this population.

Analysis

If a cohort study has been appropriately designed according to the principles given above, the analysis of the results is relatively straightforward. The first step is to estimate the incidence of the disease of interest for the cohort as a whole and, if the study was designed to make internal comparisons, in the subgroups of 'exposed' and 'non-exposed'. If the follow-up period is relatively short and there is little or no loss to follow-up due to deaths from other conditions, a simple estimate of risk is easily calculated as the number of new (incident) cases diagnosed during the study period divided by the total population at risk at the beginning of the period. Persons who already have the disease at the outset of the study (prevalent cases) are eliminated from the population at risk. For studies of longer duration, however, the risk of disease may change over the course of the study and there may be appreciable losses from the population at risk due to deaths from other causes, due to losses from follow-up, or due to the occurrence of the illness of interest itself. Then it is advantageous to divide the study period into a number of intervals (Fig. 10.1) and to estimate the incidence rate of disease as outlined in Tables 10.1 and 10.2.

Disease risk refers to the probability of developing the disease during the study period (or some sub-interval) on the assumption that the subject does not succumb to another cause of death before the study has ended. As a probability it is a unitless quantity that must range in value between zero and one. The incidence rate, on the other hand, is a measure

Table 10.1. Notation for cohort analysis

$l_j = t_j - t_{j-1}$	Length of jth interval, $j = 1, \ldots, J$
N_j	Number of subjects being followed at time t_j
D_j	Number of new disease cases diagnosed in the jth interval
T_j	Total observation time for all subjects during the jth interval
$D_+ = \Sigma D_j$	Total number of cases
$T_+ = \Sigma T_j$	Total observation time

Table 10.2. Measures of incidence and risk

Equation 10.1	I	$= D_+/T_+$	Average incidence rate over the entire study period
Equation 10.2	I_j	$= D_j/T_j$	Average incidence rate over the jth interval
Equation 10.3	CI_j	$= \sum\limits_{i=1}^{j} I_i l_i$	Cumulative incidence rate to time t_j
Equation 10.4	CI	$= CI_J$	Cumulative incidence rate over the entire study period to time $= t_J$
Equation 10.5	CR	$= 1 - \exp(-CI)$	Cumulative disease risk over the entire study period (adjusted for intercurrent mortality and loss to follow-up)
Equation 10.6	$CI(t)$	$= \sum\limits_{t_j \le t} D_j/N_j$	Cumulative incidence to time t (non-parametric estimate)*
Equation 10.7	$CR(t)$	$= 1 - \prod\limits_{t_j \le t} (1 - D_j/N_j)$	Cumulative risk to time t (non-parametric estimate)*

* Assumes the intervals are so fine that diagnoses are made *only* at times t_j.

of the frequency of the occurrence of disease per unit of time relative to the size of the population at risk. Crude incidence is the ratio of the disease risk during a time interval to the length of the interval. Instantaneous incidence, also known as the hazard rate of force of morbidity, measures the rate of diagnosis of new cases per unit of time relative to the size of the disease-free population at risk at time t. The units for incidence rates are 1/time and they have no upper limit quantitatively. Due to limitations on the available data, it is not possible to estimate precisely the incidence rate at each time t. Instead, estimates are made of the average rates over the study period or over each sub-interval by dividing the number of new cases diagnosed in the interval by the total person-years of observation time accumulated during the interval (Table 10.2, equations 10.1 and 10.2). Accurate estimation of the person-years denominators requires, for each individual in the study, knowledge of the exact duration of follow-up from the start of the study until diagnosis of the disease of interest, death from a competing cause, or loss from further observation. The contributions from each individual at risk during the jth interval are summed to yield the totals T_j shown in Table 10.1. If such data are lacking, various methods are available for approximating the person-years of observation. For example, the estimated size of the population at the mid-point of the interval maybe multiplied by the interval length.

Another useful measure of disease occurrence, known as the *cumulative incidence rate*, is obtained by summing up the products of incidence rate times interval length over a series of intervals (equations 10.3 and 10.4). The cumulative incidence rate over a specified interval, a unitless quantity with

no upper limit, is related via the exponential function to the disease risk over that interval (equation 10.5). If the disease is rare or the study period is short, so that the risk is no more than 5–10 per cent cumulative incidence and risk are nearly equal. Both may be estimated *non-parametrically* as a function of time t by choosing the intervals to be so fine that the interval end-points occur exactly at the times of disease diagnosis (equations 10.6 and 10.7). Plots of cumulative incidence over time provide a powerful graphical tool for examining the evolution of disease risk in the exposed and non-exposed subgroups. As an example, Figure 10.2 shows that breast cancer incidence in a cohort of women treated with radiation for post-partum mastitis paralleled the incidence in the control population until some 16–20 years after treatment, but then increased to substantially higher levels.

Although the preceding definitions used time on study as the basic time-scale for estimation of instantaneous incidence, other choices may be more appropriate in some circumstances. The possibilities include age, calendar time, and, for studies where exposure starts before entry into the study, time since initial exposure. With these other time-scales the population at risk changes not only due to the *loss* from observation of subjects who die or develop the disease of interest, but also due to the *entry* into the cohort of other subjects, depending for example on their age or the calendar year at the time they join the study. All the definitions and formulae continue to apply with these alternative time-scales. More advanced statistical analyses often consider several time-scales simultaneously, using a multi-dimensional classification of incident cases and person-years denominators

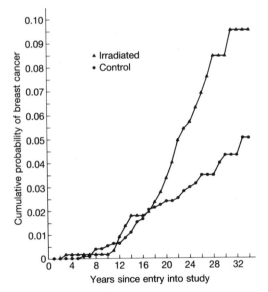

Fig. 10.2. Cumulative breast cancer morbidity curves for women treated with X-rays for post-partum mastitis (▲) and a control group (●), adjusted to the age distribution of the control group. (Source: Shore, R.E., Hempelmann, L.H., Kowaluk, E., *et al.* (1977). Breast neoplasms in women treated with X-rays for acute postpartum mastitis. *Journal of the National Cancer Institute* **59**, 813).

Table 10.3. Measures of association†

Equation 10.8	SMR	$= D_+/\Sigma T_j\,I_j^*$	Standardized morbidity ratio
Equation 10.9	RR_j	$= I_j^E/I_j^0$	Rate ratio in the jth interval
Equation 10.10	RD_j	$= I_j^E - I_j^0$	Rate difference in the jth interval
Equation 10.11	RR_{MH}	$= \dfrac{\Sigma D_j^E T_j^0/(T_j^E + T_j^0)}{\Sigma D_j^0 T_j^E/(T_j^E + T_j^0)}$	Mantel–Haenszel summary rate ratio
Equation 10.12	RD	$= CR^E - CR^0$	Cumulative risk difference (attributable risk)
Equation 10.13	RR^E	$= CR^E/CR^0$	Relative risk (crude)
Equation 10.14	AF	$= (CR - CR^0)/CR$	Aetiological fraction
		$= \dfrac{P^E\,(RP^E - 1)}{P^E\,(RR^E - 1) + 1}$	

† The superscripts refer to the exposed subcohort (E), the non-exposed subcohort (0), and the external standard population (*), respectively. Non-superscripted quantities refer to the entire cohort. P^E denotes the proportion of the population that is exposed.

according to age, calendar year, time on study, and other fixed and time-varying factors.

Cohort studies also facilitate the estimation of various measures of association between the exposure of interest and the occurrence of disease. The *standardized morbidity ratio* (SMR) is frequently used in occupational cohort studies to estimate the ratio of cohort rates to standard rates obtained from national health statistics registers or other standard sources. As shown in equation 10.8 (Table 10.3), the SMR is simply the ratio of the number of cases of disease *observed* to the number of cases *expected* from the standard rates as applied to the age/time-specific person-years of observation. Dose–response trends may be evident from SMRs that are estimated separately for subcohorts defined by levels of cumulative exposure (Table 10.4). However, doubts about the comparability of the cohort and standard population, coupled with the fact that the ratio of SMRs for two or more subcohorts may not adequately summarize the ratios of age/time-specific rates, have led many investigators to discard the SMR in favour of measures of association that do not depend on external rates. The Mantel–Haenszel (MH) rate ratio (equation 10.11) summarizes the ratios of the age/time-specific rates for the exposed versus the non exposed members of the cohort. It is closely related to the Mantel–Haenszel relative risk measure that is widely used to summarize tables of exposure/disease odds ratios in case–control studies (Chapter 9, this volume). This is the preferred measure of association when, as is often the case, the rate ratios are relatively constant over time but the rate differences are not. The cumulative risk difference (equation 10.12), known also as the attributable risk or excess risk, provides an absolute measures of the effect of exposure that is useful for public health workers.

Data from cohort studies can also be used to measure the potential impact of the removal of a suspected aetiological factor. This is measured either in terms of the estimated effect of removal on disease incidence or on the cumulative risk over the study period. The most direct measure of potential impact is known as the aetiological fraction, defined here using risk (equation 10.3) rather than incidence. It represents the proportion of all new cases of disease that can be con-

sidered due to the exposure and that are therefore potentially preventable if the exposure were to be completely removed. Equation 10.14 shows how it may be represented using the two parameters P^E, the proportion of the total population with the exposure, and RR^E, the risk ratio. Although useful for studies of short duration involving a single risk factor, serious conceptual difficulties arise when one attempts to extend the definition of use with multiple, interacting risk factors or in situations where a long study period is needed in order to ascertain the temporal aspects of the exposure/disease association.

Example 10—Artificial menopause and breast cancer

A portion of the results of this study is shown in Table 10.5. For the four 'exposure' categories of women who were less than 40 years old at time of admission into the cohort, 37 cases of breast cancer were discovered to have occurred during the follow-up period. Several difficulties in applying the usual estimates of incidence and risk are readily apparent:

1. Because of migration and incompleteness of follow-up, the observed cases are known to be an undercount.

2. The time of onset for each malignancy was not usually known (those ascertained from death certificates did not usually state age at onset).

3. The duration of follow-up for most of the women was not precisely known.

However, by making several assumptions it is possible to obtain some reasonable estimates of the association of breast cancer occurrence to the extent of pelvic surgery. The basic assumption is that whatever inadequacies there were in the follow-up procedures, they occurred uniformly in each of the exposure groups, for example the women with natural menopause were no more likely to have migrated than those with surgical menopause, and any cases of breast cancer were equally likely to be ascertained in each group. Another problem is that during the 21 years of potential admission to the study, the frequency with which the various procedures were performed varied considerably, for example pelvic irradiation was more frequent in the 1920s than later on. Thus, the women in the radiation group tended to have longer potential periods of follow-up than those in the surgical groups. This was handled by examining the specific dates of entry into the study for each women.

The fourth column of Table 10.5 gives an estimate of the crude risk of developing breast cancer. For the reasons given above, this is a very poor estimate. The next column gives an estimate of the cumulative incidence rate of breast cancer over the average 30-year follow-up period that was obtained by estimating the person-years contribution of each women. Because of the inadequacies in follow-up mentioned above, the estimates shown are undoubtedly lower than the true rates. However, provided that the underascertainment was approximately equal in the different exposure groups, there should be less bias in the estimates of relative risk shown in the last column. The cholecystectomy group was considered

Table 10.4. Lung cancer mortality by cumulative radiation exposure among Canadian fluorspar miners*

Cumulative WLM†	No. of person-years at risk	No. of lung cancer deaths		SMR (O/E ratio)‡
		Observed	Expected	
0	13 657.8	7	7.00	1.00
1–9	3 045.5	3	2.02	1.49
10–239	9 510.5	13	7.22	1.80
240–599	5 105.5	10	3.87	2.58
600–1979	7 107.0	6	1.71	3.51
1980–2039	2 415.5	25	1.54	16.23
≥ 2040	2 889.0	40	1.07	37.38

* Committee on the Biological Effects of Ionizing Radiations (1988). In *Health risks of radon and other internally deposited alpha-emitters*, National Academy Press, p. 471.
† WLM, working level months of random daughter exposure.
‡ SMR, standardized mortality ratio; O, observed; E, expected.

Table 10.5. Breast cancer in patients with and without artificial menopause

Exposure group	Number in group (No)	Number of cases (D_+)	Crude rate (D_+/No)	Estimated cumulative risk	RR (relative risk)
Cholecystectomy	400	6	0.0150	0.0198	1.00
Unilateral oophorectomy	1635	20	0.0122	0.0210	1.06
Hysterectomy and bilateral oophorectomy	1278	6	0.0047	0.0054	0.27
Radiation	468	5	0.0107	0.0106	0.54

to be the 'unexposed' or control group. Using equation 10.13, the relative risk for the women with unilateral oophorectomy is 1.06, which is not statistically different from the standard group. For the women with hysterectomy and bilateral oophorectomy, the relative risk is 0.27, which is significantly less than the standard group. The women receiving irradiation also had a low relative risk for developing breast cancer, 0.54, but because this group is small the risk is not statistically significant.

Because of the problems in estimation of the cumulative incidence rates in this study, no attempt had been made to obtain estimates of the aetiological fraction.

Example 11—Coronary heart disease and smoking among British doctors

Table 10.6 shows the numbers of deaths from coronary heart disease (CHD) and corresponding person-years denominators for smokers and non-smokers observed during the first ten years of follow-up of the British Doctors Study. The CHD rates increase markedly with age, but less so for smokers than for non-smokers. Since the rate ratios for smokers versus non-smokers decline sharply with age whereas the rate differences generally increase, this is an example where neither the Mantel–Haenszel rate ratio nor the cumulative rate difference is very useful in summarizing the age-specific quantities. Either the age-specific rates themselves, or else the results of fitting a statistical model that allows for variations in the rate ratios or differences with age, are needed to describe the results of the study adequately. Nevertheless, using the age-specific rates, one readily calcu-

lates that the cumulative mortality rate in the 35–64 years age group is 17.0 per cent for non-smokers and 24.9 per cent for smokers. The corresponding cumulative risks are $CR^E = 1 - \exp(-0.170) = 15.6$ per cent and $CR^0 = 1 - \exp(-0.249) = 22.0$ per cent for a risk ratio of $RR^E = 1.41$ and an attributable risk of $RD = 6.4$ per cent. Assuming that $P^E = 83$ per cent of British doctors were smokers at the beginning of the study period, the aetiological fraction is $AF = (0.83 \times 0.41) / (0.83 \times 0.41 + 1) = 25$ per cent. However, this number should be interpreted cautiously for the reasons mentioned earlier. The aetiological fraction is much smaller when the CHD deaths occurring at ages 75–84 years are also taken into account. The Mantel–Haenszel rate ratio for the entire 50-year age span is $RR_{MH} = 1.42$; the attributable risk is $RD = 3.9$ per cent; and the aetiological fraction is $AF = 9$ per cent.

Example 12—The Framingham Study

During the first three decades of its existence, the Framingham Study has generated more than 300 publications. Many have involved quite sophisticated methodological applications which are beyond the scope of this presentation. An example of the relation between the occurrence of CHD and serum cholesterol based on six years of follow-up is shown in Table 10.7. Data are shown for men who were aged between 40 and 59 years and were free from CHD at entry. There were 1333 men with measured cholesterol levels and follow-up was complete for six years for nearly all of them. These men were classified into tertiles on the basis of their initial serum cholesterol levels as shown in the first column of the table. Person-years of observation were estimated for each tertile based on the assumption that each of the men who developed CHD was followed on average for half of the study period, whereas the other men were followed for the entire six years. (It would be better to count those who developed CHD plus those who died from other causes as contributing three years each.) Thus, for example, the average annual incidence for the entire cohort was estimated from equation 10.1 as $I = 96/[1333 - 48] \times 6] = 0.0125$. The cumulative risks determined from equation 10.5 are virtually identical in this instance to the crude risks (number of cases divided by number of persons at risk at the start of the period). The relative risks associated with high cholesterol levels are shown in the next column, where the men with cholesterol levels lower than 210 mg per 100 ml are taken as the standard or unexposed group. Men with cholesterol levels between 210 and 244 mg per 100 ml have 1.81 times the risk of developing CHD compared with men with lower cholesterol levels, and men with cholesterol levels above 244 mg per 100 ml have risks 3.43 times greater. The attributable risks (RD) associated with higher cholesterol levels are shown in the last column.

If men could be prevented from having cholesterol levels above 245 mg per 100 ml, the potential impact upon the incidence of CHD can be estimated from the aetiological fraction (equation 10.14). The combined group of men with cholesterol levels below 245 mg per 100 ml is considered to be the unexposed group with a (crude) risk of $CR^0 = (16 + 29)/(454 + 455) = 0.0495$. Then $AF = (0.0720 - 0.0495)/0.0720 = 0.31$. That is, the risk of CHD among men potentially could be lowered by 31 per cent if none of them had cholesterol levels over 245 mg per 100 ml. A similar calculation showed that if all the men had cholesterol levels below 210 mg per 100 ml, the aetiological fraction would be 51 per cent, that is, half of the cases of CHD potentially could be prevented. This illustrates the strong dependence of the aetiological fraction on the rather arbitrary specification of the baseline level for a continuous valued risk factor. Furthermore, since whatever intervention was undertaken to reduce the serum cholesterol levels might have unpredictable effects on the CHD rates, it is clear that 'potential impact' as used here must be interpreted in terms of statistical association rather than causation.

Types of bias and their resolution

In this section the different types of bias that may occur in cohort studies are presented and discussed. Because all the different types of bias are not necessarily present in the same study we draw examples from additional studies as well as the two presented earlier.

Factors related to the selection of population, response rate, collection of information, methodologies used, and analytical strategies employed often introduce bias which, if not anticipated, can lead to incorrect conclusion concerning a possible relationship between an exposure (independent variable) and a disease (outcome variable). Such biases are inherent in all types of epidemiological studies and are discussed in Chapter 9 for case–control studies, in Chapter 8 for prevalence studies, and in Chapter 12 for experimental studies. In this section we shall confine our discussion to the type of biases that effect cohort studies.

There are five broad categories of bias that are operative in cohort studies. These are selection bias, follow-up bias, information bias, confounding bias, and *post hoc* bias. Each of these is discussed separately below. These biases can cause systematic errors which affect the internal validity of a study. This is in contrast to random errors which may not affect the internal validity of the study, but will reduce the probability of observing a true relationship. A true bias, that is, a systematic error that is introduced into one group or subgroup to a greater extent than in other subgroups, often leads to the observation of a relationship that is not a true relationship, or vice versa, leads to the conclusion that there is no relationship when, in fact, there is a true relationship between the independent and outcome variables. Random errors occur with equal frequency in all subgroups and, thus, do not usually affect the validity of a relationship. On the other hand, because a certain proportion of all the subgroups will contain an error, the probability of observing a true relationship is diminished.

While internal validity is paramount, it is often important to have external validity in cohort studies as well. External validity refers to the degree to which an association observed in the study populations holds true in the general population as well. In order to ensure external validity the population

Table 10.6. Deaths from coronary heart disease (CHD) among British male doctors*

Age group (years) j	No. of person-years (1000s)		No. of CHD deaths		CHD rates†		Rate ratio RR_j	Rate difference† RD_j
	Non-smokers T_j^0	Smokers T_j^E	Non-smokers D_j^0	Smokers D_j^E	Non-smokers I_j^0	Smokers I_j^E		
35–44	18.790	52.407	2	32	0.11	0.61	5.73	0.50
45–54	10.673	43.248	12	104	1.12	2.40	2.14	1.28
55–64	5.710	28.612	28	206	4.90	7.20	1.47	2.30
65–74	2.585	12.663	28	186	10.83	14.69	1.36	3.86
75–84	1.462	5.317	31	102	21.20	19.18	0.90	− 2.02
Total	39.220	142.247	101	630	2.58‡	4.43‡	1.72	1.85

* From Doll and Hill (1966) as quoted by Breslow and Day (1987).
† Per 1000 person-years.
‡ Average rates I^0 and I^E over entire age range.

studied must be representative of the general population to which the results of the study will be generalized. In many cohort studies, it is necessary for various reasons to study some subpopulation of the general population. This subpopulation may represent a non-random sample of the general population, such as an occupational group, a group selected from a particular health plan, etc. If such subpopulations are used, external validity may be reduced.

Selection bias

Selection bias may occur when the group actually studied does not reflect the same distribution of factors such as age, smoking, race, etc. as occurs in the general population. This may be because some of the originally selected members of the cohort refuse to participate or in a non-current cohort study because records on some individuals are missing or incomplete. The response rates among the various subgroups invited to participate in the study may, therefore, differ. In some studies particular subgroups may be used for convenience or for other reasons are not representative of the general population.

Example 13—Effects of volunteering

An example of selective non-response to recruitment was observed and documented in the Framingham studies cited earlier in this chapter (Example 2). It was found that individuals who agreed to participate in these cohort studies were healthier than individuals who did not agree to participate. While this would not affect the internal validity of the study, since the groups to be followed were characterized on the basis of factors present at baseline, it would be likely to reduce the incidence, particularly in the first few years of the study, of the disease of interest. Thus, the external validity would be diminished, but the internal validity should not be affected for those independent variables defined at baseline. However, because the incidence of disease might be lower in this healthier group that is being followed up, the probability of finding significant relationships would be somewhat diminished. In occupational cohort studies, this type of selection bias has been termed 'the healthy worker effect'.

A second problem with selection non-response has to do with the extent to which the population that agrees to participate in the study actually represents the true spectrum of the independent variable.

Example 14—Representativeness of study group

Early studies of the relationship of dietary cholesterol and saturated fats intake to CHD in the US gave inconclusive results. This may have been due, in part, to the fact that very few Americans have dietary cholesterol and saturated fat intakes in the lower ranges, whereas residents of less affluent countries have a higher proportion of individuals with these low levels of intake. If there exists a threshold level of the independent variable necessary to produce disease and the respondents include only individuals with levels of the independent variable that are above that threshold level, then no

relationship will be seen between the variable and the disease under study. Thus, some of the comparisons of the incidence of CHD among Americans with higher levels of cholesterol and fat in the diet may not have shown a relationship because the threshold level of dietary fats was below the levels consumed in the study population. Even in situations where there is no threshold level, inclusion of individuals at only one end of the spectrum of the independent variable will reduce the likelihood that a dose–response relationship will be observed. Thus, non-response of non-inclusion of participants in the cohort who represent one or the other extreme of the independent variable may affect the internal validity of the study and lead to a false observation that there is no relationship.

Another problem of selection bias occurs when individuals who have incipient disease are included in the cohort. Individuals with the disease should be excluded from the study population at the time of recruitment. However, with many chronic diseases which have a long induction period such as cancer and heart disease, it is difficult to identify individuals with incipient disease. Their inclusion in the study population may lead to an observation of associations that are, in fact, a result of the disease process rather than a risk factor for the disease.

Example 15—Presence of incipient disease

An association between low cholesterol and risk of cancer has been observed in several cohort studies. The induction period for cancer is probably one or more decades. Individuals who develop cancer in a follow-up period of less than the induction period probably had incipient disease at the time of the formation of the cohort. Thus, a low cholesterol level in these individuals may have been a result of the cancer process rather than a risk factor for it.

A final example of selection bias occurs when the distribution of co-variables which may be related to disease incidence is not equally represented in the study cohorts.

Example 16—Distribution of co-variables

Smoking is related to a number of diseases. In some, it is the probable major cause whereas in others, such as CHD, it represents only one of several risk factors that increase the probability of developing disease. Thus, if the non-respondents include a higher proportion of smokers than non-smokers, the total incidence of CHD in the study cohort would be lower than if smokers were appropriately represented in the cohort. The effect, however, would not only be to make the observed incidence of CHD lower in the study population than in the general population but would also have the effect of leading to a false estimate of the proportion of CHD that is associated with smoking (the aetiological fraction). Specifically, the incidence of CHD among the smokers would be correct, but the proportion of the total numbers of cases that were associated with smoking would be smaller than actually exists because the proportion of smokers in the study population would be lower than in the general population. Thus,

Table 10.7. Six-year incidence of coronary heart disease according to initial serum cholesterol in men aged 40–59 years

Serum cholesterol (mg/100 ml)	Number in group	Number of cases	I (average annual incidence)	CR (cumulative risk)	RR^E (relative risk)	RD (attributable risk)
< 210	455	16	0.0060	0.0352	1.00	0.000
210–244	455	29	0.0107	0.0637	1.81	0.0285
≥ 245	423	51	0.0214	0.1207	3.43	0.0855
Total	1333	96	0.0125	0.0720		

any estimates of the aetiological fraction of CHD due to smoking would be underestimated.

Follow-up bias

One of the major problems in cohort studies is to accomplish the successful follow-up of all members of the cohort. If the loss to follow-up occurs equally in the exposed and unexposed groups the internal validity should not be affected assuming, of course, that the rate of disease occurrence is the same among those lost to follow-up as among those not lost to follow-up within each exposure group. If, however, the rate of disease is different among those lost to follow-up, then the internal validity of the study may be affected; that is, the relationship between exposure and outcome may be changed.

Example 17—Bias resulting from differential incidence in those lost to follow-up

If the rate of lung cancer is higher in those smokers who are lost to follow-up than in those who remain in the study, the observed incidence of lung cancer in those smokers who remain in the study will be lower than the actual incidence of lung cancer in the entire cohort of smokers. The effect will be to observe a lower association between lung cancer and smoking than actually exists (provided that the incidence of lung cancer is the same in non-smokers who are or who are not followed up). If the lung cancer incidence rate is lower in smokers who are not followed than in those who are, the reverse effect would occur, that is, the observed association would be greater than the true association.

Usually, the incidence of disease is not known among those lost to follow-up, making it difficult to look for this type of bias. If possible, the occurrence and cause of death should be sought in those who are lost to follow-up. This will be easier in the US now that there is a National Death Registry. If the death-rate is similar between those lost and not lost to follow-up within each group, the occurrence of a different incidence of disease in the two groups is less likely.

Another strategy is to compare the known characteristics at baseline of those lost and not lost to follow-up. The more similar they are, the less likely it is that a different incidence of disease occurred in them.

Neither of these strategies guarantees that the incidence was the same in both those followed and those not followed. Therefore, the best strategy is to reduce the number lost to follow-up to the lowest level possible.

Another possible source of bias may be observed in studies in which the independent variable is being documented concurrently with the development of the outcome variable, presenting the opportunity for misclassification resulting from loss to follow-up.

Example 18—Bias resulting from loss to follow-up of individuals under observation for the independent variable

In evaluating the relationship between a decline in lung function test results and concurrent levels of exposure to photochemical oxidants at place of residence, a problem arises in considering how to evaluate individuals who have moved from the study area to other areas. In some instances, they will have moved to areas with lower levels of exposure to photochemical oxidants and in other instances to areas with higher levels. It is not feasible to maintain constant monitoring for levels of photochemical oxidants in all the areas to which these individuals have moved. If there is indeed a relationship between levels of exposure to photochemical oxidants and decreasing lung test performance, the inclusion in analyses of individuals who have moved to a cleaner area, as if they had remained in the area of high exposures, will lead to misclassification bias and, thus, to an underestimate of the relationship, whereas inclusion of individuals who moved to a dirtier area will lead to misclassification bias with the reverse effect.

On the other hand, if the individuals who have moved are excluded from the analysis to avoid misclassification bias, another potential bias is introduced. Individuals who have moved out of the study area may have done so because of the high level of exposure to photochemical oxidants and their awareness of declining respiratory performance. This would result in an observed relationship in those not moving which is lower than the true relationship.

While this type of bias is almost impossible to prevent, there are several pieces of information that can assist the investigator in evaluating the magnitude of the bias that may be introduced. First, the investigator may compare lung function test results among those who remained and those who moved away at baseline. Any differences between those re-tested and those not re-tested would provide information about the direction of the bias and possibly about the magnitude of the bias that occurred.

Second, it is often possible to send a mail questionnaire to individuals who have moved away from the study area which should include questions regarding reasons for moving. If it is found, for example, that many of the respondents moved because of the development of respiratory symptoms, the probability of potential bias can be recognized. In addition, the ascertainment of *diagnosed* respiratory impairment among those not re-tested would also indicate the presence of bias.

Although there is no completely satisfactory solution to this problem resulting from loss to follow-up, awareness of the potential for bias will enable the investigator to explore various methods to evaluate its effect.

Example 19—Unequal observation

Smoking is associated with a wide range of adverse health outcomes. Any one of these adverse health outcomes is more likely to result in smokers being seen by a physician, thus increasing the likelihood that the disease of interest may also be diagnosed at that time. That is to say, there would be an earlier diagnosis of disease in the smoking individual than in a comparable non-smoking individual who would be less likely to come under medical care. As a result, there would be an overestimate of the association of the disease of interest with the smoking variable, both when a straight incidence analysis is used (since cohort studies usually have a defined follow-up period) and when an analytical strategy such as person-year is used (since the individual would appear as a case after fewer years of follow-up than would normally occur if he or she were not brought to medical attention earlier as a result of smoking).

Information (misclassification) bias

Information bias occurs when there is an error in the classification of individuals with respect to the outcome variable. This may result from measurement errors, imprecise measurement, and misdiagnosis for whatever reason. Information bias is also termed misclassification bias. If the misclassification occurs equally in all the subgroups of the study population, the internal validity of the study will not be affected, but the precision or probability of being able to demonstrate a true relationship is reduced.

Example 20

If the proportion of cases under- or over-reported in a cohort study of the risk of CHD is equal among smokers and non-smokers, no change in the observed risk ratio for smoking would occur and the internal validity of the study would be unaffected. If, however, the misclassification occurs to a greater extent among either smokers or non-smokers the observed risk will be altered thereby affecting the relative risk and incidence difference and, as a result, the internal validity of the study.

Confounding bias

Confounding occurs when other factors that are associated with the outcome and exposure variables do not have the same distribution in the exposed and unexposed groups. Two common confounders in cohort studies are smoking and age. The risk of disease varies with age for almost all diseases. Likewise, smoking increases the risk of acquiring a wide range of diseases.

Example 21

In a cohort study to determine the risk of CHD among individuals who drink and do not drink, the prevalence of individuals who smoke is likely to be higher among those who drink than those who do not drink. If one does not take into account the prevalence of smoking in the two groups, there will be a higher incidence of CHD in the drinking group than in the non-drinking group which is, in fact, ascribable to smoking rather than to drinking. A false association or non-association also might be observed if the age distributions were not the same in the drinking and non-drinking groups since the incidence of CHD increases with age.

Confounding bias can result in either an overestimate or an underestimate of the relative risk of an independent variable with disease. Estimates of the effect of confounding variables in a cohort study usually require primarily the use of the investigator's judgement, although the application of specific statistical procedures can help in reducing the effects of recognized confounders.

Post hoc bias

Another source of potential bias is the use of data from a cohort study to make observations which were not part of the original study intent. Thus, interesting relationships are often observed in cohort studies which were not originally anticipated. These findings should be treated as interesting hypotheses which are an appropriate subject for additional studies. Such fortuitous findings should not be considered to have established the validity of a relationship and in no circumstance should the same data be analysed to test hypotheses arising from that data.

Resolution of bias

There are various strategies for reducing the presence of bias in cohort studies. Selection bias can be reduced by careful selection of individuals for inclusion in the study, and by making every attempt to characterize differences that may exist between respondents and non-respondents. Although consideration of characteristics that may be more frequent in non-respondents will not eliminate bias, it may permit the investigator to assess the directionality and degree of bias that may have resulted from specific selection procedures. Information bias can be reduced by using well-defined, precise measurements and classification criteria for which the sensitivity and specificity have been determined. Follow-up bias can be reduced by intensive follow-up of all study participants and by establishing criteria for follow-up that will assure that all members of the cohort have an equal opportunity for being diagnosed as having the outcome variable. Comparison of the characteristics present at baseline among those lost to follow-up and those successfully followed up may provide information upon which estimates of the nature and degree of bias that may have been introduced through loss to follow-up may be based.

Confounding bias can be reduced in the analysis stage by careful stratification and/or adjustment procedures. However, controlling for multiple confounders reduces the likelihood of observing a significant difference. Thus, careful consideration should be given to whether specific factors not equally represented in all groups are, in fact, related to the probability of observing the outcome variable. If they are not, it is usually better not to attempt to restrict the selection of participants and/or stratify or adjust during analysis.

The identification and resolution of bias is primarily a matter of epidemiological judgement. Although statistical techniques and analytical strategies can be used to reduce bias, they can be applied only to factors which in the judgement of the investigators are a potential source of bias. We have discussed the major sources of bias in a cohort study. This list, however, is far from exhaustive and additional types of bias will surely be described in the future for which investigators should be alert. Some references are given at the end of this chapter which will assist individuals requiring a more detailed discussion of the problems of cohort studies.

Summary

Cohort studies are usually the best type of studies for demonstrating the association between an exposure and a disease because it is possible to derive relative and attributable risks and often incidence measures from them. They are, however, usually expensive to carry out and large cohorts are required for rare diseases. In addition, there are very significant problems associated with selection of appropriate groups to be studied and with complete ascertainment of disease occurrence in them. Usually it is necessary to compromise the ideal, so providing the opportunity for various types of bias to occur which can result in incorrect conclusions. The success of a cohort study often depends on the care of the investigator in recognizing and correcting for these biases.

References

Breslow, N.E. and Day, N.E. (1987). *Statistical methods in cancer research. Volume II—The design and analysis of cohort studies.* International Agency for Research on Cancer, Lyon.

Dawber, T.R., Kannel, W.B., and Lyell, L.P. (1963). An approach to longitudinal studies in a community: the Framingham Study. *Annals of the New York Academy of Sciences.* **107**, 539.

Doll, R. and Peto, R. (1976). Mortality and relation to smoking: 20 years' observation on male British doctors. *British Medical Journal* **ii**, 1525.

Feinleib, M. (1968). Breast cancer and artificial menopause: A cohort study. *Journal of the National Cancer Institute* **41**, 315.

Kannel, W.B., Dawber, T.R., Kagan, A., Revotskie, N., and Stokes, J. III (1961). Factors of risk in development of coronary heart disease—six year follow-up experience. The Framingham Study. *Annals of Internal Medicine* **55**, 33.

Kleinbaum, D.G., Kupper, L.L., and Morgenstern, H. (1982). *Epidemiologic research.* Lifetime Learning Publications, Belmont, California.

MacMahon, B. and Pugh, T.F. (1970). Epidemiology. *Principles and methods.* Little Brown, Boston, Massachusetts.

Mausner, J.S. and Bahn, A.K. (1974). *Epidemiology. An introductory text.* Saunders, Philadelphia, Pennsylvania.

11

Surveillance

RUTH L. BERKELMAN and JAMES W. BUEHLER

Surveillance is the epidemiological foundation for modern public health. It serves as 'the brain and the nervous system' for programmes to prevent and control disease (Henderson 1976). The term surveillance is derived from the French word meaning 'to watch over' and, as applied to public health, means the close monitoring of the occurrence of selected health events in the population. In contrast to archival health information, surveillance systems are dynamic. Although surveillance methods were originally developed as part of efforts to control infectious diseases, basic concepts of surveillance have been applied to all areas of public health, beginning with infectious diseases but now including occupational and environmental health, injuries, maternal and child health, and chronic non-infectious conditions. Surveillance has been expanded to include not only information on the occurrence and distribution of health events but also information on the prevalence of risk factors, both personal and environmental, and on the use of preventive health practices and medical services.

Definition

In 1963, Alexander D. Langmuir defined disease surveillance as 'the continued watchfulness over the distribution and trends of incidence through the systematic collection, consolidation, and evaluation of morbidity and mortality reports and other relevant data' together with timely and regular dissemination to those who need to know. In 1968, the 21st World Health Assembly described surveillance as the systematic collection and use of epidemiological information for the planning, implementation, and assessment of disease control; in short, surveillance implied 'information for action' (World Health Organization 1968). In 1974, Richard Doll introduced the concept of cost–benefit analysis as part of the surveillance function. In 1986, the US Centers for Disease Control (CDC) reached a consensus on the definition of surveillance that was similar to that of the World Health Organization but focused increased emphasis on the application of data to prevention and control as a necessary link in the surveillance chain (CDC 1986a). Issues surrounding surveillance continue to evolve as the scope of surveillance broadens and as increasingly sophisticated methods are applied (Berkelman 1989).

Surveillance should be distinguished from health information systems that lack direct application to both epidemiological investigations and prevention and control programmes. Thus, health information systems, such as registration of births and deaths, routine abstraction of hospital records, or general health surveys in a population, do not, by themselves, constitute surveillance. However, data collected from such health information systems can be used for surveillance when systematically analysed on a timely basis and applied for prevention and control of specific health events. In turn, a surveillance programme is less likely to be sustained if no effective control or prevention measures are available.

A critical component in the definition of surveillance is that surveillance systems include the ongoing collection and use of health data. Thus, one-time or sporadic surveys do not constitute surveillance. In addition, there should be regular analysis and dissemination of surveillance data. The frequency of analysis and dissemination—weekly, monthly, annually—depends on information needs. Certainly surveillance may have a starting and an ending point, but these are not usually pre-defined. Surveillance should begin when a public health problem exists or is likely to occur and a programme for prevention and control of a health event is initiated. It may be stopped when that particular health event is no longer considered a public health priority. Likewise, case ascertainment in an epidemiological study, or periodic screening of individuals in a population, should not be misinterpreted as surveillance.

History

Recording of deaths is the oldest form of monitoring public health, and John Graunt, in his treatise Natural and Political Observations on the Bills of Mortality (1662), is generally recognized as the first person to describe use of numerical methods for this purpose. In 1776, Johann Peter Frank advocated a more extensive monitoring of health in Germany that would support public health efforts related to school health, prevention of injuries, maternal and child health, and public water and sewage disposal.

William Farr, of the UK, is recognized as the founder of modern concepts of surveillance. As Superintendent of the Statistical Department of the General Registrar's Office from 1839–79, he collected, analysed, and interpreted vital statistics, and he published weekly, quarterly, and annual reports. He did not stop with publication of official reports but regularly contributed papers to medical journals and even used the public press to achieve effective action (Langmuir 1976). Thus, he took the responsibility of seeing that action was taken on the basis of his analyses.

In the nineteenth century, Farr's efforts at health monitoring were extended by Chadwick, who investigated the relationship between environmental conditions and disease, followed by Villerme who analysed the relationship between poverty and mortality in Paris. Lemuel Shattuck in the US also published data that related deaths, infant and maternal mortality, and infectious diseases to living conditions. He further recommended the standardization of nomenclature of cause of disease and deaths, and the collection of health data that included sex, age, locality, and other demographic factors. In 1893, the first international list of causes of death was developed (Eylenbosch and Noah 1988).

Increasingly, elements of surveillance were applied to aid in detecting epidemics and in the prevention and control of infectious diseases. In 1899, the UK began compulsory notification of selected infectious diseases. In the US, national morbidity data collection on plague, smallpox, and yellow fever was initiated in 1878, and by 1925 all states were reporting weekly to the US Public Health Service on the occurrence of selected diseases. In a public health context, the term surveillance was increasingly applied to programmes of reporting selected infectious diseases in a population, with less emphasis on its application to quarantine of individuals (Langmuir 1963).

Similar reporting activities were occurring in Europe at about the same time. In 1907, the Office International d'Hygiene Publique, predominantly composed of European member states, was created (World Health Organization 1958). The office was to disseminate information in a monthly bulletin on the occurrence of selected diseases, most notably cholera, plague, and yellow fever. In the succeeding decades, other diseases were recommended for surveillance in step with the International Sanitary Regulations. However, many of the morbidity and mortality reporting systems were not systematic, and were still largely conceived to have long-term archival functions.

Since the early 1950s, the critical importance of surveillance to public health efforts has been demonstrated often. In 1955, acute poliomyelitis among recipients of the polio vaccine in the US threatened the national polio vaccination programmes that had just begun. In collaboration with State health departments, the Centers for Disease Control (CDC) developed an intensive national surveillance system and, at one point, a daily polio report was being issued regarding poliomyelitis cases. The collected information demonstrated that the problem was limited to a single manufacturer of the vaccine and allowed the vaccination programme to continue with a resulting dramatic decline in cases of acute poliomyelitis in the US in successive years (Langmuir 1963).

Surveillance also became an integral part of the world-wide malaria eradication programme. It was used to document continuation of transmission, to focus spraying efforts, and to substantiate achievement of eradication within an area. After the withdrawal of all spraying activities, surveillance constituted the sole activity of the eradication campaign (Raska 1966).

Surveillance was also the foundation for the successful global campaign to eradicate smallpox. In 1967, when the campaign began, efforts were focused on achieving a high vaccination level in countries with endemic smallpox; however, it was soon evident that a programme based on surveillance to target vaccinations in limited areas would be more efficient. Smallpox reporting sources, usually medical facilities, were contacted on a routine basis and a reporting network was thus firmly established in most countries. In addition, other reporting sources were often established including markets, schools, police, agricultural extension workers, and others. In 1973, as the goal of eradication neared, systematic house-to-house search for cases was established in India, and subsequently used widely in Pakistan and Bangladesh (Henderson 1976). Well-designed surveillance systems of data collection, tabulation, and routine feedback were vital to the success of the programme.

In 1981, shortly after a disease later named acquired immune deficiency syndrome (AIDS) was recognized, national surveillance was begun in the US and other countries. Even before the aetiological agent, human immunodeficiency virus (HIV), was identified, surveillance data had identified modes of transmission, population groups at risk for infection, and, equally important, population groups not at risk for infection. These data have been instrumental in directing public health resources to programmes preventing further spread of HIV and averting widespread public hysteria (Jaffe *et al.* 1983).

The potential usefulness of surveillance as a public health tool beyond infectious disease was emphasized in 1968 when the 21st World Health Assembly recommended the application of surveillance principles to a wider scope of problems including cancer, atherosclerosis, and social problems, such as drug addiction (World Health Organization 1968). Many of the principles of surveillance traditionally applied to acute infectious diseases have also been applied to chronic diseases and conditions, although some differences in surveillance techniques have been observed (Table 11.1). Even though chronic disease may have long latency periods, trends in their incidence may change relatively quickly, and surveillance can play a key role in detecting these changes (Berkelman and Buehler 1990). For example, the mortality rate from such chronic non-infectious conditions as stroke and cirrhosis can be shown to fall within 2–3 years after smoking cessation or

Table 11.1. Acute and chronic disease surveillance

	Common characteristics	Acute disease surveillance	Chronic disease surveillance
Purpose	Monitor trends	Emphasis on weekly or monthly variations to detect outbreaks	Emphasis on year-to-year trends
	Describe problem and estimate health burden Direct/evaluate programmes for prevention and control		
Data collection	Regular	Reliance on notification by health care providers/laboratories	Greater use of existing databases (e.g. vital statistics, hospital discharges)
Data analysis	Descriptive statistics for time, place, person	Emphasis usually on case counts	Emphasis usually on rates
Data dissemination	Regular; frequency reflects data collection Audience targeted	More frequent	Less frequent

declines in alcohol consumption, respectively (Skog 1984; Wolf *et al.* 1988). Epidemics of chronic disease also can be detected through timely analysis of appropriate data. For example, an epidemic of endometrial cancer occurred in the US in the early 1970s with an estimated excess of 15 000 cases; in this particular instance, the epidemic could have been detected several years earlier if available data had been analysed regularly to detect changes in disease occurrence. With recognition of the epidemic and its association with oestrogen use in post-menopausal women, disease incidence declined by 46 per cent within 2 years following a drop in oestrogen sales to post-menopausal women (Jick *et al.* 1980).

In addition to an increased scope of health problems under surveillance, the methods of surveillance have expanded from general disease notification systems to include survey techniques, sentinel health provider systems, and other approaches to data collection (Thacker *et al.* 1988). The assimilation of microcomputers into the workplace has also made possible both more efficient data collection and more rapid and sophisticated analyses (Graitcer and Burton 1987; Valleron *et al.* 1986).

Purpose of surveillance

A surveillance system should be designed to meet the needs of a prevention and control programme (Table 11.2) Usually these needs will include a description of the temporal and geographical trends in the occurrence of a health event in a particular population. The data should be able to substantiate patterns of both endemic and epidemic disease.

Foremost, surveillance systems should identify changes in disease occurrence. Surveillance may be initiated to identify risk factors associated with disease and to suggest hypotheses for further investigation; cases identified through surveillance are sometimes used in case–control studies as in the early studies on toxic-shock syndrome and on the epidemic of AIDS (Jaffe *et al.* 1983; Shands *et al.* 1980). Effective preventive actions were known through such research even before the aetiological agents, a toxigenic strain of *Staphylococcus aureus* and HIV, were discovered.

Evaluation of prevention and control programmes requires monitoring of changes in the incidence of a disease or its complications, associated risk factors, and/or the use of preventive health services. The rapid decline in morbidity from many infectious diseases has been related directly to vaccination campaigns in populations based on surveillance data. Direct relation of a single intervention to a specific chronic disease outcome may be difficult when the aetiology of the disease is multi-factorial, but the impact of an intervention on disease outcome remains the ultimate test of policy.

Assessment of the burden of disease including its incidence (i.e. the number of people newly affected each year) and its current and projected prevalence (i.e. the number of people affected by the disease at any point in time) is essential to planning public health programmes. For example, surveillance for AIDS has been critical to the forecasting of the future burden of that disease in the US (Gail and Brookmeyer 1988; Institute of Medicine 1988). With the ageing of the population and the higher prevalence of chronic diseases among the elderly, projections of chronic disease prevalence are needed to determine health care needs.

The role of surveillance in guiding public health programmes is illustrated by the first major national activity initiated by CDC, the Malaria Eradication Programme. Surveys in the mid-1930s had established malaria to be an endemic problem deeply rooted in the south-eastern part of the US. After the Second World War, an extensive DDT spraying programme was launched, with institution of surveillance in 1947. Quite rapidly it was established that endemic malaria had essentially disappeared, probably even before the DDT programme got underway (Langmuir 1963). In this case, surveillance was used as the basis for dismantling a public health programme and redirecting public health resources to problems of higher priority.

Criteria for establishing a surveillance system

Surveillance systems should address important public health problems, which may be defined in various ways. The main determinant of the public health importance of a disease is its burden on a population. The frequency and severity of the disease are key considerations, including the number of people affected, the case–fatality rate, and the direct

Table 11.2. Purposes of public health surveillance

To characterize disease patterns by time, place, and person
To detect epidemics
To suggest hypotheses
To identify cases for epidemiological research
To evaluate prevention and control programmes
To project future health care needs

(medical) and indirect (e.g. lost productivity) costs. In addition, the potential burden, if public health programmes are not maintained or not initiated, must be considered. Thus, surveillance for diseases controlled through use of vaccines is generally considered a high priority even though the burden of these diseases in developed countries has been greatly reduced. The segment of the population affected may also be key. For example, injuries and diseases affecting younger persons will have a higher importance if years of potential life lost are used to rank priorities (CDC 1986*b*). Health problems affecting predominantly minority groups will be of concern if reducing disparities in health is a priority (Heckler 1985).

The likelihood that interventions can be found, or are known to prevent the occurrence of disease or to alleviate the course of existing disease, is an important consideration. A surveillance programme is unlikely to succeed if no effective control or prevention measures are defined. Generally, greater emphasis has been given to programmes that prevent the occurrence of disease than to those that cure or ameliorate the condition.

Once the health priorities for surveillance are established, surveillance for associated risk factors and prevention services must be considered.

Establishing a surveillance system

Establishing a surveillance system requires a statement of objectives, definition of the disease or condition under surveillance, and implementation of procedures for collecting, interpreting, and disseminating information. Surveillance systems can be considered information loops, or cycles, that involve health care providers, public health agencies, and the public (Fig. 11.1). The cycle begins when cases occur and is completed when information about these cases is made available and is used for prevention and control. This process may involve multiple cycles, ranging from the local response to individual cases to the development of national policies based on information aggregated from many cases. Essential to the completion of the surveillance cycle is the return of information to those who 'need to know' (Langmuir 1963), and thus attention must be directed not only to procedures for collecting data but also to procedures for ensuring that useful information is returned to constituents.

System objectives

Defining the objectives of a surveillance system depends on what information is needed, who needs it, and how it will be used. Implementing a system will require a balance of competing interests, and a clear statement of objectives will provide a framework for subsequent decisions. For example, the desire to collect detailed information about cases may compete with needs to assess the number of cases rapidly. Thus, if the primary objective is to obtain rapid case counts, then less information would be collected about each case to avoid delays in reporting.

The objectives of a surveillance system will be shaped by its target population, constituents, the nature of prevention and control programmes, and the health problem under surveillance.

Target population

A surveillance system seeks to identify health events within a specified population. This population may be defined on the basis of where persons live, work, attend school, use health care services, etc. Alternatively, the population may be defined on the basis of where health events occur. For example, a surveillance system that monitors new-born health as a measure of prenatal care services in a community would focus on deliveries to women who live within the community and not on women who live elsewhere but deliver in community hospitals. In contrast, surveillance of traffic injuries aimed at identifying roadway hazards could include all injuries that occur in a community, regardless of whether affected persons are community residents.

Constituents

Surveillance systems are likely to have many constituents with diverse perspectives and uses for surveillance data, including clinicians, health agencies, academicians, politicians, media reporters, and others. Because these diverse needs cannot all be satisfied, the primary or most important constituents should be identified.

Public health programmes

The objectives of surveillance systems will be shaped by the objectives and capabilities of the public health programmes they serve. For example, a programme to eliminate a rare infectious disease requires intensive surveillance that emphasizes identification of all persons with the disease, permitting follow-up of others who have been in contact with case patients. This strategy was used in the smallpox eradication programme and has also been employed in the measles elimination programme in the US (CDC 1983; Foege *et al.* 1975). In contrast, an educational programme to influence behaviours associated with chronic disease may depend on a surveillance system that describes the practices of a sample of persons in a community (Remington *et al.* 1988).

Health problems

It is necessary to decide exactly what disease or health problem will be under surveillance. Many diseases have a spectrum of manifestations, ranging from asymptomatic to severe, and chronic diseases may progress through various

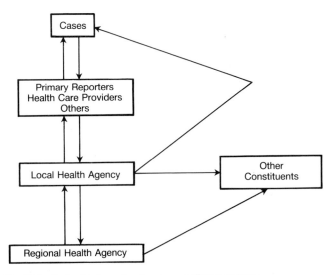

Fig. 11.1. Cycle of information flow in surveillance systems.

stages with multiple possible outcomes. One should consider which manifestation(s) or stage(s) of a disease should be under surveillance. For example, for ischaemic heart disease, manifestations include abnormal diagnostic tests in the absence of symptoms, angina pectoris, acute myocardial infarction, and sudden death. If the goal of surveillance is to assess the burden of the disease on health care systems, then a broad definition that encompasses various manifestations may be appropriate. If the purpose is to monitor trends in the disease, a more limited and severe manifestation, such as myocardial infarction, may be the appropriate target for surveillance. If resources are limited, surveillance based on analyses of death certificates may be most feasible; however, interpretation of trends may be complicated by independent trends in the occurrence of the disease, advances in treatment, or coding of vital records. Alternatively, attention may be focused on risk factors for cardiovascular disease, such as hypertension, smoking, cholesterol levels, or physical activity.

Case definition

The case definition is fundamental to any surveillance system, since it is the formal answer to the question of what manifestations of a disease or condition are under surveillance. It is both a yes/no criterion for determining who gets counted and a guide to local health departments for case investigations and follow-up. It ensures that the same measure is used over time and across geographical areas. The case definition must be sufficiently inclusive (sensitive) to identify persons who require public health attention but sufficiently exclusive (specific) to avoid unnecessary diversion of that attention. In addition, the case definition must be usable by all persons on whom the system depends for case reporting. For any particular disease or condition, there is no ideal case definition. Rather, the definition must be geared to the circumstances that govern each surveillance system. For

example, in conducting surveillance for hepatitis A, the following are two possible case definitions.

Definition 1

Illness characterized by jaundice, elevated liver enzymes, and serological detection of immunoglobulin M (IgM) antibodies against hepatitis A. This definition presumes that affected persons will have access to health care services, including diagnostic testing. It does not include persons with non-icteric disease or persons with asymptomatic hepatitis A infection. An alternative may be to limit the definition to persons with a positive test for IgM antibodies against the hepatitis A. This approach, however, would exclude persons who have an epidemiological and clinical picture consistent with hepatitis A but who lack serological testing. To accommodate these possibilities, the case definition may be subdivided into symptomatic versus asymptomatic cases or allow for gradations based on certainty of diagnosis (confirmed case, presumptive case, possible case).

Definition 2

Yellow eyes. This definition is simple and may be appropriate in a setting of limited access to diagnostic services, where field staff have little formal training and where there is an emergent need to assess a population rapidly (e.g. a refugee camp). With this definition, hepatitis case counts may include some persons with jaundice due to other conditions, but this lack of specificity may not substantially affect the usefulness of the overall information.

Neither of these two definitions or possible variations is inherently better, and each may be appropriate in a given setting.

A similar range of possible case definitions also exists for surveillance systems that focus on adverse health exposures rather than disease outcomes. For example, in a surveillance system that addresses occupational hazards, exposure to a harmful substance may be monitored by self-report of workers; company log books of manufacturing procedures; or routine measurement of substances in the work environment, on workers' clothing, or in specimens collected from workers. Each of these possible case definitions would require different levels of co-operation from the company or workers, and each may be subject to unique limitations that could bias surveillance.

The flow of information

Surveillance systems depend on the sequential flow of information through the full surveillance cycle. Each facet of this process should be carefully planned, as described below.

Reporters

Persons responsible for reporting cases may be all health care providers in a defined area, selected providers, or persons at specific institutions (clinics, hospitals, schools, factories, etc.). Some surveillance systems draw on survey techniques, particularly those that monitor health-related behaviours. Here, the question is not who is responsible for reporting but rather how the survey sample will be selected.

Forms

The desire to collect detailed information must be tempered by the need to limit data to items that can be reliably and consistently collected over the long term. Forms that are too detailed and too complicated will not be welcomed by those on whom the surveillance system must depend.

Timing

Surveillance systems provide data on a regular basis, ranging from daily to annually. Whatever periodicity is used should be specified and adhered to by participants in all phases of the surveillance loop.

Aggregation of data

Surveillance data may be in the form of individual patient records or aggregate counts and tabulations. For example, at the local level, there may be a need to maintain records on individual persons to direct follow-up services. However, at a regional level, there may be no need to maintain individual-level records, and aggregate counts or tabulations may be sufficient. While transmission of aggregate data is usually simpler than transmission of individual records, aggregate data do not permit the same flexibility of analysis as individual data.

Data transmission

The mode of data transmission will depend on both the need for timeliness and communications resources. Increasingly, computers, and particularly microcomputers, are being used in surveillance systems, permitting transmission of data in electronic formats, such as telecommunication or mailing data storage devices (Graitcer and Burton 1987; Valleron et al. 1986). This rapidity of data transmission afforded by computers may not be necessary, however, and postage of forms may be sufficient. If very limited volumes of data are being sent, such as simple case counts, voice communication by telephone may be sufficient.

Data management and dissemination

The following issues in data management and dissemination should be considered in planning for storage, analysis, and dissemination of surveillance information.

Updating records

Surveillance data are continually subject to change. Errors in reporting may be identified and corrected; duplicate case reports may be recognized and culled; information that was initially unattainable may become available; follow-up investigations yield supplemental information; persons initially classified as meeting or not meeting a case definition may be

reclassified. One approach to handling these and other changes is to maintain both provisional and final records, including separate publications for provisional and final data. Thus, provisional reports may satisfy immediate information needs, while final and more delayed reports can accommodate corrections and updates to a reasonable limit and serve an archival function.

Selecting measures for time and place

The sequence of events in reporting a case may include the onset of disease, the diagnosis, the report to local health authorites, the report to regional or national health authorities, and others. Tabulations of surveillance data may be based on the date of any of these events. However, if there are long delays between dates of diagnosis and report, for example, analyses of trends based on date of diagnosis will be unreliable for the most recent periods. Similarly, surveillance data may be tabulated on the basis of the site of occurrence of the health event, site of diagnosis, or residence of persons reported. The use of these measures for time and place may also differ for provisional and final surveillance reports.

Detail in published reports

Surveillance data may be published using either tables or graphs and maps. The advantages of tables are that more detail can be provided and that numbers are available for readers to use in further calculations and analyses. The disadvantage is that more time and effort is required of the reader to extract summary points from a table than from a good figure. The advantage of figures is that one or a few key points can be visualized by a reader almost instantly; the disadvantage is a loss of detail. Thus, the intended audience must be considered in selecting either tables, figures, or a combination for published surveillance reports.

Confidentiality

Preventing inappropriate disclosure of surveillance data is essential both to the privacy of persons with reported cases of disease and to the trust of participants in the surveillance system. The protection of confidentiality begins with limiting the collection and transmission of data to a minimum and includes ensuring the physical security of surveillance records, the discretion of surveillance staff, and legal safeguards. In addition, surveillance reports should not present data in a manner that would permit identification of individuals. This may include not reporting data in cross-tabulations when the population represented is small.

Initiating and maintaining participation

In most surveillance systems, public health agencies depend on the ongoing co-operation of others to identify and report cases. Regardless of whether reporting is required by law, is voluntary, or is financially rewarded, it takes time and effort. With any approach, personal contact and dissemination of reports that document the usefulness of surveillance data are likely to be key in initiating and maintaining participation in the system.

For certain diseases, reporting is often required by law. While legal mandates may not guarantee reporting, they establish the authority under which health agencies conduct surveillance. In addition, reporting laws may identify not only those who are required to report cases but also those who may report cases without fear of liability for violation of privacy. Statutes may also protect health agencies from forced disclosure of the identity of persons with particular diseases.

Organizational structure

For the surveillance loop of data collection, analysis, interpretation, and feedback to function as a continuous process, an organizational structure is required. Such structure depends on resources available, including the number of skilled and unskilled personnel, technology available for communication and data management (e.g. computers), financial constraints, as well as the number and type of diseases brought under surveillance. In its simplest form, it is reporting of a single disease or health event by health care providers on a regular basis to a co-ordinating public health authority. A more complex form would include a network of reporting units, dealing concurrently with problems related to many diseases. In any case, the structure must allow data to be gathered from various sources and evaluated by epidemiologists in time for appropriate action to be taken. These data must be routinely disseminated to a targeted audience; therefore, reproduction or telecommunication equipment is usually needed.

The structure should provide support for training of key personnel in surveillance through seminars or other meetings that give field and central staff the opportunity to review procedures and to resolve operational problems. The need for regular evaluation of the surveillance systems with its potential requirement for additional resources should also be recognized.

Delegating tasks between national, regional, and local health authorities should depend on information needs and resources at each level. Particular attention should be directed to the local level because primary responsibility for information collection and public health responsibilities are local. Central agencies are responsible for co-ordinating data collection procedures to ensure that surveillance data can be reliably compared from one area to another and aggregated into regional or national summaries. Also, because many monitoring efforts (e.g. traffic injuries and water pollution) are dealt with by other governmental agencies (e.g. police authorities and environmental protection agencies), there needs to be effective co-ordination between appropriate authorities and health authorities; for this purpose, procedures may need to be established to assure appropriate communication.

Data collection

Public health surveillance data are collected in many ways, depending on the nature of the health event under surveillance, potential methods for identifying the disease, the population involved, the resources available, and the goals of the programme. Some surveillance systems may rely on a single source of data with alternate data sources being used periodically to evaluate or enhance the completeness of routine surveillance data.

Notification systems

Notifiable disease reporting is the surveillance approach traditionally used by public health programmes. A system of notification is based on laws or regulations by health authorities that require reporting of selected diseases, usually infectious, to the health department to support and direct prevention/control programmes. Notification reporting may be instituted at many levels (international, national, local). Ultimately, under a system of notification, the reporting will be most useful and most accurate for diseases emphasized at a local level. Persons or institutions with responsibility for reporting to the public health authority often include physicians, other health care providers, coroners and medical examiners, laboratories, hospitals, and others. Historically, physicians have been most integral to systems of notification, but more recently, reliance on the laboratory has been increasing.

In any country, the extent of notification activities depends on the existing facilities and resources—laboratories, epidemiological services, and, of course, on the public health priority of the disease and method of diagnosis. Reports are often initiated by the health care provider or other reporting source; with diseases of high public health priority, public health professionals may contact the major reporting sources at regular intervals and inquire about the occurrence of disease to ensure that most cases are identified. These systems of reporting have been described as passive and active, respectively, but the distinctions are not always clear. Data of many surveillance programmes represent a mixture of both reports elicited by public health professionals contacting health care providers and reports submitted by health care providers to public health officials without direct solicitation.

Reporting is generally incomplete for most notifiable diseases (Hinman 1977; Vogt et al. 1983). If persons are asymptomatic or have only mild symptoms they will usually not seek health care. Health care providers may fail to report for many reasons: they may be unaware of regulations; they may treat the symptoms without a complete laboratory investigation; in addition, many laboratory tests are not exact. Patients and physicians may conceal diseases that carry social stigma, such as sexually transmitted diseases. Completeness of reporting may also be significantly influenced by factors such as medical community interest, intensity of surveillance efforts, and publicity (Davis and Vergeront 1982). Their incompleteness does not imply that these surveillance data cannot serve their purpose, however. Epidemics, as well as general temporal and geographical trends, can be determined as long as the proportion of cases detected remains consistent over time and across geographical areas.

A comparison in Israel between cases of viral hepatitis reported by practitioners in private practice and cases reported in a population covered by an insurance plan demonstrated that although completeness of reporting by the physicians was only 37 per cent, the distribution of reported cases by season and by age was similar to that recorded in the insured population (Brachott and Mosley 1972). However, all reported characteristics may not be representative; a study of under-reporting of acute viral hepatitis in the US demonstrated that homosexual men with hepatitis B and blood transfusion recipients with non-A, non-B hepatitis were less likely than other risk groups to be reported (Alter *et al.* 1987). Thus, surveillance data acquired through reports by health care providers may not accurately reflect the risk for specific populations.

Although infectious conditions have dominated the list of notifiable diseases in most countries, other diseases and conditions may also have to be notified. Adverse drug reactions, occupational injuries, poisonings and specified malignancies, and other diseases or conditions may be required to be reported, particularly in developed countries (De Bock 1988; Faich *et al.* 1987; Freund *et al.* 1989). A crude but inexpensive surveillance system for diseases of high morbidity and for which notification may not be appropriate (e.g. gastro-intestinal illnesses or influenza) may be based on absenteeism from schools or industry, depending on the ages of the affected populations.

Health care provider networks

Networks of health care providers have been organized in recent years primarily to gather information on selected health events. Most have been organized by practising physicians on a voluntary basis; in many European countries, these networks have formed firm relationships with both public health authorities and academic centres, and often form the basis for morbidity surveillance. The strengths of such a system include the commitment of the participants, the possibility of collecting longitudinal data, the flexibility of the system to address a changing set of conditions, and the ability to gain information on all patient–provider encounters, regardless of severity of illness. The most severe limitation of this type of system is that the population served by these physicians may not be representative of the general population. In addition, the illness must be fairly common to provide accurate incidence data from a small sample of physician contacts.

Example: In Belgium, a voluntary network of general practitioners was initiated in 1978 (Stroobant *et al.* 1988). Practitioners have been selected to represent Belgian general practitioners according to age and sex, and have been geographically distributed to ensure a homogeneous coverage of the country. Participants report weekly, and the results are sent to the participants on a quarterly basis. The list of health problems includes selected vaccine-preventable diseases, respiratory conditions, and suicide attempts, with some health problems such as mumps and measles reported continuously and others on a less frequent basis. In Belgium, an excellent level of participation has been documented using the degree of form completion and continuity of reporting as criteria for assessment. The network has been evaluated in terms of its possible biases, such as non-participation of practitioners and difficulties in estimating the population at risk for the health problems under study; methods have been developed to reduce these biases (Lobet *et al.* 1987).

Laboratory surveillance

Surveillance of routinely collected laboratory reports has been particularly useful in the surveillance of certain infectious conditions. In England, nearly all microbiology laboratories report positive identifications of specified infections each week to the Communicable Disease Surveillance Centre. The advantages of the system are its specificity, its flexibility in adding new diseases, its rapidity, and the amount of detail that may be provided. Reports indicate trends or the appearance of rare infections originating from a common source that could not be identified by a single laboratory. Its disadvantages are that the number of persons from whom specimens are tested is usually not reported; therefore, there is no denominator. In addition, the persons tested may not be representative of the population at risk. For some infections, such as toxic-shock syndrome, there is no suitable laboratory test, and for many common illnesses a specimen may not be taken (e.g. influenza).

Nosocomial infection surveillance is often based on review of laboratory records by an infection control nurse or other designated staff (Brachman 1982). In 1970, the National Nosocomial Infection Study was initiated to monitor the frequency and trends of nosocomial infection in US hospitals. Over 60 hospitals participate in what is now a voluntary national surveillance system, with microbiology studies reported on 90 per cent of infected patients (Hughes *et al.* 1983).

Disease registries

Registries were initially established primarily for epidemiological research on individual diseases or conditions to develop aetiological hypotheses and to identify cases for further research (Weddell 1973). Registries have also been used to ensure the provision of appropriate care and to evaluate changing patterns of medical care; unlike other disease information systems, they cut across the different levels of severity of illness, and may provide information over time about individual persons. Recently, the value of registries for monitoring the disease incidence and its distribution has been more widely recognized.

In focusing on selected diseases or conditions, registries often develop a constituency that promotes participation and reporting. Most registries rely on numerous sources of data for case detection including, but not limited to, hospitals, laboratories, and death records; few registries rely primarily on physician notification. Public health professionals probably have the most experience with cancer registries and registries for congenital malformations.

Cancer registries generally have relied on multiple sources of data, including notification by health care providers, hospital discharge abstracts, treatment records (especially from oncology or radiotherapy units), death certificates, and pathology reports. Elaborate population-based registries for all cancers, such as registries developed by the US National Cancer Institute in discrete geographical areas, have been particularly useful for national trend estimates and epidemiological research (Horm *et al.* 1984). Less sophisticated registries have been developed by State and local governments or by physician initiatives (Parkin 1988).

Surveillance of congenital malformations was first initiated with registries to establish a reliable baseline and to detect increases in the number of cases, leading to rapid identification of teratogens. Even when persons using registries have not detected changes in incidence, they have used these registries to investigate suspected problems and to confirm or deny a change alleged by others (Edmonds *et al.* 1978, 1979). An increase in concern about environmental exposures and the safety of medications taken by pregnant women have heightened interest in the surveillance of congenital malformations. Two main types of registries are used (Edmonds *et al.* 1981; Weatherall 1988). One method concentrates on major congenital malformations noted on physical examination within a few days of birth. Data gathering is relatively straightforward if information is regularly provided on all infants born, as in the Scandinavian countries and in the UK. Also, many countries have joined the International Clearing House for Birth Defects Monitoring Systems, which monitor more than 3 million births per year from over 22 countries.

The second type of system records defects no matter what age the defect is discovered, is more resource-intensive, and generally needs to be conducted in a small geographical area. Information on visual, auditory, and intellectual deficits as well as some congenital heart diseases with late manifestations may be gathered through this latter system. In addition, the effects of treatment of handicapping conditions, such as operative procedures on infants with spina bifida, can be gained (Weatherall 1988).

Hazard exposure surveillance

In addition to monitoring health outcomes, surveillance systems may focus on environmental and occupational hazards, such as chemical substances, electromagnetic radiation exposure, and others, such as vibration, noise levels, or other physical agents. For chemical substances, three types of exposure may be recorded: actual exposures, potential exposures, and inferred exposures (i.e. substances present in air in the immediate area of workers). The primary purposes

of these systems are to form a basis for regulations and preventive measures and to ensure that levels stay below regulatory standard. In theory, optimum hazard surveillance systems would have information on the chemical identification and exposure levels for every worker (Froines *et al.* 1986). Air and water pollutants have also been measured on a regular basis in many developed countries, with attempts to compare these to the health of the general population.

Health information systems

Surveillance systems may, and should, use existing health data collection systems when possible. Lack of both detail and timeliness have limited the usefulness of these data for surveillance purposes, but computerization may offer the potential for more complete and timely data collection than currently existing data systems.

Vital records

Mortality statistics serve as the most accessible and reliable source of data for international comparisons of many health problems, and for surveillance of many diseases at the local and national level. In most developed countries, registration of deaths is compulsory and largely complete. Records include basic demographic information, the cause or causes of death, and other descriptive information of the circumstance of death. In other countries, registration may be conducted only in major cities, or not at all. In all countries, the accuracy for many diagnoses is often inadequate, and changes in use of diagnostic categories and codes over time, together with variation in the quality of information, are limiting factors. For example, a study by the US National Cancer Institute revealed that seven countries in Europe and North America coded the same underlying cause of death for only 53 per cent of a sample of 1246 death certificates sent to these countries (Percy and Dolman 1978). In spite of these limitations, vital statistics, particularly mortality statistics, are used to support many surveillance activities.

Example: Death certificates have been used as a primary source of data for maternal mortality surveillance, in demonstrating progress towards reduction in maternal mortality in association with increased use of prenatal care and other factors. Analyses of death certificates have highlighted racial differences in mortality rates over time, and differences in maternal mortality rates for women aged over 35 years. Because maternal mortality rates are often based on number of live births, this surveillance system also depends on birth certificate information (Kaunitz *et al.* 1984).

There is frequently a lengthy interval between death and collection and analysis of death certificates, which may make such vital statistics less useful for surveillance purposes when more current data are needed. Summary vital data can be rapidly collected, however. As an example, weekly telephonic reporting of deaths from 121 US cities to CDC has been integral to the surveillance of influenza epidemics in that country (Choi and Thacker 1981).

Medical examiner/coroner reports

For a more detailed description of circumstances surrounding deaths, including autopsy reports, toxicology studies, and police reports, medical examiner and coroner records may be useful. In the US, these reports are most representative of deaths caused by intentional and unintentional injury, or other unnatural causes. These records have been particularly useful in surveillance of heat-wave-related mortality, sudden unexplained death syndrome in South-East Asian refugees, and alcohol-related injuries (Jones *et al.* 1982; Berkelman *et al.* 1985; Parrish *et al.* 1987).

Medical care records

Hospital records are useful sources of information on diagnoses, surgical procedures, and patient demographic characteristics. The medical record has increased in volume and complexity with an expanded use of laboratory and operative procedures and therapeutic measures. Thus, retrieval of information is often difficult and time consuming. Computerization of parts of these records, particularly the discharge summary, has allowed their use for routine surveillance. A major limitation, however, has existed when no identifiers are recorded since repeat admissions and discharges by individual patients usually cannot be identified. Also, lack of timeliness has been a limiting factor, and use of these records for surveillance has been limited primarily to chronic diseases and procedures (e.g. hysterectomies) (Sattin *et al.* 1983). To advance efforts to standardize data collection, the European community has developed a minimum basic data-set recommended for recording of hospital morbidity, including a patient number (Paterson 1988).

The Birth Defects Monitoring Program is a national programme in the US initiated in 1974 to monitor and analyse hospital discharge data on newborns for birth defects and other conditions. The approximately 1000 participating hospitals, mainly mid-sized community hospitals, are not population based. Nevertheless, trends in incidence of birth defects from this large sample of new-borns, accounting for 20 per cent of the births in the US, have been useful for surveillance (Edmonds and James 1985).

Hospital discharge records have also been useful for surveillance of many medical care technologies, such as trends in the use of hysterectomies in the US particularly by geographical region, in the rate of coronary artery bypass graft procedures by sex, and in the assessment of outcome as with carotid endarterectomies (Caper 1987; Sattin *et al.* 1983; Thacker and Berkelman 1986). More recently, hospital discharge record systems are being used as an alternative data source to evaluate surveillance data-sets, as with AIDS surveillance (Lafferty *et al.* 1988).

Insurance records and workers' compensation

Insurance records and workers' compensation claims have been useful for surveillance of injuries in specific geographical locales. Because regulations governing completion and submission of forms differ both between and within countries, data derived from these systems cannot easily be compared.

The severity of reported injury varies, being influenced by legislation related to compensation, medical care, and rehabilitation of those injured at work, and the degree of fear of job loss resulting from absence from work.

Example: In an evaluation of claims for workers' compensation as an adjunct to an occupational lead surveillance system, the usefulness of claims was demonstrated with the likelihood that a company had a case of lead poisoning strongly correlated with the number of claims against the company (Seligman *et al.* 1986).

Surveys of health behaviours and physician utilization

National household surveys of the general population such as the National Health Interview Survey, conducted in the US (National Center for Health Statistics 1985) or the General Household Survey in England and Wales (Fraser *et al.* 1978) have provided information at the national level on personal health practices, such as alcohol use and smoking, on disabilities, and on physician encounters.

Although national estimates may be gained more efficiently from such surveys, local programmes may benefit from involvement in data collection and the flexibility to adapt data collection to their particular needs. Currently nearly 40 States in the US are conducting telephone surveys on an ongoing basis, using randomly sampled telephone numbers (Remington *et al.* 1988). Interview surveys conducted in the US can obtain personal health-related information with only minor differences in the measured prevalence of various health conditions when conducted by telephone or in person. Telephone interviews have the advantages of lower cost and ease of supervising interviewers (Thornberry 1987).

Analysis of surveillance data

Surveillance data permit a description of disease in terms of the basic epidemiological parameters of time, place, and person, as well as the manifestations of disease. In addition, surveillance data can be used to make initial comparisons of disease at one time versus another, in one place versus another, or among one group versus another. When combined with appropriate population information, morbidity or mortality rates can be calculated to compare risks of disease by these parameters. Often rates are examined in broad age groups selected to reflect the different sets of conditions affecting mortality in each group (Doll 1974). Proper analysis of surveillance data can provide insight into aetiology, modes of transmission, risk factors associated with disease, and opportunities for prevention or control, in addition to detecting epidemics, monitoring long-term trends, following seasonal patterns, or making projections of future disease occurrence.

In addition to basic descriptive information, more sophisticated analyses have been used on a limited basis, such as the application of regression and time series analyses to mortality data for the surveillance of influenza (Choi and Thacker 1981; Lui and Kendal 1987; Serfling 1963) and the use of environmental data to predict the occurrence of Rocky Mountain spotted fever (Newhouse *et al.* 1986).

Epidemic detection and cluster analysis

Many epidemics are detected by astute health care providers who note or suspect an increase in disease occurrence often before disease reports are received, assembled, and reviewed by health departments. The surveillance process links practitioners to health departments and increases the likelihood that providers would contact the health department when they suspect abberations in disease occurrence.

Surveillance is most likely to detect epidemics in situations where cases, despite their aetiological link, are occurring over a wide geographical area or over a relatively gradual period, and thus time–place–person links among cases may be unrecognized by individual practitioners.

Example: Laboratory-based surveillance of *Salmonella* serotypes has identified outbreaks in which unusual serotypes and/or antimicrobial patterns identify an outbreak of diarrhoeal disease, which might otherwise have gone undetected, such as the outbreak of drug-resistant *Salmonella newport* in a large geographical area of the US that originated from animals fed antimicrobials (Holmberg *et al.* 1984).

Example: Age-adjusted oesophageal cancer mortality rates in white men and women have remained fairly steady during 1950–80, but have nearly doubled for blacks during the same period. Oesophageal cancer has gradually become one of the most common malignancies in black men aged under 55 years, while still a relatively rare cancer in similarly aged whites (Blot and Fraumeni 1987).

A frequent concern of the analysis of surveillance data is whether an apparent cluster of health events in time is significant and unlikely to have occurred by chance alone. The scan statistic, first investigated by Naus (1965), was developed to detect a sudden increase in risk. The statistic, as employed by Wallenstein (1980) uses a moving window of a defined width in time and finds the maximum number of cases revealed through the window as it 'scans' over the entire period under consideration. The probability of observing a certain size cluster is then calculated, based on the Poisson distribution. The statistic can be used in surveillance prospectively by defining the number of events in a window that will exceed a critical value. The statistic is most useful in evaluating clusters of diseases or conditions that do not have a seasonal distribution, such as clusters of leukaemia (Ederer *et al.* 1964).

Other statistical approaches to detection of rapid increases in the number of events, especially for rare events, include the 'cusum' techniques and the 'sets methods'. These methods again define parameters to set a critical value for an alarm, and have been applied to surveillance of rare events, such as congenital malformations (Gallus *et al.* 1986).

Statistical limitations of surveillance data

Surveillance data have traditionally had limitations that have made application of standard statistical techniques difficult. Reporting biases are often present, and data may not be representative of the population. For example, severe or otherwise noteworthy cases are more likely to be reported than minor illnesses. Rubella in a woman of childbearing age is more likely to be reported than rubella in a man; and a patient in a public health clinic may be more likely to be reported than a patient seeing a private physician, particularly if the disease is socially stigmatizing. Health care provider networks may have a biased sample of physicians (Lobet *et al.* 1987). In addition, descriptive data may be incomplete.

Under-reporting may also be considerable, especially in a system of notification. Using other databases, such as hospital discharge and vital statistics, the total number of cases can be established through application of methods, such as the Chandra–Sekar–Deming method and the Lincoln Petersen capture–recapture method (Chandra-Sekar and Deming 1949; Cormack 1963). For example, the Chandra–Sekar–Deming method has been applied to estimate the sensitivity of two systems for detecting vaccine-preventable diseases (Orenstein *et al.* 1986).

Provisional data increase the timeliness and hence may increase the usefulness of public health surveillance data to epidemiologists; however, provisional data may differ markedly from final data that have been confirmed. To enhance the usefulness of provisional data, these data may be compared retrospectively with confirmed data to estimate final data (Thacker *et al.* 1989). For example, for selected diseases, the final number may be consistently greater than the provisional data. A model can incorporate this consistent under-reporting to permit more accurate estimation of the final data from provisional data. In addition, when provisional data are used to examine temporal trends, current provisional data should be compared to historical provisional data rather than to final data to avoid bias.

Role of surveillance data in evaluation of community interventions

The ease with which trends in disease occurrence can be linked to interventions depends on both the disease and the intervention. The success of an immunization campaign can usually be easily inferred from surveillance data; however, such inferences become difficult when several factors contribute to a change in disease occurrence. Analyses are also difficult because of constraints, such as migration and variable acceptance of interventions in the community. Programme evaluation can be improved through conducting surveillance by monitoring risk factors as well as various stages of morbidity. In addition, combining data from several communities with similar public health programmes will strengthen the assessment of their effectiveness.

Mathematical models can be used to elucidate the

complexities of evaluating community interventions. Such models have been used most extensively for infectious diseases; however, models for predicting the decline of mortality given changes in risk factors have also been developed for cirrhosis mortality using population changes in levels of consumption of alcohol (Skog 1984) and for cardiovascular disease using changes in cigarette consumption in a population (Kullback and Cornfield 1976).

Linkage of surveillance data to other information sources

There is an increasing interest in computerized searches of large files of records to link individuals in different data systems or within the same data system to determine their mortality and health experience. Techniques involved in data linkage are complex and based on matching records by comparison of key data fields (Newcombe 1988). In some countries, a unique number may be assigned at birth that may serve as a reference number for any contact with health care services (Paterson 1988). In other countries, a number may only be assigned for use at a single health care facility or hospital. When record systems are linked, the probability that the record linkage is correct must be determined with the degree of certainty of a correct linkage depending on the comparisons of the individual identifiers, such as name or initials, date or year of birth, sex, and race/ethnicity. Names may be converted to a Soundex or similar code to allow for errors in spelling. If all identifiers match exactly, the degree of assurance that the linkage is correct is high. Few similarities argue against a correct linkage. Any linked set will normally contain a small number of pairs that should not have been linked, and conversely will have missed a few pairs that should have been linked.

Linkage of data-sets has facilitated calculation of rates, such as birth-weight specific death-rates that can be calculated following linkage of birth and death certificates (McCarthy et al. 1980).

Dissemination of data

Communication of the data is an essential step in the surveillance chain. The purpose of the communication and the audience targeted must be defined. Appropriate feedback must be given to those providing the data to demonstrate their usefulness and to stimulate further reporting. Persons providing the data should be identified by name both to credit them for the contributions and to acknowledge their responsibilities as primary reporters for the accuracy and completeness of the data. Public health professionals, policy-makers, or others who may be responsible to take action or set direction of public health programmes in response to surveillance data must receive needed information from the surveillance system on a timely basis and in an appropriate format for their use.

The data must be provided on a regular basis, with the frequency of surveillance reports dependent on the nature of the surveillance system and characteristics of the disease process (e.g. surveillance reports on measles are required more frequently than reports on cancer), and on the public health impact of the disease. For diseases requiring major policy decisions (e.g. seat belt legislation), it may be useful to provide frequent updates to remind policy-makers of the potential for prevention. In general, reports need not be issued at more frequent intervals than the data are collected from reporting sites. Provisional data should be accepted for dissemination, since rapid turnaround of data is usually more important than absolute accuracy and completeness; rarely have provisional data driven major public health decisions in directions different from those that would have been based on final data.

The format for dissemination again varies with the target audience, but in any case the design of the communications should be as creative as possible without losing needed information. A creative design will help make the information stand apart from other documents and receive greater attention. Most policy-makers and clinicians would prefer to see the data interpreted with graphics accompanied by abbreviated summary text; the important role that graphs can play in visually decoding large quantities of data has been clearly demonstrated, with graphic displays giving the reader an understanding of large and complex data-sets not conveyed easily in other ways (Cleveland 1985; Tukey 1977; Tufte 1983). Microcomputer graphics, in particular, have made the results of data analysis far more useful to private and public policy-makers in their planning and management of health care resources (Caper 1987). However, many epidemiologists and other scientists, including mathematicians projecting the future course of diseases, such as HIV infection/AIDS, find the more detailed raw data in tabular format most useful. Comparison with previous years or previous periods (e.g. experience of last 12 months compared with experience of previous 12 months) is often helpful.

Maps are useful in providing rapid insight into the geographical occurrence of diseases, and there is currently a strong interest in computer mapping and graphic displays; mapping both absolute counts of disease occurrence and rates of disease for more common conditions may be useful. Methods of mapping geographical areas in which the map shows a physical area proportional to its populations have been developed and applied to surveillance data, as illustrated by the 'exploded map' (Olson 1976). Such techniques may be useful when displaying disease burden to policy-makers if a large geographical size provides misleading impressions that the burden of disease is large, when these areas may be relatively unpopulated.

Evaluation of surveillance systems

Surveillance systems should be periodically evaluated to ensure that important public health problems are under surveillance and that useful information for disease prevention and control is collected. An evaluation of a surveillance sys-

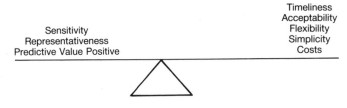

Sensitivity		Timeliness
Representativeness		Acceptability
Predictive Value Positive		Flexibility
		Simplicity
		Costs

Fig. 11.2. Balance of attributes in surveillance systems.

tem should include a review of its objectives, a detailed description of its operation, an assessment of its performance, and recommendations (CDC 1988).

The performance of surveillance systems can be judged using a series of attributes, including sensitivity, timeliness, representativeness, predictive value positive, acceptability, flexibility, simplicity, and costs. The importance of individual attributes will vary among systems, and efforts to improve one may compete with efforts to improve another. Thus, the evaluation of surveillance systems should not focus solely on the extent to which each attribute is achieved but rather on the attainment of the appropriate balance of attributes (Fig. 11.2). The ultimate impact of improvements in surveillance should be assessed in terms of improvements in health (Thacker *et al.* 1986).

Sensitivity

The sensitivity of a surveillance system is its completeness of case reporting. If all persons with the condition under surveillance in the target population are detected by a surveillance system, then its sensitivity is 100 per cent. Sensitivity of surveillance systems can be measured by comparing routinely collected case reports to data obtained by special case-finding methods. For example, the sensitivity of AIDS surveillance has been assessed through detailed review of death certificates and various hospital records, such as laboratory data, patient log books, and computerized discharge diagnoses (Chamberland *et al.* 1985; Hardy *et al.* 1987; Lafferty *et al.* 1988).

Timeliness

Timeliness refers to the entire surveillance cycle, ranging from how quickly cases are reported to the distribution of surveillance reports.

Representativeness

Surveillance reporting is rarely complete, and cases that are reported may differ from unreported cases in terms of demographic characteristics, site or use of health care services, or risk exposures (Alter *et al.* 1987). In addition to its dependence on case ascertainment, representativeness of surveillance data is also affected by the quality of descriptive data that accompanies case reports. Incomplete or incorrect data on surveillance forms limit representativeness.

Predictive value positive

Persons with reported cases of disease may not actually have the disease in question. This may reflect incorrect diagnoses (false-positives), a lack of specificity in the case definition, or errors in the interpretation of the case definition. If all persons with reported cases indeed had the disease in question, then predictive value positive would be 100 per cent. Predictive value depends both on the specificity of diagnostic tests and the case definition and on prevalence of the condition under surveillance. Evaluation of predictive value positive requires a careful review of cases detected through routine methods. For example, hospital-based stroke surveillance using readily available admission diagnoses was found to include a substantial proportion of persons without cerebrovascular disease when more stringent diagnostic criteria were applied (Barker *et al.* 1984).

The concept of predictive value positive can be extended to the detection of epidemics in a surveillance system. If changes in disease occurrence are used as a bias for triggering investigations, then a high frequency of 'false alarms' would indicate a low predictive value for epidemic detection.

Acceptability

Surveillance systems depend on the co-operation on many people over a long period. If procedures are easy to follow and useful information is returned to participants, then acceptability is likely to remain high.

Flexibility

The circumstances under which surveillance systems operate are subject to change, ranging from logistic constraints to information needs; surveillance systems should have sufficient flexibility to accommodate these changes. For example, surveillance for AIDS has been ongoing during a period of rapid evolution in the understanding of the disease, the introduction of new diagnostic tests, HIV infection, and changing diagnostic and treatment practices as a once rare disease becomes more common. As a result, the surveillance for AIDS has been flexible in accommodating revisions to the case definition (CDC 1987).

Simplicity

Simplicity is desirable throughout the entire cycle in surveillance systems and is closely tied to other attributes such as acceptability, flexibility, and costs.

Costs

Surveillance systems incur costs in time, equipment, and supplies, which may be difficult to judge relative to their public health value. Evaluation of the costs and benefits of aggressive versus less aggressive case-finding methods in surveillance of selected notifiable diseases has yielded different conclusions based on specific local circumstances (Hinds *et al.* 1985; Vogt *et al.* 1986). In the absence of a formal attempt to weigh the benefits and costs of a surveillance system, a description of its time requirements and costs remains useful to judging the value of a surveillance system (Hinds *et al*. 1985; Stroobant *et al*. 1988; Vogt *et al*. 1986).

The evaluation of a surveillance system should conclude with an assessment of its structure and usefulness, considering its mix of attributes in relation to its objectives. Recommendations should consider whether the system should be continued and what specific changes, if any, should be made.

Conclusion

For high-priority public health problems, surveillance has historically galvanized prevention and control programmes, ranging from smallpox eradication, immunization campaigns for childhood diseases, to HIV infection and AIDS. Surveillance systems represent information loops, with data flowing from local to central agencies and back. Surveillance provides a stimulus to keep prevention and control activities moving rapidly and in the right direction, guiding the response to individual cases as well as public policy. As public health priorities increasingly include non-infectious diseases and conditions, surveillance of these other health problems will take on increased visibility and importance in evaluating and directing the prevention and control efforts.

Effective public health interventions depend upon a continuing and reliable source of information. The data must be timely and representative of the population; they must be analysed and interpreted with feedback to the reporters and dissemination to those formulating and implementing public health policy. Resources necessary to the maintenance of the surveillance systems and for their regular evaluation should be allocated, balancing needs for data to direct prevention activities with needs for resources to implement those activities.

References

Alter, M.J., Mares, A., Hadler, S.C., and Maynard, J.E. (1987). The effect of underreporting on the apparent incidence and epidemiology of acute viral hepatitis. *American Journal of Epidemiology* **125**, 133.

Barker, W.H., Feldt, K.S., and Feibel, J. (1984). Assessment of hospital admission surveillance of stroke in a metropolitan community. *Journal of Chronic Diseases* **37**, 609.

Berkelman, R.L. (1989). Summary remarks: Symposium on statistics in surveillance, May 1988. *Statistics in Medicine* **8**, 393.

Berkelman, R.L. and Buchler, J.W. (1990). Public health surveillance of non-infectious chronic diseases: the potential to detect rapid change in disease burden. *International Journal of Epidemiology*. (In press.)

Berkelman, R.L., Herndon, J.L., Callaway, J.L., *et al*. (1985). A surveillance system for alcohol- and drug-related fatal injuries. *American Journal of Preventive Medicine* **1**, 21.

Blot, W.J. and Fraumeni, J.F., Jr. (1987). Trends in oesophageal cancer mortality among U.S. blacks and whites. *American Journal of Public Health* **77**, 296.

Brachman, P.S. (1982). Surveillance. In: *Bacterial infections of humans* (ed. A.S. Evans and H.H. Feldman p. 49. Plenum Medical, New York.

Brachott, D. and Mosley, J.W. (1972). Viral hepatitis in Israel: The effect of canvassing physicians on notifications and the apparent epidemiological pattern. *Bulletin of the World Health Organization* **46**, 457.

Caper, P. (1987). The epidemiologic surveillance of medical care. *American Journal of Public Health* **77**, 668.

Centers for Disease Control (1983). Classification of measles cases and categorization of measles elimination programs. *Morbidity and Mortality Weekly Report* **31**, 707.

Centers for Disease Control (1986*a*). *Comprehensive plan for epidemiologic surveillance*. Atlanta, Georgia.

Centers for Disease Control (1986*b*). Premature mortality in the United States: Public health issues in the use of years of potential life lost. *Morbidity and Mortality Weekly Report* **35** (Supplement No. 2S), 1S.

Centers for Disease Control (1987). Revision of the CDC surveillance case definition for acquired immunodeficiency syndrome. *Morbidity and Mortality Weekly Report* **36** (No. 1S), 3S.

Centers for Disease Control (1988). Guidelines for evaluating surveillance systems. *Morbidity and Mortality Weekly Report* (Supplement No. S–5), **37**, 1.

Chamberland, M.E., Allen, J.R., Monroe, J.M., *et al*. (1985). Acquired immunodeficiency syndrome, New York City: Evaluation of an active surveillance system. *Journal of the American Medical Association* **253**, 383.

Chandra-Sekar, C. and Deming, W.E. (1949). On a method for estimating birth and death rates and the extent of registration. *Journal of the American Statistical Association* **44**, 101.

Choi, K. and Thacker, S.B. (1981). An evaluation of influenza mortality surveillance, 1961–1979. I. Time series forecasts of expected pneumonia and influenza deaths. *American Journal of Epidemiology* **113**, 215.

Cleveland, W.S. (1985). *The elements of graphing data*. Bell telephone Laboratories, Murray Hill, New Jersey.

Cormack, R.M. (1963). The statistics of capture–recapture. *Ocean Marine Biology Annual Review* **6**, 455.

Davis, J.P. and Vergeront, J.M. (1982). The effect of publicity on the reporting of toxic-shock syndrome in Wisconsin. *Journal of Infectious Diseases* **145**, 449.

De Bock, A. (1988). Surveillance for accidents at work. In *Surveillance in health and disease* (ed. W.J. Eylenbosch and N.D. Noah) p. 191. Oxford University Press, New York.

Doll, R. (1974). Surveillance and monitoring. *International Journal of Epidemiology* **3**, 305.

Ederer, F., Meyers, M., and Mantel, N. (1964). A statistical problem in space and time: Do leukemia cases come in clusters? *Biometrics* **20**, 626.

Edmonds, L.D. and James, L.M. (1985). Temporal trends in the incidence of malformation in the United States, selected years, 1970–1971, 1982–83. *Centers for Disease Control Surveillance Summaries* **34** (No. 2SS), 1SS.

Edmonds, L.D., Layde, P.M., and Erickson, J.D. (1979). Airport noise and teratogenesis: A negative study. *Archives of Environmental Health* **34**, 243.

Edmonds, L.D., Anderson, C.D., Glynt, J.W., and James, L.M. (1978). Congenital central nervous system malformations and vinyl chloride exposure: A community study. *Teratology* **17**, 137.

Edmonds, L.D., Layde, P.M., James, L.M., *et al*. (1981). Congenital malformations surveillance: Two American systems. *International Journal of Epidemiology* **10**, 247.

Eylenbosch, W.J. and Noah, N.D. (ed.) (1988). *Surveillance in health and disease*. Oxford University Press.

Faich, G.A., Knapp, D., Dreis, M. and Turner, W. (1987). National adverse drug reaction surveillance: 1985. *Journal of the American Medical Association* **257**, 2068.

Foege, W.H., Millar, J.D., and Henderson, D.A. (1975). Smallpox eradication in West and Central Africa. *Bulletin of the World Health Organization* **52**, 209.

Fraser, P., Beral, V., and Chilvers, C. (1978). Monitoring disease in England and Wales: Methods applicable to routine data-collecting systems. *Journal of Epidemiology and Community Health* **32**, 294.

Freund, E., Seligman, P.J., Chorba, T.L., *et al.* (1989). Mandatory reporting of occupational diseases by clinicians. *Journal of the American Medical Association* **262**, 3041.

Froines, J.R., Dellenbaugh, C.A., and Wegman, D.H. (1986). Occupational health surveillance: A means to identify work-related risks. *American Journal of Public Health* **76**, 1089.

Gail, M.H. and Brookmeyer, R. (1988). Methods for projecting course of Acquired Immunodeficiency Syndrome Epidemic. *Journal of the National Cancer Institute* **80**, 900.

Gallus, G., Mandelli, C., Marchi, M., and Radaelli, G. (1986). On surveillance methods for congenital malformations. *Statistics in Medicine* **5**, 567.

Graitcer, P.L. and Burton, A.H. (1987). The epidemiologic surveillance project: A computer-based system for disease surveillance. *American Journal of Preventive Medicine* **3**, 123.

Hardy, A.M., Starcher, E.T., Morgan W.M., *et al.* (1987). Review of death certificates to assess completeness of reporting. *Public Health Reports* **102**, 386.

Heckler, M.D. (1985). *Report of the secretary's task force on black and minority health*. US Department of Health and Human Services, Washington, DC.

Henderson, D.A. (1976). Surveillance of smallpox. *International Journal of Epidemiology* **5**, 19.

Hinds, M.W., Skaggs, J.W., and Bergeisen, G.H. (1985). Benefit–cost analysis of active surveillance of primary care physicians for hepatitis A. *American Journal of Public Health* **75**, 176.

Hinman, A.R. (1977). Analysis, interpretation, use and dissemination of surveillance information. *Pan American Health Organization Bulletin* **11**, 338.

Holmberg, S.D., Osterholm, M.T., Senger, K.A., and Cohen, M.L. (1984). Drug-resistant *Salmonella* from animals fed antimicrobials. *New England Journal of Medicine* **311**, 617.

Horm, J.W., Asire, A.J., Young, J.L. Jr., *et al.* (1984). *SEER program: Cancer incidence and mortality in the United States, 1973–1981*. NIH publication No. 85–1837. Department of Health and Human Services, Bethesda, Maryland.

Hughes, J.M., Culver, D.H., White, J.W., Jarvis, W.R., Morgan, W.M., and Munn, V.P. (1983). Nosocomial infection surveillance 1980–1982. *Centers for Disease Control Surveillance Summaries* **32** (No. 4SS), 1SS.

Institute of Medicine (1988). *Approaches to modeling disease spread and impact. Report of a workshop on mathematical modeling of the spread of human immunodeficiency virus and the demographic impact of acquired immune deficiency syndrome October 15–17, 1987*. National Academy Press, Washington, DC.

Jaffe, H.W., Choi, K., Thomas, P.A., *et al.* (1983). National case–control study of Kaposi's sarcoma and *Pneumocystis carinii* pneumonia in homosexual men: Epidemiologic results. *Annals of Internal Medicine* **99**, 293.

Jick, H., Walker, A.M., and Rothman, K.J. (1980). The epidemic of endometrial cancer in the United States. *New England Journal of Medicine* **70**, 264.

Jones, T.S., Liang, A.P., Kilbourne, E.M., *et al.* (1982). Morbidity and mortality associated with the July 1980 heat wave in St.

Louis and Kansas City, Missouri. *Journal of the American Association* **247**, 3327.

Kaunitz, A.M., Rochat, R.W., Hughes, J.M., Smith, J.C., and Grimes, D.A. (1984). Maternal mortality surveillance, 1974–1978. *Centers for Disease Control Surveillance Summaries* **33** (No. 1SS), 5SS.

Kullback, S. and Cornfield, J. (1976). An information theoretic contingency table analysis of the Dorn study of smoking and mortality. *Computers and Biomedical Research* **9**, 409.

Lafferty, W.E., Hopkins, S.G., Honey, J., Howell, J.D., Shoemaker, P.C., and Kobayaski, J.M. (1988). Hospital charges for people with AIDS in Washington state: Utilization of a statewide hospital discharge data base. *American Journal of Public Health* **78**, 949.

Langmuir, A.D. (1963). The surveillance of communicable diseases of national importance. *New England Journal of Medicine* **268**, 182.

Langmuir, A.D. (1976). William Farr: Founder of modern concepts of surveillance. *International Journal of Epidemiology* **5**, 13.

Lobet, M.P., Stroobant, A., Mertens, R., *et al.* (1987). Tool for validation of the network of sentinel general practitioners in the Belgian health care system. *International Journal of Epidemiology* **16**, 612.

Lui, K.-J. and Kendal, A.P. (1987). Impact of influenza epidemics on mortality in the United States from October 1972 to May 1985. *American Journal of Public Health* **77**, 712.

McCarthy, B.J., Terry, J., Rochat, R.W., *et al.* (1980). The underregistration of neonatal deaths: Georgia, 1974–1977. *American Journal of Public Health* **70**, 977.

National Center for Health Statistics (1985). The National Health Interview Survey design, 1973–84, and procedures, 1975–83. Vital and Health Statistics. Series 1, No. 18, DHHS publication No. PHS 85–1320. US GPO, Washington, DC.

Naus, J. (1965). The distribution of the size of the size of the maximum cluster of points on a line. *Journal of the American Statistical Association* **60**, 532.

Newcombe, H.B. (1988). *Handbook of record linkage*. Oxford University Press.

Newhouse, V.F., Choi, K., D'Angelo, L.J., *et al.* (1986). Analysis of social and environmental factors affecting the occurrence of Rocky Mountain spotted fever in Georgia, 1961–75. *Public Health Reports* **101**, 419.

Olson, J.M. (1976). Noncontiguous area cartograms. *Professional Geographer* **28**, 371.

Orenstein, W., Bart, S.W., Bart, K.J., Sirotkin, B., and Hinman A.R. (1986). Epidemiology of rubella and its complications. In *Vaccinating against brain syndromes: The campaign against measles and rubella. Monographs in epidemiology and biostatistics* (ed. E.M. Grunberg, C. Louis, and S.E. Goldson) p. 49. Oxford University Press, New York.

Parkin, D. (1988). Surveillance of cancer. In *Surveillance in health and disease* (ed. W.J. Eylenbosch and N.D. Noah) p. 143. Oxford University Press.

Parrish, R.G., Tucker, M., Ing, R., and Encarnacion, C. (1987). Sudden unexplained death syndrome in southeast Asian refugees: A review of CDC surveillance. *Centers for Disease Control Surveillance Summaries* **36** (No. 1SS), 43SS.

Paterson, J.G. (1988). Surveillance systems from hospital data. In *Surveillance in health and disease* (ed. W.J. Eylenbosch and N.D. Noah) p. 49. Oxford University Press.

Percy, C. and Dolman, A. (1978). Comparison of the coding of death certificates related to cancer in seven countries. *Public Health Reports* **93**, 335.

Raska, K. (1966). National and international surveillance of communicable diseases. *World Health Organization Chronicle* **20**, 315.

Remingtion, P.L.S., Smith, M.Y., Williamson, D.F., *et al.* (1988). Design, characteristics and usefulness of state-based behavioral risk factor surveillance: 1981–1987. *Public Health Reports* **103**. 366.

Sattin, R.W., Rubin G.L., and Hughes, J.M. (1983). Hysterectomy among women of reproductive age, United States, update for 1979–1980. *Centers for Disease Control Surveillance Summaries* **32**, 1SS.

Seligman, P.J., Halperin, W.E., Mullan, R.J., and Frazier, T.M. (1986). Occupational lead poisoning in Ohio: Surveillance using worker's compensation data. *American Journal of Public Health* **76**, 1299.

Serfling, R.E. (1963). Methods for current statistical analysis of excess pneumonia-influenza deaths. *Public Health Reports* **78**, 494.

Shands, K.N., Schmid, G.P., Dan, B.B., *et al.* (1980). Toxic-shock syndrome in menstruating women. *New England Journal of Medicine* **303**, 1436.

Skog, O. (1984). The risk function for liver cirrhosis from lifetime alcohol consumption. *Journal of Studies on Alcohol* **45**, 199.

Stroobant, A.W., Van Casteren, V., and Thiers, G. (1988) Surveillance systems from primary-care data: Surveillance through a network of sentinel general practitioners. *Surveillance in health and disease* (ed. W.J. Eylenbosch and N.D. Noah) p. 62. Oxford University Press.

Thacker, S.B. and Berkelman, R.L. (1986). Surveillance of medical technologies. *Journal of Public Health Policy* **7**, 363.

Thacker, S.B. and Berkelman, R.L. (1988). Public health surveillance in the United States. *Epidemiologic Reviews* **10**, 164.

Thacker, S.B., Berkelman, R.L., and Stroup, D.S. (1989). The science of public health surveillance. *Journal of Public Policy* **10**, 187–203.

Thacker, S.B., Redmon, S., Rothenberg, R.B., *et al.* (1986). A controlled trial of disease surveillance strategies. *American Journal of Preventive Medicine* **2**, 345.

Thornberry, O.T., Jr. (1987). An experimental comparison of telephone and personal health interview surveys. *National Center for Health Statistics*: *Data Evaluation and Methods Research* Series 2, No. 106, 1.

Tufte, E.R. (1983). *The visual display of quantitative information.* Graphics Press, Cheshire, Connecticut.

Tukey, J.W. (1977). *Exploratory data analysis.* Addison-Wesley, Reading, Massachusetts.

Valleron, A., Bouvet, E., Garnerin, P., *et al.* (1986). Computer network for the surveillance of communicable diseases: The French experiment. *American Journal for Public Health* **76**, 1289.

Vogt, R.L., Clark, S.W., and Kappel, S. (1986). Evaluation of the state surveillance system using hospital discharge diagnoses, 1982–1983. *American Journal of Epidemilogy* **123**, 197.

Vogt, R.L., LaRue, D., Klaucke, D.N., and Jillson, D.A. (1983). Comparison of active and passive surveillance systems of primary care providers for hepatitis, measles, rubella and salmonellosis in Vermont. *American Journal of Public Health* **73**, 795.

Wallenstein, S. (1980). A test for detection of clustering over time. *American Journal of Epidemiology* **111**, 367.

Weatherall, J.A.C. (1988). Surveillance of congenital malformations and birth defects. In *Surveillance in health and disease* (ed. W.J. Eylenbosch and N.D. Noah) p. 143. Oxford University Press.

Weddell, J.M. (1973). Registers and registries: A review. *International Journal of Epidemiology* **2**, 221.

Wolf, P.A., D'Agostino R.B., Kannel, W.B., Bonita R., and Belanger, A.J. (1988). Cigarette smoking as a risk factor for stroke. *Journal of the American Medical Association* **259**, 1025.

World Health Organization (1958). *The first ten years of the World Health Organization.* WHO, Geneva.

World Health Organization (1968). Report of the technical discussions at the twenty-first World Health Assembly on 'national and global surveillance of communicable diseases', 18 May 1968. A21. WHO, Geneva.

12

Intervention and experimental studies

PEKKA PUSKA

Introduction

There are two major approaches to investigations: observational studies or surveys (the various forms of which have been dealt with in earlier chapters in this volume) and experimental studies. In an experiment the researcher decides which subjects are to be exposed to or deprived of a particular factor(s) (the 'intervention' or 'experimental' group). If at all feasible, this group is compared with a control or reference group where again allocation is directed by the researcher. In contrast, in an observational study the investigator has no control over which subjects are exposed to or deprived of a particular factor.

In applying the experimental method to human populations, it may not be possible, for a variety of reasons, to achieve all the requirements of a true experiment. The researcher may not have absolute power to decide which subjects will be exposed to a factor; it may not be possible to set up a control group; or the two groups—intervention and control—may not be strictly comparable. Study designs to accommodate such problems are often referred to as quasi-experimental. On occasions two or several groups in a population may have different levels of exposures to a factor in their environment. Although commonly referred to as 'natural experiments' and often offering great potential for study, they can only be treated as subjects of an observational study since there is no control over which subjects experience which exposure level and no assurance that the groups are comparable.

The classical application of the experimental design to studies of human populations is the clinical trial of a drug, which usually takes the form of a randomized controlled trial in which patients are randomly allocated to receive the new drug or to receive either the best available therapy or a placebo treatment. The same basic design can be applied to investigate approaches to prevention of disease including both individual measures, such as physical activity as a means of reducing an individual's risk profile for coronary heart disease, and multi-factorial population-based preventive programmes to reduce the overall incidence of a disease in a community.

A further application of the experimental study is the testing of hypotheses about the aetiology of a disease generated from various forms of observational studies. Where it is possible to manipulate exposure to one or more factors, then the natural progression is to an experiment. Any change observed in the intervention group following manipulation of its exposure to a factor or factors under study, while not automatically proving causality, provides strong evidence for such an association. To confirm causality it must be shown that no confounding exists, that there is a biologically feasible mechanism of an action, and that there is supporting evidence from observational and animal studies.

Since the early classical experimental studies conducted by the Medical Research Council (1948) on prevention of tuberculosis, the effectiveness, efficiency, and acceptability of many public health programmes have been subjected to rigorous controlled experimental trials (Cochrane 1972). Many of the diseases of current public health concern have very complex aetiologies requiring multi-factorial preventive programmes often based on a community-wide campaign. The evaluation of such a programme through an experimental study presents unique problems, and will be discussed later in this chapter.

Experimental studies of human populations and as applied to public health, follow general rules common to all scientific studies. In the following sections the basic design planning, and organization of an experimental study are outlined. This is followed by a discussion of the conduct of a specific form of experimental study, the community-based intervention trial, which is of particular relevance to present-day public health problems and will also provide a general discussion of practical issues in applying the experimental design to public health.

Basic principles

General study design

The general principle of an experiment is to achieve, through purposive allocation of subjects—individuals, groups, even whole communities—two or more groups that are as similar as possible. The factor under study is then manipulated in one group (the intervention group) and the end-points in both groups are compared. Given that the two or more groups are comparable in all aspects other than exposure to the factor under study, it can be concluded that any observed differences in the end-point for example mortality from a particular disease is due to this factor.

The method of choice for achieving comparability between groups is by random allocation, and this should be adopted if at all possible. Assuming that no bias is introduced, following random allocation any initial differences between intervention and control groups will be due to chance. Where random allocation is not possible, great effort must be made to ensure that there is no bias in allocation and that the study and control groups are as comparable as possible. Where the study unit is a whole community it is rarely possible to study a sufficient number to allocate communities randomly. In such situations, it remains only to select a reference community or communities that resemble the intervention community as closely as possible. Whatever method is adopted, the comparability of study groups must be checked and described.

Before subjects are recruited to a study, their consent to participate must be obtained. Final recruitment may be preceded by a screening procedure to achieve a homogeneous group, although for studies to evaluate the performance of a programme in a service mode, a criterion of the study may be to have a heterogeneous group representing a cross-section of the population who would use the service. There may also be a 'run-in' period with repeated visits to check the compliance of subjects, for example monitoring adherence to a drug regimen by undertaking urine analyses. Only those subjects who could comply successfully with the regimen would be included in the actual study and allocated randomly to intervention or control group. Such a procedure would reduce the drop-out rate during the period of the trial. Again, however, for evaluating the success of a programme in practice, the extent to which subjects dropped out would represent an important outcome of the study.

After allocation of study subjects, the experimental component is applied, usually for a fixed period of time. End-points, for example, morbidity, mortality, and blood pressure, are assessed after the experimental period is terminated and on occasions at intervals throughout the study period. The control group is essentially either deprived of something that may be beneficial, or remains exposed to something that may be harmful; thus the study design will often include rules that would allow the termination of the experiment earlier than planned if a significant difference is observed between the groups. There may, for example, be sequential, that is continuous, analysis of data, with termination of the study once a predetermined difference is obtained.

In certain circumstances, for example when evaluating a drug whose principal action is symptom relief, a cross-over trial may be performed in which each subject acts as his or her own control. In such a study a patient would be given the test drug for a fixed period of time and the outcome compared with that following a similar period during which he or she received a different or placebo drug. Such a design was adopted in a study of diet and coronary heart disease in a mental hospital in Finland (Miettinen *et al.* 1972).

Study population

The choice of study population is very important and should be clearly defined and stated. In medical studies the subjects are usually human, although they need not be, and this discussion will be restricted to considerations when selecting human populations. Often the nature of the problem to be studied, for example interest in a particular community or health programme, will dictate and define the population to be studied, but in many it must be purposively selected.

In selecting a study population, the suitability of the population for attaining the objectives of the study must be considered. For example, should the subjects be drawn from a random sample of a whole population or would it be more appropriate to select a specific homogeneous subgroup? A random population sample has the advantage that the results would be applicable to the general population. However, in other cases, for example when attempting to determine a causal relationship, a more homogeneous population would be more appropriate.

It must also be decided whether to restrict the study population to a particular section of the population defined by, for example, age, sex, ethnicity, or occupation. If, for example, the disease under study is associated with a certain age-group then the study can be restricted to individuals in that age range. In this way the study size can be reduced since the findings within that age range would not be diluted by those of other age-groups who are not affected by the disease. Similarly, where a condition is known to be related to a particular occupation it is appropriate to confine the study to that group; and when evaluating a preventive measure only asymptomatic individuals, possibly also from a specific age range, would be included.

Often economic and practical limitations will dictate the selection of the study population, for example the decision to study a random sample rather than every individual within a community. The study population must be of a manageable size—not so large that it is administratively impossible to handle, but not so small that the findings, even if positive, do not reach statistical significance.

Experimental component and study end-points

The experimental component applied to the intervention group will be determined by the objective of the study which must, as in all investigations, be clearly defined at the planning stage. If the objective is to confirm a causal relationship between a single factor and a specific disease then the experimental component would be a limit exposure of the intervention group to that factor and observe the number of end-points, for example positive diagnoses of the disease or mortality in each group.

However, the decision is not always so clear-cut. There are good indications that may diseases have a multi-factorial origin, and in addition some risk factors are synergistic, for example smoking and asbestos exposure in development of chronic bronchitis. Thus it must be decided whether to concentrate on one or several factors. It may be of interest to demonstrate the role of individual factors but, on the other hand, the potential for preventive actions may be determined only by examining a combination of factors in the environment.

The selection of end-points must also be considered carefully. These must be measurable and selected to complement the stated objectives. Mortality from a particular disease could be the end-point of an experimental study with the objective of assessing the effectiveness of a new therapy for a fatal condition. However, it may not be appropriate to confine the outcome measurement to a single disease. An environmental factor may not be specific to the disease under study. If low serum cholesterol prevents coronary heart disease but also increases the risk of, for example, colon cancer, then both these disease outcomes should be considered. A major World Health Organization co-ordinated study (Committee of Principal Investigators 1980) found that while clofibrate reduced the risk of coronary heart disease, the overall health outcome in the intervention and control groups was similar because the former group had a greater incidence of a number of other problems, particularly hepatic disorders, perhaps as side-effects of the drug.

To study the natural course of disease it is sufficient to observe the occurrence of disease in experimental and control groups. But from a general and public health point of view, a broader approach to outcome assessment should be considered. A drug trial should include adverse side-affects as an end-point, and major preventive programmes should attempt to measure the overall impact of intervention on health.

Often age-specific total mortality can be used as the ultimate end-point of a study and this is an outcome indicator applicable to the individual and the community. However, it is very unsensitive and can be used only for a powerful risk factor for a major fatal disease, for example for the study of the importance of treatment of blood pressure. Elevated blood pressure is a major risk factor for cardiovascular diseases that are responsible for approximately every second death in the middle-aged population of the US for example. A randomized controlled trial in the US involving 10 000 persons with raised blood pressure randomly allocated to systematic treatment or referral to community medical therapy, found a 17 per cent lower total mortality rate in the actively treated group compared with the control group over the five years of the study (Hypertension Detection and Follow-up Program Co-operative Group 1979).

Sample size and study period

The ultimate objective of an experimental study is to compare predetermined, measurable end-points in the intervention and control groups. Even where strict randomization of subjects has been observed there will always be random variation due to chance and a sufficient number of subjects must therefore be studied such that it is possible to determine that observed differences are statistically significant, i.e. that the probability of such a result occurring by chance is small.

It is not always easy to determine how large the study population should be; however, a calculation of the necessary sample size must be attempted at the planning stage. To do this a certain amount of information is required about: the objectives and design of the study, including characteristics of each study group; the expected disease rates and the magnitude of the change in disease rate or risk-factor level resulting from intervention; the accuracy of any measurements to be made; and the level of significance required to draw conclusions.

The nature of an intervention study may also result in a change in the control group, for example improvement in risk-factor score resulting from a change in behaviour brought about by an awareness of participation in a study. This was observed in the hypertension detection and follow-up study (HDFP) where at the end of 5 years 78 per cent of subjects in the intervention group were under antihypertensive therapy but there was also an increase in the control group to 58 per cent under therapy (HDFP Group 1979). Similarly another commonly experienced problem is an improvement in outcome in the control group because ethical considerations dictate that they should receive the best possible care. This may also follow from merely keeping the control group under regular observation. All these factors must be considered when planning study sample sizes since they may reduce the difference in end-points between study and control groups and thus necessitate a larger study population to achieve statistical significance.

The length of time over which subjects will be recruited into the study must also be carefully considered. If extended over too long a period, those entering the study early may not be comparable with later recruits owing to changes in other circumstances, such as introduction of new forms of treatment, publicity for a particular programme, etc. However, extending the study period is one means by which the sample size can be increased, and where a condition is rare it may be the only practical means of obtaining an appropriate number of subjects.

The time over which the subjects are studied following entry into the experiment is dictated by many practical problems, not least the commitment of subjects and investigators to co-operate with the study over long periods of time, and it is rare for experimental studies to extend over more than 5–10 years. Increasing the length of time for observation does, on the other hand, reduce the required sample size since this affords more time for differences between intervention and control groups to show up. And in the case of prevention programmes for chronic diseases, there is often a considerable time lag between the time that is optimal for initiating preventive action and positive diagnosis of disease.

Methods of intervention

The nature of the factor to be manipulated within the experiment will dictate the method of intervention. Where the study concerns the impact of a drug, the intervention is simply the administering of the drug to the intervention group and of a placebo or established treatment to the control group. Appropriate measures should, however, be taken to ensure that the subjects take the drugs as advised, and these range from counts of pills to assessment of serum drug concentrations.

For environmental and behavioural factors the selection of an approach to intervention is more problematic since individuals have great trouble changing established lifestyles such as smoking, diet, and physical activity. Furthermore, in long-term studies, while only a small proportion of the intervention group may achieve the desired change in exposure, an additional small proportion of the control group may make similar changes on their own initiative.

These problems may be so great, for example where studying chronic disease and weaker risk factors, that it is impossible in practice to conduct a controlled experimental study. The only really successful experimental study of the effect of dietary changes on risk of coronary heart disease was conducted in a psychiatric hospital using a captive population (Miettinen *et al.* 1972).

Thus the choice of intervention technique and the intensity of the intervention may be critical factors in the success or failure of an experimental study. Before answering the question 'Did the intervention prevent the disease?' one must address the question 'Did the intervention procedure succeed in manipulating the factor under study and to what extent?'. In the HDFP study mentioned above, although a drop in diastolic pressure of 17 mmHg was achieved in the actively treated group, that of the reference group also fell by 12 mmHg, and the net difference after 5 years was only 5 mmHg.

Assessment of end-points

This requires appropriate epidemiological methods for assessing the disease rate, for example repeated examination of subjects or recording and evaluating disease attacks or deaths of subjects.

It is important that all individuals originally included in the study are followed. Drop-outs should be traced where possible and included in any analysis, because they are often a highly selected group. The assessment of end-points must be identical for the intervention and control group, and every effort made to avoid bias. This is best achieved by a double-blind study design, in which neither the subjects nor the investigator knows whether the subject was allocated to an intervention or control group. This is commonly applied to clinical trials of drugs; however, often it is obvious to which group the subject has been allocated, either to the investigator (a single-blind experiment) or to both investigator and subject (open study). In such trials it is vital to apply fixed, quantitative criteria against which to assess end-points. Bias from knowledge of allocation may be overcome by having the end-points reviewed by an expert group unaware of the allocation. This would, in particular, exclude false-positive cases in an unbiased way. However, attention should also be paid to the elimination of bias concerning false-negatives, for example cases that are more likely to remain undetected in the reference or control group.

Community-based experimental studies

The classical experimental trial based on random allocation of individuals to experimental and control groups has limitations for studying certain currently important public health problems that focus on lifestyles and chronic diseases where the absolute risk is relatively low. Clinical intervention is expensive to implement on a large scale, treats individuals outside the context of their natural environment, and is unrealistic to consider for nation-wide application. Furthermore, as discussed earlier, where diseases have multi-factorial aetiologies and individual factors may affect more than one disease state it is not possible to restrict changes in the intervention group to specific risk factors without considering concurrent changes in other factors and disease outcomes.

A community-based approach overcomes many of these problems. This strategy aims at achieving overall change within the whole community rather than concentrating on individual persons, and tests the effects of a comprehensive package of interventions aimed at a whole range of aetiological factors within a natural community setting. Intervention is implemented through existing channels of influence and organizations within the community and takes advantage of natural interactions. Such a strategy may reduce costs significantly and, in some cases,, obviate ethical problems that might otherwise arise. The disadvantage of a community study is that the epidemiological inferences that can be drawn may be more restricted than those of a clinical trial because the use of large units can reduce the statistical power of a comparative study. However, despite this limitation, community studies can contribute significantly to clarifying the causality of chronic disease. Furthermore, a community study often has other aims such as achieving better use of existing health and other community services or providing a demonstration model (see below).

In the fight against the present epidemic of coronary heart disease, research has proceeded from basic and descriptive epidemiological studies to large-scale intervention studies that were started during the 1970s. The first major community-based study was the North Karelia Project in Finland that was launched in response to a petition by the local people for governmental action to counter the extremely high prevalence of heart disease in that region. Recently, several community studies have been launched in other countries. The following text makes reference to some experiences from the North Karelia Project (Puska *et al.* 1981, 1985).

Community-based intervention—study design

Such a study uses a quasi-experimental design in which one or several communities are allocated to receive the experimental intensified intervention programme and one or several communities are selected to serve as reference areas which represent the 'natural' development in the country. In the intervention community an innovative programme is implemented to apply the best possible approaches to changing the population level of one or more factors in the whole community or in a major segment of it. The reference community is not deprived of any new developments in health care, etc. that might occur other than those represented by the experimental programme.

The observation unit is a community. Several communities may be used to increase the population but it is usually not realistic to include a sufficient number to use the community as a unit of the statistical analysis. In addition, the use of the two or more communities creates interpretational complications in the event of a positive result in one community and a negative one in another. Such a situation may arise through transfer of knowledge and experience between intervention teams.

In all quasi-experimental designs, where the assignment into experimental and control units is not random, there is the possibility of both biased selection of experimental and reference units and of biased sampling (selection of study units). In the case of the North Karelia Project the experimental unit was already given before 'sampling'. North Karelia is a mainly rural county in Eastern Finland with approximately 180 000 inhabitants. The only choice for the evaluation was in the selection of a suitable reference unit. Another county in Eastern Finland was chosen as the reference area.

The main design for the assessment of the effect of intervention on risk-factor levels is the 'separate-sampling pre-test–post-test control group' design as presented by Campbell and Stanley (1963). Separate cross-sectional samples are drawn from the same populations, one in the intervention and another in the reference area, before and after the study period. The net reduction in disease and risk-factor levels in the intervention community (the reduction in the intervention area minus the reduction in the reference area) is considered to be the effect of the intervention.

Selection of intervention and reference communities

The intervention community ideally should be typical of the larger area or the whole country to which the results are to be applied. Often, however, this choice is guided by historical or practical factors, as was the case in the decision to make North Karelia the subject of an intervention trial. In selecting a community it is particularly important that the area chosen does not have exceptionally good resources; North Karelia, for example, had the lowest level of service resources and was the least developed in socio-economic terms of all the counties of Finland. If the intervention is successful in a community of average or below average resources, it can be concluded that the introduction of similar programmes would be feasible in other parts of a country.

It is usually easier to establish an intervention programme in a small community, and the evaluation of the intervention process would also be simpler. However, increasing the community size provides a greater number of disease events and usually provides a setting more typical of the region or nation as a whole for which the intervention programme is ultimately to be applied. Where one intervention and one reference community are to be compared, each community should be large enough to provide a sufficient number of disease events of interest and to enable relatively independent samples at subsequent time points and of sufficient size, so that the significance of the net differences in the disease rates in the two communities can be tested.

Where, for example, a risk factor is very prevalent, change in that risk factor will be the only end-point in assessing the effect of intervention activities, and in such cases only a relatively small community would be required. However, where chronic diseases are included as end-points of the study, considerably larger communities are required. Depending on the disease rates and the length of follow-up, a community of 250 000–500 000 would be necessary for a community intervention study aimed at coronary heart disease (World Health Organization 1981).

A reference community is essential because changes may occur 'spontaneously' as societies change their lifestyle, for example through increased popular awareness of the risk factors, technological or fashion changes, or increasing or improved treatment of risk factors by health professionals. In order to separate the effect of intervention from general trends of change, the intervention must be compared with a reference community. If risk-factor levels are decreasing nationally, then the 'net' reduction in the intervention community, i.e. the impact of the intervention programme, will be lowered. Where the national trend is for risk-factor levels to increase then there may not be sufficient time to reverse that trend although even a slowing-down of the increase would suggest that the programme had had some positive effect on lifestyles.

Changes in an intervention area can be compared with national trends. This, however, may be misleading since often there is considerable within-country variation and the 'natural' trends occurring in an area in the absence of intervention might not coincide with the national pattern. Thus it is preferable to select a reference community that is similar in all respects to the intervention community.

In the case of the North Karelia project, the reference unit

Table 12.1. Main target health behaviours and risk factors in North Karelia (NK) and the reference area (REF) according to cross-sectional population surveys in 1972, 1977, and 1982 among men and women aged 30–59 years

	Percentage of smokers		Amount of smoking		Fat from milk and on bread (g)		Serum cholesterol		Systolic blood pressure (mmHg)	
	NK	REF	NK	REF	NK	REF	NK	REF	NK	REF
Men										
1972	52	50	10.0	8.6	83	72	7.1	6.8	149	146
1977	44	45	8.6	8.5	59	62	6.7	6.7	143	146
1982	38	45	6.6	7.9	54	61	6.3	6.3	145	147
Significance[2]	***	**	***	n.s.	***	***	***	***	***	n.s.
Net change in NK										
1972–82	−17%		−28%		−22%		−3%		−4%	
Significance[3]	***		***		***		***		***	
Women										
1972	10	11	1.1	1.2	45	39	7.0	6.8	153	147
1977	10	12	1.2	1.4	31	33	6.6	6.5	142	144
1982	17	18	1.7	1.9	28	29	6.2	6.0	141	143
Significance[2]	***	***	***	***	***	***	***	***	***	***
Net change in NK										
1972–82	+1%		−14%		−16%		−1%		−5%	
Signficance[3]	n.s.		n.s.		***		n.s.		***	

[1] Number of cigarettes, cigars, and pipes per day per subject (smokers and non-smokers taken together).
[2] Statistical significance of the difference between 1972 and 1982:
*** $P < 0.01$; ** $P < 0.01$; n.s., not significant.
[3] Statistical significance of the net change in North Karelia between 1972 and 1982.

selected was a neighbouring county in Eastern Finland, Kuopio. This was the county that most resembled North Karelia in its cardiovascular disease mortality and morbidity, and in its geographical, occupational, economical, and social features. The reference unit was in this sense 'matched'. Bias in the selection of the control unit, if any, was against the study hypotheses, because the reference county was not amongst the most unfavourable in view of its future prospects of improvement in economy, social conditions, living habits of the population, and mortality and morbidity. For example, a new university including a medical school was opened in the reference area in 1972.

The available results from the North Karelia project surveys confirm that North Karelia had slightly higher risk-factor levels than the reference area. The significant net reduction observed during the five-year period was to a great extent due to North Karelia catching up with the reference area. The results might have been due to a saturation effect or to regression towards the mean; but this is not the case for all risk factors. For example, for systolic blood pressure and also for smoking in men, North Karelia overtook the reference area (Table 12.1).

There were, however, also some favourable changes in the reference area (for example, reduction of smoking). It is difficult to determine whether these changes represent national trends independent of the intervention, whether the intervention programme contributed to national interest and lifestyle changes and risk-factor control, or whether there was direct 'spill over' or contamination from North Karelia to the neighbouring county.

Study period

The length of time over which the intervention study continues is very important since a short study period may not provide sufficient time for permanent changes to occur while a long study period may result in a levelling-off in the differences between areas. Furthermore, different end-points may have different optimum time periods. Changes in health behaviour, for example, can be detected quite quickly, changes in levels of risk factors somewhat later, while changes in incidence and finally mortality will be detected only after a considerably longer time period.

In the North Karelia project, for example, most of the reduction in cigarette smoking took place in the first year of the programme; most hypertensives who brought their blood pressure under control, achieved this by the end of the third year; dietary changes took place gradually over a five-year period; and, as noted earlier, at the end of five years, a net reduction in risk-factor levels was observed. Five years, however, was too short a period of time to assess changes in mortality. Coronary heart disease incidence and mortality rates started to decline surprisingly quickly after the start of the intervention in North Karelia. In the rest of the country similar decline started several years later. Thus a significant net change in favour of North Karelia was observed, especially in 1974–9 (Salonen *et al.* 1983). Thereafter, although the decline in North Karelia continued, the net decline was gradually reduced (Fig. 12.1). For cancer mortality, a net reduction in favour of North Karelia could be observed much later, i.e. 5–10 years after start of the intervention.

Samples for population surveys

The aim of an intervention programme is to effect changes in risk factors that will in turn lead to changes in disease rates. The success of a programme in achieving such changes is assessed by comparing risk-factor levels and disease rates in a cross-sectional population sample at the outset with the findings for an independent cross-sectional sample at the end of the intervention.

Independent cross-sectional samples are used in preference to a longitudinal follow-up of a cohort because disadvantages of the cohort design severely limit its application to estimating the impact of intervention. Involvement in a survey may in itself affect the behaviour of the subjects, and those who participate in the survey preceding implementation of an intervention programme subsequently may be more sensitive to the programme activities. Among this group, any observed change may be due in part to participation in the pre-test survey and not to the intervention. Thus the true magnitude of the effect in the intervention programme can be measured only by examining a new random sample of the population at the end of the trial period.

Findings from the North Karelia project confirm that longitudinal follow-up inflates changes compared with the findings from independent cross-sectional surveys. For example, North Karelian men followed up longitudinally decreased their smoking by 11 per cent from that of the baseline survey over 3 years, while in an independent representative random sample surveyed 3 years into the study, a drop of only 7 per cent was recorded.

For these reasons the main assessment of the impact of intervention is based on repeat cross-sectional samples. Longitudinal follow-up of baseline survey samples, however, can provide useful supplementary information, for example about the characteristics of individuals who change their behaviour and lifestyles compared with those who do not. The cohort design also has certain analytical advantages such as adjustment for differences in baseline levels of risk factors and more efficient testing procedures.

The choice of sample size usually depends on the magnitude of changes that are to be detected, the required confidence, and the cross-sectional intra- and interpersonal variation. Change in the reference area must also be taken into consideration. The detection of risk factors *per se* does not usually need a very large sample size. However, detection of net changes that, while small, are important if in the same direction for all risk factors implies a larger sample size. This is also the case if several subgroups are to be analysed separately.

The sample age range is also an important issue. Obviously the whole population is the target of the intervention but often, as in North Karelia, the intervention programme, although comprehensive, emphasizes persons of certain ages because of the nature of the problem.

At the end of the intervention period either a second independent sample of the same age-groups is examined, or an

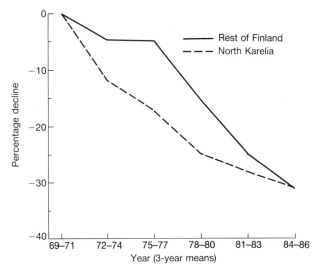

Fig. 12.1 Percentage decline in CHD mortality of 35–64-year-old men in North Karelia and the rest of Finland.

independent cross-section of the same birth cohort. Use of the same birth cohort increases the comparability of the baseline and terminal measurements because it avoids any possible unrecognized birth cohort effects (for example due to wars or famines). However, this means that the sample at the end is, for example, five years older, which may bias the observed absolute changes (that is, intervention-related changes in risk factors are countered by increases due to ageing). This effect is controlled for when the change in the intervention area is compared with the change in the reference area to describe the net change. Obviously, the analyses can, if necessary, be restricted to the same age-group at both time points.

Survey implementation

The pre-and post-test survey conducted in intervention and reference areas should be strictly comparable. The measurements should be well standardized and tested. Often self-administered and pre-coded questionnaires are used. The measurements are carried out by personnel (often nurses) who are carefully instructed and trained before the surveys to use standardized and often internationally accepted measurement techniques. Strenuous efforts must be made to use identical procedures in the two areas and in the two surveys. The time of the year should also be considered because of possible seasonal variation.

High participation rates are of vital importance to avoid bias in the results. In the North Karelia project, the data quality was strengthened by the high participation rates of both areas in the two surveys. The participation rate did vary between the two areas, although the tendency was small. At the outset the participation was higher in the intervention area, presumably because people were interested in participating in the programme. At the end, however, the response rate was smaller in the intervening area, probably due to a

fall-off of interest following exposure to the numerous intervention activities organized during the several years of the
programme.

Proper laboratory standardization of survey measurement
is important if results are to be comparable, and ideally samples should be sent to a central laboratory for analysis by
technicians who do not know whether samples originated
from the intervention or reference area.

Disease and mortality surveillance

Monitoring disease and mortality rates at the community
level has many problems. Examinations of cross-sectional
representative samples do not give much information about
the incidence of new cases.

A register, even with the most complete coverage and
rigorous criteria, is dependent on individuals seeking health
services or being identified in other ways. It is possible and
even likely that an intensive community intervention stimulates people to seek medical aid more actively and with
milder symptoms, which would tend spuriously to increase
the incidence rates. In the North Karelia register it was found
that incidence of 'definite' acute myocardial infarction
decreased more than 'possible' acute myocardial infarction
during the intervention. This might have been because, in
response to the intervention, persons with milder symptoms
were more eager to seek medical help. The actual decrease in
the acute myocardial infarction incidence rate in North
Karelia may thus have been greater than indicated by the
register.

A further problem encountered with a community-based
disease register is the maintenance of the same diagnostic criteria and coverage. A blind reclassification of cases may be
done after the study period to confirm the consistency of the
diagnostic criteria. To ensure the completeness of coverage,
death certificates, hospital records, and other available
sources of notification should be checked continuously.

The launching of a new permanent disease register in the
reference area can represent a substantial intervention which
may minimize the impact of the programme in the intervention area. And because a better health information system
(including registers) can be part of the comprehensive intervention programme, its contribution cannot be assessed if a
register is also established in the reference area. To avoid
contamination due to the introduction of an *ad hoc* register,
the ideal solution would be to monitor disease and mortality
rates based on national routine statistics. This is adequate
when a comprehensive centralized hospital data system is
available. In some countries hospital discharge data achieve
complete coverage although the reliability of the diagnostic
data is less satisfactory. In other countries, where hospital
discharge data are less complete and reliable, it is necessary
to establish disease registers in both the intervention and
reference areas.

The hardest end-point is mortality, although there are limitations with regard to cause-specific mortality, especially in
areas where autopsy rates are relatively low. The observed
rates are dependent on physician customs of completing the
death certificates and these may change along with the programme. Age- and sex-specific total mortality rates are, of
course, the more reliable indices; however, they lack sensitivity because mortality is the far end-point in the course of
the disease and, for example, only part (although a major
one) of the total mortality is due to cardiovascular disease.

Interpreting the results of a community study

If the effect of the intervention programme on a risk-factor
level is negative this may be due to two reasons: (i) the effort
was not great enough; or (ii) the method was ineffective. A
major problem is that often the intensity of the intervention
is too limited compared with the magnitude of the task and
the limited success achievable through curative medicine. It
is difficult to separate these two alternative explanations.

In the case of the North Karelia project, the results indicate that the population risk-factor levels were changed as a
consequence of the intervention. Five years was considered a
sufficiently long period to conclude that the changes on the
individual level were not short-term fluctuations. The community-based disease registers showed in North Karelia a
reduction first in the incidence of stroke and, later, in the
incidence of acute myocardial infarction and cardiovascular
disease. Since this was reflected in all-causes mortality rates it
was not due to a shift in diagnostic habits. However, firmer
conclusions needed a longer period of time: a significant net
reduction in coronary heart disease mortality could be
observed only in the analysis of the ten-year experience
(Salonen et al. 1983).

If disease rates do not follow changes in risk-factor levels
of the population, there are several possible explanations for
this: (i) the risk factors are not causal; (ii) the risk-factor
changes should be initiated earlier in life; (iii) the risk-factor
changes were not large enough; or (iv) the risk-factor
changes should be sustained for a longer time for disease
changes to occur.

The evaluation of a community-based intervention can
ultimately only assess the effect of the whole package that is
applied to the intervention community. The relevant contributions of the different components to the success or failure
can be evaluated only to a limited extent; separate study
designs would be needed for that. It is therefore important
that the whole package be designed for possible application
on a larger scale in other areas or the whole country.

The results concerning the effects of community intervention on risk factors are bound by time, place, and situation.
However, through careful and comprehensive evaluation,
the meaningfulness of the situation as well as the effects of
different intervention components can be discussed and
interpreted. The results concerning the impact of risk-factor
changes on disease rates should be more universal, although
even here there may be differences between populations.

In follow-up studies it has been shown that the relative

impact of some risk factors can vary in different populations. In this field, as many others, comparing experiences from different community studies in different cultural and community settings is valuable.

Evaluation of a community-based intervention programme

As stated earlier, a community-based intervention study can usually be seen from a broader perspective than just an experiment to test the possible relationship between a given exposure (risk factor) and a disease. It aims at assessing to what extent and by what means present knowledge can be applied to the community to tackle a health problem. Thus it is often a demonstration of the effectiveness of a given intervention programme, and a prerequisite is a considerable amount of prior epidemiological knowledge about the health problem concerned. The intervention is usually carried out as a systematic programme, often integrated with the community service structure and social organization. Depending on the needs and resources, a broad range of evaluation objectives and principles can be applied to the assessment of a community-based intervention programme.

General concepts

Evaluation is an essential and integral part of any programme, for example to determine whether the programme is achieving its objectives and to improve the programme (for decisions concerning its continuation and wider application). The aims and priorities of the particular programme concerned obviously have a great influence on the details of the evaluation.

The intervention aims, through certain inputs and processes in the community (health services and other activities), to achieve certain desired effects, output. The programme involves planning, implementation, and finally evaluation of the results. Often this is a continuous process, with all these activities occurring at the same time, and evaluation results continuously fed back into the planning.

Different kinds of evaluation are needed, the proper balance of which is determined by the nature and priorities of the programme. Some of these different (and overlapping) concepts are listed below.

1. Evaluation concerns whether the programme has reached the stated objectives, while evaluation research involves scientific inference of the basis hypotheses on changes which occur.

2. Formative evaluation is continuously carried out in connection with the different activities in order to formulate and improve the programme during its course. It often takes the form of simple interviews of small groups of people or follow-up of such things as utilization or sales statistics. Summative evaluation is conducted after a set period of the programme, to obtain a summary picture of the programme results.

3. Every programme should include both continuous evaluation and final or periodic evaluation. Continuous evaluation concerns the follow-up of the indicators of different levels of objectives (health services use, habits, risk factors, disease rates, etc.). Monitoring of these trends is a tool of this continuous follow-up. Final or periodic evaluation is carried out after a certain, usually predetermined, period of programme operation. The same data sources as for the continuous evaluation (e.g. disease registers) are normally also used for final evaluation, as are specific terminal surveys, etc.

4. Internal evaluation deals with the simple evaluative measures built into as many of the programme activities as possible and usually carried out by the same people that implement the activities. External evaluation is carried out by a specific evaluation team to assess the overall programme results. The former activity usually deals with easily obtainable forms of information on the activities, while the latter concerns the high-quality data related to the central programme aims.

Evaluation priorities and components

Intervention programmes aim at the prevention and control of certain diseases in an area. The aims of the programme may have a different emphasis, which leads in turn to a different emphasis in the evaluation. Major long-term programmes in larger communities are interested in assessing, in epidemiological terms, whether the programme has influenced the level of risk factors in the target population and whether this in turn has resulted in decreased disease rates and improved health ('experimental' component). Other programmes are restricted only to demonstrating that certain activities lead to such changes in habits, risk factors, or environmental factors that are generally considered beneficial to prevention or control of certain chronic conditions ('demonstration' component). Still other programmes are primarily health service oriented, that is, the development of better ways to deliver the health services to control these diseases ('health service operational' component).

If the programme is health service oriented (for example, better detection and treatment), then the evaluation naturally emphasizes health service-oriented research. But in programmes of intervention against chronic disease risk factors, much of the work goes far beyond the health services (various strategies in promoting lifestyles changes in the community, etc.) and evaluation focuses on other aspects, such as epidemiology and sociology. Even in a health service-oriented programme in a larger community it is advisable, in the long run, to evaluate performance also in terms of effects on mortality and morbidity.

Since in practice most programmes will combine different aims, the evaluation should accordingly be comprehensive with emphasis dependent upon that of the programme and upon local conditions.

The evaluation of any pilot programme should consider the

following related, and on occasions overlapping, evaluation components.

1. *Feasibility*. This is the assessment of whether the planned intervention activities could be implemented and to what extent, and of what proportion of the target population was covered or reached by the programme. In feasibility evaluation, the final performance of the programme is compared with the initial list of planned activities, and this forms the basis for understanding the possible effects of the programme. If this part of the evaluation shows that the programme was not feasible under the local conditions, it is not necessary then to conduct an effect evaluation.

2. *Effects*. This part of the evaluation is concerned with assessing whether the programme reached its stated objectives. Depending on the local programmes (as mentioned above), this may concern health-related habits, risk factors, and/or health service utilization, plus, possibly, mortality, morbidity, and/or disability rates (representing different levels of objectives).

   After definition of the objectives the respective indicators of the objectives have to be defined and appropriate data sources and measurements (mortality statistics, disease registers, random sample surveys, health service utilization data, etc.) selected. A standardized system for comparing changes in these indicators will be required because the changes observed may be due to factors other than the programme ('spontaneous change', 'national development', etc.). For this purpose, a reference community is needed that is, ideally, matched with the programme community and even decided by randomization. In practice, however, choice of both programme and reference communities is often determined by a number of practical aspects. The changes in the intervention community should also be compared with whatever statistics are available on a regional and national basis. This is important, especially in those cases where a specific reference area is not feasible.

3. *Process*. Process evaluation aims at assessing how the different programme components in the local community (integrated with the local health services and social organization of the community) achieve, with time, the programme objectives. This evaluation concerns both detailed assessment of the different steps and measures in the intervention, and the occurrence of risk factor and disease changes with time. In the latter aspect, the systems for monitoring these trends form the evaluation tool. In order to facilitate process evaluation, it is advisable at the planning stage to have a detailed flow-chart to illustrate the plan of action.

4. *Other consequences*. A major intervention in the community is likely to lead to consequences other than those specified in the objectives. These may include health-related, social, or psychological changes, and can be both positive or negative. Such consequences might be, for example, increased side effects or widespread medication, increased anxiety, or increased feelings of security, satisfaction about the services, and better quality of life. A major pilot programme should pay attention to this part of evaluation because of its implications for more widespread nation-wide application of the programme.

5. *Costs*. Assessment of programme costs in relation to the observed effects, that is the cost–effect ratio of the programme (the efficiency), may be expressed, for instance, as costs per saved life. This information can be used to compare costs of different strategies leading to the same effect or to compare programmes leading to different effects but which cost a similar amount to run. The essence of a cost–benefit analysis is to quantify and value all benefits and costs associated with the programme and to express them in a common denominator.

The first step in the evaluation of costs is to assess the direct costs of the project, that is the extra input (training, materials, co-ordination) that lead to the implementation of the intergrated programme in the community. Thereafter, the direct community costs should be assessed. They include the costs of the health services and other activities in the community for prevention and control of the diseases under study. It may, however, be advisable to differentiate between (the usually very large) costs that would have occurred in the absence of a specific programme and the extra costs (or savings) resulting from the intensified system. Comparison with the respective costs in the reference area may be used in such an evaluation.

Other aspects

The evaluation of a programme draws on a number of different professional skills, preferably in an intergrated way, although the emphasis will again be guided by the priorities of the programme. Epidemiological research (disease rates, risk-factor levels, etc.) and health service research (operations research) are especially important. But attention should also be paid to research concerning nutritional, economical, psychological, social, cultural, and anthropological aspects. In addition to conventional statistical-type studies, other approaches should be considered, such as in-depth, unstructured interviews, participatory observations, and other unobtrusive measurements.

The suitable time period for a summative evaluation depends very much on the nature of the programme and the aims of the evaluation. It is obvious that a long period would be required to assess possible morbidity and mortality changes, while a shorter period would be sufficient for such intermediate objectives as changes in health service use, environmental factors, lifestyles, and, possibly, risk factors.

Even if a programme ultimately concerns a whole community, for evaluation purposes practical priority decisions have to be made concerning the age-groups to be assessed. Again, these obviously depend on the nature of the programme. A major emphasis on mortality and disease rates, invalidity, health services, and disease-related costs would emphasize older age-groups, while a major emphasis on future community development and social processes would stress younger age-groups in the community.

Finally, it is important that appropriate decisions be made on the basis of the evaluation results. They may concern either the strengthening and/or continuation of a programme or possible national or other large-scale applications. It is important to realize that evaluations seldom give a straightforward answer to the problems concerned. Thus, it is highly recommended that the programme team, throughout the process of evaluation, stays in close contact with the decision-makers for continuous feedback of information and ideas to contribute both to the ongoing evaluation and health policy decision-making.

References

Campbell, D.T. and Stanley, J.C. (1963). *Experimental and quasi-experimental designs for research*. Rand McNally, Chicago.

Cochrane, A.L. (1972). *Effectiveness and efficiency. Random reflections on health services*. The Nuffield Provincial Hospitals Trust, Oxford.

Committee of Principal Investigators (1980). WHO co-operative trial on primary prevention of ischaemic heart disease using clofibrate to lower serum cholesterol: mortality follow-up. *Lancet* i, 379.

Hypertension Detection and Follow-up Program Co-operative Group (1979). Five-year findings of the Hypertension Detection and Follow-up Program. Reduction in mortality of persons with high blood pressure, including mild hypertension. *Journal of the American Medical Association* **272** (23), 25.

Medical Research Council (1948). Streptomycin treatment of pulmonary tuberculosis. *British Medical Journal* ii, 769.

Miettinen, M., Turpeinen, O., Karvonen, M., Elosuo, R., and Paavilainen, E. (1972). Effect of cholesterol-lowering diet on mortality from coronary heart disease and other causes: a 12-year clinical trial in men and women. *Lancet* ii, 835.

Puska, P., Tuomilehto, J., Salonen, J., *et al.* (1981). *The North Karelia Project: evaluation of a comprehensive community programme for control of cardiovascular diseases in North Karelia, Finland 1972–77*. WHO, Copenhagen.

Puska, P., Nissinen, A., Tuomilehto, J., *et al.* (1985). The community based strategy to prevent coronary heart disease: Conclusions from the ten years of the North Karelia Project. *Annual Review of Public Health* **6**, 177–193.

Salonen, J.T., Puska, P., Kottke, T.E. *et al.* (1983). Decline in mortality from coronary heart disease in Finland from 1969 to 1979. *British Medical Journal* **286**, 1957.

World Health Organization (1981). *Proposal for multinational monitoring of trends and determinants in cardiovascular disease and provisional protocol (MONICA project)*. WHO, Geneva.

13

Statistical methods

A.V. SWAN

Introduction

Statistical methods are required whenever groups of individuals, not entirely predictable in their characteristics or behaviour, need to be described or compared. A wide range of techniques has been developed and this text cannot cover them all. The most useful are known as regression techniques. This chapter concentrates on statistical methods that can be considered under that heading. The material on case–control studies is dealt with more fully in a separate chapter.

The chapter is divided into six sections which can be regarded as a course in three parts. The first two sections after this introduction, 'Design and sampling' and 'Practical questions from a statistical point of view', provide a general survey of statistical ideas and practical problems. This is followed by a more advanced discussion of the underlying principles of statistical reasoning under the heading, 'Basic concepts'. The final part discusses in some detail the application of the statistical methods appropriate to the problems presented in the 'Practical questions' under the heading 'Types of analyses'.

The aim of the chapter as a whole is to give a clear picture of the range of problems that can be handled by statistical methods with a general appreciation of the methods and the underlying statistical concepts necessary for analyses using computer packages such as Minitab (Ryan *et al.* 1981) and GLIM (Payne 1983). In addition, more comprehensive details are provided for the special cases where the problem can be simplified sufficiently for the analysis to be done by hand.

Design and sampling

The statistical design of an investigation is determined by the form of analysis appropriate to answering the question of interest, the precision required, and the assumptions that appear justified. The practical question should indicate quite clearly what analyses are necessary. The design is then determined by what is necessary to make these analyses possible and sufficiently precise.

Circumstances and ethics may mean an *observational* study has to be used. That is, a study where subjects exposed to a variety of influences are simply observed. In such cases sampling is restricted to a process of choosing who enters the study. If the investigation is using previously collected data, for example, cardiovascular mortality figures from areas of differing water hardness, there is no sampling involved. If an *experiment* or *clinical trial* is possible then the sampling process also includes allocation to the groups to be compared. A comprehensive discussion of this and other aspects of clinical trials is given by Pocock (1984). Because it is often important to determine the most effective treatment as quickly as possible and to minimize the number of patients receiving the other treatments, what are known as sequential trials with special allocation rules may be very useful. The subject is well covered by Armitage (1960).

In general, sampling does not need to be complicated. Although it may be awkward in an administrative sense, the process is in principle straightforward. The first necessity is to decide on the population of interest and identify that part of the population that is accessible in practice. The sampling process is then applied to that subgroup of the population with due regard to the possibility that it is not properly representative of the whole. For the sampling process it is usually sufficient in both selection and allocation to use what is known as simple random sampling. This means the use of any technique, such as 'tossing coins', 'drawing names from a hat', 'taking every tenth name on a list assumed to be in random order', etc., that gives every subject an equal chance of selection.

In practice the problem tends to be that of obtaining a list—called a sampling frame—of all the candidates for inclusion. In community health surveys in the UK a commonly used sampling frame is the electoral register.

For complex sampling problems it may be necessary to use random sampling numbers which are included in most sets of statistical tables (Lindley and Miller 1964). They usually consist of sequences of two-digit numbers, for example 87, 31, 47, 55, 56 . . . generated in such a way that in any reasonably long sequence the number of 0s, 1s, 2s, . . . 9s and consequently 00s, 01s, . . . 99s are the same. Each single digit will

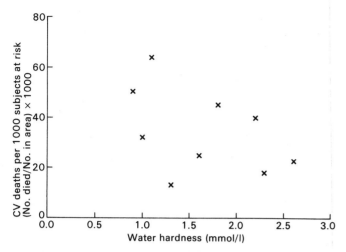

Fig. 13.1. Cardiovascular (CV) mortality by water hardness. (Source: Pocock *et al.* 1980.)

occur on average once in every ten digits and each double-digit combination once in every hundred pairs. For example, to allocate subjects to three groups in a trial a sequence of digits, ignoring the zeros, would be assigned, one at a time, to each of the subjects. Then allocation is determined by taking the digits 1, 2, and 3 to represent group 1; the digits 4, 5, and 6 to represent group 2; and the digits 7, 8, and 9 to represent group 3. Sometimes particular subgroups (essentially subpopulations) of individuals are of specific interest, for example sex or age groups. In that case it may be appropriate to identify the individuals belonging to the groups—known as strata—and sample separately within each. This is known as stratified random sampling. This process can also be performed using 'pseudo' random numbers generated with a computer.

Practical questions from a statistical point of view

Trends—routinely collected grouped data

It has been suggested that the 'hardness' of drinking water may affect the risk of an individual developing heart disease. The cardiovascular death rates and some measure of the water 'hardness' in each of a number of geographical areas (for a particular year) might give a picture as in Figure 13.1. This is known as a scatter diagram.

The question is: 'Does the risk of cardiovascular disease change according to the 'hardness' of the water you drink?' In terms of the pattern in the diagram the question becomes: 'What is the trend in this sample of points and does it reflect a real trend in all the points from other subjects, other years, and other areas that these points represent?'

The 'truth', were it possible to see it, might be as in Figure 13.2 (a) or (b). The analysis is a process of determining which is the most likely 'truth'.

Trends—individuals in an observational study

In an investigation of blood pressure and salt intake the question could be: 'Does an increase in salt intake increase systolic blood pressure?'

If it were possible to determine salt intake for a sample of individuals a plot of each individual might give a scatter diagram such as that in Figure 13.3.

As before the question becomes: 'Is there a trend in these points consistent enough to suggest a real trend in the large population of points that this sample is taken to represent?'.

If data were available from men and women (Fig. 13.4) then the question would be more complicated. The relationship between systolic blood pressure and salt intake, such as it might be, could differ between the sexes. The initial question becomes several separate questions: (i) Are the trends in the populations of men and women represented by these samples different? (ii) Is either one, if they are different, non-horizontal? (iii) What are the trends?

There is also information with which to answer the questions: (a) At a given salt intake do the male and female populations differ on the systolic blood pressure scale? That is, at a given salt intake are the population points for the males

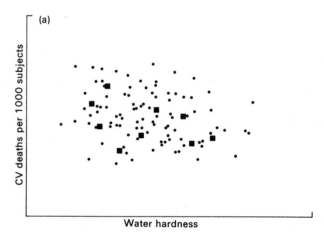

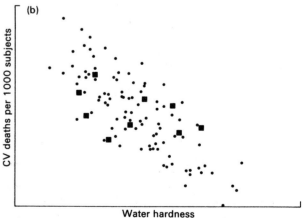

Fig. 13.2. Possible population patterns for cardiovascular (CV) death rates by water hardness. (a) No trend. (b) A negative trend.

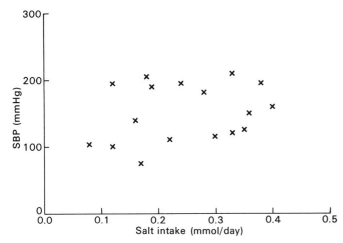

Fig. 13.3. Systolic blood pressure (SBP) against salt intake (simulated data).

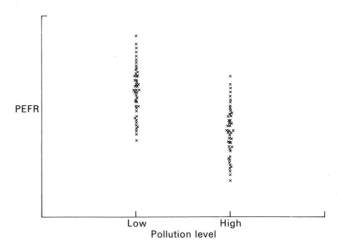

Fig. 13.5. Peak Expiratory Flow Rate (PEFR) in children by air pollution levels (simulated data).

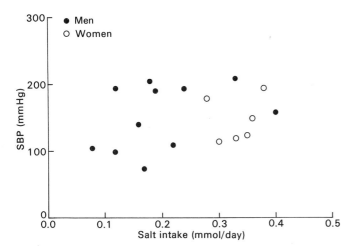

Fig. 13.4. Systolic blood pressure against salt intake in men and women (simulated data).

higher or lower, on average, than those of the females? (b) Do the male and female populations differ with respect to salt intake? Or, in the population are the male and female points positioned differently on the salt intake scale?

Simply introducing another factor, sex, has considerably complicated the problem.

In practice it may be necessary to consider weight, exercise, age, and possibly other factors related to systolic blood pressure, and the complications multiply rapidly.

Comparing groups of individuals

Investigations of air pollution and respiratory health encounter considerable problems in determining an individual's exposure. Often the design has to be a simple two-group comparison of individuals living in an area of high pollution with those living in one of low pollution. Such a study might produce peak expiratory flow rate (PEFR) values from children in high and low pollution areas as shown in Figure 13.5.

The question is: 'Does air pollution affect lung function in children?'

With some assumptions this can be phrased as: 'Do children in areas with different pollution levels have different lung function values on average?'.

On the plot this becomes: 'Do the heights of these two samples of points differ enough to suggest a difference between the populations they represent?'

Proportions from two categories of outcome

Frequently an all-or-nothing response such as presence or absence of a symptom will be the outcome measure of interest. The appropriate response scale is the proportion with the symptom, i.e. the prevalence of the symptom. An investigation of pollution as above may be interested in the prevalence of morning cough in children from areas with different levels of pollution.

The question is: 'Does exposure to pollution increase the risk of a child developing morning cough?'. A plot of the prevalences against pollution level with the points joined by straight lines will generally give a line with a number of bends in it (Fig. 13.6).

The question becomes: 'Do these sample proportions differ enough to suggest that differences would be seen if the populations of all such children exposed to these pollution levels could be observed?'. In other words, is the equivalent population line horizontal?

Information on parents' smoking habits will give a plot of two lines (Fig. 13.7). There are now several questions: (i) Are the differences between the pollution groups in the population the same in the two parental smoking groups? That is, are the population lines parallel to each other? (ii) Are the population differences between the pollution groups non-zero in either parental smoking group? That is, is either population line non-horizontal? (iii) At a fixed pollution level is there a difference between the parental smoking groups in

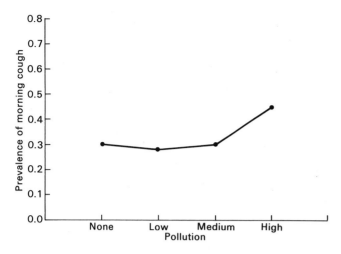

Fig. 13.6. Prevalence of morning cough in children against pollution level (simulated data).

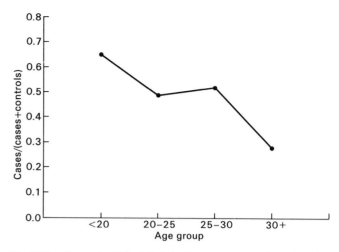

Fig. 13.8. Cases of sudden infant death as proportions (cases/(cases + controls)) by mother's age group.

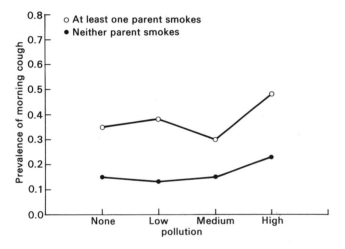

Fig. 13.7. Prevalence of morning cough in children against pollution level by parents' cigarette smoking habits (simulated data).

Table 13.1. Cases of sudden infant death and controls by mother's age

	Mother's age (years)			
	<20	20–	25–	30+
Cases	17	22	17	5
Controls	9	23	16	13
Cases and controls	26	45	33	18
P	0.65	0.49	0.52	0.28

the population? That is, is there a vertical separation between the population lines?

Proportions: case–control studies

In a study of sudden infant death syndrome various information was collected on all the infants dying this way in a particular area and period of time—these were the cases. The same information was collected on the same number of randomly selected control children born in the same period and still alive. Such studies are known as case–control or retrospective studies. The latter name arises because the study starts after the event of interest and consists of looking back in time at what preceded the event. The data are given in Table 13.1.

The question here is: 'Does the risk of sudden infant death change with the age of the mother?' It is not possible to obtain a direct estimate of this risk, which is that of an infant becoming a case (risk = no. of cases/no. at risk), because the number at risk is not known. However, with some assumptions—mainly that the sampling fraction for the controls (the proportion the sample is of the population) is the same for all the age groups—it is possible to manage without it. A proxy measure, the proportion, $P =$ cases/(cases + controls), is constructed and used to investigate the patterns in the data. These proportions are given in the last row of Table 13.1.

Plotted against the mid-point of the mothers' age group (Fig. 13.8) these proportions show a slightly negative trend with increasing mother's age.

If the controls were known to be 1 per cent of all possible controls, that is, a sampling fraction of 0.01, then the usual estimates of risk would be:

$$\frac{17}{17 + 900} \qquad \frac{22}{22 + 2300} \qquad \frac{17}{17 + 1600} \qquad \frac{5}{5 \times 1300}$$

or   0.0185   0.0095   0.0105   0.0038

These are much smaller than the proportions, but plotted give a line with practically the same shape.

The question then becomes: 'Does the line deviate from the horizontal more than could easily arise by chance if the equivalent population line is horizontal?'

Table 13.2. Classification of women by marital status and answers to questions related to depression

Marital status	No. of positive responses (per cent)				
	0	1	2	3 or 4	Total
Single	17(45)	12(32)	7(18)	2(5)	38(100)
Married	51(58)	17(19)	12(14)	8(9)	88(100)
Widowed or divorced	8(13)	25(42)	11(18)	16(27)	60(100)

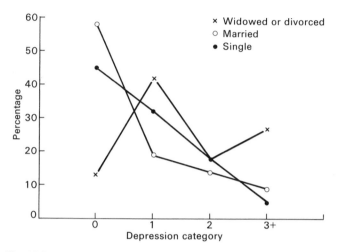

Fig. 13.9. Percentage of women in each marital status group plotted against depression category.

Three or more categories of outcome

A study of illness behaviour in women used, as a measure of depression, the number of positive responses to questions on 'loss of appetite', 'nerves', 'depression or irritability', 'sleeplessness', and 'undue tiredness'.

The numbers in the various depression categories by their marital state are given in Table 13.2. The question is: 'Does the pattern of depression in women differ according to their marital status?'

Statistically, as a deduction from the sample to the population, the question becomes: 'Is the proportional distribution among the depression categories in the population of all such women the same for each marital status?'. That is, are the 38 single women distributed among the depression categories in the same proportions, apart from differences that could arise from sampling variation, as the 88 married and the 60 widowed or divorced? The proportional allocation can be seen more easily if the frequencies are expressed as percentages of the total in that particular marital status group (Table 13.2).

Figure 13.9 gives a plot of the percentage frequencies as a separate line for each marital status group. If the proportional allocation to the depression categories were the same, these lines would fall on top of one another.

So the question becomes: 'Are these lines, obtained from a sample, more different than could easily arise by chance if the three equivalent population lines are really identical?'.

Survival data

This general title is used to cover data that arise when the interest is in the risk of some event occurring. The event may be death, but it need not be, and survival data techniques are appropriate for comparing the chances of remission for patients on different treatments and other such events.

The first example considered annual mortality in relatively large groups of people. The usual analyses for data of that type assume that all subjects were at risk of the event for the whole period. However, if someone dies half-way through the year they have been at risk only for half a year. With large numbers and small risks the numbers observed for only part of the period are small, and ignoring the problem does little harm. When the numbers in the groups to be compared are relatively small, the times at risk must be taken into account.

If it can be assumed that, within each group of subjects to be compared, each individual is running the same risk throughout the chosen time period, this is relatively straightforward. The period of time each individual was observed are summed over all the individuals in the group. If the period of time is a year and the times at risk are expressed as fractions of a year this gives the number of 'person years at risk' (obviously it may at other times be necessary to use 'woman years at risk', 'patient weeks at risk' and so on).

The estimate of the annual risk of death (or other event) is then simply:

The no. of deaths/The no. of person years at risk.

In an occupational health study of cancer, groups of current and ex-workers in two industrial environments were observed for a number of years. The overall mortality is as given in Table 13.3.

The initial question is: 'Is the risk of death from any cause different in the two environments?'.

The estimated risks per person per year are given in Table 13.3, and Figure 13.10 shows the estimated risks plotted against age.

The question becomes: 'Are these lines surprisingly far apart if the equivalent lines in the population represented by this sample of individuals are identical?'.

The techniques for answering this question are essentially the same as those for the previous examples.

Table 13.3. Mortality from all causes and estimated risk of death from any cause per person per year by age group and factory

Age at start of study (years)	Factory 1			Factory 2		
	Numbers dying	(Person years at risk)	Estimated risk of death	Numbers dying	(Person years at risk)	Estimated risk of death
50–59.9	7	(4045)	0.0017	7	(3701)	0.0019
60–69.9	27	(3571)	0.0076	37	(3702)	0.0100
70–79.9	30	(1771)	0.0169	35	(1818)	0.0193
80–89.9	8	(381)	0.0210	9	(350)	0.0257

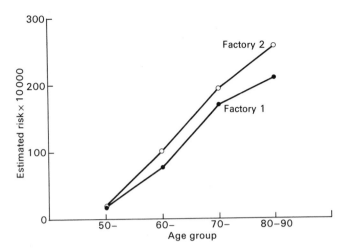

Fig. 13.10. Estimated risks of death against age by factory.

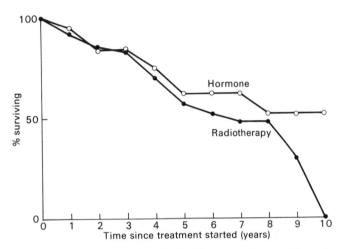

Fig. 13.11. Survival curves for patients with cancer of the prostate receiving radiotherapy or hormone treatment.

When risks are changing during the period of observation, techniques assuming one overall risk are inappropriate. To illustrate survival data of this sort, survival curves are used. These are plots of the percentage in a group still surviving against time elapsed since some defined starting point such as date of entry to the study.

Data of this type arose from a clinical trail comparing radiotherapy with hormone treatment for cancer of the prostate. The question was: 'Which is the better treatment?'.

This is not at all a simple question. The treatment giving the greater chance of survival at five years (say) is not necessarily the treatment that gives the longest life expectancy. It is necessary to look at the relative shapes of the survival curves. During the trial 44 patients received radiotherapy and 48 received hormone treatment. The survival curves for the two groups are shown in Figure 13.11.

There is some divergence of the curves and the question becomes: 'Do the curves for the population represented by these samples differ?'. Or in other words: 'Are these sample curves too different to have easily arisen by chance sampling from populations with identical survival curves?'.

Basic concepts

Because individuals subject to biological variation are to some extent unpredictable in their characteristics and behaviour, factual statements concerning them take the form: Individuals like 'this', treated like 'this', tend to respond like 'this'.

The *individuals* are the items being observed. They may, for example, be patients, tissue samples, rats, or geographical areas. They may even be the same patient observed at different times, for example when a sequence of separate blood samples is taken or results are obtained from a repeatedly administered psychiatric test.

The 'individuals like this' make up the *population* in the statistical sense, i.e. any strictly defined group of individuals.

The human population of a country may be the statistical population of interest, but it will usually be a sample of the general population of all such humans who might exist now or in the future. Researchers in London will want to be able to generalize their findings beyond Londoners and will generally wish to conclude that the results apply to all such patients in the UK now and possibly in the future.

The population should be defined before the study begins, and the sample chosen accordingly. Practical circumstances, however, can have a considerable effect on the sampling. This makes it necessary for the researcher to indicate how and with what qualification results from a sample apply to a particular population. In addition most studies will involve individuals from identifiable subpopulations. For example, studies involving physical characteristics such as lung function which may differ systematically between the sexes require that results are obtained separately for the two samples representing the subpopulations male and female. If that is not done, and the findings in the two samples not shown to be consistent, it is not safe to combine all the data to arrive at a general conclusion.

This implies that if there are two subpopulations which may differ in respects important to the study in question then data must be collected to identify from which subpopulation each individual arose. These subpopulation-defining characteristics or variables, such as sex, age, social class, and so on, are often a nuisance. It is not usually necessary, for example, to study how the sexes differ—this is mostly known. It is necessary in order to confirm that the effects of interest are consistently found in all the subpopulations. If this can be assumed at the beginning, studies can be simplified by restricting them to, for example, one sex or one age group, an option that is always worth considering at the design stage. Generally, however, data must be collected on individuals from a number of subpopulations.

Clinical trials are studies of subjects who have been 'treated' in some way which must be defined and possibly measured, for example as a dose. In most trials two or more groups receiving different treatments will be compared and so from one individual to another the treatment, type, or dosage may vary. The treatment is therefore a variable and will be represented in the data collected during the trial by numerical values identifying either or both the type of treatment and the dosage.

Observational studies of naturally occurring 'experiments' may be concerned with assessing the effect on individuals of some experience or exposure to some aspect of the environment. As for a treatment, it is necessary to define and measure the experience or exposure. Again, it must be represented by a variable with numerical values.

To assess the manner in which individuals treated in some way or undergoing some experience 'tend to respond', a variable is required to measure or classify the response. This is the response or outcome variable. Furthermore, because they will vary in a random way even among individuals from a homogeneous population in which all are treated in the same way, these variables are known as random variables. The way in which they vary is described by the distribution of their values in the sample and by inference in the populations.

Types of variables

Which analytical approach to use largely depends on the nature of the response variable. There are two main categories *quantitative* and *qualitative*.

Quantitative variables have an obvious associated scale on which distances have a clear interpretation. They can arise in two forms, continuous or discrete. Continuous variables are generally measurements and every value within some sensible range is possible, for example serum cholesterol or PEFR. Discrete variables are generally counts where only certain values on the scale are possible, such as number of previous heart attacks or number of schistosome eggs in a stool sample.

Qualitative variables are those associated with a set of categories. The categories may have a natural ordering as in severity of a condition, category of remission, social class, and so on. These are known as ordinal variables and, although they have to be treated as categories, they are usually interpretable as sections of an underlying scale. Categories such as blood group and type of health services contact do not have an order and are termed nominal variables.

Functions of variables

The function of response variables is clear. They provide the 'yardstick' on which the effects of treatment or experience are measured. However, variables are also used to identify the subpopulation to which an individual belongs so that an analysis can assess and allow for differences between subpopulations. They are also used to measure or classify treatment or experience.

There is no obvious general term for these last two classes of variable. They are sometimes called 'predictor' variables

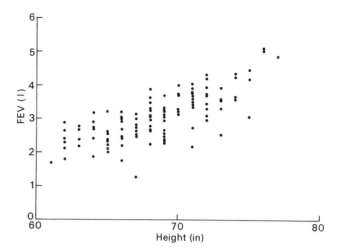

Fig. 13.12. Forced expiratory volume (FEV, litres/min) in young adults by height in inches.

because studies are often concerned with how well they predict outcome or response. They may also be referred to as independent variables because the analysis is essentially assessing how response or outcome depends on their value. Obviously in this terminology the response variable is known as the dependent variable.

Perhaps the simplest terminology of all, following mathematical convention, classes the treatment, exposure, and subpopulation variables, all of which tend to appear on the horizontal axis of graphical plots, as *x*-variables. The response variables generally appear on the vertical axis of such plots and are known as *y*-variables.

Describing data patterns—models

A model is simply an algebraic formula for describing some pattern or structure in variable data.

Consider the relationship between forced expiratory volume in one second (FEV_1) and height in a sample of students (Fig. 13.12). There is a tendency for the taller individuals to have the larger FEVs. If the individuals are grouped on the height scale, for example all individuals with heights between 62 and 64 inches are treated as if they were 63 inches, those between 65 and 67 as if they were 66 inches, and so on, the plot becomes a set of separate groups of points (Fig. 13.13).

For each height group there is an array or distribution of FEV values. These are known as conditional distributions. They are the distributions of FEV values conditional on the individuals concerned all having the same height (or in this case being in the same height group). It can now be seen that the trend in FEV values is a systematic tendency for these conditional distributions to move up as height increases. The distributions overlap, but their centres are located at a generally higher position in the taller height groups. It is not easy to describe the behaviour of the individual points, but it is possible to be reasonably precise about the relative positioning or location of the conditional distributions. The models to describe this sort of pattern have the form:

Location of distribution = Reference value + Horizontal position effect

The behaviour of individual points has to be described in terms of the location of their particular distribution and the way in which they are dispersed or spread about the location.

In Figure 13.13 the data were grouped to show the conditional distributions, but this is wasteful of information. Some method is needed to describe the trend in the raw data.

The simplest model is a straight-line trend. This model assumes that the height or *x*-variable effect is a steady increase (or decrease) in the FEV or *y*-variable distribution locations usually represented by the arithmetic mean defined below. That is to say that for each unit change in height there is the same change in mean FEV at all points on the height axis. The model for this is:

Mean FEV = Reference value + Change/unit ht × Height

This is known as a regression line model. In more general terms it becomes:

$$y = a + bx$$

which is the equation representing a straight line on a plot of *y* against *x* (in this case FEV against height) shown in Figure 13.14, where *a* is known as the intercept which is the height of the line when *x* = 0 and *b* is the slope of the line called the regression coefficient. FEV at zero height makes no practical sense but it makes the algebra simpler to represent the regression line by:

$$FEV = Intercept + b × Height$$

Finding the 'best' straight line to describe a trend means estimating the appropriate intercept and slope, and is known as fitting the regression line. The possibilities are infinite. Any values of *a* and *b* are, in theory, possible. The problem is to

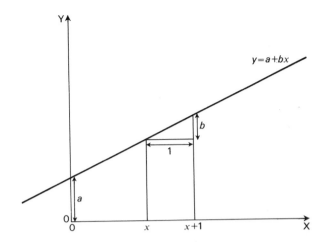

Fig. 13.14. Diagrammatic representation of the straight-line regression model $y = a + bx$.

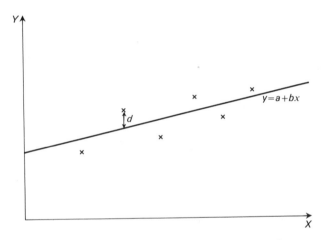

Fig. 13.15. Deviations of data points from the model $y = a + bx$.

define what is the 'best' line to describe a set of data and then to deduce the values of *a* and *b* that give this 'best' line as functions of the data values.

In practice the 'best' fitting line is defined by the 'principle of least squares'. This principle can be illustrated by considering a general line on a scatter diagram with *a* and *b* unspecified (Fig. 13.15). Each point deviates from the line (in the *y* direction) by some amount. The principle of least squares states that, the 'best' line is the one that makes the 'sum of all the deviations squared' as small as possible. This means that *a* and *b* must be chosen to make the sum of all the d^2s a minimum. Using this principle it is possible to deduce mathematically how to calculate the appropriate values from the data. The mathematics provides standard formulae with which the appropriate *a* and *b* can be calculated for any set of data.

Before further discussion of how models are defined and used it is necessary to consider how to define and measure the important characteristics of the distributions whose relative positioning they are used to describe and test.

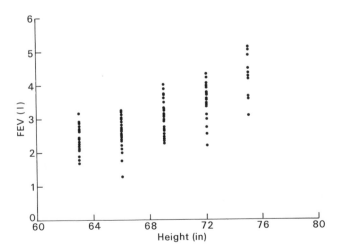

Fig. 13.13. Forced expiratory volume (FEV, litres/min) in young adults by height groups.

The characteristics of distributions

The characteristics of the distributions represented by the arrays of points in Figure 13.13 are not at all clear. Their location can be judged to some extent, but how the individuals are dispersed about the central location cannot be seen so easily. Single distributions are better displayed as histograms.

To represent a sample as a histogram the scale is divided into intervals of equal length and the points within an interval are drawn on top of one another as shown in Figure 13.16. Points falling where two intervals join could go in either, but for graphical techniques it does not really matter. Usually they are allocated to the higher interval.

The shape is rather 'ragged' but one can see, in addition to its location and spread, that the distribution is rather asymmetrical. There is a longer 'tail' on the high-value or positive side. Such distributions are called skew and since skewness can affect the analysis some method is needed to detect and measure it. There are specific measures of skewness, but they are rarely necessary in practice. Graphical techniques are usually adequate to detect whether it is present to an extent that matters. Location and dispersion on the other hand require precise measures and on occasion a complete numerical specification of the distribution is required, for which centiles (sometimes referred to as percentiles) are used.

Measures of location

There are three commonly used measures of location:

1. *The arithmetic mean.* This is the sum of all the values in a distribution or a sample divided by the number of values. It is by far the most useful measure of location and is usually referred to simply as the mean. For a sample it is:

   Mean = Sum of all values/Number of values

2. *The median.* This is the variable value such that half of the values in a distribution are above it and half are below. It is easy to find and useful for simple descriptions of asymmetrical distributions. It is not, however, very easy to handle in more complex analyses.

3. *The mode.* This is the value in a distribution that occurs most frequently. It is of occasional use for describing single distributions.

Measures of dispersion

There are a multitude of measures of dispersion in the literature, but only a few are commonly used.

The range. This is the distance, on the scale of measurement, between the highest and lowest values. It is occasionally used for describing the spread of small samples. Obviously, as sample sizes increase the range will automatically increase and, since it depends on the least characteristic pair of individuals in the sample, its usefulness is limited.

The variance and standard deviation. The variance is a direct measure of spread about the arithmetic mean. For a sample it is defined as:

Fig. 13.16.   Representing a sample of values as a histogram

Sum for all values of (value − mean value)^2/(No. of values − 1)

Notice that 'squaring' the deviations of values from their mean ensures that negative deviations from low values do not cancel out with those from high values.

For a population, usually defined so generally that the size is infinite, the variance cannot be calculated in this way. It is defined as the value approached by the calculated value as the sample size increases indefinitely.

Because the variance is an average squared deviation it is not in the original units of the variable in question. For this reason its square root, the standard deviation (s.d.), is frequently more useful. This gives quite an accurate measure of how far from the mean individuals are likely to occur in a given distribution. Unless a distribution is very asymmetrical, individuals more than three s.d.s from the mean will occur only very rarely and most (approximately 95 per cent) individuals will fall within two s.d.s of the mean.

Centiles

These are points on the variable scale that exceed a specified percentage of the distribution. Consider a sample of ten systolic blood pressure values with the distribution in Table 13.4. Ten per cent (one value) of the sample is below 110 mmHg and that is therefore the 10th centile. Seven values (70 per cent) are below 130 mmHg so that is the 70th centile, and so on. However, this approach only gives the centiles that occur at the ends of the intervals. Considering the sample as a representative of a much larger population it is possible to go a bit further by plotting the cumulative frequency, as a percentage, against the upper ends of the intervals (Fig. 13.17). It is far from smooth, but various centiles may be deduced by estimating where the 'curve' reaches the equivalent height on the per cent cumulative frequency scale. This method estimates the 50th centile, which is the median, as 125 mmHg. For more precise estimates a smooth curve could be drawn through the points. Alternatively, assumptions about the shape of the population curve may be made and the appropriate estimates deduced mathematically. To understand that process it is necessary to consider in detail how sample distributions and their characteristics relate to those of the population.

Inferring the truth—sample to population

Distributions

Consider a sequence of histograms obtained from larger and larger samples of an infinite population (Fig. 13.18). As the sample size increases, the distribution becomes smoother

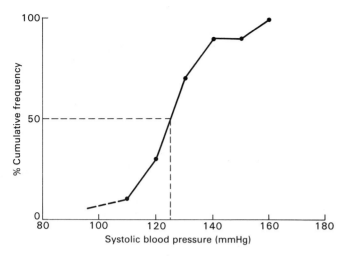

Fig. 13.17. Cumulative frequency plot of a sample of systolic blood pressure values.

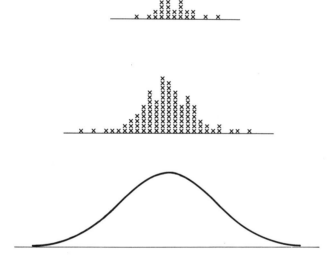

Fig. 13.18. Histograms from samples of increasing size approaching a smooth population curve.

with a more clearly defined shape. As the sample size is increased indefinitely (with the vertical scale reduced appropriately) the shape tends to a smooth curve. For many biological variables this will be close to a symmetrical 'bell-shaped' distribution known as the normal distribution, which can be defined in precise mathematical terms.

Probabilities

In the sample of ten systolic blood pressures (Table 13.4), two out of ten were in the range 110–120 mmHg. The proportion of the sample between 110 and 120 mmHg is therefore 2/10, that is 0.2. If the names of the ten individuals were written on pieces of paper, mixed up in a 'hat', and then one was picked at random, the probability that it was an individual in the range 110–120 would be 2/10 (0.2). There are two of the sort required and ten in total.

Table 13.4. Frequency distribution of a sample of 10 adult males according to their systolic blood pressures

Systolic blood pressure (mmHg)	Frequency of values	Cumulative total
from 100 to just under 110	1	1
from 110 to just under 120	2	3
from 120 to just under 130	4	7
from 130 to just under 140	2	9
from 140 to just under 150	0	9
from 160 to just under 170	1	10

The probability of selecting an individual with a systolic blood pressure less than 100 mmHg from this 'hat' is 0/10, that is, zero or impossible. The probability of one with a pressure less than 180 is 10/10, that is 1.0 or certain. Notice that 120 is the 30th centile and the probability of an individual below 120 is 0.3 or 30 per cent. The probability of an individual below the 15th centile is 0.15 or 15 per cent, and so on.

From the probability of selecting an individual below a particular point it is possible to deduce the probability of an individual on or above that point. There are three of the ten individuals below 120, which gives the probability of such an individual as 0.3. There are seven of the ten equal to or greater than 120, that is a probability of 0.7. Thus probability (value ≥ 120) is:

$$\frac{\text{No.} \geq 120}{\text{Total no.}} = \frac{10 - \text{No.} < 120}{\text{Total no.}} = \frac{10 - 3}{10}$$

$$= 7/10 \text{ or } (10 - 3)/10$$

The probability of an individual being greater than or equal to some value is one minus the probability of being less than that value:

$$p \text{ (value} \geq 120) = 1 - p \text{ (value} < 120)$$

For example:

$$p \, (< 110) = 1/10 = 0.1$$
$$p \, (\geq 110) = 9/10 = 1 - p \, (< 110) = 0.9$$

and so on.

As sample sizes increase and the histogram approaches the smooth population curve, the numbers become indefinitely large. However, the area above an interval and beneath the curve is in the same ratio to the total area as the number in that interval to the total number. In practice, population distributions are scaled so the curve encloses an area exactly equal to one. This means that the area above any interval is the proportion of individuals with values in that interval. This in turn is the probability that an individual picked at random has a value in that particular interval. So, if the area of the population distribution between two values 110 and 120 is 0.15 then the probability of an individual picked at random from this population having a systolic blood pressure between 110 and 120 is 0.15 (Fig. 13.19).

The probability that an individual selected at random has a value > 150 is the area A under the curve from 150 upwards.

Since A is 0.1 it means that 10 per cent of the population will have systolic blood pressures of 150 or above.

The area below 150 is $1 - A$ so the probability of an individual below 150 is $1 - A$ or 0.9. This means that the remaining 90 per cent of individuals have systolic blood pressures below 150. Consequently 150 is the 90th centile of the population distribution.

The centiles of the population can be obtained as long as there is some way of obtaining the areas under the population distribution curve. Representing the 10th centile of the variable y as $y_{0.1}$ the area below $y_{0.1}$ is 0.1, the area below $y_{0.2}$ is 0.2, and so on.

The normal distribution

For various reasons random variables that appear to follow a normal distribution are common in nature—for example, human height, haemoglobin levels, systolic blood pressure. Others can be made to follow a normal distribution by a transformation of their scale. Possibly even more important is the fact that for all but the smallest samples averages are themselves variables from approximately normal distributions whatever the distributions of the original values. Because of this and its amiable mathematical nature the normal distribution holds a central place in statistical theory and practice.

The theoretical normal distribution is completely specified by its mean and variance. The distribution is symmetrical about the mean and areas beneath the curve can be obtained mathematically. It is known for example that if the variable is y then the 2.5th centile:

$$y_{0.025} = \text{mean} - 1.96 \text{ s.d.}$$

This implies that in a normal distribution only 2.5 per cent of the individuals will be more than 1.96 s.d.s below the mean. Similarly only 2.5 per cent will be more than 1.96 s.d.s above it. Combining these it further implies that the range of values:

$$\text{mean} - 1.96 \text{ s.d. to mean} + 1.96 \text{ s.d.}$$

will include 95 per cent of individuals in the population and exclude only 5 per cent.

The normal distribution is so useful that the percentage cumulative frequencies for the standard distribution, which has mean = 0 and variance = 1, have been widely tabulated (Lindley and Miller 1964). From these the equivalent values

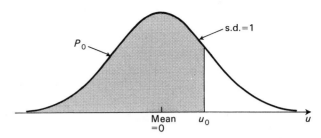

Fig. 13.20. Probabilities and the standardized normal distribution curve.

for a normal variable with any mean and variance can be deduced.

The tabled values are the areas under the mathematically defined population distribution curve below the given values of the standard normal variable u (say) as in Figure 13.20.

The area P_0, in Figure 13.20 is the percentage cumulative frequency at u_0 which is the probability of a value below u_0, that is:

$$p \text{ (value} < u_0) = P_0$$

The probability of a value at or above u_0 is:

$$p \text{ (value} > u_0) = 1 - P_0$$

and the probability of a value between u_0 and u_1 is;

$$p \text{ } (u_0 < \text{value} < u_1) = P_1 - P_0$$

where P_1 is the area under the curve below u_1.

The variable u is effectively the number of s.d.s a value is from the mean. For variables with a different mean and variance, equivalent probabilities are deduced by calculating how many s.d.s each value is from its mean and treating these as values of u.

For example, suppose systolic blood pressure in adult males is normally distributed with mean 130 and s.d. 15. To calculate the probability of an individual value less than 160 mmHg, note that this is 30 mmHg, i.e. two s.d.s above the mean. This implies that:

$$p \text{ (value} < 160) = p \text{ } (u < 2)$$

and since from tables:

$$p \text{ } (u < 2) = 0.977$$

it can be deduced that:

$$p \text{ (value} < 160) = 0.977$$

and the required probability is 97.7 per cent.

With knowledge of the population distribution, these techniques allow the probabilities of individual sample values to be deduced.

However, the problem is really in the opposite direction. The need is to deduce the population characteristics from the sample values. A method is required to assess how close sample values such as the mean are to the unknown population equivalents. This is done by considering how the values of a sample estimate might vary if further samples were taken

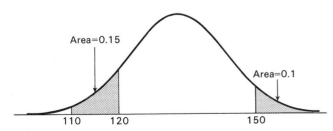

Fig. 13.19. Areas of the population distribution as probabilities.

from the same population. Each sample will differ and estimates from them will vary.

Estimation and sampling distributions

It is simplest to start by considering how the sample mean varies in a sequence of samples all taken the same way from the same population. The result can be generalized to any estimate, for example a regression coefficient, a difference between means, etc., calculated from a sample. Suppose a sequence of samples of four values ($n = 4$) is taken from the systolic blood pressure (SBP) distribution and the mean is calculated for each. These sample means will themselves have a distribution which, because it can only arise as a result of repeated sampling, is known as a *sampling distribution*. In this distribution of means, samples with all values at one extreme of the population distribution will be rare. As a result the sample means will be more tightly clustered about the population mean than the individual observations. In addition, as long as the sampling was properly random, the mean of the sampling distribution will be the same as the population mean. Because of this the sample mean is said to be an unbiased estimate of the population mean. It can be shown in fact that if the sampling is repeated indefinitely the sampling distribution of the mean will tend to a smooth curve with the same mean as the population of individual values, but a smaller standard deviation. In fact, it can be shown mathematically that this standard deviation of the mean is:

$$\text{s.d. (mean SBP)} = \text{s.d. (SBP)}/\surd \text{(sample size } n)$$

So for samples of size 4 and s.d. = 15:

$$\text{s.d. (mean SBP)} = 15/\surd 4 = 15/2 = 7.5$$

half the spread of the population.

Remember the complete sampling distribution exists only in theory. In practice there is only a single sample mean which is one observation from the distribution.

The narrower the spread of this distribution the nearer the single value available is likely to be to the truth. This spread represented by the standard deviation of the sampling distribution indicates how precise a single sample mean is as an estimate. It indicates the likely error in the estimate. Largely because of this the standard deviations of sampling distributions are almost universally referred to as standard errors:

$$\text{s.e. (mean)} = \text{s.d. (mean)} = \text{s.d. (original variable)}/\surd n$$

Thus the standard error of a mean (or of any other estimate) indicates its precision as an estimate. For samples of 100 values ($n = 100$) the sample means would have had a standard deviation of 15/10 or 1.5, a very small spread. The estimate can be made as precise as required by increasing the sample size.

So the standard error of a sample mean is:

$$\begin{aligned}\text{s.e. (mean)} &= \text{s.d. (original variable)}/\surd \text{(sample size)}\\ &= \text{s.d.}/\surd n \end{aligned} \qquad (13.1)$$

which can be deduced mathematically by the application of

some fairly general rules. The same techniques can be used to deduce the standard errors of regression coefficients, differences between means, and so on. In fact for any value calculated from a sample, appropriate standard errors can be obtained.

Confidence intervals

Although the estimate is chosen as the 'best guess' at the population value it will not usually be exactly right. In practice the standard error and the estimate are used to obtain a range of values defining an interval on the original variable scale which will contain the unknown population value with some chosen probability.

Fortunately, for reasonably large samples ($n > 30$) and some fairly mild assumptions, the sampling distributions of many common estimates are close to normal. This means, for example, that the sample value will only fall outside the range:

$$\text{true value} - 1.96 \text{ s.e.s to true value} + 1.96 \text{ s.e.s}$$

for 5 per cent of such samples. With some algebra it follows that the interval:

$$\text{estimate} - 1.96 \text{ s.e.s to estimate} + 1.96 \text{ s.e.s}$$

will include the true population value in 95 per cent of such samples. This means that one can be 95 per cent confident that an interval calculated in this way will include the true population value. Other confidence intervals such as 99 per cent are obtained by taking, instead of 1.96, the appropriate centile of the standard normal distribution.

These calculations need the population standard deviation of the original variable. Since this is rarely known it is necessary to use the sample value as an estimate. This does not matter very much for large samples, but if the sample size falls below about $n = 30$ it may. As a result for small samples the standard normal distribution cannot be used. It is necessary to use what is known as the 't' distribution. The centiles of the normal distribution, such as $u_{0.975} = 1.96$, have to be replaced by the equivalent centiles of the t distribution appropriate to the sample size. None the less the appropriate 97.5 centiles are all close to 2. These centiles are commonly denoted by $t_{f,0.975}$ where f is known as the degrees of freedom and represents the amount of information available to estimate the population standard deviation used in the calculation of the standard error. The degrees of freedom are invariably the total sample size less the number of values estimated from the sample. From tables (Lindley and Miller 1964) it can be seen that:

$$t_{20,\,0.975} = 2.09$$
$$t_{30,\,0.975} = 2.04$$
$$t_{40,\,0.975} = 2.02$$
$$t_{60,\,0.975} = 2.00$$

and

$$t_{240,\,0.975} = 1.96 = u_{0.975}$$

so as the degrees of freedom increases beyond 240, the t distribution becomes the same as the normal distribution.

Sample sizes for estimation

The sample size for an estimation is determined by the precision required. The usual approach to defining precision is to stipulate that there should be a high probability (often 95 per cent) that the estimate is close to the true value. Close is usually defined as within some small percentage such as 5 per cent of the correct value. Clearly this means that the approximate size of the value being estimated must be known.

Consider the problem of estimating a population mean from a sample. Ninety-five per cent of means from unbiased samples will be within 1.96 standard errors of the true value. This implies that if $1.96 \times$ s.e. is less than 5 per cent of the true value the precision will be as specified.

If M represents the value to be estimated then the sample will be large enough if:

$$1.96 \times \text{s.d.}/\sqrt{n} \le 0.05 \times M$$

i.e. if:

$$n \ge (1.96)^2 \times \text{s.d.}^2/(0.05 \times M)^2 \qquad (13.2)$$

With this sample size one can be 95 per cent confident that the estimate will be within 5 per cent of the truth.

Example

To estimate the mean systolic blood pressure for men between 30 and 60 years old to within 5 per cent with 95 per cent confidence. From previous work it is known that the true value (M) is about 150 mmHg and that the standard deviation is about 15.

Using equation 13.2 this means that the sample size needs to be:

$$n \ge (1.96 \times 15)^2/(0.05 \times 150)^2 = 16$$

This is very small, so 5 per cent in this context is not a very stringent requirement. Requiring a higher degree of precision, for example that the estimate must be accurate to 1 per cent, implies a sample size of:

$$n \ge (1.96 \times 15)^2/(0.01 \times 150)^2 = 385$$

Significance tests

The probability of a sample 95 per cent confidence interval not containing the true population value is less than or equal to 5 per cent. This means that for any hypothetical population value it is possible to judge whether the sample estimate is surprisingly far from it by seeing whether it falls outside the 95 per cent confidence interval. If a value hypothesized for the population falls outside the interval then the probability of the sample giving this interval and hence the estimate—assuming the population value is correct—is less than 5 per cent so:

$$p \text{ (estimate | population value)} \le 0.05$$

or as it is frequently written ($p < 0.05$).

A significance test is a procedure which can be applied, assuming that the hypothesis to be tested is true, to the calculation of the probability (p) of sample values as far or further from those expected. Some value, called a test statistic, which has a known sampling distribution is calculated from the sample. This is frequently the number of standard errors an estimate is from the population value expected from the hypothesis. The p value can then be obtained from the sampling distribution of the test statistic. If the p value is less than 5 per cent then the sample value is taken as evidence significant at the 5 per cent level that the hypothesis is untrue.

In 5 per cent of tests using a 5 per cent significance level, surprising values of the test statistic will occur even though the hypothesis is correct. This error, of concluding that an hypothesis is not true when it is, is known as a Type I error. Therefore, if a test uses a 5 per cent level of significance there will be a 5 per cent risk in the test of a Type I error. Notice that the risk may be set as required in a test by altering the level of significance to be used. Finally there is a risk that even when the hypothesis is false, samples may be obtained which produce unsurprising values of the test statistic. This leads to the error of accepting as true a hypothesis which is actually false. This is a Type II error. The size of sample needed to give a reliable test of some hypothesis must take both types of error into account. Sample size calculations for significance testing are discussed in a later section.

There is a close relationship between confidence interval calculations and significance tests. Any hypothesis that predicts a population value can be tested by inspecting whether the value falls inside the appropriate confidence interval. Using the 99 per cent confidence interval this is equivalent to $p < 0.01$.

Suppose a study is estimating the population mean systolic blood pressure for adult males. For a large sample ($n > 30$) with mean 120 mmHg and standard error 2.0 the 95 per cent confidence interval is:

$$120 \pm 1.96 \times 2.0 \text{ or } 116.1 \text{ to } 123.9$$

The probability of results as extreme as 120 if the true value is 130 is less than 5 per cent since 130 is well outside the interval. The estimated 120 mmHg is significantly different from 130 mmHg at the 5 per cent level, it is surprisingly far from the population value of 130 ($p < 0.05$). It is reasonably clear that the sample is from a population with a lower mean value than 130 mmHg. However, it must be remembered that by definition 5 per cent of the confidence intervals will exclude the true value. This is the Type I error above. If this system is used to take decisions they will be wrong 5 per cent of the time.

Although the use of confidence intervals is quite sufficient, it is common to calculate the p values directly. As implied above this is most frequently done by calculating, as a test statistic, how many standard errors separate the estimate and the hypothetical value:

$$\frac{\text{estimate} - \text{hypothetical value}}{\text{s.e. of the estimate}}$$

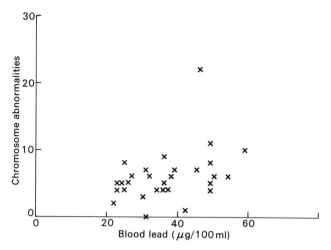

Fig. 13.21. Chromosome abnormalities per 100 cells and blood lead levels. (Source: Forni and Sciame 1975.)

What centile this is of the *t* or standard normal distribution is then determined from tables which for a given value of the variable give the probability of values below it, say *P*.

The probability of values equal to or above that obtained is then $1 - P$ which is the probability of surprisingly high values. Surprisingly low values will also be possible so the probability of values as surprising as that observed will be twice $1 - P$ or:

$$p = 2(1 - P)$$

These are the basic concepts of statistics. It is now necessary to consider how these techniques are used to answer practical questions from real data.

Types of analyses

Measured outcomes

Simple regression

Figure 13.21 shows some data on chromosome abnormalities and blood lead levels from female workers in a battery factory. The research question was: 'Does lead exposure damage chromosomes?'.

If it is assumed that current lead levels in the blood reflect overall exposure then the question can be expressed as: 'In the population of all such subjects as these is the number of chromosome abnormalities higher, on average, in those with high blood lead levels than it is in those with low levels?'.

This is the same as asking: 'Is the trend in these points sufficiently non-horizontal to indicate a real trend in the population?'.

To answer the question it is necessary to fit the least squares regression line and test whether the sample regression coefficient is consistent with a true value of zero. Alternatively it is simply necessary to show that a non-zero trend model fits the data significantly better than a horizontal

line model to demonstrate that the apparent trend is unlikely to be due to chance.

To demonstrate how a regression line is fitted a number of algebraic definitions are needed. The variance of a set of values has already been introduced as:

Sum for all values (value − mean value)^2/(no. of values − 1)

which for values of a variable *y* becomes:

$$\text{Sum } (y - \bar{y})^2/(n - 1) \tag{13.3}$$

where $\bar{y}$ (pronounced *y* bar) is the conventional abbreviation for the mean of a sample of *y* values. The sum of squared deviations construction is so common in statistical calculations that it is convenient to have a special notation. This text will use:

$$Syy = \text{Sum } (y - \bar{y})^2 \tag{13.4}$$

and for an *x* variable:

$$Sxx = \text{Sum } (x - \bar{x})^2 \tag{13.5}$$

and finally:

$$Sxy = \text{Sum } (y - \bar{y})(x - \bar{x}) \tag{13.6}$$

The sum of squared deviations minimized when we fit a regression line is essentially *Sdd* but it is more often called the residual sum of squares (*RSS*) or *Sr*.

In general:

$$Sr = \text{Sum } (y - \text{value predicted by the model})^2$$

which for a simple regression line becomes:

$$Sr = \text{Sum } (y - \text{height of line})^2$$

Now to fit the regression line of the form:

$$y = a + bx$$

to describe the trend in chromosome abnormalities as blood lead increases means choosing the most appropriate values for *a* and *b*. The principle of least squares means that *a* and *b* are required to make the sum of the squared deviations *Sr* a minimum. For this line the residual sum of squares is:

$$Sr = \text{Sum } (y - a - bx)^2 \tag{13.7}$$

It can be shown mathematically that this is a minimum when the slope:

$$b = Sxy/Sxx \tag{13.8}$$

and the intercept:

$$a = \bar{y} - b\bar{x} \tag{13.9}$$

The minimum value of *Sr* when *a* and *b* take these values is:

$$Sr = Syy - Sxy^2/Sxx \tag{13.10}$$

The residual sum of squares is the basic measure of how well a model, in this case $y = a + bx$, fits the data. If points are scattered about the line, *Sr* will be large indicating a bad fit. If the points fall more or less on a straight line *Sr* will be small indicating that a straight-line model fits the data well.

The variance of the deviations about the line is called the residual variance and denoted by:

$$s^2 = Sr/(n - 2) \qquad (13.11)$$

where $n - 2$ is known as the degrees of freedom and is the amount of information available to estimate the residual variance. This is the number of observations (or points on the plot) less the number of parameters in the model that need to be estimated. Here it was necessary to estimate two parameters, the intercept and the slope.

The residual variance can be used to calculate the standard error of b which is:

$$\text{s.e.}(b) = \sqrt{(s^2/Sxx)} \qquad (13.12)$$

For the Italian data on women battery factory workers, with y = chromosome abnormalities/100 cells and x = blood lead levels in μg/100 ml, the values required in the analysis to fit a simple regression line are:

$$n = 30 \qquad \bar{y} = 5.97 \qquad \bar{x} = 36.37$$

with:

$$Syy = 432.97 \qquad Sxx = 3302.96 \qquad Sxy = 460.37$$

The estimates of a and b are therefore:

$$b = Sxy/Sxx = 460.37/3302.96 = 0.14$$

$$a = \bar{y} - b\bar{x} = 5.97 - 0.14 \times 36.37 = 0.90$$

from equation 13.10:

$$Sr = 432.97 - 460.37^2/3302.96 = 368.80$$

so:

$$s^2 = Sr/(n - 2) = 368.80/28 = 13.17$$

and:

$$\text{s.e.}(b) = \sqrt{(s^2/Sxx)} = 0.06$$

The line representing the apparent trend 'best' in the least squares sense is therefore:

$$\text{ABS} = 0.90 + 0.14 \times \text{BL}$$

where ABS stands for chromosome abnormalities and BL for blood lead levels. Assuming that the sampling distribution of b is near enough normal, the 95 per cent confidence interval for the 'true' value of the slope is:

$$0.14 - 1.96 \, (0.06) \text{ to } 0.14 + 1.96 \, (0.06)$$

or   0.02   to   0.16

Since this does not include zero it is a reasonably strong indication that the population trend is not zero, i.e. the observed trend is significantly different from a population value of zero at the 5 per cent level. A sample regression slope could not easily occur this far from zero by chance ($p < 0.05$).

The significance test of the zero slope hypothesis is performed by calculating how many standard errors the slope 0.14 is from the expected zero, that is:

$$\frac{\text{observed slope} - \text{expected slope}}{\text{s.e. (slope)}} = \frac{0.14 - 0}{0.06} = 2.33$$

which is, as the confidence interval also showed, well above 1.96. There is therefore strong evidence of a trend in the population represented by this sample. Higher numbers of chromosome abnormalities will be found in individuals such as these with the higher blood lead levels.

There was one 'outlying' individual on the plot with 22 abnormalities. If the results had been less conclusive it might have been wise to repeat the analysis excluding this individual. Her effect on the analysis could then be assessed and the interpretation modified appropriately. Note that for samples this size the t distribution should be used. The t distribution centile equivalent to the standard normal 1.96 for 28 d.f. is 2.05 (about 5 per cent higher), but the conclusions are exactly the same.

Whether these results really mean that lead in the blood damages chromosomes is still not clear. Partly for this reason it is customary to describe significant trends as indicating a real association between the variables to make it clear that the existence of a trend in the population in no way proves a 'causal relationship'. This one fact must be added to the whole body of knowledge on the subject before conclusions can be carried that far.

Apart from tests of the assumptions that the population line is straight and not a curve, that there is a similar residual variance at all x values, and that the underlying distribution is normal, this analysis is complete.

In fact since any trend is indicative of the relationship, in this case between lead exposure and chromosome abnormalities, whether or not it is straight hardly matters. Whether the residuals have a normal distribution is more crucial since inferences from confidence intervals and significance tests all assume that they do. How to test these assumptions will be discussed later. Meanwhile it is useful to consider how this analysis can be approached in a more general way applicable to most common problems. In some circumstances the question cannot be phrased so it becomes a test of a single parameter. In those cases the question has to be expressed as a comparison of different models.

In the lead data example the equivalent model comparison is between a model of the form:

$$y = a + bx$$

which is a sloping straight line and the model:

$$y = a$$

which represents a horizontal line.

If there is a significant trend the second model will fit the data significantly worse than the first. How well a model describes data is measured by the residual sum of squares—the smaller Sr is, the better the fit. Comparisons of the fit of two models are made by comparing their residual sums of squares.

As should be clear from this example the more parameters in a model the smaller the residual sum of squares Sr can be

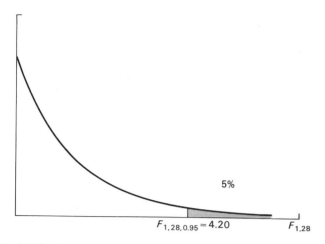

Fig. 13.22. *F* distribution with 1 and 28 degrees of freedom.

made. Choosing the intercept and the slope, two parameters for a sloping line makes it possible to fit the model closer to the data points than when the fitting process is restricted to choosing one parameter, the intercept, for a horizontal line. From the model:

$$y = a + bx$$

the residual sum of squares is:

$$Sr = 368.75$$

and the residual variance is:

$$s^2 = 13.17$$

From the horizontal line model when the intercept (height of the line) is simply the mean of *y*, the residual sum of squares is:

$$Sr = 433.0$$

that is, an increase of 64.25.

The increase has associated with it one degree of freedom since one parameter has been dropped from the more complicated model. On the hypothesis that the trend is a chance event in this particular sample 64.25 is a further estimate of the residual variance. This means that if the hypothesis is true $64.25/s^2$ should be about one since it is the ratio of two estimates of the same variance: the value in this example is 64.25/13.17 or 4.88.

The numerator which is the increase in *Sr* resulting from simplifying the model has one degree of freedom. The denominator which is the residual variance from the most complicated model has $n - 2$ which is 28 degrees of freedom. If the 'no-trend' hypothesis is true this variance ratio can be shown to have a sampling distribution with a particular mathematical form called the *F* distribution with in this case 1 and 28 degrees of freedom. This can be used to assess how surprising an observed variance ratio is to give what is known as an *F* test.

The *F* distribution of 1 d.f. and 28 d.f. is illustrated in Figure 13.22. The 95th centile of the *F* distribution is 4.20 less than 5 per cent of *F* values calculated in this way from a

sample should occur this far above one. This implies that the value 4.88 is surprisingly large ($p < 0.05$) if the population has no trend. It suggests that the horizontal-line model is incorrect. Surprisingly low values can occur, but they have little meaning. If anything they imply that the model fits the data surprisingly well given the amount of variation present. They are generally treated as insignificant.

The interpretation of this analysis is that it is either necessary to assume a freak sample or that the non-zero trend model was necessary to describe this data. In practice it would be taken as reasonably strong evidence that there was a trend in the population. The estimated regression coefficient with a confidence interval should then be obtained to demonstrate the magnitude of the effect.

This model comparison approach to the analysis is best presented in tabular form giving what is known as an 'analysis of variance' but is actually an analysis of residual sums of squares. Denoting the residual sum of squares from model 1 as S_1 and the sum of squares divided by the degrees of freedom for a model as the mean square (*MS*) the analysis is given in Table 13.5.

To construct such analysis of variance tables it is only necessary to obtain the residual sum of squares for the two models. A test of the parameters omitted when model 1 is changed to the simpler model 2 is then immediately available as:

$$F_{1.28} = (S_2 - S_1)/s^2 \qquad (13.13)$$

Since all that is needed for this approach is the residual sum of squares for each model, it is not even necessary to fit them. In practice they are fitted because it is necessary to know the likely magnitudes of the effects, not just whether they exist or not.

Non-linear regression

In a given analysis a straight-line trend may well be inadequate, a curved-line trend might fit the data better. To test this a model representing a curved trend is fitted and the fit is compared with that of a straight-line model. The simplest model for a curved trend is known as a quadratic and takes the form:

$$y = a + bx + cx^2$$

Table 13.5. Analysis of variance to test for a linear trend in chromosome abnormalities by blood lead (BL) levels

Model	SS	d.f.	MS	F
1. Intercept and trend on BL— sloping line	$S_1 = 368.75$	28	$s^2 = 13.17$	
2. Intercept, but no trend— horizontal line	$S_2 = 433.00$	29		
(2 − 1) Due to trend	64.25	1	64.25	$F_{1.28} = 64.25/13.17 = 4.88$

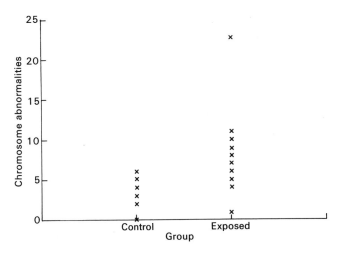

Fig. 13.23. Chromosome abnormalities and exposure group.

In the lead data example this is equivalent to:

$$ABS = Intercept + b.BL + c.BL^2$$

with ABS and BL standing for chromosome abnormalities and blood lead levels respectively. The model is fitted exactly as before with a, b, and c chosen so that the residual sum of squares is minimized.

The formulae are rather more complicated than for the straight-line model so although this model can be fitted by hand it is tedious. With a professionally produced package program on a computer it is very easy and it is not necessary to know the formulae in detail.

The hypothesis that there is no curvature in the population can be tested very easily by calculating:

$$t = \frac{c - \text{expected } c}{\text{s.e. } (c)} = \frac{c - 0}{\text{s.e. } (c)}$$

and since we have fitted three parameters, this has $n - 3$ degrees of freedom where n is the number of data points. The fitted model for the lead data gives:

Intercept	$a = 2.77$
Linear coefficient	$b = 0.03$
Quadratic coefficient	$c = 0.0014$ with s.e. $(c) = 0.0066$

so for the t test of curvature:

$$t = \frac{0.0014 - 0}{0.0066} = 0.2 \quad \text{with 27 d.f.}$$

This is very far from the value needed for significance which is a little above 2.0 ($t_{0.975} = 2.05$ on 27 d.f.). Obviously there is little evidence that the population trend is not a straight line.

Comparing two groups

The above analysis of the lead study data looked at the association between chromosome abnormalities and blood lead levels. However, current blood lead may not reflect long-term exposure. Figure 13.23 compares the individuals

Table 13.6. Chromosome abnormalities (ABS) in female battery factory workers by exposure group. (Findings when one woman with ABS = 22 was excluded are shown in parentheses)

	Not exposed	Exposed	All
Number	12(12)	18(17)	30(29)
Mean	4.00(4.00)	7.28(6.41)	5.97(5.41)
s.d.	1.71(1.71)	4.36(2.43)	3.87(2.44)

from lead-free areas in the factory with those exposed to lead in their work. The values for the exposed individuals appear to be slightly higher with a wider spread. To investigate this more precisely it is necessary to calculate the mean and standard deviations for the two groups (Table 13.6).

The question becomes: 'Are the means surprisingly different?' This is tested by fitting a model of the form:

$$\text{mean ABS} = \text{reference value} + \text{group effect}$$

Taking the 'reference value' as the mean for the first group (not exposed), the group effect is zero for that group and equal to the difference between the two means for the second group. Fitting this model by least squares is exactly the same as allowing each group to have its own mean. The model is:

$$\text{mean ABS} = 4.00 + 0 \quad \text{for group 1}$$
$$\text{mean ABS} = 4.00 + (7.28 - 4.00) \quad \text{for group 2}$$

Each mean minimizes the sum of squared deviations within its own group Syy_1 and Syy_2 (say) so the total residual sum of squares $Sr = Syy_1 + Syy_2$ is a minimum.

The variances for the groups separately are calculated as $Syy_1/(n_1 - 1)$ for group 1 and $Syy_2/(n_2 - 1)$ for group 2. To compare the group means it is necessary to assume one underlying residual variance. On this assumption both samples contain information on the residual variance, and the information is pooled as the overall residual sum of squares Sr to estimate it. The pooled estimate of variance is therefore Sr divided by its degrees of freedom.

For the above model the number of degrees of freedom is $n_1 + n_2 - 2$ because there are $n_1 + n_2$ values in total and two parameters, the two means, have been estimated. Now:

$$Syy_1 = 32.00 \quad \text{and} \quad Syy_2 = 323.61$$

So:

$$Sr = Syy_1 + Syy_2 = 355.6$$

and since $n_1 = 12$ and $n_2 = 18$:

$$s^2 = Sr/(n_1 + n_2 - 2) = 355.6/28 = 12.7 \quad (13.14)$$

The difference between the means, which is the group 2 effect, since the reference value is the group 1 mean, is:

$$\text{Difference} = 7.28 - 4.00 = 3.28$$

this has the standard error:

$$\text{s.e. (Difference)} = \sqrt{[s^2(1/n_1 + 1/n_2)]} \quad (13.15)$$
$$= \sqrt{[12.7(1/12 + 1/18)]} = 1.33$$

A significance test of the hypothesis that the population difference is zero is obtained as:

$$t = \frac{\text{Difference} - \text{Hypothesized Difference}}{\text{s.e. (Difference)}}$$

$$= \frac{3.28 - 0}{1.33} = 2.47$$

and the probability of values as far from zero as 2.47 occurring by chance, obtained from tables of the t distribution with 28 degrees of freedom, is about 0.02, so $p < 0.05$. The difference is significant at the 5 per cent level ($p < 0.05$) on the assumption that the two samples came from populations with identical variances.

However, the two groups actually have very different variances (Table 13.6) and the assumption should be tested. This is done using a form of the F test. If the population variances are the same, the variances from the two groups are estimating the same thing. Their ratio in that case will be an observation from an F distribution and should have a value close to one. If the hypothesis of equal population variances is false, the variance ratio will deviate from one. Whether this gives large or small values depends on which way the variance ratio is calculated: large/small or small/large. Since this is arbitrary, suprisingly small values are just as interesting, in this context, as surprisingly large values.

It is usual to calculate the ratio with the larger variance as the numerator. The probability of F values as high as that is then obtained from tables and doubled to give p, the probability of variances as different as observed occurring by chance. This means that the probability of values as high as the ratio calculated must be less than 0.025 for the probability of obtaining variances this different by chance to be less than 0.05.

The standard deviations of the two groups were 1.71 and 4.36 so the variance ratio (large/small) is:

$$F = (4.36/1.71)^2 = 6.50$$

with $18 - 1 = 17$ and $12 - 1 = 11$ degrees of freedom. The tables show that only 2.5 per cent of values from the $F_{17, 11}$ distribution should occur above 3.33. This implies that our variances are significantly different at the 5 per cent level ($p < 0.05$). The assumption of equal variances is clearly unsafe and some corrective action is required.

In these data the difference in the variances is largely due to one very high value of ABS (22) in the exposed group. If she is removed (Table 13.6), then, although the standard deviations are still different, the F value:

$$F_{16, 11} = (2.43/1.71)^2 = 2.02$$

is less than the 95th centile of $F_{16, 11}$ which is 2.37. Thus the probability of variances this different by chance is at least 2×0.05 or 0.10. There is now no real evidence of unequal variances and no reason not to perform the standard analysis.

The residual sum of squares reduces to 126.12 with 27 degrees of freedom so the residual variance becomes 4.67.

Table 13.7. Analysis of variance for comparing the mean number of chromosome abnormalities in two groups

Model	RSS	d.f.	MS	Variance ratio F
1. Groups different	126.12	27	4.67	
2. Groups same	167.03	28		
(2−1) Due to difference	40.91	1	40.91	$F_{1, 27} = 40.91/4.67 = 8.76$

The difference between the means is $6.41 - 4.00 = 2.41$ and the standard error of this difference is now:

$$\text{s.e. (difference)} = \sqrt{[4.67(1/12 + 1/17)]} = 0.81$$

The t test becomes, $t = 2.41/0.81 = 2.98$, which is in fact even more significant than before because removing the extreme point has reduced the standard error more than it has the difference.

The above analysis fitting a single model is possible and sufficient because the question of interest can be reduced to a test of a single parameter. However, when several groups have to be compared this is not really possible and the model comparison approach is required.

In a model comparison analysis of these data there are two models. The first is the one assuming the population means to be different.

(1) mean ABS = reference value + group effect
where the reference value is the group 1 mean. As before the residual sum of squares for this model is Syy for group 1 plus Syy for group 2, and calling this S_1:

$$S_1 = 126.12$$

(omitting the individual with ABS = 22). The residual variance from this model is $s^2 = 4.67$.

The second model, assuming no difference, is:

(2) mean ABS = reference value
where the reference value is the overall mean of the two groups estimating the supposedly constant population mean. The residual sum of squares for this model is the value of Syy calculated from both groups combined into one sample. Calling this S_2:

$$S_2 = 167.03$$

and the analysis of variance table is as given in Table 13.7. Since $F_{1, 27, 0.99} = 7.68$ this gives a probability of a difference of this magnitude, given the hypothesis of no difference between the populations, of $p < 0.01$. This is the same p value as for $t = 2.98$ obtained using the alternative t test of the relevant parameter in a single fitted model. They are the same test in two different forms.

Comparing three or more groups

Consider data from an Italian study designed to assess the effects of well-publicized hypertension clinics on blood pressure levels in the community. There is some evidence that systolic blood pressure increases with age in a non-linear way. For this reason the analyses were performed using age groups

Table 13.8. Mean systolic blood pressure by age in males from a study of the effect of hypertension clinics

	Age group (years)					
	20–29.9	30–39.9	40–49.9	50–59.9	60+	All
Number	340	303	302	207	288	1440
Mean	141.9	146.0	150.1	149.2	156.9	148.5
s.d.	20.4	19.9	22.7	21.4	24.7	22.4

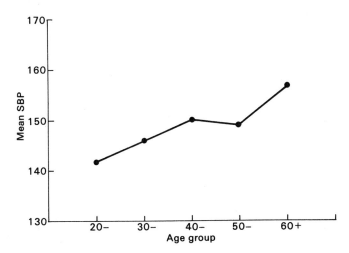

Fig. 13.24. Mean systolic blood pressure (SBP) by age in males, from a study of hypertension clinics.

rather than exact ages. At the beginning of the study the mean systolic blood pressure (SBP) for males in the control district that had no clinic was as shown in Table 13.8.

There is a tendency for the mean value to increase with age (Fig. 13.24). The s.d.s are all about the same, so an assumption of equal variances is reasonable. The appropriate model is:

(1) mean SBP = reference value + group effect

and because it simplifies the interpretation of the fitted parameters the group 1 mean (men aged 20–29 years) is taken as the 'reference value'. The 'group effects' are then the differences between the mean of that particular group and the group 1 mean, that is, the reference value. The model then has five parameters:

the reference value (the mean for group 1)

the group 2 effect G_2 (the group 2 mean − reference value)

the group 3 effect G_3 (mean for group 3 − reference value)

the group 4 effect G_4 (mean for group 4 − reference value)

and group 5 effect G_5 (mean for group 5 − reference value).

To fit the model, estimates for these five parameters must be obtained as the means and differences implied above with the residual sum of squares (Sr) and hence the s.e.s of the parameter estimates. Since the model essentially allows each group to have its own mean, Sr is the sum of the Syys

obtained separately from each of the groups. Calling this S_1 and its degrees of freedom f_1 where:

$$f_1 = (\text{total number of values} - \text{the no. of parameters})$$

gives:

$$S_1 = 685\,211 \text{ with } f_1 = (1440 - 5) = 1435$$

The residual variance is thus:

$$s^2 = 685\,211/1435 = 477.5$$

The simpler model where differences are assumed to be chance is simply one overall mean:

(2) mean SBP = reference value

with the overall mean as the reference value which represents a single horizontal line. How well this fits the data is measured by the residual sum of squares about the overall mean. This is simply Syy from all 1440 individuals treated as one sample which is $S_2 = 723\,110$. The analysis of variance is given in Table 13.9.

F values of 19.8 with 4 and 1435 d.f. are extremely unlikely to occur by chance ($p \ll 0.01$). The sample sizes are so large that the small but consistent changes in SBP are highly significant. A linear trend model:

3. mean SBP = intercept + b age

taking individuals in the same group as having the same age gives a test of whether the population trend is in any way curved.

Treating the data as one large sample and fitting a simple regression line the residual sum of squares from equation 13.10 is:

$$S_3 = Syy - Sxy^2/Sxx$$

and for these data:

$$S_3 = 688\,123$$

Since two parameters have been fitted this has:

$$1440 - 2 = 1438 \text{ d.f.}$$

The analysis of variance is given in Table 13.10. Since F with 3 and 1435 d.f. needs to exceed 2.60 for significance at the 5 per cent level there is little evidence that a linear trend is not adequate. The estimated slope is:

$$b = Sxy/Sxx = 0.68 \text{ mmHg/year}$$

and from equation 13.12:

$$\text{s.e. } (b) = 0.08$$

Table 13.9. Analysis of variance comparing mean systolic blood pressures in five age groups

Model	SS	d.f.	MS	F
1. Groups different	$S_1 = 685\,211$	1435	$s^2 = 477.5$	
2. One overall mean	$S_2 = 723\,110$	1439		
(2 − 1) Due to group differences	$S_2 - S_1 = 37\,899$	4	9475	19.8

Table 13.10. Analysis of variance testing whether mean systolic blood pressure follows a linear trend over five age groups

Model	SS	d.f.	MS	F
1. Groups different (any trend)	685 211	1435	$s^2 = 477.5$	
3. Linear trend	688 123	1438		
(3−1) Due to non-linearity	2899	3	970.7	2.03

The analysis shows that the means differ significantly and that the way they differ is adequately represented by a linear trend. This implies that the slope of the linear trend is significantly non-zero and the direct test is:

$$t = 0.68/0.08 = 8.5 \ (p < 0.01)$$

Comparing groups allowing for trends

Consider the problem of comparing lung function, in particular FEV, between smokers and non-smokers. The underlying question is: 'Does smoking affect lung function?' It appears that the study simply requires samples of smokers and non-smokers to test whether their respective populations differ. However, for an efficient analysis, the samples need to be as similar as possible in other respects so that, for example, a preponderance of males in the smokers will not bias the comparison. It is possible to avoid this bias by restricting the study to one sex. Unfortunately this approach cannot be used to avoid bias due to such things as height. Some height differences will inevitably occur. It is possible to ensure that the groups have the same range of heights which avoids the problem of systematic bias. None the less the variation in lung function is increased as a result of height differences. This decreases the sensitivity with which group differences can be detected.

Quantitative variables interfering with group comparisons in this way are known as co-variates, and analyses involving them are known as analyses of co-variance. The analysis of co-variance uses regression techniques to deduce from the data in this example: (i) whether FEV changes with height in the same way for both groups; and if it does (ii) what FEV differences would be seen between smokers and non-smokers if they all had the same height. Four main circumstances can arise illustrated in Figure 13.25.

The biases illustrated in Figure 13.25 can be avoided by designing the study to keep the height distributions of the groups similar. In practice it is more important to remove the variation due to co-variate differences in order to increase the sensitivity with which the analysis estimates and tests the group differences.

If the trend lines in two groups are parallel the groups differ by the same amount at all heights. As long as both groups have the same range of height it is possible to predict how they will differ without specifying a height. To investigate this it is necessary to compare the fit of parallel and non-parallel models. Assuming straight-line trends, an assumption that would need testing in practice, the models are:

(1) *Non-parallel*

Mean FEV = Group intercept + Group slope × Height

That is, each group has its own slope and intercept.

(2) *Parallel*

Mean FEV = Group intercept + Common slope × Height

That is, each group has its own intercept, but the same slope.

If there are $n = n_1 + n_2$ points in total, the residual sum of squares for model 1 is:

$$S_1 = \text{Sum of both groups } (Syy - Sxy^2/Sxx)$$

with $n - 4$ d.f. because four parameters have been fitted, two slopes and two intercepts. The residual variance about this model is $s^2 = S_1/(n - 4)$.

The residual sum of squares from model 2 is $S_2 = \text{Sum } Syy - (\text{Sum } Sxy)^2/\text{Sum } Sxx$ where 'Sum' implies that the Syy, Sxx and Sxy are calculated for each group separately and then summed. S_2 has $n - 3$ d.f. since in this model only three parameters are required, two intercepts and one slope. The test of parallelism, from equation 13.13 is:

$$F_{1, \ n - 4} = (S_2 - S_1)/s^2$$

which is compared with the appropriate F distribution centile.

If it seems reasonable to accept that the population trends are parallel, then the estimated slop is from equation 13.8:

$$b = \text{Sum } Sxy/\text{Sum } Sxx$$

and from equation 13.12:

$$\text{s.e.}(b) = \sqrt{(s^2/\text{Sum } Sxx)}$$

The group difference allowing for height is then constant for all heights and most easily calculated as the difference between the intercepts. This is:

Mean FEV difference − Slope × Mean height difference

Although the algebra is beyond the scope of this text it can be shown that the variance of this is:

var(diff) =
$$s^2(1/n_1 + 1/n_2 + (\text{mean height difference})^2/\text{Sum } Sxx)$$

from which we can obtain;

$$\text{s.e.}(\text{diff}) = \sqrt{(\text{var}(\text{diff}))}$$

which gives us a t test of the group difference allowing for height.

For more complex situations with several groups and possibly more than one co-variate, the group differences are more easily tested by model comparison. Considering the single co-variate case to avoid excessive complications, a third model is fitted assuming that all intercepts and slopes are the same in the population. This means that only one intercept and one slope need to be estimated. To fit them the groups are ignored and all the data combined as one. The model is:

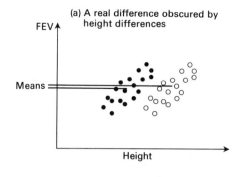

(a) A real difference obscured by height differences

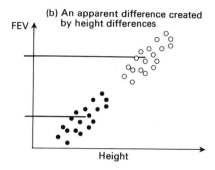

(b) An apparent difference created by height differences

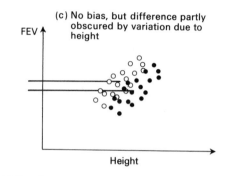

(c) No bias, but difference partly obscured by variation due to height

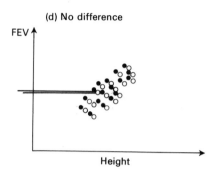

(d) No difference

Fig. 13.25. Two-group scatter diagrams to illustrate the ways in which a co-variate may influence a group comparison.

Table 13.11. Analysis of co-variance table construction for comparing groups allowing for the interfering effect of a co-variate

Model	SS	d.f.	MS	F
1. Non-parallel	S_1	$n-4$	$s^2 = S_1(n-4)$	
2. Parallel	S_2	$n-3$		
(2 − 1) Due to non-parallelism	$S_2 - S_1$	1	$S_2 - S_1$	$F_{1,n-4} = (S_2 - S_1)/s^2$
3. Same intercept and slope	S_3	$n-2$		
(3 − 2) Due to group differences	$S_3 - S_2$	1	$S_3 - S_2$	$F_{1,n-4} = (S_3 - S_2)/s^2$

(3) *Coincident trend lines*

Mean FEV = Intercept + Slope × Height

The residual sum of squares from this model is:

$$S_3 = Syy \text{ all data} - (Sxy \text{ all data})^2/Sxx \text{ all data}$$

with $n-2$ d.f. The test for group differences is again an F test performed by obtaining the difference between the Sr for the two models and dividing by the residual variance obtained from the most complex model, in this case model 1 as in equation 13.13:

$$F_{1,n-4} = (S_3 - S_2)/s^2$$

Expressed as an analysis of variance (Table 13.11) it is easier to see how this approach may be generalized to more than two groups.

Table 13.12. Mean FEV values for 50 males by smoking behaviour together with mean height in inches

	Non-smokers	Smokers
Number	45	5
Mean	3.34	3.23
s.d.	2.33	2.83
Mean height	70.6	69.0

Table 13.13. Analysis of co-variance comparing FEV in cigarette smokers and non-smokers allowing for height differences

Model	SS	d.f.	MS	F
1. Non-parallel trends with height	12.91	46	$s^2 = 0.28$	
2. Parallel trends	12.99	47		
(2 − 1) Due to non-parallelism	0.08	1	0.08	$F \ll 1$
3. Same intercept	13.03	48		
(3 − 2) Due to group difference	0.04	1	0.04	$F \ll 1$

Example

Mean FEV values were obtained for 50 males of whom five were regular smokers (Table 13.12). The smokers have slightly lower FEV values than the non-smokers, but the comparison makes no allowance for differences in height. The mean heights are in fact different, the smokers being slightly shorter on average. The analysis is given in Table 13.13.

Table 13.14. Mean systolic blood pressures by sex and intervention group of subjects in a hypertension study

	Area	
	Control	Intervention
Male		
Number	200	223
Mean	158.5	150.1
s.d.	24.2	21.0
Female		
Number	191	283
Mean	167.4	154.7
s.d.	27.5	21.5

Since F values will exceed one for surprising values, and both these F values are less than one, there is no evidence that the trends have different slopes or the smokers and non-smokers in general differ on the FEV scale.

Groups in a cross classification

The effects of intervention in an Italian hypertension study were assessed at the end of a five-year period. Samples of males and females were obtained from the control and intervention areas. The mean systolic blood pressure values for those aged 60 years and above given in Table 13.14. This is a 2×2 classification because the table has two rows and two columns. The groups are defined by the categorical variables sex in one direction and treatment in the other. For historical reasons variables used to classify individuals are known as factors and their values, the actual categories, are known as levels.

To assess the effects of intervention it is necessary to investigate whether the difference between the intervention and control areas is the same in both sexes. If it is, an overall estimate of the effect using data from both sexes is required. This must then be tested against zero to assess whether the hypothetical population of such individuals treated in this way would show a non-zero effect. If the effects are different for the two sexes they have to be estimated and tested for each sex separately.

Figure 13.26 shows that the intervention means are well below those of the equivalent control groups. The females have consistently higher values than the males and there is a slight suggestion that the intervention/control difference is greater in the females. This makes the lines joining the sample means non-parallel.

To test this non-parallelism or interaction between the effects of treatment and sex, two models must be fitted. These are a non-parallel model where each group has its own mean and a model where the population means are constrained to fall on parallel lines.

The parameter representing the deviation from parallelism is the difference, in Figure 13.26, between the control/intervention interval for the females and that for the males. Consequently if T is the treatment effect in the males (that is, the

control mean—the intervention group mean) and $T + I$ is the equivalent effect in the females then I is the deviation from parallelism. Because it represents an influence of the factor sex on the treatment effect (cf. synergy and antagonism in pharmacology) it is often referred to as an interaction. The sex effect (S say) is taken as the difference between the sexes in the intervention group. The model is:

(1) Mean SBP = Reference value + Sex effect + Treatment effect + Sex × Treatment interaction

which represents four equations for the four sex/treatment combinations:

male intervention	mean = rv
male control	mean = $rv + T$
female intervention	mean = $rv + S$
female control	mean = $rv + S + T + I$

The estimate of the interaction may be calculated from:

$I =$ (control mean – intervention mean) females – (control mean – intervention mean) males

which from these data gives:

$$(167.4 - 154.7) - (158.5 - 150.1) = 12.7 - 8.4 = 4.3$$

The Sr from this model is:

$$S_1 = \text{Sum of } Syy \text{ from the four groups} = 488\,486.2$$

with $n - 4$ d.f., where n is the total number of subjects in all four groups and four is subtracted because four parameters need to be estimated. The residual variance is therefore:

$$s^2 = S_1/(n - 4) = 488\,486.2/893 = 547.0$$

which can be used to obtain a standard error for I which is:

$$\text{s.e.}(d) = \sqrt{(s^2(1/n_1 + 1/n_2 + 1/n_4))} = 3.16$$

So a test of the hypothesis that the true deviation from parallelism is zero is:

$$t = (d - 0)/\text{s.e.}(d) = 4.3/3.16 = 1.36$$

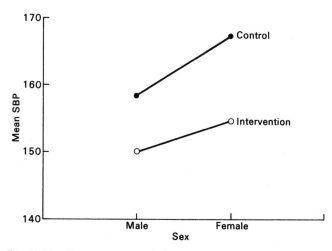

Fig. 13.26. Mean systolic blood pressure (SBP) by sex and treatment group from a study of hypertension.

Table 13.15. Analysis of variance comparing means in a two way classification

Model	SS	d.f.	MS	F
1. With interaction	S_1	$n-4$	$s^2 = S_1/(n-4)$	
2. No interaction	S_2	$n-3$		
(2 − 1) Due to interaction	$S_2 - S_1$	1	$S_2 - S_1$	$F_{1,n-4} = (S_2 - S_1)/s^2$
3. No treatment effect	S_3	$n-2$		
(3 − 2) Due to treatment	$S_3 - S_2$	1	$S_3 - S_2$	$F_{1,n-4} = (S_3 - S_2)/s^2$
4. No sex effect	S_4	$n-2$		
(4 − 2) Due to sex	$S_4 - S_2$	1	$S_4 - S_2$	$F_{1,n-4} = (S_4 - S_2)/s^2$

Since there are 893 d.f. the t distribution is the same as the standard normal distribution (Fig. 13.20). This means the value would have to exceed 1.96 for significance at the 5 per cent level. Since it does not there is little evidence of an interaction. The effect of intervention if any is the same for both sexes.

The parallel-line model is:

(2) Mean SBP = Reference value + Sex effect + Treatment effect

Unfortunately there is no simple formula for Sr from this model and the analysis cannot easily be done without a statistical package on a computer. When there are only two treatment groups a technique due to Yates (described very thoroughly by Snedecor and Cochran 1967) using averages of the mean differences weighted according to the group sizes can be used. It is, however, laborious and not very general. The Sr from this model S_2 (say) will have $n-3$ df. The three fitted parameters will be the reference mean, the average difference between the sexes, and one treatment effect averaged over both sexes. The effects can be tested using:

$$t = \text{estimate}/\text{s.e.(estimate)}$$

or, for the generality required for comparing several groups, by fitting models without them. To test the treatment effect it is necessary to fit:

(3) Mean SBP = rv + Sex effect
with $Sr = S_3$ and $n-2$ d.f.

The F test for the treatment effect is then:

$$F_{1, n-4} = (S_3 - S_2)/s^2$$

Exactly the same procedure can be used to test the sex effect. The full analysis of variance is given in Table 13.15.

Example

The actual analysis for the Italian SBP data is given in Table 13.16. Since the 95th centile $F_{1,893}$ is 3.84, there is no evidence of interaction. The other two F values are highly significant, indicating that both the differences between the

treatment groups and between the sexes are highly significant ($p < 0.001$).

The treatment effect is estimated using Model 2 and was found to be:

10.6 with s.e. 1.58

which means that, all else being equal, individuals receiving the second treatment, the intervention, will be on average, 10.6 mmHg below the control group. The approximate 95 per cent confidence interval for the 'true' effect of intervention is:

estimate ± s.e.s which gives −7.44 to −13.76

The 'true' difference between the sexes is estimated in a similar way showing females to be higher on average by 6.5 mmHg with an approximate 95 per cent confidence interval of 6.5 ± 1.57 mmHg.

The general principles of the above analyses apply to much more complicated situations. A study design may lead to cross-classifications with many more than two factors. Any of the factors may have more than two levels. As well as a cross-classification there may be one or more co-variates to allow for in the analysis. Even so the pattern of the analysis is more or less the same. With a computer the appropriate residual sums of squares can be obtained together with estimated parameters representing the group differences of interest and their standard errors.

Repeated or paired measurements

Repeated observations are made on the same subjects when changes over time are of interest. In addition subjects may be matched in pairs or larger groups to increase the precision of an investigation.

An important reason for performing analyses using co-variates or interfering factors is to explain and remove some of the variation observed in the response variable and thus increase the precision of the analysis. In comparisons of a treatment with a control the main source of variation is between the individuals within the groups. There may be a clear difference between the means of the control and treatment groups, but the degree of overlap among individual values may well mean that the observed difference cannot be distinguished from possible chance effects.

Table 13.16. Analysis of variance comparing mean systolic blood pressure in groups classified by sex and treatment

Model	SS	d.f.	MS	F
1. With interaction	488 486.2	893	$s^2 = 547.0$	
2. No interaction	489 499.2	894		
(2−1) Due to interaction	1013.0	1	1013.0	$F_{1,893} = 1.85$
3. No treatment	514 316.2	895		
(3−2) Due to treatment	24 817.0	1	24 817.0	$F_{1,893} = 45.4$
4. No sex	498 866.2	895		
(4−2) due to sex	9367.0	1	9367.0	$F_{1,893} = 17.1$

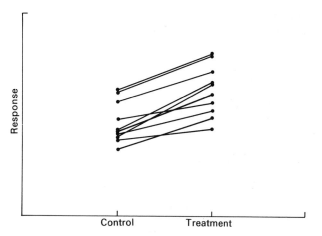

Fig. 13.27. Paired observations plotted against treatment category.

Table 13.17. Analysis of variance for paired measurements

Model	SS	d.f.	MS	F
1. Parallel lines	S_1	$n-1$	s^2	
2. Horizontal lines	S_2	n		
(2 − 1) Due to slope of lines	S_2-S_1	1	S_2-S_1	$F_{1,n-1}= (S_2-S_1)/s^2$

Table 13.18. Means of paired blood pressure readings

	1st reading	2nd reading
Number	10	10
Mean	157.6	152.1
s.d.	17.0	17.0

In certain circumstances it is possible to use each individual as his or her own control. Or, as in studies of twins, there may be a very well defined pairing. This means that the part of the response unique to the individual (or the pair) appears in both the control and the treatment measurement and cancels out when the difference is calculated. As Figure 13.27 shows, the consistent way in which the lines joining paired points slope upwards gives a much stronger impression of a genuine treatment effect than the points alone.

The questions dictating which models need to be fitted and compared are: (i) Is the effect of treatment the same for all subjects? That is, are the population lines parallel? (ii) Assuming the effect is the same for all is it non-zero? That is, are the population lines non-horizontal?

The first question cannot be answered if, as is usual, there is only one measurement for each subject within a group. In a non-parallel-line model every pair of points has its own line which fits them exactly and there are no deviations. This means there is no information about the underlying variation and inferences cannot therefore be made about the population. To address this particular question it is necessary to replicate some of the measurements.

To answer the second question it is necessary to fit a set of parallel lines to the pairs of points. The model is:

1. Mean $Y = rv$ + Subject effect + Treatment effect

where the 'treatment effect' is actually the mean of the treatment/control differences. The subject effects represent the height of the lines in Figure 13.27. It is not necessary to estimate these effects they are included in the model to keep them out of the residual sum of squares and hence reduce the residual variance. There are $2n$ points from n subjects; the model requires n parameters for the subjects and 1 for the treatment group difference. Thus the residual sum of squares S_1, has $2n - (n + 1) = n - 1$ d.f.

The model appropriate to the hypothesis that the treatment has no effect is:

2. Mean $Y = rv$ + Subject effect

which is simply a set of horizontal lines. The model requires one parameter for each subject so the residual sum of squares S_2 has $2n - n = n$ d.f. The analysis of variance is given in Table 13.17.

This is a very simple example. In practice there may well be subpopulations such as sex represented in the sample and it will be necessary to check that effects are the same in each. There will frequently be more than two treatment groups.

The model comparison approach can be extended quite simply to handle both situations. It does, however, need quite sophisticated computing facilities.

In this simplest of cases, the test of the treatment effect can be expressed as what is known as the paired t test. Denoting the treatment/control difference for each subject by d, then:

$$\text{Treatment effect} = \text{Mean } d$$

and the test of the estimate against zero is:

$$t = \text{Mean } d/\sqrt{(s_d^2/n)}$$

where s_d is the standard deviation of the n differences.

Example

In an investigation of the accuracy of blood pressure measurements, two measurements of systolic blood pressure were taken about five minutes apart. These replicated readings ought to differ in a purely random way unless the process of measurement affects the value. The comparison here is not between two treatments but between the first and second reading. The mean values are given in Table 13.18. If these are treated as two independent groups the apparent decrease is far from significant.

Table 13.19 gives the analysis of variance obtained by fitting the models with subject parameters to take account of the pairing. The 95th centile of this F distribution is $F_{1, 9, 0.95} = 5.12$, the 90th centile is 3.35, and hence $0.05 < p \ll 0.10$. The model with a reading effect fits better, but not startlingly so. The evidence is ambiguous. It is probably necessary to repeat the investigation with a larger sample before coming to a conclusion on the existence or otherwise of a reading effect.

Table 13.19. Analysis of variance to test for a change over time in duplicate systolic blood pressure readings

Model	SS	d.f.	MS	F
1. Reading effect	299.3	9	33.25	
2. No reading effect	450.5	10		
(2 − 1) Due to reading effect	151.2	1	151.2	$F_{1,9} = 4.55$

Treating the analysis as a straight paired t test the mean difference is:

$$d = 5.55 \text{ (1st − 2nd reading)}$$

with the residual sum of squares of the differences about their mean of:

$$S_{dd} = 598.6 \text{ with 9 d.f.}$$

the variance of the differences therefore is:

$$s_d^2 = 598.6/9 = 66.5$$

and:

$$t = d/\text{s.e.}(d) = 5.5/\sqrt{(66.5/10)} = 2.13$$

This must be compared with the 97.5th centile of the t distribution with 9 d.f. which is 2.26. Thus, as for the F test, p obtained in this way is greater than 0.05. In fact this t value is the square root of the above F value ($2.26^2 = 5.12$) illustrating that F and t here are effectively the same test and will always produce the same answer.

Sample size calculations for comparing means

Sample sizes for significance tests of means are calculated in much the same way as they are for estimation (see p. 119) with one extra complication. The probability of accepting an incorrect hypothesis, that is a Type II error, must be taken into account. Since the standard error of the test statistic is a function of the sample size n and the residual variance, which must be known or estimated, the calculation amounts to solving, for n:

$$\frac{\text{smallest interesting value of the test statistic} - \text{value expected if hypothesis true}}{\text{standard error of test statistic}} \geqslant Za + Zb$$

The values Za and Zb are centiles of the standard normal distribution which are determined by the size of the Type I and II errors considered acceptable. The t distribution might be more appropriate. However, since the t distribution centiles can be determined only if the degrees of freedom and hence the sample size are known, the argument becomes circular. It is simpler to use the normal distribution and remember that the calculated sample sizes may be slightly smaller than is necessary.

Za will usually be 1.96 for a Type I error of 5 per cent. Demanding an 80 per cent chance of detecting when the hypothesis is false, which implies a Type II error of 20 per cent, means that Zb must be taken as 0.84, the 80th centile of the standard normal distribution.

For a paired t test, as above, the expected difference, given the usual hypothesis of no effect, is zero. If the smallest interesting difference is D, a 5 per cent significance level is to be used ($Za = 1.96$), and an 80 per cent chance of detecting a real difference as large as D is required ($Zb = 0.84$), then the sample size n (the number of pairs) must satisfy:

$$(D - 0)/(s_d/\sqrt{n}) \geqslant 1.96 + 0.84$$

Note that D is assumed positive which makes no difference and simplifies the calculations. This requires that:

$$n \geqslant (1.96 + 0.84)^2 \times 2s_d^2/D^2$$

Example

Suppose it is of importance to detect a difference of 5 mmHg between paired blood pressure measurements. With Type I and II errors of 5 and 20 per cent, this requires a sample size of:

$$n \geqslant (1.96 + 0.84)^2 \times 2 \times 66.5/25 = 41.7$$

Thus at least 42 pairs of measurements are required.

If two independent groups are to be compared the calculations are similar except that the form of the standard error becomes that given in equation 13.14. Using D as the smallest interesting difference, the sample size is obtained by solving:

$$(D - 0)/\sqrt{[s^2(1/n_1 + 1/n_2)]} \geqslant Za + Zb$$

for n_1 and n_2.

Since analyses are most efficient if the sample sizes are equal, assuming them both equal to n the number in each group must satisfy:

$$n \geqslant (Za + Zb)^2 \, 2s^2/D^2$$

Example

In a study to assess the effect of jogging on systolic blood pressure, two independent groups, one of joggers and one of non-joggers, would be compared. The appropriate test statistic would be the difference between the means. This is known as a two group t test. If D is set at 5 mmHg, the two error levels I and II taken as 5 and 20 per cent, and the residual variance taken from previous studies as 225, then the number in each group must be:

$$n \geqslant (1.96 + 0.84)^2 \times 2 \times 225/25 = 141.1$$

That is, 142 subjects are required in each group.

The analyses in the preceding sections cover most of the situation that arise in practice when responses are measured.

Qualitative responses—two category outcomes
Maximum likelihood estimation

Exactly the same questions arise for qualitative as for quantitative responses, and the same general approach is used as covered above. There are, however, a few essential differences which must be discussed.

Typically two category outcomes arise when the effects of some treatment or experience are judged according to whether an individual responds or not, has a symptom or not, and so on. The outcome variables are responses with the values yes or no. It is conventional to represent such variables numerically as 0 and 1. Obviously they cannot have normal distributions and the principle of least squares cannot be used, but apart from this the conceptual approach to the analysis is identical to that for measured outcome variables.

In fact the underlying distribution is binomial and instead of mean values it is the proportions of individuals in each category that are of interest. In the population these proportions are the probabilities or 'risks' of individuals picked at random belonging to the specified category. An algebraic model is needed to describe how the 'risk' of an individual being in category 1 (responding, having a symptom, etc.) changes from one group to another or according to values of some measured x variable or co-variate.

To fit these models (that is, to estimate the parameters) the 'maximum likelihood principle' is used instead of least squares. Essentially this means assuming not only that the sample obtained is not peculiar, but also that it is the most likely sample to occur. Once the form of the conditional distributions is determined—in this case binomial—the probability of the sample can be calculated for any algebraic model describing the pattern in the population 'risks', for example:

population risk = some function of $(a + bx)$

The values of the parameters a and b are then chosen to maximize this probability or likelihood. Apart from this difference the process follows much the same lines as the 'least squares' approach.

To illustrate the analysis of this type of data, consider several groups of similar men exposed to differing levels of pollution classified according to whether they have persistent cough or not. If there is a positive relationship, the proportion with cough will start close to some minimum (it cannot go below zero) for low pollution levels, start to rise at some point, and then level off as it approaches some upper limit (it cannot exceed 1.0) for high levels of pollution. This produces an 'S' or sigmoid-shaped curved which is characteristic of data where the response variable is constrained between two limits—in this case 0 and 1. From a sample it may appear that there is some sort of relationship even when there is none. The curve may be a chance deviation from a horizontal line. A test is needed to assess whether there really is an effect of exposure on the risk of response. This is obtained by using an algebraic model to describe the curve. The parameters representing how increasing the dose affects response can then be estimated and tested.

If subpopulations are represented in the sample, it will be necessary to compare two or more such dose–response curves. For example, the effect of pollution on respiratory symptoms might have to be assessed using a sample containing smokers and non-smokers. Smokers might well have a consistently higher risk. It is necessary to allow for this when assessing the apparent dose–response relationship. There will be two sigmoid curves representing how risk changes with pollution, one for each of the smoking groups. This pair of curves will have to be described with algebraic models in order to allow for the smoking effect while assessing the effect of pollution. At the tails of such curves the risks are very close to zero or one. Small changes in these may be, proportionally, very important. It is necessary to take into account that a change in risk from 0.01 to 0.02 (1 to 2 per cent) is in fact a doubling of the risk while the same absolute change in the middle of the curve, say 0.50 to 0.51 (50 to 51 per cent), is a relatively trivial proportional change in risk.

In practice this is allowed for by converting the values to the 'logistic scale'. Using the variable y to represent values on this scale, the y values for the proportion with cough, i.e. p (cough), are calculated as:

$$y = \log[p(\text{cough})/(1 - p(\text{cough}))]$$

where the function $\log(\ )$ is the 'natural' logarithm to the base 'e' and the function $p/(1 - p)$ is known as the odds. So:

if $p(\text{cough}) = 0.01$ then $y = \log(0.01/0.99) = -4.6$
if $p(\text{cough}) = 0.50$ then $y = \log(0.5/0.5) = 0.0$

and

if $p(\text{cough}) = 0.99$ then $y = \log(0.99/0.01) = +4.6$

This pulls the lower tail of the curve down and pushes the upper tail up with an overall effect of straightening the sigmoid curves.

The order of points on the curve is unchanged by moving from the p to the y scale. An increase on the logistic scale automatically implies an increase on the risk scale.

Transforming proportions to the logistic scale:

$$y = \log(p/(1 - p))$$

turns a sigmoid curve into a straight line. The reverse transformation which is:

$$p = 1/(1 - \exp(-y))$$

turns a straight line to a sigmoid curve. The exponential function $\exp(\ )$ is the 'anti-log' for logarithms to the base 'e'. This means that trends in proportions can be assessed and compared by fitting straight-line models:

$$y = a + bx$$

on the logistic scale and using the reverse transformation specific risks can be estimated or compared.

All the analyses of the section on measured outcomes can be applied to proportions using this approach. Any proportion can be considered as a point on a sigmoid curve and the same approach can be used for comparing two or more groups, possibly in some cross-classification, at the same time as allowing for the effects of co-variates.

Because maximum likelihood is used to fit the models and not least squares, the residual sum of squares cannot be used

Table 13.20. Prevalence of respiratory symptoms (cough) in children by pollution level and smoking in the home (simulated data). Proportion with the symptom (number at risk)

	SO$_2$ pollution level		
	Low (100 μg/m³)	Medium (200 μg/m³)	High (300 μg/m³)
Smoking in the home			
No	0.05(37)	0.20(25)	0.33(12)
Yes	0.09(32)	0.29(28)	0.53(17)

to indicate how well a model fits the data. It is necessary to use what Nelder and Wedderburn (1972) have called the 'deviance' defined as:

$$\text{Deviance} = -2 \times \log(\text{likelihood})$$

where the 'likelihood' is the probability of the sample for the given model.

The analysis of variance to compare two models produces a ratio which is an observation from an F distribution if the two models fit the data equally well. In the analysis of deviance for qualitative response variables, the difference between two deviances is, near enough, an observation from a distribution known as the chi-squared distribution. This is a widely used distribution, so tables (Lindley and Miller 1964) are available from which centiles and p values can be obtained.

Unfortunately, most practical problems involving proportions can be handled only with the aid of a computer and a good statistical package. Only if the problem is greatly simplified can the analysis be performed by hand. The following sections cover the general approaches first and then indicate how simpler analyses can be performed at the cost of a few assumptions.

Comparing two groups allowing for a co-variate

Table 13.20 gives some simulated data on the prevalence of cough in children exposed to different levels in pollution. The proportions increase steadily with increasing pollution and they might well arise from the tails of sigmoid curves.

The questions are:

1. Is the effect of pollution the same in both groups? That is, on the logistic scale are the equivalent population lines parallel?

If parallelism can be assumed:

2. Is there an effect of pollution? That is, are the equivalent population lines non-horizontal?

and

3. Is there an effect of exposure to smoking? That is, are the equivalent population lines separated vertically?

For this analysis it is necessary to fit the models:

1. *Non-parallel*—a different pollution effect in each smoking group

$$y = rv + \text{smoking effect} + \text{smoking group pollution effect}$$

rv is the reference value which is the predicted proportion for the low pollution/no smoking reference group transformed to a value on the logistic scale.

2. *Parallel*

$$y = rv + \text{smoking effect} + \text{pollution effect}$$

3. *Horizontal*—no pollution effect

$$y = rv + \text{smoking effect}$$

and

4. *Coincident lines*—no smoking effect

$$y = rv + \text{pollution effect}$$

The analysis was performed using a computer and the statistical package GLIM (Payne 1983). Other packages such as BMD-P (Dixon 1981) can be used to perform the same analyses, but not so easily. GLIM requires as input the numbers at risk and the numbers responding (that is those with symptoms) for each cell of the two-way table (Table 13.20) to fit the models and obtain their deviances. From these the analyses of deviance in Table 13.21 can be constructed.

From line $(2 - 1)$, there is no evidence of non-parallelism. The difference between the two deviances gives a chi-squared value of 0.04 which is trivial compared to the 95th centile of the chi-squared distribution with one degree of freedom, which is 3.84. Whatever the association it is the same whether or not the child is exposed to cigarette smoking in the home.

From line $(3 - 2)$, omitting the pollution effect from the parallel-line model gives a much worse fitting model. Values as large as the chi-squared value testing the slope of the pollution trend against zero, that is 17.2, should occur, by chance, in less than 0.1 per cent of such investigations. The 99.9th centile of the chi-square distribution on 1 d.f. is only 10.83. These simulated data very strongly indicate that at these levels there is an association between increasing pollution levels and increasing risk of having a symptom.

The effect of smoking in the home is not so clear. The smoking effect, that is, the separation of the two lines, appears to be no more than could easily occur by chance. The

Table 13.21. Analysis of deviance for comparing prevalences of respiratory symptoms in children classified by exposure to air pollution and cigarette smoking

Model	Deviance	d.f.	Approximate chi-squared
1. Different pollution effects non-parallel	0.40	2	
2. Identical effects parallel	0.44	3	
(2 − 1) Due to different effects	0.04	1	0.04 (cf. ChiS$_{1,0.95}$ = 3.84)
3. No pollution effect horizontal	17.64	4	
(3 − 2) Due to pollution effect	17.20	1	17.20 ($p < 0.01$)
4. No smoking effect coincident lines	2.38	4	
(4 − 2) Due to smoking effects	1.94	1	1.94 not significant

chi-squared value of 1.94 on 1 d.f. is well below the 3.84 required for significance.

The parallel-line model (2) is adequate to describe that data and it provides estimates of the effects. The model is, on the logistic scale:

$$y = a + smk + b \times \text{pollution}$$

where a is the intercept for the no smoking group; smk, the smoking effect, is the vertical separation of the lines; and b is the slope or regression coefficient of the pollution trends.

The computed analysis gives estimates with approximate standards errors as:

$$a = -3.87 \qquad \text{with s.e. } (a) = 0.69$$
$$smk = 0.61 \qquad \text{s.e. } (smk) = 0.44$$
$$b = 0.0113 \qquad \text{s.e. } (b) = 0.0029$$

The pollution effect can be tested by comparing b with 0 using the standardized normal distribution (u):

$$u = 0.0113/0.0029 = 3.90 \qquad (\text{cf. } 1.96)$$

This gives an equally significant result to that of the chi-squared test. They are essentially two ways of doing the same thing.

The test of the smoking effect, $0.61/0.44 = 1.39$, is consistent with this chi-squared result, well below 1.96. The confidence interval which therefore includes zero is:

$$0.61 \pm 1.96 \times 0.44$$

or $\qquad\qquad -0.25$ to 1.47

However, the data are consistent with quite large values, up to 1.47. It is wise to investigate what these effects on the logistic scale mean in terms of estimated risks.

Taking pollution at 100 $\mu g/m^3$, the low category on the logistic scale, the model predicts that for the no smoking (ns) group:

$$y_{ns} = -3.87 + 0.0113 \times 100 = -2.74$$

and for the smoking (s) group:

$$y_s = -3.87 + 0.61 + 0.0113 \times 100 = -2.13$$

If $y = \log [p/(1-p)]$ then the value of p is:

$$p = 1/[1 + \exp(-y)]$$

so:

$$p_{ns} = 1/[1 + \exp(+2.74)] = 0.061$$

and

$$p_s = 1/[1 + \exp(+2.13)] = 0.106$$

This means that the estimated risks of having the symptom 'cough' are 6.1 and 10.6 per cent respectively. And although it does not reach significance in this data-set the effect of exposure to smoking is estimated as an increase in risk of 4.5 per cent. The risk of 6.1 per cent has nearly doubled, going from children not exposed to smoking to those who are. If this represents a real effect it is obviously important.

The ratio of the two risks is known as the relative risk—that is the risk of the exposed relative to that of those not exposed. This is calculated as:

$$\text{relative risk} = 0.106/0.061 = 1.74$$

which is nearly 2 indicating that the risk is nearly doubled.

An approximation of the relative risk is provided by the odds ratio which, for the smoking factor, is:

$$[p_s/(1 - p_s)]/[p_{ns}/(1 - p_{ns})]$$

Because the analysis uses the logistic scale this is in fact:

$$\exp(smk) = \exp(0.61) = 1.84$$

which is rather larger than the relative risk. None the less for small risks (about 5 per cent or less) the odds ratio is a good approximation to the relative risk.

There are a number of tests specifically designed for comparing such estimates of relative risk with the value expected if the true risks are identical—that is 1. (For further discussion see Chapter 9 on case–control studies.) Apart from slightly differing assumptions they are effectively the same test as that derived from the model comparison in the analysis of deviance. In addition, fitting and comparing models has allowed the use of all the data from several pollution groups. The model comparison approach is much more general. It can be extended to cover very complicated data-sets and it allows estimates of all the various effects to be obtained in a relatively straightforward way.

From the parallel-line model, approximate confidence intervals for the estimates of risk and relative risk can be obtained. However, this can be done only indirectly by first calculating the confidence limits on the logistic scale as:

$$\text{estimate} + 1.96 \text{ s.e. (estimate)}$$

and then converting the limits of the confidence interval back to the risk scale. The predicted y value for the smoking group is:

$$y = a + b \times 100$$

The standard error of y, s.e. (y), which is the square root of the sampling variance of y, is a function of the sampling variances of a and b with what is known as their co-variance. The interdependence of estimates has not been discussed, but they must be taken into account when standard errors of functions of estimates are required. For a full discussion see Armitage (1971). All that is needed are the co-variances of the estimates and these are readily obtained during computer analysis. For $y = a + bx$ the variance of y (var(y)) is then calculated by:

$$\text{var}(y) = \text{var}(a) + 2x \times \text{co-variance }(a,b) + x^2 \times \text{var}(b)$$

For this analysis, the computer produced the values:

$$\text{var } (a) = 0.48$$
$$\text{var } (b) = 8.3 \times 10^{-6}$$
$$\text{cov } (a, b) = -0.0017$$

So var $(y) = 0.48 + 200 (-0.0017) + 100^2 \times 10^{-6} \times 8.30$
$\qquad\qquad = 0.22$

and s.e. $(y) = \sqrt{[\text{var } (y)]} = 0.47$

Table 13.22. Numbers of children with cough according to whether they were exposed to smoking in the home and, in parentheses, the numbers expected if there was no association between exposure and risk of cough

| | Smoking in the home | | |
	No	Yes	Total
Cough			
Yes	4(5.4)	9(7.6)	13
No	8(6.6)	8(9.4)	16
Total	12	17	29

The 95 per cent confidence interval for y in the no smoking group is therefore:

$$-2.74 \pm 1.96 \times 0.47$$

or

$$-3.66 \text{ to } -1.82$$

Converting back to the risks scale, the equivalent confidence interval for the risk in this group is 0.025 to 0.139. In the group exposed to smoking the equivalent figures are 10.6 per cent for the best estimate of risk with a confidence interval of 5.0 to 21.3 per cent. This gives an estimated relative risk of $10.6/6.1 = 1.75$.

It is not simple to obtain a confidence interval for the true relative risk, but is quite straightforward for the odds ratio. On the logistic scale, the effect of smoking is 0.61 with a confidence interval of $0.61 \pm 1.96 \times 0.44$ or -0.25 to 1.47. The odds ratio is estimated as $\exp(0.61)$, and $\exp(-0.25)$ to $\exp(1.47)$, that is 0.77 to 4.35, are approximate confidence limits.

The data are reasonably consistent with a true odds ratio as low as 0.77, that is, the no smoking group running less risk than the smoking group, or as high as 4.35, that is, the group exposed to smoking running more than four times the risk of those not exposed. This means that the odds ratio is not significantly different from 1.

The conclusions from these simulated data are first that there is strong evidence of a pollution effect—going from low to medium to high pollution in these individuals at least doubles then trebles the risk in both the smoking and non-smoking groups. Second, there is some suggestion that a larger study might identify an important effect of exposure to smoking since these results are not inconsistent with a four fold increase in risk of cough in those exposed to smoking compared with those not so exposed.

Comparing two proportions

If data were available only for the high pollution group as in Table 13.22, the problem reduces to comparing two proportions. The prevalence of cough is 4/12 or 33.3 per cent in the non-smoking homes and 9/17 or 52.9 per cent in the smoking homes. The question is: 'Are these two proportions more different than could easily occur by change?'

The hypothesis to test is that there is a single 'true' risk which is the same for both groups and that the differences arose by chance. This single risk, assuming the hypothesis true, is best estimated by using all the data, i.e. by 13/29 = 44.8 per cent.

This means that on average in studies like this 44.8 per cent of each group would be expected to fall in the cough category. On this basis the expected numbers can be calculated for each cell of the table: 5.4 is 44.8 per cent of 12, 7.6 is 44.8 per cent of 17, and so on. The problem becomes one of assessing whether the cell frequencies are surprisingly far from those expected. For each cell the measure of how far apart the observed and expected frequencies are is taken as:

$$\frac{(\text{observed frequency} - \text{expected frequency})^2}{\text{expected frequency}}$$

or more simply:

$$(O - E)^2/E$$

This summed over all the cells of the table gives a measure of how the table as a whole deviates from what should be expected if the hypothesis were true, that is:

$$\text{Sum for all cells of } (O-E)^2/E$$

which can be shown to have a chi-squared distribution on one degree of freedom. This is because the analysis is essentially comparing a model with two different proportions or parameters, with a model with one proportion or parameter. Measures of the difference between these two models have one degree of freedom. In this case:

$$\text{chi-squared} = (4 - 5.4)^2/5.4 + (9 - 7.6)^2/7.6$$
$$+ (8 - 6.6)^2/6.6 + (8 - 9.4)^2/9.4 = 1.09$$

which is a long way from significance at the 5 per cent level (cf. 3.84).

Practical problems rarely reduce to the comparison of two proportions without gross simplification. None the less, a number of alternatives to this test have been developed over the years. There is also a lot of discussion as to which is right. Fortunately it rarely matters. Yates (1934) pointed out that Sum $(O - E)^2/E$ was more nearly a chi-squared variable if the test was slightly modified by subtracting 0.5 from the frequencies on the diagonal with the largest product, in this case the 8,9 diagonal. This is probably the most sensible test to use in this context. If a computer is available and it still seems appropriate to treat the analysis as nothing more than the comparison of two proportions then a test known as Fisher's exact test is more precise.

Fisher's test calculates the exact probability of this or more surprising tables occurring given the marginal totals and the hypothesis. It therefore gives p values directly. Those requiring the algebraic details should consult Armitage (1971).

Sample sizes for comparing proportions

The exact solution to this problem leads to a very complicated formula which can be found in a paper by Casagrande *et al.* (1978). For most practical purposes it will be sufficient to use a calculation analogous to that for comparing means. The sample sizes obtained will be slightly smaller than optimum, but this can be regarded as an increase in the risk of a Type II error. If this matters the risk can be set lower and the calculations repeated. The calculations are based on expressing the chi-squared test above in the form:

$$\frac{\text{estimate} - \text{expected value}}{\text{s.e. (estimate)}}$$

where the estimate is the difference between two proportions and usually the expected value is zero. The test statistic is:

$$(r_1/n_1 - r_2/n_2)/\sqrt{[p(1 - p)(1/n_1 + 1/n_2)]}$$

where p is $(r_1 + r_2)/(n_1 + n_2)$. For quite small proportions and values of n_1 and n_2 this has a sampling distribution close to normal. This means that the statistic may be tested against 1.96 for a 5 per cent significance level. As for comparing means, it is efficient to keep the groups the same size. The number in each group must then satisfy:

$$D/\sqrt{[p(1 - p)2/n]} \geq Za + Zb$$

where Za and Zb are centiles of the standard normal distribution determined by the choice of what risks of type I and II errors are acceptable. In this calculation p is taken as the average of the two proportions expected if the hypothesis is false. The number must satisfy:

$$n \geq (Za + Zb)^2 \times 2p(1 - p)/D^2$$

Example

What sample size is needed to detect a difference in the prevalence of respiratory symptoms between children in two towns with differing air pollution levels? If the prevalence is about 20 per cent and the smallest interesting difference is 5 per cent then $p = 0.20 + 0.05/2 = 0.225$. If the acceptable risks of Type I and II errors are 5 per cent 20 per cent, then the number in each group must satisfy:

$$n \geq (1.96 + 0.84)^2 \times 2 \times 0.225 (1 - 0.225)/(0.05)^2$$

which gives $n \geq 1093.7$. Thus 1094 subjects are needed in each group and 2188 in total.

This and the section on sample size for comparing means (p. 201) give a simplified guide to sample size calculations. Lachin (1981) gives a very comprehensive guide.

Repeated observations

An individual may be classified as having a symptom or not several times during a course of treatment. This produces an analysis problem analogous to that of repeated measurements. Using a subject parameter in the model allows it to be analysed in much the same way using an analysis of deviance instead of the analysis of variance. Odds ratio estimates for risks before and after some treatment or exposure may then be obtained as the square root of the equivalent estimate in

Table 13.23. Distribution of subjects according to errors made on their regimen before and after provision of a patient-held treatment record

	After record	
	Errors	**No errors**
Before record		
Errors	4	4
No errors	0	2

the section above on comparing two groups allowing for a covariate (p. 203). In practice, a modified form of model fitting is used for this type of data which also arises in matched case–control studies discussed below and in a separate chapter. A comprehensive discussion of the model fitting approach in this context is given by Breslow and Day (1980), and the method of analysis with the computer package GLIM is fully described by Adena and Wilson (1982).

In the simplest case analogous to the paired *t* test problem a simple technique known as McNemar's test may be used.

Example using McNemar's test

Geriatric patients in the community often have complex drug regimens to follow. Table 13.23 gives some data from a study to assess whether a specially designed, patient held, treatment record affects the error rate when the patients are questioned on their regimen.

It is possible to analyse these data using a model fitting approach, but in this simple case it is not necessary. On the hypothesis that the subjects are equally likely to make errors before and after, as many should improve as get worse. This means that those who changed should be equally divided between the two types of change possible.

In these data four changed—they all became better at identifying their drug dosages. The probability of a result as surprising as this if there was only a 50 per cent chance of getting better is required. This is exactly the same as the probability of getting 'heads' in all four tosses of a coin which is $0.5 \times 0.5 \times 0.5 \times 0.5 = 0.0625$. An equally surprising result would have been all four getting worse, i.e. the equivalent of four 'tails'. Since this also has probability $p = 0.0625$, the probability of a result this surprising is $p = 0.0625 \times 2 = 0.125$. So although the result seems interesting the probability of such extreme results is certainly not less than 0.05.

If the effect was genuinely of this magnitude a larger sample would be needed to show it as significant.

Comparing proportions from separate samples— case–control or retrospective studies

If cases with a particular disease or condition and controls are not matched to have the same values of potentially interfering factors such as age and sex, the data from case–control studies are analysed in exactly the same way, using the model-fitting approach, as any other set of proportions taking cases/(cases + controls) as the proportions. Odds ratios, assessing the effects on risk of factors in the model, are obtained from the fitted model as described earlier (p. 204).

If the cases and controls have been matched in some way according to factors that might be related to the risk of becoming a case, then the correct analysis is more complicated. Further discussion may be found in Chapter 9, Breslow and Day (1980), and Adena and Wilson (1982).

More than two categories of outcome

Categorical outcomes arise when individuals can respond to treatment or experience in several ways not easily measured. For example, a treatment for Hodgkin's disease may be assessed according to whether the patients 'died', 'got worse', 'stayed the same', or 'showed signs of remission'. The clinician's overall classification is preferred in this case to a single measure of well-being such as white cell count.

It is always possible to combine some of the categories so that there are only two, for example 'no remission' and 'remission'. This permits the response to be treated as a proportion and allows the use of the methods in the previous section. In practice this is often the most sensible thing to do although it does waste information. Alternatively, some way must be found for describing responses in many categories by models.

Consider data from a trial comparing two treatments, A and B, for Hodgkin's disease. The outcomes, assessed after a fixed period of treatment, were 'no response or died', 'partial remission', and 'complete remission'. The results are given in Table 13.24.

It is necessary to compare how the two groups are distributed among the response categories. This is best done by plotting the frequencies against the remission category for each of the treatment groups. Joining the points within each treatment group makes the diagram easier to interpret (Fig. 13.28).

The true responses are represented on the horizontal axis. However, if the frequencies on the y axis are considered as the response variable then the pattern to be described by an algebraic model has exactly the same form as discussed in previous sections for means and proportions. The y variable in this context is a 'count' or frequency. It is usual to assume that they arise from a theoretical distribution called the Poisson distribution.

For proportions obtained from counts in two categories of

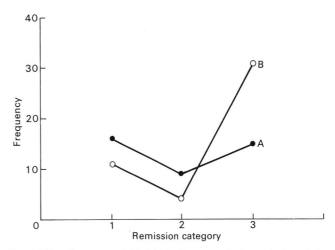

Fig. 13.28. Frequency of Hodgkin's disease patients against remission category by treatment group. (Source: British National Lymphoma Investigation 1975)

response it was necessary to use the logistic scale to fit the models. For frequencies it is necessary to use another scale, and the most useful and appropriate is the logarithmic scale. This is logical because changes in frequencies seen when moving to groups at higher risk are likely to be proportional. On the frequency scale, distances representing the same proportional difference will change with the frequency. On the log (frequency) scale these distances will stay constant which means that constant proportional effects will give parallel-line models. For this reason this analysis has come to be known as 'log linear modelling'.

Notice that if there had been many more subjects in treatment group A the frequencies would all have been larger. The A line would have been much higher than the B line, even if the treatments were equally effective. Because of this it is the parallelism of the lines that is the main interest not their separation.

To test whether there are significant differences in the shapes of the distribution it is necessary to compare models with and without interaction parameters. Because the height of the lines does not matter it is not necessary to treat it as a random variable and waste information estimating some 'true' height. This is avoided by forcing the model-fitting process to choose estimates for the 'height' parameters so the model fits or predicts the observed total frequencies exactly.

Apart from the use of the log scale, the constraints on the models and the interactions being the model parameters of prime interest, the analysis is much the same as for proportions. Maximum likelihood and the analysis of deviance are used to produce approximate chi-squared tests of the interaction parameters quantifying the hypothesis of interest.

Comparing two groups with three outcomes

In the above example the models to be compared are:

1. *Non-parallel lines*
   $y = rv$ + treatment effect + remission category effect + treatment × remission category interaction

Table 13.24. Distribution of Hodgkin's disease patients by treatment and remission category (treatment group percentages in parentheses)

| Treatment | Remission category | | | Total |
	1. None/died	2. Partial remission	3. Complete remission	
A	16(40)	9(23)	15(37)	40(100)
B	11(24)	4(9)	31(67)	46(100)

Table 13.25. Analysis of deviance comparing the frequency distribution of Hodgkin's disease patients among three remission categories for two treatments groups

Model	Deviance	d.f.	Chi-squared
1. Non-parallel	7	0	
2. Parallel	8.17	2	
(2 − 1) Due to non-parallelism	8.17	2	8.17

and

  2. *Parallel lines*
    $y = rv + \text{treatment effect} + \text{remission category effect}$

Model 1 requires six parameters. Since there are only six points it will fit the data exactly and there are no degrees of freedom and no deviance.

A computer analysis gave the deviance from the parallel-line model as 8.17 on 2 d.f. so the analysis of deviance is as in Table 13.25.

Since the 95th centile of the chi-squared distribution with 2 d.f. is 5.99, the outcome distributions of the two treatment groups are significantly different ($p < 0.05$). It only remains to make quantitative statements about how the 'relative risk' of ending up in the various remission categories varies according to treatment using the estimated frequencies from the non-parallel model and the treatment group totals, which give, in this simple case, estimates of risk equal to the percentages in Table 13.24.

This approach easily extends to more complex problems, for example if initial severity subgroups had to be taken into account.

Comparing frequencies in a two-way classification

If the data are as simple as in the above example, the frequencies can be compared directly with a chi-squared test using expected frequencies calculated assuming the hypothesis to be tested as true (Table 13.26).

On the hypothesis that the treatments are equally effective, 31.4, 15.1, and 53.5 per cent of individuals would be expected to fall into the three response categories whatever the treatment (Table 13.26). This means that 31.4 per cent of the 40 patients receiving treatment A are expected to fall into response category 1:

$$\frac{31.4}{100} \times 40 = \frac{27}{86} \times 40 = 12.6$$

in category 2, 15.1 per cent of 13 = 6, and so on.

As when comparing two proportions (pp. 205–6), the test of whether the frequencies are surprisingly far from those expected on the hypothesis of equally effective treatments is:

$$\text{chi-squared} = \text{Sum } (O - E)^2/E = 8.03.$$

This is not quite the same as the analysis of deviance chi-squared value because the analyses make slightly different assumptions. However, the conclusion is much the same and unless 5 per cent is inappropriately regarded as a 'magic' cut-off point these two approaches will usually lead to the same conclusions.

Survival or event data

These data arise when individuals observed over time are at risk of some events. The event may be death, menarche, coronary heart attack, and so on. After death or menarche, the individual is no longer at risk. The event cannot occur again. A coronary heart attack could in fact be the first of many, but if observation of the subject is discontinued after the event it can be treated as terminal and standard analyses used.

If these data consist of numbers dying in a fixed period then the analysis is simply a comparison of proportions. The techniques of the section on qualitative two-category responses (pp. 201–7) will usually suffice. Annual mortality rates from different geographical areas can be analysed in this way, treating them as proportions, but there may be complications as discussed by Pocock *et al.* (1981). None the less, the analysis of mortality data of this sort by a model-fitting approach avoids the use of standardizing techniques whereby mortality figures were traditionally adjusted for age and sex effects before group comparisons were made, notably using standardized mortality ratios (SMRs). Such adjustments make possibly unwarranted and certainly untested assumptions about the effects of age and sex being the same in all the groups to be compared, and for that reason must be used with care.

When the individuals are at risk of the event for varying periods of time the problem cannot easily be considered as an analysis of proportions. There are two main types of problem. In the first it is assumed that within the groups to be compared the risk is constant over time. In the second case the risk changes with time. The remaining discussion will be restricted to these two types of survival data.

Constant risks

From the number experiencing the event in a group of individuals observed for varying lengths of time, the risk per unit time for an individual in the group can be estimated by dividing the number of deaths by the accumulated time at risk. The modelling approach to analysing such data requires that the number of deaths in each group is taken as the response

Table 13.26. Distribution of Hodgkin's disease patients by treatments and remission category with, in parentheses, frequencies expected assuming equally effective treatments

	Response			
	1	2	3	Total
A	16(12.6)	9(6.0)	15(21.4)	40
B	11(14.4)	4(7.0)	31(24.6)	46
Total	27	13	46	86
&	31.4	15.1	53.5	100

Table 13.27. Number of deaths by age group and factory from a study of occupational mortality with person years at risk in parentheses

Age at start of study (years)	Factory	
	1	2
50–59.9	7(4045)	7(3701)
60–69.9	27(3571)	37(3702)
70–79.9	30(1777)	35(1818)
80–89.9	8(381)	9(350)

variable and the accumulated time at risk used as a covariate.

In the occupational health study data in Table 13.27 the deaths were classified according to factory and age at the start of the observation period. The table also gives the total person years at risk (PYR) within each group. Since the response variable is a count, the appropriate analysis requires using the number of deaths as a Poisson response variable and fitting the models on the log scale.

The questions to be answered are: 'Is the effect of age the same in both factories?' If it is: 'Are the risks, allowing for age, different in the two factories?' The models to answer these are:

1. *Non-parallel age trends*
   $y = \log (\text{PYR}) + rv + \text{factory effect} + \text{factory specific slope} \times \text{age}$

assuming straight line age trends.

2. *Parallel age trends*
   $y = \log (\text{PYR}) + rv + \text{factory effect} + \text{slope} \times \text{age}$

3. *No factory effect*
   $y = \log (\text{PYR}) + rv + \text{slope} \times \text{age}$

and for completeness:

4. *No age effect*
   $y = \log (\text{PYR}) + rv + \text{factory effect}$

When these models are fitted, the analysis of deviance in Table 13.28 is obtained.

The 95th centile of the chi-squared distribution with one degree of freedom is 3.84 and values less than this could well occur by chance. Obviously the age effect producing a chi-squared value of 89.19 is very marked and real. The factory effect gives a chi-squared of 1.73 which does not appear to indicate much. However, the initial deviance of 12.70 on four degrees of freedom indicates that the first model is a long way from the data points. The linear age trends do not fit the data well so the subsequent analysis must be treated with caution. In fact fitting an age-squared term to allow for curvature gives a much smaller deviance and the model with such a term, but assuming a zero difference between the two factories:

5. $y = \log (\text{PYR}) + rv + b \times \text{age} + c \times \text{age}^2$

fits the data well with a deviance of 2.09 on five degrees

of freedom. There is therefore very little evidence of a factory effect.

Observed and expected deaths

In simple cases a hypothesis may be used to deduce the number of deaths expected and these can then be used to calculate $(O - E)^2/E$ for appropriate groups. This is then used to obtain a simple chi-squared test.

In the above example the hypothesis to be tested was that the risks were equal in the two factories. The risks for each age group per person year is calculated as:

sum of deaths/sum of PYRs

The expected deaths in each factory are obtained from this multiplied by the appropriate PYR. For the 50-year-olds in factory 1 this gives:

$$\text{expected deaths} = 4045 \times \frac{(7 + 7)}{(4045 + 3701)}$$
$$= 4045 \times 0.0018 = 7.3$$

Similarly for factor 2:

$$\text{expected deaths} = 3701 \times 0.0018 = 6.7$$

and so on for all age groups.

The overall chi-squared to test the hypothesis of no factory effect on risk is the sum of the eight $(O - E)^2/E$ values, which is:

$$(7 - 7.3)^2/7.3 + (7 - 6.7)^2/6.7$$
$$+ (27 - 31.4)^2/31.4 + (37 - 32.6)^2/32.6$$
$$+ (30 - 32.1)^2/32.1 + (35 - 32.9)^2/32.9$$
$$+ (8 - 8.9)^2/8.9 + (9 - 8.1)^2/8.1$$
$$= 1.70$$

This has four degrees of freedom because there were eight observations and to calculate the chi-squared value it was necessary to estimate four parameters (the risks for the four age groups). The conclusions are the same, but this approach is impossibly tedious if there are many more than two groups to compare. The modelling approach is almost inevitable for a thorough analysis which estimates the magnitude of the various effects.

Table 13.28. Analysis of deviance comparing mortality in two factories allowing for different age structures

Model	Deviance	d.f.	Chi-squared
1. Non-parallel linear age trends	12.70	4	
2. Parallel trends	12.71	5	
(2 − 1) Due to non-parallelism	0.01	1	0.01
3. No factory effect	14.44	6	
(3 − 2) Due to factory effect	1.73	1	1.73
4. No age effect	101.90	6	
(4 − 2) Due to age effect	89.19	1	89.19

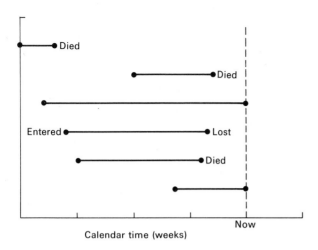

Fig. 13.29. Observation periods of subjects in a trial.

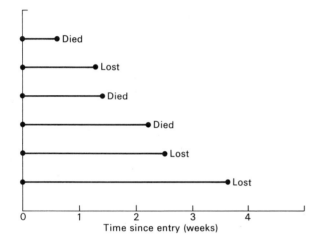

Fig. 13.30. Observation periods of subjects in a trial from time of entry.

Table 13.29. Life-table for patients with cancer of the prostate in the radiotherapy group of a clinical trial

Time since entry (years)	No. died	No. at risk	No. lost	Adj. no. at risk	Estimated probability of Death	Estimated probability of Survival	Estimated proportion of survivors at end of interval
0–1	3	44	3	42.5	0.071	0.929	0.929
1–2	2	38	10	33.0	0.061	0.939	0.873
2–3	1	26	3	24.5	0.041	0.959	0.837
3–4	3	22	3	20.5	0.146	0.854	0.715
4–5	2	16	4	14.0	0.143	0.857	0.613
5–6	1	10	1	9.5	0.105	0.895	0.548
6–7	1	8	1	7.5	0.133	0.867	0.475
7–8	0	6	3	4.5	0.000	1.000	0.475
8–9	1	3	1	2.5	0.400	0.600	0.285
9–10	1	1	0	1.0	1.000	0.000	0.000

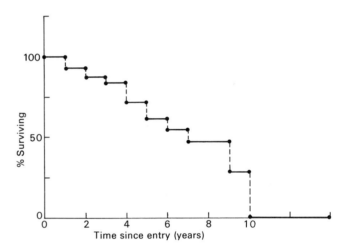

Fig. 13.31. Life-table survival curve for the radiotherapy group in a trial of treatments for cancer of the prostate.

Changing risks—survival curves

For each individual, data of this type will consist of the length of time observed and reason lost to observation. The period of observation may start at birth, diagnosis, start of treatment, or some other appropriate point in time. An individual who dies is then lost to observation. When a study is concluded observation stops on the survivors and some may drop out during the course of the study.

To investigate survival in a group of individuals observed for differing periods of time, life-table survival curves are used. The survival curve is a plot against time elapsed from diagnosis, or other appropriate points, of the percentage of individuals surviving. The life-table calculations to obtain these percentages from individuals observed for varying lengths of time are moderately simple, but not obvious.

Consider a number of subjects entering and leaving a trial at different times as in Fig. 13.29. Using time since entry as the horizontal scale gives a modified plot (Fig. 13.30) from which the number of subjects at risk and dying in each week since entry can be seen. In the first week after entry six sub-

jects were at risk and one had died. This gives an estimated death rate of 0.17 and hence a survival rate of 0.83. In week two there were four at risk all the time, one for half a week, and one died giving a death rate of $1/4.5 = 0.22$ and a survival rate for individuals reaching the second week of 0.78.

The chance of an individual surviving the second week, which requires he or she survives both the first and the second, is $0.83 \times 0.78 = 0.65$. This is the estimated cumulative survival rate. Plotting it against time from treatment gives the survival curve. Conventionally, individuals lost in any interval are treated as having been at risk for half the interval.

The life-table for cancer of the prostate patients in the radiotherapy group of a trial comparing radiotherapy and hormone treatments is given in Table 13.29. The adjusted number at risk is the number entering the interval minus half the number lost to observation during it. Plotting the percentage surviving (S) against the upper end of the time intervals gives the survival curve shown in Figure 13.31. It is not appropriate to draw a smooth curve through the points since such curves give a spurious impression of accuracy.

Comparisons of a small number of survival curves can be made by hand using what is known as the 'log rank' test. On the hypothesis of no difference between two curves the number of deaths expected in each group can be calculated. By calculating $(O - E)^2/E$ for each group a chi-squared test on one degree of freedom comparing the curves can be obtained. The method can be used to compare survival curves within a cross-classification of treatment or other groups but it is not simple. The method is described in detail by Peto *et al.* (1977).

At the cost of a few testable and often reasonable assumptions a full modelling approach can be used analogous to those discussed in earlier sections. This approach was introduced by Cox (1972) and is discussed by Anderson *et al.* (1980) and Kalbfleisch and Prentice (1980). Although these texts are rather mathematical, computer programs are now available that make it possible to fit sequences of models to survival data with relative ease. This approach provides quantitative estimates with standard errors thus aiding considerably the interpretation and presentation aspects of the analysis.

Conclusion

Obviously special problems require special techniques which cannot be discussed here. None the less, the range of problems discussed and the appropriate analytical approaches cover the major part of statistical activity in public health research and epidemiology. With reasonable computing facilities most practical problems you meet can be tackled using these methods.

References

Adena, M.A. and Wilson, S.R. (1982). *Generalised linear models in epidemiological research: case–control studies.* The Instat Foundation, Sydney.

Anderson, S., Auquier, A., Hauck, W.W., Oakes, D., Vandaele, W., and Weisberg, H.I. (1980). *Statistical methods for comparative studies.* Wiley, New York.

Armitage, P. (1960). *Sequential medical trials.* Blackwell Scientific Publications, Oxford.

Armitage, P. (1971). *Statistical methods in medical research*, Blackwell Scientific Publications, Oxford.

Breslow, N.E. and Day, N.E. (1980). *Statistical methods in cancer research, Vol. 1. The analysis of case-control studies.* IARC Scientific Publications No. 32, Internation Agency for Research on Cancer, Lyon.

British National Lymphoma Investigation (1975). Value of prednisone in combination chemotherapy of stage IV Hodgkin's disease. *British Medical Journal* **iii**, 413.

Casagrande, J.T., Pike, M.C., and Smith, P.G. (1978). An improved approximate formula for calculating sample sizes for comparing two binomial distributions. *Biometrics* **34**, 483.

Cox, D.R. (1972). Regression models and life tables (with discussion) *Journal of the Royal Statistical Society B* **34**, 187.

Dixon, W.J. (ed.) (1981). *BMDP statistical software.* University of California Press, Berkeley.

Forni, A. and Sciame, A. (1975). Chromosome and biochemical studies in women occupationally exposed to lead. *Archives of Environmental Health* **35**, 139.

Kalbfleisch, J. and Prentice, R.L. (1980). *The statistical analysis of failure time data.* Wiley, New York.

Lachin, J.M. (1981). Introduction to sample size determination and power analysis for clinical trials. *Clinical Trials* **1**, 93.

Lindley, D.V. and Miller, J.C.P. (1964). *Cambridge elementary statistical tables.* Cambridge University Press.

Nelder, J.A. and Wedderburn, R.W.M. (1972). Generalised linear models. *Journal of the Royal Statistical Society A* **135**, 370.

Payne, C. (ed.) (1983). *The GLIM system manual for release 3.77.* Numerical Algorithms Group, Oxford.

Peto, R., Pike, M.C., Armitage, P., *et al.* (1977). Design and analysis of randomised clinical trials requiring prolonged observation of each patient: II Analysis and examples. *British Journal of Cancer* **35**, 1.

Pocock, S.J. (1984). *Clinical trials: a practical approach.* Wiley, New York.

Pocock, S.J., Cook, D.G., and Beresford, S.A.A. (1981). Regression of area mortality rates on explanatory variables: what weighting is appropriate. *Journal of Applied Statistics* **30**, 286.

Pocock, S.J., Shaper, A.G., Cook, D.G., *et al.* (1980). British Regional Heart Study—geographical variation in cardiovascular mortality and the role of water quality. *British Medical Journal* **280**, 1243.

Ryan, T.A., Joiner, B.L., and Ryan, B.F. (1981). *Minitab reference manual.* Minitab Project, Pennsylvania.

Snedecor, G.W. and Cochran, W.G. (1967). *Statistical methods* (6th edn). Iowa State Press, Ames.

Yates, F. (1934). Contingency tables involving smaller numbers and the chi-squared test. *Journal of the Royal Statistical Society, Supplement* **1**, 217.

14

Mathematical models of transmission and control

R.M. ANDERSON and D.J. NOKES

Introduction

The aim of this chapter is to show how simple mathematical models of the transmission of infectious agents within human communities can help to interpret observed epidemiological trends, to guide the collection of data towards further understanding, and to design programmes for the control of infection and disease. A central theme is improvement of understanding of the interplay between the variables that determine the course of infection within an individual and the variables that control the pattern of infection and disease within communities. This theme hinges on an understanding of the basic similarities and differences between infections in terms of: the number of population variables (and consequent equations) needed for a sensible characterization of the system; the typical relations between the various rate parameters (such as birth, death, recovery, and transmission rates); and the form of expression that captures the essence of the transmission process.

Model construction, whether mathematical, verbal, or diagrammatic, is, in principle, the conceptual reduction of a complex biological or population-based process into a simple, idealized, and easily understandable sequence of events. Consequently, the use of mathematical modelling as a descriptive and interpretive tool is a common exercise in scientific study. Its use, therefore, in epidemiological study should not be viewed as intrinsically difficult or beyond the comprehension of those trained in medical or biological disciplines. The reductionist approach, inherent to model construction, which helps to define processes clearly and to identify the most important components of a system, is employed in many areas of public health research and practice. The following situations, for example, are all likely to involve, at the very least, the implicit use of models to simplify and aid understanding: (i) the assessment of the cause and severity of sporadic epidemics of *Salmonella* or hepatitis A virus food poisoning, or legionnaires' disease; (ii) the cost–benefit analysis of various measures used to combat an infection within a hospital, within a community, a nation or globally; and (iii) the identification of the factors that control the maintenance of an endemic infection within a community.

Most epidemiological problems, by definition, are concerned with the study of populations and, so, involve quantitative scores of, for example, abundances and rates of spread. Thus, it is invariably necessary to convert any descriptive model of process into a more formal mathematical framework so that we work with numbers and not words. The use of a more formal structure enables us to incorporate quantitative estimates of abundances or rates, derived from experiment or field observation, into the model and to make predictions of the likely behaviour of the system under varying conditions, particularly when the introduction or alteration of measures to control infection or disease is involved.

It is the step of translation from verbal or diagrammatic description into a formal mathematical framework that arouses the deepest suspicions amongst medical or public health workers. Quite naturally, this response is in part a consequence of the use of, what is to many, a strange symbolism to describe familiar verbal or conceptual identities. It must be remembered, however, that mathematics is the most precise language we have available for scientific study and, once a problem is formulated in mathematical terms, many techniques are available to pursue the logical consequences of the stated assumptions. The clear and unambiguous statement of assumptions is, of course, a particular attribute of mathematical, as opposed to verbal description. Excessive use of symbolism or formal methods of analysis can confuse rather than clarify, and it must be admitted that some sections of the mathematical epidemiological literature have tended towards the abstract, free from the constraints of data or relevance. But, to jump from this observation to the belief that mathematical models have nothing to contribute in practice to the design of public health programmes is a mistake. If sensibly used, mathematical models are no more, and no less, than tools for thinking about things in a precise way.

The second area of suspicion, aside from symbolism, concerns simplification. A frequent criticism of mathematical work in epidemiology is that model formulation involves too many simplifying assumptions despite known biological complexity. This is often true, and needs to be remedied, but it is, in part, a consequence of the youth of the discipline and, in some cases, is a result of inadequate quantitative understanding of a particular problem. There are, however, two important counter-arguments to the criticism of simplification. First, and most importantly, it is often the case in biological study that a few processes dominate the generation of observed pattern despite the fact that many more can, to a lesser degree, influence the outcome. The identification of the dominant processes is an important facet of model construction and of what is termed sensitivity analysis. The second point concerns scientific method. The process of understanding the consequences of a series of simple assumptions and of building upon this by slowly adding complexity is directly analogous to the laboratory scientist's approach of carefully controlling most variables and allowing a few to vary in a planned design. The careful building of complexity on a simple framework can greatly facilitate our understanding of the major factors that influence or control a particular process or pattern.

The chapter is organized as follows. The section following this introduction provides a brief review of the historical development of mathematical epidemiology and outlines the types of infection that are considered in later sections. The third section addresses the problems of model construction, design, and application. The fourth section examines the major concepts in quantitative epidemiology that have been derived from mathematical study, such as threshold host densities for the persistence of an infection, the basic reproductive rate, and herd immunity. In the fifth section, methods are explored by which to obtain some of the basic epidemiological parameters from empirical observation. The sixth section turns to applied problems and considers the use of models in the design of control strategies for infection and disease, and the final section is reserved for concluding thoughts. Throughout, mathematical details are kept to a bare minimum and the reader interested in technical details of model construction and analysis is referred to papers in specialist journals.

Historical perspective

The application of mathematics to the study of infectious disease appears to have been initiated by Daniel Bernoulli in 1760 when he used a mathematical method to evaluate the effectiveness of the techniques of variolation against smallpox (Bernoulli 1760). Further interest did not occur until the middle of the nineteenth century when William Farr in 1840 effectively fitted a normal curve to smoothed quarterly data on deaths from smallpox in England and Wales over the period 1837–9 (Farr 1840). This empirical approach was further developed by John Brownlee who considered in detail the 'geometry' of epidemic curves (Brownlee 1906). The origin of modern mathematical epidemiology owes much to the work of En'ko, Hamer, Ross, Soper, Kermack, and McKendrick who, in different ways, began to formulate specific theories abut the transmission of infectious disease in simple, but precise, mathematical statements and to investigate the properties of the resulting models (e.g. Kermack and McKendrick 1927; Ross 1911; Soper 1929). The work of Hamer (1906), Ross (1911), Soper (1929), and Kermack and McKendrick (1927) led to one of the corner-stones of modern mathematical epidemiology via the hypothesis that the course of an epidemic depends on the rate of contact between susceptible and infectious individuals. This was the so-called 'mass-action' principle in which the net rate of spread of infection is assumed to be proportional to the density of susceptible people multiplied by the density of infectious individuals. In turn, this principle generated the celebrated threshold theory according to which the introduction of a few infectious individuals into a community of susceptibles will not give rise to an epidemic outbreak unless the density or number of susceptibles is above a certain critical value.

Since these early beginnings, the growth in the literature has been very rapid and recent reviews have been published by Dietz (1987), Anderson and May (1985a), Becker (1979), and Bailey (1975). In recent work there has been an emphasis on the application of control theory to epidemic models (Wickwire 1977), the study of the spatial spread of the disease (Cliff et al. 1981), the investigation of the mechanisms underlying recurrent epidemic behaviour (Anderson and May 1982), the importance of heterogeneity in transmission (Anderson and May 1985b), the formulation of stochastic (i.e. probabilistic) models (Ball 1983), the formulation of models for indirectly transmitted infections with complex life-cycles (Anderson and May 1985c; Rodgers 1988), the study of sexually transmitted infections, such as gonorrhoea and the human immunodeficiency virus (HIV) (Anderson et al. 1986; Hethcote and Yorke 1984; May and Anderson 1987), and the development of models for infectious agent transmission in developing countries with positive net population growth rates (Anderson et al. 1988; McLean and Anderson 1988). Such theoretical work is beginning to play a role in the formulation of public health policy and the design of control programmes but there is a need, in future work, for greater emphasis on data-oriented studies that link theory with observation.

In the following sections, we attempt to give a flavour of recent work and to distil the major conclusions that have emerged in particular areas. We have deliberately chosen to concentrate on directly transmitted viral and bacterial infections that constitute the major infectious diseases of children in developed countries and, as a consequence of the recent pandemic of the acquired immune deficiency syndrome (AIDS), sexually transmitted infections. Our reasons are simply that the mathematical models are more highly developed in these fields compared with others (e.g. vector-borne infections), that theory has close contact with empirical epidemiological data in these areas, and that model structure is somewhat simpler than for other infections such as metazoan parasites.

Model construction

Definition of terms

Epidemiology

Epidemiology as a subject is concerned with the study of the 'behaviour' of an infection or disease within a population or populations of hosts (i.e. humans). 'Behaviour' refers to observed patterns such as the incidence (i.e. the rate at which new cases arise or are reported) of infection or disease. Examples of 'behaviour' are epidemics (each a rise and subsequent fall in incidence) and endemicity (the stable maintenance of infection within the human community). The aim of the discipline is to determine the underlying processes and to understand the interactions between them, that generate observed patterns (e.g. the rate of spread of infection and the pattern of susceptibility to infection). Epidemiology is a quantitative discipline that draws on statistical techniques for parameter estimation and on mathematical methods for delineating the dynamic changes that occur through time, across age classes, or over different spatial locations. The discipline also makes use of modern molecular (e.g. DNA probes) and immunological (measures of the abundances of antibodies specific to an infectious agent's antigens) techniques for the detection and quantification of current and past infection or disease.

Populations

The definition and description of the host and parasite populations is of obvious importance in epidemiological study. A population is an assemblage of organisms of the same species (or genetic type, etc.) which occupy a defined point or points in the plane created by the dimensions of space and time. The basic unit of such populations is the individual organism (i.e. parasite or human host). Populations may be divided (i.e. stratified) into a series of categories or classes, the members of which posses a unifying character or characters such as age, sex, or their stage of development. Such subdivisions may be made on spatial or temporal criteria to distinguish a local population from a larger assemblage. The boundaries in space, time, and genetic constitution between different populations are often vague, but it is important to define what constitutes the 'study population' as clearly as possible.

The natural history of infection

Mathematical models are often used to depict the rate of spread or transmission of an infectious agent through a defined human community. For their formulation, three broad classes of information are required:

(1) the modes and rates of transmission of the agent;

(2) the typical course of events within an individual following infection;

(3) the demographic and social characteristics of the human community.

The mode of transmission (i.e. direct, indirect, horizontal, vertical, etc.) is of obvious importance (see Table 14.1), but, if there is more than one route, the relative efficiency of each in determining overall transmission must be understood. When considering microparasitic infections (e.g. viruses, bacteria, and protozoa that multiply directly within the host), it is generally not possible to measure pathogen abundance within the host (i.e. the burden or intensity of infection). However, following invasion it is important to obtain quantitative information on the typical durations of the latent and infectious periods of the infection and the incubation period of the disease it induces. As depicted in Figure 14.1, the latent period is defined as the average period of time from the point of infection to the point when an individual becomes infectious to others, the infectious period denotes the average period over which an infected person is infectious to others, and the incubation period defines the average period from infection to the appearance of symptoms of disease. In practice, all these periods are variable between individuals, depending on factors such as the size of the inoculum of the infectious agent that initiates infection, the genetic background of host and parasite, past experience of infections, and the nutritional status of the host. The use of an average is an oversimplification, and where knowledge permits models should be based on distributed latent and infectious periods. In some instances, the infectious period may be influenced by patient management practices such as the confinement of an infected person once symptoms of infection are diagnosed (e.g. measles).

There are instances in the case of viral and bacterial infections when a knowledge of pathogen abundance within blood, excretions, secretions, other tissues, or organs of the host can be of importance in determining the infectivity of an infected person to susceptible contacts. A good example is provided by HIV-1. Current evidence suggests that the infectiousness of an infected person varies greatly over the long and variable incubation period of the disease, AIDS, that the virus induces (Fig. 14.2). It is believed, on the basis of recorded fluctuations in HIV antigenaemia, that a short period of high infectiousness occurs shortly after infection, followed by a long period of low to negligible infectiousness (perhaps many years) before infectiousness again increases as the infected patient develops symptoms of AIDS-related complex (ARC) and AIDS (Anderson and May 1988). In these cases, rather complex models are required to mirror the natural history of infection (Anderson 1988).

The human immune response to infection, its ability to confer protection against reinfection, and the duration of this protection have important implications for model construction. For the majority of childhood viral infections the assumption of lifelong immunity following recovery appears

Table 14.1. Epidemiological classification of infectious diseases of public health importance in developed countries

Mode of transmission	Type of parasite	Examples
VERTICAL*	Micro†	
	Viruses	Rubella, hepatitis B virus, cytomegalovirus, retroviruses
	Protozoa	*Toxoplasma gondii*
HORIZONTAL		
Direct		
Close contact	Micro	
	Viruses	Measles, mumps, rubella, Epstein–Barr virus, herpes simplex-1, respiratory syncytial virus, influenza-2, varicella, common cold
	Bacteria	Diphtheria, pertussis, bacterial meningitis
	Macro	
	Nematodes	*Enterobius vermicularis* (pinworm)
Environmental	Micro	
	Viruses	Hepatitis A, polio, Coxsackie
	Bacteria	Tetanus, shigella, salmonella, typhoid, cholera, legionnaires' disease
	Protozoa	*Giardia intestinalis*, amoebiasis
	Macro‡	
	Nematodes	Pinworm
Sexual	Micro	
	Viruses	Hepatitis B, human immunodeficiency virus, herpes simplex-2, cytomegalovirus
	Bacteria	*Neisseria gonorrhoeae*, syphilis
	Protozoa	*Trichomonas vaginalis*
Not direct		
Via other host species (zoonoses)	Micro	
	Virus	Rabies
	Protozoa	*Toxoplasma gondii*
	Macro	
	Nematodes	*Toxocara* species
	Cestodes	*Taenia solium, T. saginata, Echinococcus granulosus* (hydatid)
Vector-borne§ (including needles)	Micro	
	Viruses	Hepatitis B, human immunodeficiency virus
	Bacteria	*Yersinia* species (plague)
	Protozoa	*Plasmodium* species (malaria)

* Inclusive of transplacental and perinatal infection.
† Microparasites are those that multiply directly within the host individual, usually resulting in acute infections and subsequent durable immunity to reinfection.
‡ Macroparasites are larger parasites whose reproductive stages pass out of the host. Infection intensity is thus a process of accumulation, and can be measured as worm burdens.
§ Needle transmission is included.

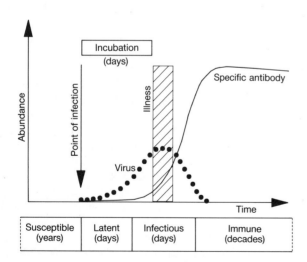

Fig. 14.1. Schematic representation of the typical time-course of an acute viral or bacterial infection in a host individual and the corresponding progression through infection classes (note the different time durations within each of these classes). (Source: Nokes and Anderson 1988, with permission from the editors of *Epidemiology and Infection*.)

to be correct. However, as one moves up the scale of parasite structural (antigenic) complexity from viruses to bacteria to protozoa, in general, the duration of acquired immunity decreases. For certain infections, such as gonorrhoea, acquired immunity is absent while for many protozoan infections it is of short duration (e.g. *Plasmodium* species). The ability to develop effective immunity is often related to the genetic diversity of the infectious agent population (antigenic diversity) such that infection with one genetic strain fails to protect against invasions by another (e.g. *Neisseria gonorrhoea*, *N. meningitidis*, and influenza viruses). The question of immunity can be complicated by a degree of cross-immunity (non-specific in character) resulting from infection by dissimilar organisms (e.g. many bacterial infections of the respiratory tract).

Demographic and behavioural characteristics of the human community are usually important in the study of transmission dynamics. For infections that confer lifelong immunity on host recovery, the rate of input, by births, of new susceptibles will influence the overall pattern of infection in a community. Similarly, the rate of transmission of 'close contact' infections (see Table 14.1) will depend upon the degree of mixing between individuals and the density and age distribution of susceptibles and infecteds. Heterogeneity in behaviour within a community is of particular importance in the study of sexually transmitted infections since rates of sexual-partner change vary greatly between individuals. More generally, heterogeneity in any behaviour, whether sexual or social mixing, must be captured in model formulation.

It will be clear from the preceding comments that much quantitative detail about the natural history of infection must be understood for accurate model formulation. In many instances such detail is not available, but model formulation can greatly facilitate our knowledge of what needs to be understood to define the transmission dynamics of a given infection. With respect to many childhood viral and bacterial infections, such as measles, rubella, mumps, pertussis, and diphtheria, a great deal is understood about the natural history and hence much of the work on mathematical models has focused on those infections. Their direct route of transmission, their tendency to induce lifelong immunity plus, in most cases, the availability of serological or virological techniques to detect past or current infection facilitate the acquisition of quantitative data.

Units of measurement

The unit of measurement employed in epidemiological study depends on the type of infection. The most basic unit is that of the individual parasite. As already discussed, in most cases this unit is not a practicable option for microparasitic organisms due to difficulties in detection and quantification (advances in molecular biology and biochemistry, however, are generating new techniques that may be of value in the near future). As such, the most useful unit is that of the infected host which allows the human community to be stratified on the basis of whether individuals are susceptible, infected but not yet infectious (i.e. latent), infectious, and recovered (i.e. immune in the case of many viral infections). Infection may be detected directly (e.g. DNA probes, virus or bacterial culture) or indirectly by the presence of antibodies specific to pathogen antigens (serological tests). Seropositivity does not necessarily discriminate between infected and recovered individuals, but for many viral and bacterial infections serological surveys of a population, perhaps stratified by age, sex, and other variables carried out longitudinally (over time by cohort monitoring) or horizontally (across age classes) provide a key measure of transmission and the broad epidemiological characteristics of the infection.

What models describe

At any point in time a population may be classified by the density or number of susceptible, infected, and immune individuals. Over time, and concomitantly as individuals age, people may move from one infection class to the next. As such, with the recruitment of new susceptibles by birth, and, in some cases, the loss of immunity, the population structure is dynamic, with individuals flowing from one class to the

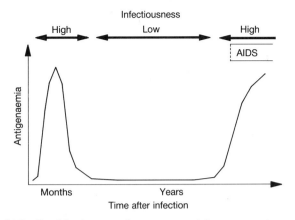

Fig. 14.2. Possible changes in human immunodeficiency virus 1 concentration in the blood of an infected individual (antigenaemia) and in the associated degree of infectiousness during the long incubation period of acquired immune deficiency syndrome (AIDS).

next. Mathematical models of transmission attempt to capture the dynamic nature of these changes in the form of difference (discrete time-steps) or differential (continuous time) equations. With respect to microparasitic infections where the population is stratified or compartmentalized by infection status, the resulting models are often referred to as compartmental models. The types and numbers of compartments will depend upon the type of infectious agent and the details of its natural or life history. A number of examples are given in Figure 14.3 in the form of flow diagrams. These diagrams form a useful intermediary step between biological comprehension and mathematical formulation.

Population rates of flow

Following the introduction of an infection into a stable population, the number or density of individuals within the various infection compartments will depend on the rates of flow between compartments, such as infection and recovery rates. The size of a population in a specific compartment will depend on the magnitude of those rates that determine entry and duration of stay. In general, the shorter the duration of stay (the higher the rate of leaving) in a particular compartment the smaller the size of the population in that category (the inverse relationship between 'standing crop' and 'rate of turnover'). If the infection attains a stable endemic equilibrium in the human community, net input into each compartment will exactly balance net output. The relative numbers in each compartment will be directly related to the duration of stay. Thus, for example, in the case of endemic measles in a developed country where immunity is lifelong (many decades), individuals remain in the susceptible class for an average of 4–5 years and in the latent and infectious classes for a few days (say 7 days on average in each). As such, most people are in the immune class, followed by the susceptible class, and few individuals at any point in time are in the latent and infectious classes. Figure 14.4 provides a diagrammatic representation of this point.

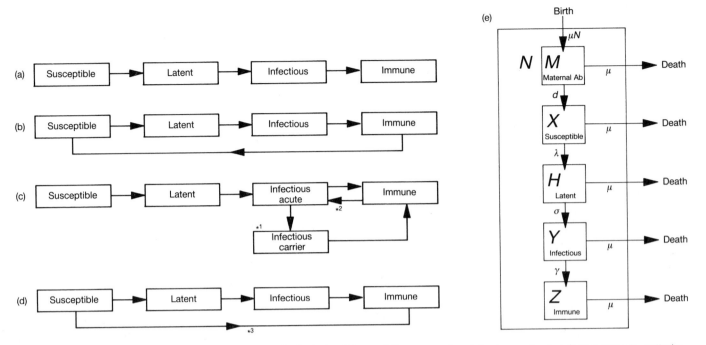

Fig. 14.3. Flow diagrams used to describe the movement of individuals within populations compartmentalized according to infection status to particular parasitic agents. (a) Simple model for infections inducing lasting immunity (e.g. measles, mumps, rubella, yellow fever, and poliomyelitis) or (b) in which immunity is transient and individuals subsequently return to the susceptible pool (e.g. *Neisseria gonorrhoea*, typhoid, cholera, *Trichomonas vaginalis*). (c) Many infections persist within the host for long periods of time, during which the infected individual may remain infectious (*1), as is the case for carriers of hepatitis B virus, gonorrhoea, *Salmonella typhi*, and *Treponema pallidum* (syphilis), chronic tuberculosis patients, or during recrudescence of herpes viruses and malaria. The epidemiological importance of this characteristic is that it enables the perpetuation of such infections in low density communities (see discussion of the mass-action principle in text). For other infections immunity is defence against disease but not asymptomatic reinfection (*2) from which new infectious individuals arise (e.g. *Haemophilis influenzae* and *Neisseria meningitidis*). (d) Vaccination (*3) has the effect of transferring individuals directly from the susceptible to the immune class. (e) More detailed description of the transmission dynamics of an acute microparasitic infection which explicitly accounts for births and deaths in the population. All neonates are born possessing maternally derived protective antibody. The net birth-rate is assumed to equal the sum of the net death-rates for each subpopulation (compartment), i.e. births = μN, where $N = M + X + H + Y + Z$ = constant population size. The per caput rates defining movement between infection classes are described in the text.

A formal demonstration of the influence of rates of flow (or durations of stay) on the proportion of susceptibles, infecteds, and immunes in a population is made possible by the translation of the flow diagram of movement between compartments (see Fig. 14.3) into a set of coupled differential equations. Typically, these describe the rates of change with respect to time (or age, or both) of the densities of infants with maternally derived immunity (due to maternal antibodies), susceptibles, infecteds not yet infectious, infectious individuals, and immunes, denoted respectively by $M(t)$, $X(t)$, $H(t)$, $Y(t)$ and $Z(t)$ at time t (see Fig. 14.3e). In writing down these equations we need to define the rates of flow between compartments by a series of symbols. A frequently used notation defines d as the per caput rate of loss of maternally derived immunity (average duration of stay $1/d$), μN as the net birth-rate of the community (all assumed to enter class M presupposing that all females of reproductive age have experienced the infection at a younger age), β as the transmission coefficient that defines the probability of contact and infection between a susceptible and infectious person, σ as the per caput rate of leaving the latent class (average latent period $1/\sigma$), γ as the per caput rate of leaving the infectious

class (average infectious period $1/\gamma$), and μ as the natural per caput mortality rate ($1/\mu$ is average life expectancy). Note that the determination of absolute rates of movement of individuals from one class to another requires that the per caput rate (i.e. the rate acting on each person) be multiplied by the size of the subpopulation in that particular class (as for example γY). For developed countries it is commonly assumed that population size is approximately constant such that net births exactly balance net deaths (hence the birth term μN for a community of size N where $N = M + X + H + Y + Z$) and that infection does not induce an extra case-mortality rate over and above natural mortality. With this notion we can define the equations for M, X, H, Y, and Z as:

$$dM/dt = \mu N - (d + \mu)M \qquad (1)$$

$$dX/dt = dM - (\beta Y + \mu)X \qquad (2)$$

$$dH/dt = \beta XY - (\sigma + \mu)H \qquad (3)$$

$$dY/dt = \sigma H - (\gamma + \mu)Y \qquad (4)$$

$$dZ/dt = \gamma Y - \mu Z \qquad (5)$$

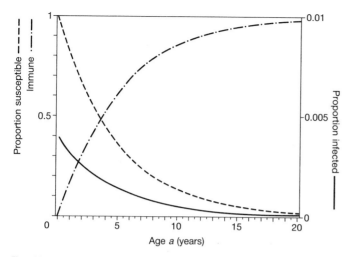

Fig. 14.4. The proportions of a population who are in the susceptible, infected (either latent or infectious), and immune classes for a typical childhood viral infection. In this example, which is based on measles, the force of infection, $\lambda = 0.2$ per year (corresponding to an average age at infection of 5 years) and the rate of movement from the latent class, σ, and recovery from infectiousness, γ, is 52 per year (corresponding to an average duration of stay in each of these infected classes of 1 week). Note that the proportion of the population in the infected classes is always much less than that in the susceptible or the immune classes (Fig. 14.1). (Source: May 1986.)

These equations constitute a simple model of transmission. The major assumptions incorporated in them are: (1) the net rate of infection βXY is proportional to the density of susceptibles multiplied by the density of infectious individuals; (2) individuals leave each compartment at a constant per caput rate; and (3) net births exactly balance net deaths (reasonably accurate for developed countries). To explore what these assumptions imply in terms of the dynamics of transmission and the numbers or proportions of individuals in each class we need to solve these equations either analytically to obtain explicit expressions for $M(t)$, $X(t)$, $H(t)$, $Y(t)$, and $Z(t)$ in terms of the rate parameters and the variable time t or numerically to generate projections of changes in the numbers in each compartment over time. In the case of simple models, we can often obtain exact analytical solutions as is the case for the equation for $M(t)$ in the model defined by equations 1–5. The solution for the number of infants with maternally derived protection at time t, $M(t)$, is:

$$M(t) = \{(\mu N)/(\mu + d)[1 - e^{-(\mu + d)t}]\} + M(0)e^{-(\mu + d)t} \quad (6)$$

where $M(0)$ is the number protected at time $t=0$.

More generally, the complexity of the life histories of many infections makes analytical solution difficult or impossible and numerical methods are required. Modern computers make light work of very complex systems of equations describing disease transmission and many software packages are available for the solution of sets of differential equations. In these cases some general insights can be obtained by examining the equilibrium properties of the model by setting the

time derivatives equal to zero ($dM/dt=dX/dt=dH/dt=dY/dt=dZ/dt=0$, i.e. there are no changes in the number of individuals within each infection class because the flows into and out of any one category are equal) and solving for the equilibrium numbers in each class (i.e. M^*, X^*, H^*, Y^*, Z^*). For example, in the simple model of equations 1–5, by simple algebraic manipulations, we obtain:

$$M^* = \mu N/(d + \mu) \quad (7)$$

$$X^* = (\sigma + \mu)(\gamma + \mu)/(\beta\sigma) \quad (8)$$

$$H^* = (\gamma + \mu)Y^*/\sigma \quad (9)$$

$$Y^* = (dM^* - \mu X^*)/\beta X^* \quad (10)$$

$$Z^* = \gamma Y^*/\mu \quad (11)$$

where N is the constant representing total population size. These equilibrium solutions illustrate how the various rate parameters that determine flow between compartments influence the numbers of individuals in each compartment when the infection is at an endemic steady state.

Parameter estimation

The preceding section provided a clear illustration of the numerous parameters that are necessary to define even the simplest model of direct transmission within a human community. To make the best use of a model it is desirable to have available estimates for each of the parameters for a given infection. Some, such as the demographic rates of birth and mortality, and total population size, can be easily obtained via national census databases (usually finely stratified by age and sex in developed countries). Others, such as the average latent and infectious periods, must be determined either by clinical studies of the course of infection in individual patients (e.g. measures of change in viral abundance during the course of infection) or by detailed household studies of case-to-case transmission. Statistical methodology plays an important role in this instance since, as noted earlier, latent, infectious, and incubation periods are rarely constant from one individual to the next. Statistical estimation procedures have been developed to help derive summary statistics of these distributions (e.g. means and variances) (Bailey 1973).

Invariably, the most difficult parameter to estimate is the transmission coefficient β (see equation 2), which is a measure of the rate of contact between members of a population plus the likelihood of infection resulting from contact. In some cases, such as certain sexually transmitted infections (e.g. gonorrhoea), direct estimates can be obtained via contact-tracing methods (see Hethcote and Yorke 1984). More commonly, indirect methods must be employed, often themselves based on model formulation and analysis. A simple example employs the model defined in the previous section by equations 1–5. We may define the component βY of the transmission term as the per caput rate at which susceptibles (X) acquire infection. This rate is commonly referred to as the 'force of infection' and denoted by the symbol λ. Analysis of the model reveals that the average age at which

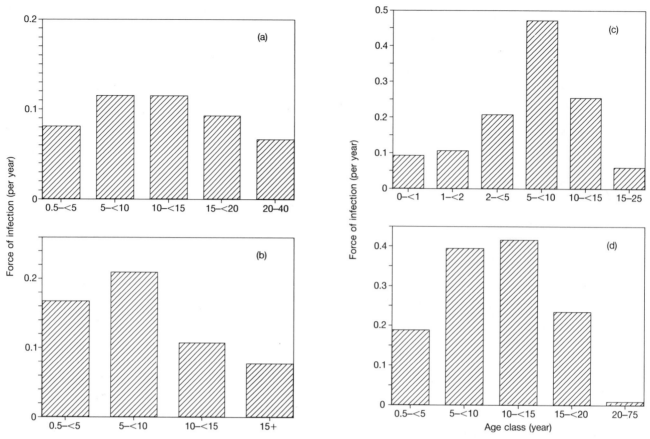

Fig. 14.5. Examples of the age-dependent nature of the rate of transmission for common childhood viral and bacterial infections. Graphs (a) and (b) derive from horizontal cross-sectional serological surveys in the UK, of rubella (Nokes *et al.* 1986) and mumps (Anderson *et al.* 1987), respectively. Graphs (c) and (d) provide estimates based on case-notification data for England and Wales, for whooping cough (Anderson and May 1985*b*) and measles (Grenfell and Anderson 1985), respectively.

an individual typically acquires infection, A, is approximately related to the force of infection by the expression:

$$A \cong 1/\lambda \qquad (12)$$

Hence, if we can estimate A from an age-stratified serological profile or from age-defined case-notification records we can, via equation 12, estimate λ. More generally, this rate often varies with age and more complex methods of estimation must be employed given good age-stratified serological data (see Grenfell and Anderson 1985). Put in simple terms, if the proportion susceptible at age $a+1$ is $X(a+1)$ then the force of infection over the age interval $a \rightarrow a+1$ (defined per unit of age) is simply:

$$\lambda_{(a \rightarrow a + 1)} = -\ln\left[X(a + 1)/X(a)\right] \qquad (13)$$

With serological data finely stratified by age, under the assumption that the infection confers lifelong immunity upon recovery, equation 13 can be used to estimate how λ changes with age in a given community. For most childhood viral and bacterial infections, λ is a function of age, changing from low values in infant classes to high in child to young teenage classes back to low in adult classes (Fig. 14.5). This is thought

to reflect patterns of intimate contact via attendance at school and play activities.

Further complications may arise if rates of contact or transmission vary through time, perhaps due to seasonal factors such as the aggregation and dispersal of children at term and school holiday periods (Anderson 1982*b*; Yorke *et al.* 1979). The problems of parameter estimation are considered in more detail in a later section.

Concepts in quantitative epidemiology

Incidence of infection and disease

Transmission by direct contact and the law of mass action

When close contact between infectious and susceptible individuals is necessary for transmission, the net rate at which new cases arise in a population (i.e. the incidence of infection) is often assumed to be approximately given by the density (or number) of susceptibles, X, multiplied by the density (or number) of infectious persons, Y, multiplied by the probability of transmission arising from a contact between susceptible and infectious persons, β (i.e. βXY). This relationship is commonly referred to as the 'law of mass

Box 14.1. The law of mass action and the incidence of infection

Imagine susceptible and infectious individuals behaving as ideal gas particles within a closed system with no immigration or emigration and occupying a defined space, where: X = number of particles of one gas (i.e. susceptibles); Y = number of particles of a second gas (i.e. infectious people); and β = collision coefficient for the formation of molecules of a new gas from one molecule each of the original gases (i.e. new cases of infection) (a).

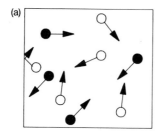
(a)

Gas particles (individuals) are mixing in a homogeneous manner such that collisions (contacts) occur at random. The law of mass action states that the net rate of production of new molecules (i.e. cases), I, is simply:

$$I = \beta XY$$

The coefficient β is a measure of (i) the rate at which collisions (contacts) occur and (ii) the probability that the repellent forces of the gas particles can be overcome to produce new molecules, or, in the case of infection, the likelihood that a contact between a susceptible and an infectious person results in the transmission of infection. Under these assumptions, the incidence of infection will be increased by larger numbers (or densities) of infectious and susceptible persons and/or high probabilities (β) of transmission (b).

(b)

Increase X and/or Y Increase β

Larger I

action' by analogy with particles colliding within an ideal gas system (see Box 14.1). The basic assumption implicit in this concept is that the population mixes in a random manner (often referred to as homogeneous mixing). The term βXY which describes net transmission is the major non-linear expression in most compartmental models of directly transmitted viral and bacterial infections. It is, of course, a crude approximation of what actually occurs in human communities, and more realistic refinements of this assumption are discussed in later sections. However, it provides a convenient point of departure for model construction and analysis.

The transmission coefficient β

The probability of transmission is made up of two components, namely, the rate at which contacts occur between susceptible and infectious persons and the likelihood that transmission will occur as a result of contact. Consequently, β

is dependent on sociological and behavioural factors within the host population (i.e. rate of mixing) and the biological properties that determine the infectiousness of an infected person and the susceptibility of an uninfected individual. These biological properties involve factors such as the virulence of the infectious agent and the genetic background plus the nutritional status of the human host.

Incidence estimates

The incidence of infection I, can be measured by direct observation of new cases, such as notifications of measles or pertussis. Unfortunately, however, measures of incidence tell us nothing about the respective densities of susceptible or infectious people, nor the magnitude of the transmission coefficient β. It is common practice in epidemiology for I to be expressed as the number of cases per unit of population (usually 100 000 people) in a defined class (e.g. age or sex)

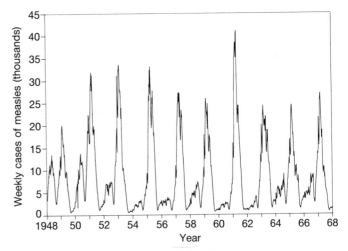

Fig. 14.6. The number of cases of measles reported each week in England and Wales between 1948 and 1968. (Source: Office of Population Censuses and Surveys, London.)

Box 14.2. Interpreting attack rates

Care should be exercised when interpreting attack rates in the absence of information on the proportion of individuals within the population who are immune as a result of previous infection (assuming we are considering an infection such as measles that induces lasting immunity on recovery). A simple illustrative example is given below based on case notification for measles.

Age (years)	Attack rate per head of population in that age class	Percentage immune in the age class	Modified attack rate based on infection per head of the susceptible population
2	180/100 000	10	180/90 000
10	20/100 000	90	180/90 000

At a first glance at column two, the attack rate suggests that infants aged 2 years have a much greater chance of acquiring infection than children aged 10 years. However, if we adjust the denominator of the attack rate from per head of population in that age class to per head of susceptible population in the age class, we see from the fourth column that the rate of infection is identical in both age classes.

over a defined period of time such as one year (e.g. 5 per 100 000 per annum). Such measures are often referred to as attack rates. However, they are a rather poor measure of the intensity of transmission within a population since they take no account of the proportion of the community (or age or sex class) that is susceptible to infection (see Box 14.2). A better measure of the rate at which susceptibles acquire infection is provided by the 'force of infection', λ. It simply defines the probability that a susceptible individual will acquire infection over a short period of time (i.e. a per caput (susceptible) rate of infection) and, in the terminology of the mass-action principle, is defined as $\lambda = \beta Y$. Estimates of this rate can be derived from age-stratified serological profiles or case notifications (Anderson and May 1983; Grenfell and Anderson 1985).

Validity of the mass-action principle

Despite the simplicity of the notion of homogeneous mixing implicit in the mass-action principle of transmission, the predictions of simple compartmental models based on this assumption often mirror observed epidemiological patterns surprisingly well (Anderson and May 1982). In part, this is a consequence of increased travel, movement, and mixing within many societies in developed countries. Measles epidemics, for example, are often synchronous in England and

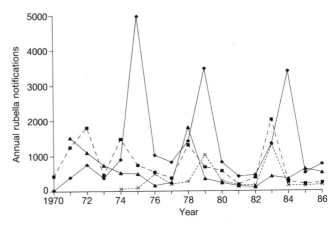

Fig. 14.7. Annual rubella case notifications reported by four city health authorities in England: Leeds (◆), Bristol (■), Manchester (▲), and Newcastle (X). The dominant interepidemic period is roughly 4–5 years with peak incidence often slightly out of phase between cities (compare with 14.6). (Source: Communicable Disease Surveillance Centre, London.)

Wales, with a clear distinction in all parts of these regions between years of high incidence and years of low incidence (Fig. 14.6). However, the less able an infection is to spread through a particular population (lower R_0) then the more important are slight deviations from homogeneous mixing, resulting in a lower degree of synchronicity of epidemics in a country (Fig. 14.7). The assumption is most appropriate for infections that are spread by close contact between individuals such as respiratory infections transmitted by contaminated droplets and nasopharyngeal secretions. In such cases, the survival of the infectious agent in the external environment is of very short duration (i.e. minutes). As such, there is no significant reservoir of infectious stages to maintain transmission in the absence of infectious persons.

Many kinds of heterogeneities can invalidate the mass-

Fig. 14.8. Variation in the numbers of different sexual partners per year revealed from surveys of the male homosexual and the heterosexual communities in the UK, 1986 (Anderson 1988). The skewed distribution observed in each instance (an indication that although the majority of individuals have few partners, a few have very many), and the mean rate of sexual-partner change (indicated), are both of significance to the perpetuation and rate of spread of sexually transmitted diseases in the community.

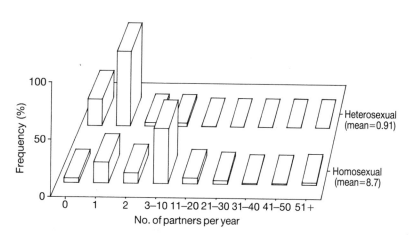

action principle and much attention in recent years has been devoted to their inclusion in compartmental models. The major sources are heterogeneities arising from age-related factors that determine contact and mixing patterns (i.e. 'who mixes with whom'), and spatial factors such as differences in population densities in urban and rural areas of a country (Anderson and May 1984; May and Anderson 1984). Such sources of heterogeneity are very important in the design of control policies based, for example, on mass vaccination, and models have been developed to assess their impact.

Heterogeneity in behaviour is of particular importance in the study of sexually transmitted infections such as gonorrhoea and HIV. One of the major determinants of the rate of spread of such infections is the distribution of the rate of sexual-partner change within a defined community (Fig. 14.8). These distributions are typically highly heterogeneous in character (i.e. the variance in the rate of partner change is much greater in value than the mean rate of partner change) where most people have few different sexual partners in a lifetime (or over a defined period of time) and a few have very many. The activities of individuals in the 'tail' of the distribution (the highly sexually active) are clearly important for the persistence and spread of infection since those with many partners are both more likely to acquire infection and more likely to transmit it to others.

Simple theory based on compartmental models of the transmission of infections such as gonorrhoea and HIV assumes that the net rate at which infection is spread in, for example, a male homosexual community, is determined by the product of the proportion of infectious persons (Y/N where N is the total size of the sexually active population) multiplied by the density of susceptibles (X) multiplied by a transmission coefficient β. This coefficient is defined as the probability that a sexual contact (per partner) results in transmission B, multiplied by the effective rate of sexual-partner change, c (which determines contacts). If the population mixed homogeneously, this effective rate would simply be the mean rate of sexual-partner change, m. When great heterogeneity in rates of partner change is present within a

population the effective rate must be defined in terms of this variability as well as the mean rate of activity. If we assume that the population is divided into classes with different rates of partner change and that partners are chosen (from any class) in proportion to their representation in the population multiplied by the rate of partner change in each group (an assumption of 'proportional mixing'; see Anderson *et al.* 1986; May and Anderson 1987), then the effective rate of partner change, c, is given by:

$$c = m + (\sigma^2/m) \tag{14}$$

where σ^2 is the variance in the rate of partner change. The importance of variability in contact is clear from this simple equation. For example, suppose the mean rate per year is unity but the variance is five times greater. If we assume that homogeneous mixing occurred our estimate of the effective rate would be one, but if we take account of heterogeneity the effective rate is six. The influence of the small proportion of highly sexually active individuals on the overall transmission rate is very significant.

Transmission thresholds and the basic reproductive rate of infection

The basic reproductive rate of infection, R_0

A key measure of the transmissibility of an infectious agent is provided by a parameter termed the basic reproductive rate and denoted by the symbol R_0. It measures the average number of secondary cases of infection generated by one primary case in a susceptible population. Its value is defined by the number of susceptibles with which the primary case can come into contact (X) multiplied by the length of time the primary case is infectious to others, D, multiplied by the transmission coefficient, β (rate of mixing and innate contagiousness or virulence of the infectious agent):

$$R_0 = \beta X D \tag{15}$$

Note that R_0 is a dimensionless quantity (i.e. the units of measurement cancel out) that defines the potential to produce secondary cases (in a totally susceptible population) per generation time (i.e. the average duration of the infection).

The basic reproductive rate is of major epidemiological significance since the condition $R_0=1$ defines a transmission threshold below which the generation of secondary cases is insufficient to maintain the infection within the human community. For values above unity the infection will trigger an epidemic, and, with a continual input of susceptibles, will result in endemic persistence. A further quantity of interest is the effective reproductive rate, R, which defines the generation of secondary cases in a population containing susceptibles and immunes (as opposed to just susceptible individuals). If the prevalence or incidence of infection is stable through time, R must equal unity in value which is a situation in which each primary case gives rise, on average, to a single secondary infectious individual.

Factors that influence R_0

The simple expression $R_0 = \beta XD$ (appropriate for directly transmitted infections under the mass-action assumption) provides a framework for assessing how different epidemiological factors influence transmission success. Clearly, high transmission coefficients, long periods of infectiousness, and high densities of susceptibles enhance the generation of secondary cases. Note that the value of R_0 depends not only on the properties that define the course of infection in an individual (i.e. the duration of infectiousness, D), but also on attributes of the host population such as the density of susceptibles, X, and the component of β that determines the rate of contact or mixing. A good example of the influence of population level characteristics is provided by the rate of transmission of the measles virus in urban centres in developed and developing countries. A more rapid rise in the proportion of children who have experienced infection, with age, in developing countries by comparison with developed regions is in part a consequence of higher population densities and poorer living conditions (McLean and Anderson 1987, 1988).

Principles of control

The threshold condition for persistence of an infection, defined by $R_0 = 1$, captures the essence of the problem of control. To eradicate an infection we must reduce the value of the basic reproductive rate below unity. Similarly, to reduce incidence the value of R_0 must be reduced below the level that pertains prior to the introduction of control measures. Reductions can be achieved by: (a) reducing the infectious period, D, by, for example, the isolation of infectious persons (perhaps recognized by clinical symptoms of disease); (b) reducing the number or density of susceptibles, usually by immunization; and (c) altering the social and behavioural factors that determine transmission such as improving living conditions to reduce overcrowding (in the case of sexually transmitted infections education can serve to reduce rates of sexual-partner change or promote the use of condoms to lower the probability of transmission).

The threshold density of susceptibles

It is clear from the definition of R_0 given above that, to maintain the value of the basic reproductive rate above unity, the density of susceptibles in the population must exceed a critical value. More precisely, this critical level, X_T is (for the mass-action assumption) obtained by setting $R_0=1$ in equation 15 and rearranging it thus:

$$X_T = 1/(\beta D) \tag{16}$$

The aim of mass vaccination, aside from protecting the individual, is to lower the density of susceptible people in the population. If eradication is the aim of control, the density of susceptibles must be reduced to less than X_T in value.

Critical community size

The magnitude of R_0 and, concomitantly, the size of the threshold density of susceptibles determines whether or not an epidemic of an infection will occur when introduced into a given community. In practice, however, for infections that induce lasting immunity in those who recover, the long-term endemic persistence of infection will depend on the renewal of the supply of susceptibles by new births or, to a lesser extent, by immigration. As such, the net birth-rate in a community, which is itself dependent on total population size (or density), will influence the likelihood of persistence. There is, therefore, a critical community size for the endemic persistence of a given infection. In certain island communities, immigration of susceptibles and infecteds may also play a role in the long-term persistence of a given infection (Anderson and May 1986; Black 1966; Cox *et al.* 1988). These factors are of growing significance as rates of population movement increase as a result of, for example, improved air transport services. Table 14.2 provides an example of the relationship between community size and the likelihood of the endemic persistence of the measles virus.

The concepts of a threshold density of susceptibles and a critical community size are most relevant for directly transmitted viral and bacterial infections that induce lasting (i.e. lifelong) immunity. The production of long-lived infective stages or the use of vectors (such as mosquitoes) lessen the importance of human population density for the persistence of an infection. In the case of sexually transmitted infections, simple models suggest that there is no critical density of susceptibles for persistence since the magnitude of R_0 can be approximately given by:

$$R_0 = BcD \tag{17}$$

where c is the effective rate of sexual-partner change, D is the average duration of infectiousness, and B is the transmission probability per partner contact (Anderson *et al.* 1986). This is simple to arrive at theoretically. If, as stated earlier, the incidence of cases of a sexually transmitted disease is defined as:

$$I = BcXY/N \tag{18}$$

Table 14.2. Island community size and endemic persistence of measles

Island	Population size (units of 100 000)	Percentage of months in which no cases were reported
Hawaii	5.50	0
Fiji	3.46	36
Iceland	1.60	39
Samoa	1.18	72
Solomon	1.10	68
Fr. Polynesia	0.75	92
New Caledonia	0.68	68
Guam	0.63	20
Tonga	0.57	88
New Hebrides	0.52	70
Gilbert and Ellice	0.40	85
Greenland	0.28	76
Bermuda	0.41	49
Faroe	0.34	68
Cook	0.16	94
Niue	0.05	95
Nauru	0.03	95
St. Helena	0.05	96
Falkland	0.02	100

Source: Anderson (1982*b*).

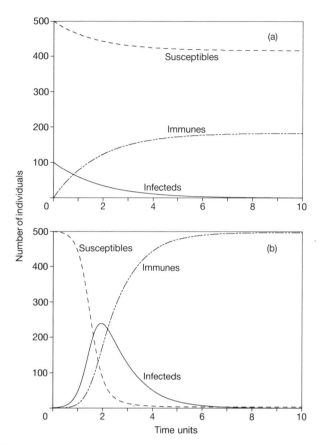

Fig. 14.9. Conditions for an epidemic: (a) host density (susceptibles) below the threshold level (at time 0, $X=500$, $Y=100$, $Z=0$, $\beta=0.0001$, $\gamma=1$); (b) host density above the threshold level (at time 0, $X=500$, $Y=1$, $Z=0$, $\beta=0.01$, $\gamma=1$). (See text for definition of threshold criteria. Anderson 1982*a*.)

then, following the introduction of a single infectious person ($Y=1$), infectious over a period D, into a totally susceptible population ($N=X$), the number of secondary cases will be represented by equation 17.

The dependence upon the number of susceptibles is lost. Biologically, this is more difficult to grasp, but it does seem reasonable that the rate of sexual-partner change should be more important to the potential for spread of a sexually transmitted disease than the number of susceptibles in the population.

Regulation of infection within human communities

The regulation (i.e. modulation or control) of the incidence or prevalence of a particular infection within a human community is largely determined by the level of herd immunity (i.e. the proportion of the population immune to infection) and the net rate of input of new susceptible individuals. A simple example serves to illustrate this point. Consider a closed population with no inflow or outflow of susceptible, infected, or immune individuals. If the densities of susceptibles, infecteds, and immunes at time t are defined by $X(t)$, $Y(t)$, and $Z(t)$ respectively then, under the mass-action assumption of transmission, the rates of change in the densities with respect to time can be captured by three coupled differential equations:

$$dX/dt = -\beta XY \qquad (19)$$

$$dY/dt = \beta XY - \gamma Y \qquad (20)$$

$$dZ/dt = \gamma Y \qquad (21)$$

It is here assumed that there is no latent period of infection (individuals are infectious once infected), that the average duration of infectiousness, D, is given by $D = 1/\gamma$ where γ is

the rate of recovery from infection, that immunity is lifelong, and that no losses occur due to death. If we start with a totally susceptible population and introduce a few infecteds, the occurrence of an epidemic will depend upon the magnitude of the basic reproductive rate R_0 ($R_0 = \beta XD$) and, concomitantly, whether or not the density of susceptibles exceeds the critical threshold value X_T ($X_T = 1/\beta D$) (Fig. 14.9). Assuming that R_0 is greater than unity then an epidemic will occur, but as time progresses the density of susceptibles will decline ($X \rightarrow Y \rightarrow Z$) until the effective reproductive rate, R is less than unity and the infection dies out. Note that at this point of elimination (where $Y=0$) the density of the remaining susceptibles is exactly equal to the critical density, X_T, required to maintain the infection and the density of immunes, Z, is $N-X$ where N denotes total population size ($N=X+Y+Z$).

For persistence of the infection, one of two things must happen. First, suppose susceptibles are continually introduced into the population at a net rate bN where b is the per caput birth-rate, and that natural deaths occur in each class at a per caput rate μ. For simplicity we further assume that net births exactly balance net deaths ($bN=\mu N$) to maintain the total population at constant size. With these assumptions,

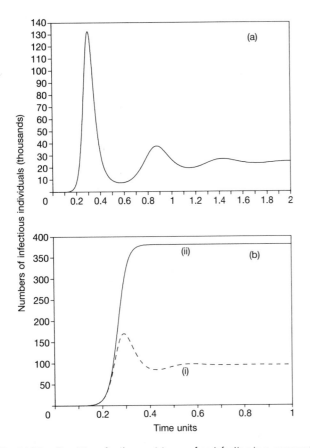

Fig. 14.10. Conditions for the persistence of an infection in a community: (a) renewal of susceptibles by births (initial conditions: $X=1\times10^6$, $Y=1$, $Z=0$, $\gamma=52$, $\beta=0.0001$, $b=\mu=3$, and immunity is lifelong); (b) renewal of susceptibles because there is no immunity (solid line), or because immunity is short lived (dashed line) (initial conditions as for (a) except $b=\mu=0$, $\alpha(i)=13$, and $\alpha(ii)=200$).

and provided $R_0 \geqslant 1$, we find that the infection persists in the population (see Fig. 14.10a) with an endemic equilibrium density of susceptibles again equal to X_T and equilibrium densities of infecteds, Y^*, and immunes, Z^*, given by:

$$Y^* = [\mu/(\mu + \gamma)](N - X_T) \qquad (22)$$

$$Z^* = (\gamma/\mu)Y^* \qquad (23)$$

Second, suppose there are no new births and no deaths but that immunity is of short duration such that individuals leave the immune class Z to re-enter the susceptible class X at a per caput rate α where $1/\alpha$ is the average duration of immunity. We again find that the infection can persist (Fig. 14.10b) (provided $R_0 \geqslant 1$) with equilibrium densities of infecteds and immunes of:

$$Y^* = [\sigma/(\alpha + \gamma)](N - X_T) \qquad (24)$$

and

$$Z^* = (\gamma/\sigma)Y^* \qquad (25)$$

Note that the faster the loss of immunity (i.e. α is large) the higher the equilibrium density of infecteds and the lower Z^*.

These two examples show how the net input of susceptibles and the degree of herd immunity (as controlled by the duration of immunity to reinfection following recovery) influence the likelihood that an infection will persist endemically after the initial epidemic has swept through a susceptible population following the introduction of an infection. In these simple models of the transmission of direct-contact infections, the density of infecteds tends to oscillate after the introduction of infection due to the rise and fall in the density of susceptibles taking the effective reproductive rate above and below unity in value. This propensity to oscillate is more apparent if the infection induces lasting immunity since it takes some time, under these circumstances, for new births or loss of immunity to replenish the supply of susceptibles such that R is again above unity in value.

Other factors that can promote long-term persistence include the production of infective stages that are able to survive for long periods in the external environment, sexual transmission, transmission from mother to unborn offspring, vector transmission, and the carrier state in which some individuals (for genetic or other reasons) atypically harbour the infection for long periods of time (see Table 14.1 and Fig. 14.3 for examples).

Herd immunity and mass vaccination

When an infection is persisting endemically in a community such that the net rate at which new cases of infection arise is approximately equal to the net rate at which individuals recover and acquire immunity, the effective reproductive rate, R, is equal to unity in value. This is known as endemic equilibrium. In practice, for many common viral and bacterial infections the incidence of infection fluctuates both on seasonal and longer-term cycles. The effective reproductive rate, therefore, fluctuates below and above unity in value as the incidence and density of susceptibles change. However, the average value over a series of incidence cycles (both seasonal and longer-term) will be approximately equal to unity in the absence of control intervention or changing social and demographic patterns. The effective reproductive rate is related to the basic reproductive rate (the potential to generate secondary cases in a totally susceptible population) by the simple equation:

$$R = R_0 x^* \qquad (26)$$

Here, x^* is the equilibrium fraction susceptible in the population. The magnitude of x^* can be determined from cross-sectional serological surveys given data on the age structure of the population (if x_i is the proportion susceptible in age class i and p_i is the proportion of the population in the same age class then

$$x^* = \sum_{i=1}^{n} x_i p_i$$

in a population with n age classes). At equilibrium where $R = 1$, equation 26 reveals that the proportion susceptible, x^*, is

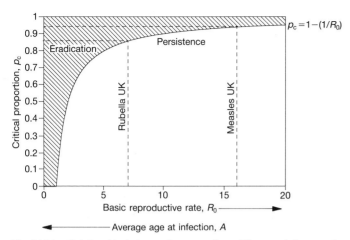

Fig. 14.11. Relationship between the proportion of the population vaccinated at or near birth and the likelihood of an infection persisting or, alternatively, being eliminated. Infectious agents with higher basic reproductive rates in defined communities will be more difficult to control by mass vaccination as illustrated by the example of measles and rubella in the UK. (Source: Nokes and Anderson 1988.)

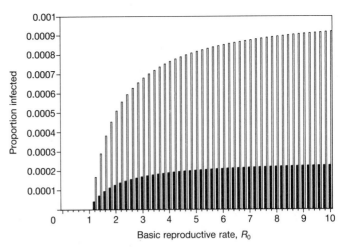

Fig. 14.12. Predicted changes in the equilibrium proportion of a population infected (i.e. the stable endemic prevalence of an infection) as the transmission potential of the microparasitic agent varies. For infections where there is no loss of immunity, the level of the plateau of prevalence is dependent upon the rate of input of new susceptibles (i.e. the birth-rate, b) and the duration of infectiousness, $1/\gamma$. In the figure, $b=1/75$ per year and $\gamma=52$ per year (i.e. a 1-week infectious period) (closed bars) or $\gamma=13$ per year (i.e. a 4-week period) (open bars). An important point to observe is that the greatest changes in the proportion infected occur over the first few increments of R_0 (irrespective of the magnitude of b or γ.)

equal to the reciprocal of the basic reproductive rate R_0 (i.e. $R_0 \cong 1/x^*$).

To block transmission and eradicate an infection it is necessary to raise the level of herd immunity by mass vaccination such that the magnitude of the effective reproductive rate is less than unity in value. If we immunize a proportion p of the population then equation 26 tells us that the new reproductive rate is simply $R_0(1 - p)$. Since this quantity must be less than unity in value in order to block transmission, the critical proportion to be immunized, p_c to achieve this is simply:

$$p_c = 1 - 1/R_0 \qquad (27)$$

The relationship between p_c and R_0 is depicted diagrammatically in Figure 14.11; the larger the magnitude of the infection's transmission potential (as measured by R_0) the greater the proportion of the population that must be immunized to block transmission. Note that it is not necessary to vaccinate everyone in the community to prevent the spread of infection. The principle of herd immunity implies the protection of the individual by the protection (i.e. vaccination) of the population. The mechanism underlying this concept is that of the critical density of susceptibles required to maintain the magnitude of the reproductive rate above unity in value.

Age at vaccination

In practice, immunization programmes are introduced by focusing on cohorts of children such that the level of immunization coverage is built up over many years of cohort vaccination. In these circumstances, the p_c of equation 27 must be interpreted as the proportion of each cohort vaccinated as soon after birth as is practically feasible, taking account of the need to immunize after the decay in maternally derived specific antibody. For most viral infections the average

duration of protection against infection provided by maternal antibodies is approximately 6 months. It will clearly take many years of cohort immunization to achieve the desired level of artificially induced herd immunity. A further complication is introduced by the practice of vaccinating a series of age classes. In this case, simple mathematical models suggest that the level of vaccination coverage required to eradicate the infection under a policy which vaccinates at an average age of V years is:

$$p > [1 + (V/L)]/[1 + (A/L)] \qquad (28)$$

Here, L is human life expectancy and A is the average age at which the infection was acquired before the introduction of vaccination (Anderson and May 1983). It is clear from this expression that transmission cannot be interrupted unless the average age at vaccination, V, is less than the average age at infection, A, prior to control.

Prevalence of infection and the basic reproductive rate

A further epidemiological feature arising from the existence of a critical density of susceptibles to maintain infection concerns the relationship between the magnitude of the basic reproductive rate and the prevalence of infection in a population in which the infectious agent persists endemically. As depicted in Fig. 14.12, simple models predict that the relationship is non-linear such that a marked reduction in the endemic prevalence or incidence will only occur as the transmission potential is reduced to an extent where it approaches the threshold level $R_0 = 1$. The practical implication is that we should not expect the decline in the incidence of infection induced by mass vaccination to be directly proportional to the level of vaccination coverage. The greatest changes are predicted to occur when coverage attains high levels.

Inter-epidemic period T

Many viral and bacterial infections that induce lasting immunity to reinfection and which have high transmission potentials (R_0 is large) tend to oscillate in incidence. A good example is that of measles which, in the UK before mass vaccination, oscillated on a seasonal basis (due to the aggregation and disaggregation of children for school term and holiday periods) and a longer-term 2-year cycle with years of high incidence separated by years of low incidence (Anderson *et al.* 1984). Time-series analyses reveal that these longer-term cycles for infection, such as measles, mumps, rubella, and pertussis, are not due to chance fluctuations but arise as a result of the dynamic interaction between the net rates of acquisition of infection and immunity on recovery.

Simple models based on the mass-action assumption suggest that the inter-epidemic period, T, of the longer-term cycles is determined by the generation time of the infection, K, defined as the sum of the latent and infectious periods, and the transmission potential of the infection inversely measured by the average age at infection, A, where:

$$T = 2\pi(AK)^{\frac{1}{2}} \tag{29}$$

This simple prediction well matches observation for a variety of common childhood infections before mass vaccination (i.e. the 2-year cycles of measles, the 3-year cycles of mumps, the 4–5-year cycles of rubella, and the 3–4-year cycles of pertussis). Non-seasonal oscillation arises as a consequence of the exhaustion of a supply of susceptibles, as an epidemic passes through a population, plus the time-lag that arises before new births replenish the pool to trigger the next epidemic. As such, the inter-epidemic period is also influenced by the birth-rate of the community (which influences the average age of infection, A, in equation 29). For example, in developing countries such as Kenya with high birth-rates, measles tends to cycle on a 1-year time-scale in urban centres as opposed to the 2-year cycle seen in the UK prior to control (McLean and Anderson 1987).

Parameter estimation

Survey data

Survey data on the incidence or prevalence of infection (past or current) can be obtained in a variety of ways: (a) longitudinal (i.e. over time) data can be acquired by monitoring a cohort of people over time and recording infection as it occurs; (b) horizontal (i.e. one point in time) cross-sectional (across age and sex classes) data can be acquired by a survey at one point in time, or over a short interval of time, by the examination of different age classes within the population. Such surveys are of most use when based on serological

examinations to determine the proportion of individuals in a given age class who have antibodies specific to the antigens of a particular infectious agent. These cross-sectional serological profiles reflect the proportion in each age class who have, at some time in the past, experienced infection; (c) case-notification data stratified by age and sex and recorded over a set interval of time, such as a year, can be accumulated to indicate what proportion of the cases occur by any given age. This may then be used to infer changes in the proportion who experience infection as a function of age. Such data are less reliable than serological information since they are dependent on lack of bias in reporting efficiency by age class. Bias is to be expected if the seriousness of the disease induced by infection changes with age (e.g. rubella in women and mumps in men) or where the incidence of subclinical (i.e. undetectable) infections is age dependent.

When conducting surveys a number of points should be borne in mind. First, sample sizes should be as large as is practically possible, finely stratified by age (preferably infants to elderly people). How large will depend upon what we wish the accuracy or power (see Sokal and Rholf 1981) of subsequent analyses to be, but 25–50 per yearly age class is a rough working estimate. Second, the incidence of infection may oscillate on a seasonal or longer-term basis. As such, it is good practice to carry out surveys that span epidemic and inter-epidemic years. Third, systematic changes over time may occur in a given population due to social, behavioural, economic, or other changes. Examples include the observed reduction in the incidence of hepatitis A in northern European countries over the past few decades due to improved standards of hygiene, and the rise in the incidence of gonorrhoea in certain developed countries during the 1960s and 1970s due to changes in sexual behaviour (e.g. increased rates of sexual-partner change). Basic reproductive rates and rates of infection may therefore change over time irrespective of the impact of control measures.

The basic reproductive rate of infection

Estimating individually the component parameters that determine the magnitude of the basic reproductive rate, R_0, is fraught with many problems. In the case of directly transmitted viral and bacterial infections, we require a knowledge of the transmission coefficient, β, the density of susceptibles, X, and the average duration of infectiousness. In practice, it is often easier to use indirect methods to arrive at estimates of R_0 employing serological data finely stratified by age. As discussed earlier, the rate of decay with age in the proportion susceptible to infection provides a measure of the age-dependent forces of infection (the $\lambda(a)$). This in turn can be used to obtain an estimate of the average age, A, at which an individual typically acquires infection. Mathematical models can be used to define a relationship between the magnitude of R_0 and the average age at infection. In the simplest case, the relationship is of the form:

$$R_0 = Q/A \tag{30}$$

where Q denotes the reciprocal of the net birth-rate of the community. In developed countries where net births are approximately equal to net deaths the quantity Q is equal to average life expectancy (from birth), L (Anderson and May 1985b). More generally, if maternally derived antibodies provide protection for an average of F years, R_0 is related to A by the expression:

$$R_0 = Q/(A - F) \tag{31}$$

A simple example of the use of this equation is provided by the transmission of the measles virus in the UK before the introduction of mass vaccination. In this case, the values of A, L, and F were 5, 75, and 0.5 years respectively, leading to an R_0 estimate of between 16 and 17. The inverse relationship between R_0 and A makes good intuitive sense; infections with high transmission potentials will tend to have low average ages at infection and vice versa. These notions are depicted diagrammatically in Box 14.3, and Table 14.3 lists some estimates of R_0, A, L, and the critical level of vaccination coverage to block transmission p_c for a variety of common infectious agents in defined localities.

An alternative method to that outlined above is based on the prediction of simple models that the magnitude of R_0 is related to the fraction of the population susceptible to infection, x^*, when the infection has attained its endemic equilibrium. The relationship is simply:

$$R_0 = 1/x^* \tag{32}$$

and arises from the fact that at equilibrium the effective reproductive rate is equal to unity in value (see equation 25). Note that equations 31 and 32 imply that the average age at infection, A, is directly related to the equilibrium fraction of susceptibles in a population, x^*, required to ensure each primary case gives rise, on average, to at least one secondary case (see Box 14.3). In general, however, the method based on estimating the average age at infection is the better one given good age-stratified serological data.

Latent and infectious periods

Two sources of data are available with which to estimate latent and infectious periods. The first derives from clinical, virological, and immunological studies of the course of infection in individual patients. For some common microparasitic infections, presence of the infectious agent in host tissues, excretions, and secretions can be directly assessed. Durations of antigenaemia in body fluids and secretions, or of infective particles in specific cells, will, in many instances, reflect the period over which an infected person is infectious to others (although this is, of course, not always the case, for example the latent herpes viruses).

Alternatively, statistical methods can be employed in the study of transmission within small groups of individuals. The classic data on measles, collected by Hope Simpson in the Cirencester area of England during the years 1946–52, record the distribution of the observed time interval between two cases of measles in 219 families with two children under the age of 15 years. The bulk of these observations represent case-to-case transmission within a family. However, in a small number of families, where the observed interval is only a few days, it may be assumed that these cases are double primaries, both children having been simultaneously infected from some outside source. Statistical methods, based on chain binomial models, can be used to derive estimates of the latent, infectious, and incubation periods (see Bailey 1973). A rough guide to these periods for various common viral and bacterial infections is presented in Table 14.4. Some of these estimates are based on detailed analyses of case-to-case data, while others are more speculative.

Sexually transmitted infections

Rather different problems in parameter estimation, compared with those outlined above, are presented by sexually transmitted infections. By way of an illustration and given the topicality of the infection, we focus on HIV.

The characteristics of most sexually transmitted diseases cause their epidemiology to differ from that of common childhood viral and bacterial infections. First, the rate at which new infections are produced is not dependent on population density. Second, the carrier phenomenon in which certain individuals harbour asymptomatic infection is often important. Third, many sexually transmitted diseases induce little or no acquired immunity on recovery, and, fourth net transmission depends on the degree of heterogeneity in sexual activity prevailing in the population. The basic reproductive rate, R_0, in its simplest form is determined by the product of the transmission probability, B, multiplied by the effective rate of sexual-partner change, c, multiplied by the average duration of infectiousness, D. In the case of HIV, each component is difficult to measure due to the sensitivity and the practical difficulties associated with the study of sexual behaviour, and the long and variable incubation period of the disease AIDS induced by the infection. Over the long incubation period infectiousness appears to vary widely for an individual and between individuals. In these circumstances some indirect measure of transmission potential is required. Mathematical models of transmission suggest that the doubling time, t_d (the average time over which the number of cases of infection doubles), of an epidemic of HIV in a defined risk group (e.g. homosexuals), during the early stages of the epidemic, is related to the magnitude of R_0 by the equation:

$$t_d = D\ln(2)/(R_0 - 1) \tag{33}$$

where D denotes the average duration of infectiousness. Current estimates of the incubation period of AIDS suggest a mean period of around 8 years. It is possible that the average infectious period is much shorter, perhaps of the order of 2 years or so, although this is uncertain at present (see Anderson and May 1988). If we assume the value of D lies between 2 and 6 years, equation 33 gives estimates of R_0 in the range of 2.7–6, based on an observed t_d of around 10 months in homosexual communities in the US during the early 1980s

Box 14.3. Surveillance profiles

Two infections, (i) and (ii), are at endemic equilibrium (i.e. roughly constant incidence in time) in a stationary host population (i.e. births are equal to deaths). The changes, with time or increasing age, in the proportion of the population that has experienced each infection may be estimated from longitudinal cohort or horizontal cross-sectional surveys (serological or case notifications) of individuals from birth to life expectancy, L, as shown below.

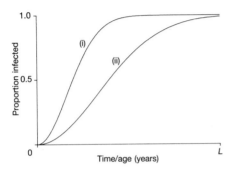

The steeper profile of (i) compared with (ii) is an indication that the basic reproductive potential, R_0, of infection (i) exceeds that for infection (ii) such that:

$$R_0(i) > R_0(ii)$$

Assume that each infection induces lifelong immunity, then, from the above profiles, changes with age/time in the proportion susceptible, x, to each infection are:

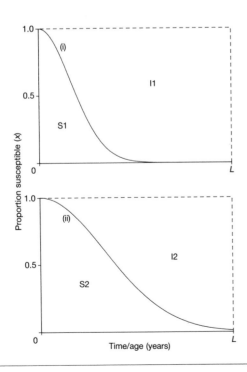

The (equilibrium) proportion of the total population susceptible to infection (i) is:

$$x^*(i) = \text{area S1}/(\text{area S1} + \text{area I1})$$

and infection (ii):

$$x^*(ii) = \text{area S2}/(\text{area S2} + \text{area I2})$$

Note that the equilibrium proportion susceptible to infection (i) (with the higher reproductive rate) is smaller than that for infection (ii) (with a lower rate of reproduction) i.e.:

$$x^*(i) < x^*(ii)$$

and the relationship between these two epidemiological parameters may be usefully expressed as:

$$R_0(i) = 1/x^*(i)$$

and

$$R_0(ii) = 1/x^*(ii)$$

Summing the proportion susceptible, $x(a)$, in the above graphs for each age class from age 0 years (time 0) to L years, we can determine the average age at infection, A, thus:

$$x(a) = x(0) + x(1) + x(2) + \ldots + x(L) = A = S$$

from which it can be seen:

$$A(i) < A(ii)$$

Note also that:

$$R_0(i) = L/A(i) = L/S1$$

and

$$R_0(ii) = L/A(ii) = L/S2$$

(See also equations 12 and 13, and Anderson and May (1983), for estimation of the force of infection from surveillance profiles.)

Summary examples

Assume infection (i) is measles and (ii) is rubella in the UK with average life-span, L, of 75 years. If S1 = 5 and S2 = 10, then $A(i) = 5$ years and $A(ii) = 10$ years, and:

$$x^*(i) = 5/75 = 0.066'$$
$$x^*(ii) = 10/75 = 0.133'.$$

Therefore:

$$R_0(i) = 1/0.066' = 15$$
or
$$= 75/5 = 15$$

and

$$R_0(ii) = 1/0.133' = 7.5$$
or
$$= 75/10 = 7.5$$

The implications of this difference in the basic reproductive rate of infection to the proportion of the population that must be vaccinated in order to eradicate each infection can be seen in Figure 14.11.

Table 14.3. Epidemiological parameters for a variety of childhood infections in developed countries in the absence of mass vaccination. Parameter definitions given in text (data from a variety of sources)

Infection	Average age at infection, A (years)	Location and date	Data type	Life expectancy, L	R_0*	p_c (%)
Measles	5.0	England and Wales, 1948–68	Case notifications	70	15.6	94
	5.5	US, large families, 1957	Serology	70	14.0	93
	8.0	US, small families, 1957	Serology	70	9.3	89
Whooping cough	4.5	England and Wales, 1944–78†	Case notifications	70	17.5	94
	4.9	US, urban, 1908–17	Case notifications	60	13.6	93
	6.5	US, rural, 1908–17	Case notifications	60	10.0	90
Chickenpox	8.6	US, urban, 1913–17	Case notifications	60	7.4	86
	6.8	US, urban, 1943	Case notifications	70	11.1	91
Mumps	7.0	UK, urban, 1977	Serology	75	11.5	91
	5.7	Netherlands, urban, 1980	Serology	75	14.4	93
	9.9	US, urban, 1943	Case notifications	70	7.4	86
Diphtheria	10.4	US, 1912–28	Case notifications	60	6.1	84
Rubella	10.8	England, urban, 1980–84‡	Serology	75	7.3	86
	10.2	GDR, 1972	Serology	70	7.2	86
Scarlet fever	8.0	US, urban, 1908–17	Case notifications	60	8.0	88
	12.3	US, rural, 1918–19	Case notifications	60	5.1	80

* $R_0 = L/(A - F)$ where F is duration of maternally derived protection, assumed to last for 6 months in all cases. Note that no consideration of age-dependent forces of infection is given (see text).

† Encroaches on to vaccination era.

‡ Male serology—only females vaccinated under selective immunization policy.

(May and Anderson 1987). Of course, this method of estimation is very crude, but it provides a rough guide to the degree to which sexual habits must change in order to reduce the magnitude of R_0 below unity in value (i.e. by a factor 3–6).

More generally, certain of the parameters that determine the magnitude of R_0 may vary between the sexes. This is certainly the case for gonorrhoea (see Hethcote and Yorke 1984) and it may be true for HIV. In these circumstances, when considering transmission via heterosexual contact, the basic reproductive rate adopts the form:

$$R_0 = (B_1 B_2 c_1 c_2 D_1 D_2) \qquad (34)$$

where the subscripts 1 and 2 denote males and females respectively.

Models and the design of control programmes

Mathematical models can be of help in defining the targets for a control programme, in interpreting observed epidemiological changes under the impact of control, and in discriminating between different approaches (Nokes and Anderson 1987, 1988). In this section we consider two examples, namely, the design of mass vaccination programmes to control childhood viral and bacterial infections and education to induce changes in sexual behaviour to control sexually transmitted diseases.

Impact of mass vaccination

In practical terms, the level of vaccination coverage in a given community or country is determined by a variety of economic and logistic factors (developing countries) or motivational and legislative issues (industrialized countries). However, models can define the ideal goal of a given programme. We have already outlined the relationship between the critical level of vaccination coverage required to block transmission, p_c, and various epidemiological (R_0), demographic (net birth-rate and life expectancy, Q and L) and logistical (V, the average age at vaccination) parameters (see equation 28 and Table 14.3). In many instances, the high transmission potentials of common childhood viral and bacterial infections imply very high levels of vaccination coverage if transmission is to be interrupted. If vaccine efficacy is less than 100 per cent (e.g. the current pertussis vaccines) then problems may arise in attaining these targets even if legislation enforces vaccination of all children before entry to school (as in the US). Models simply emphasize the point that, to obtain the best effects, very high levels of coverage should be aimed at with vaccination at as young an age as is practically feasible given the complications presented by the presence of maternally derived antibodies in infants.

Aside from defining targets for vaccination coverage, models can assist in interpreting the impact of a given programme on epidemiological parameters, such as the incidence of infection, the average age at infection, and the inter-epidemic period.

Incidence of infection

Immunization has the direct effect of reducing the number of cases of infection as a result of the protection of the vaccinated individuals ($X \rightarrow Z$, see Fig. 14.3d). Since this reduces the number of infectious persons in the vaccinated

Table 14.4. Average duration of infection classes for a variety of microparasites

Infectious disease	Latent period, $1/\sigma$ (days)	Infectious period, $1/\gamma$ (days)	Incubation period (time to appearance of symptoms; days)
Measles	6–9	6–7	11–14
Chickenpox	8–12	10–11	13–17
Rubella	7–14	11–12	16–20
Hepatitis A	13–17	19–22	30–37
Mumps	12–18	4–8	12–26
Polio	1–3	14–20	7–12
Smallpox	8–11	2–3	10–12
Influenza	1–3	2–3	1–3
Scarlet fever	1–2	14–21	2–3
Whooping cough	6–7	21–23	7–10
Diphtheria	2–5	14–21	2–5

Source: Anderson (1982b).

population, an indirect effect is a reduction in the net rate of transmission of the virus or bacterium. This is the principle of herd immunity where susceptibles gain protection from the vaccinated proportion of the population. Provided the infection is able to persist endemically (i.e. the level of coverage is less than that required for eradication), models suggest that the equilibrium proportion of susceptibles in the population will remain constant irrespective of the level of coverage below the critical point for eradication. This prediction is illustrated diagrammatically in Figure 14.13. The level of coverage simply reduces the proportion of seropositive individuals who acquire immunity via infection as opposed to via vaccination. As mentioned earlier (see Fig. 14.12), the manner in which the incidence declines as the level of coverage rises is non-linear in form, with the most dramatic reductions occurring as the proportion vaccinated approaches the critical point for the interruption of transmission. As the level of coverage approaches the critical point, the proportion of immune persons who possess vaccine-induced immunity approaches unity.

The average age at infection

As a direct result of reducing the net rate of transmission, vaccination acts to increase the average age at which susceptibles acquire infection over that pertaining prior to control (i.e. by reducing the probability of coming into contact with an infectious person). A number of examples of vaccination changing the age distribution of susceptibility in a population are documented in Fig. 14.14. As discussed later, this change in the age distribution of the incidence of infection can influence the incidence of disease arising from infection if older people differ in their vulnerability to complications and concomitant morbidity when compared with younger people.

Inter-epidemic period

Simple models also predict that a reduction in the transmission rate in a vaccinated population will lengthen the inter-epidemic period over that pertaining prior to control (Anderson and May 1983). This pattern has been observed in various vaccinated communities (e.g. Fig. 14.15).

Cautionary notes

The changes in epidemiological patterns of infection induced by vaccination are not always beneficial. An increased inter-epidemic period, for example, can induce complacency in the community with respect to the need to maintain high levels of vaccination coverage. Motivation of parents to ensure that their children are vaccinated during long periods of low incidence (the troughs in the epidemic cycle) can be problematic particularly if there is some small but measurable risk associated with vaccination. At the start of a mass immunization programme the probability of serious disease arising from vaccination is usually orders of magnitude smaller than the risk of serious disease arising from natural infection. As the point of eradication is approached, the relative magnitude of these two probabilities must inevitably be reversed. The optimum strategy for the individual (not to be vaccinated) therefore becomes at odds with the needs of society (to maintain herd immunity). This issue—which was central to the decline in uptake of pertussis vaccine in the UK during the mid to late 1970s—could be influenced by legislation to enforce vaccination (as in the US), but its final resolution is achievable only by global eradication of the disease agent so that routine vaccination can cease.

Other problems concern doubts over the role played by exposure to natural infection in boosting vaccine-induced immunity, and, in some cases, worries over the duration of protection provided by vaccination. If enough is understood about these problems, mathematical models could be used to decide whether or not to revaccinate a proportion of the immunized population and, if so, what is the best age at which to revaccinate.

Variation in vaccine uptake

Ideally, vaccination coverage should be high and constant both over time and in different regions of a country. In practice, however, this is rarely the case. With respect to time, once incidence is reduced to a low level, problems can arise in stimulating public health workers to maintain coverage at high levels. More importantly, after introduction, most immunization programmes show only a slow increase in rates of coverage. This obviously results in a delay in experiencing

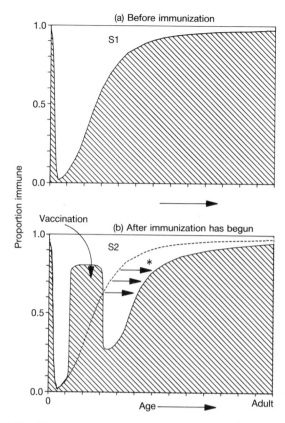

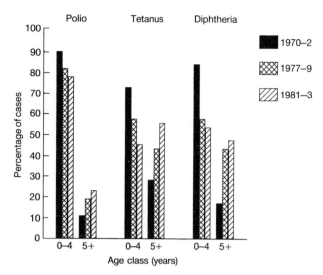

Fig. 14.14. Observed changes in the age distribution of cases of three vaccine-preventable infections in Bangkok, Thailand, following the World Health Organization's Expanded Programme on Immunization (EPI) initiative. Note that although the *proportions* of cases in the older age class increased, actual *numbers* declined significantly for all ages over the period indicated. (Source: Nokes and Anderson 1988, with permission from the editors of *Epidemiology and Infection*.)

Fig. 14.13. Diagrammatic representation of the predicted impact of mass immunization (against a typical childhood viral or bacterial infection) on the age distribution of susceptibility in a population. Before immunization (a) there is a 'valley' of susceptibles (S1) in the young age classes. Attempts to fill in this valley by vaccination (b) reduces the rate of transmission of the infection thus lowering the probability of unvaccinated individuals being infected. As a consequence there is an upward shift in the ages of susceptibles (*) from that pertaining before vaccination (dotted line). Two points are important: (i) the number or proportion of susceptibles after immunization has begun (area S2) is roughly unchanged from that which existed before immunization (area S1); and (ii) the average age of susceptibles increases. (Source: Nokes and Anderson 1988, with permission from the editors of *Epidemiology and Infection*.)

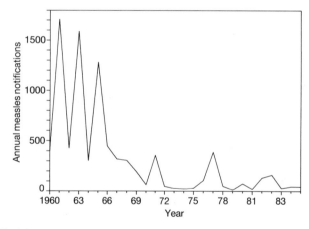

Fig. 14.15. Annual measles notifications for the city of Oxford, England, for the period 1960–85. The introduction of measles vaccination in 1966 has resulted in a significant increase in the period between epidemics. (Source: Office of Population Censuses and Surveys, London.)

Non-uniformity in human population density

the full benefits and must be recognized in assessing the impact of a given policy. It takes many decades before the full benefits of a cohort immunization programme are manifest. Model simulations of the impact of such programmes on the incidence of infection and disease clearly illustrates this point (Anderson and May 1983, 1985*a*, *b*). Of greater concern, however, is variation in vaccine uptake in different regions of a country. Levels of measles and rubella vaccine coverage in the UK, for example, vary widely between different regions (Fig. 14.16). To block transmission effectively country-wide it is necessary to ensure that the targets laid out in Table 14.3 are attained in each area. Otherwise, pockets of infection in regions of low uptake will continue to trigger small epidemics in other areas. The upsurge of mumps in certain States in the US (Fig. 14.17) is an example of the potential hazards of spatial variation in vaccine uptake.

Non-uniformity in the spatial distribution of humans, with some people living in dense aggregates and others living in isolated or small groups, can lead to heterogeneity in transmission rates. Models suggest that this can result in the transmission potential of an infection (R_0) being greater, on average, than suggested by estimation procedures which assume spatial homogeneity (Anderson and May 1984; May and Anderson 1984). Under these circumstances, theory suggests that the optimal solution appears to involve 'targeting' vaccination coverage in relation to group size, with dense

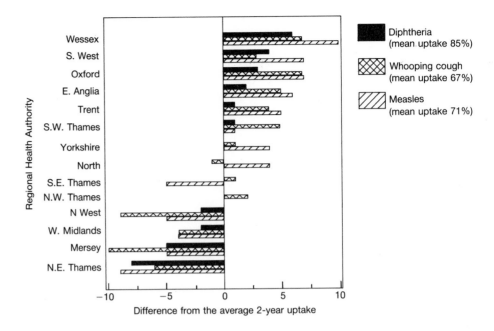

Fig. 14.16. Regional variation in immunization uptake for sentinel agents in England, 1986. (Source: Nokes and Anderson 1988, with permission from the editors of *Epidemiology and Infection*.)

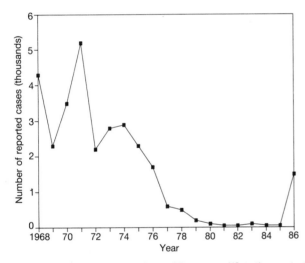

Fig. 14.17. Mumps cases in the State of Tennessee, US, in the vaccination era. The recent rise in incidence is a symptom of variation in vaccine uptake across the country. (Source: Tennessee Department of Health and Environment.)

groups receiving the highest levels of coverage. The optimal programme is defined as that minimizing the total, community-wide number of immunizations needed for elimination or for a defined level of control. This strategy reduces the overall proportion that must be vaccinated to block transmission, compared with that estimated on the assumption of spatial homogeneity. This conclusion has practical significance for the control of infections, such as measles and pertussis, in some developing countries, where rural/urban differences in population density tend to be much more marked than in developed countries. It is probable that, in

many regions of Africa and Asia, diseases such as measles cannot persist endemically in rural areas without frequent movement of people between low-density (rural) and high-density (urban) populations. Under these circumstances, transmission might be blocked in both regions by high levels of mass immunization in the urban centres alone.

Age-dependent factors

Analyses of case-notification records and serological profiles suggest that, for many common infections (measles, rubella, and pertussis), the per caput rate of infection ($\lambda(a)$) depends on the ages of susceptible individuals, changing from a low level in the 0–5-year-old classes, via a high level in the 5–15-year-old classes, back to a lower level in the adult classes (see Fig. 14.5). This is of interest both because it reflects behavioural attributes of human communities, and because of its impact on the predicted level of vaccination required to eliminate transmission. The high levels of the force of infection in the 5–15-year-old classes are thought to arise as a consequence of frequent and intimate contacts within school environments (Anderson and May 1985b; Anderson et al. 1987; Nokes et al. 1986). Theoretical studies which take account of age dependence in the force of infection predict somewhat lower rates of vaccination than those arrived at under the simple mass-action assumption (see Table 14.3). However, it should be emphasized that the values listed in Table 14.3 provide a good first approximation of the targets to be obtained in a vaccination programme. The reason why the observed age-related changes in the force of infection influence the predicted level of coverage relates to the tendency for mass vaccination to shift the age distribution of susceptibility (Fig. 14.13). Susceptibles who avoid infection and

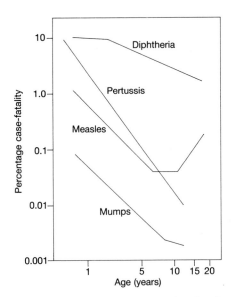

Fig. 14.18. Age-dependent mortality associated with infection from a variety of childhood viruses and bacteria. (Source: Mims 1987.)

vaccination may move from an age class with a high force of infection into an older class with a lower rate.

Does mass vaccination always reduce disease incidence?

The risk of complications arising from infection is often dependent upon the age at which exposure occurs. The newborn are especially vulnerable due to their immunological immaturity, and are therefore more likely to suffer morbidity and even mortality (Fig. 14.18). Protection by maternally derived antibody moderates the risk during this time of great vulnerability but, in developing countries, factors such as malnutrition and high incidences of secondary 'opportunist' infections can result in high mortality rates as a result of infant and childhood viral and bacterial infection. In general, where the risk of serious disease is higher in the young than old people, mass vaccination will always act to reduce the incidence of disease.

In developed countries, case fatalities are much less common and the greatest problem is morbidity and the risk of serious disease. Of particular concern are infections where the probability of severe complications increases with age (Fig. 14.19). Whether this trend is important depends on the quantitative details of such factors as how risk changes with age, the average age at which the vaccine is administered, the average age at infection, and how the rate (or force) of infection changes with age before the introduction of immunization (Anderson and May 1983, 1985b; Anderson *et al.* 1987; Knox 1980) (see Fig. 14.5). Rubella and mumps are clear examples because of the problem of congenital rubella syndrome in infants born to mothers who contracted rubella in their first trimester of pregnancy, and the occurrence of orchitis and the associated risk of sterility in post-pubertal males plus infection of the central nervous system following mumps infection. The crux of the problem relates to how

mass vaccination changes the age profile of the incidence of infection. Any level of coverage will reduce the incidence of infection, but, by increasing the average age at which those still susceptible acquire infection, certain levels of coverage may increase the incidence of disease. The important question is whether the increase in the proportion of cases in older people will result in an increase in the absolute numbers of cases of serious disease.

This problem has resulted in the adoption of different vaccination programmes against rubella (to control congenital rubella syndrome) in different countries (Table 14.5). Until recently in the UK the main thrust of the immunization policy was to vaccinate girls at an average age of around 12 years, so as to allow rubella virus to circulate in males and young females and create naturally acquired immunity in the

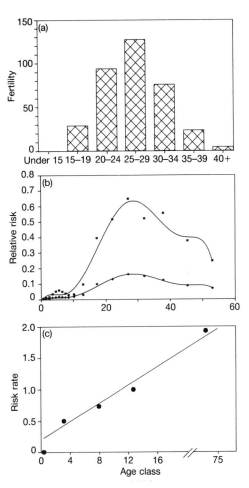

Fig. 14.19. Age-dependent risk of complications from infection: (a) the likelihood of fetal transmission of rubella virus with concomitant risk of congenital rubella syndrome is directly related to age-specific fertility of women (data for England and Wales, 1985; source OPCS Monitor FMI 86/2); (b) changes in the risk of complications from mumps infection in the UK relative to age and sex. In addition to meningitis and encephalitis, males (■) may suffer orchitis. (Source: Anderson *et al.* 1987); (c) measles encephalitis per 100 000 cases in the US. (Source: Anderson and May 1983.)

Table 14.5. Strategies of rubella immunization

	Selective	Mass cohort
Aim	Eliminate congenital rubella, not rubella infection	Eliminate rubella infection, and so congenital rubella
Age at vaccination	Pre-pubertal girls (10–15 years)	Boys and girls of 1–2 years
Philosophy	(i) Build upon levels of herd immunity attained through childhood (ii) Reduce the proportion of susceptible women of childbearing age (iii) Allow continued circulation of virus in male and young female segments of the population	(i) Reduce circulation of wild virus in community, especially children (ii) Lower the probability of susceptible women catching infection via the action of herd immunity
Overall incidence of infection	Very little impact at any level of coverage	(i) Reduction in cases in a non-linear manner as vaccine level increases (see Fig. 14.11) (ii) Increase in average age at infection
Other concerns	(i) Cannot eradicate congenital rubella unless 100 per cent of women 'at risk' are immune (via infection in childhood or immunization) (ii) Herd immunity largely natural with continued re-exposure to infection and boosting of antibody response	(i) Proportion of remaining cases increases in older age classes, hence possible to increase congenital rubella at certain levels of immunization (ii) Herd immunity ultimately all vaccine induced. Less solid? No boosting of immunity by re-exposure to virus
Which policy?	Suitable for lower levels of vaccination coverage (see Fig. 14.20)	Suitable if high levels of uptake can be achieved (see Fig. 14.20)
Country (as example)	UK	US

early years. By contrast, in the US, both boys and girls are vaccinated at around 2 years of age, with the aim of blocking rubella virus transmission. Mathematical models predict that the US policy is best if very high levels of vaccination (80–85 per cent of each yearly cohort) can be achieved at a young age, while the UK policy is better if this cannot be guaranteed (see Fig. 14.20). A mixed US and UK policy is predicted to be of additional benefit over the UK single-stage policy if moderate to high levels of vaccine uptake among boys and girls can be achieved at a young age (> 60 per cent) (Anderson and Grenfell 1986). However, ironically, introducing mass vaccination will tend to lower levels of herd immunity in all age classes other than the very young (Fig. 14.21).

In the UK, a change in vaccination policy took place in October 1988 with the introduction of a triple measles, mumps, and rubella (MMR) vaccine administered to young children. Mathematical models have been employed to assess the impact of this change on the incidence of serious disease resulting from mumps infection (Box 14.4). The conclusion arrived at was that, provided moderate levels of coverage are attained (60–65 per cent), mass vaccination is unlikely to increase the incidence of serious disease. More problematic, however, is the question of how long should a two-stage rubella vaccination policy (with young children immunized via the use of MMR and girls vaccinated at around 12 years of age) be run before the vaccination of girls can cease. Recent work suggests that unless levels of coverage with the MMR vaccine can be maintained in excess of 80–85 per cent, the vaccination of teenage girls should continue into the foresee-

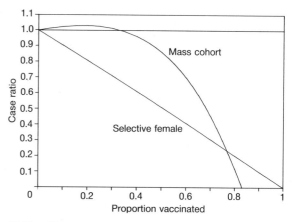

Fig. 14.20. Effectiveness of different rubella immunization programmes. Changes in the predicted case ratio (i.e. the average number of rubella infections in pregnant women after the introduction of immunization divided by the average pre-vaccination number) under increasing levels of coverage, for two types of policy, namely, selective immunization of girls of average age 12 years or mass vaccination of children (aged 2 years). Low to medium levels of uptake favour adoption of a selective immunization programme compared with mass vaccination which has the undesirable effect of increasing the average age at infection.

able future to keep the incidence of congenital rubella syndrome to a very low level (Nokes and Anderson 1987). This particular problem, with the impacts of different policies depending on the quantitative details of the transmission dynamics of the virus, levels of vaccination coverage, and age-related changes in case complication rates, illustrates well the use of mathematical models in public health planning and research.

Changes in sexual behaviour and the transmission of sexually transmitted infections

The current pandemic of AIDS, and the absence of effective drugs and a vaccine to combat infection, has focused much attention in recent years on how to induce changes in sexual behaviour via education and media publicity campaigns to slow the spread of infection. The most important behaviour relevant to the rate of spread is the distribution of the rates of acquiring new sexual partners within a defined population (see Fig. 14.8). A major characteristic of this behaviour is the heterogeneity between individuals within a given community. A central question in this problem is whether it is best to aim health education programmes at the whole population, with the aim of reducing average rates of sexual-partner change, or whether it is best to target education at high-risk groups, such as those with very high rates of sexual-partner change (in either homosexual or heterosexual communities). This is a complicated question and its resolution depends, in part, on a detailed quantitative knowledge of the pattern of sexual behaviour within a given population. However, simple mathematical models can help provide some clues to the resolution of this issue. Of particular importance in understanding the dynamics of transmission of HIV is determining how sexual behaviour influences the magnitude of the basic reproductive rate, R_0. As discussed earlier, for a sexually transmitted disease such as HIV, the magnitude of R_0 is (in simple terms) defined by the product of the probability of transmission per partner contact, B, multiplied by the effective rate of sexual-partner change, c, multiplied by the average duration of infectiousness, D, of an infected person. Under the assumption of proportional mixing (see section on validity of the mass-action principle above), the effective rate of partner change is defined as the mean rate of partner change, m, plus the variance to mean ratio, (σ^2/m) (i.e. $c = m + (\sigma^2/m)$ and $R_0 = BcD$). This definition makes clear the relative contribution of average behaviour and variation in sexual behaviour to the magnitude of R_0. The variance is typically much larger in value than the mean and, hence, those with high rates of partner change play a disproportionate role (relative to their proportional representation within a sexually active population) to the spread of infection. This simple theoretical result suggests that greater benefit is to be gained (in terms of reducing R_0) by targeting education at those with higher than average rates of sexual-partner change. In practice, the identification of such individuals is problematic in the absence of detailed survey data that relate this behaviour to other characteristics. The surveys of sexual behaviour that have been completed to date show a strong age dependency (with young adults having the highest rates of sexual-partner change) but little else of help in identifying correlates (Anderson 1988; Anderson and May 1988). However, attendees at sexually transmitted disease clinics are an important target group, since diseases other than HIV are more frequently present among those with high rates of partner change. This area of research is in its infancy at present but models are beginning to play a role in defining how changes in behaviour can influence the spread of infection.

Conclusions

We have glossed over much detail and ignored many complications in model formulation and analysis in this chapter. The

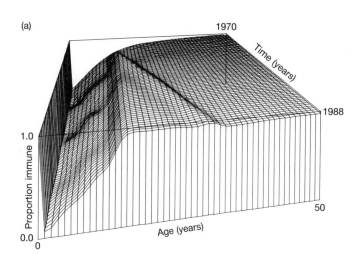

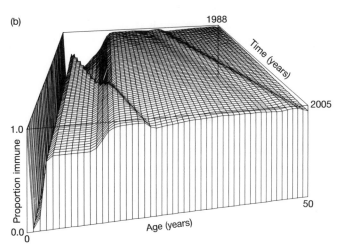

Fig. 14.21. The impact of immunization on population levels of herd immunity. Predicted time-dependent changes in the proportion of the female population in England and Wales immune to rubella virus infection: (a) between 1970 and 1988 under the programme of selective immunization of girls (10–15 years of age); and (b) between 1988 and 2005 under a dual policy of mass childhood immunization (mumps/measles/rubella vaccination initiated in October 1988) and continued selective schoolgirl vaccination. A coverage of 72 per cent of boys and girls by the age of 2 years is assumed (consistent with current levels of measles uptake). The interesting point to observe from these surfaces is that under selective immunization alone (a) herd immunity levels in the years of childbearing are appreciably raised, whereas following the introduction of childhood vaccination (b) there is an apparent decline in such levels of immunity (a consequence of reduced transmission rates—see text).

Box 14.4. Epidemiology and control of mumps virus infection

Incidence of infection

Mumps is typical of the childhood viral infections with peak incidence in the young age classes and relatively few cases occurring in adulthood (a).

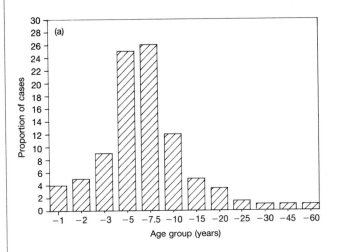

Information of this sort, obtained from age-specific case notification data or age-serological profiles, is used to derive age-dependent rates of transmission as shown in Figure 14.5b.

Incidence of disease

Various types of complications are associated with mumps virus infection (b). Mumps is the most common origin of viral meningitis in the UK, and is also a significant cause of encephalitis and, in post-pubertal males, of orchitis.

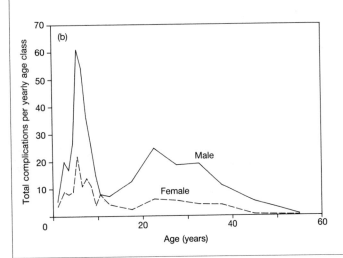

Scaling these age-complication data by the proportion of cases of *infection* in the corresponding age classes (shown above), it is possible to derive the relative risk of complications from infection, as shown in Figure 14.19b in the main text. What becomes apparent from these data analyses is that although fewest cases of infection occur in the older age classes, there remain substantial numbers of cases of complications, such that infection in older persons runs a considerably greater risk of resulting in complications when compared with infection in the young.

Mass vaccination and the incidence of infection

The figure below shows the predicted numbers of cases of mumps infection across a wide range of age classes, through time, before and after the introduction of a programme of mass cohort immunization (60 per cent of 2-year-olds).

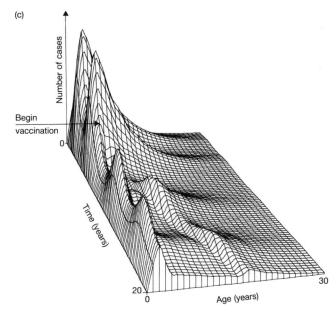

The force of infection is assumed to remain constant with age at 0.15 per year (corresponding to an average age at infection before immunization of 6.7 years). The epidemic peaks in the prevaccine period show the majority of cases occurring in the youngest age classes. Subsequent to the initiation of immunization two changes should be noted: (a) the obvious and expected decline in infection incidence (particularly in the young); and (b) an increase in the age at which the remaining cases occur, indicated by the wave of infections migrating, in time, into the older age classes. The implications of this shift in the age distribution of cases on the incidence of disease are addressed below.

Mass vaccination and the incidence of disease

The effect of a rise in the proportion of cases in the older age classes (predicted above) on the incidence of complications is dependent upon two things: (a) the level of cohort immunization that can be attained; and (b) the age-dependent nature of the risk of complications seen in Figure 14.19. Simulations that help to unravel this problem are shown below (d) (adapted from Anderson *et al.* 1987), recording the change in the predicted risk ratio (i.e. the average number of complications after immunization has begun divided by the average number of complications occurring before) over various levels of childhood immunization (note that the risk ratio is unity for no benefit from immunization).

Obviously there is little benefit to be obtained by vaccination at less than 60 per cent, and indeed vaccination at anything less than 70 per cent is potentially hazardous

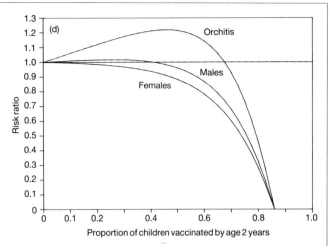

when considering orchitis alone. Such a phenomenon is a direct result of the combination of increased average age at infection and of the risk of complications with age.

interested reader is therefore urged to consult the source references. Our aim has been to define, as simply as possible, the central concepts underpinning the study of the transmission dynamics of infectious diseases and the major conclusions that have emerged from the development and analysis of mathematical models of transmission and control.

The recent convergence of mathematical theory and observation in epidemiology has created a powerful set of tools for the study of the population biology of infectious disease agents. At present, the potential value of these techniques is not widely appreciated by public health scientists and medical personnel. Many people have rightly criticized models that pursue the mathematics in isolation, making only perfunctory attempts to relate the findings to epidemiological data. But there is a converse danger which is less widely understood. The complexities of the course of infection within an individual and its spread between people are such that years of clinical experience and the most refined intuition will not always yield reliable insights into the factors that control the transmission dynamics of a given infectious agent and how these are influenced by perturbations introduced by control measures. Moreover, insensitive use of a computer will not always help us to understand these problems, for, if a computer is given inappropriate instruction, it will usually give inappropriate answers. What is needed, in our view, is increased collaboration between epidemiologists and mathematicians, with the models being founded on data (and with their predictions being tested against available facts), and with verbal hypotheses being founded on clear mathematical statements of the assumptions. We hope that the contents of this chapter stimulate interest in this goal.

References

Anderson, R.M. (1982*a*). Epidemiology. In *Modern parasitology* (ed. F.E.G. Cox) p. 204. Blackwell Scientific Publications, London.

Anderson, R.M. (1982*b*). Directly transmitted viral and bacterial infections of man. In *Population dynamics of infectious diseases agents: Theory and applications* (ed. R.M. Anderson) p. 1 Chapman and Hall, London.

Anderson, R.M. (1988). Epidemiology of HIV infection: Variable incubation plus infectious periods and heterogeneity in sexual activity. *Journal of the Royal Statistical Society, Series A* **151**, 66.

Anderson, R.M. and Grenfell, B.T. (1986). Quantitative investigations of different vaccination policies for the control of congenital rubella syndrome (CRS) in the United Kingdom. *Journal of Hygiene (Cambridge)* **96**, 305.

Anderson, R.M. and May, R.M. (1982). Directly transmitted infectious diseases: Control by vaccination. *Science* **215**, 1053.

Anderson, R.M. and May, R.M. (1983). Vaccination against rubella and measles: Quantitative investigations of different policies. *Journal of Hygiene* (*Cambridge*) **90**, 259.

Anderson, R.M. and May, R.M. (1984). Spatial, temporal and genetic heterogeneity in host populations and the design of immunization programmes. *IMA Journal of Mathematics Applied in Medicine and Biology* **1**, 233.

Anderson, R.M. and May, R.M. (1985*a*) Vaccination and herd immunity to infectious disease. *Nature (London)* **318**, 323.

Anderson, R.M. and May, R.M. (1985*b*). Age-related changes in the rate of disease transmission: Implications for the design of vaccination programmes. *Journal of Hygiene Cambridge* **94**, 365.

Anderson, R.M. and May, R.M. (1985*c*). Herd immunity to helminth infection and implications for parasite control. *Nature (London)* **315**, 493.

Anderson, R.M. and May, R.M. (1986). The invasion, persistence and spread of infectious diseases within animal and plant communities. *Philosophical Transactions of the Royal Society of London* **314**, 533.

Anderson, R.M. and May, R.M. (1988). Epidemiological parameters of HIV transmission *Nature (London)* **333**, 514.

Anderson, R.M., Crombie, J.A., and Grenfell, B.T. (1987). The epidemiology of mumps in the U.K.; a preliminary study of virus transmission, herd immunity and the potential impact of immunisation. *Epidemiology and Infection* **99**, 65.

Anderson, R.M., Grenfell, B.T., and May, R.M. (1984). Oscillatory fluctuations in the incidence of infectious disease and the impact of vaccination; time series analysis. *Journal of Hygiene (Cambridge)* **93**, 587.

Anderson, R.M., May, R.M., and McLean, A.R.; (1988). Possible demographic consequences of AIDS in developing countries. *Nature* **332**, 191.

Anderson, R.M., May, R.M., Medley, G.F., and Johnson, A. (1986). A preliminary study of the transmission dynamics of the human immunodeficiency virus (HIV), the causative agent of AIDS. *IMA Journal of Mathematics Applied in Medicine and Biology* **3**, 229.

Bailey, N.T.J. (1973). Estimation of parameters from epidemic models. In *Mathematical theory of the dynamics of biological populations* (ed. M.S. Bartlett and R.W. Hiorns) p. 253. Academic Press, London.

Bailey, N.T.J. (1975). *The mathematical theory of infectious diseases and its implications*. Griffin, London.

Ball, F. (1983). The threshold behaviour of epidemic models. *Journal of Applied Probability* **20**, 227.

Becker, N. (1979). The uses of epidemic models. *Biometrics* **35**, 295.

Bernoulli, D. (1760). Essai d'une nouvelle analyse de la mortalité causée pour la verole et des avantages de l'incubation pour la prevenir. *Mem. Math. Phys. Acad. Roy. Sci (Paris)* **1**.

Black, F.L. (1966). Measles endemicity in insular populations: Critical community size and its evolutionary implications. *Journal of Theoretical Biology* **II**, 207.

Brownlee, J. (1906). Statistical studies in immunity: The theory of an epidemic. *Proceedings of the Royal Society (Edinburgh)* **26**, 484.

Cliff, A.D., Haggett, P., Ord, J.K., and Versey, G.R. (1981). *Spatial diffusion: An historical geography of epidemics in an island community*. Cambridge University Press, London.

Cox, M.J., Anderson, R.M., Bundy, D.A.P., *et al.* (1988). Seroepidemiological study of the mumps virus in St. Lucia, West Indies. *Epidemiology and Infection* **102**, 147–60.

Dietz, K. (1987). Mathematical models for the control of malaria. In *Malaria* (ed. W.H. Wensdorfe and J.A. MacGregor) p. 1087. Churchill Livingstone, Edinburgh.

Farr, W. (1840). Progress of epidemics. *Second report of The Registrar General of England and Wales*. p. 91.

Grenfell, B.T. and Anderson, R.M. (1985). The estimation of age-related rates of infection from case notifications and serological data. *Journal of Hygiene (Cambridge)* **95**, 419.

Hamer, W.H. (1906). Epidemic disease in England. *Lancet* **i**, 733.

Hethcote, H.W. and Yorke, J.A (1984). Gonorrhoea: Transmission dynamics and control. *Lecture Notes in Biomathematics* **56**, 1.

Kermack, W.O. and McKendrick, A.G. (1927). A contribution to the mathematical theory of epidemics. *Proceedings of the Royal Society of London, Series A* **115**, 700.

Knox, E.G. (1980). Strategy for rubella vaccination. *International Journal of Epidemiology* **9**, 13.

May, R.M. (1986). Population biology of microparasitic infections. *Biomathematics* **17**, 405.

May, R.M. and Anderson, R.M. (1984). Spatial heterogeneity and the design of immunization programs. *Mathematical Biosciences* **72**, 83.

May, R.M. and Anderson, R.M. (1987). The transmission dynamics of HIV infection. *Nature (London)* **326**, 137.

Mims, C.A. (1987). *The pathogenesis of infectious disease* (3rd edn). Academic Press, London.

McLean, A.R. and Anderson, R.M. (1987). Measles in developing countries. Part I. Epidemiological parameters and patterns. *Epidemiology and Infection* **100**, 111.

McLean, A.R. and Anderson, R.M. (1988). Measles in developing countries. Part II. The predicted impact of mass vaccination. *Epidemiology and Infection* **100**, 419.

Nokes, D.J. and Anderson R.M. (1987). Rubella vaccination policy: A note of caution. *Lancet* **i**, 1441.

Nokes, D.J. and Anderson, R.M. (1988). The use of mathematical models in the epidemiological study of infectious diseases and in the design of mass immunization programmes. *Epidemiology and Infection* **101**, 1.

Nokes, D.J., Anderson, R.M., and Anderson, M.J. (1986). Rubella epidemiology in South East England: 1980–1984. *Journal of Hygiene (Cambridge)* **96**, 291.

Rodgers, D.J. (1988). A general model for the African trypanosomiases. *Parasitology* **97**, 193.

Ross, R. (1911). *The prevention of malaria* (2nd edn). Murray, London.

Sokal, R.R. and Rohlf, F.J. (1981). *Biometry* (2nd edn). W.H. Freeman and Company, San Francisco, California.

Soper, H.E. (1929). Interpretation of periodicity in disease prevalence. *Journal of the Royal Society* **92**, 34.

Wickwire, K. (1977). Mathematical models for the control of pests and infectious diseases; a survey. *Theoretical Population Biology* **II**, 181.

Yorke, J.A., Nathanson, N., Pianigiani, G., and Martia, J. (1979). Seasonality and the requirements for perpetuation and eradication of viruses in population. *American Journal of Epidemiology* **109**, 103.

15

Concepts of validity in epidemiological research

SANDER GREENLAND

Introduction

Epidemiological inference is the process of drawing conclusions from epidemiological data, such as prediction of disease patterns or identification of causes of diseases or epidemics. These inferences must often be made without the benefits of direct experimental evidence or established theory about disease aetiology. Consider the problem of predicting the risk and incubation (induction) time for acquired immunedeficiency syndrome (AIDS) among persons infected with type 1 human immunodeficiency virus (HIV-1). Unlike an experiment, in which the exposure is administered by the investigator, the date of HIV-1 infection cannot be accurately estimated in most cases; furthermore, the mechanism by which *silent* HIV-1 infection progresses to AIDS is unknown at this time (1988). Nevertheless, some prediction must be made from the available data if future healthcare needs are to be prepared for effectively.

As another example, consider the problem of estimating how much excess risk of coronary heart disease (if any) is produced by coffee drinking. Unlike an experimental exposure, coffee drinking is self-selected; it appears that persons who use coffee are more likely to smoke than non-users, and probably tend to differ in many other behaviours as well (Greenland 1987). As a result, *even if coffee use is harmless*, we should not expect to observe the same pattern of heart disease in users and non-users. Small coffee effects should thus be very difficult to disentangle from the effects of other behaviours. Nevertheless, because of the high prevalence of coffee use and the high incidence of heart disease, determination of the effect of coffee on heart disease risk may be of considerable public health importance.

In both of these examples, and in general, inference will depend on evaluating the *validity* of the available studies, that is, the degree to which the studies meet basic logical criteria for absence of bias. This chapter outlines some of the major validity concepts in epidemiological research. The contents are organized around three major headings: validity in prediction problems, validity in causal inference, and special validity problems in case–control studies. The chapter ends with a summary and references to further topics.

Several textbooks provide more background and depth than can be given here: Checkoway *et al.* (1989), Kelsey *et al.* (1986), Rothman (1986) and Schlesselman (1982) present good introductory treatments, while Kleinbaum *et al.* (1982) and Miettinen (1985) provide more algebraic and abstract approaches. There is considerable diversity and conflict among the classification schemes and terminology employed in these and other books. This diversity reflects the fact that there is no unique way of classifying the various validity conditions and biases that have been described in the literature. In particular, the scheme employed here should not be regarded as anything more than a convenient framework for organizing the present discussion.

This chapter assumes that the reader is familiar with the basics of epidemiological study design and is comfortable with a number of terms of epidemiological theory, among them 'risk', 'competing risks', 'average risk', 'population at risk', and 'rate'. An appendix is provided to review briefly the meanings of these terms as they are used here.

Validity in prediction problems

The following prediction problem will be used to illustrate basic concepts of validity in epidemiological inference: a gay men's health clinic is about to begin enrolling HIV-1 negative men in an unrestricted programme that will involve retesting each participant for HIV-1 antibodies at 6-month intervals. We can expect that, in the course of the programme, many participants will seroconvert to positive HIV-1 status. Such participants will invariably ask difficult questions, such as 'what are my chances of developing AIDS over the next 5 years?', or 'How many years do I have before I develop AIDS?' In attempting to answer these questions, it will be convenient to refer to such participants (i.e. those who seroconvert) as the 'target cohort': even though membership in this cohort is not determined in advance, this cohort will be the target of our predictions. It will also be convenient to refer to the time from HIV-1 infection until the onset of clinical AIDS as the 'AIDS incubation time'. We could provide reasonable answers to a participant's questions if we could accurately predict AIDS incubation times, although we would also have to estimate the time elapsed between infection and the first positive test.

There might be someone who responds to the questions posed above with the following anecdote: 'I've known several men just like the ones in this cohort, and they all developed AIDS within 5 years after a positive HIV-1 test'. No trained scientist would conclude from this anecdote that all or most of the target cohort will develop AIDS within 5 years of seroconversion. One reason is, of course, that the men in the anecdote cannot be 'just like' men in our cohort in every respect: they may have been older or younger when they were infected, they may have experienced a greater degree of stress following their infection, they may have been heavier smokers, drinkers, or drug users, etc. In other words, we know that the anecdotal men and their post-infection life events could not have been exactly the same as the men in our target cohort with respect to all factors that affect AIDS incubation time. Furthermore, it may be that some or all of the men in the anecdote had been infected long before they were first tested, so that (unlike men in our target cohort) the time from their first positive test to AIDS onset was much shorter than the time from seroconversion to AIDS onset.

Any reasonable predictions must be based on observing the distribution of AIDS incubation times in another cohort. Suppose that we obtain data from a study of gay men who underwent regular HIV-1 testing, and assemble from these data a study cohort of men who were observed to seroconvert. Suppose also that most of these men were followed for at least 5 years after seroconversion. We cannot expect any member of this study cohort to be 'just like' any member of our target cohort in every respect. Nevertheless, if there were only random differences the two cohorts, we could argue that the study cohort could serve as a valid point of reference for predicting incubation times in the target cohort. Thus we will henceforth refer to the study cohort as our *reference cohort*. Note that our reference and target cohorts may have originated from different populations; for example, the clinic generating the target cohort could be in New York, but the study that generated the reference cohort may have been in San Francisco. For both the target and reference cohorts, the actual times of HIV-1 infection will of course have to be imputed, based on the dates of last negative and first positive tests.

Suppose our statistical analysis of data from the reference cohort produces estimates of 0.05, 0.25, and 0.45 for the average risk of contracting AIDS within 2, 5, and 8 years of HIV-1 infection. What conditions would be sufficient to guarantee the validity of these figures as estimates or predictions of the proportion of the target cohort that would develop AIDS within 2, 5 , and 8 years of infection? If by 'valid' we

mean that any discrepancy between our predictions and the true target proportions are purely random, the following conditions would be sufficient:

C Comparison validity. The distribution of incubation times in the target cohort will be approximately the same as the distribution in the references cohort.

F Follow-up validity. Within the reference cohort, risk of censoring (i.e. follow-up ended by an event other than AIDS) is not associated with risk of AIDS.

Sp Specification validity. The distribution of incubation times in the reference cohort can be closely approximated by the statistical model used to compute the estimates. For example, if one employs a log normal distribution to model the distribution of incubation times in the reference cohort, this model should be approximately correct.

M Measurements validity. All measurements of variables used in the analysis closely approximate the true values of the variables. In particular, each imputed time of HIV-1 infection closely approximates the true infection time, and each reported time of AIDS onset closely approximates a clinical event defined as AIDS onset.

The first conditions concerns the 'external' validity of making predictions about the target cohort based on the reference cohort. The remaining conditions concern the 'internal' validity of the predictions as estimates of average risk in the reference cohort. The following sections will explore the meaning of these conditions in prediction problems.

Comparison validity

Comparison validity is probably the easiest condition to describe, although it is difficult to evaluate. Intuitively, it simply means that the distribution of incubation times in the target cohort could be almost perfectly predicted from the distribution of incubation times in the reference cohort, *if* the incubation times were observed without error. Other ways of stating this condition is that the two cohorts are *comparable* or *exchangeable* with respect to incubation times, or that the AIDS experience of the target cohort can be predicted from the experience of the reference cohort.

Confounding

If the two cohorts are not comparable, some or all of our risk estimates for the target cohort based on the reference cohort will be biased* as a result. This bias is sometimes called *confounding* or *comparison bias*. Much research has been done on methods for identifying and adjusting for such bias; see the textbooks referenced above for overviews of such methods.

To evaluate comparison validity, we must investigate whether the two cohorts differ in any factors that influence

---

* Readers with some background in mathematical statistics may substitute *inconsistent* for *biased* throughout this chapter.

incubation time. If so, we cannot reasonably expect the incubation-time distributions of the two cohorts to be comparable. A factor responsible for some or all of the confounding in an estimate is called a *confounder* or *confounding variable*; the estimate is said to be *confounded* by the factor, and the factor is said to *confound* the estimate.

To illustrate these concepts, suppose that men infected at younger ages tend to have longer incubation times, and that the members of the reference cohort are on average younger than members of the target cohort. If there were no other differences to counterbalance this age difference, we should then expect that members of the reference cohort will on average have longer incubation times than members of the target cohort. Consequently, unadjusted predictions of risk for the target cohort derived from the reference cohort would be biased (confounded) by age in a downward direction. In other words, age would be a confounder for estimating risk in the target cohort, and confounding by age would result in underestimation of the proportion of men in the target cohort who will develop AIDS within 5 years.

Now, suppose that we can compute the age at infection of men in the reference cohort, and that within (say) 1-year strata of age the target and reference cohorts had virtually identical distributions of incubation times. The age-specific estimates of risk would then be free of age confounding, and so could be used as unconfounded estimates of age-specific risk for men in the target cohort. If we wished to construct unconfounded estimates of average risk in the entire target cohort, we could do so using a method for *age adjustment*.

Methods for removing bias in estimates by taking account of a third variable (such as age) responsible for some or all of the bias are known as *adjustment* or *covariate control* methods. Standardization is perhaps the oldest and simplest example of such a method; more complex methods are based on multivariate models, which are discussed in the textbooks referenced in this chapter.

Unmeasured confounders

Were all confounders measured accurately, one could achieve comparison validity simply by adjusting for these confounders (although in attempting to do so various technical problems might arise). Nevertheless, in any non-randomized study we would ordinarily be able to think of a number of possible confounders that had not been measured, or had been measured only in a very poor fashion. In such cases, one may still be able to predict the direction of uncontrolled confounding by examining the manner in which persons were selected into the target and reference cohorts from the population at large. If the cohorts are derived from populations with different distributions of predictors of the outcome, or the predictors themselves are associated with admission differentially across the cohorts, these predictors will become confounders in the analysis.

To illustrate this problem, suppose that HIV-1 infection via an intravenous route (e.g. through needle sharing) leads to shorter incubation times than HIV-1 infection through sex-

ual activity. Suppose also that the reference cohort had excluded all or most intravenous drug users, whereas the target cohort was unselective in this regard. Then incubation times in the target cohort will on average be shorter than times in the reference cohort, due to the presence of intravenously infected persons in the target cohort. We should thus expect the results from the reference cohort to underestimate average risks of AIDS onset in the target cohort.

Random sampling and confounding

Suppose, for the moment, that our reference cohort had been formed by taking a random sample of the target cohort. Can predictions about the target made from such a random sample still be confounded? By our definition of confounding, the answer is yes. To see this, note for example that by chance alone men in our sample reference cohort could be younger on average than the total target; this age difference would in turn downwardly bias the unadjusted risk predictions if men had longer incubation times at younger ages.

Nevertheless, random sampling can help ensure that the distribution of the reference cohort is not too far from the distribution of the target cohort. In essence, the probability of severe confounding can be made as small as necessary by increasing the sample size. Furthermore, if random sampling is used, any confounding left after adjustment will be accounted for by the standard errors of the estimates, provided that the correct statistical model is used to compute the estimates and standard errors. We will examine the latter condition under the heading of specification validity.

Follow-up validity

In any cohort study covering an extended period of risk, subjects will be followed for different lengths of time. Some subjects will be lost to follow-up before the study ends. Others will be removed from study by competing risks, which in this setting are causes of death occurring before AIDS onset: fatal accidents, fatal myocardial infarctions, etc. And because subjects come under study at different times, those who are not lost to follow-up or removed by competing risks will still have experienced different lengths of follow-up when the study ends; traditionally, a subject still under follow-up at study end is said to have been 'withdrawn from study' at the time of study end.

Suppose we wish to estimate the average risk of AIDS onset within 5 years of infection. The data from a member of the references cohort who is observed not to develop AIDS but is also not followed the full 5 years from infection are said to be *censored* for the outcome of interest (AIDS within 5 years of infection). Consider, for example, a subject killed in a car crash 2 years after infection who had not contracted AIDS: the incubation time of this subject was censored at 2 years of follow-up.

All common methods for estimating risk from situations in which censoring occurs (e.g. person-years, life-table, and

Kaplan–Meier methods) are based on the assumption of follow-up validity. Follow-up validity means that over any span of follow-up time, risk of censoring is unassociated with risk of the outcome of interest. In our example, follow-up validity means that over any span of time following infection, risk of censoring (loss, withdrawal, or death before AIDS) is unassociated with risk of AIDS. Given follow-up validity, we can expect that, at any time t after infection, the distribution of incubation times will be the same for subjects lost or withdrawn at t and subjects whose follow-up continues beyond t.

Violations of follow-up validity can result in biased estimates of risk; such violations can be referred to as *follow-up bias* or *biased censoring*. To illustrate, suppose that younger reference subjects tend to have longer incubation times (i.e. lower risks) and are lost to follow-up at a higher rate than older reference subjects. In other words, lower-risk subjects are lost at a higher rate than higher-risk subjects. Then, after enough time, the average risk of AIDS in the observed portion of the reference cohort will tend to be higher than (overestimate) the average risk occurring in the full reference cohort (as the latter includes both censored and uncensored subject experience).

Note that the follow-up bias in the last illustration would not affect the age-specific estimates of risk (where 'age' refers to 'age at infection'). Consequently, the age bias in follow-up would not produce bias in age-standardized estimates of risk. More generally, if follow-up bias can be traced to a particular variable that is a predictor of both the outcome of interest and censoring, bias in the estimates can be removed by adjusting for that variable. Thus, some forms of follow-up bias can be dealt with in the same manner as confounding.

Specification validity

All statistical techniques (including so-called 'distribution-free' or 'non-parametric' methods) are derived by assuming the validity of a *statistical model* for the probabilities of observing the various possible data patterns. An estimate may be said to have *specification validity* if it is derived using a statistical model that is correct or nearly so. If the statistical model used for analysis is incorrect, the resulting estimates may be biased. Such bias is sometimes called *specification bias*, while the use of an incorrect model is known as *model misspecification* or *specification error*. Even when misspecifications does not lead to bias, it can lead to errors in statistical tests and confidence intervals. Minimizing such error consists largely of contrasting the statistical model against the data and against any available information about the processes generating the data (Greenland 1989).

Further details regarding specification validity are beyond the technical level of this chapter. Advanced treatments of specification validity in prediction problems can be found in Leamer (1978) and White (1989). Breslow and Day (1980, 1987) and Kleinbaum *et al.* (1982) provide introductory discussions of modelling in epidemiological analysis.

Measurement validity

An estimate from a study may be said to have measurement validity if it suffers from no bias due to errors in measuring the study variables. Unfortunately, there are sources of measurement error in nearly all studies, and nearly all sources of measurement error will contribute to bias in estimates. Thus evaluation of measurement validity focuses primarily on identifying sources of measurement error, and attempting to deduce the direction and magnitude of bias produced by these sources.

To aid in the task of identifying sources of measurement error, it may be useful to classify such errors according to their source. Errors from specific sources can then be classified further according to characteristics that are predictive of the direction of the resulting bias. One classification scheme divides errors into three major categories, according to their source:

1. Procedural error: errors arising from mistakes or defects in measurement procedures.
2. Proxy-variable error: errors arising from using a 'proxy' variable as a substitute for an actual variable of interest.
3. Construct error: errors arising from ambiguities in the definition of the variables.

Regardless of their source, errors may be divided into two basic types, *differential* and *non-differential*, according to whether or not the direction or magnitude of error depends on the true values of the study variables. Two different sources of error may be classified *dependent* or *independent*, according to whether or not the direction or magnitude of the error from one source depends on the direction or magnitude of the error from the other source. Finally, errors in continuous measurements may be factored into *systematic* and *random* components. As described in the following sections, these classifications have important implications for bias.

Procedural error

Procedural error is the most straightforward to imagine. It includes errors in recall when variables are measured through retrospective interview (e.g. mistakes in remembering all medications taken during pregnancy). It also includes coding errors, errors in calibration of instruments, and all other errors in which the target of measurement is well defined and the attempts at measurement are direct, but method of measurement is faulty. In our example, one target of measurement is HIV-1 antibody presence in blood. All available tests for antibody presence are subject to error (false negatives and false positives), and these may be considered procedural errors of measurement.

Proxy-variable error

Proxy-variable error is distinguished from procedural error in that use of proxies necessitates imputation, and hence virtually guarantees that there will be imputation error contributing to measurement error. In our example, we must impute the time of HIV-1 infection. For example, we might take as a proxy the 'infection time' computed as (say) 6 weeks before the midpoint between the last negative test and the first positive test for HIV-1 antibodies. Even if our HIV-1 tests are perfect, this measurement incorporates error if (as is certainly the case) time of infection does not always occur 6 weeks before the midpoint between the last negative and first positive tests.

Construct error

Construct error is often overlooked, although it may be a major source of error. Consider our example, in which the ultimate target of measurement is the time between HIV-1 infection and onset of AIDS. Before attempting to measure this time-span, we must unambiguously define the events that mark the beginning and end of the span. While infection may be reasonable to think of as a point event, AIDS onset is not. Various symptoms and signs may gradually accumulate, and then it is only by convention that some point in time is declared the start of the disease. If this convention cannot be translated into reasonably precise clinical criteria for diagnosing the onset of AIDS, the construct of 'incubation time' (the time-span between infection and AIDS onset) will not be well defined, let alone accurately measurable. In such situations it may be left to various clinicians to improvise answers to the question of time of AIDS onset—and this will introduce another source of extraneous variation into the final 'measurement' of incubation time.

Differential and non-differential error

Errors in measuring a variable are said to be differential when the direction or magnitude of the errors tends to vary across the true values of the study variables. Suppose, for example, that recall of drug use during pregnancy is enhanced among mothers of children with birth defects. Then a retrospective interview about drug use during pregnancy will yield results with differential error, since false-negative error will occur more frequently among mothers whose children have no birth defects.

Another type of differential error occurs in the measurement of continuous variables when the distribution of errors varies with the true value of the variable. Suppose, for example, that women more accurately recall the date of a recent Pap smear than the date of a more distant Pap smear. Then a retrospective interview to determine length of time since a woman's last Pap smear would tend to suffer from larger errors when measuring longer times.

Errors in measuring a variable are said to be non-differential with respect to another variable if the magnitude of errors does not vary with the true values of the other variable. As we will see below, procedures that yield only non-differential errors with respect to other study variables have important advantages over other procedures.

Dependent and independent error

Errors in measuring two variables are said to be *dependent* if the direction or magnitude of the errors made in measuring one of the variables is associated with the direction or magnitude of the errors made in measuring the other variable. If there is no association of errors, the errors are said to be *independent*.

In our example, errors in measuring age at HIV-1 infection and AIDS incubation time are dependent. Our measure of incubation time is equal to our measure of age at AIDS onset minus our measure of age at infection; hence overestimation of age at infection will contribute to underestimation of incubation time, and underestimation of age at infection will contribute to overestimation of incubation time. In contrast, in the same example it is plausible that the errors in measuring age at infection and age at onset are independent.

Independent non-differential errors have an important special property: the direction of bias produced by such errors is predictable. In particular, independent non-differential error in measuring two binary variables cannot inflate the association observed between the variables. In other words, any bias produced by independent non-differential error in measuring the variables can only be towards the null value of the association (which is 1 for a relative-risk measure). In situations involving more than two variables or variables with more than two levels, the effects of independent non-differential error are not always as straightforward; nevertheless, the direction of the bias produced by such error is still mathematically predictable.

To illustrate the two-variable situation, suppose that errors in measuring AIDS incubation time using a certain measurement protocol are non-differential with respect to sex (i.e. the direction and degree of error in the protocol is not affected by the sex of the subject). Then a study that uses the protocol will tend to underestimate the association of sex and incubation time, i.e. the non-differential errors will produce bias towards the null. We will later give an example involving three variables in which non-differential errors produce bias away from the null.

There is one important situation in which the assumption of independent non-differential measurement error has especially high plausibility: in a double-blind trial, successful blinding of treatment status during outcome evaluation should guarantee independence and non-differentiality of treatment and outcome measurement errors. Successful blinding thus helps insure that any bias produced by measurement error contributes to underestimation of treatment effects ('conservative bias').

Systematic and random components of error

For well-defined measurement procedures on continuous variables, measurement errors may be subdivided into *systematic* and *random* components. The systematic component (sometimes called the 'bias' of the measurement) measures the degree to which the procedure tends to underestimate or overestimate the true value on repeated application. The random component is the 'residual' error one is left with after subtracting the systematic component from the total error.

To illustrate, suppose that in our study HIV-1 infection time was unrelated to time of antibody testing, and that the average time of HIV-1 seroconversion was 8 weeks after infection. Then, even if one used a perfect HIV-1 test, a procedure that estimated infection time as 6 weeks before the midpoint between the last negative and first positive test would on average yield an estimated infection time that was 2 weeks later than the true time. Thus the systematic component of the error of this procedure would be +2 weeks. Since incubation time is onset time minus infection time, use of this procedure would add −2 weeks (i.e. a 2-week underestimation) to the systematic component of error in estimating incubation time.

Each of the components of an error, systematic and random, may be differential (i.e. may vary with other variable values) or non-differential, and may or may not be independent of errors (or components of errors) in other variables. We will not explore the consequences of the numerous possibilities, However, one important (but semantically confusing) fact is that, for certain quantities, independent and non-differential systematic components of error will not harm measurement validity, in that they will produce no bias in estimation.

To illustrate this, suppose in our example we wish to estimate the degree to which AIDS incubation time depends on age at HIV-1 infection. Suppose also that the systematic components of the measurements of incubation time and age of infection are −2 weeks and +2 weeks (as above), and do not vary with true incubation time and age at measurement (i.e. the systematic components are non-differential). Then the systematic components, being equal, will cancel out when one computes *changes* in incubation time and *changes* in age at infection. Since only these changes are used to estimate an association, the observed dependence of incubation time on age at infection will not be affected by the systematic components of error (although it may be biased towards the null by the random components if these are also independent and non-differential).

Summary of example

The example of this section provides an illustration of the most common threats to the validity of predictions. The unadjusted estimates of AIDS risk may be confounded if the target and reference cohorts differ in composition, and may also be biased by losses to follow-up or by use of an incorrect statistical model. Finally, our predictions are likely to be compromised by errors in measurements. These sources of error should be borne in mind if one attempts to predict AIDS incidence in populations newly infected by HIV-1.

Validity in causal inference

Concepts of valid prediction are applicable in evaluating studies of causation: comparison validity, follow-up validity, specification validity, and measurement validity must each be considered. In fact, as we shall see, problems of causal inference can be viewed as a special type of prediction problem, namely, prediction of what would happen (or what would have happened) to a population if certain characteristics of the population were (or had been) altered.

To illustrate validity issues in causal inference, we will consider the hypothesis that coffee drinking causes acute myocardial infarction (MI). This hypothesis can be operationally interpreted in a number of ways, for example:

1. There are people for whom the consumption of coffee results in their experiencing an MI sooner than they might have, had they avoided coffee.

While this is appealingly precise, it offers little practical guidance to an epidemiological researcher. The problem lies in our inability to recognize an individual whose MI was caused by coffee drinking. It is quite possible that MIs precipitated by coffee use are clinically and pathologically indistinguishable from other MIs. If so, the prospect of finding convincing physiological evidence concerning the hypothesis is not good.

We could overcome this impasse by examining a related epidemiological hypothesis, that is, a hypothesis that refers to the distribution of disease in populations. Here is one of many such hypotheses:

2. Among five-cups-a-day coffee drinkers, cessation of coffee use will lower the frequency of MI.

This form not only involves a population (five-cups-a-day coffee drinkers), it asserts that a mass action (coffee cessation) will reduce the frequency of the study disease. Thus the form of the hypothesis immediately suggests a strong test of the hypothesis: conduct a randomized intervention trial to examine the impact of coffee cessation on MI frequency.

This solution has some profound practical limitations, not the least of which would be persuading anyone to give up or take up coffee drinking to test a speculative hypothesis. Having ruled out intervention, we might consider an observational cohort study. In this case our epidemiological hypothesis should refer to natural conditions, rather than intervention. One such hypothesis is:

3. Among five-cups-a-day coffee drinkers, coffee use has elevated the frequency of MI.

There have in fact been a number of conflicting cohort and case–control studies of coffee and MI. The present discussion will be confined to the issues arising in the analysis of a *single* study. For a review of issues arising in the meta-analysis of multiple studies, using the coffee–MI literature as an example, see Greenland (1987). Additional references to the coffee–MI literature are given by Rosenberg *et al.* (1988).

Consider a 'classical' cohort study of coffee and first MI. At baseline, a cohort of people with no history of MI is assembled and classified into subcohorts according to coffee use, for example never-drinkers, ex-drinkers, occasional drinkers, one-cup-a-day drinkers, two-cups-a-day drinkers, etc. Other variables are measured as well: age, sex, smoking habits, blood pressure, serum cholesterol. Suppose that, at the end of (say) 10 years of monitoring this cohort for MI events, we compare the five-cups-a-day and never-drinker subcohorts, and obtain an unadjusted estimate of 1.22 for the ratio of the rates of first MI among five-cups-a-day drinkers and never-drinkers, with 95 per cent confidence limits of 1.00 and 1.49. In other words, it appears that the rate of MI among five-cups-a-day drinkers was 1.22 times higher than the rate among never-drinkers. (Hereafter, we will use the abbreviation 'MI' to mean 'first MI'.)

The estimated rate ratio of 1.22 may not seem large. Nevertheless, if it accurately reflects the impact of coffee use on the five-cups-a-day subcohort, this estimate implies that persons drinking five cups a day at baseline suffered a 22 per cent increase in their MI rate as a result of their coffee use. Given the high frequency of both coffee use and MI in many populations, this could represent a substantial health impact. We should therefore want to evaluate the validity of the estimate carefully.

As in the AIDS example, we may proceed by examining a series of conditions sufficient for validity of the estimate as a measure of coffee effect.

C Comparison validity. If the members of the five-cups-a-day subcohort had instead never drunk coffee, their distribution of MI events over time would have been approximately the same as the distribution among the never-drinkers.

F Follow-up validity. Within each subcohort, the risk of censoring (i.e. follow-up ended by an event other than MI) is not associated with the risk of MI.

Sp Specification validity. The distribution of MI events over time in the subcohorts can be closely approximated by the statistical model on which the estimates are based.

M Measurement validity. All measurements of variables used in the analysis closely approximate the true values of the variables.

These four conditions are sometimes called 'internal' validity conditions because they pertain only to estimating effects within the study, rather than to generalizing results to other studies. The following sections will explore the meaning of these conditions for an observational cohort study of a causal hypothesis. We will also discuss an important phenomenon known as *effect modification*, which is relevant to both internal validity and generalizability.

Comparison validity

In our example, comparison validity simply means that the distribution of MI times (or rates) among never-drinkers accurately predicts what would have happened in the coffee-drinking groups had the members of these groups never drunk coffee. Another way of stating condition C is that the five-cups and never-drinker subcohorts would be comparable or exchangeable with respect to MI times if no one had ever drunk coffee.

Despite this simplicity, note that the comparison validity condition depends on the hypothesis of interest in a very precise way. In particular, the research hypothesis (no. 3 above) is a statement about the impact of coffee among five-cups-a-day drinkers. Five-cups-a-day drinkers are thus the target cohort, while never-drinkers serve as the reference cohort for making predictions about this target.

To illustrate further the correspondence between comparison validity and the hypothesis at issue, suppose for the moment that our research hypothesis was:

4. Among never-drinkers, five-cups-a-day coffee use would elevate the frequency of MI.

In examining this hypothesis, the never-drinkers would be the target cohort and the coffee drinkers would be the reference cohort. Thus the comparison validity condition would have to be replaced by a condition such as:

C′ If the never-drinkers had drunk five cups of coffee per day, their distribution of MI times would have been approximately the same as the distribution among five-cups-a-day drinkers.

Other ways of stating condition C′ are that the five-cups and never-drinker subcohorts would be comparable or exchangeable with respect to MI times if everyone had been five-cups-a-day drinkers, and that the MI experience of five-cups drinkers accurately predicts what would have happened to the never-drinkers if the latter had drunk five cups a day.

Confounding

Failure to meet condition C results in a biased estimate of the effect of five-cups-a-day coffee drinking on five-cups-a-day drinkers, a condition sometimes referred to as confounding of the estimate. Similarly, failure to meet condition C′ results in a biased estimate of the effect that five-cups-a-day drinking would have had on never-drinkers.

To evaluate comparison validity, we must check whether the subcohorts differed at baseline in any factors that influence MI time. If so, we could not reasonably expect the MI distributions of the subcohorts to be comparable, even if the subcohorts had the same level of coffee use. In other words, we could not expect condition C (or C′) to hold, and so our estimates would suffer from confounding.

For our example, we may note that several studies have found a positive association between cigarette smoking (an established risk factor for MI) and coffee use (Greenland 1987). It also seems *a priori* sensible that a person habituated to a stimulant such as nicotine would be attracted to coffee use as well. Thus we should expect to see a higher prevalence of smoking among coffee users in our study.

Suppose then that, in our cohort, smoking is more prevalent among five-cups-a-day subjects than never-drinkers.

This elevated smoking prevalence should have led to elevated MI rates among five-cups-a-day drinkers, *even if coffee had no effect*. More generally, we should expect the MI rate among never-drinkers to underestimate the MI rate that five-cups-a-day drinkers would have had if they had never drunk coffee. The result would be an inflated estimate of the impact of coffee on the MI rate of five-cups-a-day drinkers. Similarly, we should expect the MI rate among five-cups-a-day drinkers to overestimate the MI rate that never-drinkers would have had if they had drunk five cups a day.

Adjustment for measured confounders

As in the prediction problem, we may stratify the data on potential confounders, with the objective of creating strata within which confounding is minimal or absent. We can also employ adjustment methods, such as standardization, to remove confounding from estimates of overall effect.

Standardization is appealingly simple in both justification and computation. Unfortunately, if the number of cases occurring within the confounder categories tends to be small (less than five or so), the technique will be subject to various technical problems, including possible bias. One remedy is to broaden confounder categories, but this is likely to result in incomplete control of confounding. To avoid these problems, many, if not most, researchers attempt to control confounding using a multivariate model. This remedy has problems of its own, chiefly in that it involves much greater risks of misinterpretation and specification bias (Greenland 1989).

Unmeasured confounders

Among the possible confounders not measured in our hypothetical study are diet and exercise. Suppose 'health conscious' subjects who exercise regularly and eat low-fat diets also avoid coffee and have lower rates of MI. The result will be a concentration of these low-risk subjects among coffee non-users, and a consequent overestimation of coffee's effect on risk.

Confounding by unmeasured confounders can sometimes be minimized by controlling variables along pathways of the confounders' effect. If, for example, exercise and low-fat diet lowered MI risk only by lowering serum cholesterol and blood pressure, control of serum cholesterol and blood pressure would remove confounding by exercise and dietary fat. Unfortunately, such control may generate bias if the controlled variables are also intermediates between our study variable and our outcome variable.

Intermediate variables

In effect estimation, we must take care to distinguish *intermediate* variables from confounding variables. Intermediate variables represent steps in the causal pathway from the study exposure to the outcome event. The distinction is essential, for control of intermediate variables can *increase* the bias of estimates.

To illustrate, suppose that coffee use affects serum cholesterol levels (as suggested by the results of Curb *et al.* 1986).

Then, given that serum cholesterol affects MI risk, serum cholesterol is an intermediate variable for the study of coffee effects on MI risk. Now, suppose that we stratify our cohort data on serum cholesterol levels. Some coffee drinkers will be in elevated cholesterol categories because of coffee use, and so will be at elevated MI risk because of coffee effects. Yet these subjects will be compared with never-drinkers in the same stratum, who are also at elevated risk due to their elevated cholesterol. Therefore, the effect of coffee on MI risk via the cholesterol pathway will not be apparent within the cholesterol strata, and so cholesterol adjustment will contribute to underestimation of the coffee effect on MI risk. Analogously, if coffee affected MI risk by elevating blood pressure, blood-pressure adjustment will contribute to underestimation of the coffee effect. Such underestimation may be termed 'over-adjustment bias'.

Intermediate variables may also be confounders, and thus present the investigator with a severe dilemma. Consider that most of the variation in serum cholesterol levels is *not* due to coffee use, and that much (perhaps most) of the association between coffee use and cholesterol is not due to coffee effects, but rather to factors associated with both coffee and cholesterol (such as exercise and dietary fat). This means that serum cholesterol may also be viewed as a confounder for the coffee–MI study, and that estimates unadjusted for serum cholesterol will be biased unless they are also adjusted for the factors contributing to the coffee–cholesterol association.

Suppose a variable is both an intermediate and a confounder. It will usually be impossible to determine how much of the change in the effect estimate produced by adjusting for the variable is due to introduction of over-adjustment bias and how much is due to removal of confounding. Nevertheless, a qualitative assessment may be possible in some situations. For example, if we know that the effects of coffee on serum cholesterol are weak, and that most of the association between coffee and serum cholesterol is due to confounding of this association by uncontrolled factors (such as exercise and diet), we can conclude that the cholesterol-adjusted estimate is the less biased of the two. Alternatively, if we have accurately measured all the factors that confound the coffee–cholesterol association, we may control these factors instead of cholesterol to obtain an estimate free from both over-adjustment bias and confounding by cholesterol.

Randomization and confounding

Suppose, for the moment, that level of coffee use in our cohort had been assigned by randomization, and that the participants diligently consumed only their assigned amount of coffee. Could our estimates of coffee effects from such a randomized trial still be confounded? By our definition of confounding, the answer is yes. To see this, note for example that by chance alone the five-cups-a-day drinkers could be older on average than the never-drinkers; this difference would in turn result in an upward bias in the unadjusted estimate of the effect of five cups a day, since age is an important risk factor for MI.

Nevertheless, randomization can help to ensure that the distributions of confounders in the different exposure groups are not too far apart. In essence, the probability of severe confounding can be made as small as necessary by increasing the size of the randomized groups. Furthermore, if randomization is used, and confounding left after adjustment will be accounted for by the standard errors of the estimates, provided that the correct statistical model is used to compute the effect estimates and their standard errors.

Follow-up validity

In our example, follow-up validity means that follow-up is valid within every subcohort being compared. In other words, over any span of time during follow-up, MI risk within a subcohort is unassociated with censoring risk in the subcohort. Given follow-up validity, we can expect that at any follow-up time t the MI rates in a subcohort will be the same for subjects lost or withdrawn at t and subjects whose follow-up continues beyond t.

We should in fact expect follow-up up to be biased by cigarette smoking: smoking is associated with mortality from MI and many causes besides MI; the association of smoking with socio-economic status might also produce an association between smoking and loss-to-follow-up. The result would be elevated censoring among high-risk (smoking) subjects. As a consequence, the unadjusted estimates of MI risks will underestimate the MI risks in the complete subcohorts (as the latter includes both censored and uncensored subject experience). If the degree of underestimation varies across subcohorts, bias in the relative-risk estimates will result.

In fact, the degree of underestimation should vary in this example because of the variation in smoking prevalence across subcohorts. Nevertheless, variation in smoking prevalence is *not* necessary for smoking-related censoring to produce biased estimates of absolute effect. If, for example, smoking-related censoring produced a uniform 15 per cent underestimation of the MI rate in each subcohort, all rate differences would also be underestimated by 15 per cent.

Analogously to control of confounding, any bias produced by the association of smoking with MI and censoring can be removed by smoking adjustment. As before, if adjustment is by standardization, the standard distribution should be chosen from the target subcohort.

Because the same correction methods can sometimes be applied, some authors (e.g. Miettinen 1985) classify follow-up bias as a form of confounding. Nevertheless, the two phenomena are reversed with respect to the causal ordering of the third variable responsible for the bias: confounding arises from an association of the study exposure (coffee use) with other exposures (such as smoking) that affect outcome risk; in contrast, follow-up bias arises from an association between the risk of the study outcome (MI) and risks of other endpoints (such as other-cause mortality or loss to follow-up) that are affected by exposure. Furthermore, certain forms of follow-up bias cannot be removed by adjustment. These problems are discussed under the heading of *dependent competing risks* in the statistics literature; see Kalbfleisch and Prentice (1980) and Slud and Byar (1988) for discussions of this issue.

Some authors (e.g. Kelsey *et al.* 1986) classify follow-up bias as a form of selection bias. Here, we reserve the latter term for a special problem of case–control studies (discussed below).

Specification validity

As noted earlier, use of a statistical method based on an incorrect model (specification error) can lead to bias in estimates and improper performance of statistical tests and interval estimates. All statistical techniques, including 'non-parametric' methods, must assume some sort of model for the process generating the data; yet, in the absence of randomization or random sampling, one will rarely be able to identify a 'correct' model. Thus we should expect some degree of specification error in typical epidemiological studies. Minimization of specification error must rely on checking the model against the data and against background information about the processes generating the data. For further discussion of these points, see Greenland (1989).

Measurement validity

The continuous variables of coffee use, cigarette use, blood pressure, cholesterol, and age are *time-dependent covariates*. With the exception of age (whose value at any time can be computed from birth-date), this fact adds considerable complexity to measuring these variables and estimating their effects.

Consider that we cannot reasonably expect a single baseline measurement, no matter how accurate, to summarize adequately a subject's entire history of coffee drinking, smoking, blood pressure, or cholesterol. Even if the effect of a subject's history could be largely captured by using a single summary number (for example, total number of cigarettes smoked), the baseline measurement may well be a poor proxy for this ideal and unknown summary. For these reasons, we should expect that proxy-variable errors would be very large in our example.

Proxy-variable error in the study factor

The degree of proxy-variable error in measuring the study variable depends on the exact definition of the variable that we wish to study. This definition should, in turn, reflect the hypothesized effect we wish to study. To illustrate, consider the following 'acute effect' hypotheses:

Drinking a cup of coffee produces an immediate rise in short-term MI risk. In other words, coffee consumption is an acute risk factor.

Note that this hypothesis does not exclude the possibility that coffee use also elevates long-term risk of MI, perhaps through some other mechanism; it simply does not address the issue of chronic effects.

One way to examine the hypothesis would be to compare the MI rates among person-days in which one, two, three, etc. cups were drunk with the MI rate among person-days in which no cups were drunk (adjusting, of course, for confounding and follow-up bias). If we had only baseline data, however, baseline daily consumption would have to serve as the proxy for consumption on every day of follow-up. This would probably be a poor proxy for daily consumption at later follow-up times, where more outcome events occur. It turns out that a 'standard' analysis, which simply examines the association of baseline coffee use with MI rates, is equivalent to an analysis that uses baseline consumption as a proxy for consumption on all later days. Thus, estimates from a 'standard' analysis would suffer large bias if considered as estimates of acute coffee effect.

Note that the proxy-variable error in the last example could easily be differential with respect to the outcome: person-days accumulate more rapidly in early follow-up, where the error from using baseline consumption as the proxy is relatively low; in contrast, MI events accumulate more rapidly in late follow-up, where the error is probably higher. This illustrates an important general point: errors in variables can be differential, even if the variables are measured before the outcome event. Such phenomena occur when errors are associated with risk factors for the outcome; in our example, the error is associated with follow-up time and, hence, age. Such associations are in turn likely to occur when measurements are based on proxy variables.

Suppose now that we examine the following 'chronic effect' hypothesis:

Each cup of coffee drunk eventually results in long-term elevation of MI risk.

This hypothesis was suggested by reports that coffee drinking produces a rise in serum lipid levels (see Curb *et al.* 1986), and so may accelerate coronary stenosis. Note that this hypothesis does not address the issue of acute effects.

One way to examine the preceding hypothesis would be to compare the MI rates among person-months with different cumulative doses of coffee (perhaps using a lag period in calculating dose, e.g. one might ignore the most recent month of consumption). If we had only baseline data, however, baseline daily consumption would have to be used to construct a proxy for cumulative consumption at every month of follow-up. This could be done in several different ways: for example, one could estimate a subject's cumulative dose up to a particular date by multiplying their baseline daily consumption by the number of days they lived between the age of 18 years and the date in question. This estimate assumes that coffee drinking began at age 18 years, and the baseline daily consumption is the average daily consumption since that age. We should expect considerable error in such a crude measure of cumulative consumption.

The degree of bias in estimating chronic effects could be quite different from the degree of bias in estimating acute effects. Furthermore, as discussed below, the errors in each

proxy will make it virtually impossible to discriminate between acute and chronic effects.

Measurement error and confounding

If a variable is measured with error, estimates adjusted for the variable as measured will still be somewhat confounded by the variable. This residual confounding arises because measurement error prevents us from constructing strata that are internally homogeneous with respect to the true confounding variable.

To illustrate, consider baseline daily cigarette consumption. This variable may be considered a proxy for consumption on each day of follow-up, or may be used to construct an estimate of cumulative cigarettes consumption (analogous to the cumulative coffee variable discussed above).

Suppose that we stratify the data on a cumulative smoking index constructed from the baseline smoking measurement. Within any stratum of the index, there would remain a broad range of cumulative cigarette consumption. For example, two subjects who were aged 40 years and smoked one pack a day at baseline would receive the same value for the smoking index and so end up in the same stratum. Yet, if one of them quit smoking immediately after baseline, while the other continued to smoke a pack a day, after 10 years of follow-up the quitter would have 10 less pack-years of cigarette consumption than the continuing smoker.

Suppose now that cumulative cigarette consumption is positively associated with cumulative coffee consumption. Then, even within strata of the smoking index, we should expect subjects with high coffee consumption to exhibit elevated MI rates simply by virtue of having higher levels of cigarette consumption. As a consequence, the estimate of coffee effect adjusted for the smoking index would still contain some confounding by cumulative cigarette consumption.

In some cases, a study variable may appear to have an effect only because of poor measurement of an apparently unimportant confounder. This can occur, for example, when an important confounding variable is measured with a large amount of non-differential error. Such error reduces the apparent association of the variable with the exposure, and makes the variable appear to be a weak risk factor, perhaps weaker than the study exposure. This in turn makes the variable appear to be only weakly confounding, in that adjustment for the variable as measured produces little change in the result. What is actually occurring, however, is that adjustment for the variable as measured eliminates little of the confounding by the variable.

To illustrate this phenomenon, suppose that coronary proneness of personality is measured only by the baseline yes/no question, 'Do you consider yourself a hard-driving person?' Such a crude measure of the original construct would be unlikely to show more than a weak association with either coffee use or MI, and adjusting for it would produce little change in our estimate of coffee effect. Suppose, however, that coronary-prone personalities have an elevated preference for coffee. Such a phenomenon would lead to a concentration of coronary-prone persons (and hence a spuriously elevated MI rate) among coffee drinkers, even after stratification on response to above question.

Measurement error and separation of effects

Because of their impact on the effectiveness of adjustment procedures, measurement errors can severely reduce our ability to separate different effects of the study variable. Suppose that, in our example, we wished to estimate the relative strength of acute and chronic coffee effects. To do so, we must take account of the fact that acute and chronic effects will be confounded. Thus, when examining acute effects, person-days with high coffee consumption will occur most frequently among persons with high cumulative coffee consumption. As a consequence, if cumulative coffee consumption is a risk factor, it will be a confounder for estimating the acute effects of coffee consumption. By similar arguments, if coffee consumption has acute effects, these will confound estimates of the chronic effects of cumulative consumption.

Unfortunately, both cumulative and daily consumption are measured with considerable error. As a result, any effect observed for one may be wholly or partially due to the other, even if the other has little or no apparent effect.

Repeated measures

One costly but effective method for reducing the degree of proxy-variable error in measuring time-dependent variables is to take repeated (serial) measurements over the follow-up period, and ask subjects to report their pre-baseline history of such variables at the baseline interview. In our example, subjects could be asked about their age at first use and level of consumption at different ages for coffee and cigarettes, and be recontacted every year or two to assess their current consumption. Of course, not all subjects may be willing to co-operate with such active follow-up, but the penalties of some extra loss may be far outweighed by the benefit of improved measurement accuracy.

Errors in assessing incidence

An especially important form of measurement error in assessing incidence is misdiagnosis of the outcome event. In the AIDS example, a false-positive diagnosis would result in underestimation of incubation time, while a false-negative diagnosis would result in overestimation. In the present example, false-positive errors would result in overestimation of MI rates, while false-negative errors would result in underestimation of MI rates. These errors will be of particular concern when the study depends on existing surveillance systems or records for detection of outcome events. There are some special cases, however, in which the errors will induce little or no bias in estimates (Poole 1985).

If the only form of misdiagnosis is false-negative error, the proportion of outcome events missed in this fashion is the same across cohorts, and there is no follow-up bias, the relative-risk estimates will not be distorted by the underdiagnosis. Suppose in our example that all recorded MI events are true MIs, but in each subcohort 10 per cent of MIs are missed. The MI rates in each subcohort will then be underestimated by 10 per cent; nevertheless, if we consider any two of these rates, say R_0 and R_5, the observed rate ratio will be:

$$0.9R_5/(0.9R_0) = R_5/R_0$$

which is undistorted by the underdiagnosis of MI. On the other hand, if coffee primarily induced 'silent' MIs and these were most frequently undiagnosed, the coffee effect would be underestimated.

In an analogous fashion, if the only form of misdiagnosis is false-positive error, the rate of false positives is the same across cohorts, and there is no follow-up bias, rate differences will not be distorted by the overdiagnosis. Suppose in our example that the rate of false positives is R_f in all subcohorts; then if we consider any two true rates, say R_0 and R_5, the observed rate difference will be:

$$(R_5 + R_f) - (R_0 + R_f) = R_5 - R_0$$

which is undistorted by the overdiagnosis of MI. On the other hand, if (as is likely the case in our example) there is non-differential underdiagnosis of MI, the rate difference will be underestimated.

Effect modification (heterogeneity of effect)

Estimation of effects usually requires consideration of *effect modification*, also known as effect variation or heterogeneity of effect. As an example, suppose that drinking five cups of coffee a day elevated the MI rate of men in our cohort by a factor of 1.40 (i.e. a 40 per cent increase), but elevated the MI rate of women by a factor of only 1.10 (i.e. a 10 per cent increase). This situation would be termed modification (or variation, or heterogeneity) of the rate ratio by sex, and sex would be called a *modifier* of the coffee–MI rate ratio.

As another example, suppose that drinking five cups of coffee a day elevated the MI rate in men in our cohort by a factor of 400 cases per 100 000 person-years, but elevated the MI rate in women by a factor of only 40 cases per 100 000 person-years. This situation would be termed modification of the rate difference by sex, and sex would be called a *modifier* of the coffee–MI rate difference.

As a final example, suppose that drinking five cups of coffee per day elevated the MI rate in our cohort by a factor of 1.22 in both men and women. This situation would be termed *homogeneity* of the rate ratios across sex.

Note that effect modification and homogeneity are *not* absolute properties of an effect, but are instead properties of the way one measures the effect. For example, suppose that drinking five cups of coffee per day elevated the MI rate in men from 1000 cases per 100 000 person-years to 1220 cases per 100 000 person-years, but elevated the MI rate in women from 400 cases per 100 000 person-years to 488 cases per 100 000 person-years. Then the sex-specific rate ratios would both be 1.22, homogenous across sex, whereas the sex-specific rate differences of 220 cases per 100 000 person-years for men and 88 cases per 100 000 person-years for women would be heterogeneous or 'modified' by sex. Examples such as this show that one should not equate 'effect modification' with biological concepts of interaction, such as synergy or antagonism.

Effect modification can be analysed by stratifying the data on the potential effect modifier under study, estimating the effect within each stratum, and comparing the estimates across strata. There are several potential problems with this approach: the number of subjects in each stratum may be too small to produce stable estimates of stratum-specific effects, especially after adjustment for confounder effects. Estimates may fluctuate wildly from stratum to stratum due to random error. A related problem is that statistical tests for heterogeneity in stratified data have extremely low power in many situations, and so are likely to miss much if not most of the heterogeneity when used with conventional significance levels (such as 0.05). These problems can be partially addressed by modelling effect modification (see Greenland 1989). Unfortunately, the amount of bias from confounding, measurement error, etc. may vary from stratum to stratum, in which case both the observed and modelled pattern of modification will be biased.

Effect modification and generalizability

Suppose that we succeed in obtaining approximately unbiased estimates from our study. We may then confront issues of *generalizability* (external validity) of our results. For example, we may ask whether they accurately reflect the effect of coffee on MI rates in a new target population. We may view such a question as a prediction problem, one in which the objective is to predict the strength of coffee effects in the new target population. From this perspective, generalizability of an effect estimate involves just one validity issue in addition to those discussed so far, namely confounding of the predicted effect by effect modifiers.

Suppose that, among five-cups-a-day drinkers in both our study cohort and the new target, the rate increase (in cases per 100 000 person-years) produced by coffee use is 400 for men and 40 for women. If, however, our study cohort is 70 per cent male while the new target is only 30 per cent male, the *average* increase among five-cups-a-day drinkers in our study cohort would be $0.7(400) + 0.3(4) = 292$, whereas the average increase in the new target would be only $0.3(400) + 0.7(40) = 148$. Thus any valid estimate of the average increase in our study cohort will tend greatly to overestimate the average increase in the new target. In other words, modification of the effect of coffee by sex confounds the prediction of its effect in the new target. This bias can be avoided by making only sex-specific predictions of effect, or by standardizing the study results to the sex distribution of the new target population.

Summary of example

The example of this section provides an illustration of the most common threats of the validity of effect estimates from cohort studies. The unadjusted estimates of coffee effect on MI will be confounded by many other variables (such as smoking), and there will be follow-up-bias. As a result, the number of variables that must be controlled is too large to allow adequate control using only stratification. The true functional dependence of MI rates on coffee and the confounder is unknown, so that estimates based on multivariate models are likely to be biased. Even if this bias is unimportant, our estimates will remain confounded because of our inability to measure the key confounders acurately. Finally, our inability to summarize coffee consumption accurately would further bias our estimates, and make it impossible to reliably separate acute and chronic effects of coffee use.

Given that there are several sources of bias of unknown magnitude and different directions, it would appear that no conclusions about coffee effect could be drawn from a study like the one described above, other than perhaps that coffee does not have a large effect. This type of result—inconclusive, other than to rule out very large effects—is a common result in thorough epidemiological analyses of observational data. In particular, it is an exceedingly common result when the data being analysed were collected for purposes other than to address the hypothesis at issue, for in such cases the data often lack accurate measurements of key variables.

Case–control validity

The practical difficulties of mounting cohort studies have led to extensive development of case–control study designs. The distinguishing feature of such designs is that sampling is intentionally based on the outcome of individuals.

In a population-based or *population-initiated* case–control study, one first identifies a *population at risk* of the outcome of interest, which is to be studied over a specified period of time or *risk period*. As in a follow-up study (such as a traditional cohort study), one attempts to ascertain outcome events in the population at risk; but, unlike a follow-up study, one selects only persons experiencing the outcome event (cases) and a 'control' sample of the entire population at risk for acertainment of exposure and covariate status.

In a case-initiated case–control study, one starts by identifying a source of study cases (for example, a hospital emergency room is a source of MI cases). One then attempts to identify a population at risk such that the source of cases provides a random or complete sample of all cases occurring in this population. Study cases recruited from the source occur over a risk period; controls are selected in order to ascertain the distribution of exposure in the population at risk over that period.

Case–control studies may also begin with an existing series of controls (Greenland 1985). Regardless of how a case–control study is initiated, evaluation of validity must ultimately refer to a population at risk that represents the target of inference for the study.

Relative-risk estimation in case–control studies

The control sample may or may not be selected in a manner that excludes cases. If persons who become cases over the risk period are ineligible from inclusion in the control group (as in traditional case–control designs) a 'rare-disease' assumption may be needed to estimate relative risks from the case–control data. If, however, persons who become cases over the risk period are also ineligible for inclusion in the control group (as in newer case–control designs), the rare-disease assumption may be discarded. These points are discussed in several textbooks, including Kleinbaum *et al.* (1982), Miettinen (1985), and Rothman (1986).

The basics of case–control estimation will be illustrated with the following example. We wish to study the effect of coffee drinking on rates of first MI, and we have selected a population for study (for example, all residents aged 40–64 years in a particular town) over a 1-year risk period. At any point during the risk period, the population at risk comprises persons in this selected population who have not yet had an MI.

Suppose that, over the risk period, the average number of never-drinkers in the population at risk was 20 000; the average number of five-cups-a-day drinkers was 10 000; there were 120 first MIs among never-drinkers; and there were 90 first MIs among five-cups-a-day drinkers. Then, if one observed the entire population without error, the estimated rates among never-drinkers and five-cups-a-day drinkers would be:

$$\frac{120}{20\,000 \text{ person-years}} \text{ and } \frac{90}{10\,000 \text{ person-years}}$$

Thus, if we observed the entire population, the estimated relative risk would be:

$$\frac{90/10\,000 \text{ person-years}}{120/20\,000 \text{ person-years}} = \frac{90/120}{10\,000/20\,000} = 1.50$$

This estimate depends on only two figures: the relative prevalence of five-cups-a-day versus never-drinkers among cases (90/120), and the same relative prevalence in the person-years at risk (10 000/20 000).

The first (numerator) relative prevalence could be estimated by interviewing an unbiased sample of all the new MI cases that occur over the risk period, and the second (denominator) relative prevalence could be estimated by interviewing an unbiased sample of the population at risk over the risk period. The ratio of relative prevalence from the case and control sample interviews would then be an unbiased estimate of the population rate ratio of 1.50.

Note well three points about the preceding argument: first, no rare-disease assumption was made. Second, the control sample of the population at risk is accumulated over the entire risk period (rather than at the end of the risk period); such sampling has come to be called 'density sampling'.

Third, because of the density sampling, someone may be selected for the control sample, and later in the risk period have an MI and become part of the case sample as well. Specific methods for carrying out density sampling may be found in the textbooks referenced earlier.

Validity conditions in case–control studies

The primary advantages of case–control studies are their short time-frame and large reduction in the number of subjects needed to achieve the same statistical power as a cohort study. The primary disadvantage is that more conditions must be meet to ensure the validity of case–control studies (in addition to the four listed in the cohort study example).

Suppose that our case–control study data yield an unadjusted relative-risk estimate (odds ratio) of 1.50, with 95 per cent confidence limits of 1.00 and 2.25. In other words, it appears that the rate of MI in the subpopulation of five-cups-a-day drinkers was 1.50 times higher than the rate in the subpopulation of never-drinkers. The following series of conditions would be sufficient for the validity of this figure as an estimate of the effect of drinking five cups a day (versus none) on the MI rate:

C Comparison validity. If five-cups-a-day drinkers in the population at risk had instead drunk no coffee, their MI rate would have been approximately the same as the MI rate among never-drinkers.

F Follow-up validity. Censoring risk (risk of emigration from the population plus the risk of death from other causes) is either the same across coffee-use subpopulations, or else is unassociated with MI risk within subpopulations.

Sp Specification validity. The MI rates in the subpopulations can be closely approximated by the statistical model on which the estimates are based.

M Measurement validity. All measurements of variables used in the analysis closely approximate the true values of the variables.

Se Selection validity. This has two components:

1. Case-selection validity. If one studies only a subset of the MI cases occurring in the population over the risk period (for example, because of failure to detect all cases), this subset provides unbiased estimates of the prevalence of different levels of coffee use among all cases occurring in the population over the risk period.

2. Control-selection validity. The control sample provides unbiased estimates of the prevalence of different levels of coffee use in the population at risk over the risk period.

Issues of comparison validity, follow-up validity, specification validity, effect modification, and generalizability in case–control studies parallel those in follow-up studies, and so will not be discussed here. Since, however, case–control estimation is usually limited to relative risk, only censoring that is associated with both coffee use and the MI rate need be of concern when evaluating follow-up validity.

Case–control studies are vulnerable to certain problems of measurement error that are less severe or do not exist in prospective cohort studies. We will discuss these problems first, and then examine selection validity. Finally, we will briefly discuss analogous issues in retrospective cohort studies.

Retrospective ascertainment

A special class of measurement errors arises from *retrospective ascertainment* of time-dependent variables, i.e. attempting to measure past values of the variables. Retrospective ascertainment must be based on individual memories, existing records of past values, or some combination of the two. Therefore such ascertainment usually suffers from faulty recall, missing or mistaken records, or lack of direct measurements in existing records.

Retrospective ascertainment may be an important component of a cohort study. For example, the cohort study of coffee and MI discussed above could have been improved by asking subjects about their coffee use and smoking before the start of follow-up. This information would allow one to construct better cumulative indices than could be constructed from baseline consumption alone, although the resulting indices would still incorporate error due to faulty recall.

Unless records of past measurements are available for all subjects, measurements on cases and controls *must* be made after the time period under study, since subjects are not selected for study until after that period. Thus, unlike cohort studies, most case–control studies of time-dependent variables depend on retrospective ascertainment.

Considering our example, there may be much more error in determining daily coffee consumption 10 years before interview than 1 month before interview. Given this, one might then expect that case–control studies are more accurate for studying acute effects than chronic effects. If, however, acute and chronic effects are heavily confounded, the elevated inaccuracies of long-term recall will make it impossible to disentangle short-term from long-term effects. As illustrated earlier, this confounding can arise in a cohort study. Nevertheless, in a cohort study such confounding can be minimized by taking repeated measurements. In contrast, such confounding would be unavoidable in a case–control study based on recall, even if detailed longitudinal histories were requested from the subjects.

The preceding observations should be tempered by observing that some case–control studies have access to exposure measurements of the same quality as found in cohort studies, and that the exposure measurements in some cohort studies may be no better than those used in some case–control studies. For example, a cohort study in which measurements are derived by abstracting routine medical records would suffer from no less measurement error than a case–control study in which measurements are derived by abstracting the same records.

Outcome-affected measurements

One problem that occurs in case–control but not in cohort studies is *outcome-affected recall*, often termed recall bias. These terms refer to the differential measurement error that originates when the outcome event affects recall of past events. Examples occur in case–control studies of birth defects, in which it may happen that the trauma of having an affected child either enhances recall of prenatal exposures among case mothers, or increases the frequency of false-positive reports among case mothers. In either situation, estimates of relative risk will be upwardly biased by effects of the outcome on recall.

One commonly proposed method for preventing bias due to outcome-affected recall is to restrict controls to a group believed to have recall similar to the cases. Unfortunately, one usually cannot tell to what degree this restricted selection corrects the bias from outcome-affected recall. Even more unfortunately, one usually cannot tell if the selection bias produced by such restriction is worse than the recall bias one is attempting to correct.

A problem similar to outcome-affected recall occurs when the outcome event affects a psychological or physiological measurement. This is of particular concern in case–control studies of nutrient levels and chronic disease. For example, if colon cancer leads to a drop in serum retinol levels, the relative risk for the effect of serum retinol will be underestimated if serum retinol is measured after the cancer develops. Errors of this type may be viewed as proxy-variable errors, in which the post-outcome value is a poor proxy for the pre-outcome value of interest.

Selection validity

Selection validity is straightforward to understand but can be extraordinarily difficult to verify. A violation of the selection-validity conditions is known as *selection bias*; many case–control design and field methods are devoted to avoiding such bias (see Schlesselman 1982). In some instances, it may be possible to identify a factor or factors that affect the chance of selection into the study. If in such instances we have accurate measurements of one of these factors, we can stratify on (or otherwise adjust for) the factor and thereby remove the selection bias due to the factor. Because of this possibility, some authors classify selection bias as a form of confounding. Nevertheless, there are some forms of selection bias that cannot be removed by adjustment. These points will be illustrated in the following subsections.

Case-selection validity

Unbiased selection of cases can be best assured if one can identify every case that occurs in the population at risk over the risk period. This requires a *surveillance system* for the outcome of interest, such as a population-based disease registry. In our example, we would probably have to construct our own MI surveillance system from existing resources, such as emergency-room admission records, ambulance-service records, and paramedical records.

Whether or not we can identify all cases of interest, selection bias may arise from failure to obtain information on all cases. In our example, many patients would be dead before interview was possible. For such cases, there are only two alternatives: (1) attempt to obtain information from some other source, such as next-of-kin or co-workers; or (2) exclude such cases from the study. The first alternative increases measurement error and the resulting bias in the study.

The second alternative will introduce bias if coffee affects risk of fatal Mi and risk of non-fatal MI differently, or if coffee affects risk of MI *survivorship*. To illustrate, suppose that coffee drinking reduced one's chance of reaching the hospital alive when an MI occurred. Then the prevalence of coffee use among MI survivors would under-represent the prevalence among all MI cases. Underestimation of the relative risk would result if only MI survivors were included in the study.

Note that, in the previous example, we could redefine the study so that the case selection bias did not exist by redefining the study outcome as 'non-fatal myocardial infarction'. Unfortunately, this does *not* remove the bias, but leads only to its reclassification as a bias due to dependent competing risks (here classified as a form of follow-up bias). In a study of non-fatal MI, fatal MI is a censoring event that is associated with risk of non-fatal MI; if fatal MI is also associated with coffee use, the result will be underestimation of the relative risks for non-fatal MI. More generally, it is usually not possible to remove bias by placing restrictions on admissible outcomes.

Unfortunately, exclusion is the only alternative for patients that refuse to participate or cannot be located. In our example, if such cases tend to be heavier coffee users than others, underestimation of the relative risk would result. Suppose however that, within levels of cigarette use, such cases were no different from other cases with respect to coffee use. Then adjustment for smoking would remove the selection bias induced by refusals and locating failures. (Such adjustment would of course require accurate smoking measurement, which is a problem in itself.)

Bias that arises from failure to detect certain cases is sometimes called *detection bias*. If our surveillance system used only hospital admissions, many out-of-hospital MI deaths would be excluded, and a detection bias of the sort described above could result.

Control-selection validity

Unbiased selection of controls can be best assured if, at every time during the risk period, one can potentially identify every member of the population at risk. In such a situation one could select controls with one of many available probability sampling techniques, using the entire population at risk as the sampling frame. Unfortunately, such situations are exceptional.

Many studies attempt to approximate the ideal sampling situation through use of existing population lists. An example is control selection by random-digit dialling. Here, the list (of residential phone numbers) is not used directly but nevertheless serves as a partial enumeration of the population at risk. This list excludes persons without phone numbers. In our example, if persons without telephones drink less coffee than persons with telephones, a control group selected by random-digit dialling would over-represent coffee use in the population at risk. The result would be underestimation of the relative risk.

Note that, in the previous example, one could redefine the population-at-risk so that the telephone-related selection bias did not exist by restricting the study to persons with telephones. This would require excluding persons without telephones from the case series. The resulting relative risk estimate would suffer no selection bias. The only important penalty from this restriction is that the resulting estimate might apply only to the population of phone users; this, however, is a problem of generalizability rather than of selection validity. More generally, it is often possible to prevent confounding or selection bias by placing restrictions on the population at risk (and hence the control group); in such instances, however, one must take care to apply the same restrictions to the case series.

Whether or not we can identify all members of the population at risk, selection bias may arise from failure to obtain information on all person selected as controls. The implications are the reverse of those for case-selection bias. In our example, if controls who refuse to participate or cannot be located tend to be heavier coffee users than other controls, overestimation of the relative risk would result. This should be contrasted to the underestimation that results from the same tendency among cases.

More generally, we might expect an association of selection probabilities with the study variable to be in the same direction for both cases and controls. If so, the resulting case-selection bias and control-selection bias would be in opposite directions, and so to some extent they would cancel one another out, although not completely. To illustrate, suppose that among cases the proportions who refuse to participate are 0.05 for five-cups-a-day drinkers and 0.02 for never-drinkers, and among controls the analogous proportions are 0.20 and 0.10. Refusal will thus result in the relative prevalence of five-cups versus never-use among cases being underestimated by a factor of $0.95/0.98 = 0.97$; this in turn results in a 3 per cent underestimation of relative risk. Among controls, the relative prevalence will be underestimated by a factor of $0.80/0.90 = 0.89$; this results in a $1/0.89 = 1.12$ or 12 per cent overestimation of relative risk. The net selection bias in the relative-risk estimate will then be $0.97/0.89 = 1.09$, or 9 per cent overestimation.

Case–control matching

In cohort studies, *matching* refers to selection of exposure subcohorts in a manner that forces the matched factors to have similar distributions across the subcohorts. If the matched factors are accurately measured and the proportion lost to follow-up does not depend on the matched factors, cohort matching will prevent confounding by the matched factors (Kleinbaum *et al.* 1982).

In case–control studies, matching refers to selection of subjects in a manner that forces the distribution of certain factors to be similar in cases and controls. Because the population at risk is not changed by case–control matching, such matching does *not* prevent confounding by the matched factors. In fact, it is now widely recognized that case–control matching is a form of selection bias that can be removed by adjusting for the matching factor; this adjustment also controls for any confounding by the factor. The textbooks by Miettinen (1985) and Rothman (1986) further discuss these points.

As an example, suppose that our population at risk is half male, the men tend to drink less coffee than women, and that about 75 per cent of our cases are men. Unbiased control selection should yield about 50 per cent men in the control group. If, however, we matched controls to cases on sex, about 75 per cent of our controls would be men. Since men drink less coffee than women and men would be over-represented in the matched control group, the matched control group would under-represent coffee use in the population at risk. As a result, the crude (unadjusted) relative risk would be overestimated. Note, however, that matching does not affect the sex-specific prevalence of coffee use among controls, and so the sex-specific and sex-adjusted estimates would be unaffected by matching. In other words, the selection bias produced by matching could be removed by adjustment for the matching factor.

The general conclusion to be drawn is that matching can necessitate control of the matching factors. Thus, in order to avoid unnecessarily increasing the number of factors requiring control, one should limit matching to factors for which control would probably be necessary anyway. In particular, matching is usually best limited to known strong confounders, such as age and sex in the above example (Schlesselman 1982).

More generally, the primary theoretical value of matching is that it can sometimes reduce the variance of adjusted estimators; see Kleinbaum *et al.* (1982) and Miettinen (1985) for discussions of this point. There are, however, circumstances in which matching can facilitate control selection and so is justified on practical grounds. For example, neighbourhood controls may be far easier to obtain than unmatched general-population controls. In addition, although neighbourhood matching would necessitate use of a matched analysis method, the neighbourhood-matched results would incorporate some control of confounding by factors associated with neighbourhood (such as socio-economic status and air pollution).

Special control groups

It is not unusual for investigators to select a special control group that is clearly not representative of the population at risk if they can argue that (1) the group will adequately reflect the distribution of the study factor in the population at risk, or (2) that the selection bias in the control group is of the same magnitude of (and so will cancel with) the selection bias in the case group. The first rationale is common in case–control studies of mortality, in which persons dying of other selected causes of death are used as controls: in such studies, selection validity can be assured only if the control causes of death are unrelated to the study factor. The second rationale is common in studies using hospital cases and controls; in particular, selection validity can be assured in such studies if the control conditions are unrelated to the study factor, and the study disease and the control conditions have proportional exposure-specific hospitalization rates.

Selection into a special control group usually requires membership in a small and highly select subset of the population at risk. Use of a special control group thus requires careful scrutiny for mechanisms by which the study factor may influence entry into the subset. See Schlesselman (1982) and Kelsey *et al.* (1986) for discussions of practical issues in evaluating special control groups, and Miettinen (1985) and Rothman (1986) for validity principles in mortality case–control studies (so-called 'proportionate mortality studies').

Summary of example

The example of this section provides an illustration of the most common threats to validity in case–control studies, beyond those already discussed for cohort studies. After adjustments for possible confounding and follow-up bias (along the lines described for the cohort study), there still may be irremediable selection bias, especially if we use only select case groups (e.g. MI survivors) or control groups (e.g. hospital controls). In addition, retrospective ascertainment will lead to greater measurement error than in the cohort example, and some of this additional error may be differential.

Given the even greater number of potential biases of unknown magnitude and different directions, it would appear that (as in the cohort example) no conclusions about coffee effect could be drawn from a study such as the one described above, other than that coffee does not have a large effect. Again, this is a common result in thorough epidemiological analyses of observational data.

Retrospective cohort studies

Two major types of cohort studies may be distinguished according to whether members of the study cohort are identified before or after the follow-up period under study. Studies in which all members are identified before their follow-up period are called concurrent or *prospective* cohort studies, while studies in which all members are identified after their follow-up period are called historical or *retrospective* cohort studies. Like case–control studies, retrospective cohort studies often require consideration of retrospective ascertainment and of selection validity.

In particular, retrospective cohort studies that obtain exposure history from post-event reconstructions are vulnerable to bias from outcome-affected measurements. Suppose, for example, that a study of cancer incidence at an industrial facility had to rely on company personnel to determine the location and nature of various exposures in the plant during the relevant exposure periods. If these personnel were aware of the locations at which cases worked (as when a publicized 'cluster' of cases has occurred), biased exposure assessment could result.

Retrospective cohort studies can also suffer from selection biases analogous to those found in case–control studies. Suppose, for example, that a retrospective cohort study relied on company records to identify members of the cohort of plant employees. If retention of an employee's records (and hence identification of the employee as a cohort member) was associated with both the exposure and outcome status of the employee, the exposure–outcome association observed in the incomplete study cohort could poorly represent the exposure–outcome association in the complete cohort of plant employees.

Conclusion

Consistency versus validity

Uncertainty about validity conditions is responsible for most of the inconclusiveness inherent in epidemiological studies. This problem can be partially overcome when multiple complementary studies are conducted, that is, when new studies are conducted under conditions that effectively limit bias from one or more sources present in earlier studies. Ideally, after enough complementary studies have been completed, each known or suspected source of bias will have been rendered unimportant in at least one study. If at this point all the study results appear consistent with one another (not the case for coffee and MI, although the studies of smoking and lung cancer provide a good example), some consensus may be reached by the epidemiological community about the existence and strength of an effect.

Even in such ideal situations, however, one should bear in mind that consistency is not validity. There may, for example, be some unsuspected source of bias present in all the studies, so that all the studies are consistently biased in the same direction. Or it may be that all the known sources of bias are in the same direction, so that all the studies remain biased in the same direction if no one study eliminates all known sources of bias. For these and other reasons, many authors warn that all causal inferences should be considered tentative, at least if drawn from observational epidemiological data alone. See Schlesselman (1982) and Rothman (1986, 1988) for further discussion and references on this point.

Summary

This chapter has outlined and illustrated major concepts of validity in epidemiological research, as applied in three settings: prediction from one population to another; causal inference from observational cohort studies; and causal inference from case–control studies. Parallel aspects of each application have been emphasized. In particular, each problem requires consideration of comparison validity, follow-up validity, specification validity, and measurement validity. Case–control studies require the additional consideration of case and control selection validity, and are often subject to additional sources of measurement error beyond those occurring in prospective cohort studies. Similar problems arise in retrospective cohort studies.

Further topics

We have not covered several important study designs, including prevalence studies and ecological studies. These studies require consideration of the above validity conditions, and also require special considerations of their own. Further details on these and other designs may be found in the general textbooks referenced in the introduction of this chapter.

We have also not covered a number of central topics of epidemiological inference, including epidemiological measures of occurrence and of effect, and problems of statistical inference. Oakes (1986) presents an extensive introduction to competing schools of statistical inference; he emphasizes the shortcomings of the prevailing frequentist approach to statistics, but critically discusses Bayesian and pure likelihood approaches as well. Berger and Berry (1988) and Goodman and Royall (1988) criticize frequentist approaches, respectively emphasizing Bayesian and pure likelihood approaches as alternatives. Miettinen (1985), Poole (1987), and Rothman (1986) discuss the use and misuse of statistical inference in epidemiological research.

Appendix: chapter terminology

A person is *at risk* of an outcome event at a given time if it is logically possible for the person to experience the event at that time. For example, any women is at risk of cervical cancer until she dies or has her cervix removed. A *population at risk* is a population that at any given time consists only of people at risk of the outcome of interest.

A *competing risk* for an outcome is any other event that removes a person from being at risk of the outcome. For example, total hysterectomy is a competing risk for cervical cancer. Death from a cause other than the outcome is a competing risk for every outcome.

A person's (conditional) *risk* of an outcome over an interval is the probability that the person experiences the outcome, given (1) the person is at risk at the start of the interval, and (2) no competing risks occur over the interval. Throughout, we use the term 'risk' for this concept. Thus a woman's risk of cervical cancer between the ages of 51 and 60 years is her probability of cervical cancer by age 60 years given that (1) she is still at risk at age 51 years, and (2) she experiences no competing risks (such as total hysterectomy) by the age of 60 years.

Consider a cohort in which every member is at risk at the start of an interval. The *average risk* of the cohort over an interval is just the arithmetical mean of the risks of all the cohort members over that interval. It is equal to the proportion of the cohort we should expect to experience the outcome if no competing risks occurred. This is the quantity estimated by acturial methods of risk analysis, such as lifetable and Kaplan–Meier methods, if the competing risks are unassociated with risk of the outcome of interest. Average risk is sometimes called the cumulative incidence, incidence proportion, or risk of the cohort over the interval.

The observed *person-time incidence rate* in a population at risk over an interval is the number of cases that occur in the population over the interval, divided by all the time spent at risk by population members over the interval (the person-time at risk). This quantity is also known as the empirical incidence density or the person-time index. Throughout, we use the term 'estimated rate' to refer to this quantity, and 'rate' to refer to its statistical expectation.

For further discussion of the above and related concepts, see the epidemiology texts referenced in the introduction of this chapter.

Acknowledgements

The author wishes to thank Jennifer Kelsey, James Robins, James Schesselman, and Alexander Walker for their comments on this chapter.

References

Berger, J.O. and Berry, D.A. (1988). Statistical analysis and the illusion of objectivity. *American Scientist* **76**, 159.

Breslow, N.E. and Day, N.E. (1980). *Statistical methods in cancer research. I: The analysis of case-control studies.* IRAC, Lyon.

Breslow, N.E. and Day, N.E. (1988). *Statistical methods in cancer research. II: the analysis of cohort data.* IRAC, Lyon.

Checkoway, H., Pearce, N., and Crawford-Brown, D. (1989). *Research methods in occupational epidemiology.* Oxford University Press, New York.

Curb, J.D., Reed, D.M., Kautz, J.A., *et al.* (1986). Coffee, caffeine, and serum cholesterol in Japanese men in Hawaii. *American Journal of Epidemiology* **123**, 648.

Goodman, S.N. and Royall, R.M. (1988). Evidence and scientific research. *American Journal of Public Health* **78**, 1568.

Greenland, S. (1985). Control initiated case-control studies. *International Journal of Epidemiology* **14**, 130.

Greenland, S. (1987). Quantitative methods in the review of epidemiologic literature. *Epidemiologic Reviews* **9**, 1–30.

Greenland, S. (1989). Modeling and variable selection in epidemiologic analysis. *American Journal of Public Health* **79**, 340.

Kalbfleisch, J.D. and Prentice, R.L. (1980). *The statistical analysis of failure time data.* Wiley, New York.

Kelsey, J.L., Thompson, W.D., and Evans, A.S. (1986). *Methods in observational epidemiology.* Oxford University Press, New York.

Kleinbaum, D.G., Kupper, L.L., and Morgenstern, H. (1982). *Epidemiologic research: Principles and quantitative methods.* Lifetime Learning Publications, Belmont, California.

Leamer, E.E. (1978). *Specification searches.* Wiley, New York.

Miettinen, O.S. (1985). *Theoretical epidemiology.* Wiley, New York.

Oakes, M. (1986). *Statistical inference: A commentary for the social and behavioural sciences.* Wiley, New York.

Poole, C. (1985). Exceptions to the rule about nondifferential misclassification (abstract). *American Journal of Epidemiology* **122**, 508.

Poole, C. (1987). Beyond the confidence interval. *American Journal of Public Health* **77**, 197.

Rosenberg, L., Palmer, J.R., Kelley, J.P., Kaufman, D.W., and Shapiro, S. (1988). Coffee drinking and nonfatal myocardial infarction in men under 55 years of age. *American Journal of Epidemiology* **128**, 570.

Rothman, K.J. (1986). *Modern epidemiology.* Little, Brown and Company, Boston.

Rothman, K.J. (1988). *Causal inference.* Epidemiology Resources Incorporated, Chestnut Hill, Massachusetts.

Schlesselman, J.J. (1982). *Case-control studies: Design, conduct, analysis.* Oxford University Press, New York.

Slud, E. and Byar, D. (1988). How dependent causes of death can make risk factors appear protective. *Biometrics* **44**, 265.

White, H. (1990). *Estimation, inference, and specification analysis.* Cambridge Univerisity Press, New York.

16

Microcomputer applications in epidemiology

RALPH R. FRERICHS and BEATRICE J. SELWYN

Introduction

The microcomputer has revolutionized the approach epidemiologists take to the investigation of diseases in populations. The revolution, however, is rooted in the continuing evolution of epidemiological methods and in the dramatic advancements in computer technology. Historically, epidemiologists have always focused on reasons for differential patterns of disease in communities. Yet these reflections then, as now, required detailed measurements before causal mechanisms could be adequately evaluated. Relatively crude methods of analysis were developed for investigations of outbreaks with few subjects and limited numbers of variables. Data were usually analysed by hand, with separate piles of paper used to categorize persons into the four exposure–disease groups necessary to construct 2×2 tables. Larger studies were more troublesome.

Then, after the Second World War, came the computer. At first, access to computers was limited to persons skilled in the binary language of the machine. Professionals such as computer programmers and systems analysts developed very quickly and soon became the gatekeepers to the machines. Computers were housed in large, air-conditioned rooms with access permitted only to those with the special programming skills. During the same period of time, epidemiology was undergoing an evolution of its own. The principle causes of mortality in many societies shifted from the infectious diseases to degenerative and malignant conditions, often termed chronic diseases. The basic model relating exposure factors to disease still held, yet the time period was greatly extended. Equally important, the web of causation was found to be more complex for the degenerative and malignant diseases.

In viewing the evolution of contemporary epidemiology, Greenland has divided the years from the Second World War to the early 1980s into two developmental time periods (Greenland 1987). During the first period, from the end of the war to the mid-1960s, the foundation was established for modern epidemiological analyses, including basic methods for the estimation of relative rate and risk. As noted by Rothman, there was still considerable confusion at that time over even the most basic concept to epidemiological measure-ment, the definition of rates (Rothman 1986). From the mid-1960s to the early 1980s, the theoretical framework for epidemiological methods was firmly established with the clarification of concepts such as confounding, selection bias, effect-modification, and the like. This period also saw the emergence of the case–control design as the most appropriate method for investigating determinants of numerous chronic diseases. Articles in epidemiology journals became much more quantitative in content as statisticians and mathematicians increasingly focused their attention on issues of bias and measurement. Finally, new epidemiology texts of a more quantitative nature appeared at the end of this second period, reflecting the rapid evolution of the field (Kleinbaum et al. 1982).

While it is difficult to view developments in isolation, it appears doubtful that the more quantitative methods would have been developed independent of computers. In many ways, the use of computers created a demand for clarification in basic epidemiological methods since data could now be readily analysed in increasingly sophisticated ways. At first, access to the computers was limited to programmers or the few epidemiologists with programming abilities. This soon changed, however, as new software made communicating with the computer all the more easy. In technologically developed countries, epidemiologists increasingly had to become experts in relating to computers as well as being able to develop hypotheses and design field studies. Unfortunately, because of the expense of the machines, access to computers was limited to epidemiologists employed at academic or large governmental centres. Excluded were epidemiologists who did not work for large governmental or research agencies and those working in less developed regions of the world. The latter group often found information and teaching on the new quantitative methods difficult to acquire and, since they had only calculators, inappropriate for establishing local needs and evaluating programmatic efforts. Now, with the advent of the microcomputer, all this has changed.

In many ways, the microcomputer has empowered epidemiologists around the world to serve the interests of science and their society. Of course, different disease patterns prevail in the various countries. Heart disease, cancer, and strokes are of primary importance in the more developed countries, while infectious conditions such as malaria, tuberculosis, and onchocerciasis prevail in less developed regions. Yet the methods necessary to unravel the local determinants of these diseases are the same. As Rothman has stated: 'Although some specialized methods have been developed solely to study the spread of infectious illness, whatever distinctions exist between traditional and modern areas of epidemiology are certainly less important than the broad base of concepts that are shared' (Rothman 1986). The microcomputer both permits and encourages these methods to be used by epidemiologists throughout the world who are interested in unravelling the complexity of disease occurrence in their society.

In the following sections of this chapter, we first describe the microcomputer hardware and software that has fostered the process of empowerment for epidemiologists. Next, we present four applications of microcomputer use in epidemiology, with one focusing specifically on developing countries. Finally, there is a concluding section with a view of the future.

Categories of hardware

The components necessary for a functioning microcomputer are often divided into two groups, hardware and software. Hardware is a term used to describe the parts of a computer that have a physical presence. Thus the microcomputer, monitor, memory boards, and printer are all hardware. Software, on the other hand, is the programs or instructions that tell a computer what to do. In this section, we first describe the typical microcomputer hardware and then focus on several categories of software currently being used by epidemiologists.

Microcomputers have traditionally been categorized as desktop computers and portable or laptop computers. The desktop machines typically are less expensive per unit of computing power than portable units. They tend to be much larger than portable computers and have a separate central processing unit, monitor, and keyboard. Conversely the portable microcomputers are smaller and lighter, with all the components joined in a single unit. They tend to be much more rugged than desktop machines and in many ways are ideal for field use and for general use in developing countries (Frerichs and Tar Tar 1989).

Central processing unit

The brain of the microcomputer lies in the central processing unit or CPU. Here are a series of small silicon chips which are mounted on a motherboard on the floor of the unit. The chips are the microprocessor; they understand and process a series of binary numbers with values of either one or zero. These binary digits are termed *bits* and are the basic units of a computer. All communication with the microcomputer has to be translated into bits before the computer can work with the information. The bits are combined to create numeric, alphabetic, and graphic characters termed *bytes*. An example of a byte is the letter Q, which in bits is written as 10101001. Usually eight bits are needed by the computer to create one character or byte. Most computers can process more than one byte at a time; this ability defines the speed of the computer. Thus a 32-bit computer processes four characters at a time while a 64-bit computer can process eight characters at a time.

There are various ways to communicate with the computer. First, the message can be written in binary code, often termed *machine language*, which speaks directly to the computer. Second, a programming language can be used which uses alphabetic or numeric terms common to human communication and translates them into bits or machine language. Both of these approaches require the help of computer programmers who have specialized knowledge in the respective languages. The third mode of communication uses words, sentences, or symbols to instruct the computer. It is this third mode that has helped revolutionize the way computers are being used by epidemiologists and other scientists. Here special software is written which non-programmers can readily understand and use. Thus the epidemiologist is able to interact directly with the microcomputer, using it to do a wide variety of tasks.

Besides the microprocessor, the CPU also contains the memory of the computer. Memory is measured in kilobytes (approximately one thousand bytes) or megabytes (approximately one million bytes) using the symbols K and MB, respectively. There are actually two kinds of memory internal to the microcomputer, both maintained in small chips attached to the motherboard. The first is random access memory or RAM. This memory can be filled with words or data entered either by typing on the keyboard or from disks or other modes of storage. The information can easily be altered or erased, depending on the wishes of the computer operator. Once the machine is turned off, however, all information in RAM is usually cleared. Thus it must first be stored on disks or some other storage device. The only exception is if the RAM chips are of the complementary metal oxide semiconductor type, known as CMOS. These CMOS chips use very little power. A small battery in the microcomputer maintains just enough power for the CMOS chips so that the information is retained in RAM even when the machine is turned off.

The size of RAM in a microcomputer is a major determinant of how fast information can be analysed. If all the data are in the internal memory of the machine, the computer can quickly fulfil the requests of the operator. Conversely, if RAM is too small, the computer can only process some of the data at one time and must constantly go back and forth to the disk where the complete data-set is stored. While the software program also takes up internal memory, a microcomputer with 8 MB of RAM can easily hold in memory a data-set comprised of 5000 subjects, 200 variables per subject, and 5–6 characters per variable.

Besides RAM, the microcomputer also has read-only memory or ROM. This second form of memory contains instructions written in machine language and cannot be erased or altered by the operator. ROM often contains a set of instructions that allows the computer to use words when interacting with the operator. ROM chips may also hold a programming language. Finally, various forms of ROM are available which contain favourite software programs or reference data-sets such as a dictionary or a thesaurus that do not need to be changed.

Input devices

The most common input device for the microcomputer is the keyboard. Often there is a separate set of numeric keys off to one side for entering numbers. Information can also be entered with voice recognition devices so that memos or letters can be dictated rather than typed into the machine. Another common input device is a mouse. These small devices control a cursor or arrow on the computer screen which either points and activates an instruction statement or is used to draw graphs or other symbols.

Information from other cities or locations within a city can be entered into the computer using a modem. The words or data arrive over a telephone line as sound from another computer. The modem converts the sound signals to the binary machine language used by the microcomputer.

Disk drives, to be mentioned in the next section, serve as both input and output devices. The magnetic disk is the main storage mechanism in microcomputers for information and is used to send both programs and data from one investigator to another. Commercial software is most often distributed on magnetic disks.

A more recent source of extensive data is the compact disc or CD, the same kind of disc on which stereo music is recorded. (Note that the spelling is with a 'c' rather than a 'k' as used with magnetic disks.) The latter is an abbreviation of diskette while the former refers to a round, disc-shaped object. Compact discs hold more than 500 MB of information and are thus ideal for storing and providing access to large data-sets from health surveys or population censuses, reference material such as dictionaries or encyclopedias, and book chapters and journal articles.

The final input mechanism of use to epidemiologists is the optical scanner. This device can read questionnaire responses made with a special pencil directly into the computer rather than relying on data-entry persons. Scanners can also read both text and graphs from books, journals, or manuscripts for entry into the microcomputer. When used with compact discs, the investigator can save appropriate book chapters and articles and create an extensive reference library requiring very little physical space. More important, the epidemiologist can use simple search statements to access specific text, topics, or authors in his or her personal reference library.

Output devices

The three main output devices are the monitor, printer, and magnetic disk. The monitor may be either colour or black and white, and is by far the most important output device for most epidemiologists using microcomputers. Since much time is spent reading text or viewing graphs on the screen, the monitor must be of a high quality. Contributing to the sharpness of the screen image is the number of dots or pixels that appear on the screen. Quality monitors usually can display 640×400 pixels, while high resolution monitors can display several thousand pixels on each axis of the screen. Most monitors resemble a television set and use a cathode ray tube (CRT). Low energy monitors with a flat, liquid crystal display (LCD) are often found on portable computers.

For epidemiologists and other scientists, the written word remains the most common form of communication. Whether reports, memos, letters, or manuscripts, all require output to a printer. Figures are often included in the documents. Thus the printer must be of sufficient quality to produce both text and graphs. The highest quality of print is derived with a laser printer, an output device that uses technology similar to some photocopying machines. Also commonly used to produce characters and graphs are ink jet printers which spray a jet of ink onto a page through small pin-holes, and dot matrix printers which use small pins to produce dots in various patterns on the paper.

The final form of output is to a magnetic disk. Some are termed 'floppy' disks while others are referred to as 'hard' or fixed disks. Most floppy disks are now encased in a non-bendable plastic cover with a small spring-loaded door which opens to the computer. Earlier versions of the disks had thin external covers that were easy to bend, accounting for the name 'floppy'. Floppy disks hold up to two MB of information and serve as an inexpensive storage and software distribution mechanism. Unlike floppy disks, a hard disk is housed in a protective cartridge within the CPU of the microcomputer. The hard disk provides the epidemiologist with mass storage for up to two gigabytes (2 billion bytes) of information. In addition, getting information on and off a hard disk is much faster than with a floppy disk.

Categories of software

Without software, most epidemiologists would find computers to be nothing more than complicated machines. Software allows the operator to dictate what the microcomputer should do. In this section we briefly mention several common categories of software, which may or may not be sold as separate packages. Software programs are constantly changing. Thus we will refrain from mentioning specific packages in other than general descriptive terms.

Word processing

For epidemiologists, one of the most common applications of microcomputers is likely to be word processing. Whether designing questionnaires, preparing lecture notes, writing reports, or drafting manuscripts, the microcomputer greatly reduces the necessary time from beginning to completion. This category of software enables the user to type, edit, and change documents while still on the screen and then to save the revised document on a disk. Text and figures or graphs can be printed in hard copy with any sized margins or line spacing, and with various categories of type depending on the printer. Spelling is checked with an electronic dictionary, while grammar programs check for typographical errors, writing style, and sentence level. Once completed, the text can be sent via a modem and telephone lines to the microcomputers of collaborators in other cities. Using word-processing software in battery-powered portable computers, reference material can easily be abstracted in a library or other facility. Finally, word-processing programs can create individualized form letters for sending messages to the field staff or to colleagues in other agencies.

Data-base management

Data-base management programs have been developed to transform the raw data collected by epidemiologists quickly into useful information for unravelling disease aetiology, assisting administrators, or formulating public policy. McNichols and Rushinek (1988) have described seven data management functions which have historically been done manually but are now more efficiently performed by this category of program: (1) recording; (2) storing; (3) retrieving; (4) selecting or classifying; (5) sorting; (6) computing; and (7) displaying.

Recording consists of capturing individual facts in a form that allows later reference, either from questionnaries, interview, or examination forms, or by direct entry into the microcomputer. *Storing* is the transfer of data to floppy or hard disks, or to laser discs for later retrieval. *Retrieving* is getting specific information quickly back out of the computer, such as a person's name, or disease status. *Selecting or classifying* is separating information into different categories such as never smoked, current smoker, or ex-smoker for subsequent retrieval. *Sorting* is arranging data in numerical or alphabetical order. *Computing* is summarizing, transforming, or cross-tabulating data. The last function, *displaying*, is showing the results in an interpretable form in tables or possibly graphs. Most data-base management programs allow the user to ask for particular kinds of information in words and use a display screen to set up the format in which the data will be presented.

Statistical

Statistical software has become essential to most practising epidemiologists. The software helps epidemiologists do two kinds of statistical analyses: descriptive and inferential.

Descriptive analysis summarizes common data properties such as the mean value of a given variable and the frequency of various responses. In contrast, inferential statistics determine if study findings are based on fact or are the result of chance. Typical functions done by statistical software packages are data handling, contingency tables, non-parametric tests, *t* tests, correlation, regression, analysis of variance, log-linear models, survival analysis, cluster analysis, and time–series analysis. Not all, however, do functions or tests favoured by epidemiologists such as direct and indirect standardization, Mantel–Haenszel adjusted risk and odds ratios, or logistic regression.

Spreadsheet

Spreadsheet programs are among the most useful for epidemiologists. The concept was originated on the microcomputer, although spreadsheet programs are now available for use in mainframe computers as well. The term 'spreadsheet' comes from the practice of accountants who spread large sheets of paper on a table to look at columns and rows of financial data. These programs divide the computer screen into a series of columns and rows with individual cells containing data, text, or formulae that relate one cell to another. When a cell entry is changed, it may affect the numeric or text values in many other cells, depending on how the variables are related. This procedure is often referred to as doing a 'what-if' or 'sensitivity' analysis. Spreadsheet programs are most often used for organizing, calculating, and presenting financial, statistical, and other numerical data that serve as the basis for decision-making. Spreadsheets are also used to model disease processes and to understand the consequences in studies of factors such as selection and misclassification bias.

Graphics

Graphic software permits the epidemiologist and others quickly to visualize relationships that may not be as apparent when viewing a table. Unfortunately, epidemiologists have been notoriously dull in their graphic presentations, using histograms and the plain line graphs almost exclusively to describe their findings. Typical are the findings of Selwyn (1985) who reviewed during a one-year period the graphics content of the *American Journal of Epidemiology*. Of 121 graphs, 98 per cent were either histograms or line graphs. Only 2 per cent were pie charts and none were three-dimensional graphs, a category that is ideal when considering the association between a risk factor and disease for illustrating the modifying or intervening effect or a third variable. For example, the relationship between number of antenatal care visits and neonatal mortality might differ by birth-weight. This could easily be visualized in a graph with number of antenatal care visits on the horizonal or x-axis, neonatal mortality on the vertical or y-axis, and birth weight categories on the third axis.

Many data-base management, statistics, and spreadsheet

programs include graphics of acceptable quality. Special graphics software, however, will create a wide variety of graphs including three-dimensional bar and line graphs, pie charts with separable slices, graphs with superimposed symbols, detailed geographical maps, and word charts. The graphs can be printed on paper, transparencies, or converted to slides. Alternatively, the graphic images can be shown as they appear on the computer monitor using a real-time overhead display. These displays are flat panels containing a transparent liquid crystal display (LCD), the same monitor found on many portable computers. The display is placed on an overhead projector with the projected image on a wall or screen showing the text, graph, or figure as it is created by the microcomputer.

Communication

Communications programs use telephone lines to link one microcomputer with either another microcomputer or a mainframe computer. Required is a modem which converts the machine language to sound and back to machine language, and software which tells the computer when to send and receive the data. Most communication programs will work with a variety of modems and hardware configurations. Standard features are auto-dialling, auto-answering, and the maintenance of a directory with telephone numbers and other information. The transfer of data or text from one computer to another is often referred to as 'electronic mail'. This form of communication is near instant and may be less expensive than conventional mail, depending on the telephone charges.

Software is also available for epidemiologists and others who need to communicate with information services using mainframe computers. Typical of these are electronic encyclopaedias, newspapers and magazine articles, and abstracting services such as the Medical Literature Analysis and Retrieval System (MEDLARS) maintained by the US National Library of Medicine.

Project management

To do an epidemiological study usually requires careful balancing of time, money, and resources. Project management software requires that the investigator separate the proposed study into various components and estimate both the order of completion and the time and resources necessary for completion of the individual components. Such items as unexpected sick days, reductions in funds, or the availability of extra interviewers can quickly be entered into the program to determine what is the most efficient course to take. Reporting and monitoring features of this category of software can also be used for progress reports required by the funding agency.

Special interest

Some software cannot easily be categorized. Various groups are constantly developing programs of special interest to epi-

demiologists or others interested in population-based research. Notable have been two programs developed at the Centers for Disease Control (CDC) in Atlanta, Georgia. The first, *Epi Info*, is intended for outbreak investigations or other epidemiological studies, and consists of a series of computer programs used to create and analyse questionnaires, design interview or examination schedules, and perform other common epidemiological tasks (Dean *et al.* 1988). The second, *CASP*, uses a modification of the National Center for Health Statistics (NCHS) growth curves to do anthropometric analyses of growth in children from birth through to 18 years of age (Jordan 1987). The NCHS growth curves have been recommended by the World Health Organization as the reference set for international comparisons. Included are measures of weight-for-age, height-for-age, and weight-for-height. CDC has smoothed and extended the standard anthropometric curves into lower growth ranges more appropriate for children in developing countries.

Applications

Microcomputers are being used by epidemiologists to do a wide variety of activities. Four examples which have relevance to epidemiologists in many different settings are presented in this section.

Case–control studies

In the US, those epidemiologists who plan to conduct case–control studies find that the microcomputer is increasingly being viewed as an essential tool. Most recent graduates of epidemiology programs bring to the profession special expertise in quantitative methodology, data management, and the use of microcomputers. These skills, coupled with an understanding of the natural history of diseases in populations and the web of causation which links risk factors with disease, permit the current wave of epidemiologists to be more productive than ever before.

Assume the epidemiologist is planning to conduct a case–control study. The cases would most likely be derived from a hospital or clinic, while the controls would be selected from the source population of the cases. The epidemiologist, either alone or with a small staff, would use a microcomputer to accelerate and ease all steps necessary for conducting such an investigation (see Table 16.1).

The first step would be to review the literature to determine what is known about the disease. Using a modem and a microcomputer, a search would be conducted of bibliographical data bases such as MEDLINE maintained by the US National Library of Medicine. Identified information and abstracts would be retrieved from appropriate articles and maintained in a text data base in the computer. The most important articles would be obtained from a local library. The text and graphs would be read by an optical scanner and stored on compact discs for later review.

The next step would be to determine the number of cases and controls necessary for estimating the odds ratio with a

Table 16.1. Use of microcomputer for case–control studies

Procedure	Necessary hardware	Necessary software
Review of the public health and medical literature	Microcomputer Modem Compact disc drive Optical scanner	Communication Data base Word processing
Determination of necessary sample size	Microcomputer	Spreadsheet
Plan steps necessary to complete the project	Microcomputer Printer	Project management
Development of preliminary interview schedule and protocol	Microcomputer Printer	Word processing
Preparation of proposal for funding	Microcomputer Printer Spreadsheet	Word processing Project management
Development of final version of interview schedule and protocol	Microcomputer Printer	Word processing
Development of control forms for managing the study	Microcomputer Printer	Form preparation or spreadsheet
Selection and interview of hospital- or clinic-based cases	Portable microcomputer	Data base
Selection and interview of controls with exposure history representative of persons in the source population of cases	Microcomputer Modem	Communication Data base
Data processing and initial analysis (frequency distributions)	Microcomputer Printer	Data base
Final analysis of data	Microcomputer Printer	Statistical Spreadsheet Graphics
Preparation of manuscripts of final report to funding agency	Microcomputer Printer Modem	Word processing
Preparation for oral presentations of findings	Microcomputer Printer Overhead display	Word processing Graphics

given level of precision. The formulae for the minimum sample size are obtained from appropriate epidemiology texts such as Schlesselman (1982) or Kelsey *et al.* (1986) and entered into a spreadsheet program. The same program is used to assess the effect on precision of multiple rather than single controls.

At this stage, a project management program would be used to organize and plan the various steps necessary to complete the study. Once this plan has been written, the program makes it easy to enter changes, determine their effect, and keep track of how well various phases of the study are going. This software is all the more important if the funding agency requires periodic progress reports, the study is overly complex, or if it involves multiple centres.

A preliminary interview or examination schedule is constructed along with a consent form and study protocol using a word-processing program. If necessary, this information is sent with a brief description of the study objectives and methods to the local ethics committee for approval. Once approved, the final proposal is written and submitted to the funding agency, again using a word-processing program. A time-line graph is generated using a project management program and the budget is created using spreadsheet software.

If the project is funded, the final version of the interview schedule or questionnaire is created along with the protocol using word-processing software. There may be several revisions at this stage, depending on how many other investigators are involved in the study. Control forms are also created for managing the study using a form production program or a spreadsheet program. These forms keep track of the persons contacted and their willingness to participate in the study, the activities of the staff, and the final disposition of the interview schedules.

The next step is to interview the cases either in the hospital or clinic where the diagnosis was made, or later in their home. If it does not disturb the patient, the data are entered directly into a portable, battery-powered computer while the interview is taking place. Data entry screens are created using data-base management software, including skip patterns and checks for non-eligible and illogical entries. For confidential information, the epidemiologist may want the patient to enter the data directly into the computer without being asked by an interviewer. If the computer is viewed by the patient as being intrusive, the interviewer can simply write the responses down on a paper form and transfer them later to the computer.

Controls may be selected in several different ways with the necessary software and hardware dependent on the method.

The intention of all the methods is to select persons without the disease in the source population of the cases. First, controls may be selected from patients in the same hospital or visiting the same clinic, but with conditions unrelated to the exposure factor of interest. For this type of institution-based control, the same procedure would be followed as for the cases. Second, controls may be selected from the same neighbourhood as the cases. These neighbourhood-matched controls are interviewed in their home, again following the same procedure as the cases. Finally, when all cases in a defined geographical area are included in the study, the controls are selected from a random sample of the population. A common technique for obtaining this latter category of controls is to contact and interview them by telephone using random-digit dialling. For this method, the epidemiologist needs a microcomputer, and both data-base management and communication software.

Once all the information has been entered into the microcomputer, the data are edited and frequency distributions are printed for all variables. The epidemiologist would scan the information for outliers or other unusual values. These would then be corrected and re-entered into the microcomputer. A category of data-base management software is used which generates frequency distributions.

The final analysis involves more detailed statistical manipulations to control for confounding and to determine the presence of effect modification. Usually studies are analysed first by stratification and then by multi-variate analysis. The crude and adjusted odds ratios are often presented with confidence intervals showing how precisely the variable was measured. For the final analysis, the epidemiologist would use a microcomputer and graphics, statistical, and spread-sheet software to generate tables and graphs of the most important findings.

Once the tables and graphs have been prepared, the epidemiologist would use a word-processing program to prepare manuscripts and write the final report to the funding agency. If the epidemiologist is working with investigators in other institutions, the text could be transferred for editing from one to the other using a modem. The second investigator would edit the text and, if necessary, immediately send it back via the telephone line and modem to the original author. To ease the editing task, the collaborators could use 'text comparing' software which would show the original author all the places where changes had been made.

For oral presentations, the graphics program would be used to create simple word charts and graphs for slides or overhead projection. Alternatively, an overhead projector and overhead display panel can be used to show real-time images from a computer onto a screen. Thus the epidemiologist might prepare a set of graphs or word charts using a portable computer, save the images, and then show them in the desired order using the overhead display panel as the monitor. Conversely, different views of the data might be requested by members of the audience—a request that could easily be satisfied with real time viewing of the analysed results.

Cohort studies

Microcomputers are especially useful for conducting large-scale cohort studies involving a large number of records with many variables gathered over a long period of time. With up to 8 megabytes of RAM and either large laser discs or hard disks to store data, microcomputers are now powerful enough to handle such large data-sets and do most statistical manipulations in a reasonable amount of time. The software used for such projects is commercially available with required features much like those for any epidemiological research project. There is a need to keep up with bibliographical reference work, to plan and manage the project of several years, to prepare questionnaires and forms, to enter data, edit data, manipulate files and variables, link files, generate periodic reports, do data analysis, prepare graphics, and write up the results.

Typical might be a three-year cohort study of 300 children describing the natural history of paediatric infectious diseases. Individual data would be tallied on nearly 50 000 visits over the three-year span. Included would be demographic and environmental data, specimens for microbiological assessments, and information on the signs and symptoms of the diagnosed illnesses. The measure of effect is the incidence rate with new illnesses in the numerator and child-weeks at risk in the denominator. The data from the weekly visits would be used to keep track of time-at-risk and the occurrence of new episodes of illness, and would serve as a link to the laboratory data.

With microcomputer-based systems, investigators can maintain up-to-date data entry, so that within a day of the visit, the information would be stored in the computer. Data entry of the weekly data would take less than a minute per child, depending on the number of variables. Thus, during a typical week, it would take up to 4–5 hours to enter the data on the 300 children. A preliminary analysis could easily be done on a weekly basis, serving both to monitor the field activities and to provide an advanced view of the study findings. In large long-term studies such as the one being described, errors must be checked in a timely manner since some of the information changes from week to week and incorrect data would be permanently lost. By reviewing the data as they are being collected, the investigator can keep a watchful eye on the quality of the information and have rapid turnaround for correcting mistakes and omissions.

In contrast to the spread and convenience of microcomputers, investigators using mainframe computers might take longer to complete their study even though the mainframe computer's processing speed is considerably faster than the microcomputer. Since a programmer and data entry personnel would have to be hired, the mainframe-bound investigator might be tempted to wait until the end of the data collection period before entering and analysing the data.

With a large study of nearly 50 000 records such as the one being described, it could take up to six months to complete the data-entry phase. In addition, omitted or incorrect data of a seasonal nature would be lost to the study with no opportunity for retrieval.

Laboratory work usually entails keeping detailed log books on specimens, results of laboratory findings, and the like. Often, microbiologists prepare their analyses by hand, especially when working with a fairly small set of specimens. Using a microcomputer, the laboratory data can easily be abstracted from the log books and merged with other data derived from the field. Having the data readily available allows the investigator to check on the laboratory work quickly and ensure that no specimens are missing.

The advanced summary view of the data may be especially important for the many progress reports required by typical funding agencies. Using spreadsheet software, standard reporting tables and graphs can easily be created for simple updating for each progress report.

The microcomputer is all the more important in the analysis phrase of our hypothetical cohort study. In the ideal study, all hypotheses are clearly written with a detailed list of the necessary analyses. Most epidemiologists find that this ideal is hard to match. Instead, they analyse the data in waves, providing descriptive information in the first phase and then looking for possible and plausible associations in the second phase. If done on the mainframe computer, the epidemiologist usually works with a programmer and submits jobs on a daily basis for analysis. Since everything is formal and billing is on a per-run basis, the investigator feels the pressure to be exact and not to spend too much time viewing the data in different ways or testing the applicability of other statistical approaches. The consequence may be an incomplete, or even erroneous, view of the data. As often happens in many epidemiological investigations, the findings are not in accord with prior hypotheses and are suggestive, instead, of other possible associations. The unusual findings may be real or may be a consequence of errors in the data, the categories used in the analyses, or an inappropriate statistical model. Using a microcomputer with 'user-friendly' statistical software, the epidemiologist can avoid the psychological restrictions imposed by the mainframe computer. The data can be slowly and carefully explored using both frequency distributions and cross-tabulations. If the identity of the interviewer or the day of the week when the data were collected appears to be important, the epidemiologist can explore the issue. If the laboratory specimens were all positive during two weeks in July, it would be easy to determine if they were processed by the same technician who may have cross-contaminated the specimens or made errors in data entry. If other investigators state a logarithmic transformation is necessary before analysing some of the blood variables, the epidemiologist can easily do the analyses both ways and see if it makes a difference. Finally, the investigator can experiment with various ways to show his or her major findings clearly, using different types of graph or figure.

Differential and non-differential misclassification

In this section, a spreadsheet program will be described which assesses how misclassification might have affected the findings of epidemiological studies. The program can also be used by epidemiology graduate students to understand more clearly this often complex problem. First, we will discuss what misclassification is and then show a spreadsheet program which compares true values for selected measures of effect with those observed in a typical epidemiological study.

Explanation of misclassification

Misclassification affects most studies conducted by epidemiologists. Whether the cause is laboratory errors, problems with interview questions, or inexact diagnostic algorithms, the effect is that persons in the study are not classified correctly as to their exposure or disease status. If the variables are dichotomous, such as disease and no disease or exposed and unexposed, measurement error is described in terms of 'sensitivity' and 'specificity'. For a given disease, sensitivity is the probability that a person with the disease will be classified in the study as having the disease. Conversely, specificity is the probability that a person not having the disease is correctly classified in the study as not having the disease. The two terms are used the same way when considering exposure to risk factors.

To illustrate the effects of misclassification, consider the causal association between a risk factor and disease. The relationship is diagrammed as:

$$\text{Risk factor} \xrightarrow{\;+\;} \text{Disease}$$

where we assume that exposure to the factor is positively associated with the disease. If a cohort study had been conducted in which the population was classified as to exposure and followed over time to determine the onset of disease, the fourfold table and the incidence among the exposed and unexposed would look like this:

		Onset of disease during time period of interest			
		Yes	No		
Risk factor	Exposed	A	B	$A + B$	$I_e = A/(A + B)$
	Unexposed	C	D	$C + D$	$I_u = C/(C + D)$

The risk ratio (RR) comparing the exposed to unexposed is:

$$RR = \frac{I_e}{I_u} = \frac{A/(A + B)}{C/(C + D)}$$

Misclassification of either the exposure factor or the disease status can affect the observed value of the risk ratio. The relationship between the truth and what is observed in the study is shown for the two variables as follows:

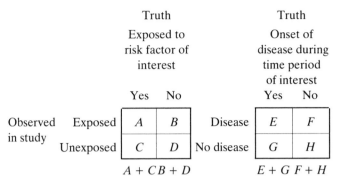

		Truth Exposed to risk factor of interest			Truth Onset of disease during time period of interest	
		Yes	No		Yes	No
Observed in study	Exposed	A	B	Disease	E	F
	Unexposed	C	D	No disease	G	H
		$A+C$	$B+D$		$E+G$	$F+H$

		Exposed to risk factor		Unexposed to risk factor	
		Truth Onset of disease during time period of interest			
		Yes	No	Yes	No
Observed in study	Disease	E_e	F_e	E_u	F_u
	No disease	G_e	H_e	G_u	H_u
		E_e+G_e	F_e+H_e	E_u+G_u	F_u+H_u

In the two tables, the sensitivity of the measurement is $A/(A+C)$ and $E/(E+G)$, respectively, and the specificity is $D/(B+D)$ and $H/(F+H)$, respectively.

Exposure status is measured in two groups: persons who will develop the disease and persons who will not develop the disease during the time period of observation. Thus, for *exposure status* we can divide the one fourfold table into two tables reflecting the two disease strata:

		Onset of disease during study interval		No onset of disease during study interval	
		Truth Exposed to risk factor of interest			
		Yes	No	Yes	No
Observed in study	Exposed	A_d	B_d	A_n	B_n
	Unexposed	C_d	D_d	C_n	D_n
		A_d+C_d	B_d+D_d	A_n+C_n	B_n+D_n

The subscripts 'd' and 'n' refer to the 'disease' and 'no disease' strata.

The question arises: 'Is the level of misclassification of exposure status the same among those who will and will not develop the disease of interest?' If the answer is 'yes', then our observation of exposure status is affected by *non-differential* misclassification. If the answer is 'no', then our observation is affected by *differential* misclassification. Note that if the study is a regular cohort investigation in which the exposure variable is measured before the disease variable, most likely the misclassification would be non-differential. On the other hand if the study follows a historical cohort design, as is often the case in acute outbreak investigations, misclassification may be differential since having the disease could affect exposure recall.

The measure of *disease onset* can be viewed in a similar manner, but this time we consider the measurement among those exposed and not exposed to the risk factor of interest:

The subscripts 'e' and 'u' refer to the 'exposed' and 'unexposed' strata.

As when considering the measurement of the exposure status, misclassification of the disease status can be the same in both strata resulting in *non-differential* misclassification, or different in the two strata resulting in *differential* misclassification. If the hypothesized link between exposure and disease is firmly embedded in the mind of the investigator or the subjects and there are few controls in the study for observation bias, it is likely that the degree of misclassification would be different for the exposed and unexposed groups, resulting in differential misclassification of the disease variable.

In most epidemiological studies relating an exposure variable to a disease, the measure of effect is the risk ratio or an estimate of the risk ratio, the odds ratio. For those interested in the policy implications of the research, two additional measures are often used: the attributable fraction among the exposed (AF_e) and the attributable fraction in the population (AF_p). In terms of incidence measures (I), the attributable fractions are defined as follows:

$$AF_e = \frac{I_e - I_u}{I_e} \qquad AF_p = \frac{I_p - I_u}{I_p}$$

where I_e = incidence among exposed persons
I_u = incidence among unexposed persons
I_p = incidence in the total group of exposed and unexposed persons.

AF_e measures the fraction or proportion of new cases among persons with the risk factor which is attributed to the risk factor. Thus, it indicates the proportion of disease cases among exposed persons which would theoretically be prevented if they were not exposed to the risk factor of interest. AF_p measures the fraction or proportion of new cases in the total population which is attributed to the risk factor. It shows the proportion of disease cases in the total population which could theoretically be prevented if the exposure factor was removed from the population.

The usefulness of the two attributable fraction measures is dependent on several assumptions. First we assume that the risk factor is causally associated with the disease of interest, and second that there is no confounding or bias in the measures of incidence among the exposed and unexposed groups.

Before showing how the computer can help, we need to summarize what has been said so far. If there is non-differential misclassification we need four values of sensitivity or specificity to describe the relationship between the 'true' and the 'observed' RR, AF_e, and AF_p in the study population. These are:

1. Sensitivity for the measure of the risk factor:
$$A / (A + C).$$

2. Specificity for the measure of the risk factor:
$$D / (B + D).$$

3. Sensitivity for the measure of the disease:
$$E / (E + G).$$

4. Specificity for the measure of the disease:
$$H / (F + H).$$

Non-differential misclassification affects the observed RR by attenuating the value towards 1.0, the value of no association between exposure and disease (Kleinbaum *et al*. 1982). For AF_e and AF_p, the observed values are attenuated towards zero.

For differential misclassification, we need eight values of sensitivity or specificity to describe the same relationship between the 'true' and 'observed' measures of effect.

1. Sensitivity for the measure of the risk factor among persons who develop the disease:
$$A_d / (A_d + C_d).$$

2. Specificity for the measure of the risk factor among persons who develop the disease:
$$D_d / (B_d + D_d).$$

3. Sensitivity for the measure of the risk factor among persons who do not develop the disease:
$$A_n / (A_n + C_n).$$

4. Specificity for the measure of the risk factor among persons who do not develop the disease:
$$D_n / (B_n + D_n).$$

5. Sensitivity for the measure of the disease among persons exposed to the risk factor of interest:
$$E_e / (E_e + G_e).$$

6. Specificity for the measure of the disease among persons exposed to the risk factor of interest:
$$H_e / (F_e + H_e).$$

7. Sensitivity for the measure of the disease among persons not exposed to the risk factor of interest:
$$E_u / (E_u + G_u).$$

8. Specificity for the measure of the disease among persons not exposed to the risk factor of interest:
$$H_u / (F_u + H_u).$$

Differential misclassification affects the observed RR in either direction. Thus the observed RR may be greater or less than the true RR, depending on the eight values of sensitivity and specificity (Kleinbaum *et al*. 1982). For AF_e and AF_p, the observed values can be either attenuated towards zero or increased towards one.

The computer program

While formulae for assessing the impact of misclassification are available (Copeland *et al*. 1977; Gullen *et al*. 1968), most likely they are not routinely used since they involve extensive mathematical calculations. While it might be desirable to do a 'what-if' analysis to determine how the measures of RR, AF_e and AF_p are affected by various changes in sensitivity and specificity, it would be cumbersome to do so by hand. Ideal for this form of analysis are spreadsheet programs. Here the various relationships are included as formulae which link one cell to another. Spreadsheets are reasonably easy to program and are most useful for quickly determining the effect changes in table entries have on other variables. The program to be shown was written by one of us (R.R.F.) and has been used for some while to teach epidemiology graduate students about the potential effects of misclassification. The investigator or student sits in front of the microcomputer monitor, changes the values of the various parameters, pushed the recalculation key, and immediately sees the effect on the outcome variables.

Assume that 5000 persons are to be enrolled in a two-year cohort study. Based on other investigations, we expect that the cumulative incidence (or risk) of disease during the two-year period is 20 cases per 1000 or 0.02, that the proportion exposed to the risk factor is 0.40, and that the 'true risk ratio is 20. With these values, we would expect that the AF_e would be 0.95 and the AF_p would be 0.88, suggesting that control measures aimed at the risk factor would have a considerable effect on the subsequent decline of the disease. Yet, all of these values are based on the assumption that there is no misclassification of either the exposure history or disease status. The person using the program must first consider if misclassification is differential or non-differential. For our example, we assume that, based on past experience, the misclassification will be differential.

The output displayed on the computer monitor is shown in Figure 16.1. Eight reasonably high values for sensitivity and specificity of exposure and disease status are entered by the investigator in the Differential section of the program in the column under the word 'Values'. As is seen in the box labelled 'Observed', the RR is reduced from a true value of 20 to an observed value of 1.54. The AF_e and AF_p are also greatly reduced from 0.95 and 0.88 to 0.35 and 0.18, respectively, suggesting a much lower theoretical impact of a prevention programme than would really occur.

The misclassification program divides the 'true' fourfold table into four cells: A (exposed and disease), B (exposed and no disease), C (unexposed and disease), and D (unexposed and no disease). For cell A there are 93 true cases. Since all are exposed to the risk factor, we only determine the sensitivity of the exposure measurement for this group. Furthermore, since all 93 are true cases, we only determine the sensitivity for the disease measurement. The same rationale

```
        EFFECTS OF MISCLASSIFICATION OF BOTH EXPOSURE AND DISEASE

    Enter parameter estimates for either non-differential or differential
     misclassification but not both (i.e., leave one column blank) and the
          TRUE exposure and disease values assuming no misclassification
-----------------------------------------------------------------------
 NON-DIFFERENTIAL    Values      DIFFERENTIAL      Values
-----------------------------------------------------------------------
 EXPOSED STATUS                  EXPOSURE STATUS (among disease cases)
   Sensitivity =                   Sensitivity =      .94
   Specificity =                   Specificity =      .97
 DISEASE STATUS                  EXPOSURE STATUS (among non-cases)
   Sensitivity =                   Sensitivity =      .92
   Specificity =                   Specificity =      .95
                                 DISEASE STATUS (among exposed)
                                   Sensitivity =      .98
                                   Specificity =      .96
                                 DISEASE STATUS (among unexposed)
                                   Sensitivity =      .97
                                   Specificity =      .95
-----------------------------------------------------------------------
         TRUTH                           ENTER VALUES
-----------------------------------------------------------------------
 Number of subjects in the study          =    5,000
 Proportion developing the DISEASE (total) =    .020
 Proportion EXPOSED to the factor         =    .40
 Risk Ratio                               =    20.0
-----------------------------------------------------------------------
                                 *************************************
         TRUTH                   *            OBSERVED               *
                                 *                                   *
       Disease                   *          Disease                  *
     Yes     No                  *        Yes     No                 *
                                 *       ----------------            *
 Exposed |  93 | 1,907 | 2,000   * Exposed | 164 | 1,828 | 1,992     *
 --------                        *       ----------------            *
 Unexposed|  7 | 2,993 | 3,000   * Unexposed| 160 | 2,848 | 3,008    *
                                 *                                   *
          100   4,900   5,000    *        324   4,676   5,000        *
                                 *                                   *
          Ie =  .047             *        Ie =  .082                 *
          Iu =  .002             *        Iu =  .053                 *
          RR = 20.00             *        RR = 1.54                  *
          AFe =  .95             *        AFe =  .35                 *
          AFp =  .88             *        AFp =  .18                 *
                                 *************************************
-----------------------------------------------------------------------

   TRUTH  Cell A =   93              TRUTH  Cell B = 1,907

       Disease                          Disease
     Yes     No                       Yes     No
     ----------                       ----------
 Exposed |  86 |   2 |  87        Exposed |  70 | 1,684 | 1,754
 --------                         --------
 Unexposed|  5 |   0 |   6        Unexposed|  6 |  146 |   153
          91      2     93                 76   1,831   1,907

 EXPOSURE STATUS (among disease cases)  EXPOSURE STATUS (among non-cases)
   Sensitivity =   .94                    Sensitivity =   .92

 DISEASE STATUS (among exposed)         DISEASE STATUS (among exposed)
   Sensitivity =   .98                    Specificity =   .96


   TRUTH  Cell C =    7               TRUTH  Cell D = 2,993

       Disease                          Disease
     Yes     No                       Yes     No
     ----------                       ----------
 Exposed |   0 |   0 |   0        Exposed |   7 |  142 |   150
 --------                         --------
 Unexposed|  7 |   0 |   7        Unexposed| 142 | 2,701 | 2,843
           7      0     7                  150   2,843   2,993

 EXPOSURE STATUS (among disease cases)  EXPOSURE STATUS (among non-cases)
   Specificity =   .97                    Specificity =   .95

 DISEASE STATUS (among unexposed)       DISEASE STATUS (among unexposed)
   Sensitivity =   .97                    Specificity =   .95
```

Fig. 16.1. Image on monitor screen for spreadsheet program to assess the potential impact of misclassification.

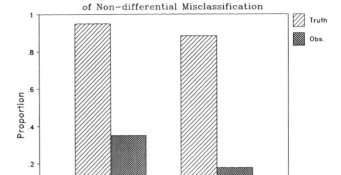

Effects on Attributable Fraction
of Non-differential Misclassification

Fig. 16.2. Computer-generated graph of the effects of measurement error due to differential misclassification on true and observed values of attributable fraction among exposed (AF_e) and attributable fraction in total population (AF_p).

The spreadsheet program also permits the immediate visualization of the study findings as a graph, either on the monitor or printed out in hard copy (see Fig. 16.2).

Rapid computer-assisted surveys in developing countries

Questions are often asked by decision-makers in developing countries, as elsewhere, about the burden that illness places on people at the community level and how resources are being allocated to address the various health problems. While the questions are easy to ask, the answers are often hard to come by. One solution is to rely more extensively on health surveys. If done correctly, trained interviewers or examiners can sample populations and obtain standardized data of high quality. Until recently, however, the processing and analysis of most survey data has been so cumbersome and has taken so long that information is old and often irrelevant before it is available for dissemination. All this is now changing with the development of Rapid Survey Methodology (RSM). Using portable battery-powered computers and contemporary software, surveys can now be done very quickly in even the poorest of countries (Frerichs 1989; Frerichs and Tar Tar 1988a, b, 1989). For example, using RSM in rural Burma, health department staff were able in less than five days to gather the appropriate survey data, complete the analysis, prepare simple tables and graphs, and present the survey findings to the local Township Medical Officer and his staff. Typical of graphs prepared in the field is the one of DPT immunization coverage shown in Figure 16.3. Another five days later, they presented a 50-page report written with a word-processing program to the Director of Public Health in Rangoon. Thus, within 10 days of going into the field, the Burmese health department staff were able to

holds for the other three cells. Note that the numbers in the cells appear as integers but are maintained as real numbers for calculation purposes. Thus hand calculations of the values in the individual cells may not produce the same exact answers.

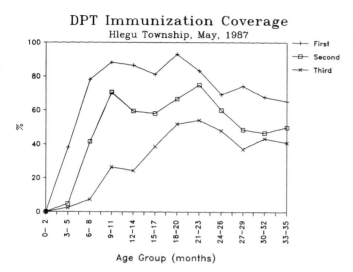

Fig. 16.3. Computer-generated graph of DTT (diphtheria/pertussis/teta-nus) immunization coverage by age group among Burmese children.

provide detailed information necessary for decision-making both at the local and national levels.

Two similar surveys were conducted six months later in a rural province of Thailand, one focusing on family planning practices and the other on antenatal care (Frerichs 1989). As in Burma, the two survey teams were in the field for three days, spent one more day on the analysis, and gave a presentation of the survey results to the Provincial Chief Medical Officer and his staff on the fifth day.

Several developments have taken place in recent years which fostered the development of RSM and enable community-based surveys to be done more quickly than ever before. First is the development of sampling methods applicable to developing countries. The Expanded Program on Immunization of the World Health Organization (EPI/WHO) has for years been using a two-stage cluster sampling scheme to assess immunization coverage throughout the world (Expanded Program on Immunization 1987; Henderson and Sundaresan 1982). Until recently, however, the validity of the sampling method has not been tested. Using computer simulation techniques, Lemeshow and Robinson (1985), have shown that the approach used by EPI/WHO does provide valid estimates, at least within the range specified by the sampling proponents. Others have successfully used the EPI/WHO sampling procedure to assess the occurrence of disease (Rothenberg *et al.* 1985) or the use of health services (Expanded Program on Immunization 1986). Second is the emergence of relatively inexpensive 'laptop' computers. These small, but powerful, computers usually weigh between 6 and 14 pounds, are battery powered, and are relatively resistant to the effects of unstable electrical power, heat, humidity, and dust. Concurrent with the development of these computers has been small, portable printers which also operate on batteries. Third is the availability of 'user-friendly' software for operating the computers.

To assemble all these components into a working unit requires an initial investment and a change in thinking. Since the price of computer hardware and software has come down dramatically, epidemiologists in developing countries are finding that both portable and desktop computers are increasingly affordable. The change in thinking, however, may take more time. In the US, most children have been introduced to microcomputers before their tenth birthday either at school or in the home. As a result, they are comfortable with computers and are willing to use them for new and varied applications. In the less developed regions of the world, however, this is clearly not the case. Few computers are available, although the introduction of microcomputers has been progressing at a more rapid pace (Bertrand 1985). While it would seem reasonable that health professionals in developing countries would be apprehensive when faced with a new computer, we have found just the opposite. Based on our experience in Bangladesh and more recently in Burma, it has become apparent that once computers are introduced it takes very little time for English-speaking professionals to master their use (Frerichs and Miller 1985; Gould and Frerichs 1986).

For doing rapid surveys, we stress that the portable computer is an instrument of the investigator, much as a stethoscope is an instrument of the clinician. He or she uses the computer to do all the steps in survey work with one exception, data entry. Here some other lower-paid person is usually asked to complete this time-consuming task. Thereafter, however, the investigator does all the editing and analysis, prepares the appropriate graphs and tables, and writes the final report.

Table 16.2 shows the necessary steps to be completed when doing a rapid survey. First, using the word-processing program, the survey instrument is constructed. Second, the investigator decides the desired level of precision for the estimate and determines the sample size. A special routine is easily written to assess the desired sample size using various spreadsheet programs. Third, revisions are rapidly made and the final survey instrument is printed. Fourth, the necessary forms to manage the study are constructed and printed. Fifth, a spreadsheet program is used to select the necessary clusters with probability proportionate to size of the population residing in the clusters (the EPI/WHO routine). Sixth, the data are collected in the field. Note that at this step, the computer is not used. We never take a computer with us into the home of a respondent. Our fear is that it would be too disruptive both for the interviewer and the respondent. Instead, the machine is maintained in a central location where the survey team assembles in the evening. Seventh, the data are entered into the computer each day and, on the final day, the analysis is completed. Since processing and editing is done on a daily basis, the investigator can review the findings and send the interviewer or examiner back to a respondent's home to check blank or unreasonable data entries. An eighth step may be necessary if more sophisticated analyses are desired. Many analysis packages are available, although all require a

Table 16.2. Steps for doing a rapid, computer-assisted survey

Description of steps	Necessary hardware	Necessary software
Development of survey instrument for feasibility testing	Microcomputer Printer	Word processing
Determination of necessary sample size	Microcomputer	Spreadsheet
Development of final survey instrument	Microcomputer Printer	Word processing
Development of control forms for managing the survey	Microcomputer Printer	Form preparation on Spreadsheet
Selection of sample clusters with probability proportionate to size	Microcomputer Printer	Spreadsheet
Gathering of data (interview and/or examination)	—	—
Processing and initial analysis of data	Portable computer Portable printer Spreadsheet	Data base Statistical
Final analysis of data (optional)	Microcomputer Printer	Statistical
Preparation of final report	Microcomputer Printer	Word processing

more detailed knowledge of statistics. Finally, the ninth step uses word-processing software to prepare the final report. All nine steps can be completed in less than three weeks.

View of the future

The computer revolution is continuing at such a rapid pace that many epidemiologists look back with feelings of nostalgia at favourite computers from the past—with the past being five years earlier. The individual investigator is now able to conduct studies using a microcomputer that previously would have required a multi-disciplinary team. This empowerment, however, is not easy to acquire. The epidemiologist must be willing to learn new approaches to gathering, processing, and analysing study data. The microcomputer has quickly become an essential component for the working epidemiologist, one that will help him or her conduct efficient investigations or assist planners or policy-makers in most countries of the world.

The applications mentioned in this chapter are intended mainly to stimulate the imagination rather than provide a detailed protocol for the respective studies. With some exceptions we focused on the future, relying on the continuing evolution of both computer hardware and software to maintain a contemporary focus. With an open mind and proper training, there is no reason the approaches we have outlined should not become reality for epidemiologists throughout the world.

We have hesitated to mention specific computers or software in this chapter because of how rapidly they are changing; they would be obsolete by the time this chapter was read. Instead we have tried to focus on general categories of hardware and software which should be available in the future as well. In addition, we have described four applications of epidemiological importance using current technology in ways

that were not possible even a few years earlier. Now it is time to look into the future.

Hardware

The next generation of desktop microcomputer will have 64 megabytes (i.e. space for 64 million characters) of RAM, a 300 megabyte hard disk drive which serves as a holding area for 1–10 gigabytes of erasable optical storage, a 20-megabyte 3.5-inch floppy disk drive, and 2000 × 2000 pixel screen resolution with the ability to create thousands of on-screen colours. An internal modem will be built in and be able routinely to transmit data at 9600 bits per second. A microphone will attach to the CPU for dictating and converting oral words to written words. Both a page scanner and electronic pencil will be common for entry of book chapters, journal articles, photographs, and hand-written notes.

Portable or laptop computers will also be more powerful, although not as powerful as desktop machines. Colour screens will soon be the norm on the more expensive laptop computers, which will weigh between 5 and 10 pounds depending on advances in battery technology. Less expensive laptop computers with black-and-white screens will become widely available in developing countries, relying on solar power to charge the internal batteries.

Printers will produce a wide variety of colours using mainly laser technology, although some high-resolution dot-matrix printers will still be popular for personal use. Portable printers will use ink-jet technology to create high-resolution black-and-white text and graphs. As with portable computers, the size and weight of portable printers will depend on advances in battery design.

Software

Most of the applications described earlier in this chapter will be available, but epidemiologists will probably not use them as single programs. Instead, they will fit together in combinations, dependent on the needs of the epidemiologist. Word-processing software will be able to integrate complex statistical formulas, graphs, photographs, and handwritten notes into documents which will print an image on paper the same as it appears on the computer monitor. Software will be available for translating written passages between English, Spanish, and French, facilitating professional communication. While analyses will be completed faster than before, the basic statistical and analytical methods will still be the same, although the programs will be easier to use once these methods are understood. Current information on specific diseases will be readily available in the US and possibly other countries using a modem and communication software to link with computers at the Centers for Disease Control or the National Library of Medicine.

Probably the most dramatic changes will occur in graphics software. Besides the conventional bar and line charts, the epidemiologist will be able to visualize complex relationships in three-dimensional forms using a wide variety of shapes and colours. Mapping programs will be common and will be routinely used by public health agencies to view disease patterns by basic geographical units.

While keeping abreast of the changes will not be easy, the epidemiologist who invests time in learning how to use the new hardware and software will find it all the easier to explain his or her study findings to other researchers, administrators, planners, or the general public. If so, the promise of this new technology will have been fulfilled.

References

Bertrand, W.E. (1985). Microcomputer applications in health population surveys: experience and potential in developing countries. *World Health Statistics Quarterly* **38**, 91.

Copeland, K.T., Checkoway, H., Holbrook, R.H., and McMichael, A.J. (1977). Bias due to misclassification in the estimate of relative risk. *American Journal Epidemiology* **105**, 488.

Dean, J.F., Dean, A.G., Burton, A., and Dicker, R. (1988). *Epi Info—computer programs for epidemiology*. Division of Surveillance and Epidemiologic Studies, Epidemiology Program Office, Centers for Disease Control, Atlanta, Georgia.

Expanded Program on Immunization (1986). Program review. *Weekly Epidemiologic Record* **4**, 21.

Expanded Program on Immunization (1987). Immunization coverage surveys. *Weekly Epidemiologic Record* **25**, 183.

Frerichs, R.R. (1989). Simple analytic procedures for rapid micro-computer-assisted cluster surveys. *Public Health Reports* **104**, 24.

Frerichs, R.R. and Miller, R.A. (1985). Introduction of a microcomputer for health research in a developing country—the Bangladesh experience. *Public Health Reports* **100**, 638.

Frerichs, R.R. and Tar Tar, K. (1988*a*). Use of rapid survey methodology to determine immunization coverage in rural Burma. *Journal of Tropical Pediatrics* **34**, 125.

Frerichs, R.R. and Tar Tar, K. (1988*b*). Breastfeeding, dietary intake and weight-for-age of children in rural Burma. *Asia-Pacific Journal of Public Health* **2**, 16.

Frerichs, R.R. and Tar Tar, K. (1989). Computer-assisted rapid surveys in developing countries. *Public Health Reports* **104**, 14.

Gould, J.B. and Frerichs, R.R. (1986). Training faculty in Bangladesh to use a microcomputer for public health: follow-up report. *Public Health Reports* **101**, 616.

Greenland, S. (1987). Preface. In *Evolution of epidemiologic ideas: Annotated readings on concepts and methods* (ed. S. Greenland). Epidemiology Resources, Chestnut Hill, Massachusetts.

Gullen, W.H., Bearman, J.E., and Johnson, E.A. (1968). Effects of misclassification in epidemiologic studies. *Public Health Reports* **53**, 1956.

Henderson, R.H. and Sundaresan, T. (1982). Cluster sampling to assess immunization coverage: a review of experience with a simplified method. *Bulletin of the World Health Organization* **60**, 253.

Jordan, M.D. (1987). *The CDC Anthropometric Software Package (CASP)*. Division of Nutrition, Center for Health Promotion and Education, Centers for Disease Control, Atlanta, Georgia.

Kelsey, J.L., Thompson, W.D., and Evans, A.S. (1982). *Methods in observational epidemiology*. Oxford University Press, New York.

Kleinbaum, D.G., Kupper, L.L., and Morgenstern, H. (1982). *Epidemiologic research—principles and quantitative methods*. Lifetime Learning Publications, Belmont, California.

Lemeshow, S. and Robinson, D. (1985). Surveys to measure program coverage and impact: a review of the methodology used by the Expanded Program on Immunization. *World Health Statistics Quarterly* **38**, 65.

McNichols, C.W. and Rushinek, S.F. (1988). *Data base management—a microcomputer approach*. Prentice Hall, Englewood Cliffs, New Jersey.

Rothenberg, R.B., Lobanov, A., Singh, K.B., and Stroh Jr, G. (1985). Observations on the application of EPI cluster survey methods for estimating disease incidence. *Bulletin of the World Health Organization* **63**, 93.

Rothman, K.J. (1986). *Modern epidemiology*. Little Brown and Company, Boston, Massachusetts.

Schlesselman, J.J. (1982). *Case–control studies*. Oxford University Press, New York.

Selwyn, B.J. (1985). Graphic packages and graphs in epidemiology. *Epidemiology Monitor* **6**, 3.

17

Epidemiology: the foundation of public health

R. DETELS

The previous chapters in this section have presented the principles and methods of epidemiology. In this chapter, I will attempt to define epidemiology, to present ways in which epidemiology is used in the advancement of public health, and, finally, to discuss the range of applications of epidemiological methodologies.

What is epidemiology?

There are probably as many definitions of epidemiology as there are epidemiologists, although every epidemiologist will know exactly what it is that he or she does. Defining epidemiology is difficult primarily because it does not represent a body of knowledge, as does, for example, anatomy, nor does it target a specific organ-system, as does cardiology. Epidemiology represents a philosophical method of studying a health problem, and can be applied to a wide range of problems, ranging from transmission of an infectious disease-agent to the design of a new strategy of health-care delivery. Furthermore, that methodology is continually changing as it is adapted to a greater range of health problems and more techniques are borrowed and adapted from other disciplines such as mathematics and statistics.

Maxcy, one of the pioneer epidemiologists of this century, offered the following definition: 'Epidemiology is that field of medical science which is concerned with the relationship of various factors and conditions which determine the frequencies and distributions of an infectious process, a disease, or a physiologic state in a human community' (Lilienfeld 1978). The word itself comes from the Greek *epi*, *demos*, and *logos*; and literally translated means the study (*logos*) of what is among (*epi*) the people (*demos*). All epidemiologists will agree that epidemiology concerns itself with populations rather than individuals, thereby separating itself from the rest of medicine and constituting the basic science of public health. Following from this therefore, is the need to describe health and disease in terms of frequencies and distributions in the population. The epidemiologist relates these frequencies and distributions of specific health parameters to the frequencies of other factors to which populations are exposed in order to identify those which may be causes of a disease or

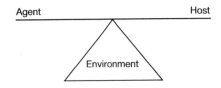

Fig. 17.1. The triadic relationship between agent, host, and environment in epidemiology.

promoters of good health. Inherent in the philosophy of epidemiology is the idea that ill health is not randomly distributed in populations, and that elucidating the reasons for this non-random distribution will provide clues regarding the risk-factors for disease and the biological mechanisms which result in loss of health. Because epidemiology usually focuses on health in *human* populations it is rarely able to provide experimental proof in the sense of Koch's postulates, as can often be done in the laboratory sciences. Epidemiology more often provides an accumulation of increasingly convincing indirect evidence of a relationship between health or disease and other factors.

Although they will differ on the exact definitions of epidemiology, most epidemiologists will agree that they try to characterize the relationship between the agent, the environment, and the host (usually man). The epidemiologist considers health to represent a balance among these three forces, as shown in Figure 17.1.

Changes in any one of these three factors may result in loss of health. For example, the host may be compromised as a result of treatment with steroids, making him/her more susceptible to agents which do not ordinarily cause disease. On the other hand, a breakdown in the water-supply system may result in an increased exposure of people to hepatitis B, as happened some years ago when the main water-supply of New Delhi, the Jumna river, was drastically reduced by drought. Finally, some agents may become more or less virulent over time, thereby disturbing the dynamic balance among agent, host, and environment.

The epidemiologist uses another triad to study the relationship of agent, host, and environment: time—place—person. Using various epidemiological techniques described in previous chapters, the epidemiologist describes the loss of health in terms of the characteristics of time (for example, trends, outbreaks, etc.), the agent (transmissibility, the usual reservoir, host-range, etc.), and the host (demographic, sociological and biological characteristics, susceptibility, immune response, etc.). By describing the agent, host, and environment in terms of time, place, and person, the epidemiologist is often able to elucidate the causative agent, the natural history of the disease, and risk-factors which increase the likelihood of the host's acquiring the disease. With this information the epidemiologist is often then able to suggest ways to intervene in the disease-process to prevent disease or death.

Epidemiology has been described as the 'art of the possible'. Because epidemiologists work with human populations, they are rarely able to manipulate events or the environment as can the laboratory scientist. They must, therefore, exploit situations as they exist naturally to advance knowledge. They must be both pragmatic and modest (realistic). They must realize both what is possible and what are the limitations of the discipline. Morris has said that the 'epidemiologic method is the only way to ask some questions . . . , one way of asking others and no way at all to ask many' (Morris 1975). The art of epidemiology is to know both when epidemiology is the method of choice and how to use it to answer the question.

Applying the epidemiological method to resolve a health question successfully can be compared to constructing a memorable Chinese banquet. It is not enough to have the best ingredients and to know the various Chinese cooking methods. The truly great Chinese chef must be able to select the appropriate ingredients and cooking methods for each individual dish, and, further, must know how to construct the correct sequence of dishes to excite the palate without overwhelming it. He creates the memorable banquet by adding his creative genius to the raw ingredients and the established cooking methods. Similarly, it is not enough for the epidemiologist to know the various strategies and methods of epidemiology; she or he must be able to apply them creatively to obtain the information needed to understand the natural history of the disease. It is not enough to know what a cohort study is; the epidemiologist must know when the cohort design is the appropriate design for the question at hand, and then must apply that design appropriately and creatively. It is this essential skill that makes epidemiology more than a methodology. It is this opportunity for creativity and innovation that provides excitement for the practitioner and makes the successful practice of epidemiology an art.

For example, Imagawa and colleagues recently demonstrated the existence of HIV-1-infected individuals who were antibody-negative by culturing men for HIV-1 who were antibody-negative, but who persisted in activities which conveyed a very high risk of infection (Imagawa *et al.* 1989). A simple cohort study of antibody–negative individuals would have required a cohort of thousands of men rather than the 133 included in the Imagawa example. The effects of passive smoking were demonstrated by doing cohort studies of non-smoking family-members of smokers. Using this population, Hirayama demonstrated that spouses of smokers had a higher lung-cancer incidence than spouses of non-smokers (Hirayama 1981). Tager *et al.* (1979) and Tashkin *et al.* (1984) demonstrated that children of smokers had lower levels of lung-function that children of non-smokers. All of these investigators used the traditional cohort-study design, but demonstrated their creativity by applying that design to specific populations which were most likely to reveal a relationship if it existed.

Uses of epidemiology in support of public health

Epidemiology is the basic science of public health because it is the health science which describes health and disease in *populations* rather than in individuals, information essential for the formulation of effective public-health initiatives to prevent disease and promote health in the community. Epidemiology can be used for the following.

1. *Describe the spectrum of disease.*
Disease represents the end-point of a process of alteration of the host's biological systems. Although many disease-agents are limited in the range of alterations they can initiate, others such as measles can cause a variety of disease end-points. For example, the majority of infections with rubeola (the measles virus) result in the classical febrile, blotchy rash-disease; but the rubeola virus can also cause generalized haemorrhagic rash and acute encephalitis. Years after initial infection, measles can also cause subacute sclerosing panencephalitis (SSPE), a fatal disease of the central nervous system. Measles virus is also suspected of being a causative factor in multiple sclerosis (Alter 1976; Sullivan *et al.* 1984).

Various types of epidemiological studies have been used to elucidate the spectrum of disease resulting from many agents and conditions. Cohort studies have been used to document the role of high blood-pressure as a major cause of stroke, myocardial infarct, and chronic kidney disease. For rare diseases such as SSPE and multiple sclerosis, case–control studies have been useful to identify the role of the rubeola virus (Alter 1976; Detels *et al.* 1973). Knowing the spectrum of disease which can result from specific infections and conditions allows the public-health professional to design more effective intervention strategies: for example, education, screening, and treatment programmes to reduce the prevalence of high blood-pressure will also reduce the incidence of myocardial infarct, stroke, and chronic kidney disease (Hypertension Detection and Follow-up Program Co-operative Group 1979).

2. *Describe the natural history of disease.*
Epidemiological studies can be used to describe the natural history of disease and to elucidate the specific alterations in the biological system in the host. For example, cohort studies of individuals who were infected with Human Immuno-

deficiency Virus (HIV), the AIDS virus, revealed that a drop in the level of T-lymphocytes having the CD4 marker was associated with infection with HIV, and that a further decline in CD4 cells was associated with developing clinical symptoms and AIDS (Detels *et al.* 1987; Polk *et al.* 1987). This observation stimulated immunologists to focus their research on the interaction of the immune system and the HIV. From a clinical perspective, clinicians can target HIV-antibody-positive individuals who have declining CD4 cells for prophylactic treatment when it is most likely to be effective. Thus describing the natural history of AIDS has assisted researchers to focus their studies and clinicians to use the limited treatment-modalities available more effectively.

3. *Identify factors which increase the risk of acquiring disease.*

Having some factors increases the probability that individuals will develop disease. These 'risk-factors' may be social (smoking, drinking), genetic (ethnicity), dietary (saturated fats, vitamin deficiencies), and so on. Knowing these risk-factors can provide public-health professionals with the necessary tools to design effective programmes to intervene before disease occurs. For example, descriptive studies, cross-sectional studies, case–control studies, cohort studies, and intervention studies have all shown that smoking is the biggest single risk-factor for ill health, because it is a major risk-factor for cardiovascular disease, chronic respiratory disease, and many cancers (for example, of the lung, nasopharynx, and bladder), the leading causes of disability and death in developed countries. Health-education campaigns to stop or reduce smoking are now a major public-health initiative in all developed countries of the world.

4. *Predict disease trends.*

The ability to predict future epidemics provides the public-health professional with the opportunity to muster the most effective forces to combat the disease. Descriptive studies of many infectious diseases, such as measles, polio, and influenza, have revealed a periodicity of pandemics and epidemics caused by them. Knowledge of these disease-patterns has, in the past, been useful to public-health officials in preparing for these epidemics.

More recently, studies of the trends of infection with human immunodeficiency virus in high-risk groups and of changing frequency of high-risk activities have permitted epidemiologists and statisticians to develop models which predict the number of cases of AIDS likely to occur in 5 to 10 years (Brookmeyer and Damiano 1989; Taylor 1989). This information is particularly useful for public-health professionals who must anticipate future health-care needs.

5. *Elucidate mechanisms of disease-transmission.*

Understanding the mechanisms of disease-transmission can suggest ways in which public-health professionals can protect the public by stopping transmission of the disease agent. Epidemiological studies of the various arboviral encephalitides have incriminated certain species of mosquitoes as the vec-

tors of disease and specific animals as the reservoirs for the virus. For example, public-health efforts in California to prevent western equine encephalitis have concentrated on eradication of the mosquito vector and vaccination of horses, which are the reservoir of the virus. Although an effective vaccine for smallpox had been available for almost two hundred years, eradication of the disease was not achieved until the recognition that the low infectivity of varicella virus and the relatively long incubation for development of smallpox could be used to develop a strategy of surveillance for cases, with identification and immediate vaccination of all susceptible contacts (containment). Using this strategy, smallpox was eradicated through a world-wide effort in less than ten years (Fenner *et al.* 1988).

6. *Test the efficacy of intervention strategies.*

A primary objective of public health is to prevent disease through intervention in the disease process. But a vaccine or other intervention programme must be proven to be effective before it is used in the community. Epidemiological studies (double-blind placebo-controlled trials) are a necessary step in developing an intervention programme, whether that programme is administration of a new vaccine or a behavioural-intervention strategy to stop smoking. Although it may be argued that injection of saline is no longer considered ethical, a proven vaccine can often be used as a placebo for a trial of a new vaccine for a different disease (Detels *et al.* 1969). Widespread use of an intervention not subjected to epidemiological studies of efficacy may result in implementation of an ineffective intervention programme at great public expense, and may actually result in greater morbidity and mortality, because of an increased reliance on the favoured but untested intervention and a reduced use of other strategies, which are thought to be less effective, but which are actually more effective.

7. *To evaluate intervention programmes.*

Although an intervention such as a vaccine may have been demonstrated to have efficacy in double-blind trials, it may fail to provide protection when used in the community. Double-blind trials may demonstrate the 'biological efficacy' of the vaccine; but if the vaccine is not acceptable to the majority of the public they will refuse to be vaccinated, and the 'public-health efficacy' of the vaccine will be very low. For example, the typhoid vaccine may provide some protection against small infecting inocula; but the frequency of unpleasant side-effects and the need for multiple injections influence many people against being vaccinated (Hornick 1982).

Another problem of inferring public-health efficacy from small vaccine trials is that volunteers for vaccine trials may not be representative of the general public which needs to be protected against a specific disease. Thus broad-based intervention trials also need to be carried out, to demonstrate the acceptability and public-health efficacy of a vaccine or other intervention to the population in need of protection.

Since there are adverse side-effects associated with any

vaccine, ongoing evaluations of the cost-benefit relationship of specific vaccines are important. By comparing the incidence of smallpox with the incidence of adverse side-effects from the vaccine, Lane *et al.* (1969) demonstrated that more disease resulted from use of the vaccine in the US than from importation of cases. This study led to revisions in the smallpox vaccination policy in the US. More recently, the report of adverse effects from the use of pertussis vaccine has led to re-evaluation of the routine use of DPT vaccines in infants (Miller *et al.* 1982; Cody *et al.* 1981). Although most public-health experts continue to recommend DPT for immunization of normal infants, there is now a much better appreciation of the changing risk–benefit relationship for the pertussis vaccine.

There are several epidemiological strategies which can be used for ongoing evaluation of intervention programmes. Serial cross-sectional studies can be used to determine if there has been a change in the prevalence of disease or of indicators of health status over time. The cohort design can be used to compare incidence of disease in comparable populations receiving and not receiving the prevention programme. The case–control design can be used to determine if there are differences in the proportion of cases and non-cases who had the intervention programme.

8. *Identify the health needs of a community.*

To be effective in promoting the health of a community or country, health agencies must know what the major health problems of that community are and which subgroups in it are most affected. Cross-sectional studies will reveal the prevalence of disease in the community as well as in specific subgroups of the population, while surveillance programmes can identify trends in disease, infection, and/or health status over time. For example, the prevalence of untreated high blood-pressure is high in most developed countries, and disproportionately high in some subgroups, such as blacks and the poor. In the last ten years the prevalence has declined; but to a lesser degree in blacks and the poor than in the rest of the population. With this information, many health departments are now focusing their education and screening programmes for high blood-pressure on blacks and the poor.

9. *Evaluate public health programmes.*

Departments of health are engaged in a variety of activities to promote the health of the community, ranging from vaccination programmes to clinics for the treatment of specific diseases in the needy. Ongoing evaluation of such programmes is necessary to assure that they continue to be cost-effective. Periodic review of routinely collected health statistics can provide information about the effectiveness of many programmes. For those programmes for which relevant statistics are not routinely available, cohort studies and serial cross-sectional studies of the incidence and prevalence of the targeted disease in the populations who are the intended recipients of these programmes can measure whether the programmes have had an impact and are cost-effective.

Applications of epidemiology

Specific epidemiological study-designs are used to achieve specific public-health goals. These goals range from identifying a suspected exposure–disease relationship to establishing that relationship, to designing an intervention to prevent it, and, finally, to assessing the effectiveness of that intervention. The usual sequence of study-designs in the identification and resolution of a disease-problem are:

- Descriptive studies
- Cross-sectional (prevalence) surveys
- Case–control studies
- Cohort studies
- Experimental studies

There are, however, many exceptions to the application of this sequence of study-designs, depending on such things as the prevalence and virulence of the agent and the nature of the human response to the agent.

The earliest suspicion that a relationship exists between a disease and a possible causative factor is frequently obtained from observing correlations between exposure and disease from existing data, often collected through prevalence surveys or surveillance programmes, or from mortality statistics. These can be correlations observed across geographical areas (ecological studies) or over time, or a combination of both. Many of the initial epidemiological investigations into chronic bronchitis used vital statistics data, particularly data on mortality (Colley 1985). Case–control studies identified smoking as a possible factor (Oswald *et al.* 1953). Subsequent prevalence studies confirmed the relationship (Higgins 1974), as have cohort studies (Fletcher *et al.* 1976; PHS 1979). Finally, a decline in respiratory symptoms of chronic bronchitis and a slower decline in lung-function have been observed in individuals who cease smoking (Fletcher *et al.* 1976).

Although this is the usual sequence in which the various epidemiological study-designs are applied, there are exceptions to this sequence. Furthermore, all study-designs are not appropriate to answer all health questions. The usual applications of each of the different epidemiological study-designs and the limitations of each are, therefore, presented below.

Descriptive studies

The use of existing statistics to correlate the prevalence or incidence of disease to the frequency or trends of suspected causal factors in specific localities or over time has often provided the first clues that a particular factor may cause a specific disease. These epidemiological strategies, however, document only the co-occurrence of disease and other factors in a population: the risk-factor and the disease may not be occurring in the same people. These types of descriptive studies are cheap and relatively easy to do, but the co-occurrence observed may be due merely to chance. For example, the incidence of both heart-disease and lung cancer has increased concurrently with the prevalence of automatic dishwashers in the US. Few people, however, would attribute the increase in these two diseases to the use of automatic dishwashers. Thus, descriptive studies usually only provide a rationale for undertaking more expensive analytic studies.

Cross-sectional/prevalence surveys

Cross-sectional/prevalence surveys establish the frequency of disease and other factors in a community. Since they require the collection of data, however, they can be expensive. They are useful to estimate the number of people in a population who have disease, and can also identify the difference in frequency of disease in different subpopulations. This information is particularly useful to health administrators who are responsible for developing appropriate and effective public-health programmes. Cross-sectional studies document the co-occurrence of disease and suspected factors not only in the population, but also in specific individuals within the population. The cross-sectional study design is useful to study chronic diseases such as multiple sclerosis and chronic bronchitis, which have a high prevalence, but an incidence that is too low to make a cohort study feasible (Detels *et al.* 1978). On the other hand, they are not useful for studying diseases which have a low prevalence, such as SSPE. Cross-sectional studies are subject to problems of respondent bias and undocumented confounders. Further, unless historical information is obtained from all the individuals surveyed, the time-relationship between the factor and the disease is not known.

Case–control studies

The case–control study compares the prevalence of suspected factors between cases and controls. If the prevalence of the factor is significantly different in cases than it is in controls, this suggests that that factor is associated with the disease. Although case–control studies can identify associations, they do not measure risk. An estimate of risk, however, can be derived by calculating the odds ratio. Case–control studies are often the analytic study-design used to investigate a suspected association. Compared to cohort and experimental studies they are relatively cheap and easy to do. Cases can often be selected from hospital patients and controls either from hospitalized patients with other diseases or by using algorithms for selecting community or other types of controls. The participants are seen only once, and no follow-up is necessary. Although time-sequences can often be established for factors elicited by interview, they usually cannot be for laboratory-test results. Thus, an elevation in factor 'B' may either be causally related, or it may be a result of the disease-process, and not a cause. Furthermore, factors elicited from interview are subject to recall bias; for example, patients are often better motivated for recall of events than controls are, because they are concerned about their disease. The case–control study is particularly useful for exploring relationships noted in observational studies. A hypothesis, however, is necessary for case–control studies. Relationships will be observed only for those factors studied. Case–control studies

are not useful for determining the spectrum of health outcomes resulting from specific exposures, since a definition of a case is required in order to do a case–control study. On the other hand, case–control studies are the method of choice for studying rare diseases. Case–control studies are often indicated when a specific health-question needs to be answered quickly.

Cohort studies

Cohort studies have the advantage of establishing the temporal relationship between an exposure and a health-outcome, and thus they measure risk directly. Because the population studied is defined on the basis of its exposure to the suspected factor, cohort studies are particularly suitable for investigating health-hazards associated with environmental or occupational exposures. Further, cohort studies will measure more than one outcome of a given exposure, and therefore are useful for defining the spectrum of disease resulting from exposure to a given factor. Occasionally, a cohort study is done to elucidate the natural history of a disease when a group can be identified which has a high incidence of disease, but in which specific risk-factors are not known (Kaslow *et al.* 1987; Chmiel *et al.* 1987). Although this cohort is not defined on the basis of a known exposure, questions are asked and biological specimens are collected from which exposure variables can be identified concurrently or in the future. Unfortunately, cohort studies are both expensive and time-consuming. Unless the investigator can define a cohort from some time in the past, and has assurance that that cohort has been completely followed up for disease-outcome in the interim, the cohort design can take years to decades to yield information about the risks of disease resulting from exposure to specific factors. Persuading participants to remain in a cohort study for such long periods of time is both difficult and expensive. For this reason, cohort studies are usually done primarily to confirm an association that has been well established on the basis of descriptive, cross-sectional, and/or case–control studies. The size of the cohort to be studied is dependent in part on the anticipated incidence of the disease resulting from the exposure. For diseases with a very low incidence, cohort studies are often not feasible, either in terms of the logistics or of the expense of following very large numbers of people, or of both. Cohort studies establish the risk of disease associated with exposure to a factor, but do not 'prove' that the factor is causal. The observed factor may merely be very closely correlated with the real causative factor, or may even be related to the participants' choice to be exposed.

Experimental studies

Experimental studies differ from cohort studies because it is the investigator who makes the decision about who will be exposed to the factor. Therefore, confounding factors which may have led to the subjects' being exposed in the cohort studies are not a problem in experimental studies. Because epidemiologists usually study human populations, there are few opportunities for an investigator to deliberately expose participants to a suspected factor. On the other hand, intervention studies of individuals randomly assigned to receive or not receive the intervention programme which demonstrates a reduction in a specific health-outcome do provide strong evidence, if not proof, of a causal relationship. Because of the serious implications of applying an intervention which may alter the biological status of an individual, intervention studies are not undertaken until the probability of a causal relationship has been well established using the other types of study-designs.

The uses of limitations of the various epidemiological study-designs have been presented to illustrate and underscore the fact that the successful application of epidemiology requires more than a knowledge of study-designs and epidemiological methods. These designs and methods must be applied both appropriately and innovatively if they are to yield the desired information. The field of epidemiology has been expanding dramatically over the last two decades, as more epidemiologists have demonstrated new uses and variations of traditional study-designs and methods. We can anticipate that the uses of epidemiology will expand even more in the future as increasing numbers of creative epidemiologists develop new strategies and techniques of epidemiology.

Summary

Epidemiology is the basic science of public health because it is that science which describes the relationship of health or disease and other factors in human populations. Furthermore, epidemiology can be used to generate much of the information required by public-health professionals to develop and implement effective intervention programmes for the prevention of disease and the promotion of health. Finally, it is the best strategy to evaluate the effectiveness of public-health programmes.

Unlike pathology, which constitutes a basic area of knowledge, and cardiology, which is the study of a specific organ, epidemiology is a medical philosophy or methodology which can be applied to learning about and resolving a very broad range of health problems. The art of epidemiology is knowing when and how to apply epidemiological methods creatively to answer specific health-questions; it is not enough to know what the various study-designs and statistical methodologies are. Used innovatively, epidemiology can be one of the most effective tools science has to combat disease.

References

Alter, M. (1976). Is multiple sclerosis an age-dependent host response to measles? *Lancet* **i**, 456.

Brookmeyer, R. and Damiano, A. (1989). Statistical methods for short-term projections of AIDS incidence. *Statistics in Medicine* **8**, 23.

Chmiel, J.S., Detels, R., Kaslow, R.A., Van Raden, M., Kingsley, L.A., Brookmeyer, R., and the Multicenter AIDS Cohort Study Group (1987). Factors associated with prevalent Human Immunodeficiency Virus (HIV) infection in the Multicenter AIDS Cohort Study. *Amercian Journal of Epidemiology* **126**, 568.

Cody, C.L., Baraff, L.J., Cherry, J.D., *et al.* (1981). Nature and rates of adverse reactions associated with DTP and DT immunizations in infants and children. *Pediatrics* **68**, 650.

Colley, J.R.T. (1985). Respiratory system. In the *Oxford textbook of public health* (1st edn) (ed. W.W. Holland, R. Detels, and G. Knox). Vol. 4, p. 145. Oxford University Press.

Detels, R., Grayston, J.T., Kim, K.S.W., *et al.* (1969). Prevention of clinical and subclinical rubella infection: Efficacy of three HPV-77 derivative vaccines. *American Journal of Diseases of Children* **118**, 295.

Detels, R., Brody, J.A., McNew, J., and Edgar, A.H. (1973). Further epidemiological studies of subacute sclerosing panencephalitis. *Lancet* **ii**, 11.

Detels, R., Visscher, B.R., Haile, R.W., Malmgren, R.M., Dudley, J.P., and Coulson, A.H. (1978). Multiple sclerosis and age at migration. *American Journal of Epidemiology* **108**, 386.

Detels, R., Visscher, B.R., Fahey, J.L., *et al.* (1987). Predictors of clinical AIDS in young homosexual men in a high-risk area. *International Journal of Epidemiology* **16**, 271.

Fenner, F., Henderson, D.A., Arita, I., Jezek, Z., and Ladnyi, I.D. (1988). *Smallpox and its eradication*. WHO, Geneva.

Fletcher, C., Peto, R., Tinker, C., and Speizer, F.E. (1976). *The natural history of chronic bronchitis and emphysema*. Oxford University Press.

Higgins, I.T.T. (1974). *Epidemiology of chronic respiratory disease: A literature review*. Environmental Health Effects Research Series. EPA 650 1-74-007. Office of Research and Development, Environmental Protection Agency, Washington, DC.

Hirayama, T. (1981). Non-smoking wives of heavy smokers have a higher risk of lung cancer: A study from Japan. *British Medical Journal* **282**, 183.

Hornick, R.B. (1982). Typhoid fever. In *Bacterial infection of humans: Epidemiology and control* (ed. A.S. Evans and H.A. Feldman), p. 673. Plenum, New York.

Hypertension Detection and Follow-up Program Co-operative Group (1979). Five-year findings of the hypertension detection and follow-up program. I. Reduction in mortality of persons with high blood pressure, including mild hypertension. *Journal of the American Medical Association* **242**, 2562.

Imagawa, D.T., Lee, M.H., Wolinsky, S.M., *et al.* (1989). Human immunodeficiency virus type 1 infection in homosexual men who remain seronegative for prolonged periods. *New England Journal of Medicine* **320**, 1458.

Kaslow, R.A., Ostrow, D.G., Detels, R., Phair, J.P., Polk, B.F., and Rinaldo, C.R., jun. (1987). The Multicenter AIDS Cohort Study: Rationale, organization and selected characteristics of the participants. *American Journal of Epidemiology* **126**, 310.

Lane, J.M., Ruben, F.L., Neff, J.M., and Millar, J.D. (1969). Complications of smallpox vaccination, 1968. National surveillance in the United States. *New England Journal of Medicine* **281**, 1201.

Lilienfeld, D.E. (1978). Definitions of epidemiology. *American Journal of Epidemiology* **107**, 87.

Miller, D.L., Alderslade, R., and Ross E.M. (1982). Whooping cough and whooping cough vaccine: The risks and benefits debate. *Epidemiologic Reviews* **4**, 1.

Morris, J.N. (1975). *Uses of epidemiology*, 3rd edn. Churchill Livingstone, London.

Oswald, N.C., Harold, J.T.P., and Martin, W.J. (1953). Clinical pattern of chronic bronchitis. *Lancet* **ii**, 639.

Public Health Service (1979). *Smoking and health: A report of the Surgeon General*, USDHEW Publication No. (PHS) 79–50066. Office of the Assistant Secretary for Health, Office on Smoking and Health, Washington, DC.

Polk, B.F., Fox, R., Brookmeyer, R., *et al.* (1987). Predictors of the acquired immunodeficiency syndrome developing in a cohort of seropositive homosexual men. *New England Journal of Medicine* **316**, 61.

Sullivan, C.B., Visscher, B.R., and Detels, R. (1984). Multiple sclerosis and age at exposure to childhood diseases and animals: Cases and their friends. *Neurology* **34**, 1144.

Tager, I.B., Weiss, S.T., Rosner, B., and Speizer, F.E. (1979). Effect of parental cigarette smoking on the pulmonary function of children. *American Journal of Epidemiology* **110**, 15.

Tashkin, D.P., Clark, V.A., Simmons, M., *et al.* (1984). The UCLA Population Studies of Chronic Obstructive Respiratory Disease. VII. Relationship between parental smoking and children's lung function. *American Review of Respiratory Disease* **129**, 891.

Taylor, J.M.G. (1989). Models for the HIV infection and the AIDS epidemic in the US. *Statistics in Medicine* **8**, 45.

C

Social Science Techniques

18

Economic studies

M. DRUMMOND

Introduction

'Is this procedure cost-effective?' is a question often over-heard in debates concerning the provision of health services these days. This increased interest in the economic aspects of health services has been mirrored by a rapid growth in recent years in the publication of economic studies of health care alternatives. In a search of the literature for the period 1966–78, Warner and Hutton (1980) found more than 500 relevant references, growing from half a dozen per year at the beginning of the period to close to 100 in each of the most recent two years. They also found that the growth had been more rapid in medical than in non-medical journals.

This chapter reviews studies that assess the relative merits and demerits of health care programmes or treatments from an economic perspective. This general approach to the analysis of alternative strategies is known as the *cost–benefit approach* or *economic appraisal* and embodies specific analytical forms such as *cost–benefit analysis* and *cost–effectiveness analysis*. First, the reasons for wishing to apply this approach to health care choices will be discussed. Second, the methodological principles underlying both the general approach and its specific forms will be outlined, illustrated by reference to published studies. Finally, the contribution of this approach to health service decision-making will be considered and the problems and prospects for its use in the future discussed.

The background to the use of economic appraisal in health care

The main reason for wishing to apply economic appraisal to choices in health care is that resources are *scarce*. That is, there are not, and never will be, enough resources to satisfy human wants completely. This means that choice between alternative programmes both within and outside the health care system is inescapable, since more resources devoted to one beneficial activity automatically means that benefits which would arise from the use of those same resources in another activity are forgone. Within the health care field scarcity often manifests itself in terms of budget restrictions

or shortages of money, but it is important to recognize that the economist's notion of cost goes deeper than money expenditures. To an economist, the cost of an activity is the benefits that the resources consumed by the activity would generate in their best alternative use. It just happens that money expenditures are often a convenient measuring rod for costs since exchanges will, under certain conditions, reflect the value of resources and, in developed economies, generate money prices. However, to emphasize the conceptual difference between cost and money expenditure, economists often use the terms 'opportunity cost' or 'resource cost'.

If one accepts the notion of scarcity, it is clear that it is no longer sufficient to compare options in health care solely on the basis of the benefits they generate; comparisons should be made on the basis of benefits *and* costs, since costs merely reflect benefits forgone elsewhere. It is not difficult to find practical examples of this; the undisciplined use of high-technology diagnostic aids may mean that there are fewer resources to devote to the development of community services for the elderly or mentally handicapped. However, at least one commentator (Williams 1974) has argued that decisions in the health care field are too often made on the basis of one option being more beneficial than another—irrespective of cost—or of being cheaper and disregarding relative benefits. Doctors are more prone to the first error, accountants and health ministries to the second. Nevertheless, even if one accepts that health care alternatives should be compared on the basis of relative costs and benefits, there are numerous problems in converting this undeniable logic into analyses whose results can be trusted and used. Economic appraisal takes on a wide brief in attempting to assess all the costs and benefits of treatment or planning options, not only to the health sector but also to patients, families, other public sector agencies, and the community at large. There are obvious problems both in assessing the impact on health of treatment and in attempting to bring all costs and benefits into a common unit for comparison. Whereas health service costs and savings may be readily converted into money terms, patient and family items such as the time taken

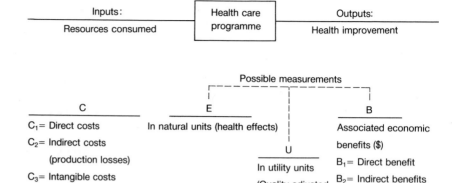

Fig. 18.1. Components of economic evaluation.

to obtain treatment or nurse sick relatives and the benefits of reducing pain and suffering are much more difficult to evaluate, as will be seen from the examples given later.

The methodology of economic appraisal in health care

In principle the methodology of economic appraisal is quite simple; namely, to select two or more alternatives for appraisal, to assess the costs and consequences of the alternatives, and to make a comparison based on relative costs and benefits. There are various forms of economic appraisal (Fig. 18.1). All forms consider costs and differ in the extent to which they measure and value consequences. The different forms of appraisal are discussed below, by reference to studies of increasing degrees of complexity.

Cost analysis

One of the simplest situations for the application of economic appraisal is that where the choice between alternatives is dependent largely on relative costs, rather than relative benefits. Consider, for example, a study of alternative methods of providing long-term domiciliary oxygen therapy (Lowson *et al.* 1981). Medical research has shown that patients with chronic bronchitis can benefit from long-term oxygen treatment in the home (up to 15 hours a day). There are three quite distinct methods of service delivery: cylinder oxygen, liquid oxygen, and the oxygen concentrator, a machine that extracts oxygen from air. The relative medical effectiveness of the treatment methods can be assumed to be similar, since

under all three the patient receives a steady supply of oxygen via a face mask. However, it is likely that the costs differ between methods, because cylinders and liquid oxygen represent a steady resource commitment through time, whereas concentrators require a large capital outlay in terms of equipment and workshop facilities but may have lower running costs in the long term. In addition, since the service facilities required for the concentrator option are shared by all the patients on the therapy in a given location, the unit cost per patient per annum is likely to fall as more patients are given the therapy. The results of the cost analysis undertaken by Lowson *et al.* are given in Figure 18.2. For all but small numbers of patients, the concentrator option is to be preferred on cost grounds. Therefore, it can be seen that in some situations merely revealing the costs of options can provide useful information for decision-making purposes.

However, even in this simple costing example a number of methodological issues arise; three in particular are worth exploring. First, *whose* costs should be considered? It is clear that health sector costs are relevant to the choice of treatment made, but what about the costs borne by the patients, such as electricity consumption in the concentrator option? The view taken by Lowson *et al.* was that all resources consumed, no matter on whose budget they fall, are relevant to the choice. That is, the appraisal was being performed from the viewpoint of society at large. This is the broadest viewpoint from which options can be evaluated and should be the approach adopted in economic appraisal. However, it may also be relevant to investigate other key viewpoints, such as those of the third party payer, individual institutions, or the patient. This helps to identify the incentives that may need to be introduced in order to encourage adoption of the socially efficient option (Drummond *et al.* 1987). For example, an option that is cost-effective when judged from society's viewpoint may not be introduced if a key group of providers would lose income as a result.

Second, how are cost flows compared through time? The concentrator option requires a large outlay at the beginning of the therapy, whereas the other options require a steady annual resource outlay. It is usually argued that, as individuals or as a society, we prefer to postpone costs (or conversely, bring benefits forward). That is, even in the absence of inflation we would prefer to receive $100 this year rather than $100 next year; in economists' language we are said to have a *positive rate of time preference*. The normal method of incorporating this notion into economic appraisals is to *discount* costs and benefits in the future *to present values*. In effect this implies treating costs and benefits in the future as though they have slightly less weight than those in the present. (Those requiring more explanation of discounting and the method of calculating present values should consult Drummond (1980).) In the example described above a discount rate of 7 per cent was used to convert the initial capital outlays to an annual charge to reflect not only the actual sums involved, but also the fact that they occur sooner rather than later. This approach gives the same result as discounting all the recurring annual outlays to present values and is more convenient in this case since the initial capital outlays for the concentrator option are the only costs that do not occur each and every year. For example, Figure 18.3 shows how the calculation would be applied in the case of a piece of equipment, such as a computed tomographic (CT) scanner, costing £140 000.

There is a large literature on the choice of discount rate in economic appraisals and the debate is too complex to reproduce here. (See Fleming (1977) for a review of the major issues.) However, most studies in the health care field have used rates between 2 and 10 per cent and it is often worthwhile investigating whether the study result is sensitive to the choice of rate. In this particular case the study result was not very sensitive to choice of rate, but other choices in health care, such as those between a curative option and a preventive one aimed at the same disease, may well be. The reason is that a preventive option requires large resource outlays *now* in return for potential benefits (in reduction of disease) *in the future*.

Finally, the cost estimates derived in the study by Lowson *et al.* incorporate a number of key assumptions. First, the costs depend on the existing facilities available, especially for the concentrator option. Therefore, in the study two assumptions were made about the availability of technician time and workshop space. (The different assumptions generated the two curves labelled A and B in Figure 18.2.) Second, it is implicitly assumed that this treatment will be given; the issue of whether it is worthwhile providing oxygen therapy to chronic bronchitics at all, given other treatment priorities, is not addressed. Usually such issues are not presented as 'all or

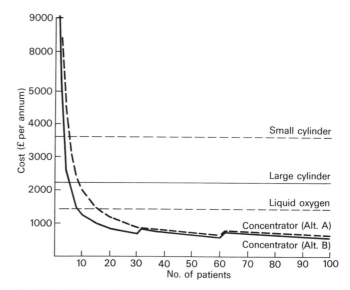

Fig. 18.2. Cost per patient per annum for all methods of providing oxygen. (Source: Lowson *et al.* 1981.)

nothing' questions, but those of *how much* one should pursue particular activities. A good example of how simple costing studies can inform this kind of debate is given by Neuhauser and Lewicki (1975). They examined the costs of pursuing a protocol of six sequential tests for detecting asymptomatic colonic cancer. Each test can detect about 92 per cent of the cancers being looked for in the screened population, so that the second test detects about 92 per cent of the 8 per cent not detected by the first test, and so on. The costs include the screening itself and barium enemas for all positive cases. After six sequential tests the mean cost (that is, total costs divided by total cases detected by all six tests) was estimated at less than US $2500. But the *marginal* cost (that is, the extra cost per case detected by the sixth test and the figure relevant to deciding whether to perform six tests or five) was over US $47 million. The difference between marginal costs and average costs is an important one to note. In the case of provision of concentrators, for example, the marginal cost of adding an extra patient to the therapy is lower than the average cost owing to the fact that the initial investment in service facilities can be shared by more than one patient.

The assessment of costs and benefits at the margin is particularly important since this is where the relevant choices in health care are made. As argued above, the question is not usually where we do '*x*' (such as develop community programmes for the mentally ill), but rather *how much* of '*x*' do we do (for example, for which type of patient should the programmes be developed). There are many such choices in health care: should CT scanning be performed when headache is the only indication, or should there also be abnormal neurological findings (Larson *et al.* 1980); should skull X-rays be given routinely to patients admitted to hospital accident and emergency units with head injury, or only when indicated by clinical diagnosis (Royal College of Radiologists 1981); should coronary artery bypass grafting be given only

Fig. 18.3. The importance of length of life and the discount rate in estimating the annual cost of capital equipment. (Source: Drummond 1984.)

If an item of equipment (such as a body scanner) has a purchase price of £140 000, one might be tempted to assume that this represents an equivalent annual cost of £14 000 over 10 years. However, this ignores the fact that we are not indifferent to the timing of resource outlays. The UK Treasury currently argues that a public sector discount rate of 5 per cent be used to compare costs ocurring at different points in time. In our example the discount rate can be used (rather like a bank interest rate) to calculate the annual amount in each of 10 years (i.e. annuity) that would be equivalent to £140 000 now.

Assuming the annuity is in arrears (i.e. paid at the end of each year), the calculation is to find the amount x in each of n years which, at discount rate r, would be equivalent to our capital outlay k. This is given by the formula:

$$k = \frac{x}{1 + r} + \frac{x}{(1 + r)^2} + \frac{x}{(1 + r)^3} + \ldots + \frac{x}{(1 + r)^n}$$

For our example:

$$140\ 000 = \frac{x}{1 + 0.05} + \frac{x}{1 + (0.05)^2} + \frac{x}{1 + (0.05)^3} + \ldots + \frac{x}{(1 + 0.05)^{10}}$$

$$140\ 000 = x \div (\text{annuity factor for 10 years at 5 per cent})$$

Using tables to obtain the annuity factor (the amount in brackets), it can be established that:

$$x = £18\ 130$$

If a useful clinical life of 5 years, rather than 10 years, were assumed, the equivalent annual cost of the scanner would be £32 336, a far cry from the £14 000 originally suggested.

to patients with severe angina, or also to those suffering from mild angina with one- or two-vessel disease (Williams 1985); should breast cancer screening be offered to women from 50–64 years of age at 3-year intervals, or more frequently and to other age groups (Department of Health and Social Security 1986)?

Economists argue that the efficient approach is to expand health care programmes up to the point where marginal costs equal marginal benefits (Fig. 18.4). If the programme expands beyond this point, the *extra* benefits from the additional utilization would be lower than the *extra* costs. Conversely, operating the programme at lower levels than that indicated by the equivalence of marginal costs and marginal benefits would also be inefficient, since by increasing the utilization slightly the *extra* benefits would be greater than the *extra* costs.

In allocating resources between competing programmes, the efficient point would be one where the marginal benefit from the last dollar spent is the same for all the programmes, the logic being that if the marginal benefits were unequal, a reallocation of resources would produce more benefits in total.

Of course, smooth curves like those drawn in Figure 18.4 do not exist in practice; investments in health care usually imply discrete packages of resource consumption, for example a movement from a targeted to a population-based screening policy will require a minimum additional investment; so would a decision to purchase an extra CT scanner for the hospital. In the main it is the *principle* of considering marginal costs and benefits that is important.

One should note that an implication of the marginal 'way of thinking' is that decisions based on average costs can be misleading. For example, in expanding a screening programme, the marginal costs of screening the final 10 per cent of the population may be considerably higher than the average costs of screening the first 90 per cent. Similarly, the marginal savings from shortening the hospital length of stay of a given category of patient by one day may be lower than the average daily cost of the hospital, since the early days of the patient's stay are more resource-intensive than the later days.

Therefore, it can be seen that even simple costing exercises require a number of methodological issues to be resolved, such as defining the *range* of costs to be considered, making allowances for differential *timing* of resource outlays, and matching the costs to the decision being considered (i.e. iden-

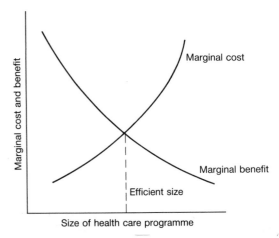

Fig. 18.4. Efficient size of a health care programme as determined by marginal cost and benefit.

Table 18.1. Results of a comparison of day-case and long-stay surgery for patients with hernia and haemorrhoids

	Planned length of stay		
	Day-case patients	Long-stay (trial patients)	Long-stay (excluded patients)
No. of patients	55	56	54
Complications (however slight)			
Hernia	12 out of 32	11 out of 35	17 out of 44
Haemorrhoids*	13 out of 23	6 out of 21	4 out of 11
Length of convalescence			
< 4 weeks	8	11	7
4–8	22	21	12
> 8 weeks	14	12	12
Special arrangements made for return home	30	30	32
Average additional expenditure	£4.67	£5.12	£4.33
Preferred length of stay			
Shorter than occurred	2	11	1
Same as occurred	27	41	41
Longer than occurred	26†	4	12
Average no. of general practitioner consultations			
Home	0.87	0.55	0.60
Surgery	1.49	1.31	1.39
Average no. of district nurse visits	5.96	1.79	2.07

Source: Russell *et al.* (1977).
* Differences almost significant at 5 per cent level.
† Including 17 who would have preferred 24 hours.

Table 18.2. Cost-effectiveness of treatments for chronic renal failure

	Cost per year of life gained (US $)
Transplantation	2 600
Centre dialysis	11 600
Home dialysis	4 200

Source: Klarman *et al.* (1968).

tifying the marginal costs). However, at the same time, the examples cited show that costing studies can, in certain situations, generate useful information.

Cost–effectiveness analysis

A slightly more complex situation is one where again it is implicitly assumed that a given treatment or health care planning objective will be pursued, but that no prior assumptions can be made about relative effectiveness of the alternative treatments or programmes. Here a cost–effectiveness study is usually carried out. There are a number of good examples of such studies from the field of elective surgery, where economic appraisals have been undertaken in conjunction with randomized controlled clinical trials. Consider, for example, the study by Russell *et al.* (1977) on day-case surgery for hernias and haemorrhoids. The results of the medical evaluation were that there was no significant difference in outcome (as assessed by number of complications) between day-case hernia patients (8-hour stay) and those treated by traditional in-patient methods (see Table 18.1). Given equivalence in medical outcome, it was possible to estimate the relative resource use in the two alternatives so as to advise on relative cost–effectiveness. This type of cost–effectiveness study, where the benefits of the two options are equal, is sometimes called *cost minimization analysis*.

Russell *et al.* found that although the day-case patients consumed more community services, in general practitioner consultations and district nurse visits, these extra costs were far outweighed by the savings in hospital care. This particular study also employed an innovative approach to estimating the marginal savings in hospital resources from day-case surgery. (As argued earlier, shortening stays will not save the equivalent of *average* hospital cost per day as merely the 'hotel' element of cost is being escaped.) Rather than basing their estimate of savings on average costs, Russell *et al.* argued that a movement towards the treatment of more day-cases would either (i) enable a five-bed ward to be closed or (ii) enable construction of a new five-bed ward to be avoided. It was estimated that the net savings from day-case surgery would be between £19 and £24 per case (1973 prices), depending upon which of these two outcomes resulted from the change in surgical policy.

There are other interesting examples of cost–effectiveness analysis undertaken when two medical procedures produce equivalent medical outcomes. Waller *et al.* (1978) considered in addition patients' views on shortened hospital stays for hernia, haemorrhoids, and varicose veins surgery and found that when patients' costs were considered, the overall cost advantage of short-stay surgery (48 hours post-operative stay) was only slight. It can be seen that in certain situations it can be highly advantageous to link economic appraisal to clinical trials (Drummond and Stoddart 1984). It should also be noted that cost–effectiveness analyses, and the other forms of economic appraisal to be discussed below, are dependent on having good evidence on the effectiveness of medical interventions. In that respect economic appraisal is a complement to, and not a substitute for, some of the other approaches to evaluation described in this volume.

But what if the alternatives being considered have different relative effectiveness? One of the pioneering cost–effectiveness studies by Klarman *et al.* (1968) addressed this issue. The study concerned the choice between different mixes of transplantation, centre dialysis, and home dialysis for patients with chronic renal failure. For the patients surveyed, transplantation gave a longer survival time (17 years on average compared with 9 years on dialysis) although it had a higher initial cost. The authors argued that since the main objective of treatment for chronic renal failure was to extend life, a legitimate comparison would be in terms of cost per year of life gained. On this basis, transplantation appears to be a good buy (see Table 18.2). The logic is that, given a fixed budget, the number of years of life gained would be maximized by selection of the maximum transplantation route. This finding has been confirmed by later studies

(Ludbrook 1981; Stange and Sumner 1978) although the differences in cost per year of life gained are smaller with older patients. The more recent studies also examine home dialysis and continuous ambulatory peritoneal dialysis (Churchill *et al.* 1984).

Cost–effectiveness analysis is useful in those situations where there is a clear objective of medical intervention, such as to extend life. For example, Gravelle *et al.* (1982) have compared different tests in breast cancer screening in terms of cost per woman-year of life saved. However, there are many health care investment decisions for which the *quality* of life gained may differ between medical procedures for the same condition, or between conditions. The treatments for chronic renal failure are a case in point; most people would argue that the quality of life gained by transplant is higher than that gained by dialysis. In the early work, Klarman *et al.* (1968) made a fairly arbitrary judgement, that a year gained by transplant is equivalent to 1.25 years gained by dialysis. However, later researchers have tackled this issue in a more sophisticated manner.

Cost–utility analysis

Torrance *et al.* (1973) and Bush *et al.* (1973) have estimated these relative valuations of health states ('utilities') empirically by talking to physicians or patients. One approach is to pose the question 'How many years on dialysis would you exchange for one by transplant?'; another is to obtain a linear ranking of health states by presenting the individual with a choice between a certainty of being in one health state and a 'gamble' of either being worse off or better off. The relative valuations of states are obtained by varying the probabilities assigned to the gamble until the respondent is indifferent between the gamble and the certainty. These two approaches are known as the *time trade-off* method and the *standard gamble* respectively (Berg 1973; Culyer 1978). The measurement of consequences in health 'utilities' is an essential component of the variant of cost-effectiveness analysis known as cost–utility analysis. This approach is gaining popularity because it has broader application than simple cost–effectiveness analysis, yet avoids the controversial issue of measuring benefits in money terms.

An example of a study embodying this approach is that by Stason and Weinstein (1977). Their purpose was to apply cost–effectiveness analysis to the management of essential hypertension, both to determine how resources can be used most efficiently within a programme to treat hypertension and to provide a yardstick for comparison with alternative health-related uses of these resources. The central benefit measurement issue concerns the fact that although blood-pressure screening and control may extend life, the medication also has certain unpleasant side-effects. Also, extension of life is not the only potential benefit of lowered blood-pressure, since non-fatal (but disabling) strokes and myocardial infarctions may also be averted. Therefore, a utility analysis was performed and the results expressed in terms of

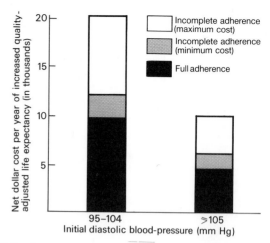

Fig. 18.5. Cost per quality-adjusted life year by management of hypertension (40-year-old subjects; discount rate of 5 per cent assumed). (Source: Stason and Weinstein 1977.)

cost per quality-adjusted life year (Fig. 18.5). As in the study by Klarman *et al.* (1968), the adjustment chosen was not based on thorough empirical analysis, but at least the authors did explore the sensitivity of their results to the assumption made. In fact, *sensitivity analysis* was a central feature of this particular study, given the uncertainties about patient compliance with therapy, medical treatment costs, and valuation of adverse subjective medication side-effects. Variations in a number of these factors were found to have a substantial impact on the cost–effectiveness ratio. Nevertheless the study was able to generate useful insights into the efficiency of different ways of managing hypertension. One of the main findings was that an intervention to improve patient compliance with antihypertensive therapy may be a better use of limited resources than maximum efforts to detect hypertension by extending screening programmes. The question of whether treatment of hypertension represents an efficient use of health resources was answered only indirectly by the analysis. The cost per quality-adjusted life-year (assuming difficulties with patient compliance) was found to be US $10 500 for patients with diastolic blood pressures above 105 mmHg and US $20 400 for those with diastolic blood-pressures between 95 and 104 mmHg. Whether or not this is a reasonable price to pay depends on the subjective valuation one places on life and the returns from other life-saving investments. (This issue is explored below.)

Other examples of cost–utility analyses include those by Boyle *et al.* (1983) on neonatal intensive care, Torrance and Zipursky (1984) on anti-D immunization, and Williams (1985) on coronary artery bypass grafting. There have also been recent methodological advances in utility measurement and assessment of the validity, reliability, and reproducibility of the various measures (Torrence 1987).

Cost–benefit analysis

Discussion of those cost-effectiveness studies that employ benefit measurement in terms of 'utilities' raises the wider issue of valuation of health benefits. The broadest kind of comparison between programmes would be one in which all the costs and benefits were valued in the same unit (e.g. money terms). In theory one should then be able to assess whether particular investments in health treatments or programmes are worth undertaking when compared to other uses for the same scarce resources.

Very few studies succeed in tackling this wide brief, but one which comes close is the study by Weisbrod *et al.* (1980) on alternatives in the treatment of psychiatric illness. This study compared conventional hospital-based treatment with a community-based alternative called 'Training in Community Living', and was linked to a controlled prospective evaluation. A wide range of costs and benefits were quantified in money terms. These included the direct costs of the alternative programmes, the costs imposed on other public sector agencies (such as social services agencies and sheltered workshops), law enforcement costs resulting from offences committed by patients, patient maintenance costs, and costs imposed on the family. The money benefits included earnings resulting from return to employment. Of course some items could not be valued in money terms, although many were quantified in physical units; these included the number of arrests, suicides, days in employment, indicators of 'improved consumer decision-making' and patient mental health (for example, the presence of clinical symptoms and patient satisfaction) (Table 18.3.).

Therefore, this study investigated fairly fully the relevant changes in a comparison of the economic efficiency of treatments (Fig. 18.6). The main finding was that although the community-based programme cost about US $800 more per annum per patient, the monetary benefits were around US $1200 per patient per annum. A number of the forms of benefits and costs that were measured in quantitative but non-monetary terms showed additional advantages of the community-based experimental programme. At the same time the authors were careful to point out that the generalizability of a single experiment is limited, that the analysis can be viewed from different perspectives (for example, society, the governmental budget, and the patient), and that those costs and benefits for which it was impossible to provide monetary values should not be ignored.

Discussion of cost–benefit analysis raises two key issues: (i) is it possible to obtain reliable money estimates for all the benefits of health services; and (ii) *whose values* are the appropriate ones to use? With regard to the first issue, it has often been argued that it is impossible to place a value on the 'intangible' elements of the benefits of health services, such as the reduction of pain and suffering or the extension of life itself. However, it should be noted that, implicitly at least, a valuation *is* being placed on these items every time a health care investment decision is made. Returning to the example of screening for cancer of the colon cited earlier, a decision to endorse a protocol of six sequential tests rather than five is implicitly valuing the extra lives saved in excess of US $47 million. Conversely, Cochrane and Holland (1971) calculated

Table 18.3. Costs and benefits per patient (12 months after admission) for alternative mental illness programmes

Category	Conventional hospital-oriented programme	Community-based programme
Money costs (C)		
Direct treatment costs	$3138	$4798
Indirect treatment costs (falling on other agencies)	$2142	$1838
Law enforcement costs	$409	$350
Patient maintenance costs	$1487	$1035
Lost family earnings	$120	$72
Money benefits (B)		
Patient earnings	$1168	$2364
Net money cost (C − B)	$6128	$5729
Non-money costs		
No. of arrests	1.2	1.0
Suicide	1.5	1.5
Non-money benefits		
Days in employment	87	216
Patient satisfaction	Significantly higher with community programme	
Clinical symptomatology	Significantly better with community programme (on 7 of 13 measures)	

Source: Weisbrod *et al.* (1980).

that the cost of saving the life of a patient with porphyria variegata, by population screening, was around £250 000, yet community screening for the disease had never been tried in the US or the UK. They pointed out that this 'suggests a conscious or unconscious recognition of the financial factor'. It is apparent that, although valuations are *implicitly* being placed on health service outputs every day, there are likely to be many inconsistencies. This has led some economists to undertake studies that reveal these implicit valuations so that decision-makers can give them more active consideration. For example, Buxton and West (1975) calculated the implicit social value of a year of life on renal dialysis, that is, the amount which the community was currently paying to keep patients alive under this particular regimen. It was found to be £2600 and £4720 per annum respectively for home and hospital dialysis in the UK. Buxton and West point out that these figures should then be compared with those implied by other life-saving health treatments. It is interesting to note that at a time when the UK had restrictions on the growth of dialysis, the cost per life implied by building safety measures implemented following the collapse of an apartment block in London was around £20 million (Mooney 1977).

From time to time economists have attempted to place a money value on life directly. The most popular approach in the early literature was to take discounted future earnings (either gross or net of the individual's consumption) as an estimate of the loss to the community from premature death of an individual. (The logic was the 'value' of the lost life-years averted related to the productive output of the

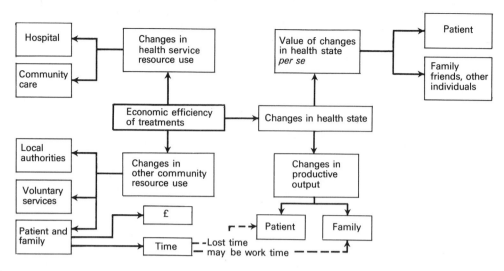

Fig. 18.6. The relevant changes in a comparison of the economic efficiency of treatments. (Source: Drummond 1980.)

individual during that time.) This *human capital* approach has always been recognized as, at best, partial and at worst, misleading (Mishan 1971). (See Drummond (1981*a*) for more discussion of this point.) Certainly, the approach can give answers that may mislead decision-makers. For example, in a study of the size and distribution of benefits from US medical research into cancer and heart disease, Holtmann (1973) found that if cancer were eliminated the average white male would gain about $561 and the average black male about $316 (present values using 5 per cent discount rate). The corresponding figures for cardiovascular disease were $1034 and $508. However, Holtmann states that 'we should note that the higher returns for white males should not be taken as a justification for investing more in treatment of whites at the expense of blacks', presumably because if blacks had equal opportunities in the labour market the figures may be different.

An alternative approach is to elicit from individuals values for their own lives or improvements in health, either by inference from their actions (for example, attitudes towards personal safety) or by asking them directly. (See, for example, the work of Jones-Lee (1976).) A recent study obtained estimates from arthritis sufferers of the proportion of their household income they would be willing to pay for a cure of their arthritis. The estimates were found to be remarkably stable and reproducible (Thompson 1986). Progress is therefore being made in overcoming the considerable measurement problems inherent in direct valuation of health improvements, but for the time being the more immediate contribution of economic analysis is likely to be in making previously hidden values explicit.

This leads on to the other major issue raised by cost–benefit analysis; that of *whose* values should be used? Economists typically start from the premise that the consumer is the best judge of the utility that he or she obtains from a particular good or service. Cost–benefit analysis is firmly based

in the theoretical traditions of Paretian welfare economics. Here, the purpose of cost–benefit analysis is *to identify potential Pareto improvements*; that is, situations where the maximum total sum of money that the gainers from a project would be prepared to pay to ensure that the project were undertaken exceeds the minimum total sum of money that the losers from it would accept as compensation to allow it to be undertaken. Therefore, this perspective requires that benefits from health services be valued in terms of consumers' willingness to pay. Not all economists would totally adhere to these principles when judging health service investments. Indeed, there is nothing in the theoretical foundations of Paretian welfare economics which says that consumers are always the best judges of their own welfare. However, most economists would argue that consumers' values are too often neglected by policy-makers (at the planning and clinical levels) who merely replace them with their own. In general, economists would prefer values to be *made explicit*, no matter whose values are used. (See Drummond (1981*a*) for more discussion of this point.) One advantage of the cost–utility approach (as described above) is that the technical and value judgements implicit in measuring health service benefits are clearly identified and can be discussed in the constructive arena of multi-disciplinary activity.

Finally, and most fundamentally, economic appraisal in health care is an *aid* to decision-making and not a substitute for it. Resource allocation decisions in health care are likely to depend on a number of considerations and the main contribution of economic appraisal is to provide an assessment of the consequences, *in terms of economic efficiency*, of various choices. The other major consideration is *equity*, and on occasions economists have explored the equity–efficiency trade-offs skilfully. A good example is in the study by Rich *et al.* (1976) of alternative screening procedures for asymptomatic bacteriuria in schoolgirls. The most cost-effective method, an unsupervised test, had a lower detection rate in

Fig. 18.7. Ten Questions to ask of any published study. (Source: Drummond *et al.* 1987.)

1. Was a well-defined question posed in answerable form?
   (a) Did the study examine both costs and effects of the service(s) or programme(s)?
   (b) Did the study involve a comparison of alternatives?
   (c) Was a viewpoint for the analysis stated or was the study placed in a particular decision-making context?
2. Was a comprehensive description of the competing alternatives given (i.e. can you tell who did what to whom, where and how often)?
   (a) Were any important alternatives omitted?
   (b) Was (should) a "do-nothing" alternative (have been) considered?
3. Was there evidence that the programme's effectiveness had been established?
   Was this done through a randomized, controlled clinical trial? If not, how strong was the evidence of effectiveness?
4. Were all important and relevant costs and consequences for each alternative identified?
   (a) Was the range wide enough for the research question at hand?
   (b) Did it cover all relevant viewpoints (e.g. those of the commnnunity or society, patients, and third-party payers)?
   (c) Were capital costs as well as operating costs included?
5. Were costs and consequences measured accurately in appropriate physical units (e.g. hours of nursing time, number of physician visits, days lost from work, or years of life gained) prior to valuation?
   (a) Were any identified items omitted from measurement? if so, does this mean that they carried no weight in the subsequent analysis?
   (b) Were there any special circumstances (e.g. joint use of resources) that made measurement difficult? Were these circumstances handled appropriately?
6. Were costs and consequences valued credibly?
   (a) Were the sources of all values (e.g. market values, patient or client preferences and views, policy-makers' views and health care professionals' judgements) clearly identified?
   (b) Were market values used for changes involving resources gained or used?
   (c) When market values were absent (e.g. when volunteers were used) or did not reflect actual values (e.g. clinic space was donated at a reduced rate) were adjustments made to approximate market values?
   (d) Was the valuation of consequences appropriate for the question posed (i.e. was the appropriate type, or types, of analysis—cost–effectiveness, cost–benefit, or cost-utility—selected)?
7. Were costs and consequences that occurred adjusted for differential timing?
   (a) Were costs and consequences that occurred in the future 'discounted' to their present values?
   (b) Was any justification given for the discount rate used?
8. Was an incremental analysis of costs and consequences of alternatives performed?
   Were the additional (incremental) costs generated by the use of one alternative over another compared with the additional effects, benefits, or utilities generated?
9. Was a sensitivity analysis performed?
   (a) Was justification provided for the ranges of values (for key parameters) used in the sensitivity analysis?
   (b) Were the study results sensitive to changes in the values (within the assumed range)?
10. Did the presentation and discussion of the results of the study include all issues of concern to users?
   (a) Were the conclusions of the analysis based on some overall index or ratio of costs to consequences (e.g. cost–effectiveness ratio)? if so, was the index interpreted intelligently or in a mechanistic fashion?
   (b) Were the results compared with those of other studies that had investigated the same questions?
   (c) Did the study discuss the generalizability of the results to other settings and patient/client groups?
   (d) Did the study allude to, or take account of, other important factors in the choice or decision under consideration (*e.g.* distribution of costs and consequences or relevant ethical issues)?
   (e) Did the study discuss issues of implementation, such as the feasibility of adopting the 'preferred' programme, given existing financial or other constraints, and whether any freed resources could be used for other worthwhile programmes?

children from low-income groups. The authors present this information to the decision-maker in a way that highlights the key issues involved in choice of test.

The contribution of economic appraisal to health service decision-making

Judging the quality of the literature

It is clear from the discussion above that undertaking an economic appraisal in health care requires numerous methodological judgements on the part of the analyst, many of which can be debated. Recent reviews of the literature (Drummond *et al.* 1986; Warner and Luce 1982) have shown published studies to be of a variable quality. How, therefore, can the non-economist critically appraise the quality of this growing body of literature? A simple checklist of questions to ask of any published study has been offered by Drummond *et al.* (1987) and is elaborated upon here (Fig. 18.7).

First, as in all fields of scientific inquiry, it is important to *be clear on the study question*. In particular, it was stressed earlier that the viewpoint(s) from which the alternatives are being compared should be clearly identified. Questions such as 'Is the new medicine worthwhile in the prevention of coronary heart disease?' beg the questions 'to whom?' and 'compared with what?'. A better specified question would be something like the following: 'From the viewpoint(s) of (a) the Ministry of Health, (b) other agencies providing care, and (c) patients and their families, would a preventive programme including the new medicine be preferable to the existing programme, which concentrates mainly on treating coronary heart disease as and when it occurs?'

It is important that studies include a *comprehensive*

description of the competing alternatives so that the decision-maker can assess the implications of study results for his or her own setting. For example, it is important to describe the treatments concerned, including the health care professionals involved, the procedures used, the frequency and length of care, and its location.

Of course, given the need to consider both the costs and consequences of interventions in an economic evaluation, it is important that the *effectiveness of the programmes or treatments is established*. This further emphasizes the need, discussed earlier, to integrate the economic evaluation of health treatments as fully as possible with their clinical evaluation.

The *identification, measurement, and valuation of all relevant costs and consequences* are also important. Obviously the range included needs to match the breadth of the viewpoint(s) being considered and the study questions being posed. In particular, broader questions demand that a wider range of costs and benefits is measured and valued, since frequently the issue of whether the treatment is worthwhile, when compared with the alternative uses of the same resources, is being explored.

If the costs and consequences of the alternatives occur at different points in time they need to be adjusted for *differential timing*. Furthermore, *sensitivity analysis* should be performed, exploring the sensitivity of study conclusions to the values of those parameters about which there may be methodological controversy or imprecision in estimation. Typically, the factors varied in a sensitivity analysis include the discount rate, the costs of (or savings from reduced) hospitalization, the medical evidence on the success of therapy, and the relative valuations of states of health. The precise selection of items for inclusion in a sensitivity analysis depends on particular circumstances, but users of evaluation results should be suspicious of a study that does not embody this general approach, as it is likely that many of the estimates used are more optimistic than would be found in practice.

Finally, in the *presentation of results*, it is normal that an *incremental analysis* be shown. That is, compared with the existing programme or treatment, what *extra* costs and *extra* benefits would result if the new intervention were used? It should be remembered that where the implicit existing programme is 'doing nothing', this rarely results in zero costs and zero benefits. In addition, the presentation of results should include a discussion of other concerns to users, such as the implications for other policy objectives (e.g. equity), the managerial costs of changing to the recommended intervention, and the extent to which the results of the particular study are confirmed by the results of other studies of the same topic.

Obviously, few economic evaluations would pass such a stringent test of their methodology. Rather, the ten-question checklist should be regarded as a methodological 'gold standard' to which analysts should aspire. In the same way that we do not abandon medicine because it occasionally fails, we should not abandon economic evaluation as an aid to

decision-making because some of the studies have methodological imperfections.

Problems and prospects of applying economic appraisal in health service decision-making

There is now a substantial body of literature relating to issues in the economic burden of disease, prevention, diagnosis, treatment, and rehabilitation. However, despite the increasing popularity of economic appraisal among health service and medical researchers, there is very little evidence of its being used in planning or clinical decisions. Warner and Luce (1982) cite the work of Eddy (1980) as a good example of economic analysis being presented in a form and quality which enabled decision-makers to appreciate its relevance. It contributed to decisions by Blue Cross–Blue Shield and the American Cancer Society to alter their recommendations on screening practices. However, Warner and Luce point out that both in terms of its technical competence and its policy impact, Eddy's work remains a distinct deviation from the norm. Whilst also recognizing the variable quality of the existing literature (a point discussed above), Drummond (1983) cites additional evidence from Europe, such as the decision by the Department of Health and Social Security in the UK to adopt a new approach (rather like economic appraisal) to the appraisal of planning options where one of these is a large capital scheme, such as building or extending a hospital (Department of Health and Social Security 1981). Perhaps, given the methodological comments made above, a promising approach may be to build an economic element into medical research design where this appears to be relevant. Requests for research monies from one of the Ministry of Health's committees in Ontario (Canada) are evaluated not only in terms of their methodological soundness, but also in terms of whether 'the proposal is likely to have an important economic impact in reducing the costs or increasing the efficiency of health services' and the grant application has to provide, where appropriate, 'an adequate cost–effectiveness, cost–benefit, or cost–utility analysis' (Ontario Ministry of Health 1982).

In recent years there have been a number of methodological developments that increase the relevance of economic appraisal to planning and clinical decisions. At the clinical level a number of analyses are based on a *clinical decision tree*. This is a convenient way of understanding a complex sequence of events in diagnosis or treatment (Weinstein and Fineberg 1980). Figure 18.8 shows a decision tree for the diagnosis and treatment of cancer of unknown primary origin. The tree flows from left to right; the squares denote 'decision nodes', outcomes over which the decision-maker has control; the circles denote chance nodes, which show the probabilities of given outcomes.

Levine *et al.* (1985) used the decision tree to estimate that comprehensive diagnosis and treatment of patients would result in five extra patients being alive at the end of the first year, when compared with a limited diagnostic strategy (find-

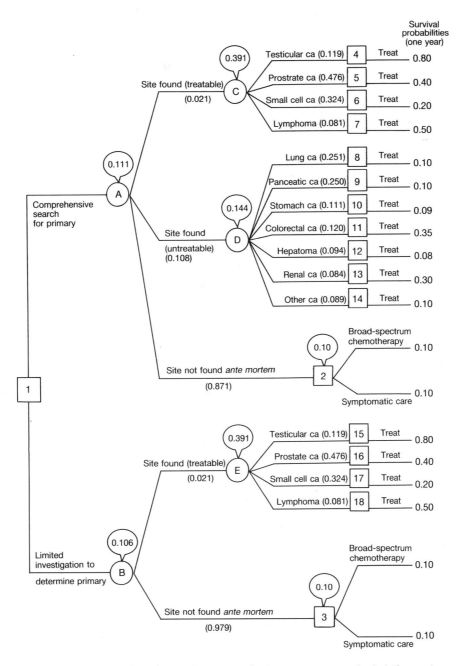

Fig. 18.8. Decision tree for diagnosis and treatment of carcinoma of unknown primary origin (males). (Source: Levine *et al.* 1985.)

ing only those cases for which there was an effective systematic therapy). Other examples of decision analysis can be found in Weinstein (1986). The decision trees can be used on an interactive basis with decision-makers. This is particularly useful when evidence on some of the outcomes is lacking or imperfect. Different decision-makers' estimates can be entered into the model and the sensitivity of results to the different estimates can be examined.

Levine *et al.* also used the tree to estimate the comparative costs of the alternative diagnostic and treatment policies. The incremental costs of the comprehensive strategy over and

above the limited strategy were more than Can.$7 000 000 per annum. Whether or not costs should enter into clinical decision-making is an important ethical issue. Certainly economic appraisal is more relevant to planning decisions, where choices are being made for the community as a whole. At the clinical level the clinician may feel that he or she needs to do as much as possible for the patient, within the physical or resource constraints imposed. However, given scarcity of resources, there is clearly an opportunity cost when more resources are consumed on a given patient. Indeed, other patients are likely to lose out. In practice it is unlikely that in

Table 18.4. 'League table' of costs and QALYs for selected health care interventions (1983–4 prices)

Intervention	Present value of extra cost per QALY gained (£)
General practitioner advice to stop smoking	170
Pacemaker implantation for heart block	700
Hip replacement	750
Coronary artery bypass graft for severe angina with left main disease	1 040
General practitioner control of total serum cholesterol	1 700
Coronary artery bypass graft for severe angina with two-vessel disease	2 280
Kidney transplantation (cadaver)	3 000
Breast cancer screening	3 500
Heart transplantation	5 000
Coronary artery bypass graft for mild angina with two-vessel disease	12 600
Hospital haemodialysis	14 000

Source: Williams (1985).
QALY, quality-adjusted life-year.

treating the individual patient the clinician would want to restrict care, but he or she may have agreed to a general clinical policy that attempts to limit the use of resources in situations where the additional benefits are small in relation to the additional costs. Two examples were cited above, in the fields of cancer screening and routine skull X-ray in accident and emergency departments (Department of Health and Social Security 1986; Royal College of Radiologists 1981).

Turning to planning decisions, the most important development is the construction of 'league tables' of health care interventions in terms of their cost per quality-adjusted life-year (QALY) (Table 18.4). The implication of the league table is that interventions near the top give relatively good value for money when compared with those near the bottom and should therefore be given higher priority for funding. Calculations have been made both in North America and Europe (Torrance and Zipursky 1984; Williams 1985).

However, despite their superficial attraction, cost per QALY league tables raise a number of questions (Drummond *et al.* 1988). It has been pointed out that the mortality and morbidity data upon which the QALY calculations are based are not sufficiently precise and that different methods of measuring utilities give different results. Also, many of the cost values are average where what is really required are marginal costs of expansion of the various programmes. These technical issues clearly require more investigation before reliable judgement can be made on the basis of cost per QALY calculation.

In addition, the production of the league tables raises a number of philosophical issues. For example, the community may wish to provide treatments or programmes with poor cost per QALY value, either on the grounds of equity or because future research may improve the results. However, most important of all, the highly summarized presentation in one cost per QALY value may obscure the complex series of

technical and value judgements from the decision-maker. This is clearly unproductive and it must be reaffirmed that economic appraisal is only an aid to decision-making.

Concluding remarks

The main focus of this chapter has been on outlining the methodology of economic appraisal and on giving examples of studies undertaken to date. It is clear that refinements in methodology can be, and are being, made. However, the major challenge, if economic appraisal is to make a lasting contribution to health service decision-making, will be in finding better ways of disseminating the 'way of thinking' implicit in economic studies and in making the studies themselves more relevant to decisions taken at the planning and clinical levels.

References

Berg, R.L. (ed.) (1973). *Health status indexes*. Hospital Research and Educational Trust, Chicago, Illinois.

Boyle, M.H., Torrance, G.W., Sinclair, J.C., and Horwood, S.P. (1983). Economic evaluation of neonatal intensive care of very low birth weight infants. *New England Journal of Medicine* **308**, 1330.

Bush, J.W., Chen, M.M., and Patrick, D.L. (1973). Cost effectiveness of a PKU program. In *Health status indexes* (ed. R.L. Berg), p. 172. Hospital Research and Educational Trust, Chicago, Illinois.

Buxton, M.J. and West, R.R. (1975). Cost–benefit analysis of long-term haemodialysis for chronic renal failure. *British Medical Journal* **ii**, 376.

Churchill, D.N., Lemon, B.C., and Torrance, G.W. (1984). Cost effectiveness analysis comparing continuous ambulatory peritoneal dialysis to hospital haemodialysis. *Medical Decision Making* **4**(4), 489.

Cochrane, A.L. and Holland, W.W. (1971). Validation of screening procedures. *British Medical Bulletin* **27**, 3.

Culyer, A.J. (1978). *Measuring health: Lessons for Ontario*. Toronto University Press for Ontario Economic Council, Toronto.

Department of Health and Social Security (1981). *Health services management: Health building procedures*. HN(81)30. DHSS, London.

Department of Health and Social Security (1986). *Breast cancer screening*. Report by a working group chaired by Professor Sir Patrick Forrest. HMSO, London.

Drummond, M.F. (1980). *Principles of economic appraisal in health care*. Oxford University Press.

Drummond, M.F. (1981*a*). Welfare economics and cost–benefit analysis in health care. *Scottish Journal of Political Economy* **28**, 125.

Drummond, M.F. (1981*b*). *Studies in economic appraisal in health care*. Oxford University Press.

Drummond, M.F. (1983). Economic appraisal and health service decision making. *Effective Health Care* **1**, 25.

Drummond, MF. (1984). *Cost-effective analysis in health care*. Nuffield/York Folios 6. Nuffield Provincial Hospitals Trust, London.

Drummond, M.F. and Mooney, G.H. (1981). Economic appraisal in health care. 1. A guide to the methodology of economic appraisal. *Hospital and Health Services Review* **77**(9), 277.

Drummond, M.F. and Stoddart, G.L. (1984). Economic analysis and clinical trials. *Controlled Clinical Trials* **5**, 115.

Drummond, M.F., Stoddart, G.L., and Torrance, G.W. (1987). *Methods for the economic evaluation of health care programmes.* Oxford University Press, Oxford.

Drummond, M.F., Teeling Smith, G., and Wells, N. (1988). *Economic evaluation in the development of medicines.* Office of Health Economics, London.

Drummond, M.F., Ludbrook, A., Lowson, K.V., and Steele, A. (1986). *Studies in economic appraisal in health care. Volume 2.* Oxford University Press, Oxford.

Eddy, D. (1980). *Screening for cancer: Theory, analysis and design.* Prentice-Hall, Englewood Cliffs, New Jersey.

Fleming, J.S. (1977). What discount rate for public expenditure? In *Public expenditure: Allocation between competing ends* (ed. M. Posner) p. 45. Cambridge University Press, Cambridge.

Gravelle, H.S.E., Simpson, P.R., and Chamberlain, J. (1982). Breast cancer screening and health service costs. *Journal of Health Economics* **1**, 185.

Holtmann, A.G. (1973). The size and distribution of benefits from U.S. medical research: The case of eliminating cancer and heart disease. *Public Finance* **28**, 354.

Jones-Lee, M.W. (1976). *The value of life: An economic analysis.* Martin Robertson, London.

Klarman, H.E., Francis, J.O'S., and Rosenthal, G.D. (1968). Cost effectiveness analysis applied to the chronic renal disease. *Medical Care* **6**, 48.

Larson, E.B., Omenn, G.S., and Lewis, H. (1980). Diagnostic evaluation of headache. Impact of computerised tomography and cost-effectiveness. *Journal of the American Medical Association* **243**(4), 359.

Levine, M.N., Drummond, M.F., and Labelle, R.J. (1985). Cost–effectiveness in the diagnosis and treatment of carcinoma of unknown primary origin. *Canadian Medical Association Journal* **133**, 977.

Lowson, K.V., Drummond, M.F., and Bishop, J.M. (1981). Costing new services: Long-term domiciliary oxygen therapy. *Lancet* **ii**, 1146.

Ludbrook, A. (1981). A cost–effectiveness analysis of the treatment of chronic renal failure. *Applied Economics* **13**, 337.

Mishan, E.J. (1971). Evaluation of life and limb: A theoretical approach. *Journal of Political Economy* **79**, 687.

Mooney, G.H. (1977). *The valuation of human life.* Macmillan, London.

Neuhauser, D. and Lewicki, A.M. (1975). What do we gain from the sixth stool guaiac? *New England Journal of Medicine* **293**, 226.

Ontario Ministry of Health (1982). *Health care systems research grant review committee: Criteria for assessment of applications.* Toronto, Canada.

Rich, G., Glass, N.J., and Selkon, J.B. (1976). Cost effectiveness of two methods of screening for asymptomatic bacteriuria. *British Journal of Preventive and Social Medicine* **30**, 54.

Royal College of Radiologists (1981). Costs and benefits of skull radiography for head injury. *Lancet* **ii**, 791.

Russell, L.T., Devlin, H.B., Fell, M., Glass, N.J., and Newell, D.J. (1977). Day case surgery for hernias and haemorrhoids: A clinical, social and economic evaluation. *Lancet* **i**, 844.

Stange, P.V. and Sumner, A.T. (1978). Predicting treatment costs and life expectancy for end-stage renal disease. *New England Journal of Medicine* **298**, 372.

Stason, W.B. and Weinstein, M.C. (1977). Allocation of resources to manage hypertension. *New England Journal of Medicine* **296**, 732.

Stoddart, G.L. (1982). Economic evaluation methods and health policy. *Evaluation and the Health Professions* **5**, 393.

Thompson, M.S. (1986). Willingness to pay and accept risks to cure chronic disease. *American Journal of Public Health* **76**(4), 392.

Torrance, G.W. (1987). Utility approach to measuring health-related quality of life. *Journal of Chronic Diseases*, **40**(6), 593.

Torrance, G.W. and Zipursky, A. (1984). Cost-effectiveness of antepartum prevention of RH immunisation. *Clinics in Perinatology* **11**(2), 267.

Torrance, G.W., Sackett, D.L., and Thomas, W.H. (1973). Utility maximisation model for program evaluation: A demonstration application. In *Health status indexes* (ed. R.L. Berg) p. 156. Hospital Research and Education Trust, Chicago.

Waller, J., Adler, M., Creese, A., and Thorne, S. (1978). *Early discharge from hospital for patients with hernia or varicose veins.* HMSO, London.

Warner, K.E. and Hutton, R.C. (1980). Cost benefit and cost effectiveness analysis in health care: Growth and composition of the literature. *Medical Care* **18**, 1069.

Warner, K.E. and Luce, B.R. (1982). *Cost benefit and cost effectiveness analysis in health care: Principles, practice and potential.* Health Administration Press, Ann Arbor, Michigan.

Weinstein, M.C. (1986). Risky choices in medical decision making: A survey. *The Geneva Papers on Risk and Insurance* **11**, 197.

Weinstein, M.C. and Fineberg, H.V. (1980). *Clinical decision analysis.* W.B. Saunders, Philadelphia.

Weisbrod, B.A., Test, M.A., and Stein, L.I. (1980). Alternative to mental hospital treatment. II Economic benefit–cost analysis. *Archives of General Psychiatry* **37**, 400.

Williams, A.H. (1974). The cost benefit approach. *British Medical Bulletin* **30**, 252.

Williams, A.H. (1985). Economics of coronary artery bypass grafting. *British Medical Journal* **291**, 326.

19

Sociological investigations

MYFANWY MORGAN

Introduction

Most issues in the field of public health have an important social dimension. These range from traditional epidemiological concerns regarding the social causes of disease and its distribution among the major social groups in the population, to the consideration of health and illness in a broader context, including the social meanings and responses to chronic and stigmatizing conditions, and the social forces that shape prevailing medical institutions and approaches to health. In addition, issues relating directly to the organization and provision of health services and their use and evaluation, pose questions of a social nature and have led to the involvement of sociologists and other social scientists in undertaking research with a view to informing health policy.

The field of health and medicine now forms a major area of specialization for sociologists in both the United States and Britain, while in several other countries there are groups of social scientists active in the field even though health and medicine has not yet developed a strong identity as a sociological specialism (Claus 1981). The variations between countries in the development of the specialism reflects the interaction of several factors. Of particular importance is the strength of sociology itself as an academic discipline, as well as the emphasis placed on applied research and the links established between sociology and medicine. The development of health and medicine as an area of sociological enquiry has also been encouraged in many countries by the government funding of research concerned with the organization and provision of health care.

As with all branches of sociology, investigations in the field of public health are characterized by considerable diversity in terms of their aims, perspectives and methods of enquiry. Many sociological studies share similar assumptions and methods of investigation to those of epidemiological research. Other studies adopt a more distinctly sociological perspective and may also employ qualitative rather than quantitative methods of investigation. This chapter describes these different approaches and their underlying assumptions, and provides examples of their application to public health issues. As a background to this discussion, the development of sociology and its links with public health is first briefly described.

Sociology and public health

The name Sociology was introduced in 1839 by Auguste Comte, a French philosopher, who believed that abstract theorizing about the nature of society should be combined with the systematic and scientific study of the social world to create a new 'science of society'. However, the theoretical origins of modern sociology are generally viewed as being in the late nineteenth century. This period saw the writings of Karl Marx, Max Weber, and Emile Durkheim among others, who, aware of the great social changes brought about by the industrial revolution, were concerned to identify the causes of these changes and the basis of social order. Each presented their own theories regarding the nature of society and the determinants of social relationships and thus laid important foundations for the development of sociology.

One common link between sociology and public health lies in the examination of health issues by both Marx and Durkheim as part of their more general theoretical analysis of society. Particularly notable is Durkheim's study, *Suicide* (1897), both because of its content and its methods of enquiry. *Suicide* was an attempt to explain a public health problem, namely the social distribution of people taking their own life. However, the sub-title of the book, a 'study of society', shows that Durkheim, like many sociologists after him, was not directly concerned with health problems in themselves, but only as a means of furthering an understanding of society. Suicide is a behaviour which is perceived as being, *par excellence*, an individual act. What Durkheim wished to show was that individuals are social products and therefore that individual self-destruction would be socially structured in important ways. The method he adopted was essentially the scientific method, involving a statistical examination of secondary data to test hypothesized relationships. He took existing suicide data from various areas and regions of France and of other European countries and analysed them in relation to such factors as levels of education, seasons of the year, and levels of insanity. He also compared the rates for religious groups, marital groups, and other social categories which he hypothesized would differ in their suicide rate. An important methodological feature of this study was Durkheim's use of three variable tables which forms one of the earliest uses of multivariate analysis. Through this statistical method he found that suicide rates were higher among single, widowed, and divorced than married people, and among Protestants compare with Catholics and Jews. This provided support for his hypothesis that the type and intensity of people's social bonds serve as a social force influencing individuals' behaviour, and specifically the likelihood of taking their own life. This early sociological study thus contributed to the identification of social risk factors in terms of the protective effects of social ties. However, it also served to define the subject matter of sociology by demonstrating the way in which individuals are social products and that their behaviours can therefore only be fully understood in terms of characteristics of the social groups of which they form a part.

A second common origin of sociology and public health lies in the development of empirical enquiries into the social conditions, health and poverty of the population as a basis for social reform. This tradition of empirical social research played a particularly important role in Britain, where an early landmark was the Report of the Royal Commission on the Poor Law of 1834. This involved extensive first-hand enquiries being undertaken into the situation of the poor and operation of poor relief as part of a government commission of investigation. It was shortly followed by a more systematic social investigation under the direction of Edwin Chadwick, then Secretary to the Poor Law Commission, which was published in 1842 as *Report on the Sanitary Conditions of the Labouring Population*. Through comparing mortality rates for different areas, Chadwick made the vital link between the social conditions of an area and the health of the population and recognized that poor health formed an important cause of pauperism. The Report's ultimate outcome was the Public Health Act of 1848 in which the British government, for the first time, charged itself with a measure of responsibility for safeguarding the health of the population.

Besides these official enquiries, social surveys were also undertaken by wealthy philanthropists. Notable examples are Charles Booth's massive survey of the social conditions and extent of poverty among the population of London, published in seventeen volumes between 1889 and 1903, and Seebohm Rowntree's subsequent survey of poverty in York, entitled *Poverty: a study of town life* (1903). The method adopted in such studies involved the investigation of a defined geographical area often through house to house enquiries, to record data in a systematic manner on poverty, housing conditions, employment, etc. However, this material was frequently supplemented by more qualitative data derived from people's own accounts of the experience of poverty and by the observations of investigators who were concerned not merely to document social conditions but also to try to determine the causes and effects of poverty.

Both sociologists and epidemiologists regard these nineteen-century social enquiries as laying foundations for the development of their disciplines by establishing the social survey as a method of investigation. In addition, people such as Chadwick in Britain, Shattuck in the US, and Virchow in Germany, placed health and disease in a social context through identifying the social determinants of disease and recognising that poor health carried economic costs and formed an important cause of pauperism.

Whereas the public health concerns of the mid-nineteenth century emphasized the importance of environmental sanitation and more general social reform in improving the health of the population, from the 1890s developments in bacteriology and the germ theory of disease led to a change in the focus of public health. Thus, greater emphasis was placed on individual measures and the role of the health services in promoting health. The collaboration of social scientists and epidemiologists in carrying out research on these questions probably dates from the 1940s in both Britain and the US. At this time the continuing high rates of infant mortality were seen as posing questions concerning the organization of maternal and child health services, and their uptake by different groups in the population. In addition, epidemiologists were becoming increasingly concerned with the distribution of non-infectious diseases, such as lung cancer, bronchitis, and coronary heart disease, and their social causes in terms of housing conditions, diet, smoking, etc. This resulted in the employment in Departments of Social Medicine of social researchers (many of whom were later identified as medical sociologists), who brought their particular skills in social survey techniques to study these questions, as well as their knowledge of social variables, such as social class, housing conditions, poverty, etc. The development of a more theoretically informed analysis of health and medicine and its emergence as a sociological specialization occurred rather later. In the US this probably dates from the early 1950s and in Britain occurred about ten years later, reflecting the earlier and more firmly established position of sociology as an academic discipline in the US. However, the growth of sociological studies in the field of health and medicine has since been rapid. This had been encouraged by the development of sociology as an academic discipline, as well as by the increasing recognition of the social dimensions of health and health care and the greater involvement of governments in the provision of health services and funding of research.

If journals might be used as disciplinary markers, then perhaps the founding of the *Journal of Health and Human Behaviour* in the US in 1960 (later amended to *Journal of Health and Social Behaviour*) identifies the emergence of medical sociology as a specialism. This was followed in 1967 with the appearance of an international journal, *Social Science and Medicine*, while the UK journal *Sociology of Health and Illness* was first published in 1979. As the title of this recent journal indicates, sociologists view their concerns as being with health and illness in its broadest sense, and to include the study of features of the social and economic system which inhibit health and produce illness and disease, the range of healing and caring professions and people's beliefs about the causes of disease and experience of illness. For these reasons many medical sociologists are not entirely happy with the title of their specialty.

Concerns of medical sociology

A distinction is commonly drawn between sociological enquiries whose primary aim is to contribute directly to the health of the population and the organization and delivery of healthcare, and sociological enquiries which have as their primary goal the furthering of sociological knowledge through the study of health and medical issues. Strauss (1957) termed these different orientations to research sociology *in* medicine and sociology *of* medicine. However, research characterized as sociology of medicine, although not primarily concerned with policy and practice, has often produced ideas and concepts which have been applied directly to the field of public health. Notable early examples include Parsons' concept of the sick role and the social nature of the doctor–patient relationship (Parsons 1951), and Goffman's analysis of stigma and of the characteristics and effects of psychiatric hospitals and other 'total' institutions organized on custodial lines (Goffman 1961, 1963). Studies concerned more directly with epidemiological and health service questions are also increasingly contributing to the more general analysis of social institutions and processes. As a result, the difference in emphasis between sociology *in* and *of* medicine is becoming more blurred. Nevertheless, it is possible to identify some differences between countries in terms of the orientation of sociological investigations. For example, in France and Poland medical sociology is strongly linked to and informed by sociology, whereas in West Germany as in Britain sociological investigations have developed alongside and been more closely linked to public health (Claus 1981).

The concerns of both sociology and public health have led to the investigation of a wide range of issues, while an important trend has been the gradual expansion of areas of research as a result of developments both in sociology and in the health care field. Several books describe what the authors identify as major areas of sociological investigation, including Illsley (1980), Susser *et al*. (1985), Morgan *et al*. (1985), and Stacey (1988). However, it is difficult to determine the precise boundaries between sociology and the other social sciences which contribute to the field of public health, including social administration, medical anthropology, health economics, health policy, and medical geography. This is largely because the concepts and methods derived from these different disciplines often became absorbed into the general body of knowledge to be applied in the study of public health issues. In addition, there is considerable overlap of interests between different disciplines, while much research is undertaken by multi-disciplinary teams (Spruit and Kromhaut 1987).

Medical sociology, like its parent discipline, encompasses a diversity of approaches in terms of its theoretical perspectives and methods of investigation. This diversity often puzzles the medical observer. This is because medicine has been characterized for the last two centuries by a single dominant approach to research based on the biomedical model. It thus seeks to identify the causes of specific disease processes that reside within the body and produce cancer, coronary occlusion, etc., and adopts for this purpose the scientific method, which involves the testing of hypotheses through experimental or statistical methods. Much sociological research shares this emphasis on quantification, while sociologists employed in medical settings often conduct research within an epidemiological framework. However, sociological enquiries may also employ qualitative rather than quantitative methods and emphasize the subjective nature of social reality, rather than its objective, measurable, characteristics.

The next section describes the assumptions and methods of quantitative research and then outlines the various sociological perspectives, or views of society, that are generally associated with this form of enquiry. This is followed by a discussion of the assumptions and methods of qualitative research. Each approach is illustrated by examples of its application to the field of public health, with particular attention being paid to the ways in which sociological research may differ from epidemiological enquiries and complements this approach to the study of public health issues.

Quantitative research

Quantitative scientific research is based on a positivist approach to the social world. This is a philosophical position that views social phenomena as objective and external to the individual and which thus take the form of social 'facts'. These social facts are therefore studied in much the same way as phenomena in the natural world through the application of the scientific method, with the aim of producing valid and generalizable findings. This approach to the study of the social world was advanced by Durkheim and adopted in his study on suicide, and forms the basis of the scientific status of sociology.

It is recognized by those advocating a positivist approach that there are some fundamental differences between the social and the natural worlds. One difference is in the way that human beings react to situations. For example, people, unlike molecules, may change their behaviour if they are aware that they are being observed, or may give what they perceive to be socially acceptable answers in interview surveys. A second important difference is that sociological research can rarely employ and experimental design. Even randomized controlled trials, which form the method of epidemiological research most closely corresponding to the experimental methods of the natural sciences, are rarely practicable in sociological studies which therefore rely on controlling for variables statistically. Although acknowledging the differences between the social and natural sciences, sociologists adopting a quantitative scientific approach regard

Table 19.1. Quantitative and qualitative research

	Quantitative	Qualitative
View of social world	Social reality exists as objective, measurable phenomena, external to the individual (Positivism)	Social reality is subjectively interpreted and experienced (Naturalism)
Logic of enquiry	Deductive based on testing formal hypotheses to establish casual relationships	Inductive reasoning with understanding of social processes derived from data
Research design	Quantitative, with sample selection, data collection and analysis based on scientific procedures and ensuring repeatable and generalizable results	Qualitative, based on detailed study of beliefs and behaviours of groups of interest to elicit their interpretations and responses
Validity	Corresponds to an objective reality	Corresponds to a subjective reality

these differences as problems to be overcome by appropriate research methods rather than requiring a different form of enquiry (Strong and McPherson 1982).

The basic features of the scientific method that provide the framework for quantitative sociological investigations are outlined in Table 19.1. These require that hypotheses are specified at the outset, or at least precise research questions formulated. The relevant variables are then 'operationalized', that is, defined in ways that are measurable. For example, the hypothesis that unemployed men with high levels of social support will experience fewer health problems than unemployed men lacking high levels of social support, requires that the concepts of unemployment, high level of social support, and health problems are each defined in ways that are measurable. Data are then collected from the entire population of interest, or from a representative sample, to test the hypothesis or examine the research question, and relationship between variables is analysed statistically. Each stage in this process is guided by the requirement to produce data that conform to an objective reality and to achieve generalizable results.

In some cases available or 'secondary' data are employed in the form of routinely collected statistics or other written records. However, many research questions can only be answered through the collection of first-hand or primary data (Table 19.2). The main method of collecting primary data in quantitative studies is through a structured questionnaire. This may be self-completed by the respondent or administered by an interviewer, and allows information to be collected fairly quickly from large numbers of people. An important feature of the structured questionnaire is that it is designed to be administered in exactly the same way to all respondents. Interviewers are therefore given no discretion as to how to ask the questions and in what order, and are required to conduct each interview in exactly the same manner. The questions are also often pre-coded and allow only a limited range of responses. In this way, subjectivity and the effects of interviewer variability are reduced, and the results obtained through the use of this standard instrument are regarded as being repeatable.

Observation is occasionally employed as a method of data collection in quantitative research. However, an important limitation is regarded as being the small number of people or situations that it is generally feasible to observe, which raises issues of the generalizability of the results. Observation when employed in quantitative research is usually tightly defined with the aim of achieving a uniform, standardized method of data collection. The observer is therefore required to record only certain types of pre-defined behaviours or comments. An example is provided by Clark and Bowling's (1989) study designed to test the hypothesis that small nursing homes would provide long-stay elderly patients with more personal attention and freedom of choice compared with hospital wards, and that the nursing home environment would therefore provide a better quality of life. The study employed observation as one of several forms of data collection. Observation required that the observers logged activities, interactions, and moods in 15-minute blocks. A total of 232 observational sessions were conducted over a three-month period, in four settings. The types of activities and interactions recorded included meal times, serving of tea/coffee, presence of visitors, playing of radio/TV/music, interaction with professionals, etc. The recording was undertaken by two observers. In order to achieve a high level of inter-observer reliability they piloted the observational schedule until they achieved consistent ratings.

A variant of direct observation which is sometimes employed to minimise the intrusive effect of an observer, is video or audio tape recording. When employed in quantitative research, the material is translated into numerical data for analysis. This is achieved by classifying data into categories based on the content of responses, or by using rating scales (Armstrong et al. 1989). One example is provided by a study of the effects of differences in the length of medical consultations on both the amount and type of communication which occurred (Roland et al. 1986). A total of 623 consultations of varying lengths were tape recorded. These consultations were then analysed by classifying statements into categories; for example, statement by doctor explaining treatment, statement interrupting patient, or question asked by patient. This produced a count of the frequency with

Table 19.2. Sources of data in sociological investigations

Secondary data	Primary data
Routine statistics (e.g. mortality data, hospital statistics)	Structured questionnaire
	Semi-structured interview
Government surveys and other *ad hoc* enquiries	Unstructured (or focused) interview
	Diary method
Written records (including historical and literary sources)	Observation

which different types of statements occurred and thus provided data in a quantitative form for analysis.

The diary method in which people record feelings or activities, usually daily, in a 'diary' or record book, is another method of data collection which is occasionally employed in quantitative research. The data recorded are usually highly structured and are analysed in numerical terms. An example is the use of health diaries as one of several forms of data collection in a study of the factors influencing women's demands for medical care. The diaries were completed every evening over a four-week period and recorded information or perceptions of symptoms, action taken, and the occurrence of special events such as stressful situations (Banks *et al.* 1975). This allowed quantitative statements to be made, concerning, for example, how many people experience particular types of symptoms, such as headaches, backpain, and feverishness, and how many consulted the doctor about their symptoms.

The methods of data collection employed in quantitative research are thus designed to follow scientific principles and to achieve results that are repeatable and generalizable, and which conform to an objective reality. The acceptance by epidemiology and to a large extent by sociology of the scientific model of research has formed an important bond promoting their collaboration in the study of public health issues. However, to understand the nature of sociological enquiries it is important not only to consider the methods of research employed but also the broader perspective or view of society that informs research. This is because the theoretical perspective adopted by the researcher influences the questions selected for study, as well as providing a framework for the interpretation of data in terms of a particular theory of society.

Two main perspectives are generally associated with the acceptance of a positivist approach and the use of quantitative methods of research. These are consensual perspectives which emphasize the consensus and shared values that exist in society, and conflict perspectives (Marxist and pluralist) which emphasize the conflict that exists between groups in society arising from differences in their interests and goals.

Consensual perspectives

A consensual view of society guided sociological analyses during the 1950s and early 1960s. This was associated with functionalist theories which regard the various institutions in society, such as the class system, the family system and the medical profession, as contributing to the smooth functioning of society as a whole, very much as the heart, lungs, etc., contribute to the functioning of the human body. For example, the class system, with its differential rewards and statuses, is regarded as promoting the well-being of society as a whole (Davis and Moore 1945). This is because the differences in status and resources associated with different occupations serve to ensure that the most able people are motivated to undergo the often lengthy training necessary to take on the jobs requiring special talents and training. In addition, they

are motivated to perform well in these positions. Social inequality is thus viewed from a functionalist perspective as an unconsciously evolved device that contributes to the smooth functioning and general well-being of society, rather than being explained in terms of conflict and the exercise of power.

Probably the most notable contribution of functionalism to medical sociology is Parsons' concept of the sick role, which refers to the special social position occupied by sick people. Parsons (1951) was concerned with the question of how society handles deviance, or non-conformity with social expectations. Parsons regarded the ill as deviant not only in biological terms but also in social terms. This is because sick people no longer perform their normal social roles, while such deviance on a large scale could threaten the smooth functioning of society. Parsons identified society's mechanism for handling sickness so that it does not place too great a strain on the social system, as being to institute a special status and social role for sick people, namely the 'sick role'. Like all social roles this carries both rights and obligations. The obligation on the sick person is that they are expected to be motivated to want to get well as soon as possible, and should seek technically competent help and co-operate with medical experts. The privileges are that they are given exemption from the performance of their normal social roles (e.g. may be allowed time off work) and are regarded as being in need of care.

Parsons viewed the sick role as a temporary social role, that has been instituted by society to ensure that sick people are returned to normal functioning as quickly a possible. It is also a universal social role, in that anyone may enter the sick role regardless of their status in other spheres. Complementing the sick role is the role of the doctor. Parsons also depicted this as involving rights and obligations designed to promote the cure of sick people and their return to normal functioning. They include the expectation that doctors will apply a high degree of knowledge and skill to the problems of illness, act for the welfare of the patient rather than for their own self-interest, be objective and emotionally detached, and be guided by the rules of professional conduct. In return for these obligations, doctors are granted the right to examine patients physically and enquire into intimate areas of their physical and personal life. They also enjoy considerable autonomy in relation to the patient and occupy a position of authority.

Parsons' theoretical analysis of society's response to illness is generally regarded as marking the beginning of health and medicine as a sociological specialism. This is because it extended the view of illness from that of a purely biological state to the social sphere, and identified a specific set of rights and obligations that surround the sick person. Similarly, the relationship between doctor and patient was portrayed as not merely a technical relationship but as a social relationship, in which each party to the relationship is guided by social expectations and obligations. These ideas have since been widely employed and developed in the analysis of illness and the medical consultation. One example is the analysis of the appropriateness of the sick role for chronically ill people. This is because expectations that encourage the permanent occupancy of the sick role, and hence continued dependency, are seen as having an adverse effect on their functioning. Instead there is a need, so far as is possible, to promote normal functioning. The often ambivalent responses of the public and medical staff to conditions such as alcoholism, overdoses, and AIDS, can also be explained in terms of their unwillingness to view such conditions purely as sickness requiring treatment and therefore to grant people with these conditions the privileges of the sick role. Instead there is a tendency to ascribe notions of responsibility and thus of punishment to people with these conditions, and possibly to their close relatives. Similarly, the notion that the relationship between doctors and patients is a social relationship and not merely a technical relationship has focused attention on ways in which this relationship can be improved, so as to promote the diagnostic and therapeutic process and increase patients' satisfaction with the consultation (Morgan 1986).

Parsons' analysis and other functionalist writing, although important in extending the medical view, do not in any way challenge medicine or serve as a critique of the *status quo*. This is because they emphasize the consensus that exists in society and the way in which the various social institutions contribute to the general well-being of society. For example, Parsons emphasized the shared interests and expectations between doctor and patient, in which the doctor acts in the best interests of the patient and patients obey the doctor, rather than the conflicts that may occur in this relationship as a result of differences in doctors' and patients' assessment of the seriousness of the presented condition, or in the doctors' failure to fulfil patients' expectations in terms of the investigations undertaken or treatment and advice given (Freidson 1970). Doctors may also experience conflicts in terms of their need to act in the best interests of an individual patient and the requirements of other patients awaiting a consultation. Besides this emphasis on the consensual nature of the doctor–patient relationship, Parsons also emphasized the positive role of the doctor in acting as a gatekeeper to the sick role and deciding who is healthy and who is sick, and thus of protecting society against malingerers who claim the privileges of the sick role unnecessarily. In contrast, other writers have drawn attention to the ways in which this power may in some situations be misused by the state. For example, doctors may be required by the State to certify political dissidents and other people whose behaviour does not conform to social norms as mentally ill and thus enable such people to be detained against their will, with the label of illness being used as a means of coercive social control (Szasz 1971).

Just functionalist theories developed an analysis of society and its institutions based on a consensual perspective, the assumption of shared interests and values has also provided a framework for the early tradition of applied social research and continues to underpin much research in the field of public health. Sociologists and others adopting this approach seek to ascertain the 'facts' through quantitative methods, with the aim promoting the health of the population and the organization and delivery of health care by contributing to a process of enlightenment and suggesting ways of achieving socially valued goals. This quantitative socio-medical approach is illustrated here by research concerned with questions of the social distribution of disease and causes of social inequalities in health, and by studies examining the reasons for the non-uptake of health services and non-compliance with medical advice.

Sociologists often adopt an epidemiological approach to questions of the social causes of disease and in the US are frequently classified as non-medical epidemiologists. Like epidemiologists they have examined the contribution of social risk factors, such as cigarette smoking, diet, housing conditions, etc., to the differences in the cause specific morbidity or mortality rates of social groups, defined in terms of class, education, gender, ethnicity, etc. One example concerns the causes of childhood accidents. In the UK this forms the most important cause of death among children aged 1–14 years, while there is also a steep class gradient with the rates being higher in the manual compared with the non-manual occupational groups (Whitehead 1987). A number of studies have examined how far this can be explained in terms of housing conditions, and especially the presence of hazards, such as steep unguarded stairs, a lack of gardens, and the poor storage of potentially dangerous substances around the home, and have evaluated the success of health education interventions designed to reduce domestic hazards. However, these approaches appear to have had limited success (Colver *et al.* 1982; Jackson 1988). Brown and Davidson (1978) hypothesized that differences in rates of accidents may also be explained by differences in the psychiatric state of mothers. This was examined in a study based on a random sample of 458 women aged between 18 and 65. Psychiatric problems were identified using a structured questionnaire and the women divided into three groups: 'cases' with definite psychiatric symptoms (mostly depressive), 'borderline cases', and no psychiatric problems. This showed that working-class mothers were more likely to have experienced psychiatric problems than middle-class women. In both social groups, rates of accidents among children were highest among mothers with psychiatric problems. Working-class children whose mother was classified as a case or borderline case had an accident rate of 19.2 accidents per 100 children, compared with 9.6 for other working-class children. For middle-class children the figures were 5.3 per 100 and 1.5 per 100 respectively. In both social classes the accident rate was highest during the weeks the mother was psychiatrically disturbed. This study therefore suggested that the greater prevalence of psychiatric problems among working-class mothers explains some of the class gradient in rates of accidents among children, and may also account for the limited success of interventions to reduce home accidents.

Whereas some studies have adopted an epidemiological perspective and have sought to identify the contribution of particular social risk factors to specific causes of morbidity and mortality with the aim of promoting an understanding of the aetiology of disease, sociological concerns have also led to a broadening of this approach. Noting the regularity in the patterning of social inequalities in health across disease categories, they have been concerned to identify the particular aspects or characteristics of social groups, defined in terms of class, gender, or ethnicity, etc., that give rise to these differences in health (MacIntyre 1986). For example, the finding that women report more morbidity and have higher rates of medical care utilization than men but have lower mortality rates and a greater expectation of life, raises questions of how far this can be explained in terms of differences in the social roles, life-styles, and behaviour, and in the social expectations and responses to men and women's health problems. A contribution of quantitative studies has been to provide statistical data relating to these issues. For example, the health and rates of medical care contact of employed and non-employed women have been compared, controlling for marital status and other intervening variables, to determine the possible effects of women's social roles on their higher reported morbidity and higher rates of utilization of medical services compared with men (Nathanson 1980). Similarly, there are questions of the contribution of differences in material conditions (e.g. housing, pollution, working conditions), life-styles (e.g. rates of smoking, dietary patterns) and of health-related mobility to the higher morbidity and mortality rates of manual compared with non-manual occupational groups. An early study quantifying the effects of health-related morbidity was undertaken by Illsley (1955) based on data collected in 1951–4 for women having a first pregnancy in Aberdeen. He showed that women who were brought up in social classes I and II but who were downwardly mobile at marriage and entered classes IV and V, differed in important ways from other women who stayed in their social class of origin. For example, they left school earlier and entered less prestigious occupations, they were also shorter in stature, had a poorer physique and a higher perinatal death rate in their first pregnancies. However, the position was reversed for those who moved out of classes IV and V and into a higher class. This analysis thus pointed to a selective exchange between classes, which would have the effect of maintaining class differentials in morbidity and mortality. A subsequent analysis for first pregnancies to married women in Aberdeen District 1969–75, identified a similar process (Illsley 1980). This suggests that a factor contributing to the continuing social class differentials in mortality is the effects of health-related mobility associated with occupational achievement and marital selection, as well as the effects of social position on risks of exposure to illness. The size of the contribution of health selection to the observed variations in the health of social groups is much debated, although differences in material conditions and behaviour are generally acknowledged as forming the major determinants of social inequalities in health (Wilkinson 1986; Townsend and Davidson 1982).

Sociological enquiries are concerned not only to identify the contribution of different aspects of the social position and behaviour of groups in society to their differential health experience but also to explain why social groups differ in their roles, behaviours, and life-styles. This analysis of the determinants of social phenomena can be viewed as paralleling at a social level the concern of the medical sciences to explain disease mechanisms and processes at the biological level. Explanations of these social phenomena are necessarily influenced by the theoretical perspective adopted. For example, a functionalistic approach explains the differences in the roles of men and women in terms of their physiological sex differences and the way in which this natural division of labour promotes the well-being of society, whereas Marxist and feminist analyses emphasize the social construction of gender differences and the dominant influence of groups with power in determining social roles (Stacey 1988). As already noted (p. 313) a consensual approach explains the existence of social class inequality in terms of the importance for society of a system of differential occupational rewards, rather than viewing the social class system as based on conflict and serving the interests of the groups with power. Similarly, explanations of differences in behaviours, such as the higher rates of smoking of manual compared with non-manual groups may emphasize the role of education and knowledge, or view this as forming a response to life stresses and other aspects of their social and economic position, while this structural position may be explained in terms of consensual or conflict theories of society.

Socio-medical enquiries have thus contributed to and broadened the traditional epidemiological approach to the study of the social causes and distribution of disease, through the quantification of social variables and by the analysis of the determinants of social risk factors from a consensual perspective. This quantitative approach to social phenomena has also been widely employed in health services research and can be illustrated by studies concerned to identify the social factors associated with non-compliant behaviour. During the 1950s and early 1960s the findings of community surveys that people frequently delayed or failed to seek professional advice for conditions which would benefit from treatment, led to the study of what Mechanic (1962) termed illness behaviour. He defined this as 'the ways in which given symptoms may be differentially perceived, evaluated, and acted (or not acted) upon by different kinds of persons'. Illness behaviour thus forms the first step towards the official legitimization of illness and entry into the sick role. A number of models have been developed to identify those variables regarded as explaining illness behaviour, and which account for the variations in responses to illness and the uptake of services (Becker *et al.* 1977; Dingwall 1976). The general approach adopted by researchers employing quantitative methods is to identify at the outset a list of variables hypothesized to influence behaviours, including, for example, a lack of knowledge of the benefits of service use, difficulties of access due to distance, a lack of transport or other commitments, problems in relationships with health care providers, etc. Data relating to these variables are then collected, using a structured questionnaire, and the characteristics of individuals or groups with inappropriate behaviours are identified, with the aim of providing information necessary to influence behaviours. Recently, this approach has been used extensively to identify the reasons for the non-uptake of screening. for example, in a study of the reasons for non-attendance for computer-managed cervical screening, women who did not attend were followed up to ascertain their reasons for non-attendance, using a mixture of structured and open questions (Maclean *et al.* 1984). This identified problems of the adequacy of records of attendance, as well as specific reasons for non-attendance. These included various organizational problems, such as the failure to receive an invitation, and the inconvenience of the appointment, and the influence of attitudes and knowledge, including beliefs regarding the appropriateness of the test for them and feelings of fear and embarrassment. Other studies have similarly used quantitative methods to examine the reasons for delays in consulting about potentially serious medical conditions (Nichols *et al.* 1981), patients' non-compliance with medication instructions (Norman *et al.* 1985), and failure to take up child health services (Morgan *et al.* 1987). The major emphasis of these studies is to promote behaviours that are regarded as conducive to health by identifying the various organizational barriers to service use and other reasons for non-compliant behaviour. As with all research employing structured questionnaires, the social factors identified as influencing service use or following medical advice depends on the researcher's prior hypothesis, and the variables selected for study.

Quantitative research conducted within a biomedical or consensual framework has provided data on a wide range of issues relating to the provision, organization, and evaluation of health care (e.g. Patrick and Peach 1989; Dowie 1983; Bowling and Cartwright 1982; Cartwright and Anderson 1981). However, quantitative research based on a consensual approach is increasingly complemented by studies deriving from a conflict perspective which offer different forms of explanation of social phenomena.

Conflict perspectives

The emphasis of functionalist sociology and health services research on the consensual nature of society was challenged in the 1960s by the emergence of 'conflict' theories of social order. Rather than depicting society as based on shared values and altruism, these theories drew attention to the conflict that exists in terms of the struggle for advantage between groups in society. An influential conflict perspective came from Marxist analysis, although pluralist theories more commonly underpin research in the field of public health.

Marxist theories

The form of social relationships in any society, in terms of the distribution of power and authority, is viewed from a Marxist perspective as being determined by its economic organiz-

ation. Under capitalism, economic production involves the participation of two major groups, namely a capitalist class who own the means of production (land, factories, etc.) and dominate and exploit a working class who are simply a source of labour in the productive process. Both parties are thus seen as having their own interests. Relationships between them are characterized by conflict, since one group can only gain at the expense of the other. Marx viewed other social institutions (law, medicine, religion, education, etc.) as reflecting this basic conflict in the economic sphere, and as being shaped and supported by an ideology that essentially reflects the beliefs and values of the ruling class and justifies its power and privilege.

Marxist theories of society have been applied by sociologists and epidemiologists to two main areas of public health, namely the analysis of the social causes and distribution of disease and the structure and organization of health services. Whereas socio-medical enquiries have been important in identifying the distribution of social risk factors that account for the relatively high rates of morbidity and mortality among manual workers and their families compared with non-manual groups, Marxist analyses have sought to explain the origins and distribution of the social risk factors in terms of the characteristics and demands of the capitalist economic system.

One example of this approach is provided by Doyal in *Political Economy of Health* (1979), in which she identifies various ways in which the organization of production in capitalist society increases the health risks for workers. For example, the 'safe' levels set for exposure to chemicals, gases, dusts, and other toxic substances is viewed as reflecting the interests of the owners and controllers of industry, who operate on the basis of 'acceptable' risks for a given return on investment and assume that work creates no danger unless there is clear evidence to the contrary. Health and safety legislation has also generally been fairly limited and fragmentary, with little priority being assigned by the government to its enforcement, as evidence by the relatively small number of factory inspectors and the limited penalties when prosecutions do occur. In these ways the economic interests of the capitalists and ruling classes are viewed as having priority over considerations of the health risks for the work-force. As a result, manual workers are exposed to increased risks of industrial accidents and injury, and to cancer, respiratory problems, skin problems, and other diseases. The pursuit of profit is also regarded as giving rise to a lack of effective controls on industrial pollution and the spreading of industrial wastes in the air or water, with the result that the health risks of production extend into the community and affect workers' families. Doyal suggests that not only are health risks inherent in many aspects of production, but that some of the commodities produced, including cigarettes and highly processed foods, are either harmful to health or have little nutritional value. However, a high level of demand is necessary for the maintenance of the capitalist economic system. As a result, demand for these products is encouraged through a considerable investment in advertising, while the State imposes only limited controls on the goods produced.

The question of the ways in which individual behaviours are determined by the broader social and economic organization of society has also been considered by Navarro (1982). He attributes the relatively high rates of cigarette smoking among manual workers to the alienation they experience at work. Alienation arises because workers lack control over their work and thus gain little in the way of self-fulfilment from their labours. As a result, they regard work as merely a means of achieving goods and services and acquire satisfaction through the rewards of work in terms of possessing the money to spend on cigarettes and other consumer goods, which in turn is encouraged by the need to generate a high level of demand to sustain the economic system.

As with other perspectives, Marxist analyses differ in the emphasis they give to particular aspects of capitalist production in the generation of ill health. For example, Eyer (1977) believes that whereas physical hazards (poor housing, occupational risks, environmental risks, etc.) formed the primary cause of ill health in Britain and the US in the nineteenth century, in this century stress forms the major factor responsible for the social class inequalities in health. He regards the cyclical nature of capitalist economies as producing high levels of stress during periods of economic boom. This is because these periods are characterized by high rates of labour-force mobility, competition for jobs, and long hours of work, which have the effect of breaking up worker solidarity, and disrupting family and community ties. In contrast, these social bonds are strong during periods of recession and high levels of unemployment. Eyer suggests that the stresses associated with economic booms have a direct effect on the health of workers and also on their families, as well as having an indirect effect on health through causing an increased consumption of tobacco and alcohol.

Questions of the relative contribution of social stress and physical hazards to the social distribution of disease thus characterize a political economy approach, as they do a traditional epidemiological approach. However, a fundamental difference between these approaches lies in their explanations of the determinants of these risk factors and of unhealthy behaviours, and hence in the scope and approaches to reducing social risk factors. The policy recommendations deriving from a consensual perspective take the form of reformist measures to improve the availability and uptake of health services, to educate the population about health matters, to improve housing conditions, reduce poverty, etc. A political economy approach regards the promotion of health as requiring more fundamental changes in the social and economic organization of society. It is also critical of the emphasis given to individual intervention as the dominant approach to health and views, this as being explained by and serving the interests of the capitalist class.

The interventionist approach to health is reflected in the dominance of curative medicine, which is concerned with the treatment of those people who become sick, rather than with the prevention of ill health and with the emphasis of preventive medicine encouraging individuals to stop smoking, change their diet, and generally adopt a 'healthy life-style', rather than with preventing ill health through attention to its social and economic causes. Writers adopting a Marxist perspective argue that individual intervention and individual responsibility for health forms the dominant ideology and approach to health, as it supports the interests of the capitalist class. This is because by focusing on the diseased individual and on individual behaviours, people are blamed for their poor health. As a result, relatively little attention is given to the health risks inherent in the social and economic organization of capitalist society and the need for more fundamental reforms (Crawford 1977). In contrast, critiques of the dominance of an interventionist approach to health, based on a consensual perspective, regard this as reflecting a mistaken belief in the contribution of curative measures to the historical decline in mortality, which has led to a neglect of the social and environmental determinants of health (McKeown 1976).

Another issue which has formed the focus of Marxist analysis is the causes of the increased government funding of health care in capitalist society. Writers differ in their explanations of this general trend, but share similar beliefs concerning the central role of class interests and class conflict as determinants of social change, rather than regarding social change in consensual terms as forming the product of progressive enlightenment and humanitarian concerns. One approach is to view the extension of government funding of health care simply as a means of buying off resistance to the economic system. In short, workers with 'free' health services will be more satisfied with their lot and with the progress of society and its institutions, and are therefore less likely to seek radical change which threatens the *status quo* than are those who are left disgruntled. This is illustrated by Navarro (1978) in his analysis of the crucial antecedents to key phases in the development of government provision of health services in Britain. For example, he attributes the 1911 National Health Insurance Act, which provided free health care to the employed population, to the fears of social unrest among the working classes in Britain and the uprisings in other parts of Europe. This alarmed the upper and middle classes and meant that concessions were made to the working population to stem any possible revolutionary movement. Similarly, the establishment of the National Health Service (NHS) in 1948, which represented an expansion of free medical care to the entire population, is viewed by Navarro as forming a response to the demands for social change following the Second World War and to the fear among the Conservative Party of the radicalization of the Labour movement. The actual nature of these changes, in terms of the structure and composition of the NHS that emerged, is also viewed as being shaped by and reflecting the class interests within British society. This includes the dominant voice given to the top echelon of consultants and their representatives in determining the structure of the NHS, and concessions they gained which reinforced their privileged position, including the possibility of having private beds and private practices, a higher level of rewards than general practitioners, and their heavy representation on the main decision-making and administrative bodies of the NHS. Navarro (1989) similarly explains the differences in the health care systems of capitalist economies in terms of differences in the strength of the capitalist and working classes. For example, the health care model characterized by the absence of a comprehensive and universal health programme, as found in the US, is attributed by Navarro to the fact that working-class organizations have traditionally been weak and divided along racial and ethnic lines. This resulted in low levels of unionization and the absence of a mass socialist party in the US. In contrast, the capitalist class was relatively strong in the US and was therefore able to determine the form of public and health policies.

Marxist analyses have been important in challenging assumptions regarding the consensual nature of society and in developing specific forms of explanation of social phenomena, particularly of the prevailing social inequalities in health and the structure and organization of health care systems. However, some form of power group of pluralist approach more commonly underpins sociological analyses of the health field.

Power group and pluralist theories

This perspective views power as being distributed among groups who differ in their interests, agendas, and goals. Thus, while groups such as the medical profession, government representatives, entrepreneurs, and managers of industry have some interests in common, they are also viewed as having opposing interests. Public and health policies, therefore, form the outcome of a complex interplay between various groups, rather than a product of social consensus. However, writers differ in the relative power they attribute to particular groups, and especially whether they focus on the interplay between the major power groups, or adopt a more pluralist approach in which even those groups with little formal power are viewed as having the potential to shape policies and their actual implementation in practice.

An example of the way in which health policies can be viewed as the outcome of the interplay between multiple interest groups is provided by an analysis of policies toward the tobacco industry in the UK (Calnan 1984). Rather than explaining the voluntary and fairly limited government controls on the tobacco industry and emphasis on self-regulation in terms of class interests, power group theories draw attention to the conflicts of interest between those with power. For example, the tobacco industry, the Department of Trade, and Treasury have opposed stricter controls on the tobacco industry, as a result of economic considerations in terms of the effects on profits and government revenue. In contrast, the Department of Health, the medical profession, and various pressure groups have been important in pressing for and achieving those restrictions in advertising and the health

warnings that do exist, reflecting their concerns with the health risks of smoking.

Analysis of the groups which have shaped the health care system and contributed to the dominance of an interventionist approach to health, frequently ascribe a central role to the medical profession, which forms an occupational group that has achieved considerable autonomy and power to influence the nature and organization of its work (Freidson 1970). Alford (1985) in an analysis of the US health care system has identified three sets of structural interests as being of importance in shaping health policy. He terms these the professional monopolists (or medical profession), whose interests are dominant in the health services, the corporate rationalisers (including health planners and hospital administrators), who frequently challenge the prevailing organization of health care, and patient or consumer groups who represent the repressed interests of the population. Starr (1982) in his analysis of American medicine is similarly concerned with the way in which different groups, and particularly the medical profession, have shaped health policy. However, whereas Alford's analysis is based on structural interests, Starr focuses on the role of specific interest groups, since occupations of organizations may not have a common interest. For example, hospital administrators do not necessarily have interests opposed to those of private physicians, while hospital planners and administrators may not always have shared interests. Starr regards the medical profession as having been able to exert a dominant control over the development of the US health care system from the 1920s. However, he identifies new trends emerging, in terms of the increasing involvement and power of the federal government and of the corporation as third party payers, and he sees these as increasingly challenging the dominance of the medical profession.

Adopting a more pluralistic approach, Klein (1983) views the establishment of a system of public provision of health care in Britain in the 1940s as being encouraged by a variety of interests. These include the population's increasing concern for social justice, the Civil Servants' desire to achieve a more rational system of health care, and the interests of the medical profession in gaining new facilities and opportunities to develop new technologies. Similarly, he regards the actual form of the NHS as representing the outcome of a compromise between the interests of family doctors, local authorities, industry, hospitals, and Conservative and Labour politicians.

Pluralist theories have been important not only in explaining the form of the health care system and of health policies but also in informing more micro-level analyses of the relationship between different groups in health settings. For example, whereas Parsons viewed the roles of doctor and patient as based on consensus and shared understandings, a pluralist approach acknowledges that groups may differ in their interests and goals. For example, the patients' desires for a prescription, a second opinion, or sick note, may not accord with the doctor's views and treatment decisions. As a result, the medical consultation may be characterized by conflict, although this generally remains latent rather than becoming overt. Thus, both doctors and patients may try to influence the consultation to attain their desired outcome, rather than the patient obeying the doctor unquestioningly as depicted by Parsons (Freidson 1970). Similarly, although hospitals and other organizations are characterized by formal duties and lines of authority, Strauss (1963) has shown that daily life in the hospital is characterized by a continuous process of negotiation and renegotiation among the various personnel. Acute staff shortages and short-term emergencies may mean that ward staff and trainees become parties to negotiation concerning the *ad hoc* allocation of duties and responsibilities, while negotiation also often occurs between the different categories of staff involved in patient care. Particular attention has been paid to this process of defining roles and the power of doctors in negotiating their position in relation to the development of multi-disciplinary teams in patient care, which involves bringing together groups who may emphasize different patient needs and give priority to different goals (Evers 1982).

This section has described a number of perspectives that inform sociological research. They differ in their assumptions concerning the distribution of power and the extent to which society is viewed as characterized by conflict or consensus. These differences influence the questions posed for research and the explanatory models developed. However, these perspectives also have a number characteristics in common. First, they adopt a structuralist position, in that beliefs and behaviours are regarded as being influenced by broader social values and expectations, with individuals acting out socially prescribed roles and responding to the demands of their position. Second, they emphasize the objective nature of social phenomena and regard the existence of social classes, the variations in rates of suicide and other causes of death, and the socially prescribed behaviours of doctors and patients etc., as social facts that can be studied and measured in much the same way a phenomena in the natural world. As a result, they are generally associated with quantitative methods of investigation.

Qualitative research

Whereas quantitative research has dominated sociological investigation in the field of health and medicine, qualitative research has a long tradition in sociology and characterized many of the early sociological studies undertaken in the US in the 1940s. However, its application to the field of public health mainly dates from the early 1970s and was associated with the development of new sociological perspectives based on a naturalist approach to the social world.

Naturalism regards the subject matter of the social sciences as being fundamentally different from the subject matter of the natural sciences. This is because human beings act toward things on the basis of the meanings that they have for them. Social reality is thus subjectively perceived and experienced,

rather than consisting of objective facts. Individuals are therefore viewed as constantly involved in making sense of and responding to situations and events, rather than acting out socially prescribed roles. The assumptions of a naturalist position thus directly contrast with those of a positivist structural approach and requires that quantitative scientific methods are replaced by methods of research that allow these subjective meanings and social processes to be studied.

Central to a naturalistic approach is the use of inductive reasoning or a 'ground-up' approach (Table 19.1). This means that rather than specifying hypotheses and defining concepts as the first stage of the research and then collecting data with which to test the hypothesis, the data themselves suggest concepts and explanations and draw attention to social processes. In other words, research is concerned with the generation rather than the testing of theory. Glaser and Strauss (1967) termed this the discovery of grounded theory, which essentially arises out of the categories that the participants themselves use to order their experience, rather than being based on categories that are defined at the outset by the researcher and may therefore fit poorly with the participants' perspectives. This aim of seeing the world through the eyes of their subjects also requires that forms of data collection are relatively open and unstructured, so as not to confine responses to inappropriate categories.

Various forms of data collection are employed in qualitative research, including both primary and secondary sources (Table 19.2). An example of the use of secondary data is provided by Atkinson's (1978) study of suicide in which he was concerned to identify how the decision to categorize a death as 'suicide' or 'accidental' was reached. To examine this he employed various forms of primary data as well as secondary data in the form of information derived from coroners' written reports. Sontag (1977) also employed secondary data in the form of poetry, letters, and other literary sources to identify the social meanings that surrounded tuberculosis during the early years of this century and which now surround cancer.

Interviews probably provide the main source of data in qualitative research, as they do in quantitative research. However, rather than involving structured questionnaires with pre-coded answers, a much freer format is adopted in qualitative research. The terms used to describe interviews with a relatively open format tend to vary, and are identified here as semi-structured and unstructured (or focused) interviews. In semi-structured interviews respondents are asked a series of open-ended questions and their answers are recorded in full, generally using a tape recorder. The interviewer is free to probe as necessary to amplify and clarify responses and is able to follow up any interesting ideas and explanations disclosed by the respondent. The interviewer also chooses the order in which questions are asked, so as to assist the flow of the interview, while the interview is conducted in a conversational manner. An even freer format is provided by the unstructured or focused interview, and is employed when the researcher is more concerned to examine the features of individual cases. A focused interview involves the interviewer in simply engaging the respondent in conversation about particular topics and following up points of interest as they develop. Examples of the use of this form of interview include Cornwall's study (1987) of the ideas and theories about health, illness, and health services held by 24 people in East London and Locker's study (1983) of the impact of severely disabling rheumatoid arthritis on people's daily life and life chances based on interviews with 24 people who had experienced rheumatoid arthritis for several years. Atkinson (1978) also employed focused interviews with coroners to assist in understanding their decision-making process. A notable feature of the study by Cornwall (1987) was that more than one interview was undertaken, with the second and third interviews being employed by the researcher to explore in greater detail some of the themes identified in the initial interview. Similarly, Locker (1983) conducted a second interview one year after the first to examine the changes in impairment, disability, and personal and social circumstances that had occurred, and took into account in this second interview their responses in the initial interview. He notes that the quantity of data produced meant that each interview frequently required two or three visits to complete so as not to overburden respondents.

Other forms of data collection occasionally employed in qualitative research are observation and the diary method. In both cases the data to be collected are less precisely defined at the outset than in quantitative research. For example, Clark and Bowling (1989), in their evaluation of the quality of life in hospitals and nursing homes, supplemented their structured observations with a less structured approach. The latter required the researchers to write an account of events, interactions, and moods as well as their overall impressions and remarks. Qualitative studies may also involve participant observation in which the researcher is part of the group being studied. This method was adopted by Goffman who observed life in a psychiatric hospital while employed as a remedial gymnast (Goffman 1961).

Differences between qualitative and quantitative research also extend to data analysis. In quantitative research this involves the statistical analysis of numerical data. In contrast, data analysis in qualitative research involves identifying themes, in terms of shared meanings, interpretations, and responses. These themes are identified through studying the interview transcript or other case materials and analysing responses and behaviours in their context. It therefore involves content analysis, with ideas and explanations being examined through the perspective of the respondents, with a view to eliciting the respondents' own understandings and reasons for their behaviours. The presentation of data in qualitative studies generally involves providing case material in terms of verbatim statements, with quantification usually being limited to identifying the numbers of people sharing similar experiences, views or behaviours. There are thus fundamental differences in both the conduct and presentation of research employing quantitative and qualitative methods

which are associated with differences in their emphasis on the objective or subjective nature of social reality.

Examples of qualitative research

The application of qualitative methods can be illustrated by studies of the behaviour of people in medical settings. This research is generally informed by pluralist notions and employs qualitative methods to identify the interests of different participants and the ways in which they try to influence events and achieve their desired goals. An example is West's (1976) study of the interactions between clinicians and parents with a child diagnosed as epileptic. His data consisted of observations of 64 consultations plus tape recorded interviews with mothers at home. He found that whereas the doctor maintained tight control over the initial consultation, at later consultations parents were often more active and frequently pressed for a diagnosis, or a reduction in the dosage of the drugs prescribed. As a result, doctors were often faced with the problem of maintaining their credibility and the legitimacy of their position as 'expert' in the face both of considerable clinical uncertainty and this pressure from parents. Thus, rather than both doctor and patient acting out their socially prescribed roles as depicted by Parsons (1951) with the doctor being dominant in the consultation, the doctor's authority was challenged as patients became more knowledgeable, and bargaining and negotiation occurred.

Another example of qualitative research in medical settings is Hall's (1977) study of the implementation of the decision to introduce play leaders for children in hospital. He collected data through observing on the wards and interviewing nursing staff. This showed that nurses viewed the introduction of play leaders as disrupting the conventional routines of the ward. They therefore tried to renegotiate the opportunities children had for play and to confine play leaders to a more marginal position within the ward than official policy decreed. This points to the importance of the acceptance of organizational change by the various groups affected and not merely by those with formal authority. Indeed, in some cases those who occupy a low position in the organization may be most resistant to change. This is especially true if the change is likely to undermine their skills, as these generally task-specific rather than more general and adaptable. For example, staff who only have skills and expertise in teaching deaf people sign language may resist attempts to replace this by lip reading (Scott 1974).

Recognizing that the various participants in an organization may have differing interests and goals, Smith and Cantley (1985) combined a pluralistic approach with qualitative methods of research in their evaluation of a psychogeriatric day hospital. This involved a rejection of the traditional epidemiological approach to evaluation, which involves specifying the criteria of success at the outset and then collecting data using standardized techniques to establish whether this criteria has been met. Instead they began by conducting focused interviews to elicit the criteria of success held by the various groups responsible for planning, implementing, and using the service, including doctors, nurses, managers, paramedical staff, and patients' relatives. This led to the identification of six criteria of success: free patient flow, clinical cure, provision of an integrated service, impact on related services, support for relatives, and the quality of care received by residents. Success was then evaluated in terms of the extent to which each of these criteria was met. The perceptions of the various groups as to the reasons for failure to achieve the criteria of success was also elicited. A variety of data were employed, including open-ended interviews with staff and patients' relatives, the recording of meetings, and observations of the activities of the day hospital, as well as the use of secondary data derived from nursing notes and medical and social work reports. An important finding of the research was of the differences between the various groups in their goals and perceptions of success. For example, in terms of patient flow the major concern of the ward staff is to avoid 'silting up' of their part of the system by enhancing patient flow out of in-patient wards through the day hospital, whereas for day hospital staff this poses problems of 'silting up' of their part of the system. For social workers, their concern to maintain patient flow is tempered with concern for the needs of the family unit, and thus in this respect their interests differ from most other staff groups. As the authors point out, their adoption of a pluralistic model of evaluation, although not conforming to the traditional scientific evaluative model, should serve to facilitate the effective implementation of the research results. This is because it takes into account the interests and views of the major groups involved whose co-operation is crucial in promoting change.

Another area in which qualitative research has made a major contribution is in relation to studies of the social meanings of medical conditions and experiences of patients. One of the earliest and classical studies is Goffman's analysis of stigma, published as *Stigma: the management of a spoiled identity* (1963). Drawing on his own observations, Goffman identified the effects of possessing conditions and behaviours that are socially devalued and regarded as something shameful or 'stigmatizing'. He identifies three categories of conditions that tend to be stigmatized, namely physically disfiguring conditions, defects of character or mind (e.g. mental illness, epilepsy, a criminal conviction), and membership of certain racial or religious groups. However, he points out that the social meanings associated with particular conditions or behaviours change over time. An example is the considerable change in attitudes to mental illness, homosexuality, and single parenthood that has occurred in many western countries over the past 50 years. Variations also occur between societies, as illustrated by Waxler's (1985) description of the differences in the social meanings and responses to leprosy between India and Sri Lanka and between different countries of Africa.

When conditions are socially devalued or stigmatized, they are frequently associated in people's minds with other devalued conditions, as with the association of physical disability with low intellectual ability, or with negatively valued personality traits, such as being an introvert or loner. Furthermore, the possession of a stigmatizing condition tends to become generalized to the whole person and forms what Goffman termed a 'master status', which replaces the stigma bearers' actual identity with a spoiled identity. The person with epilepsy, a history of mental illness, or a physically disabling condition, is thus faced with the problems of coping not only with the direct effects of their medical problem but also with the feelings of shame and differentness associated with these special meanings, and with the responses that such social meanings may evoke in terms of avoidance, and various forms of discrimination and exclusion. Goffman graphically describes this experience from the point of view of the stigma bearer and identifies various strategies they may adopt. This may involve 'passing' as normal, if the stigmatizing condition is not visible or otherwise known about. However, he describes the considerable psychological strain that passing may impose, if stigma bearers are constantly concerned to conceal their physical deficit, or to control information which if known about would be likely to give rise to stigma. For example, the information that a person has been a patient in a psychiatric hospital may be concealed and a period of unemployment explained in more socially acceptable terms, such as travelling abroad or caring for a sick relative, while contacts with other ex-patients may be avoided, as may situations that would reveal a physical disability. Another strategy Goffman identifies is the attempt to normalize the condition. This refers to when a person accepts their possession of a stigmatizing condition but attempts to show that in all other respects they are 'normal' and want to be accepted as such, rather than for their stigmatizing condition to be generalized to their whole person. Goffman further suggests that some people may cease to seek the company of 'normals' (i.e. the non-stigmatized) and withdraw from such social interaction.

Goffman's analysis of stigma was important in taking Parson's analysis of illness as a social state a step further, and showing that the social meanings surrounding particular biological states can have a major impact on people's everyday life and self-identity. Other studies have since developed this analysis to consider such issues as the role of the medical profession as stigmatizers and destigmatizers (Volinn 1983). Particular attention has also been paid to the experience and responses of people with particular types of stigmatizing illness. For example, Scambler and Hopkins (1986) examined sufferers' perceptions of epilepsy and its impact on them, based on semi-structured interviews with 108 people who had experienced at least one epileptic seizure. They found that the perceived stigma of epilepsy had a powerful effect on people's lives, and was frequently the major problem for people with epilepsy. However, the respondents' descriptions of their experiences and responses led the researchers to introduce a distinction between 'enacted' and 'felt' stigma. Enacted stigma refers to instances of discrimination on the grounds of their perceived unacceptability or inferiority, although excluding what the respondents viewed as legitimate discrimination in terms of banning them from driving or operating machinery. Instances of enacted stigma recounted by respondents were fairly rare. The most disruptive effect of stigma on their lives and a source of considerable anxiety was that of felt stigma, or the fear of enacted stigma and shame associated with being epileptic. The experience of felt stigma meant that they often avoided situations that might lead to enacted stigma and frequently tried to conceal their epilepsy even from fairly close acquaintances. However, whereas felt stigma exerted a powerful effect on the lives of this group of people with epilepsy, some groups, rather than sharing the values of the wider society and experiencing felt stigma, are increasingly challenging these social values and rejecting the stigma surrounding their condition. Examples include the Gay Liberation movement and various pressure groups formed by disabled people that seek to change social meanings and thus reduce stigmatization (Anspach 1979).

For some chronically ill and disabled people, their main problems are not so much their feelings of stigma but rather the difficulties they experience in accomplishing everyday activities and the need to adapt and accept their restricted functioning. Their problems and ways of coping at a practical and psychological level have been described in detail through studies employing qualitative methods. An example is Locker's (1983) study of the experiences of people with rheumatoid arthritis. This was based on semi-structured interviews with 24 people with rheumatoid arthritis who were each interviewed on two occasions. Their considerable difficulties in undertaking everyday tasks such as dressing and climbing stairs are described, as well as the effects of their disability on social contacts, social relationships and family life. Respondents were also shown to be engaged in a continuing search to identify the cause of their condition and to locate some factor in their family history or own experiences to explain why they had acquired the disease. Their medical contacts are described as being important in serving as a source of hope and of providing psychological and social support, although patients generally recognized the limitations of medicine in terms of the possibilities for intervention. In this way, the study identified the specific concerns and responses to services of the patients themselves, which may differ from the views and assumptions of service providers.

Questions as to how people understand and make sense of the world has also led to studies of the conceptions of 'health' held by lay people. This suggests that lay people do not view health as a single concept but identify several different dimensions. Williams (1983), on the basis of open-ended interviews with 70 respondents aged 60 years and over living in a Scottish city, identified three dimensions of health. These were: (a) health as the absence of disease; (b) health as strength, in terms of the power to come through illness; and (c) health as functional fitness, judged as the ability to undertake activities normal for one's age. These dimensions were not necessarily related. For example, people viewed it as possible to have health as strength, without functional fit-

ness. The first two dimensions were similar to those identified by Herzlich (1973) based on semi-structured interviews with 80 middle-class people in France. The main difference was that rather than defining health in terms of functional fitness, the French sample identified health as 'equilibrium', referring to a state of perfect physical and psychological well-being; a difference which may reflect cultural patterns or a generational effect. Identifying the conceptions of health held by the lay population may help to explain what has often appeared inconsistent from a medical perspective, that those with known medical problems may none the less rate themselves as having 'good' health.

Studies by Blaxter (1983a) and others, involving discussions of people's perceptions of the causes of disease, have similarly shown that although their perceived causes may not necessarily accord with professional views, they nevertheless have their own internal logic. Their explanations were also often complex and rather than identifying a single risk factor frequently involved multiple causes and might include a psychosomatic component. Blaxter also noted that when accounting for their own chronic health problems people tended to identify various aspects of their experiences, including childhood experiences, their pregnancies, their work and environment, and the major illnesses they had suffered, to explain their susceptibility or to locate the origins of their problems, and they rarely viewed diseases as striking randomly.

Just as lay people's perceptions of the cause of disease can be seen to have an internal logic (although not necessarily coinciding with professional medical views), in the same way decisions not to follow medical advice can be viewed as rational when seen from the patient's perspective. This is illustrated by a study examining the causes of non-compliance with anti-hypertensive medication. Rather than adopting a quantitative approach and specifying at the outset the variables hypothesized to explain non-compliance, semi-structured interviews were held with groups of 30 respondents of European origin and 30 respondents of West Indian origin who were being treated for hypertension by their general practitioner (Morgan and Watkins 1988). This allowed respondents' own beliefs about hypertension, their use of the medication, and their reasons for non-compliance to be examined in detail. It showed that about half the West Indian patients did not take their medication as prescribed and regularly 'left off' the drugs for a few days each week or for a week or so at a time. All these patients were aware of their doctor's expectations and also of the increased risks of a heart attack or stroke if their blood pressure was not controlled. The major factor leading to non-compliance with medication was described as being their fears of possible long-term harmful effects of regular drug taking and a dislike of feeling dependent or 'addicted' to the drugs, as well as their belief in the value of herbal remedies which many took as well as, or sometimes instead of, their prescribed medication. Thus, for this group of respondents their behaviour was unlikely to be influenced by conventional responses to non-compliance, in terms of explaining instructions more clearly or providing aids to remembering. Instead their main concern and reason for non-compliance, as identified by the respondents themselves, was their worry about the long-term harmful effects of drugs. Furthermore, although frequently 'leaving off' the prescribed drugs, they appeared to be concerned to monitor their blood pressure level, and visited their doctor regularly for this purpose.

Qualitative methods have been applied not only in the study of lay views and experiences but also to identify the social processes and interpretations that surround professional decision-making. An example is Atkinson's (1978) study of suicide. Rather than accepting suicide statistics as objective data, Atkinson studied the process by which deaths are certified as suicide, through first-hand observation of the work of the coroners' offices and inquests, unstructured interviews with officials to elaborate and clarify observed events, and the analysis of coroners' written reports and newspaper reports. He showed that although there is general agreement that 'intention to die' is in some ways a necessary distinguishing characteristic of suicide, there are important differences among officials as to how such intentions may be inferred. A range of evidence is often employed in making this assessment, including suicide notes and threats, the mode of dying, the location and circumstances of the death, and the biography of the deceased person. Such 'clues' are then interpreted in terms of the particular but varying assumptions that officials hold about what constitutes a 'typical' suicide or 'typical suicide biography'. Atkinson concluded that whether the official cause of death is 'suicide' or 'accidental' is very much a social product. This is because it reflects the complex interaction and interpretations of various officials, such as doctors and policemen, as well as the responses of the public and close relatives, who together influence the choice of the 'correct' category and the production of a 'solid' statistic. Thus, whereas Durkheim's approach to the variations in suicide rates was to focus on the social causes of differences in individuals' propensity to take their own life (p. 311), Atkinson, adopting an interpretative approach, drew attention to the social production of official statistics and the way in which the differing views of participants in the decision-making process may contribute to variations in suicide statistics and other cause specific death rates. Variations in the ways of categorizing the cause of death tend to be greatest when considerable medical uncertainty exists as to the correct category, and when a particular cause of death carries a stigma, such as death from chronic alcoholism, which is frequently recorded as a more acceptable cause of death, such as myocardial ischaemia. Another condition for which the considerable social stigma and medical uncertainty influences the cause of death certification is death from AIDS. The underlying cause of death is often either omitted or selected to avoid the stigma and distress for relatives and close friends associated with AIDS (King 1989; Bennett 1987).

The way in which professional interpretations influence official rates is also illustrated by the investigation by Bloor *et al.* (1979) to explain the variations between specialists in operation rates for adeno-tonsillectomy between regions in Scotland. Their initial analysis indicated that the differences in operation rates occurred even within regions and for individual ear, nose, and throat (ENT) specialties, and did not appear to be completely explained by differences in incidence or in general practice referral rates. They therefore investigated the actual assessment practices of surgeons, based on the study of eleven ENT surgeons. This involved observation at 493 consultations to elucidate what information the consultant used to come to his decision, focused interviews with consultants, and information derived from referrals letters and case notes. Using this material, the researchers demonstrated that although subscribing to a common body of scientific knowledge of ENT practice, phrased in general terms, there were considerable variations in the way in which this was done. These included the clinical features that surgeons considered relevant to their assessments, the importance they accorded to examination findings and to history, the procedures used in the examination, the degree to which they restricted the operation to younger children, and the importance attached to aural symptoms. They thus concluded that local differences in surgical rates are largely 'created' by local differences in the nature of specialist practice. By focusing on procedures of assessment and clinical decision-making, this study thus complements broader socio-medical enquiries which focus on the contribution of differences in morbidity and health service resources to small area variations in hospital admission rates (Wennberg 1987; McPherson 1988).

As these examples illustrate, naturalist approaches have been important in focusing attention on social meanings and processes, and on the content of lay beliefs and experiences of health and medicine. However, the use of qualitative methods is frequently criticized for their failure to satisfy the requirements of the scientific method. This is viewed as casting doubt on the validity of the findings of qualitative research. It has also been an important reason for a reluctance of organizations more familiar with the scientific method to provide funding for research employing qualitative methods.

Critiques of qualitative research

A frequent criticism of qualitative research is that the study of small numbers of people, who are not necessarily selected so as to be representative of the broader population from which they are drawn, casts doubt on the generalizability of the findings. Although accepting that qualitative methods often do not satisfy scientific principles of sampling, it can be argued that the detailed study of what may be regarded as 'typical' cases allows general social categories and social processes to be identified, and thus produces results that are generalizable in these terms. For example, Scambler and Hopkins (1986) were able to contribute to a more general understanding of the experience of epilepsy based on the study of a small group of people aged 16 years and over who were identified from general practice records as having experienced at least one epileptic seizure in the last two years and/or were on anti-convulsant drugs. Through semi-structured interviews with these respondents, the researchers developed the notions of 'felt' and 'enacted' stigma, and were able to describe their causes and how they affected the daily lives of people with epilepsy. They also suggest that felt stigma was more common than enacted stigma among people with epilepsy (p. 321). However, although identifying generalizable social processes, this piece of research could not answer questions of population prevalences, in terms of whether the strength of felt stigma and people's coping responses differ between social classes or educational groups, and between different age groups, etc. One way of approaching these questions is to employ large representative samples which allow the beliefs and responses of different groups to be compared. This approach generally requires that quantitative forms of data collection are employed and that such concepts as felt stigma and enacted stigma are operationalized and measured, using standard scales. These scales can then be incorporated into a structured questionnaire and administered in a uniform way to large numbers of people. An example of this approach is Macdonald's (1988) use of a stigma rating scale. This consisted of a number of statements about avoidance of others, avoidance by others, feelings of self-consciousness, of unattractiveness, and of being different from other people. Typical statements were, 'I feel less attractive than I used to', 'I feel self-conscious and embarrassed about myself', and 'I feel odd and different from other people'. Items were scored from 0 (complete disagreement) to 3 (complete agreement), with scores of 0 or 1 regarded as representing negligible stigma and 2 or 3 some stigma. In terms of overall scores the upper 15 per cent of the frequency distribution defined severe stigma. Although such standard instruments provide a uniform and repeatable measure, there is the question of whether the selected dimensions and the scores attached to statements accurately reflect individuals' feelings and experiences, and hence whether this forms a subjectively valid measure. For example, do those who score most highly on a stigma rating scale actually feel most stigmatized? Similarly, do those who score lowest on formal measures of social contact actually perceive themselves as having a low level of social contact and by implication do they view their level of social contact as inadequate? A second issue is whether people respond in the same way in a structured interview as they do in a more open, conversational interview. Cornwall (1987) believes that structured interviews tend to elicit people's public accounts, or the dominant cultural beliefs and values, whereas more open, in-depth interviews frequently elicit people's private accounts which are drawn more directly from their own personal experiences. For example, public accounts of health, illness, and health services tend to reflect ideas that people believe are acceptable to doctors or how they should behave, whereas private accounts emphasize their own responses to these

social and medical expectations. For this reason, there may often be problems in employing qualitative techniques as a means of extending the findings of qualitative studies.

Whereas the primary concern of qualitative research is with producing data and explanations that have subjective validity, there is a question as to whether this is achieved. In particular, there is a problem in that the close involvement of the researcher in the process of data collection and the requirement to be more active in terms of identifying and probing ideas in semi-structured and focused interviews, affords considerable scope for the researcher's beliefs and values to influence the information elicited, as well as its interpretation. Investigators also sometimes feel a strain between their role as a researcher and their acceptance by respondents as a friend, especially if several interviews are involved or the subject matter is stressful (Cannon 1989). Quantitative studies employing standard questionnaires and the statistical analysis of data, although not placing these demands on the researcher, do not entirely overcome problems of the subjective influence of the investigator. This is because the researcher not only selects the variables to be studied but is also frequently engaged in a process of interpreting responses to very specific questions, and supplies meaning in idiosyncratic or culturally patterned ways (Cicourel 1964).

One approach to increasing the validity of qualitative research is through the use of multiple sources of data to assist in developing and checking working hypotheses (e.g. Atkinson 1978; Bloor et al. 1979). Another approach is to allow respondents, or subjects, to comment on the interpretations drawn by the researcher. This often occurs fairly informally during open-ended interviews. An example of a more formal approach is described by Bloor (1976) who, in his investigation of ENT surgeons routine assessment practices, sent a written report to each surgeon. This described the surgeon's routines as identified by the research, based on observations of consultations and the analysis of case notes. Each report was then discussed with the specialist in question. Where any point in the report disagreed with the surgeon's own impressions of his or her clinical practice, the points were discussed, and, where necessary, the data were re-analysed in the light of the discussion.

It is important that assessments of the methods and validity of the data produced by qualitative studies are judged in their own terms rather than being viewed as deficient for their failure to satisfy the requirements of the scientific method. This is because qualitative and quantitative approaches derive from different philosophical traditions and differ in their aims, assumptions, and forms of data they elicit. They are thus important in focusing on different types of questions and can be viewed as complementary approaches to the field.

Combining quantitative and qualitative approaches

So far, quantitative and qualitative research have been contrasted and viewed as distinct approaches. However, in reality it is probably more accurate to regard these as polar types, for much research combines some elements of these two approaches. For example, preparatory investigations prior to setting up quantitative studies, often involve qualitative techniques of data collection to assist in the identification of variables of interest. Predominately structured questionnaires may also include some open-ended questions to give respondents an opportunity to express their own views more freely. Health services research also frequently departs from the hypothetico-deductive model of hypothesis testing. Instead it adopts a more inductive approach in which hypotheses and explanations are developed from the data, although employing quantitative methods of data collection and analysis. An example is Bowling and Cartwright's (1982) study of the needs of elderly widowed people. Rather than testing specific hypotheses concerning the effects of formal services on the quality of life of elderly widowed people, their aims were much broader. They defined these as being to examine the whole spectrum of problems that are encountered by both the elderly widowed and their supporters, and to identify what elderly people themselves and their general practitioners thought could be done to reduce these problems. Posing such broadly defined questions influenced the design of the research. Rather than adopting an experimental design involving a randomized controlled trial, or the comparison of an intervention and control group based on a non-random allocation, descriptive data were collected using a structured questionnaire administered to a random sample of elderly widowed people four to six months after widowhood. Structured interviews were also held with a close relative or friend, and a postal questionnaire sent to their general practitioners. The survey thus provided data on a range of issues, including the widowed person's role in caring for the spouse who died, information and lack of information about the prognosis, the practical problems faced in the early months of widowhood, his or her own health and the role of the general practitioner in helping to cope with the new situation, problems of emotional adjustment, isolation and loneliness, and the role of relatives and friends as sources of support. This study therefore led to the identification of a series of things that might be done by statutory and voluntary organizations and by relatives and friends to help people recently widowed.

Although quantitative sociological investigations in the field of public health often depart from a strict interpretation of the scientific model through their use of an inductive approach, and may also be supplemented by more qualitative data, few studies have attempted to integrate quantitative and qualitative approaches. A notable exception is the Brown and Harris (1978) study of the causes of depression among middle-class and working-class women in Camber-well, South London. This study aimed to investigate the influence of various vulnerability and protective factors on risks of developing depression, including the women's experience of life events and the quality of their social support. It thus focused on one of the central issues of social epidemiological research. Their approach was informed by the requirements of quantification. For example, the women were drawn from households selected at random from local authority records, depression was identified using a shortened version of the Present State Examination, and the effects of various risk and vulnerability factors were examined using statistical methods. However, their measurement of life events and social support derived from a naturalist perspective which emphasizes the meanings for individuals.

The traditional epidemiological approach to the measurement of life events is to employ the Holmes–Rahe Social Readjustment Rating Scale (Holmes and Rahe 1967) or one of its many modifications. This consists of a standard checklist of life events in which each event is assigned a score to indicate its severity, defined as the amount of readjustment required. These scores have been derived on the basis of previous scaling studies and assume that everyone experiencing a particular type of life event (e.g. widowhood, divorce, changing job, moving house, etc.) undergoes a similar amount of readjustment. In the same way, social support is usually assessed by asking respondents a series of questions about the number and types of social contacts they experience. On the basis of this information respondents are assigned a support score. However, these scores and the cut-off points defining high or low levels of support are often determined fairly arbitrarily by the researcher, with no attempt being made to gain any assessment from the respondent of the adequacy of their support (Cohen and Syme 1985).

In contrast to this approach, Brown and Harris (1978) wished to elicit the women's own views about their life events and support, recognizing that the stressfulness of life events and the quality of support depend on individuals' subjective assessments. They therefore adopted qualitative methods involving semi-structured tape-recorded interviews to elicit the women's own feelings and experiences. This material was then analysed by raters who employed carefully defined criteria by which to assess the women's experience of life events, including the perceived short-term threat of a life event, its long-term threat, the feelings of anxiety it produced, etc. The quality of their social ties was similarly elicited through interviews and rated in terms of number of predefined scales. The use by Brown and Harris of standard rating scales meant that a uniform method of assessment was applied to the women's own descriptions of their experiences. This approach was preferred to requiring the respondents themselves to make judgments of the severity of the life events they experienced, or the strength of their social support. This is because the criteria employed in making such judgements is likely to vary between respondents and may be influenced by whether or not the life event resulted in subsequent illness. These considerations emphasize the problems of measurement and meaning in social science research and difficulties of establishing causality, since human beings actively interpret and respond to situations.

Contributions of sociological investigations

Sociology forms one of the basic disciplines contributing to the field of public health, and contributes both to epidemiological questions and to issues relating to the organization, provision, and evaluation of health services. As this chapter has shown, an important feature of sociological investigations is that they are not characterized by a single disciplinary approach. Instead they encompass a number of 'sociologies' that differ in their theoretical perspective, or view of the social world, and in their methods of enquiry. Sociological investigations sometimes share similar assumptions and methods of enquiry as epidemiological research deriving from the biomedical model, and involve the quantification of social variables. Other sociological investigations, although involving quantification, are informed by a view of society that emphasizes the conflict rather than the consensus inherent in social relationships. A third approach is characterized by an emphasis on the subjective nature of social reality and employs qualitative rather than quantitative methods of investigation to elicit these subjective meanings. As we have seen, these 'sociologies' have posed different questions about health and medicine and provided different explanatory models, and thus made their distinctive contributions to the body of knowledge in the field.

The contribution of sociological research to public health can be considered in terms of two models derived from analyses of the influence of research on the policy process. These are the problem-solving or engineering model and the enlightenment model (Bulmer 1982). The problem-solving model refers to a situation in which a problem is identified, such as the low rates of uptake of preventive services or the nature and measurement of disadvantage. Research findings are then drawn on, or research commissioned, to fill the knowledge-gap. In this way, particular research studies provide information that is utilized directly in seeking a solution to a problem, or in selecting among alternative solutions by clarifying the situation and reducing uncertainty. Similarly, sociological research may contribute directly to epidemiology and health service studies by providing ways of measuring poverty, stigma, health status, or other variables. However, in many cases the contribution of sociological investigations, as with social science research more generally, is fairly diffuse, and conforms more closely to the 'enlightenment' model. This refers to the way in which sociological investigations may help to shape the way in which epidemiologists, policy makers, and others think about such issues as inequalities in health, disability, patient compliance, the care of the chronically sick, etc. As Becker (1976) notes, it is not sufficient that situations or conditions should exist for them to

be defined as a problem, but it also requires that attention is drawn to them. Thus one important contribution of sociological enquiries is to identify problems of illness and health care and to influence what are viewed as priorities for action by providing evidence on, for example, the substantial inequalities in health between social classes or the extent of non-compliance with medication.

Sociological research also identifies those variables or aspects of a problem that require to be taken into account. For example, whereas non-compliance with medication by people with hypertension and other chronic conditions has traditionally been viewed as a problem of forgetting or a lack of knowledge of medical advice, sociological investigations have identified a number of other explanations for such behaviours. These include people's desire to reject their illness and to view themselves and be viewed by others as 'normal', as well as their fears of the long-term harmful effects of drugs. Similarly, sociological investigations have influenced thinking in relation to the care of the elderly by drawing attention to the social costs that are often experienced by families in caring for an elderly relative. This includes the effects on the health and social role of the main carer and the tensions that may develop within the family unit, which require to be taken into account in assessing needs for services and evaluating this form of care.

Research serving an enlightenment function need not necessarily be compatible with prevailing values and goals, or methods of enquiry. Indeed, it may exert an important influence on thinking in the field by questioning these assumptions. For example, sociological research has questioned the ideology of individual choice and responsibility for health that underpins traditional health education approaches. Instead it suggests that behaviours such as the relatively high prevalence of cigarette smoking among the more disadvantaged groups forms a response to their socio-economic circumstances, and particularly to the problems and stresses they experience. Similarly, low uptake of preventive health services among disadvantaged groups is viewed as a product not only of problems of access to services but also of the low value placed on reducing long-term health risks among people who are involved in coping with more immediate difficulties, such as housing problems, unemployment, etc. More fundamentally, the analysis of the social determinants and distribution of disease raises questions of the priority assigned by society to the promotion of health and the reduction of inequalities in health compared with other social goals (Blaxter 1983b). It also focuses attention on questions of the extent to which inequalities in health can be reduced through health services alone, without tackling the major inequalities and sources of disadvantage in society. Such concerns led the Working Group on Inequalities in Health in the UK (Department of Health and Social Security 1980) to move beyond the medical sphere in their policy recommendations to address the social causes of ill health in terms of poverty, poor housing, and occupational risks. Conflict perspectives have also drawn attention to ways in which the medical profession, the government, industrial corporations, and other groups in society, differ in their interests and goals and to the influence they may have in shaping health policies, including the controls on industrial pollution and the tobacco industry, and the organization and funding of health services.

It is often difficult to identify the precise contributions of particular disciplines to the field of public health. This is partly because through a general process of enlightenment they contribute to the common stock of knowledge in the field. Areas of study are also often common to several disciplines, although each may bring its own perspective and methods of research. For example, official agencies and other groups now frequently undertake surveys of consumer views as a means of monitoring and assessing the quality of health services and informing health service planning. This has been facilitated and encouraged by the tradition of social science research that promoted the consumers' view as an important dimension in evaluating health services and contributed to methods of eliciting such assessments (e.g. Cartwright 1967; Locker and Dunt 1978). However, during the 1960s when sociologists were beginning to provide data on patients' views of medical care, changes were also taking place in medicine, which involved a greater emphasis being given to the importance of doctors listening to patients, both as part of the diagnostic process and as part of the therapeutic process as a means of helping people to organize their problems (Armstrong 1984). More recently, the emphasis on consumer evaluations has been encouraged by broader changes in the ideology and structure of health care systems, and has been endorsed by the various disciplines contributing to the field of public health. Sociological studies can thus be viewed as forming one of several factors contributing to the changing concerns and priorities in this area and to developments in the measurement of consumer views.

A second example of the differing forms and contributions of sociological research is provided by the field of disability and handicap. Community surveys undertaken by sociologists, epidemiologists, and statisticians among others, have been important in identifying the numbers, characteristics, and needs of disabled people and thus of contributing to policy and planning (Patrick *et al.* 1981). Another important contribution of sociological research has been provided by studies of the social consequences of disability and its effects on the individual's self identity. Drawing on Goffman's notion of stigma, the disabled person is viewed as frequently experiencing disadvantages in terms of work, social life, etc., both as a direct result of their activity limitations and from the social meanings surrounding his or her condition that may give rise to actual or feared stigmatization. In addition, the consequences of particular activity restrictions are socially determined and depend on the individual's social role and expectations. For example, arthritis causing restrictions of movement in the hands places a concert pianist at much greater disadvantage than it does a teacher or elderly retired person. The ways in which the consequences of disabling illness for the individual are determined not only by the severity of their impairment and the degree of activity restriction experienced, but also by the social meanings surrounding their condition and the extent to which it disrupts their usual social roles, has been formally acknowledged in the World Health Organization's classification of disablement (World Health Organization 1980). This identifies the social consequences of disability for the individual in terms of 'handicap', which is defined as 'a disadvantage for a given individual resulting from an impairment or disability, that limits or prevents the fulfilment of a role that is normal (depending on age, sex, social, and cultural factors) for that individual'. Such disadvantage may result directly from their activity restrictions, as well as from the social responses to their impairment or disability which may result in avoidance or exclusion. The WHO classification of disablement thus reflects sociological analyses of disability, and identifies ways in which the effects of chronic illness extend into the social sphere. It also serves to focus attention on the contribution of non-medical measures to reducing the social disadvantages experienced by disabled people. This includes the role of education in changing social attitudes and the effects of policies and provision in relation to mobility, housing, and employment in increasing the integration of disabled people into the wider society.

Sociological investigations thus contribute to both research and practice in the field of public health through their problem-solving or technical contributions, as well as their more general enlightenment function. It is likely that the well-established tradition of sociological research in this area will continue to flourish and develop, while the increasing involvement of governments in the planning and provision of health services will encourage the growth of social science research and the development of medical sociology in countries where the specialism is less firmly established. Another general trend may be that sociological investigations will increasingly be informed by a naturalist approach and employ qualitative methods of research, associated with a greater acceptance of the value of such studies and their contribution to an understanding of social behaviours and processes. Furthermore, sociological research may increasingly be conducted within a pluralist framework, which acknowledges the importance of taking account of the interests and expectations of different groups in society in understanding behaviours and developing and implementing changes in the health care field. In addition, it is likely that the growth of new medical technologies, and changes in health care systems and in the concerns of public health as well as in sociology itself, are likely to continue to raise new issues for research with important social dimensions.

Acknowledgements

I have benefited greatly in preparing this chapter from discussing many of the issues with my colleague, David Armstrong, and for his helpful comments on earlier drafts.

References

Alford, R. (1985). *Health care policies*. University of Chicago Press, Chicago.

Anspach, R. (1979). From stigma to identity politics. *Social Science and Medicine*, **13A**, 765.

Armstrong, D. (1984). The patient's view. *Social Science and Medicine*, **18**, 737.

Armstrong, D., Calnan, M., and Grace, J. (1989). *Research methods for general practitioners*. Oxford University Press.

Atkinson, J.M. (1978). *Discovering suicide: studies in the social organisation of sudden death*. Macmillan, London.

Banks, M.H., Beresford, S.A., Morrell, D.C., Watkins, C.J., and Waller, J. (1975). Factors influencing demand for primary medical care in women aged 20–44 years. *International Journal of Epidemiology*, **4**, 189.

Becker, H. (ed.) (1976). *Social problems: A modern approach*. Wiley, London.

Becker, M.H., Hoefrer, D.P., Kasl, S.V., Kirscht, J., Maiman, L., and Rosenstock, I. (1977). Selected psychological models and correlates of individual health related behaviours. *Medical Care*, **15**(5), Supplement 27.

Bennett, F.A. (1987). AIDS as a social phenomenon. *Social Science and Medicine*, **25**(6), 529.

Blaxter, M. (1983*a*). The causes of disease: Women talking. *Social Science Medicine*, **17**, 59.

Blaxter, M. (1983*b*). Health services as a defence against the consequences of poverty in industrialised societies. *Social Science and Medicine*, **17**, 1139.

Bloor, M. (1976). Bishop Berkeley and the adenotonsillectomy enigma: An exploration of variation in the social construction of medical disposals. *Sociology*, **10**(1), 43.

Bloor, M., Venters, G.A., and Samphier, L. (1979). Geographical variation in the incidence of operations on the tonsils and adenoids: An epidemiological and sociological investigation. *Journal of Laryngology and Otology*, **92**, 791.

Bowling, A. and Cartwright, A. (1982). *Life after death*. Tavistock, London.

Brown, G.W. and Davidson, S. (1978). Social class, psychiatric disorder of mother and accidents to children. *Lancet*, **1**(i), 378.

Brown, G.W. and Harris, T. (1978). *Social origins of depression: A study of psychiatric disorder in women*. Tavistock Publications, London.

Bulmer, M. (1982). *The uses of social research*. Allen & Unwin, London.

Calnan, M. (1984). The politics of health: The case of smoking control. *Journal of Social Policy*, **13**(3), 279.

Cannon, S. (1989). Social research in stressful settings: Difficulties for the sociologist in studying the treatment of breast cancer. *Sociology of Health and Illness*, **11**(1), 62.

Cartwright, A. (1967). *Patients and their doctors*. Routledge & Kegan Paul, London.

Cartwright, A. and Anderson, R. (1981). *General practice revisited*. Tavistock, London.

Chadwick, E. (1842, rep. 1965). *Report on the sanitary conditions of the labouring populations on Great Britain*. University of Edinburgh Press, Edinburgh.

Cicourel, A.V. (1964). *Method and measurement in sociology*. Free Press, New York.

Clark, P. and Bowling, A. (1989). Observational study of quality of life in NHS nursing homes and a long-stay ward for the elderly. *Ageing and Society*, **9**, 123.

Claus, L. (1981). *The growth of a sociological discipline*. Sociological Research Institute, KU, Leuven, Belgium.

Cohen, S. and Syme, S.L. (ed.) (1985). *Social support and health*. Academic Press, New York.

Colver, A.F., Hutchinson, P.J., and Judson, E.C. (1982). Promoting children's home safety. *British Medical Journal*, **285**, 1177.

Cornwall, J. (1987). *Hard-earned lives: accounts of health and illness from East London*. Tavistock, London.

Crawford, R. (1977). You are dangerous to your health: The ideology and politics of victim blaming. *International Journal of Health Services*, **7**(4), 663.

Davis, K. and Moore, W.E. (1945). Some principles of stratification. *American Sociology*, **2**, 242.

Department of Health and Social Security (1980). *Inequalities in health: Report of a research working group*, chaired by Sir Douglas Black. DHSS, London.

Dingwall, R. (1976). *Aspects of illness*. Martin Robertson, London.

Dowie, R. (1983). *General practitioners and consultants: A study of outpatient referrals*. King Edward's Hospital Fund, London.

Doyal, L. (1979). *The political economy of health*. Pluto Press, London.

Durkheim, E. (1897, trans. 1951). *Suicide*. Free Press, New York.

Evers, H. (1982). Professional practice and patient care. *Ageing and Society*, **2**, 57.

Eyer, R. (1977). Prosperity as a cause of death. *International Journal of Services*, **7**, 625.

Freidson, E. (1970). *Profession of medicine*. Aldine, New York.

Glaser, B.G. and Strauss, A.L. (1967)., *The discovery of grounded theory: Strategies for qualitative research*. Aldine, New York.

Goffman, E. (1961). *Asylums: Essays on the social situation of mental patients and other inmates*. Doubleday, New York.

Goffman, E. (1963). *Stigma: Notes on the management of a spoiled identity*. Prentice Hall, Englewood Cliffs, New Jersey.

Hall, D. (1977). *Social relations and innovation*. Routledge & Kegan Paul, London.

Herzlich, C. (1973). *Health and illness*. Academic Press, London.

Holmes, T.H. and Rahe, R.H. (1967). The social readjustment rating scale. *Journal of Psychosomatic Research*, **11**, 213.

Illsley, R. (1955). Social class and selection and class differences in relation to stillbirths and infant deaths. *British Medical Journal*, **ii**, 1520.

Illsley, R. (1980). *Professional or public health?* Nuffield Provincial Hospitals Trust, London.

Jackson, R.H. (1988). Management response to childhood accidents. *British Medical Journal*, **296**, 448.

King, M.B. (1989). AIDS on the death certificate: The final stigma. *British Medical Journal*, **298**, 734.

Klein, R. (1973). *The politics of the National Health Service*. Longman, London.

Locker, D. (1983). *Disability and disadvantage: The consequences of chronic illness*. Tavistock, London.

Locker, D. and Dunt, D. (1978). Theoretical and methodological issues in sociological studies of consumer satisfaction with health care. *Social Science and Medicine*, **12**(4), 283.

Macdonald, L. (1988). The experience of stigma: Living with rectal cancer. In *Living with chronic illness* (ed. R. Anderson and M. Bury), Chapter 8. Allen & Unwin, London.

MacIntyre, S. (1986). The patterning of health by social position in contemporary Britain. *Social Science and Medicine*, **23**, 393.

Maclean, U., Sinfield, D., Klein, S., and Hamden, B. (1984). Women who decline breast screening. *Journal of Epidemiology and Community Health*, **38**, 278.

McKeown, T. (1976). *The role of medicine: Dream, mirage or nemesis?* Nuffield Provincial Hospitals Trust, London.

McPherson, K. (1988). Variations in hospitalization rates: Why and how to study them. In *Health care variations: Assessing the evidence* (ed. C. Ham) Chapter 2. King's Fund Institute, London.

Mechanic, D. (1962). The concept of illness behaviour. *Journal of Chronic Disease*, **15**, 189.

Morgan, M. (1986). The doctor–patient relationship. In *Sociology as applied to medicine* (ed. D.L. Patrick and G. Scambler) Chapter 5. Bailliere Tindall, London.

Morgan, M. and Watkins, C. (1988). Managing hypertension: beliefs and responses to medication among cultural groups. *Sociology of Health and Illness*, **10**, 556.

Morgan, M., Calnan, M., and Manning, N. (1985). *Sociological approaches to health and medicine*. Croom Helm, London.

Morgan, M., Lakhani, A.D., Morris, R.W., and Dale, C. (1987). Parent's attitudes to measles immunization. *Journal of the Royal College of General Practitioners*, **37**, 25.

Nathanson, C.A. (1980). Social roles and health status among women: The significance of unemployment. *Social Science and Medicine*, **14A**, 463.

Navarro, V. (1978). *Class struggle, the state and medicine*. Martin Robertson, London.

Navarro, V. (1982). The labour process and health: A historical materialistic interpretation. *International Journal of Health Services*, **12**(1), 5.

Navarro, V. (1989). Why some countries have national health insurance, others have national health services, and the US has neither. *Social Science and Medicine*, **28**(9), 887.

Nichols, S., Waters, W., Fraser, J., Wheeler, M., and Ingham, S. (1981). Delay in the presentation of breast symptoms for consultant investigations. *Community Medicine*, **3**, 217.

Norman, S.A., Marconi, K., and Schezel, G. (1985). Beliefs, social normative influences and compliance with antihypertensive medication. *American Journal of Preventive Medicine*, **1**, 10.

Parsons, T. (1951). *The social system*. Free Press, New York.

Patrick, D.L. and Peach, H. (ed.) (1989). *Disablement in the community*. Oxford University Press.

Patrick, D.L., Darby, S., Green, S., Horton, G., Locker, D., and Wiggins, D. (1981). Screening for disability in the inner city. *Journal of Epidemiology and Community Health*, **35**, 64.

Roland, M., Bartholomew, J., Courtenay, M.F.J., Morris, R.W., and Morrell, D.C. (1986). The 'five minute' consultation: Effect of time constraints on verbal communication. *British Medical Journal*, **292**, 874.

Rowntree, B.S. (1903). *Poverty: A study of town life*. Macmillan, London.

Scambler, G. and Hopkins, A. (1986). Being epileptic: Coming to terms with stigma. *Sociology of Health and Illness*, **8**, 26.

Scott, R. (1974). The social construction of concepts of stigma by professional experts. In *The handicapped person in the community* (ed. D. Boswell and J. Wingrove) Chapter 12. Open University Press, Milton Keynes.

Smith, G. and Cantley, C. (1985). *Assessing health care: A study in organizational evaluation*. Open University Press, Milton Keynes.

Sontag, S. (1977). *Illness as metaphor*. Allen Lane, New York.

Spruit, I.P. and Kromhaut, K. (1987). Medical sociology and epidemiology: Convergences, divergences and legitimate boundaries. *Social Science and Medicine*, **25**(6), 579.

Stacey, M. (1988). *The sociology of health and healing*. Unwin Hyman, London.

Starr, P. (1982). *The social transformation of American medicine*. Basic Books, New York.

Strauss, R. (1957). The nature and status of medical sociology. *American Sociological Review*, **22**, 200.

Strauss, A. (1963). The hospital and its negotiated order. In *The hospital in modern society* (ed. E. Freidson) Chapter 6. Free Press, New York.

Strong, P. and McPherson, K. (1982). Natural science and medicine; Social science and medicine: Some methodological controversies. *Social Science and Medicine*, **16**(6), 643.

Susser, M., Watson, W., and Hopper, K. (1985). *Sociology in medicine* (3rd edn). Oxford University Press.

Szasz, T. (1971). *The manufacture of madness*. Routledge & Kegan Paul, London.

Townsend, P. and Davidson, N. (1982). *Inequalities in health: The Black Report*. Penguin, London.

Volinn, I.J. (1983). Health professionals as stigmatisers and destigmatisers of diseases: Alcoholism and leprosy as examples. *Social Science and Medicine*, **22**, 385.

Waxler, N.E. (1985). Learning to be a leper: A case study in the social construction of illness. (In *Social contexts of health, illness and patient care* (ed. E.G. Mishler, Amara Singham, S.T. Haurer, *et al.*). Cambridge University Press, Cambridge.

Wennberg, J. (1987). Population illness rates do not explain population hospitalization rates. *Medical Care*, **25**, 354.

West, P. (1976). The physician and the management of childhood epilepsy. In *Studies in everyday medical life* (ed. M. Wadsworth and D. Robinson). Martin Robertson, London.

Whitehead, M. (1987). *The health divide: Inequalities in health in the 1980s*. Health Education Council, London.

Williams, R. (1983). Concepts of health: An analysis of lay logic. *Sociology*, **17**, 185.

Wilkinson, R.G. (ed.) (1986). *Class and health: Research and longitudinal data*. Tavistock, London.

World Health Organization (1980). *International classification of impairments, disabilities and handicaps*. WHO, Geneva.

20

Methods of communication to influence behaviour

JOHN W. FARQUHAR, STEPHEN P. FORTMANN, JUNE A. FLORA, and NATHAN MACCOBY

Introduction

Public health is concerned with the determinants and distribution of disease in populations (epidemiology) and with control of disease in populations. Disease control encompasses both treatment (especially where complete cure is possible) and prevention. During the past century, the developed world has experienced a shift in the major causes of death from certain acute infectious diseases to chronic diseases such as cancer, hypertension, and atherosclerosis. These diseases, as well as acquired immune deficiency syndrome (AIDS) and many other infectious diseases, have important foundations in social conditions and individual behaviour and, to a great degree, are preventable. Increasingly, therefore, public health interventions must deal with changing individual behaviour. At the same time, many people in developed countries are seeking information on how to live a longer, healthier, and fuller life. Public health methods must now include those needed to communicate with large numbers of individuals (communities) in order to transmit knowledge and influence behaviour. We use a broad definition of communication, including not only face-to-face and mediated education, persuasion, and training, but also the emerging communication technologies, including personal computers, videotape players, and laser discs. The fact that about one-half of the world's population witnessed the moon-landing in 1969 is a telling example of how communication patterns have changed since electronic media came into being. As McGuire (1985) pointed out, large parts of the modern world are engulfed in a pervasive period of extensive communication which he termed an *era of persuasion*. Graubard (1982) labelled this period an *information revolution* that has altered all of our prior views on how we learn, including, we submit, how habits related to health are formed. The estimated US $2.5 billion spent annually in the US alone (data from 1985) to promote the use of cigarettes (Taylor 1984) or the US $1.5 billion expended world-wide (in 1981) on inducements to drink alcoholic beverages (Ashley and Rankin 1988) are among the salient features of this era of persuasion.

The purpose of this chapter is to describe the methods of communication useful to modern societies in counteracting factors that lead to ill-health and in promoting beneficial health practices. We emphasize the methods applicable to comprehensive community-level disease prevention programmes, reviewing work done on potential components of broad community efforts (such as projects in schools and worksites) and drawing on community studies done in various locales and on the experience of our research group at Stanford. We describe some of the various theoretical perspectives that have been found useful in planning and executing community interventions. However, community disease control cannot be achieved solely by creating educational materials and programmes: they must be implemented. We, therefore, describe the approaches to community organization that are also critical for this implementation. This approach results in some useful redundancy with certain chapters in Volume 3.

The focus of the chapter is on the use of multiple methods of communication as a means of influencing behaviour through education. Our group has generally employed a comprehensive *educational* model in formulating community prevention interventions for several reasons. First, the links between behaviour and health are scientifically established to varying degrees. Our approach is to present individuals with the evidence and to facilitate self-determination of healthful behaviour through both acquisition of knowledge and new skills. Second, regulatory and environmental approaches to influencing behaviour (e.g. restrictions on tobacco smoking and changing food supplies) are more easily achieved when a broad population consensus that these are desirable exists. Our emphasis on education, however, is not intended to imply that other approaches are to be ignored in a truly comprehensive health promotion and disease prevention programme.

Table 20.1. Characteristics of the beneficiaries and mediators of health communication

---

Beneficiaries of health communication
  Recipients of health communication (individuals who differ in various
    characteristics, such as economic status, age, education level, ethnicity,
    cultural values, health needs, health interests, use of different media
    sources)

Mediators of health communication
  Network level
    Extended social networks
    Peer groups
    Families
  Organization level
    Worksites
    Restaurants
    Grocery stores
    Schools
    Cafeterias
    Mass media organizations
    Health care organizations
    Public health organizations
  Community level
    Mass media organizations
    Health professional organizations
    Integration of all the above levels leading to change in public opinion,
      social norms, legislation, food production, and environmental changes

---

The communication process

Communication to achieve healthful behaviour change is defined in this chapter as embracing all messages that influence the health of the individual. The most fundamental requirements for effectiveness are that the messages' content and context be designed to flow through an individual's social network, be appropriate to the needs of the individual, and follow empirically devised theories of human learning. It is apparent that the most common error in health education stems from ignoring current knowledge of health communication methods. The major goal of this chapter is to present the scientific basis for and method of effective mass health communication.

The simplest conceptual framework for communication distinguishes the origin of the message (the sender) from the channel or means of communication. Channels typically include face-to-face communication (which usually provides a means for two-way communication), mediated communication (which most often is a one-way process), and environmental influences (such as the retail food outlet that influences dietary behaviour through what is available and how it is displayed).

In addition to sender and channel, health communications can be classified by the characteristics of the recipients of the intervention. Messages must be designed for *individuals*, often divided by age, sex, economic, or cultural group according to needs and interests. The *social network* of the individual (either family or peer group) is an important reinforcing mediator of health messages. *Organizations* (such as libraries, worksites, schools, or hospitals) may also become

important mediators of health communication campaigns. Lastly, *communities* themselves (as broader systems that include shared geography and common educational and governmental functions) are the logical locale for mediators of health communication when all implementation methods are integrated into a planned and phased campaign (Table 20.1).

The individual who receives a health message can often become a sender or mediator of that message. Community settings especially favour spread of ideas or innovations through their many social networks, multiplying this diffusion. As more individuals are reached and are changed by whatever means, the process accelerates (Rogers 1983). At the community level, one has the opportunity for co-ordination of the system and for use of personalized mass media linked to other campaign elements. Such co-ordination fosters the achievement of a change in public opinion and social norms and, ultimately, environmental, policy, and legislative change.

Implementation: evidence of success of health communication to change behaviour

The fields of public health and biomedical science have benefited from experiments in which interventions designed for high risk individuals, special subgroups of the population, or the entire general population were carried out. Some of these have been in the category of small group studies where their generalizability has not been tested adequately. Other studies have been carried out in the format of controlled clinical trials that have been designed primarily to test hypotheses regarding causality of risk factors and disease outcomes. Some of these have coincidentally given evidence that particular modes of intervention are likely to be feasible when applied more generally in the population at large. A growing number of studies is being done with adolescents, largely in school settings, and these are beginning to carry with them a reasonable assurance of their feasibility in general use (Best *et al.* 1988; Killen 1985). Most of the school-based studies, however, have lacked evidence that they are adequately generalizable to different cultures. A few well-evaluated regional or national mass media health campaigns have occurred, but achievement of other than short-term success seems dependent upon a broader approach. Lastly, a series of studies has been carried out in many different countries on the use of comprehensive, integrated programmes within communities themselves. Some of these studies have reported encouraging results on life-styles, on risk factors, and on some chronic disease end-points.

Small group studies in adults

These studies have generally used face-to-face communication combined with supplementary print materials in highly selected group of adults, often those at high risk for the disease under study. Studies of this kind may be used to develop the components needed for more comprehensive community-based interventions and can be of benefit to a subset of the general population if resources do not allow a broader programme. Many studies have used the principles of Bandura's social learning theory (or social cognitive theory). This model of self-directed behaviour change (Bandura 1977, 1986; Farquhar 1987) assumes that a person is able to self-regulate behaviour and to participate actively in the learning and application of behaviour change skills. The components of self-directed change include problem identification, goal setting, training in self-monitoring, and active training in skills needed both to make changes and to avoid relapses. A corner-stone of social cognitive theory is that active practice of a new skill is more likely to achieve lasting change than is written or verbal persuasion or the vicarious observation of other individuals acting as models of a new behaviour. It is generally agreed that interventions based on behavioural theories such as social learning are more effective than the more traditional health education method of providing knowledge alone. Although the specific techniques used vary somewhat, this general approach has been reasonably effective in achieving smoking cessation among adults (Glasgow 1986; Meyer et al. 1982; MRFIT Research Group 1982; Pierce et al. 1986), at least in developed countries. Future research will need to address cultural and ethnic subgroups as well as smokers who are unwilling to come to time-consuming and complex interventions offered through smoking cessation clinics. Programmes in small groups providing information, monitoring, and guided practice designed to change dietary constituents of saturated fat and cholesterol have been quite successful in lowering blood cholesterol levels (Artzenius et al. 1985; Bruno et al. 1983; Carmody et al. 1982; Ehnholm et al. 1982; Glanz 1985; Hjermann et al. 1981; Puska 1985; Wilhelmsen et al. 1986) and partly mediated education programmes requiring relatively small amounts of health professional attention have also been successful (Crouch et al. 1986). Programmes using modern behavioural techniques designed to achieve weight reduction through caloric restriction have also in some instances proved to be effective (Brownell 1982). The application of these methods of weight reduction to achieve reductions in blood pressure have also succeeded (Hovell 1982). A number of studies has been carried out with adequate comparisons to control groups to show that a wide variety of techniques ranging from specific behavioural methods to simple persuasion and careful explanation of protocol requirements has increased the exercise habits of previously sedentary people (Haskell 1984; Martin and Dubbert 1982; Wood et al. 1983, 1985). These small group studies indicate that the body of educational methods to achieve lasting change is growing, they have been applied to different cultural groups in various countries, and that these changes in life-styles are associated with decreases in postulated risk factors for various chronic diseases such as cancer, cardiovascular disease, and diabetes.

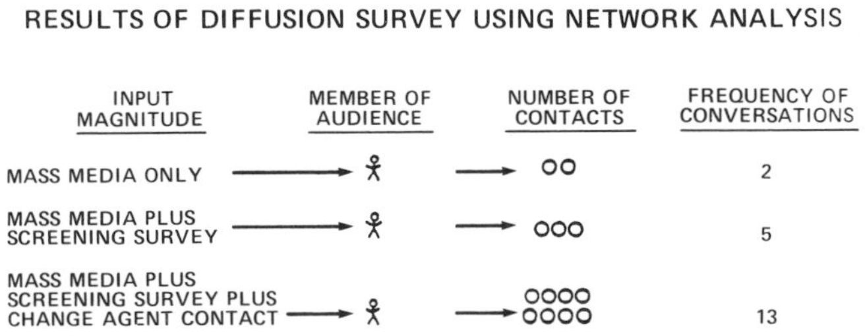

RESULTS OF DIFFUSION SURVEY USING NETWORK ANALYSIS

Fig. 20.1. Evidence for variable diffusion and opinion leader status as a result of varied input of health education during the first year of campaigning in the Three Community Study. Intervention recipients who were personally contacted (bottom) had more conversations about the health issues with a larger number of acquaintances than recipients of mass media only or of mass media plus screening survey. (From: Farquhar 1978, with permission.)

Social networks and health communication

Increasingly television acts to create diffusion of new ideas for commercial, political, or religious purposes in a setting where individuals in a new network are linked to the media source but not usually to each other. However, interactive computer networking now allows participants to communicate with one another in numerous ways (Hiltz and Turoff 1978). Future application of such computer technology in planned social change for human betterment is clearly possible. However, traditional peer and family networks that provide two-way communication are still the strongest inherent links to the individual, particularly when the safety and 'good sense' of a recommended novel approach to life-style and health is questioned (Bandura 1977). These networks can be seen as guardians of existing social norms and can clearly act as barriers to behaviour change. However, they can also be activated to promote diffusion and adoption of new and beneficial health practices, and are thus an important potential resource.

Health communication programmes can recruit adults from the general population to serve as 'change agents' within their peer networks. In the Finnish North Karelia Project, for example, such recruited individuals were found to be similar to the general adult population and after a brief 4 hour training, became a useful part of the programme. Using face-to-face channels, they helped to create diffusion of the project's goals (Puska et al. 1986).

Similarly, a group recruited from the general population on the basis of shared attributes, such as high risk status, may also become effective communicators of health messages. In the Stanford Three Community Study, a group of such adults, who had about 20 hours of intensive instruction in self-directed change, were subsequently shown to be change agents in their peer groups. They conversed with a larger number of their acquaintances, more frequently, than did other members of the adult population (Fig. 20.1) (Meyer et al. 1977).

It can also be seen in Figure 20.1 that members of the audience exposed to a mass media health eduction campaign were more likely to become amateur opinion leaders if they felt personally involved, by virtue of receiving a lengthy survey that identified their personal cardiovascular disease risk status. Given this evidence for 'spontaneous' creation of opinion leaders as a function of the degree of exposure to health education, one can see why a health communication programme applied as any part of a system (such as a school, a worksite, or a community) can lead to at least some diffusion within that system. The often debated question of a 'high risk' versus a 'public health' approach (Lewis *et al.* 1986) to disease prevention is partly resolved by considering the high risk targets of a programme as potential participants in the spread of a health message even if no formal instructions are given asking them to spread the message.

Families can also receive special attention to enhance behaviour changes for one or even all members. Such purposeful family-centred programmes have been shown to enhance success in smoking cessation (Glasgow 1986; Meyer *et al.* 1982; MRFIT Research Group 1982), weight control (Brownell 1982), and nutrition change (Artzenius *et al.* 1985; Bruno *et al.* 1983; Carmody *et al.* 1982; Ehnholm *et al.* 1982; Glanz 1985; Hjermann *et al.* 1981; Nader *et al.* 1986; Puska 1985; Wilhelmsen *et al.* 1986). This evidence buttresses the view that social support provides added benefits to individual instruction and may provide non-specific health benefits as well (Berkman 1984).

Networks are varied and numerous. For optimum effect in health communication one must first discover which natural networks to use (Rogers 1983). For example, it was earlier shown that birth control practices and agricultural innovations diffuse through different networks in the same community (Marshall 1971).

Life-style changes in adolescents

Schools are at once channels to young individuals and collections of adolescent peer networks. Schools have therefore been used in numerous health communication studies, most notably in attempts to prevent smoking adoption in adolescents (Best *et al.* 1988; Telch *et al.* 1982). The most successful programmes have used what Killen has called the 'social pressure resistance training approach' which embodies many features of Bandura's social cognitive theory (Killen 1985). These studies have reported a surprisingly long duration of impact and have generally indicated that extra benefits are derived from including instruction by non-smoking student peers as compared to instruction given entirely by adults (Best *et al.* 1988; Killen 1985; Telch *et al.* 1982). Recent research has attempted to determine also the effects of delivering such programmes by videotape with the hope that improved generalizability and lower costs of implementation would result (Telch *et al.* 1990). A smaller amount of research evidence indicates a beneficial impact on self-reported marijuana and alcohol use in adolescents using the social pressure resistance training method (Pentz *et al.* 1989, Telch *et al.* 1990). This promising approach to preventing or delaying the risk-taking behaviours among young people

needs to be extended to many cultural, ethnic, and economic subgroups and integrated into regular class-room teaching.

The extension and modification of these smoking adoption prevention methods to exercise, weight control, and diet change among adolescents is greatly needed. A few studies have recently reported that instruction in methods of self-directed change can increase exercise habits, change adiposity, produce decreases in resting pulse rate, and induce self-reported decreases in saturated fat and cholesterol intake among groups of tenth grade students in two different areas in northern California (Killen *et al.* 1988, King *et al.* 1988). Recently, Howard has reported the successful use of the social resistance training method in delaying the onset of sexual activity among young women (Howard 1985).

Studies in the workplace

In a second example of change at the level of organizations, worksites are an obvious locale to carry out health communication since they offer the opportunity to influence change through peer network and environmental channels as well as providing obvious opportunities for both mediated and face-to-face communication. Additional motivation is provided given the advantage to employers of a healthy workforce. It is not surprising, therefore, that the past few decades have seen many examples of worksite-based health promotion, focused either on single topics (such as smoking cessation or blood pressure control) or on multi-factor campaigns (Sallis *et al.* 1986).

In a recent review, Sallis and co-workers reported that about 50 worksite programmes have been evaluated and reported in the US alone in the past decade. Many of these appear to have been effective in achieving at least short-term changes in various life-styles and risk factors. They concluded that the scientific basis for the widespread belief that worksite programmes are effective for cardiovascular disease reduction is rather weak, with the exception of multiple risk factor programmes for high risk individuals. A critical lapse is the lack of demonstrated cost-effectiveness.

The largest worksite study conducted was the World Health Organization's European collaborative study in which 66 worksites in four countries were randomized to treatment or control conditions over a 6 year study of a multi-factor cardiovascular risk intervention (WHO 1983). Although treatment intensity and risk factor change outcomes varied, there was a significant overall risk factor change of 11 per cent at the end of 6 years (WHO 1983). Despite a comparatively low level of intervention during the 6 years (comprising three or four brief appointments with a company physician or nurse for high risk men and a general information campaign for all employees) there was evidence for continued benefit after 12 years in a follow-up of ten British factories (Bauer *et al.* 1985). This large study, coupled with numerous other smaller efforts, leads to guarded optimism that worksite programmes can have an important effect on health behaviours and that their use should, and undoubtedly will, increase. The major question for health planners is to know what is the most cost-effective way of achieving different behaviour changes. Information is needed on the role of regulatory and environmental change, the choice of educational method, and the potential for mediated education.

Unfortunately, many worksite interventions are incompletely described. Successful programmes usually incorporated the self-directed change aspects of social cognitive theory, more commonly for nutrition change, weight control, or smoking cessation than for increasing physical activity (Blair *et al.* 1986; Bruno *et al.* 1983; Glanz 1985; Klesges *et al.* 1986; Meyer and Henderson 1974; Sallis *et al.* 1986; Stunkard *et al.* 1985; Wilber 1983). Some of the most effective worksite programmes have also contained regulatory and environmental change features, such as use of restrictions on smoking (Wilber 1983), provision of exercise facilities (Blair *et al.* 1986; Wilber 1983), and inclusion of lower fat menus in company cafeterias (Glanz 1985). Fostering competition among linked work groups through contests has increased the success of programmes designed to achieve smoking cessation (Kleges *et al.* 1986) and increased physical fitness (Blair *et al.* 1986). Such results are testimony to the virtues of worksites as examples of systems that can allow education, peer networks, and environmental influences to be used in a co-ordinated and mutually reinforcing manner.

An important challenge is whether the sequenced pattern of self-directed change instruction can be incorporated into largely mediated nutrition education programmes. One promising study provided initial videotaped instruction followed by repeated brief dietary assessments obtained in widely dispersed sites with individual sequenced instructions for change generated from a central computer (Miller *et al.* 1988).

Health communication at points of purchase

Bandura's social cognitive theory is based on evidence for a reciprocal linkage of a person's cognitions with both their behaviour and their environment (Bandura 1977, 1986). Many recent studies illustrate Bandura's views on the role of environmental influence by showing that modifying environmental cues in school, worksite, or community restaurants and in grocery stores have produced changes toward healthier food choices (Glanz and Mullis 1988). Tactics include providing healthier food choices, altered shelf displays, easily visible labelling of healthier products, and provision of nutrient information through posters and brochures (Glanz and Mullis 1988).

The most extensive point of purchase study in health education has been created by the Minnesota Heart Health Program (Mullis *et al.* 1987) which has also been able to obtain co-operation with the meat industry to increase supplies of lean meats and to train meat department personnel to aid the consumer in identifying lean meat products (Mullis and Pirie 1988). Overall, modest gains in knowledge and slight changes in behaviour are observed as a result of point-of-purchase programmes, and both knowledge and behaviour changes are

often transitory. Point-of-purchase programmes should be part of a larger campaign to become more effective (Glanz and Mullis 1988). The full potential for the development of healthier food products is also unknown.

One recent study has shown a remarkable 50 per cent decrease in (illegal) retail sales of cigarettes to young people (under the age of 18 years) through a combined programme of newspaper, radio, and television publicity and direct face-to-face merchant education (Altman *et al.* 1989). Thus it may well be that point-of-sale health communication methods will have relevance beyond food sales to cigarette and possibly to alcohol sales.

Regional and national mass media campaigns

It would be attractive indeed if all health problems could be solved by use of mass media campaigns alone, given the potential for cost savings inherent in mass communication through print or electronic media. However, a broad consensus exists that such campaigns, directed toward alcohol use, smoking, automobile seat-belt use or multi-factor cardiovascular disease risk factors, although useful for changing knowledge and attitudes, are better linked to more comprehensive campaigns if large and lasting behaviour change is to be achieved (Atkin 1979; Farquhar *et al.* 1977, 1985a; Glanz 1985; Hewitt and Blane 1984; McGuire 1964, 1985; Meyer *et al.* 1977; Puska *et al.* 1981, 1985; Roberts and Maccoby 1985; Rootman 1985; WHO 1986).

Two successful antismoking mass media campaigns have occurred; a regional campaign in Australia (Pierce *et al.* 1986) and a national campaign in Finland (Puska *et al.* 1979). In both instances numerous ancillary activities supplemented the mass media, including use of distribution of printed self-help booklets. Recently, a national television campaign in the US was found to increase the public's awareness of dietary fibre and their self-reported use of high-fibre cereals, and to increase purchase of high-fibre cereals in one test city (Freimuth *et al.* 1988; Levy and Stokes 1987). This success is probably attributable to a combination of unusually effective television messages, prior extensive national publicity in the US linking high-fibre foods with prevention of colon cancer, and to an unusual co-operation between a credible source, the National Cancer Institute, and a large cereal manufacturer (Freimuth *et al.* 1988). Successful multi-factor cardiovascular disease prevention campaigns using mass media have also been carried out, all with broader programmes. They are reviewed in the next section.

Research studies of comprehensive, integrated, community-based health communication programmes

An increasing number of community-based health communication studies has been reported or are under way. The progress of some ten projects that began between 1972 and 1982 was recently reviewed (Farquhar *et al.* 1985a). The two best known of such projects are the Stanford Three Community

Study in northern California (Farquhar *et al.* 1977, 1978; Fortmann *et al.* 1981; Meyer *et al.* 1977; Williams *et al.* 1981) and the Finnish North Karelia Project (Puska 1985; Puska *et al.* 1979, 1981, 1986; Tuomilehto *et al.* 1986), which both began in 1972. The Three Community Study consisted of two treatment cities and one control city with a total population of 35 000. The cholesterol, blood pressure, and smoking results achieved after 2 and 3 years of education in the Three Community Study were comparable with those achieved after the first 5 years of education in the North Karelia Project. Composite cardiovascular disease risk scores in each of these two studies predicted a 20–25 per cent reduction in coronary events. The Finnish Study was extended for a total of 10 years, and favourable effects on cardiovascular disease morbidity and mortality were reported in comparison not only against an adjoining county but in respect of the trends in the remainder of Finland (Tuomilehto *et al.* 1986).

A few other community-based studies using methods similar to the Stanford Three Community Study and the North Karelia Project have reported their main findings. Only one of them, in three small rural South African towns, reported significant changes in cholesterol and blood pressure as well as in smoking (Rossouw *et al.* 1981). Two others, in four towns in Switzerland (Gutzwiller *et al.* 1985) and a three-town study in Australia (Egger *et al.* 1983), reported a 6 and a 9 per cent drop in smoking rates, respectively. In a study from the US carried out in two counties in Pennsylvania, favourable effects were reported in body-weight and dietary habits at workplaces (Stunkard *et al.* 1985), which were the major targets of this study. Another large and complex long-term study, the Stanford Five City Project (Farquhar *et al.* 1985b), has reported interim results showing favourable effects on smoking, blood pressure, exercise habits, and blood cholesterol, comparable with the changes achieved in the Three Community Study (Farquhar *et al.* 1986, 1988). At this point, the analogous results in the Stanford Three Community Study, the Stanford Five City Project, and the North Karelia Project suggest a reasonable generalizability of this extensive type of community approach. Two other extensive and well-evaluated studies are now in progress in the US: the Minnesota Heart Health Study involving six communities of total population 356 000 (Blackburn *et al.* 1984), and the Pawtucket Heart Health Study in Rhode Island involving two cities of total population of 173 000 (Lasater *et al.* 1984). Further information on the feasibility and methods of carrying out such studies will result when later reports of these studies are reported.

A major trial testing a multi-component attack on adolescent tobacco, alcohol, and drug use is under way in 15 midwestern US cities, the Midwestern Prevention Project (MPP). The programme has introduced the following activities sequentially over 6 years: mass media education, school-based education using the previously described social pressure resistance training, parent education, community organization, and health policy components. Preliminary results have shown a decreased prevalence for all three drugs in the numerous intervention schools (Pentz *et al.* 1989). This ambitious study, although restricted in its outcome goals only to youth, is an encouraging addition to the previously described comprehensive cardiovascular disease prevention studies in communities.

Recommended steps in planning, implementing and evaluating comprehensive health communication programmes in communities

Introduction

Planners must first ask whether or not comprehensive community-based public education strategies *should* be implemented. This requires that an adequate consensus exists that resources should be devoted to such strategies, in contrast with those strategies in which only high-risk individuals are identified and managed in traditional medical settings. Part of this question is whether a need for public intervention exists even for disorders where genetic susceptibility plays a large role, such as in adult-onset diabetes. Planners need also to decide if regulatory changes alone will suffice to solve an identified problem. However, if it is assumed that a consensus exists that comprehensive public intervention that combines high-risk approaches, broad educational efforts, and regulatory change is indicated, many questions are posed on how to carry out such programmes. There are such questions as when to intervene, how to carry out the intervention, how to evaluate such interventions, and how to decide on the methods to use.

Problem identification

Problem identification entails an assembly of individuals who are charged with the responsibility for a review of the scientific justification for a particular type of comprehensive public intervention. We must recognize that public intervention concerning life-style can occur at an international, national, regional, or local level. In each instance, the kind of intervention available varies considerably. The final common pathway of public intervention on life-style requires local participation so that programmes deemed useful can be implemented by health workers, educators, mass media outlets, and the citizens themselves. Of course, regional, national, and even international action can be carried out in the areas of regulatory and environmental change and, to a lesser extent, in health education delivered through mass media channels.

We will use the example of how public health officials might proceed if they were dealing with a community that is both a political and a geographical unit, with health education, transportation, regulatory, and media issues under the control of that political and geographical unit. Such an initiating group of organizers would need to have the public health and medical authorities examine the scientific basis for action. In the case of cardiovascular disease prevention, they would find a reasonably good international consensus that cardiovascular risk factors have much of their origin and maintenance determined by powerful cultural forces that lend themselves to concerted and comprehensive public intervention. The organizers would first assemble the literature referred to in the preceding section attesting to the fact that cardiovascular disease remains the major cause of death and disability in the industrialized countries in the world, that controlled clinical trials have supported risk factor hypotheses, that studies on 'components' (such as schools, worksites, and high-risk groups) have succeeded and that some comprehensive community-based studies have had favourable effects. The results of these aforementioned studies therefore furnish support for the idea that planned social change based on a comprehensive linkage of medical, educational, and media resources results in favourable effects on cardiovascular disease. This belief is shared by the World Health Organization, which has set forth guide-lines to encourage member countries to carry out analogous programmes across a broad spectrum of non-communicable diseases (Glasunov *et al.* 1983; WHO 1986). At this point, the local organizers might now conclude that support exists from international studies for comprehensive public intervention, but they may need to seek support of such activities within their own country. If the major professional groups and public health agencies within a country are against such activities, or if review bodies within that country have not yet achieved national consensus, it would be important to recognize this fact and take steps to achieve consensus.

Public health policy must both lead and follow scientific consensus. To a degree, scientific questions are never answered fully, and the opinion that 'more data are needed' is widespread in the scientific and medical communities. Policy workers must recognize that no action is also a policy and may not be the best policy even in the face of uncertainty. Nevertheless, policy cannot advance too far beyond scientific knowledge, and one potential outcome of the problem identification stage is a decision to perform additional aetiological or intervention–development studies. This occurred in the late 1960s in the US for cardiovascular disease, resulting in several major clinical trials such as the Hypertension Detection and Follow-up Program (1979), the Multiple Risk Factor Intervention Trial (MRFIT Research Group 1982), and the Coronary Primary Prevention Trial (Lipid Research Clinics' Program 1984*a*, *b*).

Achieving commitment

The problem identification steps outlined above represent an early stage of consensus that identifies the broad outlines of the problems to be addressed and recognizes that methods exist that can be used to attack the problem with a reasonable chance of success. The next phase of planning requires the existence of a more extensive consensus that recognizes local needs and interests. The initiating group, whether intrinsic or extrinsic to the community, must at this stage recruit a set of community leaders into a community council that is generally representative of various sectors of the economy and life of the community. An early task is to identify, through opinion leader surveys, one or more community residents who can facilitate the formation of a partnership of community leaders and agencies in achieving change. This partnership needs to perform a 'community needs assessment' and 'resource inventory'. A community needs assessment uses formal and informal surveys to determine the attitudes and perceived needs of the agencies and citizens of the community. It incorporates local health statistics to determine the extent of the medical problem under scrutiny. The resource inventory identifies how many organizations and agencies within the community are now working on the problem in question. This allows gaps to be determined and defines the need for additional resources. If the initiating agency is external to the community, then a joint planning group that includes community representatives can use the results of the needs assessment to increase the commitment for action and to develop a preliminary consensus on the general methods to use. Various surprises may emerge in carrying out the needs assessment and resource inventory tasks. As an example, in the field of cardiovascular disease prevention, one might find not only an adequate consensus that the problem needs to be addressed, but also that various resources already exist to carry out at least some of the educational programmes. One might, however, find that there is less consensus on such issues as substance abuse, adolescent pregnancy prevention, cancer prevention, or injury prevention. At this phase, the joint planning group might decide that pilot studies would be necessary to develop methods to carry out any proposed programmes or that an inadequate consensus exists on the need for public intervention in the health topic initially addressed.

Planning

Planning a comprehensive health communication programme requires adequate 'formative' evaluation, which may be defined as evaluation used to modify the programme as it is created. The Center for Research in Disease Prevention at Stanford University has been convinced that formative evaluation is critically important for the success of broad community intervention programmes. Our emphasis in this area follows the general criticism by Atkin that health communication campaigns usually lack adequate formative evaluation, resulting in common failure to meet campaign objectives (Atkin 1979). The community needs assessment and the resource inventory described above constitute the earliest stages of formative evaluation. There are three general categories of formative evaluation used in the actual development of materials and programmes. They are needs analysis, pre-testing of education programmes, and evaluation following the introduction of education programmes into the field. Concepts used in commercial marketing, such as product, price, and promotion, may be usefully applied to social goals, as described by Kotler (1975) and others (Lefebvre and Flora 1988), and are mentioned below.

Audience needs analysis

Considering all community residents as the 'audience', one needs an overview of its attitudes, beliefs, knowledge, and interests. It is important to identify subsections of the community that have different characteristics in age, sex, relative risk, media use, and expressed interest in various educational programmes. This allows educational programmes and materials to be developed that are best suited for each sub-audience. For example, older individuals might respond favourably to programmes that encourage walking, whereas younger age groups are more amenable to more vigorous forms of physical activity. Programmes for adolescents and younger children must be developed that are appropriate to their different stages of development. This stage of formative research also addresses such questions as the proper name, location, and time for educational activities.

Pre-testing of education programmes

If educational materials and programmes are thought of as 'products', then they must be the right products at the right price and must be properly promoted to be successful. Pre-testing determines that the target audience for a given 'product' finds it comprehensible and appropriate; that the audience finds the 'price' (usually time, effort, and other non-monetary costs) to be acceptable; and that it will be available in the right place. A variety of methods is used, such as paper and pencil tests, telephone interviews of target audience samples, and small group discussions focused on a predetermined set of critical issues (led by trained leaders). The implementers of educational programmes must deal here with source, message, channel, and receiver factors that determine the content and method of delivery of the message. The pre-test data are used to revise the message or programme before use.

Evaluation of education programmes

After the introduction of any educational programme in the field, it is important to confirm the conclusions of the pre-testing phase. Did the programme reach the intended audience, did they learn the knowledge and skills as planned, and did the appropriate behaviour change occur? This information can also be used to modify the programme, and to assess which components of the overall effort most contributed to success. The methods used are similar to the pre-testing phase and include questionnaires, face-to-face interviews, focus group discussions, and telephone surveys as well as unobtrusive measures (Webb et al. 1981). This stage of formative evaluation is often termed 'process evaluation'.

Process evaluation is applicable to various community activities such as health fairs, classes, contests, lectures, point-of-purchase programmes, etc. Mass media components of the overall programme must similarly be tracked to ensure that the expected audience is reached and affected as planned. The data obtained in process evaluation are principally concerned with short-term success in achieving the intervening process leading to behaviour change such as attendance, satisfaction, attitudes, and knowledge. These data must also contribute to evaluating the overall outcome of the health communication programme, but complete outcome evaluation requires longer-term, more complete, and more rigorous evaluation methods. Process evaluation gives valuable information to the implementer on how the earliest versions of educational programmes and messages might be altered to increase their effectiveness when repeated.

It can be helpful, when possible, to couple process evaluation with more formal tests of effectiveness by conducting 'pre-field' studies. For example, we developed some intensive instruction methods that followed Bandura's methods of self-directed change (Bandura 1977) in a workplace setting near Stanford University. These methods were subsequently used in the Three Community Study (Meyer and Henderson 1974). This pre-field study succeeded not only in training our instructors and giving us needed protocols and manuals but also in giving us confidence that Bandura's methods could be applied to cardiovascular disease in settings analogous to those we anticipated in our future field experiments. Similarly, success in a pre-field study in preventing adoption of cigarette smoking in young adolescents prepared our group for use of these methods in the Five City Project (Telch et al. 1982). In certain health topics, such as prevention of alcohol-related problems, substance abuse prevention, adolescent pregnancy prevention, and injury prevention, it is likely that more such pre-field studies are needed.

Implementation

Decisions need to be made on the content, sequence, mixture, and duration of the components of health communication programmes. To make these decisions, the implementer can use behavioural and communication theory, the experience of other implementers, and the results of pre-field and formative studies carried out in the planning phase. Health communication programmes should be implemented with a spirit of problem-solving influenced by the continued use of the process evaluation described earlier.

Our Stanford group has previously described the theoretical foundations for comprehensive health communication campaigns in community settings (Farquhar et al. 1985a), and these are briefly described here. We see public health campaigns as requiring both community organization to obtain collaboration with community institutions and also consideration of theories of learning and of mass and inter-personal communication. We termed the organizational change methods 'community organization for health' and the foundations on mass communication and learning as 'health communication–behaviour change'. We have seen the former process of organizational change as a necessary step to ensure local acceptance, collaboration, and, finally, institutionalization and maintenance of any intervention programme. The latter theories on mass communication and learning were seen as guides to the use of effective behaviour change methods within a community setting made more receptive by ongoing community organizational effort. One component of the health communication–behaviour change framework is the use of social marketing principles previously described (Kotler 1975; Lefebvre and Flora 1988). Another useful basis for constructing communications that influence personal networks comes from the work of Rogers (1983), Nader and colleagues (1986), and Puska and colleagues (1986). In respect to content, we draw on prior research, principally McGuire's (1964), which demonstrated that individuals can be inoculated against negative or health compromising acts of persuasion (Roberts and Maccoby 1985). Cartwright's pioneering research (1949) on mass persuasion principles has been found to be useful. He posited that changes must be made not only in people's knowledge and motivation but also in what he calls their 'action structure' for changes to occur. These principles are included in Table 20.2.

Bandura's social cognitive theory (1977, 1986), previously alluded to, has been an important resource for our group at Stanford in determining how to teach the behavioural skills that are often necessary prerequisites to the establishment of healthful habits. Skills can be acquired through social modelling and guided practice, which increase self-efficacy, provide incentives for healthy behaviour, and give direct feedback of one's behaviour to oneself. Table 20.2 outlines the essential components of the health communication—behaviour change formulation, combining elements of the research of both Cartwright and Bandura. The first column depicts the instructional products and events that are used, the middle column lists communication functions required to meet the behaviour change objectives listed in the right-hand column such as knowledge gained, skills acquisition, and maintenance.

Inspection of Table 20.2 leads one to the view that the sequence of communication should follow the stepwise progress from awareness through maintenance. The table, therefore, suggests how to break a large community health education task into manageable pieces such that starting points and sequence and content flow along this predetermined path. It is, of course, never possible to have a clean demarcation among the various parts of the communication and behaviour change objectives. It does, however, suggest that early in a campaign, awareness and knowledge require a greater amount of attention than enhancement of motivation. It also suggests that skills instruction generally occurs later and that maintenance and relapse prevention occur at even later time-points.

The optimal mix of community organization and health education is difficult to know with confidence given the complexity and uncertainties of community studies and the expected variability in state of readiness of individuals to

Table 20.2. The health communication–behaviour change formulation

Communication inputs	Communication functions (for the sender)	Behaviour objectives (for the receiver)
Face-to-face messages	Determine receiver's needs	Become aware
Mediated messages	Gain attention (set the agenda)	Increase knowledge
Community events	Provide information	Increase motivation and interest
Environmental cues	Provide incentives	Learn and practise skills
	Provide training	Take action, assess outcomes
	Provide cues to action, including environmental change	Maintain action, practice self-management skills
	Provide support, self-management	Become an opinion leader (exert peer group influence)
		Give feedback to sender

change. Past experience indicates that high-quality public health communication campaigns carried out for more than a year have produced community-wide changes in various risk factors as indicated in the introduction. In our own experience in the Stanford Three Community Study, and in our current Five City Project, we have developed a general measure of the minimum 'dose' of education needed (Farquhar et al. 1977, 1985a, 1985b, 1986, 1988; Meyer et al. 1977; Meyer and Henderson 1974). It appears that a yearly exposure to approximately 5 hours of a combination of mediated and face-to-face education carried out for 2 or more years will produce an adequate change in cardiovascular risk factors, at least in the populations studied. It appears from our prior experience that if as little as 10 per cent of the total exposure is devoted to the inherently more effective face-to-face communication, adequate change can still occur. Given the lower cost of mediated education, it is, of course, tempting to reduce the face-to-face component to as low an amount as possible. Its inclusion, however, seems to us to be essential for achieving and maintaining long-term behaviour change. Additional experience in the Five City Project derived from use of community-wide and worksite contests indicates that incentives may be provided on a mass basis even with programmes that are predominantly mediated (King et al. 1987). Another overview of the determinant of how much education is sufficient can be answered as follows: it is important to seek the 'critical mass' to achieve an approximate 15–25 per cent of the population adopting a new health innovation such that the diffusion of that innovation may occur more rapidly (Meyer et al. 1977; Rogers 1983).

A number of principles of field implementation of comprehensive public intervention strategies may be derived from a combination of the theoretical framework and practical field experience. Some of these principles are briefly described below.

Establish credibility of the implementer

Scientific credibility of the implementers is an important pre-requisite for success. The planners of broad public intervention programmes must therefore include credible scientific groups and professional societies as true participants in the planning and implementation phase of such interventions. In addition to the scientific credibility needed for success, it is also important that wide community support exists. Therefore, co-sponsorship of programmes by well known and trusted community organizations is a great help in ensuring public acceptance. Rogers (1983) reviewed extensively such credibility issues and Hovland (1954) has contributed to the basic research in this area.

The implementer must be an effective role model

Our Stanford group is persuaded that the would-be health educator and communicator themselves often need a period of active practice of new health habits in order to become effective. We have, therefore, advocated a 'health professional heal thyself' motto. Within a community setting, nurses, physicians, health educators, and teachers should be offered special programmes in the core health curriculum before widespread public intervention occurs. This is important both for creation of role models and for acquisition of skills. One's teaching skills can be derived partly from one's own struggles and eventual life-style change.

Use multiple channels for distributing education

Adoption of a new behaviour is more likely if the individual is encouraged to do so through multiple channels, such as radio, television, print, contests, and so forth. For example, a student's homework may be used to increase the parent's knowledge on a health topic. The most successful programmes use components that reinforce each other such as using the radio to promote a contest that includes printed educational materials that in turn promote a radio component (King et al. 1987). Such co-ordinated multi-component programmes are unusual in health education, but their use is growing and they act effectively to reach the most people.

Use mini-campaigns

Our Stanford projects have generally used 3-month cycles of separate campaigns on individual topics such as nutrition, weight control, smoking, exercise, or hypertension. Only one topic is used at a time, although a general background level of attention to all topics can occur throughout the year. These 'mini-campaigns' should be carried out in accordance with any natural pattern that might exist in that community. For example, exercise programmes in North America are most advantageously launched during the summer months. If a national programme such as one devoted to heart disease or hypertension or smoking exists, then the local campaign can have accentuated public intervention occurring during that period of time.

Promote collaboration with community organizations

The original group of planners and implementors should work collaboratively with local community groups leading toward eventual adoption of various programmes by these community groups, a process needed for institutionalization and maintenance.

Promote environmental and regulatory change

Achieving change in nutrition is facilitated if lunch programmes in hospitals, schools, worksites, and other institutions are changed to provide healthier food choices. Although the organizational challenge is greater, restaurants and food markets may be enlisted to provide healthier menu choices and labelling that allows customers to select healthier foods. Consultation should be provided to worksites and other organizations on methods of achieving phased regulatory change in relevant policies (e.g. smoking restrictions), seeking to match the rate of policy change to the increasing acceptance of the health goal. Such acceptance becomes more likely as social norms are altered by the overall health communication programme.

Evaluation

The need for use of multiple levels of evaluation for comprehensive health communication programmes cannot be over-emphasized. Sustained health promotion interventions must be continuously re-evaluated and modified if success is to occur. We described formative and process evaluation above and we have previously described many of the methods used in our earlier studies (Farquhar 1987; Farquhar et al. 1977, 1985a, b; Killen et al. 1988; King et al. 1987). It is inadequate merely to determine that a component of the programme was made available. Through formative evaluation one must determine that each component is likely to reach the intended audience with the expected impact. Process evaluation determines that the programme component actually achieved the expected impact, or that it failed, and helps to indicate why. Taken together, the various results of process evaluation enable 'mid-course' corrections to be taken. Comprehensive health communication programmes should also be evaluated for impact on outcome variables, such as knowledge, behaviour, and disease rates. Low-cost interview methods may often be useful in evaluating knowledge and self-reported behaviour (Remington et al. 1988), and existing public health data sources (mortality or incidence registers, hospital data, etc.) may be useful in tracking the impact on disease. The best available data sources should be used for evaluating changes in outcome and, if necessary, special systems can be established if resources permit (Farquhar et al. 1985a, b; Fortmann et al. 1986).

Maintenance of healthful behaviours

Success in maintenance of new health habits is dependent not only on individual factors but on system factors as well. In individual factors, the nature of the health habit undergoing change has some relationship to maintenance. For example, weight changes, especially those due to periodic use of caloric restriction, are associated with rather high relapse rates. In contrast, those who have succeeded in stopping smoking for a number of months are much less likely to return, although prior exposure to relapse prevention training increases the probability of maintenance of the non-smoking condition. Therefore, the likelihood of maintenance is dependent to a degree on the quality of the initial skills training programme. As a further example, individuals who have changed their dietary habits through a phased approach that emphasizes the pleasure of new foods will often develop a new set of food preferences and then retain the new behaviours through internal motivation.

System-wide factors are also critical in the maintenance of change at the community level. An important determinant is the initial proportion of individuals who gained knowledge and changed behaviour in certain respects. As indicated previously, if the proportion of the population adopting a new beneficial health behaviour exceeds a 15–25 per cent range, than the process of diffusion through the natural networks of the system becomes more likely. One can see the process of change as analogous to the spread of epidemics where collision frequency of infected with uninfected people is a determinant of the range of change in the ascending limb of the curve. Maintenance is thus partly dependent on the degree of initial success. The largest question in maintenance is whether or not the public intervention strategies cease completely or whether they continue at a level sufficient to maintain health benefits gained during the initiation phases of the intervention campaign. The Stanford group believes that adoption of health promotion activities by existing community orgianzations, including schools, hospitals, physicians, health agencies, and citizens groups is generally necessary to supply ongoing reinforcement and reminders and to provide new knowledge and skills as they become needed (Farquhar et al. 1985a, b). This process of adoption and continuation of a new set of technologies in health promotion represents 'institutionalization' and requires an ongoing process of community organization to ensure its success. The form of continuation is likely to vary considerably, but one model or pattern is to develop a council or consortium of all of the agencies that deal with health education, health promotion, and preventive medicine delivery in order to decrease territorial conflict and ensure inter-organizational collaboration. It is the Stanford group's belief that a continued relationship of these community agencies with an external research and development organization would be very helpful. This relationship would then allow continued acquisition by the community groups of new technologies as they become available. Furthermore, this research and development organization could carry out a periodic retraining of community individuals to ensure a high quality of ongoing health promotion. This model could also be implemented through the existing public health authority in areas where it is mandated and funded to include chronic disease prevention activities.

Extension

A nation, a state, a province, or region in a country may wish to extend lessons learned from a successful demonstration project in a community to other communities within the region or to a larger political unit, including the country as a whole. Successful projects of this sort have a way of becoming adopted by others even without official sanction. Nonetheless, co-ordinated planning to ensure maximum transfer of technology to other units is in order. A co-ordinated programme to achieve this goal should include provision of training to future community leaders and to individuals who can assist communities in that region to achieve these goals. As examples, the Stanford group, the North Karelia Project staff, and the staff of the Minnesota (Blackburn et al. 1984) and Rhode Island (Lasater et al. 1984) projects have engaged in both informal and formal training programmes of that nature. Furthermore, our group at Stanford has formed a Health Promotion Resource Centre, funded by the Kaiser Family Foundation, to expand our ability to be of assistance to organizations and to communities that wish to carry out public intervention strategies (Tarlov et al. 1987).

Conclusions

An international movement towards an increase in the amount of comprehensive community-based communication to influence behaviour has occurred. Much of this has centred on the topic of cardiovascular disease prevention provided in a co-ordinated, comprehensive set of strategies applied at the community level. Analogous projects in other non-communicable diseases, substance abuse, cancer prevention, injury prevention, and adolescent pregnancy prevention are underway in various parts of the world (Pentz et al. 1989; Tarlov et al. 1987; WHO 1986). In this chapter, we have reviewed the theoretical and scientific basis for such programmes, as well as some of the challenges facing new projects. Finally, we would argue that any problems with implementation of comprehensive life-style change programmes in communities are outweighed by the problem of not carrying them out, given the need to counter the immense social pressures that promote ill-health, including the burgeoning growth of commercially generated pressures carried out through the mass media (McGuire 1985). Fortunately a little inoculation goes a long way not only in infectious disease prevention, but, as we have seen, also in chronic disease prevention and health maintenance.

References

Altman, D.G., Foster, V., Rosenick-Douss, L., et al. (1989). Reducing illegal sales of cigarettes to minors. *Journal of the American Medical Association* **261**, 80.

Artzenius, A.C., Kromhout, D., Barth, J.D., et al. (1985). Diet lipoproteins and the progression of coronary atherosclerosis. The Leiden intervention trial. *New England Journal of Medicine* **312**, 805.

Ashley, M.J. and Rankin, J.G. (1988). A public health approach to the prevention of alcohol-related health problems. *Annual Review of Public Health* **9**, 233.

Atkin, C. (1979). Research evidence on mass mediated health communication campaigns. In *Communication yearbook III* (ed. D. Nimmo), p. 655. Transaction Books, New Brunswick, New Jersey.

Bandura, A. (1977). *Social learning theory*. Prentice-Hall, Englewood Cliffs, New Jersey.

Bandura, A. (1986). *Social foundations of thought and action: A social cognitive theory*. Prentice-Hall, Englewood Cliffs, New Jersey.

Bauer, R.L., Heller, R.F., and Challah, S. (1985). United Kingdom Heart Disease Prevention Project: 12-year follow-up of risk factors. *American Journal of Epidemiology* **121**, 563.

Berkman, L.F. (1984). Assessing the physical effects of social networks and social support. *Annual Review of Public Health* **5**, 413.

Best, J.A., Thomson, J., Santi, S., *et al.* (1988). Preventing cigarette smoking among school children. *Annual Review of Public Health* **9**, 161.

Blackburn, H., Luepker, R.V., Kline, F.G., *et al.* (1984). The Minnesota Heart Health Program: A research and demonstration project in cardiovascular disease prevention. In *Behavioural health: A handbook of health enhancement and disease prevention* (ed. J.D. Matarazzo, N.E. Miller, S.M. Weiss, *et al.*), p. 1171. John Wiley, Silver Springs, Maryland.

Blair, S.N., Piserchia, P.V., Wilber, C.S., *et al.* (1986). A public health model for work-site health promotion. *Journal of the American Medical Association* **255**, 921.

Brownell, K.D. (1982). Obesity: understanding and treating a serious prevalent, and refractory disorder. *Journal of Consulting and Clinical Psychology* **50**, 820.

Bruno, R., Arnold, C., Jacobson, L., *et al.* (1983). Randomized controlled trial of a nonpharmacological cholesterol reduction program at the worksite. *Preventive Medicine* **12**, 523.

Carmody, T.P., Fey, S.L., Pierce, D.K., *et al.* (1982). Behavioral treatment of hyperlipidemia: Techniques, results, and future directions. *Journal of Behavioural Medicine* **5**, 91.

Cartwright, D. (1949). Some principles of mass persuasion. *Human Relations* **2**, 253.

Crouch, M., Sallis, J.F., Farquhar, J.W., *et al.* (1986). Personal and mediated health counselling for sustained dietary reduction of hypercholesterolemia. *Preventive Medicine* **15**, 282.

Egger, G., Fitzgerald, W., Frape, G., *et al.* (1983). Results of a large scale media antismoking campaign: North Coast 'Quit for life' Programme. *British Medical Journal* **296**, 1125.

Enholm, C., Huttunen, J.K., Pietinen, P., *et al.* (1982). Effect of diet on serum lipoproteins in a population with a high risk of coronary heart disease. *New England Journal of Medicine* **307**, 850.

Farquhar, J.W. (1978). The community-based model of lifestyle intervention trials. *American Journal of Epidemiology* **108**, 103.

Farquhar, J.W. (1987). *The American way of life need not be hazardous to your health* (revised edn). Addison-Wesley, Reading, Massachusetts.

Farquhar, J.W., Maccoby, N., and Wood, P.D. (1985a). Education and communication studies. In *Oxford textbook of public health* (ed. W. Hollond, R. Detels, and G. Knox). Vol. 3, p. 207. Oxford University Press, Oxford.

Farquhar. J.W., Fortmann, S.P., Flora, J.A., *et al.* (1986). Interim results of the Stanford Five City Project. *CVD Epidemiology Newsletter, American Heart Association* (Abstract No. 106), **39**, 32.

Farquhar, J.W., Fortmann, S.P., Flora, J.A., *et al.* (1988). The Stanford Five City Project: Results after 5⅓ years of education. *CVD Epidemiology Newsletter, American Heart Association* (Abstract No. 77), 43, p. 27.

Farquhar, J.W., Fortmann, S.P., Maccoby, N., *et al.* (1985b). The Stanford Five-City Project: Design and methods. *American Journal of Epidemiology* **122**, 323.

Farquhar, J.W., Maccoby, N., Wood, P.D., *et al.* (1977). Community education for cardiovascular health. *Lancet* i, 1192.

Felix, M.R.J., Stunkard, A.J., Cohen, R.Y., *et al.* (1985). Health promotion at the worksite. 1. A process for establishing programs. *Preventive Medicine* **14**, 99.

Fortmann, S.P., Haskell, W.L., Williams, P. T., *et al.* (1986). Community surveillance of cardiovascular disease in the Stanford Five City Project. *American Journal of Epidemiology* **123**, 656.

Fortmann, S.P., Williams, P.T., Hulley, S.B., *et al.* (1981). Effect of health education on dietary behaviour: The Stanford Three Community Study. *American Journal of Clinical Nutrition* **34**, 2030.

Freimuth, V.S., Hammond, S.L., and Stein, J.A. (1988). Health advertising: Prevention for profit. *American Journal of Public Health* **78**, 557.

Glanz, K. (1985). Nutrition education for risk factor reduction and patient education: A review. *Preventive Medicine* **14**, 721.

Glanz, K. and Mullis, R.M. (1988). Environmental interventions to promote healthy eating: A review of models, programs, and evidence. *Health Education Quarterly* **15**, 395.

Glasgow, R.E. (1986). Smoking. In *Self-management of chronic disease: Handbook of clinical interventions and research* (ed. K. Holroyd and T. Creer) p. 21. Academic Press, New York.

Glasunov, I.S., Grabauskas, V., Holland, W.W., and Epstein, F.H. (1983). An integrated programme for the prevention and control of noncommunicable disease: A Kaunas report. *Journal of Chronic Diseases* **36**, 419.

Graubard, S.R. (1982). Print culture and video culture. *Daedalus* Preface to vol. 111, p. v.

Gutzwiller, F., Nater, B., and Martin, J. (1985). Community-based primary prevention of cardiovascular disease in Switzerland: Methods and results of the National Research Program (NRP 1A). *Preventive Medicine* **14**, 482.

Haskell, W.L. (1984). Exercise-induced changes in plasma lipids and lipoproteins. *Preventive Medicine* **13**, 23.

Hewitt, L.W. and Blane, H.T. (1984). Prevention through mass media communication. In *Prevention of alcohol abuse* (ed. P.M. Miller and T.D. Nirenberg), p. 281. Plenum, New York.

Hiltz, S.R., and Turoff, M. (1978). *The network nation: Human communication via computer*. Addison Wesley, Reading, Massachusetts.

Hjermann, I., Holme, I., Velve-Brye, K., *et al.* (1981). Effect of diet and smoking intervention on the incidence of coronary disease: Report from the Oslo Study Group of a randomized trial in healthy men. *Lancet* ii, 1303.

Hovell, M.F. (1982). The experimental evidence for weight loss treatment of essential hypertension: A critical review. *American Journal of Public Health* **72**, 359.

Hovland, C.I. (1954). Effects of the mass media of communication. In *Handbook of social psychology* (1st edn) (ed. G. Lindzey), p. 1062. Addison Wesley, Reading, Massachuetts.

Howard, M. (1985). Postponing sexual involvement among adolescents. *Journal of Adolescent Health Care* **6**, 271.

Hypertension Detection and Follow-up Program Cooperative

Group (1979). Five-year findings of the hypertension detection and follow-up program. 1. Reduction in mortality of persons with high blood pressure, including mild hypertension. *Journal of the American Medical Association* **242**, 2562.

Killen, J.D. (1985). Prevention of adolescent tobacco smoking: The social pressure resistance training approach. *Journal of Child Psychology and Psychiatry* **26**, 7.

Killen, J.D., Telch, M.J., Robinson, T.N., *et al.* (1988). Cardiovascular disease risk reduction for tenth graders: Multiple factor school-based approach. *Journal of the American Medical Association* **260**, 1728.

King, A.C., Flora, J.A., Fortmann, S.P., *et al.* (1987). Smokers' challenge: Immediate and long-term findings of a community smoking cessation contest. *American Journal of Public Health* **77**, 1340.

King, A.C., Saylor, K.E., Foster, S., *et al.* (1988). Promoting dietary change in adolescents: A school-based approach for modifying and maintaining healthful behaviour. *American Journal of Preventive Medicine* **4**, 68.

Klesges, R.C., Vasey, M.M., and Glasgow, R.E. (1986). A worksite smoking modification competition: Potential for public health impact. *American Journal of Public Health* **75**, 198.

Kotler, P. (1975). *Marketing for nonprofit organizations*. Prentice Hall, Englewood Cliffs, New Jersey.

Lasater, T., Abrams, D., Artz, L., *et al.* (1984). Lay volunteer delivery of a community-based cardiovascular risk factor change program: The Pawtucket Experiment. In *Behavioural health: A handbook of health enhancement and disease prevention* (ed. J.D. Matarazzo, N.E. Miller, S.M. Weiss, *et al.*) p. 1166. John Wiley, Silver Springs, Maryland.

Lefebvre, C.R. and Flora, J.A. (1988). Social marketing and public health interventions. *Health Education Quarterly* **15**, 299.

Levy, A.S. and Stokes, R.C. (1987). Effects of a health promotion advertising campaign on sales of ready-to-eat cereals. *Public Health Reports* **102**, 398.

Lewis, B., Mann, J.I., and Mancini, M. (1986). Reducing the risks of coronary heart disease in individuals and in the population. *Lancet* i, 956.

Lipid Research Clinics Program (1984*a*). The Lipid Research Clinics' coronary primary prevention trial results. 1. Reduction of incidence of coronary heart disease. *Journal of the American Medical Association* **251**, 351.

Lipid Research Clinics Program (1984*b*). The Lipid Research Clinics' coronary primary prevention trial results. II. The relationship of reduction in incidence of coronary heart disease to cholesterol lowering. *Journal of the American Medical Association* **251**, 365.

Marshall, J.F. (1971). Topics and networks in intravillage communication. In *Culture and population: A collection of current studies* (ed. S. Polgar), p. 160. Carolina Population Center, Monograph 9, Chapel Hill, North Carolina.

Martin, J.W. and Dubbert, P.M. (1982). Exercise applications and promotion in behaviorial medicine: Current status and future directions. *Journal of Consulting and Clinical Psychology* **50**, 1004.

McGuire, W.J. (1964). *Inducing resistance to persuasion*. In *Advances in experimental social psychology*, (ed. L. Berkowitz), Vol. I, p. 191. Academic Press, New York.

McGuire, W.J. (1985). The nature of attitudes and attitude change. In *The handbook of social psychology* (3rd edn) (ed. G. Lindzey and E. Aronson), Vol. 2., p. 233. Random House, New York.

Meyer, A.J. and Henderson, J.B. (1974). Multiple risk factor reduction in the prevention of cardiovascular disease. *Preventive Medicine* **3**, 225.

Meyer, A.L., Maccoby, N., and Farquhar, J.W. (1977). The role of opinion leadership and the diffusion of innovations in a cardiovascular health education campaign. *Communication Yearbook* **1**, 579.

Meyer, A.J., Nash, J.D., McAlister, A.L., *et al.* (1982). Skills training in a cardiovascular health education campaign. *Journal of Consulting and Clinical Psychology* **48**, 129.

Miller, N., Wagner, E., and Rogers, P. (1988). Worksite-based multifactorial risk intervention trial. *Journal of the American College of Cardiology Supplement A*, **11**(2), 207A.

MRFIT Research Group (1982). Multiple risk factor intervention trial: Risk factor changes and mortality results. *Journal of the American Medical Association* **28**, 1465.

Mullis, R.M. and Pirie, P. (1988). Lean meats make the grade: A collaborative nutrition education program. *Journal of the American Dietetic Association* **88**, 191.

Mullis, R.M., Hunt, M.K., Foster, M., *et al.* (1987). The Shop Smart For Your Heart grocery program. *Journal of Nutrition Education* **19**, 225.

Nader, R.R., Sallis, J.F., Rupp, J., *et al.* (1986). San Diego Family Health Project: Reaching families through the schools. *Journal of School Health* **56**, 227.

Pentz, M.A., Dwyer, J.H., MacKinnon, D.P., *et al.* (1989). A multi-community trial for primary prevention of adolescent drug abuse: Effects on drug use prevalence, *Journal of the American Medical Association* **261**, 3259.

Pierce, J.P., Dwyer, T., Frape, G., *et al.* (1986). Evaluation of the Sydney 'Quit for Life' anti-smoking campaign. *Medical Journal of Australia* **144**, 341.

Puska, P. (1985). Effectiveness of nutrition education strategies. *Proceedings of the Fourth European Nutrition Conference*, Amsterdam, 1983. Voorlichtingsbureau Voor de Voeding, The Hague, Netherlands.

Puska, P., Koskela, K., McAlister, A., *et al.* (1979). A comprehensive television smoking cessation program in Finland: Background, principles, implementation and evaluation. *International Journal of Health Education* **18**, 7.

Puska, P., Koskela, K., McAlister, A., *et al.* (1986). Use of lay leaders to promote diffusion of health innovations in a community program: Lessons learned from the North Karelia project. *Bulletin of the World Health Organization* **64**(3), 437.

Puska, P., Tuomilehto, J., Salonen, J., *et al.* (1981). *The North Karelia Project: Evaluation of a comprehensive community programme for control of cardiovascular diseases in 1972–77 in North Karelia, Finland*. Copenhagen, Public Health in Europe, WHO/EURO Monograph Series.

Remington, P.L, Smith, M.Y., Williamson, D.F., *et al.* (1988). Design, characteristics, and usefulness of state-based behavioral risk factor surveillance: 1981–87. *Public Health Reports* **103**, 366.

Roberts, D.F. and Maccoby, N. (1985). Effects of mass communication. In *Handbook of social psychology*. (3rd edn) (ed. G. Lindzey and E. Aronson) p. 539. Random House, New York.

Rogers, E.M. (1983). *Diffusion of innovations*, (3rd edn) Free Press, New York.

Rootman, I. (1985). Using health promotion to reduce alcohol problems. In *Alcohol policies*. (ed. M. Grant), p. 571. WHO, Copenhagen.

Rossouw, J.E., Jooste, P.L., Kotze, J.P., *et al.* (1981). The control of hypertension in two communities: An interim evaluation. *South African Medical Journal* **60**, 208.

Sallis, J.F., Hill, R.D., Fortmann, S.P., *et al.* (1986). Health behavior change at the worksite: Cardiovascular risk reduction. *Progress in Behaviour Modification* **20**, 161.

Stunkard, A.J., Felix, M.R.J., and Cohen, R.Y. (1985). Mobilizing a community to promote health: The Pennsylvania County Health Improvement Program (CHIP). In *Prevention in health psychology* (ed. J.C Rosen and L.J. Solomon), p. 143. Hanover University Press of New England.

Tarlov, A.R., Kehrer, B.H., Hall, D.P., *et al.* (1987). Foundation work: The health promotion program of the Henry J. Kaiser Family Foundation. *American Journal of Health Promotion* **2(2)**, 74.

Taylor, P. (1984). *The smoke ring—tobacco, money, and multinational politics.* Pantheon Books, New York, and Bodley Head UK.

Telch, M.J., Killen, J.D., McAlister, A.L., *et al.* (1982). Long-term follow-up of a pilot project on smoking prevention with adolescents. *Journal of Behavioural Medicine* **5**, 1.

Telch, M.J., Miller, L.M., Killen, J.D., *et al.* (1990). Social influences approach to smoking prevention: The effects of videotape delivery with and without same-age peer leader participation. *Addictive Behaviours* **15**, 21.

Tuomilehto, J., Geboers, J., Salonen, J.T., *et al.* (1986). Decline in cardiovascular mortality in North Karelia and other parts of Finland. *British Medical Journal* **293**, 1068.

Webb, E.J., Campbell, O.T., Schwartz, R.D., Sechrest, L., and Grove, J.B. (1981). *Nonreactive measures in the social sciences* (2nd edn). Houghten-Mifflin, Boston, Massachusetts.

Wilber, C.S. (1983). The Johnson and Johnson program. *Preventive Medicine* **12**, 672.

Wilhelmsen, L., Berglund, G., Elmfeldt, D., *et al.* (1986). The multifactor primary prevention trial in Goteborg, Sweden. *European Heart Journal* **7**, 270.

Williams, P.T., Fortmann, S.P., Farquhar, J.W., *et al.* (1981). A comparison of statistical methods for evaluating risk factor changes in community-based studies: An example from the Stanford Three Community Study. *Journal of Chronic Diseases* **34**, 565.

Wood, P.D., Terry, R.B., and Haskell, W.L., (1985). Metabolism of substrates: Diet, lipoprotein metabolism, and exercise. *Federation Proceedings* **44**, 358.

Wood, P.D., Haskell, W.L., Blair, S.N., *et al.* (1983). Increased exercise level and plasma lipoprotein concentrations: A one-year randomized, controlled study in sedentary, middle-aged men. *Metabolism* **32**, 31.

World Health Organization, European Collaborative Group (1983). Multifactorial trial in the prevention of coronary heart disease: 3. Incidence and mortality results. *European Heart Journal* **4**, 144.

World Health Organization (1986). Report of a WHO Expert Committee: *Community prevention and control of cardiovascular diseases.* Technical Report Series No. 732. WHO, Geneva.

21

Operational and system studies

SHAN CRETIN

Introduction

Operational research was first recognized as a distinct discipline during the Second World War. Scientists, long engaged in the design of weapons, began to apply scientific methods of analysis to the deployment of weapons and to overall military operations, with the goal of providing military commanders with a quantitative basis for decisions (Morse and Kimball 1951). A decision problem (e.g. how to search for U-boats in the North Atlantic or what size to make a convoy) was recast into a mathematical model. The key variables in the problem, including the appropriate measure of the effectiveness of the operation, were quantified, and the relationships among the variables were described using mathematical expressions. These formulae comprised a mathematical model of the real decision problem and could be manipulated to predict the impact of a given decision on the measure effectiveness. The success of operational research in the Second World War demonstrated that many apparently complex problems could be usefully analysed using simple models to approximate the critical variables and relationships.

After the Second World War operational research was extended to non-military problems. Hospitals and health care systems were quickly identified as fruitful areas for analysis, with the first published reports of health applications appearing in the early 1950s. Bailey (1951, 1952) used the mathematical theory of queues to explore various out-patient appointment systems and to analyse the use of single versus multiple occupancy rooms in hospitals. Other studies applied operational research techniques to models of epidemics (Abbey 1952; Taylor 1958), emergency admissions (Newell 1954), and inventory control in general hospitals (Flagle 1960) and blood banks (Elston and Pickrel 1963; Rockwell *et al.* 1962).

Most of these early applications of operational research used mathematical models based on probability or statistical theory. Today, operational research studies use a variety of mathematical tools, some adapted from traditional mathematical fields and some newly developed. The techniques most commonly employed are as follows: (i) probability models, which use queuing theory, Monte Carlo simulations, and statistics; (ii) optimization models, which use mathematical programming, dynamic programming, and graph and network theory; (iii) differential equation methods and dynamic systems simulation; (iv) decision analyses, which use models based on Bayesian and statistical decision theory, game theory, and utility theory.

The remainder of this chapter will be divided into three parts. The first section will review the most frequently used operational research techniques, with references to sources of more detailed information about each. The second part will review the types of public health problems that have been addressed using operational research methods, with references to some typical studies. The third section will discuss the potential and the problems surrounding applications of operational research methods in public health.

Review of operational research techniques

There are several excellent textbooks surveying the methods of operational research. Duckworth *et al.* (1977) and Eppen *et al.* (1987) provide overviews of methods and applications aimed at managers. Hillier and Lieberman (1980) and Wagner (1975) are good references for basic methods. Larson and Odoni (1981) focus on techniques useful in the analysis of urban problems. Warner *et al.* (1984) stress methods and applications in health administration. In addition, there are scores of books and hundreds of articles devoted to specific techniques, models, and problems. In the brief review of techniques which follows, reference is made to a few of these more advanced texts.

Probability models

Uncertainty is a feature of life. Many important administrative and policy decisions in health care are complicated by chance elements beyond the control of the decision-maker. Recent work by psychologists (Tversky and Kahneman 1974) has demonstrated biases and inconsistencies in our intuition about probabilities. This failure of intuition may explain why operational research models employing probability theory to analyse chance events have resulted in some of the most substantial improvements in operating systems.

Stochastic processes

Stochastic, or probabilistic, processes evolve over time in a way that is not completely predictable. Some simple probabilistic processes have proved to be useful models of real events. These models have been carefully described and well analysed over the years, so that once the model is known to apply, many results can be inferred. The Bernoulli process (which gives rise to the familiar binomial distribution), the Poisson process, and the Markov process are three models that have been used extensively in health applications.

In the simplest probability models, successive events are assumed to be independent of each other. The classic example is coin flipping, modelled by a Bernoulli process; the outcome (heads or tails) on one toss does not alter the probability of heads or tails on the next toss.

In a Bernoulli process, the outcome is a result of some discrete event. When tossing coins, for example, 'heads' or 'tails' cannot occur at any instant, but only as a result of the action of 'flipping the coin'. For this reason, the Bernoulli process cannot be used to model events which can occur at any moment in time, such as a request for emergency medical service. The Poisson process extends the notion of independent events to continuous time by assuming that the number of events in one time period is independent of the number of events in any other non-overlapping time period. Poisson arrivals (sometimes called 'random arrivals') have been used to model the arrival of women in spontaneous labour at an obstetrics service as well as requests for ambulance or other emergency medical care.

The Poisson and Bernoulli processes are called memoryless. While such memoryless processes may be adequate for many situations, there are times when probabilities of future events will depend on past outcomes. The Markov mode is a simple probability model with 'memory'. A Markovian world is assumed to consist of a set of discrete non-overlapping states. For example, a Markovian model of a person's state of health might consist four states: disease free, presymptomatic, symptomatic, dead. Starting in one of the states, the system undergoes changes of state (or state transitions) according to a set of probability rules. The Markov process is allowed to have a limited memory, i.e. the probability of transition to a particular state allowed to depend on which state the system is in just before the transition. Some Markovian systems undergo transitions only at discrete times (similar to the Bernoulli process), while others may undergo transitions at any time.

Drake (1967) and Feller (1968) provide introductions to applied theory. Other useful references on stochastic processes are Bailey (1964), Cox and Miller (1965), Howard (1960, 1971), Parzen (1960), and Ross (1970, 1972).

Queuing systems

The application of stochastic models to the analysis of queues for service has been quite fruitful, and queuing theory has become a distinct and expanding field of inquiry. A.K. Erlang did much of the early work in this field in conjunction with his studies of the Copenhagen telephone system (Brockmeyer *et al.* 1948).

The simplest queuing system has one server who handles requests for service sequentially. To simplify the analysis, both arrivals and service are assumed to be 'memoryless'. The arrival of customers is modelled by a Poisson process. Service times for each customer are assumed to be drawn independently from a negative exponential distribution, a distribution with the property that no matter how long service has been in progress, the expected remaining time in service is the same. Given the average arrival rate of customers and the mean service time, we can then calculate the distribution of the length of the queue and the distribution of customer waiting times, as well as characteristics of the server's busy and quiet periods.

More complex queuing systems have been analysed which allow multiple servers, more general arrival and service time distributions, batch arrivals of customers, balking or reneging by customers who wait too long in the queue, tandem queues or hierarchical queuing networks. Less complete mathematical results are available for these more complex systems, but many approximate results have been obtained. Many complex queuing networks are now analysed using computer simulation methods and this will be discussed in the section on Monte Carlo simulations.

Cox and Smith (1961) provide a concise introduction to queuing theory, while Cooper (1972) and Kleinrock (1975) give fuller treatment to the subject. Newell (1971) emphasizes approximate solution methods.

A special class of queuing models of particular relevance to the analyses of ambulance systems is the spatially distributed queue. In this case, chance is involved not only in the timing of the requests, but also in the location from which the requests originate. The geography of the region is important in describing the location of calls for service and in determining the distribution of response times. Servers are no longer indistinguishable, although they provide the same service, since each potential server may be at a different distance from an incoming call. Models of spatially distributed queues have been developed to analyse the effects of call volume, the number and location of servers, and dispatching strategies on the efficiency and equity of service. Larson and Odoni (1981) give a description of models and methods. The Rand Fire Project report (Walker *et al.* 1979) describes the use of these and similar models in analysing fire services.

Monte Carlo simulations

The behaviour of systems involving chance elements can often be analysed by simulating the operation of the system. Simulation is also used to analyse complex deterministic systems, as in dynamic systems simulation which is discussed below. Coin flipping could be simulated using digits from a table of random numbers by labelling odd digits as 'heads' and even digits as 'tails'. Alternatively, a computer could be programmed to simulate coin flipping in a similar way using internally generated random numbers.

Simulation is usually an analysis of last resort, which is used when a system is too complex for the simpler mathematical models with closed-form solutions. Complex queuing networks are frequently simulated. In theory a simulation can be carried out 'by hand', but in practice simulations are usually computer operations using one of several simulation languages (SIMSCRIPT, GPSS, GASP). Simulations in these languages are 'event paced'. First, the logical structure of the system is described and the beginning state of the system is specified. The simulation then focuses on events that change the state of the system (for example, the arrival of a new customer or the completion of service on a current customer). Various statistics describing the operating characteristics of the system are compiled during the simulation and reported when a predetermined span of simulated time has elapsed.

In effect, running a simulation is like collecting experimental data on an operating system. As in any experiment, the interpretation of the result depends on the design of the experiment and the sample size. Interpretation of simulation results may also depend on how the data used in constructing the simulation were obtained. With a little care, a simulation can give a reliable picture of how the model behaves. As with all operational research, the crucial question is also the most difficult to answer: how well does the model represent the real system?

Emshoff and Sisson (1970) or Banks and Carson (1984) provide a general introduction to simulation methods. Manuals for specific simulation languages are also useful guides—for example, Kiviat et al. (1969) on SIMSCRIPT II or Pritsker (1986) on SLAM II.

Competing risk and other life-table models

Historically, life-tables have been used by demographers and biostatisticians to calculate life expectancies and survival curves. With some modifications the basic life-table models can incorporate the notion of competing causes of death (Chiang 1968; Manton et al. 1976). The age-specific mortality from and particular cause of death can further be described as a function of various risk factors (such as blood pressure or smoking habits). Several operational research studies have incorporated competing risk/life-table methods in larger models designed to estimate the costs and benefits of risk factor intervention programmes (Berwick et al. 1980; Eddy 1980; Weinstein and Stason 1976).

Inventory theory

Inventory theory, like queuing theory, is organized around an area of application rather than a method. Inventory models include both deterministic models (which assume that the demand for a good is completely predictable) and probability models (which allow demand and various time lags to be stochastic). The basic components of most inventory models include ordering costs, storage costs for items on hand, shortage costs, outdating costs (for perishable goods, such as whole blood), and the cost of capital. The models are analysed to determine the least-cost inventory policy, including the size and timing of orders.

Inventory models have been developed to analyse one-time ordering decisions, multiple-cycle ordering decisions, and single and multiple product systems. When an inventory problem is too complex to fit any of the existing models, the computer simulation can provide a means of analysis. Buffa and Miller (1979) review inventory theory and practice, and Veinott (1966) reviews theory.

Optimization models

Many of the probability models just discussed are descriptive rather than prescriptive. A queuing model, for example, would be useful in describing the trade-off between customer waiting time and server idle time, but would not indicate the 'optimal' number of servers for a given workload. The value of the model is in predicting the effects of uncertain events so that a few known alternatives can be evaluated. In some decisions, uncertainty about the outcome of a strategy is not the major difficulty. The outcome for any given alternative is easy to calculate, but the decision-maker is cursed with too many choices. In this kind of problem, mathematical modelling can provide an efficient way of finding the optimal alternative from a large (perhaps infinite) number of possibilities. Optimality is defined in terms of a mathematical expression (an 'objective function') which is to be made as large (or as small) as possible.

Mathematical programming

Mathematical programming includes a variety of techniques used to determine the optimal allocation of resources. The form of these models is always the same: maximize or minimize an objective function subject to a series of constraints. The specific techniques (linear programming, quadratic programming, integer programming, or non-linear programming) rest on different assumptions about the mathematical form of the objective function and constraints, and therefore involve different solution methods. Linear programming models are usually the simplest to solve and are often used to approximate more complex integer or non-linear problems.

A classic linear programming problem is the 'product mix' decision. Suppose, for example, that a potter has 80 pounds of clay and 10 hours available to prepare pottery for a craft show. She can make vases (each uses a pound of clay and takes five minutes) or bowls (each takes a pound and a half of clay and 15 minutes). If bowls sell for 20 dollars and vases for 10 dollars, how many of each should she make to maximize her revenue.

In this example, the decision variables are the number of vases to be made V and the number of bowls B. The objective is to maximize revenue, which is a linear function of the two decision variables $(20B+10V)$. There are two constraints, time and material, which can be expressed as linear inequalities:

$$\text{Time: } \tfrac{1}{12}V + \tfrac{1}{4}B < 10$$
$$\text{Clay: } V + \tfrac{3}{2}B < 80.$$

If this problem is solved using linear programming techniques, the optimal values for the decision variables will take on fractional values. When it is important to restrict the solution to integral values, the same formulation of the objective function and constraints could be solved using integer programming methods.

In the simple example just given, the decision problem is not very difficult. Once the problem is formulated mathematically, it can be solved by hand using algebraic or graphical techniques. However, many applications of linear programming involve hundreds of decision variables and constraints. The solution of problems of this magnitude only became feasible with the development of the simplex algorithm (Dantzig 1963), which is an extremely efficient solution method when used on high-speed computers. More recently, a new ellipsoid method of solution was introduced (Bland *et al.* 1981).

Although linear programming provides a powerful tool for analysing questions of resource allocation, many practical problems violate the assumptions of the model. A common problem is the need to restrict the solution to integer values. This is demonstrated in the example above, and is even more pronounced when the variable represents the number of surgical suites. In other cases, the constraints are not linear. For example, the potter may find that each additional bowl takes a little less time than the preceding bowl, because of a learning effect. While solution methods of non-linear and integer programming problems do exist, except for some rather special cases, these algorithms are far less efficient than linear programming techniques.

Luenberger (1984) provides a valuable introduction to both linear and non-linear programming. Garfinkel and Nemhauser (1972) have produced a good text on integer programming.

Graphs and networks

A graph is a set of nodes or points with lines or branches connecting certain pairs of points. A network is a graph with some type of flow along its branches. A highway system is an example of a network, with roads as branches, intersections as nodes, and traffic flowing along the roads. Network theory provides an excellent means of analysing many problems related to transportation, such as finding the shortest route from one point to another or finding the shortest tour through a network (a path that visits every node). Network theory is also the basis for a technique widely used in co-ordinating and scheduling interrelated activities. This method has been popularized as PERT (program evaluation and review technique) or CPM (critical path method).

PERT systems begin by representing a project as a network. Each node represents an event—for example, the beginning of the project, the end of the project, or the completion of some intermediate activity. Each activity or task is represented by a directed branch (a line with an arrow) starting from the node marking completion of all necessary predecessor tasks and ending at a task completion node. If the duration of each subtask is known, network theory can be used to calculate the earliest possible completion time for the whole project, the earliest and latest starting time for each subtask, and the critical path. Activities on the critical path have no slack time, meaning that if the starting time on one of these activities is delayed, the completion of the entire project is delayed.

Although the critical path is found as a result of an optimization technique, the major use of PERT systems is as an aid to scheduling, and not as an optimization tool. Sometimes a range of times for each subtask is the basis for 'best-case' and 'worst-case' analyses. Alternatively, the cost of speeding up certain subtasks can be balanced against the effect on project completion time.

Network analyses have also been applied to assignment problems (assigning personnel to tasks) and to facility location problems (finding the best location for fire stations in a region). Assignment problems and certain transportation problems can be solved using adaptations of linear programming algorithms. Ford and Fulkerson (1962) cover network theory in general. CPM and PERT are reviewed by Moder and Phillips (1970). Larson and Odoni (1981) provide a good review of routing problems and facility location problems in the context of urban services.

Dynamic programming

Dynamic programming is a technique for maximizing the overall effectiveness of a series of interrelated decisions. In a dynamic programming problem the decisions take place in stages. The choices available and the results of a decision at each stage depend on the current state of the system. After the decision has been made, the state of the system changes (either deterministically or according to a set of known probabilities). The complexity of the problem is not the result of uncertainty but of the number of possible decision sequences which must be evaluated. If there are only four stages with four possible decisions at each stage, there are 256 distinct sequences.

Finding the optimal solution to a dynamic programming problem is not a simple matter of applying a standard algorithm as was the case for linear programming. Although there is a general approach, it must be tailored to each individual problem. The solution procedure starts by finding the best decisions for each possible state at the last decision stage and works backward. The best decision set at the nth stage is found by building on the best decisions found for stage $n + 1$.

An extension of dynamic programming to allow an infinite number of recurring decision stages is found in Markovian decision models. Some decisions with continuous outcomes can also be modelled using continuous state dynamic programming. Howard (1960, 1971) discusses both dynamic programming and Markovian decision theory. Denardo (1982) reviews theory and applications of dynamic programming.

Differential equations and difference equations

Differential equations and difference equations have been used in modelling electrical, mechanical, and fluid systems for many years. Differential equations are useful in describing systems in which key variables change continuously over time, and in which the rate of change of a variable may depend on the level of that variable and on the level and rates of change of other variables. Difference equations are simply approximations to differential equations in which time is only allowed to change in discrete units.

A simple example of a system well modelled by differential equations is a world populated by caterpillars and birds. If some external force suddenly depletes the bird population, the caterpillars will thrive. This, in turn, will allow the small remaining bird population to grow. Such a system may oscillate for some time before reaching a stable equilibrium. The size of the oscillations and the length of time taken for the system to stabilize will depend on the size of the external force and on the exact relationship between bird and caterpillar populations.

In the late 1950s, Forrester (1961) began using difference equations to model the responses of industrial firms to management policies. Since that time the applications have been extended to many other settings, including health care. The models of such systems are too complex to be solved mathematically, and analysis is usually performed using simulation. The computer language DYNAMO was developed especially for this purpose.

Decision analysis

Decision analysis is a general approach to the problem of decision-making when there is uncertainty. Embedded in this widely accepted approach to structuring the problem are several thorny and highly controversial topics. How does one estimate the degree of uncertainty in the absence of data? What criteria should be applied in finding the 'best decision'? How does one value outcome, especially when there are multiple dimensions—for example, mortality, money, impaired function?

Decision analysts structure a problem by first separating decisions, chance events, and outcomes. Decisions and chance events are then displayed using a sequential 'tree'. Branches representing alternatives and chance occurrences radiate from appropriately sequenced 'decision' and 'chance' nodes respectively. Each end-point of the tree is labelled with the appropriate outcome. Each chance branch is labelled with its probability of occurring. Decision branches are initially unlabelled. At the end of the analysis the 'best' alternative is identified at each decision node.

When the outcome is a single dimension (profit in a business decision or winnings in a gamble), the best decision is often defined as that which maximizes the expected value of the outcome. A common extension of decision analysis uses the decision-maker's 'utility' for the outcome, rather than the outcome itself. In this case the best decision is the one which maximizes expected utility.

The concept of utility makes it easier to apply decision analysis to problems with inherently qualitative outcomes and to incorporate multi-dimensional outcomes. However, the problem of eliciting a decision-maker's utilities for complex outcomes is itself quite difficult. The ability of utility theory to model the preferences people express in real decision situations has been challenged in recent years (Tversky and Kahneman 1981).

Another controversial area in the practice of decision analysis is the problem of estimating probabilities in the absence of data. It is common practice to include subjective or intuitive estimates of probabilities when no other means of estimation are possible. However, the method of eliciting these subjective probabilities may bias the estimates (Tversky and Kahneman 1974) and hence the analysis. For this reason, many applications of decision analysis include extensive sensitivity analysis to test the stability of the optimal policy. An excellent general introduction to decision analysis is given by Raiffa (1968). McNeil and Pauker (1984) provide a good introduction to public health applications.

Game theory

Game theory involves the analysis of decisions against opponents or competitors. Decisions by one party will interact with decisions by the other party to produce the outcomes. In two-person zero-sum games, the winnings of one party must equal the loss of the opponent. Other games include co-operative games, non-zero-sum games, and games involving more than two players. In game theory (as in decision analysis) the analysis specifies the decision strategy which maximizes one player's gain in the face of the opponents' best strategies. One promising application of game theory is in predicting the behaviour of an industry to changes in regulations or reimbursement methods. Luce and Raiffa (1957) offer an extensive treatment of game theory and decision-making.

In the past decade, the rapid development of personal computers has revolutionized the field of operational research. Many of the techniques reviewed above have been incorporated into user-friendly software systems that can be run on desktop or other small computers, allowing decision-makers and managers without extensive training in operational research to use and interact directly with models and data. Such 'decision support systems' have sometimes been described as gimmicks (Naylor 1982), but there is no doubt that the computer has been transformed from the exclusive property of the highly trained technician to the office machine. This transformation has aided in the diffusion of operational research applications in health care planning and management (Boldy 1987).

Review of public health applications

Over the last four decades there have been many reviews of health applications of operational research. Some reviewers focused on a particular technique or problem, such as statistics (Bailey 1956), queuing theory (Baily 1954), mathematical programming (Boldy 1976), decision analysis (Krischer 1980), or facility location (ReVelle *et al.* 1977). Others have conducted more general surveys leavened with varying degrees of philosophical comment (Boldy 1981; Boldy and O'Kane 1982; Denison *et al.* 1969; Flagle 1962; Fries 1976, 1979, 1981; Horvath 1965, 1967; McLaughlin 1970; Shuman *et al.* 1975). Several reviews have looked primarily at applications in hospitals (Bailey 1957a; Luck *et al.* 1971; Nuffield Provincial Hospitals Trust 1955, 1962, 1965; Stimson and Stimson 1972). Several critical reviews—notably, Boldy (1981), Shuman *et al.* (1975), and Stimson and Stimson (1972)—have been motivated by the apparent discrepancy between the large number of published studies and the considerably smaller number of successful implementations.

Since there are so many existing reviews I shall not attempt to be comprehensive in this chapter. Rather, for each application area, I shall highlight a few representative studies.

Health facilities management

Many early operational research studies, especially in the US, addressed problems that arise in the management of hospitals and clinics. This emphasis can be explained in part by the importance of the hospital and clinic in medical care systems. It is also true that problems in hospitals and clinics (scheduling, inventory control) often have parallels in business and industry. Many hospital applications represented efforts to transfer models and solutions from business to health settings.

Out-patient appointment systems

The design of efficient appointment systems has engaged analysts for three decades. Queuing and simulation studies abound (Bailey 1952; Blanco White and Pike 1964; Fetter and Thompson 1965; Fries and Marathe 1981; Rockart and Hoffman 1969; Walter 1973; Welch 1964; Welch and Bailey 1952) and a considerable amount of data on the performance of operating systems has been collected (Johnson and Rosenfield 1968; Nuffield Provincial Hospitals Trust 1965; United Hospital Fund of New York 1967). Fries and Marathe (1981) used a dynamic programming approach to compare the performance of block appointment systems. Frenkel and Minieka (1982) applied a linear programming model to the problem of outpatient scheduling in a solo ophthalmic practice. Nevertheless, most studies have elaborated (without fundamentally altering) Bailey's original queuing formulation. Given the demand for services and the distribution of service times, the model describes the relationship between patient waiting time and provider idle time under various appointment systems. A wide array of other variables can be added to the basic model: patient and provider punctuality, the fraction of patients who arrive without appointments, the fraction of patients who fail to keep appointments, and service times dependent on patient characteristics.

Regardless of the model used or the site of the study, analysts have come to the same general conclusions: (i) patients are quite punctual, but physicians are not; (ii) minor changes in clinic procedures will usually eliminate excessive patient delays without significantly increasing the idle time of providers.

Despite the agreement among the analysts, long delays persist in many clinics, including some of those studied. O'Keefe (1985) suggested the importance of determining which policy changes can really be implemented and of educating the participants to accept the proposed changes. Operational research analysts have been quick to assume that the staff and administration of the clinic are interested in reducing patient waiting time and in increasing their own productivity. This may not be the case. Under a system of prepaid health care (the National Health Service in the UK or health maintenance organizations in the US), increasing the number of patient visits will only increase workload and not income. Staff may consciously or unconsciously use long patient waiting time to hold down demand for services. Recommended changes, however minor, must be carried out by the staff. If the staff are not interested, the implementation is likely to falter.

When clinic staff recognize a problem in the appointment system, an operational analysis can aid in identifying and evaluating possible solutions (Henderson 1976). In one successful project, staff involvement led to a model which focused on the patient's delay in obtaining an appointment rather than on waiting time in the clinic. The mark of a good operational research analysis is that it solves the problem of interest to the manager, rather than one of interest to the analyst.

In-patient admission/discharge scheduling

Interest in the efficient scheduling of in-patients arises in overcrowded facilities and also in underutilized facilities. In hospitals with high occupancy rates (90 per cent or more), efficient scheduling can be viewed as an alternative to capital investment in new facilities. Hospitals with low average occupancy face a different problem: wide fluctuations in the size of patient population are likely to occur, making staffing difficult. In this case, scheduling of patients can result in a more stable operation. Budget cuts and long hospital waiting lists in National Health Service hospitals have stimulated renewed interest in both inpatient scheduling (Worthington 1987) and facility sizing models (Vassilacopoulos 1985 and Harris 1985, below).

Many different models have been used in the analysis of inpatient scheduling: queuing (Worthington 1987), simulation (Smith and Solomon 1966; Webb *et al.* 1977), and Markov probability (Esogbue and Singh 1976; Kolesar 1970; Shonick and Jackson 1973). Some models also employ statistical or time-series forecasting methods in predicting future demand as part of the scheduling problem (Kao and Pokladnik 1978; Shonick and Jackson 1973).

Despite the variety of models, there is common ground. Most analysts consider at least two classes of admission separately—elective and emergency. A further division into medical versus surgical admissions is frequently used. Future demand in each category is forecast from past experience. Future census is forecast from future emergency demand, future scheduled demand, and length-of-stay estimates. Measures of the effectiveness of the scheduling system include resulting in-patient census, rate of turnaway for emergencies, and length of waiting time for elective cases.

The simplest systems generate fixed rules about the maximum number of elective patients who can be scheduled on a given day. The most elaborate systems consider each potential admission separately, using individual patient data to project length of stay and then scheduling the patient so that the probability of facility overflow never exceeds a pre-set value (Rubenstein 1976).

The fixed-rule systems are most commonly used. The more complex systems are necessarily limited to hospitals with on-line computerized admission and pre-admission systems. Implementation studies carried out at two hospitals suggest that systems with heavy day-to-day data requirements in underutilized hospitals are not as likely to be successful as simpler systems implemented in overcrowded facilities (Griffith et al. 1975).

Facility sizing

Most of the scheduling and appointment models discussed assume that the size of the facility is fixed. While efficient scheduling can increase the apparent capacity of a facility, there are times when a hospital or clinic can only meet a growing demand by expanding. Many of the models used in evaluating appointment systems or admissions scheduling systems can be adapted to the problem of establishing an appropriate facility size (Esogbue and Sing 1986; Fetter and Thompson 1966; Shonick and Jackson 1973). These models are limited, however, because they focus on a single facility. A more efficient health system is achieved by fixing individual facility size as part of a regional plan, as discussed in the section on regional health planning.

Several studies have tried to establish appropriate sizes for various units within a hospital. Using a priority queuing model, Vassilacopoulos (1985) addressed the question of how best to allocate in-patient beds to various hospital departments. Operating suites have been analysed using queuing theory (Whitston 1965), dynamic programming (Esogbue 1969), and simulation (Goldman and Knappenberger 1968). Maternity services have been analysed using queuing models (Thompson et al. 1963) to determine the number of labour, delivery, and post-partum beds required. Markov and semi-Markov probability models have been used to predict the resources needed for coronary patients (Kao 1972). As an example of what might be done using a linear programming model, Dowling (1971) determined the

numbers of appendectomies, cholecystectomies, and tonsillectomies that could be performed subject to constraints in the numbers of operating suites, recovery beds, and hospital beds. Responding to National Health Service budget cuts, Harris (1985) used a simulation model to examine the inter-related problems of scheduling operating theatre time and allocating hospital beds.

Most 'size' analyses assume that the pattern of demand for services is fixed. However, reducing the fluctuations in demand through better scheduling may increase the effective capacity of a facility without increasing its size. Simulation studies have been used to evaluate facility size under various scheduling policies: for example, surgical suites (Kwak et al. 1976); radiology (O'Kane, 1981). Such comprehensive models are more realistic and therefore more appealing. However, they may be less likely to be used than simpler models. Because complex models take longer to develop and require more extensive data collection, they cannot be used when quick answers are needed.

Staffing

Staffing is a major concern in most health facilities. Nurse staffing, in particular, has been the subject of many operational research analyses—see, for example, Hershey et al. (1981). Warner et al. (1984) describe a three-level classification of staffing decisions: (i) long-range planning of the numbers and types of nursing personnel needed on each unit; (ii) deriving work schedules for each individual nurse; (iii) allocation of nurses to units at the start of each shift. In addition, modelling has been applied to a fourth area: predicting the supply of personnel.

A quantitative approach to long-range staffing requires quantification of the nursing tasks needed on a unit. The simplest formulations relate nursing needs to the number of patients. More sophisticated approaches attempt to determine the nursing needs of each patient or class of patients. Simulation or statistical concepts can be used to incorporate the inevitable fluctuation in demand for nursing services (Hershey et al. 1974; Smallwood et al. 1971).

The scheduling problem consumes considerable time and effort in most hospitals. A common solution is to have each employee rotate through a fixed set of weekly schedules. Various heuristic and mathematical programming techniques have been used to help in formulating sets of schedules that meet staffing needs (James et al. 1974; Maier-Rothe and Wolfe 1973). This approach is so simple that it can be implemented on a microcomputer (Rosenbloom and Goertzen 1987). The resulting schedules are fair in that each employee cycles through the same set of schedules; however, it does not exploit differences in personal preferences. Warner (1976) has developed and implemented a system to assign schedules based on individual preferences for long weekends, night shifts, and other characteristics. Nurses are asked to assign numerical weights to reflect their preferences, and a mathematical programming algorithm is then used to maximize the sum of these weights while meeting minimal coverage requirements.

The need to allocate nurses on a shift-by-shift basis arises because the number and type of patient assigned to each ward or floor can be expected to fluctuate. In addition, some nurses may be absent from an assigned shift. Many hospitals cope with this problem by having a pool of 'floaters', i.e. nurses who are not assigned to a particular unit until the start of the shift. Trivedi and Warner (1976) use an integer programming model to assign floaters based on the quantified needs of each unit.

Staffing must also be determined for other areas of the hospital. Queuing and simulation studies have been used in staffing operating rooms (Whitston 1965), messenger services (Gupta *et al.* 1971), and out-patient clinics (Keller and Laughhunn 1973).

The issue of physician staffing has not received much attention. In the US, most physicians have a voluntary rather than a salaried or contractual relationship with hospitals. Fetter and Thompson (1966) looked at physician staffing in the outpatient clinics of a teaching hospital.

Statistical and probability models have been used to forecast the supply of physician and other health professionals. The report of the Graduate Medical Education National Advisory Committee (GMENAC) modelled both the expected supply of physicians and the anticipated demand (Tarlov 1980*a, b*) in order to make national policy recommendations on the need for physicians and other personnel. Kane *et al.* (1980) made similar projections for geriatric manpower.

Patients flow models

Patient flow models do not, in themselves, address operational questions. However, estimates derived from such models may be useful (in conjunction with other models and data) in planning and staffing facilities. Analyses of such operational decisions often hinge on having an explicit model of how patients flow through the facility. Markov and semi-Markov models are frequently proposed as good methods of analysing patient flow. Fetter and Thompson (1965) proposed simulating Markovian model of patient transition through a facility organized around Progressive Patient Care (a patient classification system based on the nursing needs of the patient). Later, Weiss *et al.* (1982) used a semi-Markov model to predict patient flow through a hospital.

When used to focus on particular patient subgroups, patient flow models blur the distinction between clinical and operational models. Kao (1972), for example, analysed the flow and recovery of coronary patients. Kastner and Shachtman (1982) looked at the flow of patients with hospital-acquired infections.

Ancillary and support services

Ancillary departments of hospitals present special cases of the general problems of sizing, scheduling, staffing, and inventory control. Simulation has been a popular tool for studying the size and operation of radiology departments (Jeans *et al.* 1972), laboratories (Rath *et al.* 1970), out-patient pharmacies (Myers *et al.* 1972), and in-patient drug distribution (Assimakopoulos 1987*a*). A complex hierarchical queuing model was used to study the operation of telemedicine systems (Willemain 1974). A birth–death model was used to describe the problem of medical records storage, allowing an analysis of microfilm policies and development of long-term storage needs (Nimmo 1983).

Inventory theory has proved useful in evaluating drug inventory (Satir and Cengiz 1987) and hospital support services such as oxygen supplies (Kilpatrick and Freund 1967). The costs of ordering, storage, and shortage are considered in determining the best supply policy. Smith *et al.* (1975) used inventory theory to develop a stock control system for hospitals which was widely implemented in the UK.

The management of blood bank inventory presents a particularly difficult problem since blood is both a product of unique medical value and one which has a limited shelf-life. Blood banks must manage whole-blood inventories for the eight major blood types and the extraction of blood components (such as platelets, packed cells, and plasma) from whole blood. Prastacos (1984) provides an excellent review of blood inventory management theory and practice, including over 75 published and unpublished studies.

Regional health planning

In the US regional health planning has fallen in and out of favour several times in the last three decades. Despite these ups and downs, the interest in quantitative models in health planning has grown steadily. Early efforts in modelling health care systems were primarily academic exercises (Navarro 1969). As a result, Shuman *et al.* (1975) found that few operations research models had resulted in useful aids for planners. More recent reports paint a brighter picture. Health planning applications in the US have addressed resource allocation for mental health using linear programming (Leff *et al.* 1986), primary health care manpower using a semi-Markov Model (Trivedi *et al.* 1987), and a food supplement programme for mothers and children using goal programming (Tingley and Liebman 1984). An equilibrium model was used to model the consumption of 16 different medical services in Quebec Province (Delorme and Rousseau 1987). While these models have been developed using data from operating programmes, they still report hypothetical, rather than actual, use in planning decisions.

A few planning models have been successfully implemented. Pliskin and Tell (1981, 1986) developed a model for forecasting the need for dialysis in Massachusetts which was implemented by the Department of Public Health and resulted in a savings of over five million dollars between 1978 and 1981. Best *et al.* (1986) described the construction of a broadly formulated planning system for a regional health council in Ontario. Traditional operations research models were incorporated in a strategic planning framework which accommodated multiple objectives, participative decision-making, co-ordination of related decisions, and consideration of the political context. Boldy and Clayden (1979) noted that in the UK, where regional health planning has long been accepted, the number of operational research projects devoted to 'strategic studies' in health care has increased, while the number of 'hospital-based tactical studies' has declined.

Because the hospital is often the building block in regional planning, there is some interest in determining which populations are apt to be served by which hospital, especially in multi-hospital regions. Simple statistical models have been used in defining hospital service areas (Griffith 1978; Meade 1974). In the US, particularly in regions with an excess number of hospital beds, hospitals have developed 'marketing strategies' to compete for patients (and physicians) based in part on service area statistics.

In both the UK and the US, as in other industrialized countries, the emphasis in regional health planning has been on assessing the health needs of a population and then allocating resources to meet those needs most effectively. Operational research models have been used in both the needs assessment and the resources allocation phases of this process. The analysis of ambulance systems has engendered such an extensive literature that we treat it as a separate planning area. There is a growing number of examples of operational research models applicable to the planning needs of developing countries, especially in the areas of primary care and epidemic control (Blumenfeld 1984).

Population-based needs assessment

Most planning projects start with an estimate of future needs. Because the determination of health needs has proved to be exceptionally difficult, planners have used past utilization (or demand) for health services as a surrogate for needs. Many economists believe that demand, not need, is the most appropriate yardstick for measuring and allocating medical care. Williams (1974), however, finds the classical economic concept of demand inadequate and concludes that some of the issues raised by 'methodologists' have merit. Multivariate statistical and econometric methods have been widely used to project future demand for health care from past usage and a few population characteristics such as age, sex, and socio-economic status (Dove and Richie 1972; Rosenthal 1964; Wirick 1966).

Several studies have used probability models, usually Markov models, to predict the need or demand for particular medical services (Liebman and Logan 1976; Navarro 1969). Smallwood et al. (1971) proposed a semi-Markov model of 'disease dynamics' which could be used to project need for nursing or medical care. The full implementation of this model would require data on the dynamics of scores of disease groupings, however, and this limits its practicability.

Another approach to assessing needs has been the use of hybrid models, combining traditional epidemiological methods and data with probability models linking the incidence or prevalence of disease with the need for health services. The HASA Health Care Systems Modelling Team used this approach (Shigan et al. 1979). Roberts and Cretin (1980) use life-table methods to predict the need for paediatric cardiac surgery, combining epidemiological data on the incidence of congenital heart disease, clinical reports on survival after surgical intervention, and subjective estimates of the fraction of patients operated on at each age. Similar models were presented in a series of reports on the use of health outcomes (usually obtained from vital statistics) as a basis for planning (Harris et al. 1977). These techniques have been used in US Health Systems Agencies (e.g. Los Angeles County 1978).

Facility location

A wide variety of operational research models has been applied to the facility location problem. Many variants of mathematical programming and network theory have been used to specify sets of hospital locations which would minimize travel time (Calvo and Marks 1973; Elshafei 1977) or maximize consumer preferences (Parker and Srinivasan 1976). Abernathy and Hershey (1972) compare the locations resulting from four different criteria for placement.

Unfortunately, the facility location problem addressed by most existing operational research models is not the problem faced by planners in western Europe and North America. The health planner interested in where to locate facilities faces a complex problem subject to many constraints. Existing facilities are not easily moved, and so the question really only applies to new or replacement facilities. Even so, in most urban settings, the number of locations actually available for expansion is limited. An additional problem has been the failure to incorporate realistic models of patient and physician behaviour in choosing facilities.

The implementation of new medical systems in developing countries presents fewer constraints in the construction of new facilities than in developed countries, making simple network models more useful. In Ecuador, a location-set-covering algorithm was used to determine the placement of medical supply centres (Reid et al. 1986). The construction of a completely new city in Saudi Arabia presented the opportunity to plan the location and size of primary health care centres using network location theory (Berghmans et al. 1984).

Health maintenance organizations have spurred interest in the facility location question, not only for hospitals but also for out-patient facilities (Shuman et al. 1973). Dokmeci (1977) used a non-linear mathematical programming model to determine the number, size, and location of medical centres, hospitals, and health clinics. This model chose an allocation which minimized the sum of facility and transportation costs. Buchanan (1980) developed a series of algorithms for both the timing and location of expansions for health maintenance organization facilities.

Ambulance system design and operation

Ambulance services present a number of design and operational problems. How many vehicles are needed in a region? Where should the vehicles be stationed? How should they be dispatched? In the last 15 years, all these problems have been addressed with the help of quantitative models, usually simulation, queuing models, or network models (Fitzsimmons 1973; Groom 1977; Larson 1975; Savas 1969; Willemain and Larson 1977). In addition, statistical models have been used to predict demand for emergency medical care (Kamenetzky *et al.* 1982).

Analysts of ambulance systems benefited greatly from the work done in modelling other urban emergency services, mainly fire and police patrol services. Models originally developed for one service were easily adapted to another because the key elements of the systems are the same: vehicles must make their way to emergency calls which may occur at any time and from any location in the region. Although the ultimate goals of the systems are different (to reduce crime, to limit fire damage, to save lives), the operators of these systems tend to think in terms of the same intermediate objective: minimize response time. The success of operational research models in ambulance systems springs in large part from the fact that emergency medical systems administrators had an agreed-upon easily quantifiable objective which could be incorporated into a mathematical model.

In the case of ambulance systems, the good fit between the mathematical model and the real concerns of the decision-makers has borne fruit. In both the US and the UK, the recommendations resulting from ambulance system modelling have been successfully implemented (Eaton *et al.* 1985; Jarvis *et al.* 1975; Raitt 1981). Recently, the same models have been successfully applied to planning medical transportation requirements for a district in Ghana (Wright *et al.* 1984) and an ambulance system in the Dominican Republic (Eaton *et al.* 1986).

Epidemic control

Epidemiologists and biomathematicians have used mathematical models in investigating the behaviour of epidemics since the nineteenth century (Valinsky 1975). Dietz (1988) reports on a discrete-time model applied to measles epidemics which was developed by P.D. En'ko' and reported in Russian in St Petersburg in 1889. By the late 1950s, both deterministic and probabilistic models of the spread of epidemics had been developed and refined (Abbey 1952; Bailey 1957b). The models were so complex that they could not be solved for realistic situations. However, computer simulations of the models were feasible (Coffey 1973; Cvietanovic *et al.* 1972; Elveback *et al.* 1976; Garg *et al.* 1967; ReVelle *et al.* 1969; Taylor 1958).

Simulations of epidemics are closer to dynamic system simulations than to the usual Monte Carlo simulation based on a queuing model and often include an economic component. Differential equations or difference equations are often incorporated to model disease incubation periods and lags in the effectiveness of control programmes (Frerichs and Prawda 1975). Recently, Assimakopoulos (1987b) took a novel approach, modelling the spread of hospital-acquired infection as a network in which bacteria from sources of infection are transferred by carriers to susceptible patients. Several hospitals in Athens have used a decision support system incorporating this model to evaluate proposed measures to reduce the spread of nosocomial infection. The worldwide AIDS epidemic has stimulated many different modelling efforts. The dynamics of the epidemic (Hyman and Stanley 1988; May and Anderson 1987), the incubation period of the virus (Liu *et al.* 1988), the importance of different modes of transmission in different populations (Kaplan 1989), and the effectiveness of proposed interventions (Kaplan and Abramson 1989) have all been explored with the help of mathematical models. The many uncertainties surrounding AIDS, its treatment, and its prevention will only be resolved over a matter of years. In the mean time, models provide a valuable way to evaluate proposals and programmes which, imperfect though they are, may still be able to save thousands of lives if promptly adopted.

Clinical decision-making

The study of medical decision-making has blossomed in the last decade. Applications of operational research methods, especially decision analysis, to the diagnosis and treatment of disease are described in several texts (Barnoon and Wolfe 1972; Lusted 1968; Weinstein *et al.* 1980). The use of statistically based algorithms and computer-aided diagnosis are widely reported in the medical literature and are well summarized in a recent review (Kassirer *et al.* 1987). In addition, simulation studies, Fourier analysis, and other mathematical techniques are used in the interpretation of data from electrocardiograms (Wolf *et al.* 1977), X-rays (Chan and Doi 1981; Kalender 1981; Kulkarni 1981), and computer-aided tomography (CAT) scans (Wills *et al.* 1981).

The early applications of decision analysis to problems of medical management were often too simplistic to be taken seriously by clinicians. While physicians try to decide each case on its individual characteristics, the decision analyst seemed to approach clinical decision-making on an aggregate basis, as if there was only one 'best answer' which applied over the whole population. The development of interactive computer programs enabled the physician decision-maker to use more complex and therefore more realistic models (Pauker and Kassirer 1981). There is evidence that these decision models are most useful to physicians in training (Goldman *et al.* 1981; Walmsley *et al.* 1977). Clinical decision-making methods have also been extended to the assessment of quality of care (Greenfield *et al.* 1982). With the growing availability of computer systems in hospitals, the place of computer-based decision aids in medical training and medical practice seems secure (Crichton and Emerson 1987; Fryback 1986).

Programme evaluation and policy analysis

The most important public health decisions are those involving the future directions for national health policy. The way in which health care is financed, for example, has consequences for the health and finances of a nation. Decisions regarding national policy on screening or treatment for a specific disease or decisions on which types of biomedical research to fund may also have major impact on the public's health. These policy decisions are, almost by definition, complex. Decision-makers at this level often confront conflicting objectives, competing interest groups, and divergent views of justice, equity, and human behaviour.

What role, if any, can operational research play in the resolution of policy debates? Few analysts (and fewer decision-makers) believe that a mathematical model can generate optimal (or even acceptable) solutions to such complex decisions. However, many analysts (and a few decision-makers) now believe that mathematical models can be useful in organizing the problem, analysing relevant data, and evaluating the readily quantified consequences of an action.

Most examples of operational research models used in policy analysis involve interdisciplinary projects. Economic and political theories are usually incorporated, as are epidemiological and statistical models. In some of these analyses, operational research plays a relatively minor role; in others, the operational research component is the means for organizing contributions from other fields. In a few examples, the entire analysis revolves around a particular mathematical approach, such as Markov models, dynamic system simulations, or integer programming. Many economic analyses of health care financing and health insurance use simulation (e.g. Drabek *et al.* 1974; Feldman and Dowd 1982).

Whether simple or complex, the models are rarely intended to be used as decision-making machines. Rather, they are designed for use in an iterative or interactive mode as an aid to effective decision-making. While this caveat applies to every application of operational research, it is especially pertinent when reviewing policy models.

The allocation of research grants in biomedical and health services research was an early area of interest. Utility theory, economic theory, and other scaling techniques were suggested as means of developing funding priorities (Cutler *et al.* 1973; Keeler 1970; Stimson 1969). Shachtman (1980) reports a decision analysis which helped secure funding for a national survey on nosocomial infection. In the US, the continuing study of the peer review process (Carter 1974) and the use of consensus panels at the National Institute of Health bear the stamp of this earlier work.

Early programme evaluation studies tended to use 'pure' operational research models. Markov models were proposed as a general approach to programme evaluation (Navarro 1969; Ortiz and Parker 1971). Markov models were also proposed to evaluate mental health programmes (Trinkl 1974), tuberculosis control programmes (Bush *et al.* 1972), and geriatric programmes (Burton *et al.* 1975; Meredith 1971), and as a method for generating a measure of benefit for health programmes in general (Chen *et al.* 1975). Dynamic system simulation was used to model narcotics addiction and control (Levin *et al.* 1972), conversion to a health maintenance organization (Hirsch and Miller 1974), and the delivery of dental care (Hirsch and Killingsworth 1975). Mathematical programming models have been used to analyse a voucher system for financing medical care (Whipple 1973). Chen and Bush (1976) used 0–1 integer programming to specify which groups should be screened for phenylketonuria or tuberculosis.

Some of these early simple models have been the foundation for more complex interdisciplinary models. Burton's Markov model of geriatric care has subsequently evolved into a complex methodology for planning geriatric care (Burton and Dellinger 1981). Boldy *et al.* (1981) describe the 'balance of care' model used to assist the UK in allocating resources among various health and social services. Although this is not an optimization model, it grew out of a traditional mathematical programming approach to resource allocation (McDonald *et al.* 1974). Both the geriatric care model and the balance of care model have been used by decision-making agencies.

Nursing homes and long-term care have attracted considerable interest in the US. Willemain (1980) has used modelling to predict the effects of two different reimbursement strategies in nursing homes. Keeler *et al.* (1981) used simple models to explore the mix of long-stay patients in nursing homes.

Case mix in hospitals has been the subject of a series of studies by a group at Yale University (Fetter *et al.* 1980). This work resulted in a set of diagnosis-related groups which form the basis for a hospital reimbursement scheme now being used for Medicare patients in the US. Klastorin (1982) proposes a very different approach to grouping hospitals for reimbursement or cost-containment purposes. Cluster analysis is used to group hospitals on the basis of price-related variables (prices of inputs, degree of union involvement, urban versus rural setting).

Several studies have used probability and statistical models to organize a large body of inconclusive scientific data bearing on a policy issue. Cretin (1974, 1977) reviewed the clinical literature on the effectiveness of prehospital cardiac care and hospital coronary care units, and then used a hypothetical life-table to evaluate the effectiveness of pre-hospital versus in-hospital care in terms of life-years saved. Weinstein and Stason (1976) used a similar approach in comparing strategies for the screening and treatment of hypertension. They incorporate techniques from decision analysis to arrive at 'quality adjusted life-years'. Berwick *et al.* (1980) reviewed the literature linking cholesterol and heart disease, and analysing alternative policies for screening and treating hypercholesterolaemic children. Eddy (1980) developed a generic Markov model for analysing cancer screening programmes. Using data from the clinical literature to estimate the parameters for specific cancers, he then estimated optimal screening strategies for several major cancers.

Policy models that incorporate extensive reviews of the clinical and epidemiological data can serve two purposes. First, the models can help the policy-maker distil the scientific data. This is especially useful when there is an extensive and apparently contradictory body of work. Second, the models can help in identifying the strengths and weaknesses of the scientific evidence, suggesting directions for future studies. Such work has already highlighted one important fact—unanswered questions which are critical to the development of sound public policy in an area are not necessarily the questions of most interest to scientists or clinicians.

This review of health applications of operational research has been of necessity incomplete. The bibliographies mentioned at the beginning of this section are good sources of additional information. In addition, the following journals often include articles using operational research methods: *Health Services Research, Medical Care, Journal of Medical Decision Making*, and *Methods of Information in Medicine*. Operational research journals may also contain reports of health applications, especially *Interfaces* and *Management Science. Operations Research, Journal of Operational Research*, and *Operations Research Quarterly* also publish health applications. Unfortunately, the selection of reports for publications is often based on the originality of the modelling approach rather than on utility of the model to the decision-maker. Many good accounts of implementation of operational research receive limited distribution as technical reports.

Problems and prospects

Application of operational research to public health problems have enjoyed mixed success. Operational research analysts and public health practitioners have worried over the discrepancy between the many published accounts of health applications and the few successful implementations. The problem is a reflection of the gap between academic operational research (theory) and applied operational research (practice) which exists in all areas of application. However, the usual difficulties take a special form in health care systems.

One problem, partly semantic and partly practical, is the notion of 'optimization' inherent in many mathematical methods. Models that optimize are necessarily limited to a single measure of effectiveness. The concepts of 'maximize' and 'minimize' run into difficulty when more than one variable is involved. It is easy to identify the tallest person in a room (the one of maximum height) or the heaviest person (the one of maximum weight). If, however, we define 'bigness' as a two-pat concept involving weight and height, we may not be able to identify the 'biggest' person (unless the tallest is also the heaviest).

The world of public health seems inherently multidimensional. The outcomes of disease inevitably encompass mortality, morbidity, and costs. The goals of health agencies are most often in terms of balancing several outcomes, not in terms of maximizing or minimizing a single variable. We strive for complete physical, psychological, and social well-being, or for the best possible medical care at the least possible cost. Operational researchers have been quick to point out the logical impossibility of such goals and to recast the goal into logically acceptable form—minimize the cost of care subject to certain constraints. While such a formulation may, in fact, be adequate for the problem at hand, there is a fundamental difference between the concepts of optimization and balancing which the operational researcher needs to acknowledge.

The analyst also needs to acknowledge that solutions which are optimal in the model are not necessarily optimal in the world. The model is an imperfect representation of the real problem, and it is crucial to test the sensitivity of an 'optimal' solution to assumptions and parameters in the model which may well be wrong. For this reason, the best use of operational research models, including optimization models, may be in an interactive mode or, as Boldy (1981) terms it, a 'what if' mode. This may help humble the analyst who claims to have an optimal solution and raise healthy scepticism in the decision-maker anxious to believe the claim.

A second difficulty arises because different observers have different opinions about whether health services are inputs to or outputs from the medical care system. Epidemiologists, for example, tend to treat medical services as inputs (or costs) which will eventually produce improved health status as an output (or benefit). Health administrators are more likely to take personnel and facilities as inputs which produce health services as outputs. Operational research studies have tended to follow the latter perspective. Both points of view are valid. However, when differences in perspective are not explicitly recognized, conflict and poor communication may result.

A third problem is the difficulty in quantifying many important parameters. While some analysts think that the main difficulty has been too little quantification—i.e. reluctance of health practitioners to clear up fuzzy thinking and put a number on some factor—I believe that there has also been too much misleading quantification.

It is relatively easy to quantify concrete things by counting, weighing, or measuring. However, some of the concepts involved in health and the delivery of health services are quite abstract; patient satisfaction, quality of care, and health status, for example. The simplicity of the labels belie the richness of the underlying concepts. It is tempting to believe that any concept described by a single word or phrase can automatically be captured by a single number. Unfortunately, this is not the case.

When a concept, such as quality of care, can only be realistically represented by an elaborate measurement scheme, it is appropriate to look for simpler approximate representations. The problem is that approximation of complex concepts has not always been carried out responsibly. This is not solely a problem of operational research and Gould (1981) gives a thorough analysis of the misuse of measures of intelligence.

A good approximation should retain the essential while cutting away the non-essential. This requires that the essential part of a concept be identified with the application in mind. The most critical component of the quality of medical care is different in an emergency department and a nursing home. Approximations are often chosen with little consideration of their appropriateness, but rather because the data are readily available. The problem of using less than adequate approximations is compounded when the analysts fail to distinguish clearly between the underlying concept and the approximate measure. The danger is that the analyst and the decision-maker may come to believe that this measure is the concept.

The fourth problem arises in all applications of operational research. Operational research is supposed to be a service to a decision-maker. Models are meant to solve the problems perceived by the manager of the system. There is a tendency, especially among academic operational researchers, to solve problems of interest to the analyst. This problem was exacerbated in health applications of operational research in the US by the manner in which much of the early work was funded. Individual hospitals, by and large, did not see the possibilities for quantitative analysis. Major funding was through research grants to the analysts. The hope was that these research projects would demonstrate the value of operational research in health and lead operating institutions to fund their own operational research units. This has happened, but very slowly. Many of the projects were never implemented or only implemented with half-hearted support from key decision-makers in the organization studied.

Despite the problems, operational research has now become an accepted part of public health, particularly health administration. The maturation and expansion of health operational research has undoubtedly benefited from the expanding role of computers in hospitals. The presence of computers in hospitals encourages the development of more ambitious models which can only be implemented with the help of computers. In addition, the use of the computer to collect and store information in hospitals gives the analyst access to data which were previously unavailable or at least difficult to extract. The ready access to clinical information has contributed to the growth of clinical decision models and computer-aided diagnosis.

Clinicians and administrators have gradually learned the value and the limitations of operational research. Operational researchers have come to respect the uniqueness and complexity of the health field. This atmosphere of mutual respect and realistic expectations bodes well for the effectiveness of future collaborations.

References

Abbey, H. (1952). An examination of the Reed-Frost theory of epidemics. *Human Biology* **24**, 20.

Abernathy, W. and Hershey, J. (1972). A spatial-allocation model for regional health services planning. *Operations Research* **20**, 629.

Assimakopoulos, N. (1987a). A medication management system. *Applied Mathematics and Computation* **21**, 73.

Assimakopoulos, N. (1987b). A network interdiction model for hospital infection control. *Computers in Biology and Medicine* **17**(6), 411.

Bailey, N.T.J. (1951). On assessing the efficiency of single-room provision of hospital wards. *Journal of Hygiene* **49**, 452.

Bailey, N.T.J. (1952). A study of queues and appointment systems in outpatient departments, with special reference to waiting times. *Journal of the Royal Statistical Society Series B* **14**, 185.

Bailey, N.T.J. (1954). Queueing for medical care. *Applied Statistics* **3**, 137.

Bailey, N.T.J. (1956). Statistics in hospital planning and design. *Applied Statistics* **5**, 146.

Bailey, N.T.J. (1957a). Operational research in hospital planning and design. *Operational Research Quarterly* **8**, 149.

Bailey, N.T.J. (1957b) *The mathematical theory of epidemics*. Griffin, London.

Bailey, N.T.J. (1964). *The elements of stochastic processes with applications to the natural sciences*. Wiley, New York.

Banks, J. and Carson, J.S. (1984). *Discrete-event system simulation* Prentice-Hall, Englewood Cliffs, New Jersey.

Barnoon, S. and Wolfe, H. (1972). *Measuring the effectiveness of medical decisions: an operations research approach*. Thomas, Springfield, Illinois.

Berghmans, L., Schoovaerts, P., and Teghem, J. Jr. (1984). Implementation of health facilities in a new city. *Journal of the Operational Research Society* **35**, 1047.

Berwick, D.K., Cretin, S., and Keeler, E.B. (1980). *Cholesterol, children and heart disease: an analysis of alternatives*. Oxford University Press, New York.

Best, G., Parston, G., and Rosenhead, J. (1986). Robustness in practice—the regional planning of health services. *Journal of the Operational Research Society* **37**(5), 464.

Blanco White, M.F. and Pike, M.C. (1964). Appointment systems in outpatient clinics and the effect of patients' unpunctuality. *Medical Care* **2**, 133.

Bland, R.G., Goldfarb, D., and Todd, M.J. (1981). The ellipsoid method: A survey. *Operations Research* **29**, 1039.

Blumenfeld, S.N. (1984). The PRICOR Project: Applications of Operations Research in PHC Planning in Developing Countries. *Public Health Review* **12**, 279.

Boldy, D. (1976). A review of application of mathematical programming to tactical and strategic health and social service problems. *Operational Research* **27**, 439.

Boldy, D.P. (1981). *Operational research applied to health services*. St. Martin's Press, New York.

Boldy, D. (1987). The relationship between decision support systems and operational research: health care examples. *European Journal of Operational Research* **29**(2), 128.

Boldy, D., Canvin, R., Russell, J., and Royston, G. (1981). Planning the balance of care. In *Operational research applied to health services* (ed. D. Boldy), p. 84. St. Martin's Press, New York.

Boldy, D. and Clayden, D. (1979). Operational research projects in health and welfare services in the United Kingdom and Ireland. *Journal of the Operational Research Society* **30**, 505.

Boldy, D.P. and O'Kane, P.C. (1982). Health operational research—a selective overview. *European Journal of Research* **10**, 1.

Brockmeyer, E., Halstrom, H.L., and Jensen, A. (1948). *The life and works of A. K. Erlang.* Transaction, Danish Academy of Technical Science, Copenhagen, No. 2.

Buchanan, J. (1980). *Planning inpatient capacity expansion in health maintenance organisations.* Unpublished dissertation, Graduate School of Management, University of California, Los Angeles, California.

Buffa, E.S. and Miller, J.G. (1979). *Production-inventory systems: planning and control,* 3rd edn. Irwin, Homewood, Illinois.

Burton, R.M., Damon, W.W., and Dellinger, D.C. (1975). Patient states and the technology matrix. *Interfaces* **5**, 43.

Burton, R.M. and Dellinger, D.C. (1981). Planning the care of the elderly. In *Operational Research applied to health services* (ed. D. Boldy), p. 129. St. Martin's Press, New York.

Bush, J.W., Fanshel, S., and Chen, M.M. (1972). Analysis of a tuberculin testing program using a health status index. *Socio-Economic Planning Science* **7**, 49.

Calvo, A. and Marks, D.H. (1973). Location of health care facilities: an analytic approach. *Socio-Economic Planning Science* **7**, 407.

Carter, G.M. (1974). *Peer review, citations and biomedical research policy: N.I.H. grants to medical school faculty,* R-1583-HEW. Rand Corporation, Santa Monica, California.

Chan, H.P. and Doi, K. (1981). Monte Carlo simulation studies of backscatter factors in mammography. *Radiology* **139**, 195.

Chen, M.M. and Bush, J.W. (1976). A mathematical programming approach for selecting an optimal health programme case mix. *Inquiry* **13**, 215.

Chen, M.M., Bush, J.W., and Patrick, D.L. (1975). Social indicators for health planning and policy analysis. *Political Science* **6**, 71.

Chiang, C.L. (1968). *Introduction to stochastic processes in biostatistics.* Wiley, New York.

Coffey, R.J. (1973). Model of communicable diseases spread in a rural population. In *Health care delivery planning* (ed. A. Reisman and M. Kiley), p. 287. Gordon and Breach, New York.

Cooper, R.N. (1972). *Introduction to queueing theory.* Macmillan, New York.

Cox, D.R. and Miller, H.D. (1965). *Theory of stochastic processes.* Wiley, New York.

Cox, D.R. and Smith, W.L. (1961). *Queues.* Chapman and Hall, London.

Cretin, S. (1974). *A model of the risk of death from myocardial infarction,* Innovative Resource Planning Project Technical Report No. TR-09-74. MIT Operations Research Center, Cambridge, MA.

Cretin, S. (1977). Cost/benefit analysis of treatment and prevention of myocardial infarction. *Health Services Research* **12**, 174.

Crichton, N.J. and Emerson, P.A. (1987). A probability-based aid for teaching medical students a logical approach to diagnosis. *Statistics in Medicine* **6**, 805.

Cutler, R.S., Martino, V.A., and Webb, A.M. (1973). Biomedical research relevance assessment. In *Health care delivery planning* (ed. A. Reisman and M. Kiley), p. 89. Gordon and Breach, New York.

Cvietanovic, B., Grab, B., Uemura, K., and Butchenko, B. (1972). Epidemiological model of tetanus and its use in the planning of immunization programmes. *International Journal of Epidemiology* **1**, 125.

Dantzig, G.B. (1963). *Linear programming and extensions.* Princeton University Press, Princeton, New Jersey.

Delorme, L. and Rousseau, J.M. (1987). MEMRA: An equilibrium model for resource allocation in a health care system. *European Journal of Operational Research* **29**, 155.

Denardo, E.V. (1982). *Dynamic programming: models and applications.* Prentice-Hall, Englewood Cliffs, New Jersey.

Denison, R.A., Wild, R., and Martin, M.J.C. (1969). *A bibliography of operational research in hospitals and the health services.* University of Bradford Management Centre, Bradford.

Dietz, K. (1988). The first epidemic model: A historical note on P.D. En'ko'. *Australian Journal of Statistics* **30A**, 56.

Dokmeci, V.F. (1977). A quantitative model to plan regional health facility systems. *Management Science* **24**, 411.

Dove, H.G. and Richie, C.G. (1972). Predicting hospital admission by state. *Inquiry* **9**, 51.

Dowling, W.L. (1971). The application of linear programming to decision making in hospitals. *Hospital Administration* **16**, 66.

Drabek, L., Intriligator, M.D., and Kimbell, L.J. (1974). *A forecasting and policy simulation model of the health care sector.* Lexington Books, Lexington, Massachusetts.

Drake, A.W. (1967). *Fundamentals of applied probability theory.* McGraw-Hill, New York.

Duckworth, W.E., Gear, A.E., and Lockett, A.G. (1977). *A guide to operational research,* 3rd edn. Chapman and Hall, London.

Eaton, D.J., Sanchez H.M., Lantiqua, R.R., and Morgan, J. (1986). Determining Ambulance Deployment in Santo Domingo, Dominican Republic. *Journal of the Operational Research Society* **37**(2), 113.

Eaton, D.J., Daskin, M.S., Simmons, D., Bulloch, B., and Jansma, G. (1985). Determining emergency medical service vehicle deployment in Austin, Texas. *Interfaces* **15**(1), 96.

Eddy, D.M. (1980). *Screening for cancer: theory analysis and design.* Prentice Hall, Englewood Cliffs, New Jersey.

Elshafei, A.N. (1977). Hospital layout as a quadratic assignment problem. *Operational Research Quarterly* **28**, 167.

Elston, R.C. and Pickrel, J.C. (1963). A statistical approach to ordering and usage policies for a hospital blood bank. *Transfusion* **3**, 41.

Elveback, L.R., Fox, J.P., Ackerman, E., Langworthy, A., Boyd, M., and Gatewood, L. (1976). An influenza simulation model for immunization studies. *American Journal of Epidemiology* **103**, 152.

Emshoff, J.R. and Sisson, R.L. (1970). *Design and use of computer simulation models.* Macmillan, New York.

Eppen, G.D., Gould, F.J., and Schmidt, C.P. (1987). *Introductory Management Science* (2nd edn). Prentice-Hall, Englewood Cliffs, New Jersey.

Esogbue, A. (1969). Dynamic programming and optimal control of variable multichannel stochiastic service systems with applications. *Mathematical Bioscience* **5**, 133.

Esogbue, A. and Singh, A.J. (1976). A stochiastic model for an optimal priority bed distribution problem in a hospital ward. *Operations Research* **24**, 884.

Feldman, R.D. and Dowd, B.E. (1982). Simulation of a health insurance market with adverse selection. *Operations Research* **30**, 1027.

Feller, W. (1968). *An introduction to probability theory and its applications* (3rd edn), Vol. 1. Wiley, New York.

Fetter, R.B. and Thompson, J.D. (1965). The simulation of hospital systems. *Operations Research* **13**, 689.

Fetter, R.B. and Thompson, J.D. (1966). Patients' waiting time and doctors' idle time in the outpatient setting. *Health Services Research* **1**, 66.

Fetter, R., Shin, Y., Freeman, J., Averill, R., and Thompson, J. (1980). Case mix definition by diagnosis related groups. *Medical Care* **18**(2), Supplement, 1.

Fitzsimmons, J. (1973). A methodology for emergency ambulance deployment. *Management Science* **19**, 627.

Flagle, C.D. (1960). The problem of organization for inpatient care. In *Management science: models and techniques* (ed. C. W. Churchman and M. Verhulst), Vol. 2. Pergamon Press, New York.

Flagle, C.D. (1962). Operations research in the health services. *Operations Research* **10**, 591.

Flagle, C.D. (1967). A decade of operations research in health. In *New methods of thought and procedure* (ed. F. Zwicky and A. G. Wilson), p. 33. Springer, New York.

Ford, L.R. Jr. and Fulkerson, D.R. (1962). *Flows in networks*. Princeton University Press, Princeton, New Jersey.

Forrester, J. (1961). *Industrial dynamics*. MIT Press, Cambridge, Massachusetts.

Frenkel, M. and Minieka, E. (1982). Optimal patient scheduling in a solo practice: an application of linear programming. *Annals of Opthalmology* **14**(9), 782.

Frerichs, R. and Prawda, J. (1975). A computer simulation model for the control of rabies in an urban area of Colombia. *Management Science* **22**, 411.

Fries, B.E. (1976). Bibliography of operations research in health care systems. *Operations Research* **24**, 801.

Fries, B.E. (1979). Bibliography of operations research in health care systems: an update. *Operations Research* **27**, 408.

Fries, B.E. (1981). *Applications of operations research to health care delivery systems*. Springer, Berlin.

Fries, B.E. and Marathe, V.P. (1981). Determination of optimal variable-sized multiple-block appointment systems. *Operations Research* **29**, 324.

Fryback, D.G. (1986). A programme for training and feedback about probability estimating for physicians. *Computer Methods and Programs in Biomedicine* **22**, 27.

Garfinkel, R.S. and Nemhauser, G.L. (1972). *Integer programming*. Wiley, New York.

Garg, M.L., Thompson, D.J., and Gezon, H.M. (1967). Assessing the influence of treatment on the spread of staphylococci in newborn infants by simulation. *American Journal of Epidemiology* **85**, 220.

Goldman, J. and Knappenberger, H.A. (1968). How to determine the optimum number of operating rooms. *Modern Hospital* **111**, 114.

Goldman, L., Waternaux, C., Garfield, F., *et al.* (1981). Impact of a cardiology data bank on physicians' prognostic estimates—evidence that cardiology fellows change their estimates to become as accurate as the faculty. *Archives of Internal Medicine* **141**, 1631.

Gould, S.J. (1981). *The mismeasure of man*. Norton, New York.

Greenfield, S.G., Cretin, S., Worthman, L., and Dorey, F. (1982). The use of an ROC curve to express quality of care results. *Medical Decision Making* **2**, 13.

Griffith, J.R. (1978). *Measuring hospital performance*. Inquiry Book, Blue Cross Association, Chicago, Illinois.

Griffith, J.R., Munson, F.C., and Hancock, W.M. (1975). *Cost control in hospitals*. Health Administration Press, Ann Arbor, Michigan.

Groom, K.N. (1977). Planning emergency ambulance services. *Operational Research Quarterly* **28**, 641.

Gupta, I., Zareda, J., and Kramer, N. (1971). Hospital manpower planning by use of queueing theory. *Health Services Research* **6**, 76.

Harris, L.J., Keeler, E.B., Kisch, A.E., Michnich, M.E., de Sola, S.F., and Drew, D.E. (1977). *Algorithms for planners: an overview*. R-2215/1, Rand Corporation, Santa Monica, California.

Harris, R.A. (1985). Hospital bed requirements planning. *European Journal of Operations Research* **25**, 212.

Henderson, K.M. (1976). Some aspects of clinic management. In *Selected papers on operational research in the health services* (ed. B. Barber), p. 161. Operational Research Society, Birmingham.

Hershey, J.C., Abernathy, W.A., and Baloff, N. (1974). Comparison of nurse allocation policies—a Monte Carlo model. *Decision Science* **5**, 58.

Hershey, J., Pierskalla, W., and Wandel, S. (1981). Nurse staffing management. In *Operational research applied to health services* (ed. D. Boldy), p. 189. St. Martin's Press, New York.

Hillier, F.S. and Lieberman, G.J. (1980). *Operations research*, 3rd edn. Holden-Day, San Francisco, California.

Hirsch, G.B. and Killingsworth, W.R. (1975). A new framework for projecting dental manpower requirements. *Inquiry* **12**, 126.

Hirsch, G.B. and Miller, S. (1974). Evaluating HMO policies with a computer simulation model. *Medical Care* **12**, 668.

Horvath, W.J. (1965). British experience with operations research in the health services. In *Medical care research* (ed. K. L. White), p. 55. Pergamon Press, New York.

Horvath, W.J. (1967). Operations research in medical and hospital practice. In *Operations research for public systems* (ed. P. M. Morse), p. 127. MIT Press, Cambridge, Massachusetts.

Howard, R. (1960). *Dynamic programming and Markov processes*. MIT Press, Cambridge, Massachusetts.

Howard, R. (1971). *Dynamic probabilistic systems*, Vols. I and II. Wiley, New York.

Hyman, J.M. and Stanley, E.A. (1988). Using mathematical models to understand the AIDS epidemic. *Mathematical Biosciences* **156**, 189.

James, S., Outten, W., Davis, P.J., and Wands, J. (1974). House staff scheduling: A computer-aided method. *Annals of Internal Medicine* **80**, 70.

Jarvis, J.P., Stevenson, K.A., and Willemain, T.R. (1975). *A simple procedure for the allocation of ambulances in semi-rural areas*, Technical report TR-13-75. MIT Operations Research Center, Cambridge, Massachusetts.

Jeans, W.D., Berger, S.R., and Gill, R. (1972). Computer simulation model of an X-ray department. *British Medical Journal* i, 674.

Johnson, W.L. and Rosenfeld, L.S. (1968). Factors affecting waiting time in ambulatory care services. *Health Services Research* **3**, 286.

Kalender, W. (1981). Monte Carlo calculations of X-ray scatter data for diagnostic radiology. *Physics in Medicine and Biology* **26**, 835.

Kamenetzky, R.D., Shuman, L.J., and Wolf, H. (1982). Estimating need and demand for prehospital care. *Operations Research* **30**, 1148.

Kane, R., Solomon, D., Beck, J., Keeler, E., and Kane, R. (1980). The future need for geriatric manpower in the United States. *New England Journal of Medicine* **302**, 1327.

Kao, E.P.C. (1972). A semi-Markov model to predict recovery progress of coronary patients. *Health Services Research* **7**, 191.

Kao, E.P.C. and Pokladnik, F.M. (1978). Incorporating exogenous factors in adaptive forecasting of hospital census. *Management Science* **24**, 1677.

Kaplan, E.H. (1989). What are the risks of risky sex? Modeling the AIDS epidemic. *Operations Research* **37**, 198.

Kaplan, E.H. and Abramson, P.R. (1989). So what if the program ain't perfect? A mathematical model of AIDS education. *Evaluation Review* **13**, 107.

Kassirer, J.P., Moskowitz, A.J. Lau, J., and Pauker, S.G. (1987). Decision analysis: A progress report. *Annals of Internal Medicine* **106**, 275.

Kastner, G.T. and Shachtman, R.H. (1982). A stochastic model to measure patient effects stemming from hospital acquired infections. *Operations Research* **30**, 1105.

Keeler, E. (1970). *Models of disease costs and their use in medical research resource allocation*. Rand Corporation, Santa Monica, California.

Keeler, E.B., Kane, R.L., and Solomon, D.H. (1981). Short-and long-term residents of nursing homes. *Medical Care* **19**, 363.

Keller, T.F. and Laughhunn, D.J. (1973). An application of queueing theory to a congestion problem in an out-patient clinic. *Decision Science* **4**, 379.

Kilpatrick, K.E. and Freund, L.E. (1967). A simulation of oxygen tank inventory at a community general hospital. *Health Services Research* **2**, 298.

Kiviat, P.J., Villaneauva, R., and Markowitz, H. (1969). *The SIMSCRIPT II programming language*. Prentice-Hall, Englewood Cliffs, New Jersey.

Klastorin, T.D. (1982). An alternative method for hospital partition determination using hierarchical cluster analysis. *Operations Research* **30**, 1134.

Kleinrock, L. (1975). *Queueing systems*. Wiley, New York.

Kolesar, P. (1970). A Markovian model for hospital admission scheduling. *Management Science* **18**, B374.

Krischer, J.P. (1980). An annotated bibliography of decision analytic applications to health care. *Operations Research* **28**, 97.

Kulkarni, R.N. (1981). Monte Carlo calculation of the dose distribution across a plane bone-marrow interface during diagnostic X-ray examinations. *British Journal of Radiology* **54**, 875.

Kwak, N.K., Kuzdrall, P.J., and Schmitz, H.H. (1976). A GPSS simulation of scheduling policies for surgical patients. *Management Science* **22**, 982.

Larson, R.C. (1975). Approximating the performance of urban emergency service systems. *Operations Research* **22**, 845.

Larson, R.C. and Odoni, A.R. (1981). *Urban operations research*. Prentice-Hall, Englewood Cliffs, New Jersey.

Leff, H.S., Dada, M., and Graves, S.C. (1986). An LP planning model for a mental health community support system. *Management Science* **32**(2), 139.

Levin, G., Hirsch, G., and Roberts, E. (1972). Narcotics and the community: a system simulation. *American Journal of Public Health* **62**, 861.

Liebman, J.S. and Logan, E. (1976). Analysing the start-up effects of new patients on an ambulatory case programme. *Medical Care* **14**, 839.

Los Angeles County (1978) *Health systems plan component, cardiovascular surgery and cardiac catheterization services*. Health Systems Agency, Los Angeles, California.

Luce, R.D. and Raiffa, H. (1975). *Game and decisions*. Wiley, New York.

Luck, G.M., Luckman, J., Smith, B.W., and Stringer, J. (1971). *Patients, hospitals and operational research*. Tavistock Publications, London.

Luenberger, D.G. (1984). *Linear and nonlinear programming*, 2nd edn. Addison-Wesley, Reading, Massachusetts.

Lui, K.J., Darow, W.W., and Rutherford, G.W., III (1988). A model-based estimate of the mean incubation period for AIDS in homosexual men. *Science* **240**, 1333.

Lusted, L.B. (1968). *Introduction to medical decision making*. Thomas, Springfield, Illinois.

McDonald, A.G., Cuddeford, G.C., and Beale, E.M.L. (1974). Balance of care: some mathematical models of the National Health Service. *British Medical Bulletin* **30**, 262.

McLaughlin, C.P. (1970). Health operations research and systems analysis literature. In *Systems and medical care* (ed. A. Sheldon, F. Baker, and C. P. McLaughlin), p. 27. MIT Press, Cambridge, Massachusetts.

McNeil, B.J. and Pauker, S.G. (1984). Decision analysis for public health; principles and illustrations. *American Review of Public Health* **5**, 135.

Maier-Rothe, C. and Wolfe, H.B. (1973). Cyclical scheduling and allocation of nursing staffing. *Socio-Economic Planning Science* **7**, 471.

Manton, K.G., Tolley, H.D., and Poss, S.S. (1976). Life table techniques for multiple cause mortality. *Demography* **13**, 541.

May, R.M. and Anderson R.M. (1987). Transmission dynamics of HIV infection. *Nature* (*London*) **328**, 719.

Meade, J. (1974). A mathematical model for deriving hospital service areas. *International Journal of Health Services* **4**, 353.

Meredith, J. (1971). A Markovian analysis of a geriatric ward. *Management Science* **19**, 604.

Moder, J.J. and Phillips, C.R. (1970). *Project management with CPM and PERT* (2nd edn). Van Nostrand, New York.

Morse, P.M. and Kimball, G.E. (1951). *Methods of operations research*, 1st edn., revised. MIT Press, Cambridge, Massachusetts.

Myers, J.E., Johnson, R.E., and Egan, D.M. (1972). A computer simulation of outpatient pharmacy operations. *Inquiry* **9**, 40.

Navarro, V. (1969). Planning personal health services: a Markovian model. *Medical Care* **7**, 242.

Naylor, T.H. (1982). Decision support systems or whatever happened to M.I.S.? *Interfaces* **12**, 92.

Newell, D.J. (1954). Provision of emergency beds in hospitals. *British Journal of Preventive Social Medicine* **8**, 77.

Newell, G.F. (1971). *Applications of queueing theory*. Chapman and Hall, London.

Nimmo, A.W. (1983). A model of medical record storage. *Journal of the Operational Research Society* **34**(5), 391.

Nuffield Provincial Hospitals Trust (1955). *Studies in the functions and design of hospitals*. Oxford University Press, London.

Nuffield Provincial Hospitals Trust (1962). *Towards a clearer view: the organization of diagnostic X-ray departments*. Oxford University Press, London.

Nuffield Provincial Hospitals Trust (1965). *Waiting in out-patient departments*. Oxford University Press, London.

O'Kane, P.C. (1981). Hospital Studies. In *Operational research applied to health services* (ed. D. Boldy), p. 159. St. Martin's Press, New York.

O'Keefe, R.M. (1985). Investigating outpatient departments: implementable policies and qualitative approaches. *Journal of the Operational Research Society* **36**(8), 705.

Ortiz, J. and Parker, R. (1971). A birth–life–death model for planning and evaluating of health service programmes. *Health Services Research* **6**, 120.

Parker, B.R. and Srinivasan, V. (1976). A consumer preference approach to the planning of rural primary health-care facilities. *Operations Research* **24**, 991.

Parzen, E. (1960). *Modern probability theory and its application*. Wiley, New York.

Pauker, S.G. and Kassirer, J.P. (1981). Clinical decision analysis by personal computer. *Archives of Internal Medicine* **141**, 1831.

Pliskin, J.S. and Tell, E.J. (1981). Using a dialysis need-project model for health planning in Massachusetts. *Interfaces* **11**(6), 84.

Pliskin, J.S. and Tell, E.J. (1986). Health planning in Massachusetts: Revisited after four years. *Interfaces* **16**(2), 72.

Prastacos, G.P. (1984). Blood inventory management: An overview of theory and practice. *Management Science* **30**(7), 777.

Pritsker, A.A.B. (1986). *Introduction to simulation and SLAM II*, (3rd edn). Wiley, New York.

Raiffa, H. (1968). *Decision analysis*. Addison-Wesley, Reading, Massachusetts.

Raitt, R. (1981). Ambulance service planning. In *Operational research applied to health services* (ed. D. Boldy), p. 239. St. Martin's Press, New York.

Rath, G.L., Balbas, J.M.A., Ikeda, T., and Kennedy, G.O. (1970). Simulation of a haematology department. *Health Services Research* **5**, 25.

Reid, R.A., Ruffing, K.L., and Smith, H.J. (1986). Managing medical supply logistics among health workers in Ecuador. *Social Science and Medicine* **22**, 9.

ReVelle, C.S., Feldmann, F., and Lynn, W. (1969). An optimization model of tuberculosis epidemiology. *Management Science* **16**, B190.

ReVelle, C.S., Bigman, D., Schilling, D., Cohon, J., and Church, R. (1977). Facility location: a review of context free and EMS models. *Health Services Research* **12**, 129.

Roberts, N. and Cretin, S. (1980). The changing face of congenital heart disease. *Medical Care* **18**, 930.

Rockart, J.F. and Hoffman, P.B. (1969). Physician and patient behaviour under different scheduling systems in a hospital outpatient department. *Medical Care* **7**, 463.

Rockwell, T.H., Barnum, R.A., and Giffin, W.C. (1962). Inventory analysis applied to hospital whole blood supply and demand. *Journal of Industrial Engineering* **13**, 109.

Rosenbloom, E.S. and Goertzen, N.F. (1987). Cyclic nurse scheduling. *European Journal of Operational Research* **31**, 19.

Rosenthal, G.D. (1964). *The demand for general hospital facilities*. Hospital Monograph Series No. 14, American Hospital Association, Chicago, Illinois.

Ross, S. (1970). *Applied probability models with optimization applications*. Holden-Day, San Francisco, California.

Ross, S. (1972). *Introduction to probability models*. Academic Press, New York.

Rubenstein, L.S. (1976). *Computerized hospital inpatient administrations scheduling system—a model*. No. 76–9010, University Microfilms International, Ann Arbor, Michigan.

Satir, A. and Cengiz, D. (1987). Medical inventory control in a University Health Centre. *Journal of the Operational Research Society* **39**(5), 387.

Savas, E.S. (1969). Simulation and cost-effectiveness analysis of New York's emergency ambulance service. *Management Science* **15**, B608.

Shachtman, R.H. (1980). Decision analysis assessment of a national medical study. *Operations Research* **28**, 44.

Shigan, E.N., Hughes, D.J., and Kitsul, P.J. (1979). *Health care systems modeling at IIASA: a status report*, SR-79-4. International Institute for Applied systems Analysis, Laxenburg, Austria.

Shonick, W. and Jackson, J.R. (1973). An improved stochastic model for occupancy-related random variables in general-acute hospitals. *Operations Research* **21**, 952.

Shuman, L.J., Hardwick, P., and Huber, G.A. (1973). Location of ambulatory care centres in a metropolitan area. *Health Services Research* **8**, 121.

Shuman, L.J., Speas, R.D., and Young, J.P. (1975). *Operations research in health care: a critical analysis*. Johns Hopkins University Press, Baltimore, Maryland.

Smallwood, R.D., Sondik, E.J., and Offensend, F.L. (1971). Towards an integrated methodology for the analysis of health-care systems. *Operations Research* **19**, 1300.

Smith, A.G., Gregory, K., and Maguire, J.D. (1975). Operational research for the hospital supply service. *Operational Research* **2**, 375.

Smith, W.G. and Solomon, M.B. Jr (1966). A simulation of hospital admission policy. *Communications A.C.M.* **9**, 362.

Stimson, D.H. (1969). Utility measurement in public health decision making. *Management Science* **16**, B17.

Stimson, D.H. and Stimson, R.H. (1972). *Operations research in hospitals: diagnosis and prognosis*. Hospital Research and Educational Trust, Chicago, Illinois.

Tarlov, A.R. (Chairman) (1980a). *Summary report*, Vol. 1, *Graduate Medical Education National Advisory Committee*. US Government Printing Office, Washington, DC.

Tarlov, A.R. (Chairman) (1980b). *Modeling, research and data technical panel*, Vol. 2. *Graduate Medical Education National Advisory Committee*. US Government Printing Office, Washington, DC.

Taylor, W.F. (1958). Some Monte Carlo methods applied to an epidemic of acute respiratory disease, *Human Biology* **30**, 185.

Thompson, J.D., Fetter, R.B., McIntosh, C.S., and Pelletier, R.J. (1963). Use of computer simulation techniques in predicting requirements for maternity facilities. *Hospitals* **37**, 132.

Tingley, K.M. and Liebman, J.S. (1984). A goal programming example in public health resource allocation. *Management Science* **30**(3), 279.

Trinkl, F.H. (1974). A stochastic analysis of programmes for the mentally retarded. *Operations Research* **22**, 1175.

Trivedi, V.M. and Warner, D.M. (1976). A branch and bound algorithm for optimal allocation of float nurses. *Management Science* **22**, 972.

Trivedi, V., Moscovice, I., Bass, R., and Brooks, J. (1987). A semi-Markov model for primary health care manpower supply prediction. *Management Science* **32**(2), 149.

Tversky, A. and Kahneman, D. (1974). Judgement under uncertainty: heuristics and biases. *Science* **185**, 1124.

Tversky, A. and Kahneman, D. (1981). The framing of decisions and the psychology of choice. *Science* **211**, 453.

United Hospital Fund of New York (1967). *System analysis and the design of outpatient department appointment and information systems*. The Fund, Training Research and Special Studies Division, New York.

Valinsky, D. (1975). Simulation. In *Operations research in health care: a critical analysis* (ed. L. Shuman, R. Speas, and J. Young), p. 114. Johns Hopkins University Press, Baltimore, Maryland.

Vassilacopoulos, G. (1985). A simulation model for bed allocation to hospital inpatient departments. *Simulation* **45**(5), 233.

Veinott, A.F., Jr. (1966). The status of mathematical inventory theory. *Management Science* **12**, 745.

Walker, W.E., Chaikan, J.M., and Ignall, E.J. (1979). *Fire department deployment analysis*. North-Holland, New York.

Walmsley, G.L., Wilson, D.H., Gunn, A.A., Jenkins, D., Horrocks, J.C., and De Dombal, F.T. (1977). Computer aided diagnosis of lower abdominal pain in women. *British Journal of Surgery* **64**, 538.

Wagner, H.M. (1975). *Principles of operations research* (2nd edn). Prentice-Hall, Englewood Cliffs, New Jersey.

Walter, S.D. (1973). A comparison of appointment schedules in a hospital radiology department. *British Journal of Preventive Social Medicine* **27**, 160.

Warner, D.M. (1976). Scheduling nursing personnel according to nursing preference: a mathematical programming approach. *Operations Research* **24**, 842.

Warner, D.M., Holloway, D.C., and Grazier, K.L. (1984). *Decision making and control for health administration*, (2nd edn). Health Administration Press, Ann Arbor, Michigan.

Webb, M., Stevens, G., and Bramson, C. (1977). An approach to the control of bed occupancy in a general hospital. *Operational Research* **28**, 391.

Weinstein, M.C. and Stason, W.B. (1976). *Hypertension: a policy perspective*. Harvard University Press, Cambridge, Massachusetts.

Weinstein, M.C., Fineberg, H.V., Elstein, A.S., *et al.* (1980). *Clinical decision analysis*. Saunders, Philadelphia, Pennsylvania.

Weiss, E.N., Cohen, M.A., and Hershey, J.C. (1982). An iterative estimation and validation procedure for specification of semi-Markov models with applications to hospital patient flow. *Operations Research* **30**, 1082.

Welch, J.D. (1964). Appointment systems in hospital outpatient department. *Operational Research* **15**, 224.

Welch, J.D. and Bailey, N.T.J. (1952). Appointment systems in hospital outpatient departments. *Lancet* **i**, 1105.

Whipple, D. (1973). A voucher plan for financing health care delivery. *Socio-Economic Planning Science* **7**, 681.

Whitston, C.W. (1965). An analysis of the problems of scheduling surgery. *Hospital Management* **99**, 58.

Willemain, T.R. (1974). Approximate analysis of a hierarchical queueing network. *Operations Research* **22**, 522.

Willemain, T.R. (1980). A comparison of patient-centered and case-mix reimbursement for nursing home care. *Health Services* **15**, 365.

Willemain, T.R. and Larson, R.C. (ed.) (1977). *Emergency medical systems analysis*. Lexington Books, Lexington, Massachusetts.

Williams, A. (1974). 'Need' as a demand concept (with special reference to health). In *Economic policies and social goals* (ed. A. Culyer), p. 60. Martin Robertson, London.

Willis, K., du Boulay, G.H., and Teather, D. (1981). Initial findings in the computer-aided diagnosis of cerebral tumours using CT scan results. *British Journal of Radiology* **54**, 948.

Wirick, G.C. (1966). A multiple equation model of demand for health care. *Health Services* **1**, 301.

Wolf, H.K., Gregor, R.D., and Chandler, B.M. (1977). Use of computers in clinical electrocardiography: an evaluation. *Canadian Medical Association Journal* **117**, 877.

Worthington, D.J. (1987). Queueing models for hospital waiting lists. *Journal of the Operational Research Society* **38**(5), 413.

Wright, D.J., Bandurka, A., Amonoo-Lartson, R., and Lovel, H.J. (1984). Forecasting transportation requirements for district primary health care in Ghana: a simulation study. *European Journal of Operational Research* **15**, 302.

22a

Management science and planning studies

PAUL R. TORRENS

History and background

It is only in recent years that management and planning have become well recognized as an important part of public health practice. For many generations previously the more active attention was paid to the epidemiology of communicable diseases and to the control of various health hazards related to sanitation. In more recent times, the focus of attention has shifted to the social policy issue of assuring equal access for all the public to the benefits of the marvellous scientific advances of the twentieth century.

By contrast relatively less attention was paid to management science and planning techniques which would allow these scientific discoveries to be provided in a more effective or more efficient fashion. Until rather recently, professional public health work was felt to be mainly medical and epidemiological in nature and comparatively little attention was paid to the newly emergent fields of management and planning. Indeed, in many cases there was a strong feeling that these new disciplines were appropriate only to large corporations and commercial enterprises, not to a human service endeavour such as public health.

In recent years, this has changed very markedly and there has been a great surge of interest in management and planning as legitimate aspects of public health in their own right. A special commission on higher education for public health of the Milbank Memorial Fund reported in 1976 that there seemed to be three basic knowledge areas that were central to the purposes of public health: (1) the measurement and analytical sciences of epidemiology and biostatistics; (2) social policy and the history and philosophy of public health; (3) principles and practice of management and organization of public health (Milbank Memorial Fund Commission 1976).

Similar surveys or studies of various aspects of public health programmes and personnel around the world, sponsored variously by the Kellogg Foundation in the United States, the King's Fund in England, and the World Health Organization in Europe, have all stressed the great importance of improved management practices throughout all areas of health care (Kellogg Foundation Commission of Edu-

cation for Health Administration 1975; King's Fund 1977; World Health Organization 1976). John Evans (then of the World Bank) writing in 1981 in a special report for the Rockefeller Foundation on *Measurement and Management in Medicine and Health Services* suggests that organized health programmes around the world will not reach their fullest potential until they have the benefit of better trained and more efficient managers (Evans 1981).

Writing in 1986, Foege and Henderson pointed out 'Our problem is not a paucity of ideas, techniques, or effective prevention and treatment for improving health. Rather, given the embarrassment of riches in terms of things that can be done, the question is one of appropriate stewardship of scarce national and international resources' (Foege and Henderson 1986). In 1985, the annual conference of the National Council for International Health focused primarily on management issues in health programmes in the developing world and came to the general conclusion that the management and planning challenges for international public health were now clearly as important as the development of new health care techniques and procedures (National Council for International Health 1985). The Committee for the Study of the Future of Public Health of the US Institute of Medicine, writing in its landmark report, *The Future of Public Health*, identified a lack of managerial capacity as one of the serious deficits in the capacity of public health agencies to conduct effective programmes. It recommended that great emphasis should be placed on managerial and leadership skills, both in public health training and in practice, so that public health agencies might be maximally effective (Institute of Medicine 1988).

This increased awareness of the importance of management and planning in public health practice has been quickly noticed by schools of public health and other institutions responsible for the training of public health practitioners around the world. Training programmes in these schools have greatly expanded their course offerings in management and administration, and in many cases these schools have linked their training efforts with those of similar programmes in graduate schools of management and departments of economics, political science, and public administration. Indeed, in some of the leading graduate schools of public health in various parts of the world, joint programmes of training have been developed that allow students to obtain graduate training and academic degrees in both public health and parallel management disciplines.

This vigorous development of new academic directions in management and planning has been accompanied, as one might expect, by the publication of numerous textbooks on management issues in health services and public health. These texts range from the more general texts dealing with management principles (Kovner and Neuhauser 1987; Lieber 1984; Rakich *et al.* 1987; Stevens 1985) to textbooks dealing with such specialized topics as strategic planning (Flexner *et al.* 1981; Fournet 1982; Pegels and Rogers 1987), inter-personal communications (Gazda 1982), management decisions (Hardy and McWhorter 1988), personnel management (Bruce 1988; Metzer 1988; Tannenbaum *et al.* 1985), marketing (Cooper 1985; MacStravic 1986*a*, *b* 1988), and organizational development (Wieland 1981). As one might expect, there have been a large number of textbooks dealing with the financial management aspects of health care and public health (Barret and Nich 1986; Beck 1989; Berman *et al.* 1986; Cleverly 1986, 1989; Goldfield and Goldsmith 1985; Herkimer 1988, 1989; Nackel *et al.* 1987; Neuman *et al.* 1984; Suver and Neuman 1985; Ward 1987).

The field of public health practice has changed in a manner that parallels the changes in academic training. Whereas in the past, the qualifications for major positions in local, state/provincial, or national public health agencies have been the more traditional ones of biostatistics, epidemiology, environmental engineering, and maternal and child health, they are now turning more toward the fields of management and planning sciences for their primary qualifications. Potential employers of public health practitioners now are looking for skills in organizing budgetary projections and controlling finances, experience in long-range planning and strategy, and competence in organizational development and the leadership of people. The ability to recognize and appreciate the traditional purposes and objectives of public health are still important, but increasingly it is the ability to manage and plan for large-scale public health enterprises that is more highly valued.

Since there have been these major changes in attitude and approach to much of the practice of public health, it is important to identify and strengthen those management and planning disciplines that have become so central to effective functioning in the 'new' public health. What are these disciplines and areas of expertise in management and planning that are important to the modern public health practitioner? In particular, for the purposes of this chapter, what are those techniques of management and planning that need to be brought more firmly and directly into the centre of public health practice?

In this chapter we will explore those management techniques and practices which should be part of the equipment of every public health practitioner. The approach in this chapter will be first to consider those areas of knowledge, skill, or procedure that have already been identified as being important in the broader non-public health management world. Examples of their application to public health programmes will be cited wherever this has taken place, and in those cases in which little or no formal transfer has taken place, discussion will be focused on what should take place in the future. In all of the discussion, attempts will be made to show the logic and general applicability of standard management principles and practice to public health, while at the same time commenting on the unique environment (public health programmes) to which they will be applied.

In the material that follows, the discussion will focus first on the nature of managerial work, particularly within public health organizations. Then the discussion will turn to the nature of organizations, particularly health care organizations. Finally, blending together what is known and what has been discussed about managers and organizations, the third section will discuss future needs in the management science and planning area as it applies to public health.

The nature of managerial work

Writing in 1980, Longest suggests that there are three ways of looking at managerial work, whether in health care programmes or in other settings, and that the appraisal of a particular manager or managerial piece of work must include all three in some form or another. The three areas of study are: (1) the *roles* a manager plays in the course of managerial work: (2) the *functions* that a manager actually carries out in completing that work; and (3) the *skills* that a manager must employ in completing these functions.

In the sections that follow, we will review each of these three aspects of managerial work, and then show how a knowledge of these areas can be used to construct a method of evaluating managers and their work; to develop improved programmes for training and educating prospective managers; and to form the basis for the improved design of managerial positions in public health organizations. Although the discussion may seem somewhat academic in places, it is important for the reader to understand each of these three areas, in order to understand and utilize the synthesis being proposed at the end of the section.

Managerial roles

As suggested both by Longest (1980) and by Kaluzny (Kaluzny *et al.* 1982) it is possible to learn a great deal about the nature of managerial work by understanding the various roles that a manager must assume in the course of that work. They point out that, just as an actor plays a role, a manager must adopt certain patterns of behaviour when a particular managerial position is assumed, those patterns of behaviour often arising out of the necessities of the job to be done.

The classic work in this area has been done by Mintzberg,

Table 22a.1. Managerial roles

Inter-personal roles
   Figurehead
   Liaison
   Leader
Informational roles
   Monitor
   Disseminator
   Spokesperson
Decisional roles
   Entrepreneur
   Disturbance handler
   Resource allocator
   Negotiator

who reviewed the various roles of managers in general and then developed a very useful topology by which to analyse the manager's work. Through intensive case studies of a set of executives in different organizational settings, he identified ten different managerial roles and then proceeded to group them into logical clusters as summarized in Table 22a.1 (Mintzberg 1973).

Mintzberg argues that all managers have authority, either formal or informal, over the units they manage and that this authority forces them into certain *inter-personal* roles that are necessary for the successful completion of the managerial tasks. The first of these inter-personal roles is that of *figurehead* and refers to the manager's role as the symbolic head of the organization. In this role, the manager is under an obligation to perform a number of routine functions of a ceremonial, legal, or social nature, to represent the organization or unit to the outside environment whenever some official representation is required. The second inter-personal role is that of *liaison* and refers to the need for the manager to interact with peers and with other people outside the organization to exchange information and maintain contact. This is different from the *figurehead* role in the sense that *liaison* is an active, participatory, communicative role, while the role as a *figurehead* is more passive and ceremonial.

The third inter-personal role identified by Mintzberg is that of *leader* and refers to the manager's role primarily within the base organization itself. In this role, the manager is seen as the primary administrative figure in the organization, the person responsible for the direction of the organization's total activities. This is different again from the *figurehead* role, since it is active, essential, and internal.

Both reports from practitioners and from researchers in recent years have pointed out the importance of the *inter-personal* roles for the manager. Brown and McCool (1987) and Kovner (1988) have stressed the importance of the leadership role of the manager, whether in developing or developed countries. Flahault and Roemer (1986) have stressed the importance of the inspirational aspects of the leader's role in primary health care around the world, suggesting that there is a significant need for leaders who are visibly and actively committed to primary health care development. Jacobson *et*

al. (1987), writing about community health workers in Kenya, stressed the supervisory role of leaders, particularly in developing countries, both for the purpose of leading and developing future personnel. Mottaz (1988) has suggested that work satisfaction among health services personnel very often is related to their perceived degree of appreciation by their leader. Savage and Blair (1989) stress the importance of relationships in various types of negotiations in hospitals, suggesting that the manager must spend time to build adequate relationships with those various groups with whom negotiations of various kinds must take place. Robbins and Rakich (1989) surveying hospital personnel management needs in the 1990s, suggest that the inter-personal roles may well decide the manager's effectiveness in the years ahead.

The second set of roles have been designated by Mintzberg as *informational* and relate to the manager's ability to gather information, disseminate it, monitor its flow and use within the organization, and establish new sources of information as appropriate or needed.

The first informational role for the manager is that of *monitor*, and refers to the manager's ability to seek and receive current information about the organization and its functions. The manager thereby has a thorough understanding of the entire organization, hopefully more complete than anyone else with that organization.

The second informational role for the manager is that of *disseminator* and refers to the manager's ability to transmit factual and interpretational information to members of the organization. By the careful selection and dissemination of appropriate information, the manager can guide members of the organization to areas that are felt to be most productive, keep them away from areas of less interest, and encourage a greater involvement in the organization as a whole.

The third of this category of roles is that of *spokesperson*, and refers to the transmission of information to outsiders about the organization's work plans and operating status. This is obviously somewhat similar to the liaison role but it is in fact different enough to warrant a separate role designation. It is entirely possible to serve as the liaison with another organization or group, while not passing along formal information about the base group's work plans or operating status. The liaison role can be essentially neutral with regard to the communication of information, thereby making it necessary to emphasize the spokesperson role as that of the formal communicator of detailed information outside the organization or unit.

It has been noted by Austin (1988, 1989) and others that when there is a discussion of the managerial role with regard to information, that discussion very frequently centres on computers and technological assistance for information handling; because the financial investment in such equipment is so large, the financial concerns and the need to plan computer purchases carefully often obscure a much more important aspect, which is to determine first of all the organization's information needs in some detail. He points out that information technology is only as valuable as its application and that application very often will be governed by how carefully the organization has assessed its information needs.

Commenting on the spokesperson aspect of a manager's informational roles, Boscarino (1988) has stressed the important role of the manager in making the purposes and the objectives of an organization clear to outside constituencies, such as the general public or higher level officials in government; very often, organizations may be doing very appropriate or even excellent work, but their activities may be unappreciated by important outside constituencies, because the manager has not taken the time to concentrate on this important spokesperson role.

A number of writers (Hingley *et al.* 1986; Imberman 1989; Markowich and Silver 1989; Staley and Staley 1989) have stressed the importance of the manager's internal communication role as it relates to important professional groups within those organizations. Physicians, nurses, and other health professionals are particularly interested in the purposes and directions that their organizations are taking and have the ability to impede the progress and the direction of these organizations if they are not fully supportive of the purposes of those organizations. These writers point out that a manager has a very great obligation to keep these important health care professionals well informed of the directions and strategies of the organization, if they are actively and vigorously to support those strategies and directions.

These two key roles, inter-personal and informational, are really preliminary to Mintzberg's third set, which are the *decisional* roles. These roles are connected with the manager's ability to make and implement decisions within an organization and to exercise real authority within that organization.

The first of these decisional roles is that of *entrepreneur*, and refers to the manager's ability to initiate change within the organization, to develop new programmes or procedures, and to move the organization or unit into new activities or endeavours that should eventually benefit the organization and enhance its objectives. Numerous writers (Foss 1989; Kabat and Wynder 1987; Lefebvre *et al.* 1988; Makuc *et al.* 1989; Shaw *et al.* 1988) have pointed out how important this role is for public health managers and leaders. Although the term 'entrepreneur' is often used to describe managers in commercial businesses, these writers point out that it is an important role for public health leaders to develop new types of service, to explore continuously ways of making these services available and desirable to the public, and to encourage the public to use these new services actively and aggressively.

The second decisional role is that of *disturbance handler* and refers to the manager's ability (and, indeed, responsibility) to take corrective action when the organization faces unanticipated problems, either externally or internally. One of the manager's key roles is to identify disruptive patterns of activity or behaviour and, more appropriately, to prevent these disruptive patterns from damaging the fabric and functioning of the organization. Cliff (1987), Hartfield (1980), Sheard (1980), and Wright (1984) have all talked about the importance of the manager's role in helping individuals and groups within organizations to deal with change, both in the organization itself and in the external environment.

The third decisional role, according to Mintzberg, is that of *resource allocator*, and refers to the manager's ability to direct the organization's resources of money, personnel, space, and equipment towards those purposes and people that are deemed most important to the organization's objectives. This is generally considered to be the source of a manager's real internal power, although many others (such as the control of information) are felt by some to be equally important. Many authors (Ferguson and Lapsley 1988; Helmi and Burton 1988; Pollitt *et al.* 1988; Smith and Taylor 1984; Young and Saltman 1983) have written about various aspects of the resource allocation role of the manager from various points of view, all suggesting that managers in public health must not only have basic skills in financial management but very often considerably advanced skills as well.

The fourth and final decisional role is that of *negotiator*, which refers to the manager's responsibility for representing the organization at major negotiations with other organizations or groups. Again, this is a different role from that of liaison, since this is a more active posture of putting forward the organization's needs and desires, and seeking to obtain the best deal possible for that organization. It is also a different role from that of spokesperson, which is more of a passive, information-handling posture and not necessarily one of actively engaging in the bargaining and manoeuvering which characterize the negotiator role. The negotiator role seeks to obtain the best arrangement possible for the organization; it is a role that expects a trophy or an achievement of some sort will be brought home to the organization. As Brown (1983) and others (Morgan and McCann 1983; MacStravic 1986*a*; Savage and Blair 1989; Shortell 1983; Whitehead *et al.* 1989; Wright 1984) all point out, as health systems develop, negotiations with physicians in particular become exquisitely important in determining a health care organization's ability to reach its goals.

Research has shown that not every managerial position includes the same set of roles to the same degree. Each managerial position requires a different 'mix' of managerial roles, and the same position may require different combinations of roles at different times. Kaluzny (Kaluzny *et al.* 1982) refers to a paper by Forrest and Johnson in which they reviewed the different activities of managers in several different kinds of health organizations and found that there was significant differences in the roles that the managers played and in the work they actually conducted. These findings were similar to those of Kuhl, also referred to by Kaluzny *et al.* (1982), who reviewed the work of managers in different types of health care organizations and who also found some significantly different sets of roles and functions depending upon the nature of the organization itself. At the same time, both studies reported that there were general similarities in the types of roles required by similar organizations or agencies, so that although managers of individual hospitals might vary significantly from one another, managers of hospitals in general shared many similar clusters of roles. It has further been pointed out by Kindig and Lastiri-Quiros (1989) that managerial roles also change over time and change as a manager matures in a particular position.

Understanding the roles that a manager is called upon to play in the conduct of managerial work is not simply an academic exercise; the approach also has significant merit for the selection and preparation of potential candidates for managerial positions. Kurtz (1980) and Slater (1980) have separately studied the personal characteristics of two populations of physician managers and have shown that it is possible to categorize them according to certain management-related styles and attitudes. Their work suggests that it may be possible to make a better match of physician managers (and perhaps, all managers) in the future, by identifying individual manager's strengths and weaknesses and then projecting them into the various roles that they must assume in various managerial positions. Studies by Stewart *et al.* (1980) and Schulz and Harrison (1983) demonstrated how differently individual chief officers in the health care field in the United Kingdom perceived and discharged their roles. Stewart, in particular, emphasized the substantial scope for personal choice in how managers use their time and in the styles that they adopted. They suggested that managers should take clear account of the needs of their organization at the time, as well as their own personal characteristics and the characteristics of their immediate colleagues. They should then make conscious choices about the roles that they will choose to fill.

Managerial functions

A second means of understanding and improving managerial work in public health is the analysis of the manager's *functions*. This analysis looks not so much at the roles or postures that a manager must assume while managing, but rather at what the manager actually does while filling those roles. It focuses more directly on the *practice* of management than does the study of roles; it deals with those actions and activities of the manager that are generally recognized as making up the manager's daily work. As with all such classifications, it suggests a tidier and more rigid situation than exists in life. The analysis is nevertheless helpful for the range of facets of management that it describes.

A variety of categorizations of managerial functions have been developed by various writers, but in general the most useful method for looking at managerial functions describes them either as: (1) future-orientated, planning, projective types of activities; (2) present-orientated, maintenance, daily operational types of activities. Just as was the case in the review of managerial roles, a better understanding of these individual managerial functions and of the clusters of managerial functions that make up a particular management job will allow for a better understanding of managerial performance.

The *future-orientated* management functions are those

Table 22a.2. Managerial functions

Future-oriented management functions
  Planning and organizational goal-setting
  Organizational design and development
  Leadership and motivation
  Management of change

Present-oriented management functions
  Organizing
  Decision-making
  Directing
  Controlling
  Conflict resolution

functions focused more on where the organization is going (or at least where it should go in the future) than on maintaining its present status. These *future-orientated* functions are: planning and organizational goal-setting; organizational design and development; leadership and motivation; management of change.

The most important of these future-orientated functions and the one around which all others must revolve is *planning and organizational goal-setting*. This function involves considering where the organization should be going and how it should get there. It requires a thorough review of present and future resources and how they might best be organized to continue the advance of the organization. It includes the development of a strategy for moving in that direction and the role each part of the organization is to play in doing so (Luke and Begun 1987; Seidel *et al.* 1989; Smith 1988). It also certainly includes development of long-range financial forecasts and is usually linked together with the budgeting process by which financial resources are allocated within the organization.

Indeed, in more recent years, the focus of organizational planning and goal-setting has been increasingly set by the financial constraints under which the organization exists. In past years, the situation was just the reverse, with organizations very frequently determining what it was that they would like to do and how they would like to do it, and then secondarily checking to see whether there was the budget to carry out those activities. In more recent years, with the limitation of finances which has become common in all health programmes around the world, organizational planning and goal-setting generally begins with a review of financial matters and moves on from there. Unfortunately, in many settings, the existence of these tighter financial limits have often tended to dampen the ingenuity and the creative energies of managers in public health and has somewhat depressed their vision about what might be possible. It is a delicate balance in organization planning and goal-setting to live within budgetary constraints, without at the same time having plans and aspirations for new programmes and expansion of old ones in the future.

Often, also, the environment around an organization may change and what have previously been acceptable plans or proposals for an organization may now need major changes because of changes in the environment. An excellent example of the major changes in organization planning and goal-setting is seen in the recent decisions by the government in the United Kingdom to change the organization of the National Health Service by introducing new marketplace reforms into what had previously been a non-market, non-competitive service (Her Majesty's Stationery Office 1989).

The second future-orientated management function is organizational design and development. This function naturally follows the development of plans for the organization's future and involves the creation of an organizational structure and identification of its shortcomings, together with the initiation of steps to create change in all or part of the present structure. Although there have been a number of prominent researchers who have studied organizational behaviour and design in other types of settings, relatively little has been done in health care organizations until recently. Although not always grounded in formal organization theory, the literature in health care recently has been filled with description and discussions of new organizational forms and of the issues that are raised by the attempts to develop new organizational activities and units. Writers such as Clement (1988), Glandon and Morrissey (1986), Kearns and Hogg (1988); McClure (1985), Munson and D'Aunno (1989), and Nackel and Kues (1986) all talked about organizational redesign in the framework of the development of new services in health programmes. The development of new units within hospitals in the United States in response to some of the newer entrepreneurial pressures in the United States has been a very marked example of organizational redesign in response to external pressures, as have the writings of authors like Starkweather (1981) in their discussions of hospital mergers. Johnson (1988) has written very cogently on the impact of increasing health care competition on the design and function of American hospitals.

The third future-orientated management function is that of *leadership and motivation of personnel* within an organization, although this function might also be described as a more present-orientated maintenance function. In this aspect of the manager's activities, efforts are directed towards inspiring and motivating people to give their best efforts in the future, to commit themselves fully to the goals and the purposes of the organization, and to co-operate with the plans and objectives that have been laid down for the future. Unfortunately, this is not an area where there has been a great deal of extensive research studies conducted from a managerial point of view, dealing with the question of motivation and leadership of personnel, even though this is a very significant area for managers to consider for the future. For example, even though many countries are experiencing shortages in nursing personnel and even though there have been numerous reports of the stresses and pressures on nurses (Morgan and McCann 1983; Mottaz 1988; Sheard 1980; Stamps and Piedmonte 1986), there has been relatively little response to the issues raised by nurses in the managerial literature.

A fourth and final future-orientated management function is that of the *management of change*. In many ways this function is clearly part of organizational design and also of leadership and motivation, but it has become such an important function that it is worth singling out as separate. One of the major characteristics of health care and public health at the present time is the continuous stream of major changes that are taking place: new technology is being developed; new methods of financing or new regulatory restraints are being put into place; new pressures from the public are building up for different types of services. The manager plays a key role in preparing the organization and the workers within it for change so that they are both encouraged and enthused by its prospect.

The *present-orientated management functions* are those that focus more on the maintenance of the organization's ability to function in the short run and less on the direction in which the organization should be heading over the longer term. In many ways it is the more mundane, less exciting part of management but an absolutely critical part none the less.

The *organizing function* of the manager depends upon the fact that the organization's goals and directions have already been set; now it is up to the manager to bring together the personnel and the resources in such a way as to accomplish the objectives laid out by the organization's plan. Within the organizing function, the manager takes on two separate tasks: first, deciding what type of organizational structure is best to get the work done; second, gathering the people and resources together into that organizational format and assigning the specific tasks, delegating the specific responsibility and authority to the individual workers or to specific work groups—all these are included in the manager's organizing function.

The decision-making function of the manager is obviously both a long-range, future-direction function and a short-term present-orientated function. On a daily basis, there are a myriad of individual details that require decisions. For a manager these include the establishment of a system for determining which decision should be brought forward for the manager's attention and which should be handled by subordinates. Very often these decisions can be assisted by decision support systems and computer support of various kinds (Turban 1982; Taylor 1984), but often they happen more by accident than by intent.

The *directing function* of the manager involves the conveying of instructions to members of the organization concerning how work is to be carried out, how things are to be done. In this function, the manager must express his or her will about the style and substance of the work being carried out, making it clear what the instructions are and what is expected of those doing the work. Directing is making explicit how the manager wishes the job to be done; it is the communication of details about the work, together with a statement of authority of the manager's position.

The *controlling function* involves the review of what is being done within the organization and making the appropriate changes in light of what is learned. It involves the reception of information on a regular basis about the organization's operation, the evaluation of that information in comparison to pre-set goals and objectives and the issuance of corrective directions wherever appropriate. In the area of finance, for example, it involves the continuous review of financial reports, the comparison of these reports with pre-set objectives or indicators, and the initiation of corrective actions with regard to expenditures, prices, inventories, and the like. *Controlling* is perhaps the area of expression of the manager's authority that most directly impacts on people within organizations; for that reason is also very closely tied up with the manager's personality and the overall personnel atmosphere within an organization. Since it is often dependent upon information that the manager gets from information systems and computers (Worthley 1982), it can sometimes become a rather complicated and somewhat technological task.

Perhaps one of the most important functions of the manager is that of *conflict resolution*; between individuals, between subunits of the organization, or between the manager's organization and other similar organizations in the environment. Although it is usually felt that the planning, organizing, and controlling functions of a manager are most important on a long-term basis, it is usually the conflict resolution function that is the most visible in the short term; it is often the function that requires the most time and drains the most energy from the manager (Cliff 1987; Wenzel 1986). In this function, the manager is called upon to bring together opposing parties, to reconcile opposing viewpoints, and to develop consensus between opposing values (Kellerman and Hackman 1988; Morgan and McCann 1983; Sheard 1980; Wright 1984). It is the most 'people' connected function of the manager and it is the one that most usually takes the manager out of the office and into the lives of the workers in the organization.

Managerial skills

The third method suggested by Longest for looking at managerial work is to appraise managerial skills. He suggests a typology created by Kast and Rosenzweig (1974) that identifies three types of skills an effective manager must use.

Technical skills are the abilities to use the methods, processes, and techniques of a particular field in managerial work; in the case of public health work, for example, these might include the traditional areas of biostatistics, epidemiology, environmental health, maternal and child health, and the like. It might include some of the technical areas related to management, such as financial management and accounting, quantitative methods for planning or decision-making, specialized systems for conflict resolution, or information system design and control.

Kast's second area of managerial skills is described as *human skills*. These involve the ability to get along with other people, to understand them, and to motivate them towards the organizational objective (or at least the organizational objective as the manager sees it). This collection of skills is utilized by the manager not so much in the abstract aspects of management that involve technical or conceptual abilities, but in translating these aspects into terms and conditions and attitudes that the people in the organization will understand and respond to.

The third area of managerial skills described by Katz is *conceptual skills*. This is the mental ability to visualize the complex interrelationships that exist in the workplace—that is, among people, among departments or units of an organization, and even between a single organization and the environment in which it exists. Conceptual skills permit the manager to understand how the various factors in a particular situation fit together and impact on each other; it also allows the manager to know what to do about them.

Just as with the managerial roles and functions, there is an appropriate but differing 'mix' of managerial skills that is necessary for each administrative position in public health. For example, in one community at one time, the medical officer of health or the local health commissioner may be faced with a problem that is really more *technical* in nature than anything else; what type of polio vaccine to use in a programme for infants and pre-school children; what type of tuberculosis screening mechanism; what type of family planning materials or programmes. The main thrust of the managerial challenge may be a technical one, with the public health administrator functioning more as a technical expert than anything else.

In another instance, the managerial challenge may be the need to provide fluoridation for the local water supply in order to reduce the incidence of dental caries. In this case, the local public health officer may not need technical skills at all. The technical scientific work in this area has already been done and is well discussed in the literature. In this jurisdiction at this time, the need may be more for *human skills*, skills that will bring together the divergent opinions and groups in the community and that will allow a major public health benefit to be successfully implemented in that community.

Finally, in another circumstance the medical officer of health or the local public health commissioner may be called upon to use *conceptual skills* that are different from either of the other two types of skills. For example, the health officer may understand that the community needs to develop a comprehensive programme of cancer prevention, early diagnosis, early treatment, and (if necessary) comprehensive care up to and including hospice care. There will be a number of complex interrelationships that will need to be worked out, an overall system designed, and perhaps an organizational structure created either to oversee the effort or actually to carry it out. Here the manager will need to use *conceptual skills* in identifying the problem, understanding what needs to be done, and organizing a system that will accomplish the necessary objective.

Summary

More effective management of health care programmes and institutions depends upon more effective health care managers (Antle and Reid 1988; Kelliher 1985; McFarlane 1987; Rindler 1987). A better understanding of managerial roles, functions, and skills is a first step toward the improvement of the work of health care managers, as it makes possible better evaluation of managers and their work, better programmes for training and educating prospective managers, and improved design of managerial positions in public health organizations.

For example, a better understanding of managerial roles, functions, and skills will allow for a better design of managerial positions in public health organizations. A better knowledge of the contents of management work might force organizations to analyse their management needs much more carefully and draw up their management tables of organization much more precisely; it might allow for more specific assignment of responsibility and authority, as well as the clearer definition of each individual manager's part in the organization *vis-à-vis* all others. A better understanding of the nature of managerial work might force public health organizations to look at their entire structure and to design positions much more specifically. At the same time this would allow for the creation of a much more specific set of expectations for the individuals filling these positions, since they would know in much more detailed terms what the actual managerial content of those positions is supposed to be.

This in turn would allow one to design a much more exact form of managerial evaluation than exists at the present time, because it would make much more specific and pertinent what is expected of the manager in each position. If it is known that a particular position calls for a manager whose main roles are mainly that of liaison and also calls upon that person to act as figurehead and spokesperson for the organization, the incumbent can be evaluated on how those roles were actually carried out. Rather than evaluating the manager on some more general and indefinite focus of activity, the evaluation can be carried out with regard to specific management roles and functions. In a similar fashion, this much more specific definition of the roles and the functions of each managerial position would allow for a much more focused and much more relevant recruiting process for personnel to fill these management roles. Now, if the management roles, functions, and skills requirements for each managerial job are better known, recruitment can focus much more on searching for people with excellent attributes in the specific areas that are required by the job. This has always been done generally and somewhat indirectly in the past, as people search for recruits who generally fill a set of vaguely understood criteria, but now with more specific management role, function, and skill definition the recruitment can be much more exact and precise. Finally, if the mix of managerial roles, functions, and skills that are necessary for maximum success in public health management positions become better known, training programme for these types of positions can be much more carefully and exactly designed and conducted. An educational curriculum or training programme can be developed to cover the more important areas, once they are known and the linkage between educational preparation and professional practice can be much more pertinent and exact than it is at the present time.

While the entire discussion of managerial roles, functions, and skills may seem at first glance to be relatively academic and with little direct relevance to the work of the public health manager, a thoughtful consideration of these important areas of management practice can lead to a much better understanding of the nature of the public health manager's work and to a much higher degree of effectiveness in that work.

The nature of organizations

A second method of understanding the administration and planning of health service programmes is to look not so much at the manager but at the organization and the environment in which management takes place. In this approach, interest is focused more on organizational issues than managerial issues, and a rather different set of research and analytical skills and approaches are called into play to understand the issues better.

It is a rather striking irony that, although most people around the world spend considerable portions of their life either working in organizations or being affected by the work of organizations, very little actual time and energy is spent in understanding those organizations and the nature of organizations themselves. It is almost as if a fish were swimming in the sea, taking the presence of the sea for granted, and not spending any time at all understanding the impact of that sea on the fish's life.

For a potential manager of public health organizations to ignore the nature of these organizations would be disastrous. An individual manager may be the most skilled technician or may have the most detailed knowledge of the manager's roles, functions, and skills, but if there is a lack of understanding about the nature of organizations in which these roles, functions, and skills must be carried out, the chances of that manager's being maximally effective are dubious. It is exquisitely important that each manager understand that every organization is different, that a particular organization itself may differ from one period of time to another, and that the manager's function is markedly dependent upon organizational factors that may or may not be readily apparent. For a manager to manage effectively, a study of organizations and their nature is essential.

Basic organization theory

Towards the end of the last century and into the beginning of this one, workers in various parts of the world began to develop a body of general knowledge of organization theory that has served as the foundation for most modern approaches to organizations. In recent years the more classical ideas of the early years have been challenged and modified and considerably, so that present organization theory is a mixture of the older, more classical ideas that have now been tempered and adapted in light of newer findings. The major developer of classic organization theory was Max Weber, a German sociologist who developed the idea of the bureaucratic model of organizations; this model and Weber's original ideas have remained as a set of basic principles for anyone interested in understanding basic organization theory.

In Weber's ideal model of organizations, several principles predominated:

(1) *specialization and division of labour* (that is, tasks are broken down into specific pieces of work and the various specific jobs are assigned clearly and definitely to individual workers or work groups);

(2) *hierarchical organization and chain of command* (organizations are structured in a hierarchy, with each succeeding higher level having authority and control of the units beneath it on the table of organization);

(3) *consistent set of organizational rules* (the organization has a set of explicit statements that govern conduct of life within the organization and without which the organization could not function effectively);

(4) *managerial impersonality* (the manager should assume a somewhat impersonal attitude towards those working under his or her direction, so that personal attributes of either the manager or the managed do not interfere with the task-orientation of the organization);

(5) *assumption of authority on the basis of competence* (that is, managerial authority and the management positions in which that authority resides should be assigned on the basis of competence in the work; in the same fashion, tenure in a position should be dependent upon effectiveness in meeting the objectives of the organization).

Inherent in this bureaucratic, traditional model of organizations were certain principles or ideas that have become widely accepted and applied throughout organizations all over the world. The division of labour and the increasing specialization of tasks is one that is universally accepted, as is the need for co-ordinating control to bring together all these various specialized tasks. The principle of unity of command is also generally accepted; this operates on the basis that each worker in the organization has one person administratively responsible for him or her and it is clear to each worker just who that controlling person is. The distinction between 'line' and 'staff' positions (that is, the distinction between those who have the responsibility for keeping the organization operating and those who are technical experts or consultants to the operating staff) has been open to widespread discussion and challenge in recent years, but continues to be a fundamental assumption in most organizations. Limitations

on the span of control exercised by an individual manager is another principle that is inherent in many of the classic interpretations of the bureaucratic model of organizations. Finally, the delegation of management decisions and actions downward to the lowest appropriate organization level is an additional principle that has become accepted in all organizational thinking.

Objections to classical organization theory

Almost from the beginning, there have been objections to classical organization theory, generally on the grounds that it was too mechanistic in its approach, too impersonal in its application. Various workers have pointed out that organizations are not simply physical constructions that deal only in physical, tangible activities; these workers have stressed that organizations are really collections of people and it is the personal aspects of work, the individual and group interactions, attitudes, and values, that more accurately describe and define an organization. In more recent years, behavioural scientists and others have played a major part in reassessing the classical bureaucratic model and in developing alternative models of organizational analysis and behaviour (Hommans 1950; Likert 1961, 1967; Maslow 1954; McGregor 1960; Mintzberg 1979, 1983; Perow 1970; Peters and Waterman 1982).

One of the major objections of behavioural scientists to the classical theory of organization is that it ignores what is now known as the informal organization, in favour of studying only the formal organization. The formal organization is the structure as it is supposed to be, in theory or on paper or in the mind of some distance management or authority figure. The informal organization is that spontaneous interaction of people in situations that more accurately resembles how work is accomplished, how authority is really exercised and received, and how an organization actually 'lives'. The general feeling among more modern organizational thinkers is that a complete view of any organization must include both the formal structure (as outlined in the organization chart, job descriptions, formal work rules, operating policies, and the like), and the informal structure (the actual interrelationships of people, authority, and work activity).

Certainly one of the most important contributions to organizational thinking in recent years has been the application of general systems theory to organizations, since it has provided a much more useful and relevant view of organizations and a method of analysing their behaviour (Churchman 1968; Johnson *et al.* 1967; Kast and Rosenzweig 1974). In general, systems theory suggests that organizations are systems or sets of interrelated and interdependent parts that, taken together, form the complex whole. Each part of the system (that is, subsystem) has a life and a momentum and set of relationships of its own, while at the same time it is affected by and in turn affects all other parts of the complex. It suggests that systems such as organizations are living, changing, interactive complexes, and also suggests that the nature of that interaction can be delineated and perhaps even quantified. At the same time, it allows for formal and informal aspects of an organization to be combined, doing away with the necessity for an artificial distinction between these two important components of organizational life.

The importance of the open systems model of organization is apparent from a number of viewpoints. First, it more accurately displays organizations as complex interactions of many component parts and many forces and influences. Second, it allows these interactions to be more accurately described and permits a more realistic portrait of the organization to be presented. Third, it permits a more total perspective of the organization to be developed for both planning and management purposes, so that the implications of any one action in any part of the system can be traced into other related parts of the system. Fourth, it allows workers and managers in one part of the system better to understand how their actions affect the total organization and vice versa. Finally, it allows for organizational design that is more pertinent to the operating needs and realities of the organization than is possible in the usually conventional two-dimensional models.

The characteristics of health care organizations

It is virtually impossible to develop a list of characteristics of health care organizations that will apply equally well to all such organizations, since diversity among health care organizations in general and public health organizations in particular is very great. There are, however, a number of characteristics that have been identified as being somewhat common among all health care organizations and a further set of characteristics that are unique among public health agencies and organizations.

It is important for those interested in the management and planning of public health organizations to be aware of and to understand their characteristics, since they are symptomatic of forces and influences that shape these agencies and determine their existence. The characteristics themselves are perhaps not so important as what has brought them about. Knowing these characteristics allows us to understand the dynamic shaping forces better; knowing these forces better allows us to develop more appropriate and effective public health organizations in the future.

In general, public health organizations are altruistic in nature and not profit- or productivity-motivated. This means that they are seen differently from other organizations, both by those within them and those without, including researchers. Workers within these organizations frequently refuse to believe that many of the principles that apply to other organizations are relevant for them, since the purposes of their organizations are so different. In general, workers within public health organizations are intent on the delivery of a service to a public or a population that needs it, and matters such as organizational efficiency or resource constraints are not seen as legitimate reasons for altering organizational behaviour. There is a moral imperative to health

work that is seen as pre-empting many of the principles that govern the operation of other types of organizations. However, although public health organizations themselves are altruistic in purpose, they are often forced to use harsh methods to reach their altruistic aims. Sometimes these methods and approaches are directly contradictory to their normal operating mode and cause serious organizational confusion of attitude and purpose.

For example, every *governmentally-sponsored* public health organization occasionally carries out a police-like function to enforce certain ordinances and laws; in this role it must assume a somewhat formidable police-like posture. At the same time, the same organization may be trying to convince mothers to bring in their babies voluntarily for well-child examinations or try to convince people to stop smoking. On the one hand, the organization is ordering people to behave in a certain way under threat of legal action, while on the other hand it is trying to convince people to behave voluntarily in certain ways that will benefit them. The organizational confusion that can arise from this contradiction should be obvious.

Second, most public health organizations are governmentally-sponsored, or they are at least very closely connected with some social or political set of values; that is, the operations of these agencies are very much influenced and directed by a broader set of social and/or political decisions that provide the framework within which the public health agency functions. The public health agency is not free to embark on whatever course of action or whatever direction it wishes, strictly by the will of the people within the organization. The framework or the environment for its work is set outside the organization and then passed along to it.

A further subset of these social purposes is the reality that the public health agency or organization usually focuses its attention more forcibly and aggressively on certain subsets of the population being served, with the result that the agency sometimes becomes identified with the individual subsets of the population than with the population as a whole. Although public health organizations work very assiduously to improve the health of the public as a whole, they work most energetically to improve the health of the poorer and disadvantaged portions of the public. As a result, public health work is sometimes not seen as an activity that has equal relevance to all parts of the population; it is seen more as a 'poor people's programme', or one that focuses on tuberculosis or venereal disease, or some other similar subset. This means that the organization may then be able to call upon the active emotional support and involvement of a comparatively limited part of the population at large, because only a limited part of the population feels that its activities are relevant to them.

Another characteristic of public health organizations is that their operations are technically or scientifically determined; that is, the whole purpose of the organization is influenced by the technology or the scientific discoveries that have been deemed to be important for a particular population to receive. The development of a polio vaccine in a laboratory somewhere in the world meant that public health agencies all over the world had to begin to consider the addition of polio vaccination efforts to their array of products and services.

This scientific determinism has two important impacts on public health organizations. First, it may well determine the operating structure of the organization by forcing the creation of one type of programme or unit, either where none had existed before or where something rather different had been in operation. The arrival of a new birth-control method or the development of some new vaccine will frequently call forth a new organizational unit to administer that new service, occasionally drawing off staff and resources from other already-existing units in the organization. A second aspect of this technical determinism relates to the social and professional distance between those who develop the new technology or science and those who must apply it. For example, the development of the polio vaccine was the result of the efforts of a large number of scientific researchers and technical experts. The implementation of a polio vaccination effort by a public health agency depends upon the work of a large number of non-technical personnel, or at least lower-level technical personnel than those who originated the product to be delivered. The translation of very sophisticated technical ideas and products into terms and procedures that are genuinely understood by lower level workers is a major feature of public health agencies and organizations throughout the world.

Another characteristic of public health organizations is the difficulty in evaluating or appraising their work, given the long-term nature of the results. It is perhaps easy to measure the effectiveness of a public health organization in finding and removing a contamination of the drinking water supply in a particular town. It is harder to measure the impact of a public health education campaign aimed at reducing the amount of cigarette smoking in a particular population, since the effect may be stretched out over a long period of time. It is often easy to measure the effort expended (for example, the number of educational sessions on cigarette smoking), but it may not be possible to measure the results of that effort (which is obviously the more important aspect of the educational work).

A final organizational characteristic of public health organizations is their often archaic structure and their retention of organizational units that perhaps no longer have any relevance. A review of the organizational structure of many public health organizations around the world will reveal units that have existed for many years next to ones that have been created only recently; it may reveal units that were initially organized to tackle one particular disease pattern (for example, infectious disease in highly developed countries) still in existence long after other health problems (for example, alcoholism or chronic disease) have assumed a more important position but are still not represented by any formal organizational unit. Structures would most probably

Table 22a.3. Framework for appraisal of public health organizations

(A) Developmental history of the organization
(B) Environmental factors affecting the organization
(C) Formal organizational structure
(D) Organizational systems
    (1) Task completion process
    (2) Distribution of power
    (3) Planning and decision-making pattern
    (4) Resource distribution and resource allocation method
    (5) Internal management structure
    (6) Communications network
    (7) Inter-personal/inter-group/inter-organizational relationships
    (8) Evaluation and control systems
    (9) Conflict resolution systems
    (10) Organizational redesign and new products system

be found that are composed of professional personnel of one kind (for example, public health nursing) immediately next to another unit that operates a programme for services (for example, maternal and child health) and which uses a multiplicity of professional personnel. The structure of public health organizations seems to evolve by indirection rather than by direction; it seems to be more an incidental happening than a deliberate attempt to develop a more pertinent organizational form to meet the particular organizational purpose.

In the next section, we will discuss a means for analysing or reviewing public health organizations from the point of view of their organizational structure, with the hope that more pertinent and appropriate structures can be developed.

Framework for analysing organizations

It is important for managers and for others interested in health services organizations to understand the basic principles or organizations and to be able to conduct a systematic review of those organizations and their operations. It will also allow them to review the public health literature from the point of view of an organizational theorist and to interpret what would ordinarily be reports of various programme activities in a more thoughtful and analytical framework (Arnold *et al.* 1987; Binder 1983; Kaiser 1986; Kaluzny and Hurley 1987; Miller, C. *et al.* 1989). To assist managers in their analyses and to point out to researchers important aspects of health care organizations that are worth study, it is useful to have a framework for analysis of health cate organizational structures.

In general there are four parts of the analysis of any organization, health services organization or otherwise. The first three deal with the organization's history, its environment, and its formal organizational structure. The fourth and much more comprehensive portion of the organizational analysis involves an orderly review of the organization's ten major operational or functional systems. All elements of this analytical frame are given in Table 22a.3.

Development history of organizations

One of the first items to be studied in any review or analysis of an organization is its developmental history. Any organization will have gone through a series of developmental steps to bring it to its present situation. None of these steps would have been purely accidental and all of them would reflect important changes in the organization itself or the environment in which it functions. Each of these steps, however, will have an important bearing on why the organization has developed into the form and function that it has at the present time.

A careful historical review will tell a great deal about an organization's past, but more important it will also tell a great deal about its present and perhaps even its future. A review of its organizational changes over the years, the growth of the services or products it has offered, a description of the personnel involved in its leadership—all of these will provide important clues to an organization's present situation and its future opportunities. It will also identify important trends and forces that may be continuing to influence the organization at the present time and that may shape its future.

Environmental factors affecting the organization

The next area to review in investigating any organization is the environment in which it exists and the forces that presently impinge upon it. In general, the key organizational factors are social, economic, political, and physical, but any aspect of the general environment in which an organization exists should be reviewed if felt to be important.

A good place to start is to consider the economy upon which the organization is dependent. Is the economy strong and growing or is it weak and unstable? Does it hold prospects for providing increasing financial support in the future or will the organization be lucky to continue to obtain its present funding? Are there prospects that the organization can assist in the economic development of the area or will it always be a drain on economic growth?

In the same fashion, a review should be made of political, social, and other factors affecting the organization. Is it an agency of government and if so, is the governmental structure stable and supportive? What are the political imperatives for the present government? How do they impact on the public health organization? Does the organization fit in well with the social and cultural values that are currently important in the community to be served, and does it have a means of determining whether its programmes are congruent with these social and cultural values?

With regard to the physical environment, does the public health organization have an appropriate physical setting in which adequately to conduct its work? Is the physical geography and topography a factor to be considered in its work or is it a relatively neutral feature? If the organization is not appropriately located, would it take major changes to bring it into a setting in which it would work better? Are there other major physical and environmental considerations that affect

its operations, such as availability of fuel, supplies, shelter, and the like? If so, what are the implications for its structure and mode of working?

Formal organizational structure

Although the formal organization is only half the picture (the informal organization being the other half), it is important to identify clearly what the formal organizational structure is supposed to be, at least in the minds of the people responsible for managing the organization. An attempt should be made to obtain such items as organization charts, job descriptions, procedure manuals, and other written and documentary material that describe the organization.

With these in hand, it should be determined how closely the present structure actually approximates what appears in the written materials. Does the structure truly resemble the organization chart? Are the personnel listed in that chart actually still in place or have they changed positions? How often is the organization chart brought up-to-date and how much attention does anyone pay to it? Is there any attempt to review and reorganize these charts, procedure manuals, job descriptions, and other formal procedures on a regular basis? What major discrepancies are there between the formal structure and the way that things are actually done?

Organizational systems

Once these three basic background areas of developmental history, environmental forces, and formal organizational structure have been reviewed, one can look at the internal systems of the public health organization itself. These are the functioning networks, the formal or informal coming together of people and problems that *in toto* really comprise the organization.

Task process

The first organizational system to examine is the process (or processes) by which the organization's primary tasks are performed. Indeed, one of the important aspects of this part of the analysis is to identify the various tasks that the organization is actually trying to perform, since most organizations will most probably be performing many. In the usual task process analysis of a public health organization, it may be determined that the organization had ten or twenty different tasks that it is performing, some of which function separately and some of which function in an integrated fashion with others. Very often, an organization has simply added on various tasks over the years without a very clear delineation of what those tasks are or how the organization should be changed to carry them out. Too many managers and too many researchers work in organizations without carefully identifying the tasks that those organizations are called upon to complete and without carefully examining the steps by which these organizations function on a daily basis.

One of the best ways to analyse an organization's various tasks processes is actually to walk through the process by which the tasks are carried out. In a very fundamental and practical kind of way, perhaps the easiest way to understand an organization's various tasks processes is simply to participate in those processes on a step-by-step basis from beginning to end. In this way the various 'processes' are no longer vague or indefinite generalities but are very specific procedures, people, and resources at work in a specific way.

Distribution of power

One of the most important features of organizations is the distribution of power, both formal and informal. Each organization will have a formal arrangement for distributing power that is generally related to supervisory authority, ability to expend resources, power to hire or dismiss, power to approve or disapprove new projects, and power to develop new programmes or services. As much as possible, the location of these foci of power must be clearly identified and the type of power that is involved must be clearly delineated. The processes and the procedures by which power is exercised must be described as specifically as possible, so that the analysis shows not only where a particular type of power is located but how it is exercised.

In addition to the distribution of formal power, what can be learned in an organization about informal power and influence? Is there a particular person who has significant ability to speed up or to impede organizational functioning, not on the basis of a formal position of authority but rather on the basis of personal knowledge or skill? Are there structural aspects of the organization that have allowed power or influence to concentrate in one unit or person more by default than by intent? Are there forces, psychological or political, that effect the ways in which power is actually applied? If a manager is to be effective, a considerable amount of time and energy must be spent in understanding the distribution of power and how it is exercised within an organization, if that manager is really to appreciate the complexities of managing that organization.

Planning and decision-making pattern

The third important area to review is how the organization's long-range intentions are developed and implemented. Are they the results of a formal process or does a long-range direction seem to emerge by some kind of informal consensus? Is planning primarily a budgetary function or is it separate? If it is separate, at what point are the two processes joined? Is there a written guide-line that describes and delineates how planning will be carried out on a regular or routine basis within the organization, or is it something that happens on an *ad hoc* basis to deal with each situation separately as it arises?

How are decisions made within the organization and by whom? What types of decisions are made on which levels? Do individuals within the organization know accurately what power they have to make decisions or must they guess what is expected of them, each time risking some significant over-expression of their authority? Are decisions in the organ-

ization really made by default, by delay, or by being unwilling to make active, positive decisions?

Resource distribution and resource allocation methods

A fourth area to analyse is the distribution of resources and the method by which resources are allocated. Although the formal organization chart will portray the supposed relation of one organizational unit to another, it does nothing to describe the allocation of resources among all the units. It is therefore useful to obtain the financial reports and personnel rosters for the organization under study and to attempt to re-create the pattern of resource distribution throughout the organization. If it is possible to break down these reports by service programme or product line (for example, health education efforts for polio vaccine campaigns), it is also helpful since it shows whether or not an organization's priority statements are reflected in the way its resources are allocated.

In parallel with the mapping of planning and decision-making, it is also of value to learn how resources of various kinds are actually allocated through the organization. Is there a formal budgetary process by which funds, personnel, and equipment are assigned, or is the process more informal? Does it take place once a year or throughout the year? Is there a single person involved in these decisions or many people? Are financial resources handled one way, personnel another, and space/equipment/supplies a third? What policies appear to govern current decisions: for example, do resource allocations suggest some major changes in policy or a continuation of the status quo? Do people in the organization know how resources are allocated and do they think the process is fair and equitable?

In reviewing much of the resource distribution and the allocation methods, a person analysing organizations will see many direct parallels with the analysis of the distribution of power, since money is usually equated with power. It is important that these two analyses be kept somewhat separate, at least in the analytical phase, since they do deal with two different but related aspects of the organization. At this point of the analysis, a person trying to understand the organization better should resist the temptation to treat financial resources and power as if they are the same and interchangeable, since in most cases they are not.

Internal management structure

It is useful to know what the basic management structure of an organization really is: which people are part of central management and which people are part of local or unit management? How far down into the organization does top management feel the 'management' system extends? Do the employees involved agree that they are part of management or do they see themselves as mainly involved in professional or technical work within their particular unit? Do managers in one portion of the organization know who are the managers in a distant part of the organization and do they have a chance to exchange experiences? Is there a clear pattern of upward management mobility or is it unclear how one moves through the management structure? Are people in the organization who are 'managers' encouraged in that identity, given extra training and experience to help them function better as managers, and encouraged to identify themselves with the management structure? If a manager in an organization will have to work through his internal management structure and the people who are the management staff within that structure, it is important to describe that structure and identify those people very clearly.

Communication networks

Some organization theorists think that an organization can be clearly pictured by mapping its communications and the actions stemming from them. Are there clear patterns of communication from the top of the organization to the bottom? From the bottom to the top? From one organizational subunit to another organizational subunit in a side-to-side fashion? Are there formal methods of communication from the organization outward to the environment as well as methods of communicating from the environment back to the organization? Is communication felt to be important or something that simply happens on an *ad hoc* basis? Is the philosophy of the organization to keep people comprehensively informed or just to tell them about those matters in which they have direct involvement? Does anyone regularly review communication systems and networks to see if some change needs to be made?

In addition to looking at the formal communication networks within the organization and between the organization and the outside environment, it is also important to identify informal networks of communication. Do the workers in an organization really learn about the life of that organization in their daily contacts with others workers or through their personal/family networks more effectively than they do through formal communication patterns established by the organization? What type of information is communicated over the informal network as opposed to the types of information that are transmitted over more formal networks? Do people make decisions and guide their actions more on the basis of informal communication network information than they do on the basis of formal network information? Do people in the organization generally feel well informed about their organization or do they feel that information about the purposes and the practices of the organization is closed to them? What are the kinds of information that people would like to have in addition to their present sources of information, in order to enable them to function more productively and to feel more comfortable within the organization.

Inter-personal/inter-group/inter-organizational relationships

A complete analysis of any organization should include a review of the inter-personal relationships within it. Do people see themselves in the midst of colleagues and supporters or do they feel beset by adversaries on all sides? Is there an atmosphere of mutual respect and co-operation, of common purpose and objectives, or is it one of isolation and aloneness? Are the inter-personal relationships between individual workers and their supervisors and superiors positive, productive, and respectful, or are they demeaning, damaging, or in some way hindering the satisfaction of either the worker or the supervisor?

At the same time as one is looking at individual interactions, it is important to look at the interactions of groups (for example, interactions of professional workers with non-professionals) and of one profession with another (nurses with doctors, for example). Does the organizational structure encourage the creation of separate groups of personnel and how does this affect the organization's functioning? Does the existence of sub groups create a sense of close support for the individual or does it serve to pigeon-hole and isolate people even more. Do sub groups of personnel form organizations on a formal or an informal basis in order to get their work done better, in order to socialize with people that are like themselves, or in order to protect themselves (such as in labour union groupings).

The organizational analyst must also look at inter-organizational relationships to see whether the workers and the procedures in one organization are linked productively and positively with the workers and procedures in a separate organizational unit. Has there been a specific effort made to determine what inter-organizational relationships between workers and procedures are appropriate and to ensure that these relationships are established and nurtured? Do the workers and supervisors in one organizational unit know which other organizational units outside their own are important to their home unit's success and do they know enough about these other organizational units to interact with them in an appropriate fashion? Is there some person who has been given specific responsibility for being sure that the relationship between organizational units runs smoothly and productively or is it simply left to be handled by individuals on an informal basis?

Evaluation and control systems

In every effective modern organization, there should be some organized effort to measure the outcome of the organization's work, to determine whether its objectives are being met, either in total or in part. There should also be some method of feeding back this valuative information into the management and planning structure so that appropriate changes and corrections can be made.

A thorough review should attempt to identify the major evaluative systems that may be in place within the organization, to learn how they function, to learn what information is developed, and to learn how that information is actually utilized in the management of the organization. This review should include a determination of whether the organization is actually interested in evaluating its activities for internal, management, or planning reasons or whether it merely conducts its evaluations because it must produce some form of report to an outside organization. The timeliness, appropriateness, and completeness of the evaluation system needs to be determined, together with the effectiveness of whatever corrective actions the evaluation calls into play.

Conflict resolutions

One of the natural outcomes of organizational life is conflict—between individuals, between groups of individuals, and between organizational units. Sometimes the conflict is based on personal differences and variations in style of work, while at other times it is based on different beliefs as to how the work should be done, where the organization should be going or how it is getting there. In some instances, the conflict is actually the result of the way in which the organization has been structured, while in other cases the organizational structure has less to do with the conflict than the people in the organization itself. Whatever the nature of conflict, it is important that the manager has a clear idea of the types of conflict that either exists or had the potential to exist, the sources of that conflict, and the possible means of reducing or eliminating that conflict in the future.

In every organization, there will exist conflict among individual workers and between organizational units. There may also exist some formal or informal mechanisms for handling conflict and possibly for resolving it. A thorough review of conflict within an organization should attempt to document not only the individual conflict situations, but also their potential causes and their possible means of resolution. An organizational analyst should ask whether there are appropriate systems in place for identifying potential sources of conflict early and for correcting or ameliorating them. Are there ways of containing conflict so that its existence does not do major damage to the broader aspects of the organization? Are there systems in place by which each episode of conflict leads to a better understanding of the organization by its managers; by which each episode leads to action to prevent future conflict? Is the general atmosphere of this organization or of this operating unit one in which conflict is a major and continuing characteristic or are conflicts intermittent, infrequent, and of a minor nature? Since every manager's work will include a significant concern for conflict and its resolution, it is important that the manager carry out an organizational analysis that looks at the existence of conflict, its probable causes, and its potential solutions.

Organization redesign and new products

Finally, every effective organization must continually seek to renew itself, to realign its resources and organizational structure, to introduce new services, programmes, and products. No organization can afford to remain static and unchanged for long, and certainly public health organizations, with their rapidly changing environments and challenges, are not exceptions.

A careful review or analysis of any public health organization must ask itself: how is this organization systematically preparing itself for the future? What information does an organization continuously seek with regard to its future directions and what steps does it take once it has this information?

Is there an organized, continuous effort to consider the long-term plans of the organization and to shape its structure accordingly? How do new ideas and new approaches get circulated for discussion and consideration, and how does the decision get made to try some new approach? Are innovation and new product development seen as a benefit to the organization or a bother? Is everyone involved in preparing for the future or is it felt to the responsibility of a special separate planning or research/demonstration unit?

The public health organization that is not continually analysing and reanalysing its organizational structure and the various programmes or products it offers, is doomed to become archaic and inappropriate for the challenges of the future. Perhaps the greatest challenge to a public health organization's top leadership is to appraise and reappraise continuously how they are carrying out their important tasks and whether there is a better way to do it.

Summary

A great deal more could be said about organizations in general and public health organizations in particular. The important message to take away from this discussion is not merely that there is a body of technical or professional knowledge about organizations, but that the way in which public health agencies are organized and function is critically important. The organization is the vehicle by which public health work gets done and if that vehicle has significant problems, the work simply will not be done or, at least, it will not be done as well as it could/should be. The more effective and streamlined the vehicle, the more powerful the outcome.

Public health workers and managers cannot afford to be passive about the organization in which they work. They should spend as much time studying, understanding, and trying to improve it as they spend on their own individual professional work, because the two (their individual professional work and the organization in which that work is carried out) are inextricably intertwined; one depends upon the other. Therefore it is an important obligation for each public health manager to understand that interdependence of professional work and organizational function, search for ways to improve it, and invest the appropriate time and energy in making organizations as efficient and effective as the public health skills and techniques that the organization itself is trying to deliver.

Management issues for the future in public health

An analysis of the public health literature and a review of public health practice reveals that there are a number of management areas that seem to have been increasingly receiving attention from public health managers and organizations around the world. The first of these areas have received some attention in the past, but developments in the most recent years have pushed them even further forward as important disciplines for the future public health manager to know better and to practice more effectively. The areas that will most probably require special attention of public health managers in the future are: (1) organizational performance measures; (2) outcome and effectiveness measurements of medical and public health programmes and treatments; (3) cost–benefit/cost–effectiveness measurement; (4) organizational and public policy decision-making; (5) information needs/information system design and management; (6) social marketing/social entrepreneurship.

Organizational performance measurement

Increasingly around the world, there is a greater understanding that it is no longer enough simply to establish organizations or agencies (whether public health organizations of other types); it now becomes much more important to know more accurately what those organizations are producing and how well they are performing. In recent years, there has been much greater interest in organizational performance measurement and in the development of techniques to carry out these types of measurements better.

Literature in this area is still rather widespread and diffuse and ranges from articles dealing with the evaluation of the performance of an organization's chief executive officer (Ad Hoc Committee on Evaluation 1984) to analysing the cost and performance of organizations producing health care (Kelliher 1985). Several authors (Estaugh 1985; Laliberty 1984; Sahney 1982; Schemerhorn 1987; Talukdar and Mac-Laughlin 1987) have written about methods for improving or enhancing productivity, while others (Gans 1986; Long and Harrison 1985; Nicol and Gibbs 1985; Serway et al. 1987) have written about methods for better management of organizational productivity. Other writers (McDougall et al. 1989; Spain et al. 1989) have talked about the need to have clearly stated performance standards developed, so that the performance of organizations can be accurately measured against a set of previously determined standards and then can be compared with the performance of similar organizations in other settings.

In all of these discussions, the overriding message that comes through most clearly is that it is no longer acceptable for a public health organization simply to establish a programme or an operating unit and then forget about the internal efficiency of its operations. The well-managed public health organization of the future will have a clearly defined set of operating standards against which the organization's performance will be measured and against which the managers of those units will be held accountable. In an era of increasingly scarce financial resources, this type of more accurate measurement of organizational internal efficiency and performance will be the hallmark of the well-managed public health organization of the future.

Outcome and effectiveness measures of medical and public health programmes and treatments

In a parallel development to the increased interest in measuring organizational performance, there has been a marked upsurge in recent years in looking at the actual outcomes and effectiveness of various forms of medical care and public health programmes. A number of writers (Brook 1989; Chassin et al. 1989; Kanouse et al. 1989) have taken the same direction as those writers who have endorsed better methods of measuring organizational performance; in this instance, the writers have strongly advocated better measurements of individual treatment and programme results. In this approach, there is an assumption of a great need to measure much more accurately the actual results and effectiveness of individual medical treatments and public health programmes, since it is now becoming apparent that many of these treatments or public health programmes are less effective than has been widely assumed. Taken together with better methods of measuring the appropriateness of treatment and programmes in different situations (Chassin et al. 1986; Leape et al. 1989), outcome and effectiveness measures will be an important tool for the public health manager in the future.

A number of public health authors had already begun to write about the need for improved evaluation of programmes and treatments (Borden 1988; Earls et al. 1989; McCusker et al. 1988) while others have been looking at the social or community impact of various programmes as a measure of their outcomes or effectiveness (Jingtaganont et al. 1988; Renaud and Suissa 1989; Robers et al. 1988; Wokutch and Fahey 1986). Other writers (Reeder et al. 1984) have taken the outcome and effectiveness evaluations one step further and have linked them to the actual cost or expense of operating the various programmes being measured.

Cost–benefit/cost–effectiveness analyses

In parallel with the heightened interest in improved measurements of programme performance and of the outcomes/effectiveness of individual medical treatments and public health programmes, there has also been an increased interest in improved methods of cost–benefit analysis and cost–effectiveness analysis as an important tool for managers of public health programmes. Indeed, in some instances, these techniques have become so valued that they have seemed to displace more traditional values of public service and community need.

In the past, decisions concerning individual programmes were often made only on the basis of an initial assessment of that programme itself. There were relatively few attempts to review whether the returns from the investment of time, money, and personnel had been worth the effort. Similarly, there is little interest in comparing the results from one potential programme with another, in order to see which yielded a better return on society's investment of its capital. There was often no real understanding in public health circles of the fact that, by choosing to perform one service or programme, a decision was simultaneously made to not perform another. There was little attempt to weigh the merits of programmes against one another (or even against not doing anything at all) in establishing priorities for the public health organizations to attack during the next budget period.

In recent years this has changed very rapidly, with a better understanding that it is incumbent on a society to appraise the comparative benefits of the various things that it can do, so that it will do those things that are most beneficial and effective. The society may still choose to do something that is less 'productive' because it is committed to helping one area of a country or one part of a population, but at least now it is making these decisions and choices knowingly and has a better understanding of the alternative options.

In order to function in this new era, public health managers will have to have a much greater knowledge of cost-accounting and other quantitative techniques for management and planning and will have to be much more familiar with the techniques of cost–benefit/cost–effectiveness analysis. They will have to be much better able to document the real costs of health programmes and to project more accurately the real benefits of such programmes (Walsh 1986; Warner and Luce 1982).

While they are often discussed together, cost–benefit analysis and cost–effectiveness analysis differ very considerably in their results and in the use of those results. Cost-effectiveness analysis compares the cost of alternative strategies to achieve specific health outcomes: lives or years of life saved, immunization coverage, improvement in health status. Cost–benefit analysis values all of those outcomes in economic or financial terms, including lives or years of life lost or disability. Cost–effectiveness analysis describes its results in terms of outcomes and effect on health, while cost–benefit analysis describes its outcomes in terms in finances. Each type of result from those two approaches is quite appropriate for use, depending upon the purpose of the analysis.

In recent years there has been a veritable explosion in reports of the use of cost–benefit/cost–effectiveness analyses in public health programmes, with applications in areas as diverse as the use of seat belts in buses (Begley and Biddle 1988), neonatal intensive care units (Boyle *et al.* 1983), treatment for gynaecological infections (Buhaug *et al.* 1989), health promotion programmes (Hatsiandreu *et al.* 1988), reduction of infant mortality (Joyce *et al.* 1988), use of lithotriptor treatment for renal stones (Lingeman *et al.* 1986), cervical cancer screening (Mandelblatt and Fahs 1988), infection control programmes in hospitals (Miller, P. *et al.* 1989), tuberculosis prevention (Rose *et al.* 1988), prenatal maternal serum alpha-feto protein screening (Taplin and Conrad 1988), and bone marrow transplantation (Welch 1989). These techniques have also been used to evaluate the use of resources in the treatment of hypertension (Stason and Weinstein 1977), pneumococcal pneumonia vaccine (Willems *et al.* 1980), the provision of primary health care (Zakir Hussain 1983), immunization programmes (Creese *et al.* 1982;

Robertson *et al.* 1984). It is safe to say that there is almost no area of public health practice in which cost–benefit/cost–effectiveness analyses have not been considered or actually applied in recent years. It is clear, both from the volume of literature that has appeared in recent years and from the reports from actual practice of public health professionals in the field that these techniques of cost–benefit/cost–effectiveness analysis are important parts of the professional skills of the public health manager of the future.

Organizational and public policy decision-making

Along with an expanded interest in cost–benefit/cost–effectiveness analysis, there has been a greatly widened attention to decision-making in general in public health. The interest here has not been so much in the decisions that are actually made, but rather in the procedures and processes of decision-making itself. In recent years a significant interest has been focused on how important organizational and policy decisions are made, what inputs of information are used, what analyses are carried out, and how the results of decisions are evaluated after they have been made. It seems clear that this is an important and burgeoning area of interest for the public health management professional and one that will be increasingly important in the future.

A number of writers have looked at decision-making in various situations, both in management (Breindel 1988; Hardy and McWhorter 1988; Shortell 1983; Taylor 1984; Turban 1982; Warner *et al.* 1984), as well as medical decision-making and decisions in public policy (Eddy 1986; Hinman *et al.* 1988; Russell 1989; Warner and Luce 1982). While this is still a field that is in its relative infancy, both in the development of practical working methodologies as well as in the application of those methodologies, it is clear that interest in how decisions are made and more important, how they can be made better and more exactly, will be a major area of activity for public health managers in the future.

Information needs information system design and management

Underlying all of the discussion of organizational performance measures, outcome and effectiveness measures of medical treatment and public health programmes, and cost–benefit/cost–effectiveness analysis is the obvious requirement that information be available upon which these measurements can be based. There is a growing realization that the increasing need for improved decision-making and analytical capacity will continue to demand increasingly improved information systems that can produce the data upon which decisions and analyses can be based. As a result, the public health manager of the future will, by necessity, be much more competent and comfortable with advanced information systems, not only their use but also their design and management.

Numerous writers covering a wide range of topics have

commented upon the need for increased excellence in information systems and for their greater use and applicability in public health management. A number of writers have discussed both the general need for improved information systems in management and policy (Austin 1988, 1989; Taylor 1984), as well as for the need of specialized information systems for particular purposes (Allgullander 1989) and for the need in better skills in managing the new information capacity that is rapidly becoming available (Worthley 1982). All of these writers stress that the new capacity for information handling that is provided by the rapidly expanding computer technology makes available a widely different range of management decision options. Some of them (Austin 1989) point out, however, that it is very easy for the new computer technology to become more master than servant and to overwhelm and intimidate managers with their great capacity. Virtually all writers stress the fact that the new information systems are only as useful as their careful design allows them to be and that it is essential that the first step in the approach to information utilization is to determine exactly and carefully what information is needed and what specific use will be made of the information once gathered. In order to identify need, design appropriate information systems, and manage their utilization, the public health manager of the future must begin early in training to become completely confident and competent in the use of advanced information systems so that they can be the servant of management rather than the master. All of the other areas of management that have been mentioned as important (organizational performance measures; outcome and efficiency evaluations of treatments and programmes; cost–benefit/cost–effectiveness analysis) depend upon the development of appropriate information systems and the development of management skills in utilizing the results of those information systems for improved management.

Social marketing/social entrepreneurship

Although the terms 'marketing' and 'entrepreneurship' conjure up images of salesmen, of advertising, or of vigorous public relations campaigns by business concerns to sell a product, in fact the term has a very different and a very important meaning in the future of public health practice. Indeed, in recent years it has become very apparent that much of public health programmes in the future may depend upon public health practitioners being able to 'market them' better to the public that needs them. In some cases, it has been suggested that the public health manager's ability to 'sell' the programmes that have been developed to impact the health of the public may be the *single* most important skill that a public health manager in the future may need to have.

Basically, marketing consists of several aspects. On the one hand, it involves a careful appraisal of what people want and need, and also what they are capable and potentially willing to buy or to use. On the other hand, it means a careful appraisal of what the organization has to offer and what new programmes or services it can develop to meet the needs of the public. It means the development of a strategy for presenting the new programme or service in such a way that people want to utilize it, and a careful continuing assessment of what the public thinks of it in practice. In recent years, the use of terms 'social marketing' and 'social entrepreneurship' have suggested a change in the public health practitioner's view towards these important areas. Although the terminology differs in its use in different kinds of programme, in essence the terms relate to the fact that the public health manager must creatively determine ways in which a needed service can be made available to the public, whether through private enterprise or public organizations. The social entrepreneurship comes from being ingenious and clever in the development of new products or new approaches that the public will understand and respond to by using the service or product that is being proposed by the public health manager. The use of the term 'social entrepreneurship' suggests that managers of public health programmes need to be active, not passive in the vigorous development of new approaches and programmes, in a fashion that has not been encouraged before.

Various writers have discussed the development of new programmes and their marketing (Ershoff *et al.* 1989; Foss 1989; Kropf and Szafran 1988; Levy *et al.* 1989; Worden *et al.* 1989) while others have focused on various techniques of marketing health programmes and their applications to management practice (MacStravic 1986c, 1988, 1989; Sherwood *et al.* 1988; Smith and Taylor 1984). While this is still a relatively new area of public health management practice, it is clear that it will be an important area in the future for the public health manager to understand and utilize.

Summary

In the last 50 years, the nature of the public health professional's work has changed markedly, moving steadily from a medical role to a more public health technology role, and finally in recent years to a management and planning role. It is clear that the trend towards greater emphasis on management effectiveness for public health professionals will increase in the years ahead and will be the deciding factor in the advancement of many public health professional's careers in the future.

Since that is the case, it is vitally important that each public health professional view management and planning skills as central to their future work, not as a minor addition or peripheral activity. If the public health professional of the future wishes to be maximally effective in working to improve the health of the public and if that public health professional is choosing to do it through public health organizations and structures, then improved management and planning skills of the latest type and technique are essential. The issues identified in this chapter and the literature referred to briefly in the course of that identification are only a small fraction of the new ideas that are circulating and the new developments that are taking place in management within public health organizations. This chapter is meant as an introduction to those issues and those developments and is also meant to provide encouragement to each public health professional in the development of a programme of self-motivated learning and development, so that each public health professional will grow in skills and knowledge as new management and planning techniques develop. It is only an understanding of the importance of management and planning techniques in public health practice, together with a firm decision on the part of each individual public health professional to develop these important skills over the years, that will lead to the competent public health leaders that public health organizations around the world will need in the future.

References

Ad Hoc Committee on Evaluating the Performance of the Hospital Chief Executive Officer (1984). *Evaluating the performance of the Hospital Chief Executive Officer.* Foundation of the American College of Hospital Administrators, Chicago, Illinois.

Allgulander, C. (1989). Psychoactive drug use in a general population sample, Sweden: Correlates with perceived health, psychiatric diagnoses, and mortality in an automated record-linkage study. *American Journal of Public Health*, **79**, 1006.

Antle, D. and Reid, R. (1988). Managing service capacity in an ambulatory care clinic. *Hospital and Health Services Administration*, **33**, 201.

Arnold, D., *et al.* (1987). Organizational culture and the marketing concept: Diagnostics keys for hospitals. *Journal of Health Care Marketing*, **7**, 18.

Austin, C. (1988). *Information systems for health services administration* (3rd edn). Health Administration Press, Ann Arbor, Michigan.

Austin, C. (1989). Information technology and the future of health services delivery. *Hospital and Health Services Administration*, **34**, 157.

Barrett, M. and Nich, D. (1986). *Effective health care internal auditing.* Aspen Systems Corporation, Rockville, Maryland.

Beck, D. (1989). *Basic hospital financial management* (2nd edn). Aspen Publishers, Frederick, Maryland.

Begley, C. and Biddle, A. (1988). Cost–benefit analysis of safety belts in Texas school buses. *Public Health Reports*, **103**, 479.

Berman, H., Weeks, L., and Kukla, S. (1986). *The financial management of hospitals.* Health Administration Press, Ann Arbor, Michigan.

Binder, J. (1983). Value conflict in health care organizations. *Nursing Economics*, **1**, 114.

Blair, J. and Whitehead, C. (1988). Stakeholder diagnosis and management for hospitals. *Hospital and Health Service Administration*, **33**, 153.

Borden, J. (1988). An assessment of the impact of diagnosis-related group (DRG)-based reimbursement on the technical efficiency of New Jersey hospitals. *Journal of Accounting and Public Policy*, **7**, 192.

Boscarino, J. (1988). The public's rating of hospitals. *Hospital and Health Services Administration*, **33**, 189.

Boyle, M., *et al.* (1983). Economic evaluation of neonatal intensive care of very-low-birth-weight infants. *New England Journal of Medicine*, **308**, 1330.

Breindel, C. (1988). Nongrowth strategies and options in health care. *Hospital and Health Services Administration*, **33**, 37.

Brook, R. (1989). Practice guidelines and practicing medicine: Are they compatible? *Journal of the American Medical Association*, **262**, 3027.

Brook, R.H., Kamberg, K., and Mayer-Oakes, A. (1989). Appropriateness of acute medical care for the elderly. Publication R-3717, The RAND Corp., Santa Monica, California.

Brown, B. (1983). Post mortem on a class battle: Doctor *versus* hospital. *Hospital and Health Services Administration*, **28**, 59.

Brown, N. and McCool, B. (1987). High-performance managers: Leadership attributes for the 1990's. *Health Care Management Review*, **12**, 69.

Bruce, S. (ed.) (1988). *How to write your employee handbook.* Business and Legal Reports, Madison, Connecticut.

Buhaug, H., *et al.* (1989). Cost effectiveness of testing for chlamydia infections in asymptomatic women. *Medical Care*, **27**, 833.

Chassin, M.R., Brook, R.H., Park, R.E., *et al.* (1986). Variation in the use of medical and surgical services by the Medicare population. *New England Journal of Medicine*, **314**, 285.

Chassin, M.R., Kosecoff, J., Park, R.E., *et al.* (1989). *The appropriateness of use of selected medical and surgical procedures and its relationship to geographic variation in their use.* Health Administration Press, Ann Arbor, Michigan.

Churchman, C.W. (1968). *The systems approach.* Delacorte Press, New York.

Clement, J. (1988). Vertical integration and diversification of acute care hospitals. *Hospital and Health Services Administration*, **33**, 99.

Cleverley, W. (1986). *Essentials of health care finance* (2nd edn). Aspen Publishers, Frederick, Maryland.

Cleverley, W. (1989). *Handbook of health care accounting and finance* (2nd edn). Aspen Publishers, Frederick, Maryland.

Cliff, G. (1987). Managing organizational conflict. *Management Review*, **76**, 51.

Cooper, P. (1985). *Health care marketing* (2nd edn). Aspen Publications, Rockville, Maryland.

Creese, A.L., Sriyabbay, A., Casabal, G., and Wisego, G. (1982). Cost-effectiveness appraisal of immunization programs. *Bulletin of the World Health Organization*, **60**, 621.

Earls, F., *et al.* (1989). Comprehensive health care for high-risk adolescents: An evaluation study. *American Journal of Public Health*, **79**, 999.

Eddy, D. (1986). Success and challenges of medical decision making. *Health Affairs*, **5**, 108.

Ershoff, D., *et al.* (1989). A randomized trial of a serialized self-help smoking cessation program for pregnant women in an HMO. *American Journal of Public Health*, **79**, 182.

Estaugh, S. (1985). Improving hospital productivity under PPS: Managing cost reductions. *Hospital and Health Services Administration*, **30**, 97.

Evans, J.R. (1981). *Measurement and management in medicine and health services.* The Rockefeller Foundation, New York.

Ferguson, K. and Lapsley, I. (1988). Investment appraisal in the National Health Service. *Financial Accountability and Management*, **4**, 281.

Flahault, D. and Roemer, M. (1986). *Leadership for primary health care.* World Health Organization, Geneva.

Flexner, W., Berkowitz, E., and Brown, M. (ed.) (1981). *Strategic planning in health care management.* Aspen Systems Corporation, Rockville, Maryland.

Foege, W. and Henderson, D. (1986). Management priorities in primary health care. In *Strategies for primary health care: Technologies appropriate for the control of disease in the developing world* (ed. J. Walsh and K. Warren). University of Chicago Press, Chicago, Illinois.

Foss, R. (1989). Evaluation of a community-wide incentive program to promote safety restraint use. *American Journal of Public Health*, **79**, 304.

Fournet, B. (1982). *Strategic planning in health services management*. Aspen Systems Corporation, Rockville, Maryland.

Gans, N. (1986). Measuring physician productivity. *Medical Group Management*, **33**, 19.

Gazda, G., *et al.* (1982). *Interpersonal communication for health professionals*. Aspen Publications, Rockville, Maryland.

Glandon, G.L. and Morrissey, M. (1986). Redefining the hospital–physician relationship under prospective payment. *Inquiry*, **23**, 166.

Goldfield, N. and Goldsmith, S. (ed.) (1985). *Financial management of ambulatory care*. Aspen Publishers, Rockville, Maryland.

Hardy, O. and McWhorter, C. (1988). *Management decisions*, Aspen Publications, Rockville, Maryland.

Hartfield, J. (1980). Managing the conflicts of change. *Group Practice Journal*, **29**, 9.

Hatsiandreu, E., *et al.* (1988). A cost–effectiveness analysis of exercise as a health promotion activity. *American Journal of Public Health*, **78**, 1417.

Helmi, M. and Burton, S. (1988). Cost control under the DRG system. *Hospital and Health Services Administration*, **33**, 263.

Her Majesty's Stationery Office (1989). *National Health Service and Community Care Bill*. HMSO, London.

Herkimer, A. (1988). *Understanding health care budgeting*. Aspen Publishers, Frederick, Maryland.

Herkimer, A. (1989). *Understanding health care accounting*. Aspen Publishers, Frederick, Maryland.

Hingley, P., *et al.* (1986). *Stress in nurse managers*, Project Paper No. 60. King Edward's Hospital Fund for London.

Hinman, A., *et al.* (1988). Decision analysis and polio immunization policy. *American Journal of Public Health*, **78**, 301.

Hommans, G. (1950). *The human group*. Harcourt-Brace-Jovanovich, New York.

Imberman, W. (1989). Strike prevention in hospitals. *Hospital and Health Services Administration*, **34**, 195.

Institute of Medicine (1988). *The future of public health*. National Academy Press, Washington, DC.

Jacobson, M., Labbok, M., Murage, A., and Parker, R. (1987). Individual and group supervision of community health workers in Kenya: A comparison. *Health Administration Education*, **5**, 83.

Jingtaganont, P., *et al.* (1988). The impact of an oral rehydration therapy program in southern Thailand. *American Journal of Public Health*, **78**, 1302.

Johnson, E. (1988). The competitive market: Changing medical staff accountability. *Hospital and Health Services Administration*, **33**, 179.

Johnson, R., Kast, F., and Rosenzweig, J. (1967). *The theory and management of systems*. McGraw-Hill, New York.

Joyce, T., *et al.* (1988). A cost–effectiveness analysis of strategies to reduce infant mortality. *Medical Care*, **26**, 348.

Kabat, G. and Wynder, E. (1987). Determinants of quitting smoking. *American Journal of Public Health*, **77**, 1301.

Kaiser, L. (1986). Organizational mindset: Ten ways to alter your world view. *Healthcare Forum*, **29**, 50.

Kaluzny, A. and Hurley, R. (1987). The role of organizational theory in the study of multi-institutional systems. *Health Administration Education*, **5**, 407.

Kaluzny, A., Warner, D., Warren, D., and Zelman, W. (1982). *Management of health services*. Prentice-Hall, Englewood Cliffs, New Jersey.

Kaluzny, A. and Veney J. (1980). *Health care organizations: A guide to research and assessment*. McCutchan Publishers, Berkeley, California.

Kanouse, D.E., Winkler, J.D., Kosecoff, J., *et al.* (1989). *Changing medical practice through technology assessment: An evaluation of the NIA Consensus Development Program*. Health Administration Press, Ann Arbor, Michigan.

Kast, F. and Rosenzweig, J. (1974). *Organization and management: A systems approach*. McGraw-Hill, New York.

Kearns, K. and Hogg, M. (1988). Employee resistance and strategies for planned organizational change. *Hospital and Health Services Administration*, **33**, 521.

Kellerman, A. and Hackman, B. (1988). Emergency department patient 'dumping': An analysis of interhospital transfers to the Regional Medical Center at Memphis, Tennessee. *American Journal of Public Health*, **78**, 1287.

Kelliher, M. (1985). Managing production, performance, and cost of services. *Healthcare Financial Management*, **39**, 23.

Kellogg Foundation Commission on Education for Health Administration (1975). *The Report of the Commission on Education for Health Administration*. Health Administration Press, Ann Arbor.

Kindig, D. and Lastiri-Quiros, S. (1989). The changing managerial role of physician executives. *Journal of Health Administration Education*, **7**, 33.

King's Fund Working Party (1977). *The education and training of senior managers in the National Health Service. Report of a working party*. King Edward's Hospital Fund for London.

Kovner, A. (1988). *Really managing: The work of effective CEOs in large health organizations*. Health Administration Press, Ann Arbor, Michigan.

Kovner, A. and Neuhauser, D. (1987). *Health services management: Readings and commentary* (3rd edn). Health Administration Press, Ann Arbor, Michigan.

Kropf, R. and Szafran, A. (1988). Developing a competitive advantage in the market for radiology services. *Hospital and Health Services Administration*, **33**, 213.

Kurtz, M. (1980). A behavioural profile of physicians in management roles. In *The physician in management* (ed. R.S. Schenke). American Academy of Medical Directors, Falls Church, Virginia.

Laliberty, R. (1984). *Enhancing productivity in health care facilities*. National Health Publishing, Owings Mills, Maryland.

Leape, L., Park, R.E., Soloman, D., Chassin, M.R., Kosecoff, J., and Brook, R.H. (1989). Relation between surgeon's practice volumes and geographic variation in the rate of carotid endarterectomy. *New England Journal of Medicine*, **321**, 653.

Lefebvre, R., *et al.* (1988). Social marketing and public health intervention. *Health Education Quarterly*, **15**, 299.

Levy, D., *et al.* (1989). Traffic safety effects of sobriety checkpoints and other local DWI programs in New Jersey. *American Journal of Public Health*, **79**, 291.

Lieber, J., *et al.* (1984). *Management principles for health professionals*. Aspen Systems Corporation, Rockville, Maryland.

Likert, R. (1961). *New patterns of management*. McGraw-Hill, New York.

Likert, R. (1967). *The human organization*. McGraw-Hill, New York.

Lingeman, J., *et al.* (1986). Cost analysis of extracorporeal shock wave lithotripsy relative to other surgical and nonsurgical treatment alternatives for urolithiasis. *Medical Care*, **24**, 1151.

Long, A. and Harrison, S. (ed.) (1985). *Health services performance: Effectiveness and efficiency*. Croom Helm, London.

Longest, B. (1980). *Management practices for the health professions*. Reston Publishing Co., Reston, Virginia.

Luke, R. and Begun, J. (1987). Industry distinctiveness: Implications for strategic management in health care organizations. *Health Administration Education*, **5**, 387.

McClure, L. (1985). Organizational development in the healthcare setting. *Hospital and Health Services Administration*, **30**, 55.

McCusker, J., *et al.* (1988). Association of electronic fetal monitoring during labor with cesarian section rate and with neonatal morbidity and mortality. *American Journal of Public Health*, **78**, 1170.

McDougall, M., *et al.* (1989). *Productivity and performance management in health care institutions.* American Hospital Association, Chicago, Illinois.

McFarlane, D. (1987). US domestic population and family planning policies, 1970–1983: Management lessons learned and pitfalls to avoid. *Health Administration Education*, **5**, 63.

McGregor, D. (1960). *The human side of enterprise.* McGraw-Hill, New York.

MacStravic, S. (1988). The patient as partner: A competitive strategy in health care marketing. *Hospital and Health Services Administration*, **33**, 15.

MacStravic, S. (1989). A customer relation strategy for health care employee relations. *Hospital and Health Services Administration*, **34**, 397.

MacStravic, R. (1986a). Hospital–physician relations: A marketing approach. *Health Care Management Review*, **11**, 69.

MacStravic, R. (1986b). *Managing health care marketing communications.* Aspen Systems Corporation, Rockville, Maryland.

MacStravic, R. (1986c). Product line administration in hospitals. *Health Care Management Review*, **11**, 35.

Makuc, D., *et al.* (1989). National trends in the use of preventive health care by women. *American Journal of Public Health*, **79**, 21.

Mandelblatt, J. and Fahs, M. (1988). The cost-effectiveness of cervical cancer screening for low-income elderly women. *Journal of the American Medical Association*, **259**, 2409.

Markowich, M. and Silver, C. (1989). How to reduce absenteeism: A comparative analysis. *Hospital and Health Services Administration*, **34**, 213.

Maslow, A. (1954). *Motivation and personality.* Harper & Row, New York.

Metzger, N. (1988). *The Health Care Supervisors Handbook.* Aspen Publications, Rockville, Maryland.

Milbank Memorial Fund Commission on Higher Education for Public Health (1976). *Report.* Milbank Memorial Fund, New York.

Miller, C., *et al.* (1989). Barriers to implementation of a prenatal care program for low income women. *American Journal of Public Health*, **79**, 62.

Miller, P., *et al.* (1989). Economic benefits of an effective infection control program: Case study and proposal. *Reviews of Infectious Diseases*, **11**, 284.

Mintzberg, H. (1973). *The nature of managerial work.* Harper & Row, New York.

Mintzberg, H. (1979). *The structuring of organizations.* Prentice-Hall, Englewood Cliffs, New Jersey.

Mintzberg, H. (1983). *Structure in fives: Designing effective organizations.* Prentice-Hall, Englewood Cliffs, New Jersey.

Morgan, A. and McCann, J. (1983). Nurse-physician relationships: The ongoing conflict. *Nursing Administration Quarterly*, **7**, 1.

Mottaz, C. (1988). Work satisfaction among hospital nurses. *Hospital and Health Services Administration*, **33**, 57.

Munson, F. and D'Aunno, T. (1989). Structural change in academic health centers. *Hospital and Health Services Administration*, **34**, 413.

Nackel, J. and Kues, I. (1986). Product-line management: Systems and strategies. *Hospital and Health Services Administration*, **31**, 109.

Nackel, J., *et al.* (1987). *Cost management for hospitals.* Aspen Publishers, Rockville, Maryland.

National Council for International Health (1985). *Management issues in health programs in the developing world.* National Council for International Health, Washington, DC.

Neuman, B., *et al.* (1984). *Financial management: Concepts and applications for health care providers.* National Health Publishing, Owings Mills, Maryland.

Nichol, D. and Gibbs, R. (1985). Development of performance indicators in health care management. *World Hospitals*, **21**, 11.

Pegels, C. and Rogers, K. (1987). *Strategic planning for hospitals and health care service corporations.* Aspen Publications, Rockville, Maryland.

Perow, C. (1970). *Organizational analysis: A sociological view.* Wadsworth Publishing, Belmont, California.

Peters, T.J. and Waterman, R.H. (1982). *In search of excellence.* Harper & Row, New York.

Pollitt, C., *et al.* (1988). The reluctant managers: Clinicians and budgets in the NHS. *Financial Accountability and Management*, **4**, 213.

Rakich, J., Longest, B., and Darr, K. (1989). *Managing health services organizations.* Saunders, Philadelphia, Pennsylvania.

Reeder, G., *et al.* (1984). Is percutaneous coronary anagioplasty less expensive than bypass surgery? *New England Journal of Medicine*, **311**, 1157.

Renaud, E. and Suissa, S. (1989). Evaluation of the efficay of simulation games in traffic safety education of kindergarten children. *American Journal of Public Health*, **79**, 307.

Rindler, R. (1987). *Managing a hospital turnaround.* Pluribus Press, Chicago, Illinios.

Robbins, S. and Rakich, J. (1989). Hospital personnel management in the early 1990s: A follow-up analysis. *Hospital and Health Services Administration*, **34**, 385.

Robers, P., *et al.* (1988). Nurse administration of sleep medication: A comparison of Registered Nurses and Licensed Practical Nurses. *American Journal of Public Health*, **78**, 1581.

Robertson, R.L., Davis, J.H., and Jobe, K. (1984). Service volume and other factors affecting the costs of immunization in the Gambia. *Bulletin of the World Health Organization*, **62**, 729.

Rose, D., *et al.* (1988). Tuberculosis prevention: Cost-effectiveness analysis of isoniazid chemoprophylaxis. *American Journal of Preventive Medicine*, **4**, 102.

Russell, L. (1989). Some of the tough decisions required by a National Health Plan. *Science*, **246**, 892.

Sahney, V. (1982). Managing variability in demand: a strategy for productivity improvement in health care services. *Health Care Management Review*, **7**, 37.

Savage, G. and Blair, J. (1989). The importance of relationships in hospital negotiation strategies. *Hospital and Health Services Administration*, **34**, 231.

Schemerhorn, J. (1987). Improving health care productivity through high performance management development. *Health Care Management Review*, **12**, 49.

Schulz, R. and Harrison, S. (1983). *Teams and top managers in the National Health Service*, Project Paper No. 41. King Edward's Hospital Fund for London.

Seidel, L., Seavey, J., and Lewis, R. (1989). *Strategic management for health care organizations.* National Health Publishing, Owings Mills, Maryland.

Serway, G., *et al.* (1987). Alternative indicators for measuring hospital productivity. *Hospital and Health Service Administration*, **32**, 379.

Shaw, K., *et al.* (1988). Correlates of reported smoke detector usage in an inner-city population: Participants in a smoke detector give-away program. *American Journal of Public Health*, **78**, 750.

Sheard, T. (1980). The structure of conflict in nurse–physician relations. *Journal of Nursing Leadership and Management*, **11**, 14.

Sherwood, P., *et al.* (1988). Dissecting health care markets: Large versus small business. *Hospital and Health Services Administration*, **33**, 249.

Shortell, S. (1983). Physician involvement in hospital decisions making. In *The new health care for profit: Doctors and hospitals in a competitive environment* (ed. B. Gray). National Academy Press, Washington, DC.

Slater, C. (1980). The physician manager's role: Results of a survey. In *The physician in management* (ed. R.S. Schenke). American Academy of Medical Directors, Falls Church, Virginia.

Smith, C.T. (1988). Strategic planning and entrepreneurism in academic health centers. *Hospital and Health Services Administration*, **33**, 143.

Smith, S. and Taylor, J. (1984). Market identifications and hospital cost containment: A comparison of revenue contributions to utilization of strategic business units. *Journal of Health Care Marketing*, **4**, 43.

Spain, C., *et al.* (1989). Model standards impact on Local Health Department performance in California. *American Journal of Public Health*, **79**, 969.

Staley, R. and Staley, C. (1989). Physician executives and communication. *Physician Executive*, **15**, 15.

Stamps, P. and Piedmonte, E. (1986). *Nurses and work satisfaction: an index for management.* Health Administration Press, Ann Arbor, Michigan.

Starkweather, D. (1981). *Hospital mergers.* Health Administration Press, Ann Arbor, Michigan.

Stason, W. and Weinstein, M. (1977). Allocation of resources to manage hypertension. *New England Journal of Medicine*, **296**, 732.

Stevens, B. (1985). *The nurse as executive* (3rd edn). Aspen Systems Corporation, Rockville, Maryland.

Stewart, R., Smith, B., Blake, J., and Wingate, P. (1980). *The District Administrator in the National Health Service.* King Edward's Hospital Fund for London.

Suver, J. and Neumann, B. (1985). *Management accounting for healthcare organizations* (2nd edn). Pluribus Press/Healthcare Financial Management Association, Chicago, Illinois.

Talukdar, R. and McLaughlin, C. (1987). Comparing the productivity of family planning operations: An American experience. *Health Administration Education*, **5**, 45.

Tannenbaum, R., Margulies, N., and Massarik, F. (1985). *Human systems development: New perspectives on people and organizations.* Jossey-Bass, San Francisco, California.

Taplin, S. and Conrad, D. (1988). Cost-justification analysis of prenatal maternal serum apha-feto protein screening. *Medical Care*, **26**, 1185.

Taylor, T. (1984). Computer support for management decision making in family practice. *Journal of Family Practice*, **19**, 567.

Turban, E. (1982). Decision support systems in hospitals. *Health Care Management Review*, **7**, 35.

Walsh, J. (1986). Prioritizing for primary health care: Methods for data collection and analysis. In *Strategies for primary health care* (ed. J. Walsh and K. Warren). University of Chicago Press, Chicago, Illinois.

Ward, W. (1987). *An introduction to health care financial management.* National Health Publishing, Owings Mills, Maryland.

Warner, K. and Luce, B. (1982). *Cost–benefit and cost–effectiveness analysis in health care: Principles, practice, and potential.* Health Administration Press, Ann Arbor, Michigan.

Warner, D.M., Holloway, D., and Grazier, K. (1984). *Decision making and control for health administration: The management of quantitative analysis* (2nd edn). Health Administration Press, Ann Arbor, Michigan.

Welch, H. and Larson, E. (1989). Cost effectiveness of bone marrows transplantation in acute nonlymphocytic leukemia. *New England Journal of Medicine*, **321**, 807.

Wenzel, F. (1986). Conflict: An imperative for success. *Journal of Medical Practice Management*, **1**, 252.

Whitehead, C., *et al.* (1989). Stakeholder strategies for the physician executive. *Physician Executive*, **15**, 2.

Wieland, G. (ed.) (1981). *Improving health care management: Organizational development and organization change.* Health Administration Press, Ann Arbor, Michigan.

Willems, J.S., Sanderes, C.R., Raddiough, M.A., and Bell, J.C. (1980). *New England Journal of Medicine*, **303**, 553.

Wokutch, R. and Fahey, L. (1986). A value explicit approach for evaluating corporate social performance. *Journal of Accounting and Public Policy*, **5**, 243.

Worden, J., *et al.* (1989). Preventive alcohol-impaired driving through community self-regulation training. *American Journal of Public Health*, **79**, 287.

World Health Organization (1976). *The education and training of Public Health Medical Officers, Report of a working party.*

Worthley, J. (1982). *Managing computers in health care.* Health Administration Press, Ann Arbor, Michigan.

Wright, M. (1984). The elective surgeon's reaction to change and conflict. *Archives of Otolaryngology*, **110**, 318.

Young, D. and Saltman, R. (1983). Preventive medicine for hospital costs. *Harvard Business Review*, **61**, 126.

Zakir Hussain (1983). Cost analysis of a primary health care center in Bangladesh. *Bulletin of the World Health Organization*, **61**, 477.

22b

The contribution of social policy and administration

NICHOLAS MAYS

What is social policy and administration?

'. . . The term "Social Administration" is a misleading one; we are not experts in office management and social book-keeping, nor are we technicians in man manipulation.' (Titmuss, 1968, p. 20)

Defining the nature of the discipline of 'social administration' or 'social policy', as it now tends to be referred to, has never been straightforward. Definitions have changed as the discipline itself has developed. It remains to be seen whether its practitioners will ever agree on a single definition. At its simplest, the discipline comprises the academic study of human welfare and the social institutions which shape the welfare of members of societies rich and poor. These include formal structures of welfare assembled under the title of the 'welfare state'. As a separate and distinctive university discipline, social policy and administration originated primarily in universities in the UK and later in the British Commonwealth (Canada, New Zealand, Australia, etc.). In the US, for example, the study of social policy still tends to take place in university departments of urban studies, planning, politics, policy analysis, sociology, applied economics, and a range of research institutes, rather than in departments explicitly focussed on the study of social policy. The early growth of the discipline was closely tied to the development of a collective, state infrastructure of policies and institutions devoted to tackling the social problems thrown up by early twentieth century western capitalism. Courses in social science were established in the UK early this century to train students who wished to work in the newly established social services. For example, a School of Social Studies was established in Liverpool as early as 1904, followed by courses at Birmingham in 1908 and the London School of Economics in 1913.

In this period, the study of social administration was decisively shaped by members of the Fabian Society, particularly Beatrice and Sidney Webb. Fabianism was characterized by an emphasis on using the techniques of empirical social investigation to develop practical proposals aimed at alleviating social problems such as poverty. Fabians believed in gradually moving towards socialism by means of a programme of social reform implemented through state intervention in the economy and in the distribution of income, wealth and services.

The subject expanded rapidly after the Second World War linked to the emergence of welfare states in western Europe committed to using their political power to supersede, supplement or modify the workings of the free market in order to improve the quality of life and living standards of their peoples. The discipline flourished in the relative consensus which existed as to the purposes and instruments of state welfare. The role of Social Administration in this period was to train the staff who would operate the new institutions of welfare while at the same time, improving the working of the welfare state through empirical analysis. In the 1940s and 1950s the discipline was dominated by the work of Richard Titmuss (1907–73). Titmuss has subsequently been criticized for taking an atheoretical approach and for his unselfconscious adherence to the values of Fabian collectivism (Reisman 1977; Walker 1988). Yet for all the limitations of his approach, as judged with the benefit of hindsight, Titmuss ensured that the study of social administration extended beyond the formal social services and their administration to encompass the full range of mechanisms by which social needs are met and benefits distributed in society. In much of his empirical work he showed that the supposed redistributive intent of the welfare state in Britain was imperfectly realized. Perhaps his most influential piece, *The Social Division of Welfare*, first published in 1956, showed how an analysis of the impact of social policies was incomplete unless it included the fiscal (tax reliefs) and occupational (fringe benefits) systems of middle-class welfare as well as the more conventionally identified statutory social services (Titmuss 1963, pp. 34–55). While Titmuss and his younger colleagues used this material to argue for reform of the welfare state, the results were also employed later by opponents of extensive state provision to argue for radical change and the development of more individualistic approaches to welfare (see below, discussion of the New Right).

Titmuss and his followers also emphasised the normative aspects of social policy: the view that social policy decisions involve choices between different objectives based on conflicting value systems and that the means chosen to achieve policy goals are at least as important as their immediate consequences since they carry highly visible messages as to the sort of society we wish to create. In *The Gift Relationship* (1970), Titmuss contended that the system of voluntary blood donation in the NHS fostered altruism, whereas the commercial system in the US had the opposite effect, commodifying a moral transaction as well as producing an inefficient and unsafe system.

Contemporary writers in the Fabian tradition are more critical of the bureaucratic, centralizing tendencies of state provision than Titmuss's generation (Deakin 1987). While a moral commitment to social improvement has remained a feature of much academic social policy analysis ever since (distinguishing it, perhaps, from other less applied social sciences such as sociology), later analysts have increasingly recognized the need to elucidate a wider range of perspectives than Fabianism; for example, by acknowledging developments in both Marxist and neo-liberal (New Right) theory and critiques of the welfare state. A leading contemporary theorist in the field of social policy has argued that the discipline of social policy is distinguished by an attempt to develop both *normative* theory (social philosophy) and *explanatory* theory (social science) concerning welfare (Mishra 1986). Under normative theory, Mishra includes the study of the values such as liberty, equality, justice, security, integration, etc., which are either implicitly or explicitly the objectives of specific social policies. Under explanatory theory, Mishra includes the analysis of how social policy develops and with what consequences, intended or unintended. For example, the distributional impact of the welfare state would constitute an important theme for explanatory studies.

As the study of social policy has developed beyond the purely vocational training of welfare practitioners such as social workers and probation officers, so interest in knowledge-building based on theory has quickened (Pinker 1971; George and Wilding 1976, 1985; Mishra 1977, 1981; Pinker 1979; Room 1979; Taylor-Gooby and Dale 1981). A wide range of 'models of welfare' or welfare paradigms has been identified which include elements of normative and explanatory theory (see Table 22b.1). Each represents an ideological prescription for the development of social policies based on a particular set of value preferences and views about the nature of humanity and society (Mishra 1986). In addition, each embodies a set of factual propositions which ought in principle to be testable empirically. Each can be seen to occupy a different position on a continuum of social policy from the absence of any collective responsibility for individual needs to a socialist idea of 'to each according to his/her needs'. Mishra (1986) contends that these models can

Table 22b.1. Models of welfare

Author	Models of welfare identified			
Wilensky and Lebeaux (1958)	Residual			Institutional
Pinker (1971)	Residual			Institutional
Titmuss (1974)	Residual	Industrial Achievement– Performance		Institutional– Redistributive
Parker (1975)	Laissez faire	Liberal		Socialist
George and Wilding (1976 and 1985)	Anti-collectivist	Reluctant Collectivist	Fabian Socialist	Marxist
Mishra (1977 and 1981)	Residual	Institutional		Structural (normative in 1977)
Pinker (1979)	Market	Mercantilist– Collectivist		Socialist
Room (1979)	Liberal	Social Democratic		Marxist
Taylor-Gooby and Dale (1981)	Individualist	Reformist		Structural

Table 22b.2. Three ideal-typical models of welfare

	Market	**Mixed**	**Need**
Methods of securing welfare	Minimal collective responsibility for welfare	Pragmatic mix of market and need criteria	Total collective responsibility for welfare
Pre-eminent values	Liberty	Mixed values	Equality
Factual propositions (empirically testable)	Free enterprise/ market best promotes economic growth and welfare of the population	A 'mixed economy' optimizes welfare by stimulating economic activity and at the same time, mitigating the adverse effects of unfettered markets	State planning best secures the welfare of the population

Source: Mishra (1986).

be logically reduced to three basic or ideal-typical models: a market model, a mixed model and a need model of social welfare (see Table 22b.2). For Mishra, therefore:

'It is the central task of social policy analysis to tease out the normative and factual propositions or assumptions (directly relevant to welfare) associated with each model and to subject them to close scrutiny.' (Mishra 1986, p. 35)

Hence, following Titmuss (1968, pp. 20–3), the discipline is concerned with the objectives of social policy (normative), its development (positive) and its consequences (normative and positive). The model-based approach offers the basis for a social science discipline which goes beyond the study of individual services or particular social problems, to study normative and positive aspects of social policy in widely differing settings taking account systematically of the effect of different economic, ideological, political, and government structures on the precise form of social policies in different countries (Walker 1981).

Robert Walker (1988) has elaborated Mishra's very general guide to the tasks facing social policy analysts in terms more comprehensible to those working at the mid- and micro-levels on the details of particular policies or services. At these levels, social policy analysis is formulated as, 'the pluralistic evaluation of policy . . . and understanding of the outcomes of policy set against a knowledge of the aspirations and intentions of the various actors involved in the policy process' (Walker 1988). A notable example of such a study in the health field is Smith and Cantley's explicitly pluralist evaluation of the success of a new NHS psychogeriatric day-centre which incorporates the differing perspectives of the range of groups involved in the service such as patients, their relatives and carers, medical staff, nurses, paramedics, the health authority, and the local authority (Smith and Cantley 1985). The study demonstrated that a comprehensive assessment of the costs and benefits of the innovatory service had to cope with the often conflicting objectives held by different groups. For example, medical and nursing staff tended to view success in terms of 'patient flow' through the centre to avoid the service 'silting up', whereas for the relatives of confused patients, the opposite was the case. They had little or no interest in seeing patients discharged to their full-time care at home! Pluralistic evaluation is inevitably controversial since it highlights conflicts of interest and perception between those involved in any programme. It lacks the level of scientific respectability accorded to quasi-experimental designs such as cost–effectiveness analysis. None the less, its fidelity to the reality of local agency conflict which can undermine policy innovations has led to its adoption in a number of other studies (Means and Smith 1988).

Smith and Cantley's study demonstrates the growing theoretical sophistication of the discipline as it becomes less pre-occupied with descriptive accounts of statutory welfare agencies and more explicitly analytic. A major thrust is to *measure*, *explain*, and *evaluate* patterns of service distribution and outcome by building on theory derived from the other social sciences, for example, Merton's concept of manifest (recognized and intended) and latent (unrecognized and unintended) functions of action (Merton 1968), or rival elitist (Mills 1956), pluralist (Dahl 1961) and corporatist (Cawson 1982) theories of policy-making.

Merton's concepts have been fruitful in yielding insights into the operation of welfare organizations. The observation that one of the latent functions of such organizations is to provide satisfying work for the staff and that this may conflict with their manifest functions of meeting the needs of patients, clients, and users draws on a Mertonian analysis. A specific application in the health field occurs in Calnan's study of accident and emergency departments in the British National Health Service in which the professionally desired development of the service into a specialized area of medicine with a clear career structure and area of expertise was found to be in conflict with the public expectation of the ser-

vice as a convenient alternative to general practice (Calnan 1982; quoted in Forder *et al.* 1984, pp. 128–9).

Corporatism generally refers to an arrangement in which the state plays a positive role in directing the economy and in providing or subsidizing services in close co-operation with a limited number of key interest groups. Klein's interpretation of the development of the NHS in the 1970s provides an example of a broadly corporatist analysis. He argues that decision-making was the outcome of relatively stable bargaining arrangements between the State and the medical profession, the most powerful 'quasi-feudal occupational barony' in the health sector (Klein 1977, p. 179; Klein 1983 and 1989*a*). It remains to be seen whether a corporatist account of this relationship will still hold good at the end of the 1990s after the introduction of competitive market principles into the NHS following the 1989 Government White Paper *Working for Patients* (Secretaries of State for Health, Wales, Northern Ireland, and Scotland 1989).

In the 1970s there was a rapid development in theoretically informed studies of the social policy process at different levels from formal, explicit central government decision-making through to the day-to-day, discretionary decisions of professional staff working in field agencies (e.g. Donnison and Chapman 1965; Hall *et al.* 1975; Banting 1979; Allen 1979; Hunter 1980; Ham 1981). In a pioneering series of case-studies, Donnison and Chapman (1965) showed how official policy changes were frequently no more than codifications and explicit recognition of changes in practice worked out previously by local service providers. A decade later, Hall *et al.* (1975) set the analysis of social policy change on a mature footing by exposing the limitations of simplistic accounts of policy development in terms of either 'social conscience' or 'rationality'. In six case-studies they showed that policy developments were profoundly shaped by the manner in which the claims of an issue were assessed in terms of 'legitimacy', 'feasibility', and 'support', the characteristics of the issue itself (e.g. the scope of an issue and its links with other issues), the political context and the resultant conflicts which surrounded the issue. For example, in their account of the struggle which preceded the 1956 Clean Air Act in the UK they attempted to show why it was that, despite seemingly good prospects for early and effective legislation, the attitude of the Government was characterized by 'reluctance, apprehension and tardiness' (Hall *et al.* 1975, p. 371). The pressures for change appeared to have been generated entirely outside the traditional Government bureaucracies since there were other pressing political priorities and no well-established civil service interest. The medical profession as a whole showed little sense of outrage and there was no well-developed medical interest group concerned primarily with the consequences of atmospheric pollution. Precise, hard evidence of the health effects of polluted air was not available, although it was widely agreed that polluted air was harmful to health. For these and other reasons, including the Conservative government's reluctance to embark on domestic controls having been elected to remove wartime restrictions on private life, the campaign for clean air took four years to reach the statute book.

In the fields of health policy and management, there have been a series of important studies in the UK National Health Services which have applied and tested a range of conceptual schemes from sociology and political science in order to shed light on the persistent problems of implementation deficit in the Service. For example, Allen (1979) attempted to explain the origins and implementation of the 1962 Hospital Plan which was directed at building district general hospitals in under-provided areas of the country in terms of rational, incremental, and bureaucratic politics models of public policy-making derived from the work of American political scientists such as Lindblom (1959) and Allison (1971). Hunter (1980) studied the difficulties faced by authority members and officials in Scottish Health Boards in making more than incremental resource shifts between services for different care groups despite clear national guidance on 'priorities'. The decision process was problematic because it was 'complicated', involving large numbers of people at different levels all representing different interests; it was 'fragmented' by the demarcations between administrative bodies; and it was 'disjointed' in that decision-makers tended not to follow plans, but to allocate development funds in small amounts to different interest groups as part of an intuitive approach based on crisis-management and appeasement. Faced with the dilemma of either allocating development funds to change the direction of policy in line with national 'priorities', or responding to the clamour for more resources for existing services, the demands from existing services normally prevailed.

Ham (1981) explained the politics of 'conflict without change' and the pre-eminence of matters of acute, curative medicine over the needs of the so-called 'priority care groups' in the actions of the Leeds Regional Hospital Board through the 1960s and early 1970s in terms of Alford's theory of the power relations between the three key structural interests identifiable in all modern health care systems: the professional monopolists (who were traditionally 'dominant'); the corporate rationalizers (planners and managers who were normally to be found 'challenging' the professionals); and the public (who were currently a 'repressed' interest) (Alford 1975). Ham showed how the hospital professionals effectively defined the limits of practical politics and used their resources of technical knowledge and social prestige to neutralize policies articulated further up the NHS hierarchy which threatened to disturb the historic centrality of acute medicine.

The study of policy-making and implementation, of course, comprises only one part of the terrain of contemporary social policy and administration. Just as definitions of the intellectual and theoretical core of the discipline remain controversial (see below for yet further feminist and neo-liberal critiques), so, too, are accounts of the subject matter of the discipline continually debated. The primary focus for many students of social policy remains the study of the formal and visible institutions of the welfare state as the embodiment, however imperfect (Brown and Madge 1982; Ringen 1987), of social justice and notions of shared citizenship. Indeed, Brown (1983) warns of the danger of an excessive broadening of the subject matter of the discipline lest it become unmanageable and lose its distinctive identity. In the United States, social policy studies tend to be organized and defined in terms of the heterogeneous and ever-changing group of topics which American society regards as 'social problems' (e.g. unemployment, drug addiction, urban decay, etc.). The 'social problems' approach has been criticized for producing an incoherent and parochial basis for the discipline. An alternative approach is to define and organize the subject matter of the discipline of social policy around a series of key concepts such as need, stigma, universality, selectivity, citizenship, social justice, etc. (Forder 1974). Like the 'social problems' approach, the 'key concepts' approach tends to produce an arbitrary set of concepts and to isolate the study of social policy from its wider societal context.

In an address to the first meeting of the UK Social Administration Association (now the Social Policy Association), Richard Titmuss, one of the main founders of British academic social policy analysis, put forward eight areas of study for students of social administration:

(1) analysis and description of policy formation and its consequences, intended and unintended;

(2) study of the structure, function, organization, planning, and administrative processes of institutions and agencies, historical and comparative;

(3) study of social needs and of problems of access to, utilization and patterns of outcome of services, transactions, and transfers;

(4) analysis of the nature, attributes, and distribution of social costs and diswelfares;

(5) analysis of distributive and allocative patterns in command-over-resources-through-time and the particular impact of the social services;

(6) study of the roles and functions of selected representatives, professional workers, administrators, and interest groups in the operation of social welfare institutions;

(7) study of the social rights of the citizen as contributor, participant, and user of the social services;

(8) study of the role of government (local and central) as an allocator of values and of rights to social property, as expressed through social and administrative law and other rule-making channels (Titmuss 1968, pp. 22–3).

Although Titmuss's definition of the subject matter of social administration was broad by comparison with many of his contemporaries, he was at pains, none the less, to distinguish social policy clearly from economic policy as a distinctive area of government activity. Later attempts to delineate the field of study have broadened its scope to include, for example, general socio-economic changes and macro-economic policy in so far as these impinge on social welfare. Reflecting this view, Donnison's list of topics for social policy studies compiled twelve years later in 1979, is still broader and reads as follows:

(1) the social services (including the private and voluntary sectors);

(2) the impact of changes in industrial structure and employment on workers;

(3) the distribution of income, wealth, public expenditure, and incidence of taxation;

(4) changing patterns in the family;

(5) administrative, legal, economic, and political issues relating to (1) to (4) above;

(6) the interaction between social change and institutions set up to cope with social change (Donnison 1979).

Walker (1981) argues even more forcefully that by placing the explicit welfare activities of government too close to the centre of their attention, social policy specialists are in danger of missing the important welfare consequences of non-governmental agencies and social institutions such as multi-national corporations, financial enterprises, and their inter-relationships with the macro-economic policies of government. Predictably, Walker's widening of the scope of the subject to include broad questions of economic, political and social theory and the change of name from 'Social Administration' to the less dreary, more theoretical and more political 'Social Policy' have not gone unchallenged. Glennerster (1988) argues that scholars have been distracted by theorizing and failed the ordinary consumers of welfare by not devoting sufficient time to analyse in depth how welfare bureaucracies actually work at the local level in order to help them work better in the future.

The persistent preoccupation with defining the nature and scope of the discipline, is in part a reflection of its eclecticism. 'Social administrators' tend to come from many different backgrounds in the social sciences. Much of their research takes place on the bounds of welfare economics, law, sociology, politics, and public administration. Social policy and administration as currently thought of, draws on philosophy to clarify definitions of terms, to identify value positions and elucidate the objectives of policy; on statistics and methods of social investigation to establish the incidence of and trends in, social phenomena; on sociology to understand the nature of social structure, organizational behaviour and social change; on psychology to assess the impact of policies on individuals; on economics for insights into macro-economic policy, resource constraints and methods of evaluation such as cost–effectiveness and cost–benefit analysis; on history and political science to explain the origins and development of social policies; and finally, on government and public administration to describe contemporary political and governmental processes.

What are social policies?

Just as definitions of the nature and scope of the academic discipline of 'social administration' or latterly, 'social policy' have changed over time as well as provoking controversy, so too analysts of social policy have articulated a range of definitions of what it is they profess to study from administrative definitions of social policy, through more sociological definitions which extend beyond the formally recognized institutions of the welfare state, to a perspective based on Marxist political economy. At the risk of imposing consensus where none exists, most, but by no means all students of social policy today would probably accept a very general definition that social as opposed to economic policy consists in collective intervention to improve, maintain or promote the welfare of individuals or society as a whole. Beyond this high level of generality, definitions diverge. T.H. Marshall's classic view from the 1940s and 1950s stressed that social policy was distinguished from economic policy which was directed at the commonwealth, by its focus on securing *individual* welfare by means of *government* action to provide services or income through the *formal* mechanisms of the welfare state, such as the National Health Service and the social security system (Marshall 1975). The role of these institutions was to relieve citizens of their anxiety in the face of the future threats to their well-being posed by life-course events such as illness, disability, unemployment, and retirement.

Titmuss drew attention to the social values reflected in social policies. He stressed the *integrative* and stabilizing functions of social policies defining them as the 'different types of moral transactions, embodying notions of gift-exchange, of reciprocal obligations, which have developed in modern societies in institutional forms to bring about and maintain social and community relations' (Titmuss 1968, pp. 20–1). One way in which social policies could achieve social harmony was by providing a degree of compensation for the personal costs associated with economic and technological development (e.g. unemployment). This implies that social policies frequently include an element of redistribution both horizontally (e.g. from those who are well to those who are sick) and vertically (e.g. between the better-off and the poor) in society in accordance with some notion of 'need'. Indeed, an alternative to Titmuss's definition of social policy focuses on its characteristic in the meeting of human needs (Rein 1976; Mishra 1977, 1981; Donnison 1979). Thus, Mishra defines social policies as, ' . . . those *social* arrangements, patterns and mechanisms that are typically concerned with the distribution of resources in accordance with some criterion of *need*'. (Mishra 1977, and 1981, p. xi). Mishra's definition includes within the sphere of social policy actions other than those of the government (e.g. by industry, commerce, and the voluntary sector) and reflects many of the criticisms which were levelled at earlier formulations by those who took a broader, more sociological view of the nature and scope of social policy. However, it continues to limit social policy to the *distribution* of resources rather than the way in which they are *produced*.

Townsend (1976) criticized the 'traditional' view of authorities such as Marshall (1975) for: its parochialism in defining social policy in terms of the specific institutions of the social services in the United Kingdom; for its assumption that collective policies can only emanate from government; for its assumption that collective policies are inevitably beneficial in terms of welfare; for implying that social policies can operate according to different principles from the market principles dominant in capitalist societies without contradiction or conflict; for assuming that social policy is exclusively concerned with the distribution of the surplus and has no part to play in modifying the nature of relations on the production side of the economy; and for overlooking important sources of social inequality and diswelfare which lie outside the sphere of control of welfare agencies (Townsend 1976, pp. 2–9).

Townsend built on Titmuss's 'social division of welfare' thesis (Titmuss 1963, pp. 34–55) to highlight the existence of 'social policies' pursued by institutions outside the framework of the state which produced results which frequently threatened the pursuit of welfare state objectives such as social justice or equality of opportunity, furthering instead the goals of the dominant economic system. In his 'structuralist' analysis, Townsend is also sensitive to questions of the distribution of power in society and the ability of those with power to prevent certain issues reaching the realms of practical politics (Bachrach and Baratz 1970). Thus, for Townsend, social policy is concerned with the production and distribution of a wide range of social resources, including cash income, savings, property, employment benefits, services in kind such as health, and education, environment, status and power (Townsend 1974, pp. 31–3). It includes ' . . . the underlying as well as the professed rationale by which social institutions and groups are developed or created to ensure social preservation or development'. (Townsend 1976, p. 6). This more radical approach focuses on the hidden intentions as well as the explicit goals of policy; all the social institutions which determine the production and distribution of resources and opportunities; and the differential status, power, and rewards which exist along gender, class, and ethnic lines in society.

Walker summarizes his version of this position, as follows:

' "Social policy" might be defined therefore, as the rationale underlying the development and use of social institutions and groups which affect the distribution of resources, status and power between individuals and groups in society.' (Walker 1981, p. 239)

The approach exemplified by Townsend and Walker recognizes that social problems such as poverty are related to the class structure of society rather than technical malfunctions in the bureaucratic system (Townsend 1979). Yet, it has been criticized by Marxists for not making explicit the strong links between the state and the capitalist economy and the scale of the material obstacles to large-scale redistributive policies required to overcome social problems such as poverty (Gough 1978).

The Marxist perspective on the nature of social policy differs from the preceding definitions in regarding social policy as intimately bound up with the functions of the modern capitalist state in:

(1) *accumulation:* maintaining conditions favourable to the accumulation of capital;

(2) *reproduction:* ensuring a healthy, skilled work-force which is at the same time disciplined;

(3) *legitimation:* maintaining political stability, social harmony, and social control (O'Connor 1973).

However, as Ian Gough, a leading Marxist theorist of welfare points out, welfare programmes are rarely straightforwardly functional for the maintenance of capitalism (Gough 1978; Gough 1979). First, the functions of accumulation, reproduction, and legitimation may come into conflict with one another; for example, when the level of taxation and public expenditure required for investment in reproduction (e.g. through spending on health services and education) is perceived by owners of capital to threaten the current profitability of their enterprises. Second, certain aspects of the welfare state may be attributable not so much to the self-interested actions of the ruling class and its agents, as the result of working-class struggles to improve social conditions under capitalism (Navarro 1978). Navarro explains the emergence of social insurance in western European countries such as Germany from the 1880s onwards in terms of the political pressure exerted on governments by organized labour (Navarro 1989).

The contribution of the discipline of social policy and administration to the practice of public health

Almost any factor which can be identified as creating human suffering, distress, or diswelfare is also highly likely to affect health adversely. Thus, the public health implications of many social policies are so much a part of those policies that they are often taken for granted. For example, it may be assumed that improvements in housing standards, higher levels of social support, a reduction in poverty, and the minimization of unemployment will improve the health of the groups affected. However, few if any welfare programmes derive their justification exclusively from their direct impact on health. Warm, dry housing for example, is an unequivocal social benefit irrespective of the strength of the evidence linking poor housing conditions such as dampness to particular aspects of ill-health (Platt *et al.* 1989). Recent research showing that the links between housing inequalities and health inequalities have remained despite the post-war expansion of public housing in the UK only serves to reinforce the existing case for high quality modern housing (Byrne *et al.* 1986;

Blackman *et al.* 1989). None the less, social policy researchers have an important contribution to make in showing how changes in parts of the system of welfare outside the health sector (e.g. housing, income maintenance, pensions, etc.) affect the physical, social and economic circumstances of individuals and ultimately their health. For example, there is a long tradition of research in social administration and epidemiology, dating back to the 1930s looking at the living standards and health of unemployed people (Morris and Titmuss 1944; Moser *et al.* 1984). In the same vein, in the 1980s, work in the UK on social security showed how a seemingly minor change in the rules surrounding the payment of board and lodging allowances to elderly people on income support who were in residential care had accelerated the growth of private nursing homes, directly contrary to the Government's avowed policy of community care, as well as leading to a huge unplanned increase in public expenditure. Concerns were raised that elderly patients were leaving their own homes to go in to inappropriate residential settings where they risked becoming unnecessarily dependent, and that existing regulatory mechanisms were inadequate to monitor the quality of care in the private sector (Day and Klein 1987). This interest in tracing the unintended consequences of a policy change in one area on the effectiveness of policy in another is typical of an important strand in social policy research.

Contemporary social policy research with a bearing on health and health services covers a wide range of themes relevant to the practice of public health. It is impossible to summarize it here (see Volume 1, Chapter 10, for a fuller coverage of the social and economic determinants of health and disease). Instead, a number of contrasting studies will be described to give an indication of the contribution of the discipline, to add to the studies already referred to earlier in the chapter. The studies cover three characteristic concerns of the discipline: the problems of policy implementation in complex organizations; the patterning of, and explanation for, social inequalities; and the success of welfare services in meeting the needs arising from these inequalities.

The 'administrative anthropology' of organizations delivering welfare services has long represented a significant part of the research effort in social adminstration. The replacement of multi-disciplinary consensus management in the NHS with commercial-style general management in 1984 following the recommendations of the Griffiths report on the management of the service (NHS Management Enquiry 1983), was widely regarded at the time as a more fundamental change than the succession of structural reorganizations which had preceded it (Day and Klein 1983). A qualitative study was undertaken in the later 1980s to assess whether the implementation of the Griffiths 'revolution' had succeeded in changing the substance as well as the form of management in an organization dominated by professional rather than bureaucratic values (Pollitt *et al.* 1988). The Griffiths reforms included a commitment to involve clinicians more closely in the management process as budget holders. Pollitt *et al.* found that inadequate

attention had been given to convincing clinicians, who were usually suspicious of budgeting, and others who were only occasionally in favour of such schemes (e.g. nurses), of the merits of the proposed changes. Instead, the scheme had been regarded largely as a technical exercise in developing appropriate information systems to the exclusion of behavioural issues and questions of organizational politics (Hunter 1990). Few of the participants in the policy field gave management budgeting (now known in the NHS as resource management) a high priority even when they were generally in favour of it. The authors concluded that progress on management budgeting would be slow and uneven unless incentives could be incorporated into the scheme to secure the support of key actors such as consultants.

Both public health and the applied social science of social adminstration in the UK share a tradition of research describing and seeking to explain the variations in health between different geographical areas in relation to their relative prosperity (M'Gonigle and Kirby 1936; Titmuss 1938). Work on geographical inequalities since the Second World War quickened in the period immediately before and after the implementation of the Resource Allocation Working Party (RAWP) formula for distributing financial resources fairly in relation to need in the NHS in England (Department of Health and Social Security 1976) and the introduction of other similar formulae in Scotland, Wales, and Northern Ireland. The RAWP formula represented the fruitful coming together of epidemiology and Fabian social administration to produce an explicit and rational system for reallocating finance. Another related research tradition involving applied social scientists and epidemiologists has been the study of health and social class and the explanation of social class gradients in ill-health with a view to informing a wide range of policy (Stevenson 1923; Brotherston 1976; Black Report 1980; Whitehead 1987). A study undertaken in the Northern Region of England between 1984 and 1986 built on both traditions to investigate the health differences between the 678 electoral wards of the Region and ascertain the extent to which differences in health were matched by differences in material and social conditions. The differences in health and in deprivation between wards were found to be very wide. Poor health and material deprivation were found to be strongly associated with one another. Analysis of poor health at ward level (as measured by mortality, permanent sickness and low birth-weight) in relation to the ward characteristics of material deprivation explained more of the idiosyncracies in area variations in health than could be explained by an analysis based simply on occupational social class variations between wards (Townsend *et al.* 1988). The authors concluded by noting that the inequalities in health and affluence which were observed by M'Gonigle and Kirby in the 1930s within the urban areas of the Region persisted 50 years later (M'Gonigle and Kirby 1936). The results of the study not only reinforced the conclusion of the earlier Black Report that material circumstances played a predominant role in explaining inequalities in health (Black Report 1980), but also provided up-to-date evidence of how far the UK had still to go before the World Health Organization goal of equity in health status could be secured.

One of the major themes of recent studies in social policy has been the distributional impact of public expenditure on the health and social services. The belief that public spending on welfare services such as housing, education, health, and social security could promote a fairer society by overcoming the inequalities resulting from the market process was an important spur to the development of the welfare state (Crosland 1956, p. 519). In a notable study of the results of public expenditure on health care, education, housing, and transport, Le Grand showed that equality had not been achieved in the United Kingdom, whether it was defined in terms of equality of expenditure, final income, use, access, or outcome (Le Grand 1982). Le Grand's conclusions concerning the National Health Service have major implications for the pursuit of public health. Having shown that equality of use in relation to equal need, equality of access, and equality of outcome had failed to be achieved, Le Grand argued that there was little the NHS could do to reduce such inequalities. Their principal determinants lay in basic social and economic inequalities which were beyond the control of the NHS and which could only be remedied by policies which extended far beyond the delivery of health care. The daunting challenge facing public health practitioners of achieving societal change is clear from this analysis (Smith and Jacobson 1988).

Current debates

Empirical studies such as Le Grand's have tempered the early conviction within the discipline that the publicly provided, professionalized, social services of the post-war welfare state were successfully achieving society's objectives and were intrinsically beneficent in their impact. Within the study of social policy and administration, newer critical perspectives on the welfare system have been developed, often based on the experiences of users themselves who may have found services unnecessarily bureaucratic, remote, unresponsive to individual needs, sexist or racist. The emergence of these perspectives has coincided with significant changes in the economic context of welfare, the tenor of the political debate surrounding welfare and the type of innovations being made in welfare systems. Since the world economic crises of the mid-1970s, governments in the western world have been continuously preoccupied with restraining the level of resources available for state welfare. The political truce between Left and Right over the economic and social merits of the welfare state has largely broken down and the role of values in making social policy has become more explicit. A change of emphasis has occurred in the prevailing policy debate and to a lesser degree in the actual provision of welfare towards 'welfare pluralism' or 'the mixed economy of welfare' (Johnson 1987). Although such terms are imprecise, they tend to refer to a system in which services are provided by the voluntary, informal, and commercial sectors, as well as

through state agencies with the state becoming less dominant. This is accompanied in some versions of welfare pluralism by measures to decentralize resource allocation decisions and encourage wider participation by clients and employees in planning services (Hadley and Hatch 1981).

Although the anti-bureaucratic and anti-professional themes in welfare pluralism appeal to Left and Right, the most potent welfare pluralists have come in the form of conservative governments imbued with ideas derived from what is now known as 'The New Right'. The New Right combines the economic liberalism of Hayek and Friedman with the social authoritarianism of neo-conservatives such as Roger Scruton and is associated in the UK particularly with the work of the Institute of Economic Affairs and the journal, the *Salisbury Review* (Green 1987). The principal New Right critique of the welfare state highlights the lack of individual choice and control suffered by patients and clients which is exacerbated by the lack of alternatives to state provision. It is argued that the monopoly position of conventional welfare bureaucracies makes them wasteful and inefficient. They are operated by paternalistic professionals and administrators whose self-interest dictates that need is constantly redefined so that the welfare state can expand without limit. This leads to excessive taxation which stifles economic growth and overall prosperity suffers (see George and Wilding 1985, Chapter 2, for a summary of these views). Three broad categories of remedy to the ills of the post-war welfare state are proposed from the New Right: incentives to the private sector to develop a range of user-friendly rivals to state provision; privatization or contracting-out of state welfare services to the private and voluntary sectors; and the introduction into state welfare of the principles of the private market-place through competitive strategies such as voucher schemes or competition between provider organizations (Enthoven 1985).

A further challenge to the philosophy of state intervention and the academic discipline which grew up around it, has come from feminist critics of the welfare state (Wilson 1977; Ungerson 1985; Pascall 1986). Feminists from a variety of positions have drawn attention to the way in which the major institutions of the welfare state in western societies have either failed to give women the help they needed or reinforced material and ideological aspects of their subordination and dependency as users of welfare and as workers in the welfare services (Williams 1989, p. 10). In this analysis, the state is both capitalist, in that it manages the reproduction of the workforce, and patriarchal, in that it sustains family forms, a domestic division of labour, patterns of paid work, and a benefits system which contribute to women's oppression. Hilary Rose sums up the development of the welfare state in western Europe as ' . . . an accommodation between capital and a male-dominated labour movement . . . ' (Rose 1986, quoted in Williams 1989, p. 63). Feminist analysts of welfare have been particularly active in the health field in areas where the existence of biological differences has been used in the past to justify divisions of labour and power between men and women. In the area of reproductive technology, for example, the feminist critique of the new technologies has focused not so much on the technologies themselves, but on creating the political and cultural conditions in which the technologies can be used to realize women's own definitions of reproduction rather than those of predominantly male doctors (Stanworth 1987).

Another area where the feminist critique of what is biologically given and what is conditioned by material circumstances and ideology, has challenged existing welfare provision, lies in the analysis of the gendered nature of caring and the separation between paid work in the formal, visible economy and unpaid work in the informal, invisible economy of caring and how the state plays a part in the structuring of work in both sectors. These forms of analysis have led to a wide range of proposals for social policy change which challenge the existing sexual division of labour and the sharp distinction between paid (male) and unpaid (female) work, such as: schemes to promote the rights of women to receive benefits in their own right; the introduction of payments for married women caring for dependent relatives; and policies for better nursery and day-care for children. Analyses of policy cast in a broadly feminist mould have had a major impact in questioning fashionable policies for 'community care' of the elderly, the frail, and the disabled which rest in part on a greater use of informal sources of care. The recognition by feminist writers that community care relies on the unpaid work of women which can seriously undermine their own health and well-being is a crucial insight for the development of humane policies on the care of dependent groups (Finch and Groves 1983).

Conclusions

Research in the field of social policy and adminstration over the last 40 years has provided much of the raw material for the contemporary reaction in western Europe against the provision of welfare through large, monolithic, bureaucratic organizations in which professional power prevails over individual wishes. There are signs that the role of the state in welfare is changing from near-monopoly provider of many social services to that of regulator and auditor of services provided in the public, voluntary, and commercial sectors with a variety of public and private sources of funding (Klein 1987). Yet, the majority position within the discipline of social policy and administration, hitherto, has tended to be identified with conventional state provision despite its limitations and failures. As a result, the New Right political philosophers and economists appear to have succeeded for the moment, in portraying their market-style, competitive, solutions as the only feasible means to realize people's desires for choice, autonomy, and flexibility in their dealings with welfare services, while at the same time making the professionals accountable for the quality, appropriateness, and cost of their services (Harris 1988; Taylor-Gooby 1987). Questions such as who funds, who provides, and who regulates welfare services and how, which were taken for granted in the last 50 years are now up for debate as it is increasingly recognized that these roles need not be undertaken by a unitary public agency, nor indeed, exclusively by the state at all. Developments in information technology, for example, offer public authorities the potential to enforce notions of accountability on providers of services without direct, face-to-face control (Klein 1989*b*).

One of the principal tasks for social policy analysis in the 1990s will, thus, be to respond to this changing scene by evaluating the consequences for the complex range of objectives underlying each welfare system, of applying the principles of market competition and new forms of organization such as purchaser–provider contracting, to areas such as health and social care. While issues of efficiency, flexibility and responsiveness of services, together with client choice, are currently pre-eminent in the minds of policy-makers, the impact of social policy change on other enduring principles such as equity, social justice, individual rights, and democratic participation cannot and should not, be ignored. Social policy research must assess the extent to which fashionable changes in the form of welfare produce the empirical results predicted by their proponents, but must also evaluate their side-effects and wider normative implications. In the health field, the major reforms of the UK National Health Service which began with the 1989 White Paper *Working for Patients* (Secretaries of State for Health, Wales, Northern Ireland, and Scotland, 1989) are a notable example of the emerging trend towards market-style reforms of cherished welfare institutions with little rational analysis of options. The changes aim to inject the characteristics commonly associated with private commerce such as dynamism, a willingness to innovate, and a responsiveness to consumer views, into public health care through the development of an untried 'provider market' (King's Fund Institute 1989) involving public and private health care operators competing for the business of providing health services for the populations of health authorities which act as purchasing agencies. Similar proposals have been canvassed elsewhere, for example in New Zealand (Maxwell 1988) and Sweden (Saltman and von Otter 1987). These important and complex changes throw up numerous questions for social policy research; for example, is it possible to apply business methods and incentives to the NHS without damaging the essential principles underlying the Service (in so far as these continue to command popular support)? If so, what combinations of state, voluntary, and commercial funding, provision and regulation are most likely to achieve this?

Similarly challenging research agendas exist in many other fields of social policy relevant to the public health at a time of rapid socio-economic and policy change in western countries. It would be naive, however, to be too optimistic that the results of research on the consequences of market solutions will be responded to enthusiastically by western governments. There is little evidence of a commitment at present, to the use of the results of rational, applied social research by governments in countries such as the United Kingdom and the US where policy is more explicitly ideologically driven than it has been for several decades (Booth 1988). None the less, it is important that the discipline maintains its distinctive contribution to intellectual life by continuing to document and explain the impact on life chances and welfare of the interaction between socio-economic change and the social policy response.

References

Alford, R.R. (1975). *Health care politics*. University of Chicago Press, Chicago.

Allen, D. (1979). *Hospital planning*. Pitman Medical, London.

Allison, G.T. (1971). *Essence of decision*. Little Brown, Boston.

Bachrach, P. and Baratz, M.S. (1970). *Power and poverty*. Oxford University Press.

Banting, K.G. (1979). *Poverty, politics and policy*. Macmillan, London.

Blackman, T., Evason, E., Melaugh, M., and Woods, R. (1989). Housing and health: a case study of two areas in West Belfast. *Journal of Social Policy* **18**, 1.

Black Report (1980). *Inequalities in health: report of a research working group*. Department of Health and Social Security, London.

Booth, T. (1988). *Developing policy research*. Gower, Aldershot.

Brotherston, J. (1976) Inequality: is it inevitable? In *Equalities and inequalities in health* (ed. C.O. Carter and J. Peel). Academic Press, London.

Brown, M. and Madge, N. (1982). *Despite the welfare state*. Heinemann, London.

Brown, M. (1983). The development of social administration. In *Social policy and social welfare: A reader* (ed. M. Loney, D. Boswell, and J. Clarke) pp. 88–103. Open University Press, Milton Keynes.

Byrne, D., Harrison, S., Keithley, J., and McCarthy P. (1986). *Housing and health: The relationship between housing conditions and the health of council tenants*. Gower, London.

Calnan, M. (1982). The hospital Accident and Emergency Department: what is its role? *Journal of Social Policy* **11**, 483.

Cawson, A. (1982). *Corporatism and welfare*. Heinemann, London.

Crosland, C.A.R. (1956). *The future of socialism*. Cape, London.

Dahl, R.A. (1961). *Who governs? Democracy and power in the American city*. Yale University Press, New Haven, Connecticut.

Day, P. and Klein, R. (1983) The mobilisation of consent versus the management of conflict: Decoding the Griffiths report. *British Medical Journal* **287**, 1813.

Day, P. and Klein, R. (1987). Quality of institutional care and the elderly: policy issues and options. *British Medical Journal* **29**, 384.

Deakin, N. (1987). *The politics of welfare*. Methuen, London.

Department of Health and Social Security (1976). *Sharing resources for health in England: the report of the Resource Allocation Working Party*. HMSO, London.

Donnison, D.V. and Chapman, V. (1965). *Social policy and administration*. Allen & Unwin, London.

Donnison, D. (1979). Social policy since Titmuss. *Journal of Social Policy* **8**, 145.

Enthoven, A.C. (1985). *Reflections on the management of the National Health Service*. Nuffield Provincial Hospitals Trust, London.

Finch, J. and Groves, D. (ed.) (1983). *A labour of love: Women, work and caring*. Routledge & Kegan Paul, London.

Forder, A. (1974). *Concepts in social administration*. Routledge & Kegan Paul, London.

Forder, A., Caslin, T., Ponton, G., and Walklate, S. (1984). *Theories of welfare*. Routledge & Kegan Paul, London.

George, V. and Wilding, P. (1976). *Ideology and social welfare*. Routledge & Kegan Paul, London.

George, V. and Wilding, P. (1985). *Ideology and social welfare* (2nd rev. edn). Routledge & Kegan Paul, London.

Glennerster, H. (1988). Requiem for the Social Administration Association. *Journal of Social Policy* **17**, 83.

Gough, I. (1978). Theories of the welfare state: a critique. *International Journal of Health Services* **8**, 27.

Gough, I. (1979). *The political economy of the welfare state*. Macmillan, London.

Green, D.G. (1987). *The New Right: The counter-revolution in political, economic and social thought*. Harvester Wheatsheaf, Brighton.

Hadley, R. and Hatch, S. (1981). *Social welfare and the failure of the state*. Allen & Unwin, London.

Hall, P., Land, H., Parker, R., and Webb, A. (1975). *Change, choice and conflict in social policy*. Heinemann, London.

Ham, C. (1981). *Policy-making in the National Health Service: A case study of the Leeds Regional Hospital Board*. Macmillan, London.

Harris, R. (1988). *Beyond the welfare state*. IEA Occasional Paper No. 77. Institute for Economic Affairs, London.

Hunter, D. (1980). *Coping with uncertainty: Policy and politics in the National Health Service*. Research Studies Press, Wiley, Chichester.

Hunter, D. (1990). Organizing and managing health care: A challenge for medical sociology. In *Readings in medical sociology* (ed. S. Cunningham-Burley and N.P. McKeganey) pp. 213–36. Tavistock/Routledge, London.

Johnson, N. (1987). *The welfare state in transition: the theory and practice of welfare pluralism*. Harvester Wheatsheaf, Brighton.

King's Fund Institute (1989). *Managed competition: A new approach to health care in Britain*. KFI Briefing Paper No. 9. King's Fund Institute, London.

Klein, R. (1977). The corporate state, the health service and the professions. *New Universities Quarterly* **31**, 161.

Klein, R. (1983). *The politics of the National Health Service*. Longman, London.

Klein, R. (1987). Toward a new pluralism. *Health Policy* **8**, 5.

Klein, R. (1989a). *The politics of the National Health Service* (2nd rev. edn). Longman, London.

Klein, R. (1989b). *Social policy agendas of the 1980s: An overview*. Paper delivered to a plenary session of the 23rd Annual Conference of the Social Policy Association, University of Bath, 10–12 July 1989.

Le Grand, J. (1982). *The strategy of equality: Redistribution and the social services*. Allen & Unwin, London.

Lindblom, C.E. (1959). The science of 'muddling through'. *Public Administration Review* **19**, 79.

Marshall, T.H. (1975) *Social policy in the twentieth century* (4th rev. edn). Hutchinson, London.

Maxwell, R.J. (1988). New Zealand proposals to unshackle hospitals: suggestions to make regional health authorities purchasers of services. *British Medical Journal* **297**, 1214.

Means, R. and Smith, R. (1988). Implementing a pluralistic approach to evaluation in health education. *Policy and Politics* **16**, 17.

Merton, R.K. (1968). *Social theory and social structure*. Free Press, New York.

M'Gonigle, G.E.N. and Kirby, J. (1936). *Poverty and public health*. Gollancz, London.

Mills, C.W. (1956). *The power elite*. Oxford University Press, New York.

Mishra, R. (1977). *Society and social policy*. Macmillan, London.

Mishra, R. (1981). *Society and social policy* (2nd rev. edn). Macmillan, London.

Mishra, R. (1986). Social policy and the discipline of social administration. *Social Policy and Administration* **20**, 28.

Morris, J.N. and Titmuss, R.M. (1944). Health and social change: the recent history of rheumatic heart disease. *Medical Officer* **2**, 69–71; 77–79; 85–87.

Moser, K.A., Fox, A.J., and Jones, D.R. (1984). Unemployment and mortality in the OPCS longitudinal study. *Lancet* **ii**, 1324.

Navarro, V. (1978). *Class struggle, the state and medicine: An historical and contemporary analysis of the medical sector in Great Britain*. Martin Robertson, Oxford.

Navarro, V. (1989). Why some countries have national health insurance, others have national health services and the U.S. has neither. *Social Science and Medicine* **28**, 887.

NHS Management Enquiry (1983). *Report* (Chairman: Sir Roy Griffiths). Department of Health and Social Security, London.

O'Connor, J. (1973). *The fiscal crisis of the state*. St James's Press, New York.

Parker, J. (1975). *Social policy and citizenship*. Macmillan, London.

Pascall, G. (1986). *Social policy: A feminist analysis*. Tavistock, London.

Pinker, R. (1971). *Social theory and social policy*. Heinemann, London.

Pinker, R. (1979). *The idea of welfare*. Heinemann, London.

Platt, S., Martin, C.J., Hunt, S.M., and Lewis, C.W. (1989). Damp housing, mould growth and symptomatic health state. *British Medical Journal* **298**, 1673.

Pollitt, C., Harrison, S., Hunter, D., and Marnoch, G. (1988). The reluctant managers: Clinicians and budgets in the NHS. *Financial Accountability and Management* **3**, 213.

Rein, M. (1976). *Social science and public policy*. Penguin, Harmondsworth.

Reisman, D.A. (1977). *Richard Titmuss: Welfare and society*. Heinemann, London.

Ringen, S. (1987). *The possibility of politics: a study in the political economy of the welfare state*. Clarendon Press, Oxford.

Room, G. (1979). *The sociology of welfare*. Basil Blackwell, Oxford.

Rose, H. (1986). Women and the restructuring of the welfare state. In *Comparing welfare states and their futures* (ed. E. Owen). Gower, Aldershot.

Saltman, R.B. and von Otter, C. (1987). Re-vitalizing public health care systems: a proposal for public competition in Sweden. *Health Policy* **7**, 21.

Secretaries of State for Health, Wales, Northern Ireland and Scotland (1989). *Working for patients*. Cm. 555. HMSO, London.

Smith, A. and Jacobson, B. (ed.) (1988). *The nation's health: A strategy for the 1990s*. King Edward's Hospital Fund for London, London.

Smith, G. and Cantley, C. (1985). *Assessing health care: A study in organizational evaluation*. Open University Press, Milton Keynes.

Stanworth, M. (ed.) (1987). *Reproductive technologies: Gender, motherhood and medicine*. Polity Press, Cambridge.

Stevenson, T.H.C. (1923). The social distribution of mortality from different causes in England and Wales 1910–1912. *Biometrika* **15**, 382.

Taylor-Gooby, P. and Dale, J. (1981). *Social theory and social welfare*. Edward Arnold, London.

Taylor-Gooby, P. (1987). Welfare attitudes: Cleavage, consensus and citizenship. *Quarterly Journal of Social Affairs* **3**, 199.

Titmuss, R.M. (1938). *Poverty and population*: A factual study of contemporary social waste. Macmillan, London.

Titmuss, R. (1963). The social division of welfare: Some reflections on the search for equity. In *Essays on the welfare state*, pp. 34–55. Allen & Unwin, London.

Titmuss, R. (1968). The subject of social administration. In *Commitment to welfare*, pp. 13–24. Allen & Unwin, London.

Titmuss, R. (1970). *The gift relationship*. Allen & Unwin, London.

Titmuss, R. (1974). *Social policy*. Allen & Unwin, London.

Townsend, P. (1974). Poverty as relative deprivation: Resources and style of living. In *Poverty, inequality and class structure* (ed. D. Wedderburn) pp. 15–41. Cambridge University Press, London.

Townsend, P. (1976). *Sociology and social policy*. Penguin, Harmondsworth.

Townsend, P. (1979). *Poverty in the United Kingdom*. Penguin, Harmondsworth.

Townsend, P., Phillimore, P., and Beattie, A. (1988). *Health and deprivation: Inequality and the North*. Croom Helm, London.

Ungerson, C. (ed.) (1985). *Women and social policy: A reader*. Macmillan, London.

Walker, A. (1981). Social policy, social administration and the social construction of welfare. *Sociology* **15**, 225.

Walker, R. (1988). Syllogism based on licensed premises. *The Times Higher Education Supplement*, 18 March, p. 15.

Whitehead, M. (1987). *The health divide*. Health Education Council, London.

Wilensky, H.L. and Lebeaux, C.N. (1958). *Industrial society and social welfare*. Free Press, New York.

Williams, F. (1989). *Social policy: A critical introduction. Issues of race, gender and class*. Polity Press, Cambridge.

Wilson, E. (1977). *Women and the welfare state*. Tavistock, London.

D

Field Investigations of Physical, Biological, and Chemical Hazards

23

The principles of an epidemic field investigation

MICHAEL B. GREGG and JULIE PARSONNET

Introduction

This chapter contains a simple, practical, and essentially non-technical discussion of how to prepare for and perform an epidemic field investigation. We focus attention on a presumed point-source (common-source) epidemic, recognized and reported by local health authorities to a regional or state health department. This frequent and typical scenario highlights some key operational and public health issues. The discussion describes the tasks to perform, but attention is also directed towards important operational and public health policy concerns. Although the chapter centres on an acute infectious disease epidemic in a community, the epidemiological and public health principles apply equally well to non-infectious diseases.

Background considerations

Overall purposes and methodology

As mentioned in earlier chapters, the purposes of epidemiology are to determine the cause(s) of a disease, its source, its mode of transmission, who is at risk of developing disease, and what exposures predispose to disease. With answers to these questions, the epidemiologist hopes to control and prevent disease. Clearly, these purposes also apply to field investigations of infectious diseases. Fortunately, in many outbreak investigations, the clinical syndromes are easily identifiable, the agents can be readily isolated and characterized, and the source, mode of transmission, and risk factors of the disease are usually well known and understood. Therefore, the epidemiologists are often quite well prepared for their field investigations. However, when the clinical diagnosis and/or laboratory findings are unclear, the task becomes much more difficult. Careful consideration of the clinical presentation of disease is needed in order to obtain key information regarding the source, mode of spread, and population(s) at risk of disease. For example, bacterial contamina-tion of food or water is usually manifested by signs and symptoms referable to the gastrointestinal tract. Pathogenic agents transmitted in air often affect the respiratory tract and sometimes the skin, eyes or mucous membranes. Skin abrasions or lesions may suggest animal or insect transmission. So the clinical manifestations of disease may serve as critical leads for epidemiologists who may at times have no other information to guide them. Regardless of how secure the clinical diagnosis may be, the thought process must include clinical, laboratory, and epidemiological evidence. Together, these provide leads and pathways to take or reject, so that ultimately the natural history of the epidemic will be understood.

Although epidemiologists perform several separate operations (which are listed below), in broad strokes they really do two things. First, they collect information that describes the setting of the outbreak, namely, over what time period people became sick, where they acquired disease, and what the characteristics of the ill people were. These are the descriptive aspects of the investigation. Often, simply by knowing these facts (and the diagnosis), the epidemiologist can determine the source and mode of spread of the agent and can identify those primarily at risk of developing disease. Common sense will often give these answers, and relatively little, if any, further analysis is required.

However, on occasion, it will not be readily apparent where the agent resided, how it was transmitted, who was at risk of disease, and what the risk factors were. Under these circumstances, a second operation, analytical epidemiology, must be used, hopefully to provide the answers. As has been described in other chapters, epidemiological analyses require comparisons of persons—ill and well, exposed and not exposed. In epidemic situations, the epidemiologist usually compares ill and well people—both believed at risk of disease—to determine what exposures ill people had that well people did not. These comparisons are made using appropriate statistical techniques. If the differences between ill and well are greater than one would expect by chance, the epidemiologist can draw certain inferences regarding the transmission of and exposure to the disease.

The pace and commitment of a field investigation

An underlying theme woven through this chapter emphasizes the need to act quickly, to establish clear operational priorities, and to perform the investigation responsibly. This should not imply the haphazard collection and inappropriate analysis of data, but rather the use of simple and workable case definitions, case-finding methods, and analyses. If at all possible, one should collect data, perform analyses, and make recommendations in the field as part of the investigation. There is often a strong tendency to collect what is believed to be the essential information in the field and then retreat to 'home base' for analysis—particularly with the availability of computers. However, such action may be viewed as lack of interest or concern, or even of possessiveness, by the local constituency. Equally important, premature departure also makes any further collection of data or direct contact with study populations and local health officials difficult, if not impossible. Once home, the epidemiology team has lost the urgency and momentum to perform, the sense of relevancy of the epidemic, and, most of all, the totally committed time for the investigation. Every field investigation should be completed not only to the team's satisfaction, but particuly to the satisfaction of the local health officers as well. Contemplate what the local health department must face if the team leaves without providing reasonably final results and firmly based recommendations.

Recognition and response to a request for assistance

The report

The regional health officer may learn of an epidemic from a variety of sources such as the local health department, a private physician, a hospital administrator, a concerned citizen, or perhaps even the news media. Generally, the most direct and reliable source of information is a local health official. There will be a 'working' diagnosis, an estimate of the number of cases, the background expected number of cases, and the affected population. Reports of a possible or real epidemic from sources such as private physicians, hospitals, etc. may reveal only a segment of the overall picture of the epidemic or may, indeed, not reflect the existence of an epidemic at all. Therefore, when reports such as these are received at the regional level, the regional official should contact the local health officials and inform them of the reports. Local officials will usually try to verify such reports and, if verified, they will often investigate the epidemic themselves. Even if no epidemic is ultimately recognized or no request for assistance results, the regional health official has clearly discharged an important responsibility by reporting back to the local health department.

The request

However, if local health officials request assistance, the regional epidemiologist, before making any decision, should try to acquire as much information as possible regarding the diagnosis, the normal occurrence of disease, and the population primarily affected. Quite frequently, local health departments will have performed a preliminary and sometimes relatively extensive investigation before calling for assistance. They will be able to provide a considerable amount of valuable information which can be used in planning for the investigation.

It is important to find out exactly why the request for assistance is forthcoming. Does the local health department simply need an extra pair of hands to perform or complete the investigation? Has it been unable to uncover the nature or source of infection or the mode of spread, thereby limiting adequate control or prevention? Perhaps the health department wants to share the responsibility of the investigation with a more seasoned and knowledgeable health authority so as to be relieved of local political or scientific pressure. Occasionally, legal or ethical issues may have become prominent in the early investigation. Those responding to requests for help must be aware of these possibilities. Rarely, an epidemic may even be declared or announced by local authorities or citizens. Assistance is then requested in order to publicize perceived adverse health conditions, to awaken state or national health leaders, or even to secure funds. Regardless of the motivation behind a call for assistance, there must be an established official basis for such a request and official local permission for an epidemiological investigation. Many a field study has been aborted simply because either those requesting assistance had no authority to do so or state, regional, or national teams were investigating without local permission.

The response and the responsibilities

The relationships between regional and local health departments vary not only from region to region within countries but also from country to country. In general, the larger health districts help serve the smaller in times of need. Yet the sensitivities between these two authorities are frequently delicate, particularly as they relate to perceived competence, local jurisdiction, and ultimate authority. The regional health officer must decide—on the basis of prevailing local-provincial amenities and agreements and his or her best judgment— the most appropriate response. There are several important reasons why requests for a field investigation should be answered, if not encouraged:

(a) to control and prevent further disease;

(b) to provide agreed upon or statutorily mandated services;

(c) to derive more information about interactions among the human host, the agent and the environment;

(d) to assess the quality of epidemiological surveillance at the local level;

(e) to maintain or improve such epidemiological surveillance by personal and direct contact;

(f) to establish a new system of epidemiological surveillance;

(g) to provide training opportunities in practical field epidemiology.

If the regional health officer decides to provide field assistance, both he or she and the local health official should discuss and hopefully agree upon (i) what resources (including personnel) will be available locally; (ii) what resources will be provided by the regional team; (iii) who will direct the day-to-day investigation; (iv) who will provide overall supervision and ultimately be responsible for the investigation; (v) how the data will be shared and who will be responsible for their analysis; (vi) will a report of the findings be written, who will write it, to whom will it go; and (vii) who will be the senior author of a scientific paper, should one be written. These are extremely critical issues, some of which cannot be totally resolved before the investigative team arrives on the scene. However, they must be addressed, discussed openly, and agreed upon as soon as possible.

Preparation for the field investigation

No attempt is made here to describe in detail what personnel or equipment should be deployed for the field investigation. These decisions will clearly depend upon the presumed cause, magnitude, geographical extent of the epidemic, and the local and regional resources available. Rather, the emphasis focuses upon the necessary collaborative relationships between health professionals at the regional office and key instructions to the investigating team before they depart.

Collaboration and consultation

Virtually all infectious disease outbreaks require the support of a competent laboratory. Even if local laboratories are capable of processing and identifying specimens, the regional epidemiological team should immediately, upon being informed of the proposed investigation, contact their counterparts within their regional laboratories. These microbiologists should be requested to provide any needed guidance and laboratory assistance. Now is the time to obtain assurance of co-operation and commitment rather than during the field investigation or near the end when specimens have already been collected and await testing. Not only must the microbiologists schedule the processing of specimens, but they should be asked to recommend what kinds of specimens to collect and how they should be collected and processed. There also may be substantive basic or applied research questions that could be appropriately addressed and answered during the field investigation. These issues should be discussed in detail with the microbiologists and every effort made to enlist and support their interest.

Advice on statistical methods may be sought at this time as well. The same philosophy applies also to contacting other health professionals, such as veterinarians, mammalogists, or entomologists, whose expertise can be crucial to a successful field investigation. Moreover, serious consideration should be given to including such professionals on the investigative team. It is important to determine whether such scientists should be part of the initial team so that appropriate information and, particularly, specimens can be collected concomitantly with other relevant epidemiological information.

Other persons who can be extremely important in the overall management of a field investigation are information specialists. When large outbreaks of disease are to be investigated that will likely attract even moderate local or regional attention in the news media, the presence of an experienced and knowledgeable information officer who can respond to public inquiries and meet with the news media on a regular basis can be invaluable. Some consideration should be given to including secretarial and/or administrative personnel on the investigating team—not only to utilize their services but to expose them to a real-life investigation. By such experience they will return home with a better understanding of field work and an increased ability to support technical personnel in the regional offices.

Basic administrative instructions

Once the field team has been designated, certain key instructions should be emphasized.

1. Identify the team leader and the person to whom he or she should report regularly at the regional level.

2. Specify when and how communications should be established with the regional home base for information and guidance. Do not permit the investigating team to notify, as it is convenient, the regional supervisors of the progress of the investigation. Establish within reason fixed times and places for regular communication regardless of whether new facts or findings have been uncovered. There may be just as important reasons for the home base to communicate with the field team as the reverse.

3. Emphasize the need for the team to meet with appropriate local health officials immediately upon arriving in the field. If the local official has not already been identified, instruct the team leader to determine as soon as possible who at the local level will be in charge. Encourage the team to identify and meet with all persons they may need co-operation from in the investigation. Such persons include local health department directors and/or chiefs of epidemiology, laboratory services, vital statistics, nursing, and maternal and child health. Other important persons would often include the mayor, the local medical society, or hospital administrators and staff. It is highly preferable to take the day or so needed to meet these persons initially—so that key doors will be opened—than to spend valuable time later in the investigation mending bridges.

4. Have the team leader identify the appropriate local person to speak for the entire investigative team when necessary. In general, the regional team should try to avoid direct contact with the news media and should always defer to local health officials. The investigative team is usually

working at the request and under the aegis of the local health authorities. Therefore, it is the local officials who not only know and appreciate the local aspects of the epidemic but who are the appropriate persons to comment on the findings of the investigation. In the most practical sense, the less the news media make contact with the investigative team, the more can be done at the pace and discretion of that team.

5. Before leaving to conduct an investigation, the team leader or preferably his or her immediate supervisor should write a memorandum. It should summarize how and when the region was contacted, what information was provided by the local health department, what the proposed response by the region is, what the agreed upon commitments of both local and regional health authorities are, who is on the field team, and when the latter is expected to arrive in the field. This memorandum should be distributed to key supervisors in both regional and local health offices, and to others who may have need to know.

The field investigation

Before the actual field activities are discussed, it should be borne in mind that the order of the tasks should not be considered fixed or binding but rather logical in terms of field operations and epidemiological thinking. The epidemiologist may perform several of these functions simultaneously or in different order during the investigation and may even institute control and preventive measures soon after beginning the investigation on the basis of intuitive reasoning and/or common sense. No two epidemiologists will take the same pathway of investigation. Yet, in general, the data they collect, the analyses they apply, and the control and preventive measures they recommend will likely be similar.

Since, by definition, the epidemic in question has resulted from a point source and may be continuing or nearly over before the field team arrives, the investigation will be retrospective in nature. This should alert the epidemiologist to some fundamental aspects of any investigation that occurs after the fact. First of all, because many illnesses and critical events have already occurred, virtually all information acquired and related to the epidemic will be based upon memory. Health officers, physicians, and patients are likely to have different recollections, views, or perceptions of what transpired, what caused the disease, and even who or what was responsible for the epidemic. Information may conflict, may not be accurate, and certainly cannot be expected to reflect the precise recounting of past events. Yet, just as the clinician may ask patients what they think is making them sick, the epidemiologist will do well to ask members of the affected community what they think caused the epidemic.

For the young, inexperienced, medical epidemiologist steeped in the tradition of molecule and millimole determinations, the 'more-or-less' measurements of the field epidemiologist can initially be major hurdles to the successful field investigation. However lacking in accuracy these data may be, they are the only data available, and must be collected, analysed, and interpreted with care, imagination, and caution.

Determine the existence of an epidemic

In most instances, local health officials will know whether more cases of disease are occurring than would normally be expected. Since most local health departments have ongoing records of the occurrence of communicable disease, comparisons by week, month, and year can be made to determine whether the observed numbers exceed the normally expected level. Although strict laboratory confirmation may be lacking at this time, an increase in the number of cases of a disease reasonably accurately reported by local physicians should stimulate further inquiry. However, the terms 'epidemic' and 'outbreak' are quite subjective, depending not only upon how local health officials view the expected rises and falls in disease incidence, but upon whether such changes merit investigation. One must be acutely aware of artefactual causes of increases or decreases in numbers of reported cases, such as changes in local reporting practices, increased interest in certain diseases because of local or national awareness, or changes in methods of diagnosis. Even the presence of a new physician or clinic in the community may lead to a substantial increase in reported numbers of cases, yet may not represent a true increase above normal.

In certain situations, however, it may be difficult to document the existence of an epidemic rapidly. One may need to acquire information from such sources as school or factory absentee records, out-patient clinic visits, hospitalizations, laboratory records, or death certificates. Sometimes a simple survey of practising physicians will strongly support the existence of an epidemic, as would a similar rapid survey of households in the community. Frequently, such quick assessments, entail asking about signs and symptoms rather than about specific diagnoses. For example, such inquiry might involve asking physicians or clinics if they are treating more people than usual with sore throats, gastro-enteritis, fever with rash, etc., in order to obtain an index of disease incidence. Although not specific for any given disease, such surveys can often document the occurrence of an epidemic. Sometimes it is extremely difficult to establish satisfactorily the existence of an epidemic. Yet, because of local pressures, epidemiologists may be obliged to continue the investigation even if they believe that no significant health problem exists.

Confirm the diagnosis

Every effort possible should be made to confirm the clinical diagnosis by standard laboratory techniques such as serology and/or isolation and characterization of the agent. One should not attempt to apply newly introduced, experimental, or otherwise not broadly recognized confirmatory tests—at least not at this stage of the investigation. If at all possible, visit the laboratory and verify the laboratory findings in person. Not every reported case has to be confirmed. If most patients have the expected or similar clinical signs and symptoms and, perhaps, 15–20 per cent of the cases are laboratory confirmed, one does not need more confirmation at this time. This should be ample confirmatory evidence. One should try to examine several representative cases of the disease as well; clinical assumptions should not be made. The diagnosis should be verified by a physician member of the team. Nothing convinces epidemiologists and responsible health officers more than an eyewitness confirmation of clinical disease by the investigating team.

Create a case definition and determine the number of cases

Now the epidemiologist must create a workable case definition, decide how to find cases, and count cases. The simplest and most objective criteria for a case definition are usually the best (for example, fever, X-ray evidence of pneumonia, white blood cells in the spinal fluid, number of bowel movements per day, blood in the stool, skin rash, etc.). However, be guided by the accepted, usual presentation of the disease, with or without standard laboratory confirmation, in the case definition. Where time may be a critical factor in a rapidly unfolding field investigation, a simple, easily applicable definition should be used—recognizing that some cases will be missed and some non-cases included. Some factors that can help determine the levels of sensitivity and specificity of the case definition are the following:

1. What is the usual apparent-to-inapparent clinical case ratio?
2. What are the important and obvious pathognomonic or strongly clinically suggestive signs and symptoms of the disease?
3. What isolations, identification, and serological techniques are easy, practicable, and reliable?
4. How accessible are the patients or those at risk; can they be recontacted after the initial investigation for follow-up questions, examination, or serology?
5. In the event that the investigation requires long-term follow-up, can the case definitions be applied easily and consistently by individuals other than the current investigating team?
6. Is it absolutely necessary that all patients be identified during the initial investigation, or would only those seen by physicians or hospitalized suffice?

These considerations and others will likely play an important role in how cases will be defined and how intensive case investigation will be. However, no matter what criteria are used, the case definition must be applied equally and without bias to all persons under investigation.

Methods for finding cases will vary considerably according to the disease in question and the community setting. In many field investigations the techniques for identifying cases will be relatively self-evident. Most outbreaks involve certain clearly identifiable groups at risk. It is simply a matter of

intensifying reporting from physicians, hospitals, laboratories, or school and industrial contacts, or perhaps using some form of public announcement to identify most of the remaining, unreported cases. However, there may be times when more intensive efforts—such as physician, telephone, door-to-door, culture or serological surveys—may be necessary to find cases. Regardless of the method, some system(s) of case identification must be established for the duration of the investigation and perhaps for some time afterwards.

In the vast majority of instances, simply determining the number of cases does not provide adequate information. Control and prevention measures depend upon knowing the source and mode of spread of an agent as well as the characteristics of ill patients. Therefore, the process of case finding should include collecting pertinent information likely to provide clues or leads to the natural history of the epidemic and, particularly, relevant characteristics of the ill. First, one should collect basic information about each patient's age, sex, residence, occupation, date of onset, etc., to define the simple and basic descriptive aspects of the epidemic. However, if the disease under investigation is usually water- or foodborne, one should ask questions about exposure to various water and food sources. If the disease is most frequently transmitted by person-to-person contact, one should seek information that will help determine the frequency, duration, and nature of personal contacts. If the nature of the disease is not known or cannot be comfortably presumed, the epidemiologist will need to ask a variety of questions covering all possible aspects of disease transmission and risk.

Orient the data in terms of time, place, and person

Having now reasonably accurately determined the number of cases of disease and an attack rate, the field epidemiologist should record the descriptive aspects of the investigation. Characterize the epidemic in terms of when patients became ill, where patients resided or became ill, and what characteristics the patients possess. There may be a tendency to wait until the epidemic is over or until all likely cases have been reported before performing such an analysis. This tendency should be strongly avoided because further inclusion of a proportionately small number of cases will usually not affect the analysis or recommendations. Moreover, the earlier one can develop ideas of why the epidemic started, the more pertinent and accurate are the data that one can collect.

Time

In most instances, it will be valuable to describe the cases by time of onset by constructing a graph that depicts the occurrence of cases over an appropriate time interval (Fig. 23.1). This 'epidemic curve', as it is frequently called, can give a considerably deeper appreciation for the magnitude of the outbreak, its possible mode of spread, and the possible duration of the epidemic than would a simple listing of cases. A remarkable amount of information can be inferred from a

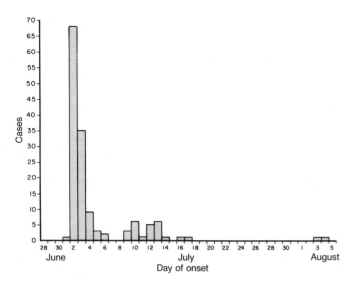

Fig. 23.1. Cases of Pontiac fever, by day of onset, Oakland County (Michigan) Health Department, 28 June to 5 August 1968.

pictorial representation of times of onset of disease. If the incubation period of the disease is known, relatively firm inferences can be made regarding the likelihood of a point-source exposure, person-to-person spread, or a mixture of the two. Also, if the epidemic is in progress, one may be able to predict, using the epidemic curve, how many more cases are likely to occur. Finally, a pictorial representation of cases over time serves as an excellent way of ready communication to non-epidemiologists, administrators, and the like who need to grasp in some fashion the nature and magnitude of the epidemic. The epidemic curve in Figure 23.1 portrays cases of Pontiac fever (subsequently confirmed legionnaires' disease) that occurred in Pontiac, Michigan, in July and August 1968, by day of onset (Glick *et al.* 1978). The epidemic was explosive in onset suggesting (a) a virtually simultaneous common-source exposure of many persons; (b) a disease with a short incubation period; and (c) a continuing exposure spanning several weeks.

Place

Not infrequently, exposures to microbial agents occur in unique locations in the community, which, if properly depicted and analysed, may provide major clues or evidence regarding the source of the agent and/or the nature of exposure. Water supplies, milk distribution routes, sewage disposal outflows, prevailing wind currents, air-flow patterns in buildings, and ecological habitats of animals may play important roles in disseminating microbial pathogens and determining who is at risk of acquiring disease. If cases are plotted geographically, a pattern of distribution may emerge that approximates these known sources, and routes of potential exposure may help identify the vehicle or mode of transmission.

Figure 23.2 illustrates the usefulness of a 'spot map' in the

investigation of an outbreak of shigellosis in Dubuque, Iowa, in 1974 (Rosenberg *et al.* 1976). Initially the investigation revealed that cases were not clustered by area of residence or by age or sex. A history of drinking water gave no useful clue as to a possible source and mode of transmission. However, it was later learned that many cases had been exposed to water by recent swimming in a camping park located on the Mississippi River. As can be seen in Figure 23.2, the river sites where 22 culture-positive cases swam within 3 days of onset of illness strongly suggested a common source of exposure. Ultimately, the epidemiologists incriminated the Mississippi River water by documenting gross contamination by the city's sewage treatment plant 5 miles upstream and by isolating *Shigella sonnei* from a sample of river water taken from the camping park beach area.

Person

Lastly, the epidemiologist should examine the character of the patients in terms of a variety of attributes, such as age,

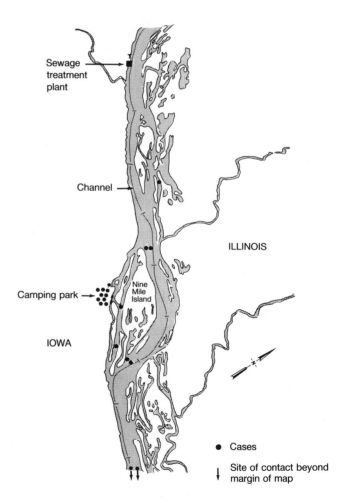

Fig. 23.2. Mississippi River sites where 2 culture-positive cases swam within 3 days of onset of illness.

sex, race, occupation, or virtually any other characteristic that may be useful in portraying the uniqueness of the case population. Some diseases primarily affect certain age groups or races; frequently, occupation is a key characteristic of people with certain infectious diseases. The list of human characteristics is nearly endless. However, the more investigators know about the infectious disease in question (the reservoir, mode(s) of spread, persons usually at greatest risk), the more specific and pertinent is the information they can seek from the cases to determine whether any of these characteristics predisposes to illness.

Determine who is at risk of becoming ill

It is at this time in the investigation that epidemiologists may begin to apply analytical techniques. They now know the number of people who are ill, when and where they were when they became ill, and what their general characteristics are, and usually they will have a firm diagnosis or a good 'working' diagnosis. These data frequently provide enough information to the field team to feel relatively sure how and why the epidemic started. For example, often this information will strongly suggest that only people in a particular community supplied by a specific water system were at risk of becoming sick, or that only certain students in school or workers in a single factory became ill. Perhaps it was only a group of people who attended a local restaurant who reported illness. In other words, the simple descriptive aspects of the epidemic frequently identify those most likely at risk of disease. However, no matter how obvious it might appear that only a single group of persons was at risk, the epidemiologist should apply analytical methods to support this conclusion.

For example, if fever, abdominal cramps, and diarrhoea occur among 35 residents of a housing subdivision 'A' (presumably caused by water contaminated with *Shigella*) and no similar illness was reported from subdivision 'B' over the same time period, one would logically conclude that only subdivision 'A' residents were at risk of developing disease. However, only after a proper survey is applied to the residents of subdivision 'B' or even elsewhere in the community, looking for the same illness and comparing illness rates in such groups, can one legitimately infer that the at-risk population were all residents of subdivision 'A'.

Develop a hypothesis that explains the specific exposure that caused disease and test this hypothesis by appropriate statistical methods

The next analytical step is often the most difficult one to perform. By now the epidemiologists should have an excellent grasp of the epidemic and an overall feel for the most likely source and mode of transmission. However, they must still determine the most likely exposure that caused disease. A classic example of what is meant here would be an investigation of an outbreak of nausea, vomiting, and diarrhoea among people who had attended a church supper presumedly caused by staphylococcal contamination of food(s) eaten at the supper. Since the disease was most likely acquired by eating or drinking something (because of the signs and symptoms) and because no other cluster of similar disease had occurred elsewhere in the community, the investigator focused attention only on those who attended the supper. The hypothesis posed was that the exposure necessary to develop nausea, vomiting, and diarrhoea was consumption of some food(s) contaminated with staphylococcal enterotoxin. Therefore, the ill people were asked what they had eaten or drunk at the church supper (i.e. what they had been exposed to) and their food histories were compared with those of people who had not become ill but who had also attended the church supper. Comparisons of food histories (eating rates) between the ill and the well participants were made and statistically analysed. The food histories were found to be very different (and unlikely to be different simply by chance alone). Therefore, the inference was drawn that a particular food or drink was the exposure that caused the illness.

Several other examples of epidemiological investigations of infectious diseases may serve to emphasize the importance of developing hypotheses and testing them.

In August 1980, a community hospital in the State of Michigan recognized seven cases of group A streptococcal post-operative wound infections which had occurred over the previous four months (Berkelman *et al.* 1982). This represented more cases than usual, and an investigation was started. Using a standard case definition the epidemiologists ultimately identified ten cases that had occurred over this time period, all of whom were patients on several surgical wards. This geographical clustering, plus the fact that the infections developed within one to two days following surgery, suggested a common source of exposure, presumably in the operating rooms. Although streptococcal disease can rarely be transmitted by inanimate objects, within the hospital setting, the most frequent source of streptococci is humans, and the most common mode of spread is person-to-person. Therefore, the epidemiologists hypothesized that the probable risk factor unique to these patients was contact with or exposure to an infected or colonized member of the hospital's professional staff sometime during surgery. After selecting appropriate non-streptococcal-infected post-surgical patients as controls, the investigators compared both ill and well patients with regard to what exposure they had with a total of 38 surgeons, anaesthetists, and nursing staff during surgery for the epidemic period. Exposure rates were not statistically different between cases and controls except for one nurse. This nurse was cultured and found to be an anal and vaginal carrier of a strain of beta-haemolytic streptococci identical to the epidemic strain. After appropriate treatment this strain of streptococcus could no longer be cultured from her. She returned to work and six months later two more cases of post-operative beta-haemolytic streptococcal infection occurred, caused, however, by a different serotype. The nurse was cultured again and found to be a vaginal and anal carrier of this different strain.

The investigation of an epidemic of *Listeria monocytogenes* is a classic example of how the epidemiological method, simple inferences, and persistent re-examination of data can point to a hitherto unknown source and mode of spread of disease. Thirty-four cases of perinatal listeriosis and seven cases of adult disease occurred between 1 March and 1 September 1981 in several maritime provinces of Canada (Schlech *et al.* 1983). These cases represented a several-fold increase over the number of cases diagnosed in previous years, suggesting some common exposure. Although *L. monocytogenes* is a common cause of abortion and nervous system diseases in cattle, sheep, and goats, the source of human infection has been obscure. The epidemiologists therefore undertook an investigation to determine if cases had contact with one another or whether there had been a common environmental source which would explain their disease. Cases could not be linked together by person-to-person contact; they shared no common water source; and food exposures, as determined from a general food history, were not different between cases and controls. However, a second, more detailed, food history and subsequent intensive interrogation of cases and controls revealed that there was a statistically significant difference between cases and controls regarding exposure to coleslaw. Even though this food had never been previously incriminated as a source of *Listeria*, it was the only food item statistically incriminated and essentially the only lead the investigators had at the time. Armed with this association, the epidemiologists subsequently found a specimen of coleslaw in the refrigerator of one of the patients which grew out the same serotype of *Listeria* found in the epidemic cases. No other food items in the refrigerator were positive for *Listeria*. The coleslaw had been prepared by a regional manufacturer who had obtained cabbages and carrots from several wholesale dealers and many local farmers. Although environmental cultures from the coleslaw plant failed to reveal *Listeria* organisms, two unopened packages of coleslaw from the plant subsequently grew *L. monocytogenes* of the same epidemic serotype. A review of the sources of the vegetable ingredients was made, and a single farmer was identified who had grown cabbages and also maintained a flock of sheep. Two of his sheep had previously died from listeriosis in 1979 and 1981. Also, he was in the habit of using sheep manure to fertilize his cabbage.

It cannot definitely be proven that this particular farm was the source of the *Listeria* organisms that caused the epidemic. However, the hypothesis that coleslaw was the source and the statistical test which supported this hypothesis provided the necessary impetus to continue the investigation. Ultimately a single highly suggestive source of the bacteria was discovered. These findings strongly implied listeriosis as a zoonotic infection transmitted from infected animals via contaminated vegetables to humans.

Lastly, a similar logic in a much more difficult situation was applied to an outbreak of legionnaries' disease among conventioneers attending meetings in Philadelphia, Pennsylva-

nia, in July 1976 (Fraser *et al.* 1977). From the very beginning to the end of the field investigation, neither the clinical presentation nor laboratory results provided the epidemiologist with a diagnosis. However, initially it appeared and was epidemiologically established that disease was not transmitted from person to person. Yet being a conventioneer who stayed at or visited the Bellevue Stratford Hotel in Philadelphia conferred an increased risk of disease. But this conclusion did not provide enough information about the source or mode of spread of the agent to be particularly useful. To put it another way, the person's presence in the Bellevue Stratford did not alone help to explain what the specific exposure was or how the disease was acquired. A legionnaire could easily have eaten a meal, consumed water from the hotel, or simply breathed in the hotel—all possible exposures to the agent that could place a person in high-risk category of becoming ill. Therefore, a series of hypothesis was proposed to determine whether eating meals, drinking water, or simply being in the hotel conferred increased risk of developing illness among the legionnaires. When the final analysis was done, there was no significant difference between ill and well legionnaires in terms of eating or drinking at the hotel. However, spending at least one hour in the hotel lobby conferred a much greater risk of disease than would have been expected by chance alone. Therefore, the investigators inferred that being in the lobby of the hotel was the necessary exposure for acquiring disease. This, coupled with the clinical features of the disease (pneumonia), implied that the agent was air borne and was transmitted through the air-conditioning system. Although the bacterium responsible for legionnaires' disease was not isolated from the Bellevue Stratford Hotel's air-conditioning system at that time, it was later recovered from lung tissue of several diseased legionnaires. Moreover, subsequent investigations of similar epidemics of legionnaires' disease elsewhere have not only confirmed the epidemiological pattern of this disease, but *Legionella* bacteria have been isolated from similar air-conditioning systems.

Again, this phase of the investigation will clearly pose the greatest challenge to epidemiologists. They must review their findings carefully, weigh the clinical, laboratory, and epidemiological features of the disease, and hypothesize possible exposures that could plausibly cause disease. In other words, they must seek, from the patient's histories, exposures that could conceivably predispose to illness. If exposure histories for ill and well people are not significantly different, the epidemiologist must develop new hypotheses. This may require imagination, perseverance, and sometimes resurveying those at risk to obtain more pertinent information.

Compare the hypothesis with the established facts

Having determined by epidemiological and statistical inference the probable exposure responsible for disease, the epidemiologist still must 'square' the hypothesis with the clinical, laboratory, and other epidemiological facts of the investigation. In other words, do the proposed exposure, mode of spread, and population affected fit well with the known facts of the disease? For example, if in the gastro-enteritis outbreak referred to above the analysis incriminated an uncooked food left at room temperature for 18–24 hours and previously known to promote growth of staphylococci, the hypothesis fits well with our understanding of staphylococcal food poisoning. However, if the analysis incriminated coffee or water—highly unlikely sources of staphylococcal enterotoxin—the epidemiologist must then reassess the findings, perhaps secure more information, reconsider the clinical diagnosis, and certainly pose and test new hypotheses.

In the field investigation, when the disease is undiagnosed, the epidemiologist will clearly find it very difficult to fit a hypothesis to the natural history of the disease in question. All that can be hoped for is that the clinical, laboratory, and epidemiological findings portray a coherent, plausible, and physiologically sound series of findings and events that make sense.

Plan a more systematic study

The actual field investigation and analyses may be completed by now, requiring only a written report (see below). However, because there may be a need to find more patients, to define better the extent of the epidemic, or because a new laboratory method or case-finding technique may need to be evaluated, the epidemiologists may want to perform more detailed and carefully executed studies. With the pressure of the investigation somewhat removed, the field team may now consider surveying the population at risk in a variety of ways to help improve the quality of data and answer particular questions. Perhaps the most important reasons to perform such studies are to improve the sensitivity and specificity of the case definition and establish more accurately the true number of persons at risk, i.e. to improve the quality of numerators and denominators. For example, serosurveys coupled with a more complete clinical history can often sharpen the accuracy of the case count and define more clearly those truly at risk of developing disease. Moreover, repeat interviews of patients with confirmed disease may allow for rough quantitation of degrees of exposure or dose responses—useful information in understanding the pathogenesis of certain diseases.

Preparation of a written report

Frequently, the final responsibility of the investigative team is to prepare a written report to document the investigation, the findings, and the recommendations. It is beyond the scope of this chapter to provide a detailed set of guide-lines on scientific report writing. However, in most instances there are several important reasons why a report should be written as soon as possible.

Administrative/operational purposes
A document for action

Not infrequently, control and prevention efforts will be taken only when a report of all relevant findings has been written. This can and should place a heavy but necessary burden on the epidemiologists to complete their work quickly. Even if all possible cases have not yet been found, or some laboratory results are still pending, reasonable written assumptions and recommendations can usually be made without fear of retraction or subsequent major change.

A record of performance

In this day of input and output measurements, programme planning, programme justifications, and performance evaluations, there is often no better record of accomplishment than a well-written report of a completed field investigation. The number of investigations performed and the time and resources expended not only document the magnitude of health problems, changes in disease trends, and the results of control and prevention efforts, but serve as concrete evidence of programme justification and needs.

A document for potential medical/legal issues

Presumably, epidemiologists investigate epidemics with objective, unbiased, and scientific purposes, and similarly prepare written reports of their findings and conclusions objectively, honestly, and fairly. Such information may prove absolutely invaluable to consumer, practising physician, or local and state health department officials in any legal action regarding health responsibilities and jurisdictions. In the long run, the health of the public is best served by simple, careful, honest documentation of events and findings made generally available for interpretation and comment.

Scientific/epidemiological purposes
Enhancement of the quality of the investigation

Although not fully explained and rarely referred to, the actual process of writing and viewing data in written form often generates new and different thought processes and associations in the mind of the epidemiologist. The discipline of committing to paper the clinical, laboratory, and epidemiological findings of an epidemic investigation almost always will bring to light not only a better understanding of the natural unfolding of events, but their importance in terms of the natural history and development of the epidemic. The actual process of creating scientific prose, summarizing data, and creating tables and figures representing the known established facts forces one to view the entire series of events in a balanced, rational, and explainable way. This is considerably more so than an oral report given to the local health department the day of departure from the field. Occasionally, previously unrecognized associations will emerge from a careful and step-by-step written analysis that may be critical in the final interpretation and recommendations. The exercise of writing what was done and what was found will sometimes uncover facts and events that were more or less assumed to be true but not specifically sought for during the investigation. This in turn may stimulate further inquiry and fact finding in order to verify these assumptions.

An instrument for teaching epidemiology

There would hardly be disagreement among epidemiologists that the exercise of writing the results of an investigation constitutes an essential building block in learning epidemiology. Much the way a lawyer prepares a brief, the epidemiologist should know how to organize and present in logical sequence the important and pertinent findings of an investigation, their quality and validity, and the scientific inferences that can be made by their written presentation. The simple, direct, and orderly array of facts and inferences will reflect not only on the quality of the investigation itself but also on the writer's basic understanding and knowledge of the epidemiological method.

Execute control and prevention measures

It is not the purpose of this chapter to elaborate on this aspect of the investigation. Nevertheless, the underlying purposes of all epidemic investigations are to control and/or prevent further disease.

Conclusion

In summary, then, the process of performing a thorough and successful epidemic field investigation has two major components. The first is the operational aspect that includes knowing why the investigation should be performed, who are the principal health officers involved, and who will assume the primary responsibility for data collection, interpretation, and implementation of preventive and control measures. The epidemiologist must also identify other health professionals who will provide the necessary laboratory and field support early in the planning stage.

Second, the field investigation is a direct application of the epidemiological method, very often with an implied relatively circumscribed timetable. This forces the investigative team (i) to establish workable case finding techniques; (ii) to collect data rapidly but carefully; and (iii) to describe these cases in a general sense regarding the time and place of occurrence and those primarily affected. Usually, the infectious agent is known and its sources and modes of transmission are well established, allowing the epidemiologist to identify the source and mode of spread rapidly. However, when the clinical disease is obscure and/or the origin of the agent is ill defined, one may be hard pressed to create a hypothesis that will not only identify the critical exposure and show statistical significance, but will logically explain the occurrence of the epidemic. Although scientific proof of causation in the strictest sense will not be established by such retrospective investigations, in most instances the careful development of epidemiological inferences coupled with persuasive clinical and laboratory evidence will almost always provide convincing evidence of the source and mode of spread of disease.

Appendix: important bacterial diseases

This appendix describes 13 bacterial diseases or conditions that are important to the industrialized world in terms of frequency of occurrence, duration of morbidity, and general impact upon the health economy. No attempt has been made to quantify their importance in any formal way; however, these choices represent the authors' best assessment following discussions with experts in the field of bacterial diseases.* Although there may be differences of opinion regarding the importance of some of the bacterial diseases discussed, an attempt was made to direct the attention of the reader to bacterial pathogens that appear to be emerging as important problems for the next several decades. For instance, *Campylobacter* infections have, within the recent past, become recognized as important agents of gastrointestinal disease in the industrialized world. Likewise, the sexually transmitted chlamydial infections of the genito-urinary tract are of increasing importance, occurring substantially more frequently even than gonorrhoea in the US and parts of Europe.

One section in the following pages discusses nosocomial bacterial infections. This is the only included disease category not caused by a single or closely related group of bacterial agents. However, nosocomial bacterial infections represent a continuing and probably slowly expanding disease problem, particularly for hospitals in the developed world that care for the elderly, the chronically ill, and those with altered or diminished immune mechanisms. As long as these kinds of patients constitute a substantial portion of medical responsibility within the hospital setting, we can expect to see an increasing burden of nosocomial infections.

The diseases are listed in alphabetical order and not by any implied order of importance. The description of these conditions is short; it is not the purpose of this chapter, nor of the text to which it belongs, to describe such diseases in any clinical, laboratory, or epidemiological detail. There are a variety of excellent texts that can provide the reader with extensive descriptions of these illnesses. Selected references are listed at the end of the chapter.

Bacterial nosocomial infections
Clinical picture

A variety of clinical presentations occurs as a result of bacterial nosocomial infections but the great majority involve urinary tract infection, surgical wound infection, and lower respiratory tract infection. Primary bacteraemia, although considerably less common than other in-hospital infections, is

---

* Exact incidence of some of these diseases is not really known for the US. However, the Centers for Disease Control in Atlanta, Georgia, US, has attempted to make estimates of certain bacterial disease occurrence using a variety of sources and techniques for its own programme and planning purposes. Some of these estimates have been used in the discussions that follow.

the most serious manifestation of nosocomial disease. Urinary tract infections comprise approximately 42 per cent of known bacterial infections; they are generally mild, often asymptomatic, and resolve spontaneously. Some patients develop ascending infection, however, resulting in pyelonephritis or secondary bacteraemia. Surgical wound infections account for 24 per cent of bacterial nosocomial infections, and vary according to surgical procedure. Presentation is relatively characteristic with low-grade fever, redness, swelling, and pain at the surgical operative site. Lower respiratory infections, characterized by pneumonia, lung abscesses, empyema, or bronchitis, equal about 10 per cent of bacterial nosocomial infections. Diagnosis may be overlooked because of underlying medical and/or surgical conditions. Primary bacteraemia (5 per cent of all bacterial nosocomial infections) causes high morbidity and mortality within the hospital and often appears as spiking fever, shaking chills, and sometimes profound hypotension.

Agent

Among the Gram-negative bacteria *Escherichia coli*, *Pseudomonas aeruginosa*, and species of *Enterobacter*, *Proteus*, and *Providencia* are the most frequently isolated organisms. Of the Gram-positive bacteria, *Staphylococcus aureus* and enterococcal group B streptococci are most frequently reported.

Epidemiology

Bacterial nosocomial infections occur world-wide. They have become recognized as a major cause of morbidity and mortality, particularly in the developed world. In the US in 1975–6, it was estimated that nosocomial bacterial infection rates were approximately 5.7 per 100 hospital discharges—about 2.1 million infections annually in acute care hospitals. More recent estimates, including all hospitals, double this figure.

In the majority of instances, the reservoir of bacteria is the patient. Although hospital personnel and medical or surgical equipment may be sources of infection, pathogenic bacteria of the genito-urinary tract, of the skin, and of the respiratory tract most frequently come from the patient. For urinary tract infections, the indwelling catheter most frequently introduces faecal organisms into the bladder. For skin infections, staphylococci from the skin of the patient enter the incision site. For most lower respiratory tract infection, aspiration of respiratory secretions provides the most frequent mode of introduction of pathogens. Incubation periods of disease vary tremendously according to the susceptibility and natural resistance of the patient, the mode of introduction of the organism, and the virulence and pathogenicity of the particular bacterium. However, because most infections tend to result from invasive procedures (indwelling catheters, surgery, mechanical methods for removing respiratory secretions), incubation periods tend to be short, usually hours to several days in length.

In general, nosocomial bacterial infections become communicable in an epidemic sense only when hospital staff inadvertently carry them from patient to patient or when equipment or instruments within the hospital are inadequately sterilized and transmit the organisms.

Methods of control and prevention

Primary preventive and control programmes are based on intensive, active surveillance of bacterial infection in the hospital. Surveillance includes an active search for infections by regular examination of patients; by regular, careful, review of hospital charts; and by a similar review of bacterial laboratory results. These duties are best performed by a specially designated full-time hospital epidemiologist who reviews these data daily, looking for clustering of disease, by time, place, or by organism, and frequently by antimicrobial susceptibility patterns of bacteria.

Careful attention should always be paid to proper sterilization and handling of hospital equipment, proper food handling, and careful housekeeping and laundry services.

Campylobacter enteritis

Clinical picture

Campylobacter enteritis is usually characterized by the abrupt onset of diarrhoea, bloody stools, fever, and abdominal pain. Most patients recover in less than a week, but up to 20 per cent of cases may relapse or have prolonged illness. Milder, non-specific disease with watery diarrhoea may also occur.

Agent

Campylobacter jejuni and *Campylobacter coli*, the two most commonly isolated agents of *Campylobacter* enteritis, are Gram-negative rods.

Epidemiology

The organism is distributed world-wide. In the US, infants and young adults are primarily affected; in developing countries, the disease occurs almost exclusively among young children. Recent estimates judge the occurrence of *Campylobacter* infection in the US to approach that of *Salmonella*, approximately 2–3 million cases per year. Common-source outbreaks have been associated with unpasteurized milk, untreated water, and contaminated foods, particularly poultry. Sporadic cases occur and may result from inadequate food preparation or from contact with infected pets or domesticated animals. *Campylobacter* is also a cause of 'traveller's diarrhoea'. Person-to-person transmission is not well documented. The incubation period is 2–10 days. Asymptomatic infection can occur.

Methods of prevention and control

Prevention rests on thorough cooking and preparation of foods derived from animal sources, particularly poultry. Pasteurization of milk and disinfection of water before drinking can also prevent transmission.

Chlamydial genito-urinary infections
Clinical picture
The bacterial genus *Chlamydia* is comprised of the species *Chlamydia psittaci* and *C. trachomatis*, and a proposed third species, *C. twar*. *C. psittaci* causes the pulmonary disease, psittacosis. *C. twar* is a newly recognized agent of pharyngitis and pneumonia. *C. trachomatis*, the most important of the three species, is responsible for a myriad of clinical illnesses, including trachoma, inclusion conjunctivitis, lymphogranuloma venereum, infant pneumonia (sometimes called chlamydial pneumonia of infancy) and genito-urinary tract infections. In the developed world, genito-urinary infection generates the highest incidence, prevalence, and morbidity among the *C. trachomatis* syndromes.

   C. trachomatis genito-urinary tract infection in males may present as urethritis, epididymitis, prostatitis, or proctitis. Symptoms of *Chlamydia* urethritis, commonly called non-gonococcal urethritis, include urethral irritation, genital itching, and frank dysuria; these are generally less severe than symptoms associated with gonococcal urethritis. Scant, mucoid urethral discharge may be seen. If untreated, symptoms can persist for months or years. In women, genito-urinary *C. trachomatis* infections resemble gonococcal infections, causing salpingitis, cervicitis, and sometimes urethritis. However, women may be completely asymptomatic. Infertility is the most serious consequence of disease in both males and females.

Agent
Chlamydia trachomatis is a small Gram-negative, spherical bacterium.

Epidemiology
Although the organism is found world-wide, the sexually transmitted genito-urinary infections have only recently been described as a problem in western Europe and the US. In both the US and the UK, chlamydial genito-urinary infections are estimated to have twice the incidence of gonorrhoea. Transmission occurs when infected secretions come into contact with mucous membranes of the genito-urinary tract. The incubation period and duration of communicability have not been well established; most patients develop symptoms 1–3 weeks after sexual contact with an infected partner. Because the infection may persist for several months, communicability can last for extended periods of time if the disease is left untreated.

Methods of prevention and control
These methods are similar to those of syphilis and gonorrhoea.

Gonorrhoea
Clinical picture
Gonococcal infections may manifest themselves in a variety of clinical syndromes, including urethritis, epididymitis, cervicitis, salpingitis, conjunctivitis of the new-born, and rarely bacteraemia with arthritis and endocarditis. The more frequent and important clinical manifestations of gonorrhoea are acute urethritis in the male with urethral discharge and pain upon urination, and pelvic inflammatory disease in the female, with fever, abdominal pain, vaginal discharge, salpingitis, and endometritis. Gonococcal conjunctivitis of the new-born appears as acute redness and swelling of the conjunctiva with a purulent discharge and usually occurs within the first 3 weeks of life.

Agent
Neisseria gonorrhoeae is a Gram-negative diplococcus.

Epidemiology
The reservoir for *N. gonorrhoeae* is exclusively humans. More than 700 000 cases were reported in the US in 1988, representing an estimated half of the true incidence. The disease is primarily transmitted by sexual contact with exudates from mucuous membranes of infected individuals. Gonococcal conjunctivitis of the new-born is transmitted by contact of the baby with the infected birth canal of a carrier of the organism. The incubation period is usually 2–7 days for sexually acquired gonorrhoea and approximately 1–5 days for gonorrhoeal conjunctivitis. Communicability of gonorrhoea varies but may last for months, particularly in asymptomatic females who can become chronic carriers. Susceptibility to the organism is general, and no immunity is developed following infection.

Methods of control and prevention
These are identical to those that apply to syphilis. In addition, prevention of gonococcal conjunctivitis in the new-born can be accomplished by immediate installation of appropriate doses of silver nitrate, tetracycline, or erythromycin into the eyes.

Haemophilus influenzae infection
Clinical picture
Haemophilus influenzae causes two important groups of clinical diseases, particularly in infants and children: invasive disease (meningitis, epiglottitis, or sepsis) and otitis media. Meningitis caused by this organism is usually sudden in onset with fever, vomiting, meningeal irritation, stiff neck, and frequently lethargy and coma. *H. influenzae* epiglottitis produces hyperacute, potentially lethal, respiratory obstruction. Sepsis can occur without associated focal disease and generally presents as fever, anorexia, and lethargy in infants. Otitis media, often a precursor of *Haemophilus* meningitis, is characterized by irritability, ear pain, and fever, usually following an upper respiratory infection.

Agent
Haemophilus influenzae is a small pleomorphic Gram-negative coccobacillus. There are six antigenically distinct

capsular types and other non-typable strains. Otitis isolates are almost always non-typable.

Epidemiology

Haemophilus influenzae infection is world-wide in distribution and occurs most frequently in the age group of from 2 months to 3 years old. It accounts for approximately 25 per cent of cases of acute suppurative otitis media, a major cause of deafness. The reservoir is humans; the source of the organism is the upper respiratory tract. Asymptomatic colonization is frequent, and person-to-person transmission occurs by droplet nuclei from nasal and pharyngeal discharge during the infectious period. The incubation period is usually 2–4 days, and persons presumably remain communicable for as long as the organism remains in the upper respiratory tract. Immunity, as reflected by circulating antibody, follows infection with the specific capsular type.

Methods of prevention and control

There are no vaccines presently available for general use in infants, the group at highest risk. However, a vaccine has been licensed for use in children aged from 18 to 60 months. The vaccines are directed against capsule type b and will not prevent infection due to other types or unencapsulated strains such as those responsible for otitis media. Prevention measures centre primarily around the practice of good, general, personal hygiene and education of parents regarding the risk of secondary cases in children less than 6 years of age. Rifampicin can be given prophylactically to children less than 4 years old who are close household contacts of patients with invasive *H. influenzae*.

Haemorrhagic colitis

Clinical picture

Escherichia coli can cause several types of gastro-intestinal illness, from mild, self-limited 'turista' to refractory, chronic diarrhoea. Most recently, some members of the species have been found to cause a severe illness called haemorrhagic colitis. Affected patients have low grade fever, cramps, and watery, followed by bloody, diarrhoea. As many as 10–20 per cent of haemorrhagic colitis patients, usually those at the extremes of age go on to develop the haemolytic uraemic syndrome or thrombotic thrombocytopenic purpura, life-threatening triads of renal failure, anaemia, and thrombocytopenia.

Agent

The family of *E. coli* contains hundreds of different serotypes. In the US, the serotype 0157:H7 is that most often associated with haemorrhagic colitis, although other serotypes have caused the disease as well.

Epidemiology

Although no national surveillance for *E. coli* 0157:H7 exists, some evidence suggests that it is the third most common bacterial cause of diarrhoea in some regions of the US, behind *Salmonella* and *Campylobacter*. In Canada, the prevalence may be even higher. Cattle serve as a reservoir for the organism, and outbreaks have been traced to both unpasteurized milk and to rare ground beef. Other animal hosts have not yet been identified. Person-to-person transmission can occur and outbreaks have been seen in nursing homes, institutions for the handicapped, and day-care centres. The incubation period is 3–8 days. The period of communicability is unknown.

Methods of control and prevention

Beef should be eaten well done and unpasteurized milk products should be avoided. In institutional outbreaks, cohorting of ill patients and strict enteric precautions may limit transmission. No vaccine is available.

Listeria monocytogenes

Clinical picture

Listeria causes a spectrum of disease ranging from a non-specific flu-like illness and spontaneous abortion in pregnant women to sepsis and fatal meningo-encephalitis in infants and immunocompromised hosts. Although infection occurs infrequently in the US, the severe morbidity and high mortality associated with *Listeria* make it an important pathogen.

Epidemiology

Approximately 1500–2000 cases of *L. monocytogenes* occur in the US each year, primarily in pregnant women, newborns, the elderly, and the immunocompromised. The organism is ubiquitous in the environment and has been found in dust, water, vegetation, animal and human waste, and a wide variety of foods. Outbreaks of listeriosis have been traced to dairy products (cheese and milk) and vegetables fertilized with animal waste. Most cases, however, are sporadic and have no clearly identifiable source, although contaminated, undercooked meats have been suggested as vehicles for some of these cases. Person-to-person spread has not been documented. The incubation period of *Listeria*-related disease is unknown. Cases in outbreaks have occurred up to 10 weeks after exposure.

Methods of control and prevention

Until risks factors for *L. monocytogenes* are better delineated, prevention will be difficult. Recognition of outbreaks and elimination of epidemiologically suspect vehicles are the only effective disease control methods to date.

Pneumococcal infections

Clinical picture

The most frequent clinical manifestations of pneumococcal disease are lobar pneumonia and otitis media. Pneumococcal pneumonia is an acute infection, most severe in infants and the elderly, characterized by the sudden onset of chills, fever, chest pain, difficulty breathing, and a productive cough.

Acute otitis media commonly complicates viral upper respiratory infections in children and is characterized by fever, irritability, and pain in the ear.

Agent

Streptococcus pneumoniae is a Gram-positive coccus which has more than 80 identifiable serotypes.

Epidemiology

The pneumococci are ubiquitous organisms frequently present in the upper respiratory tract of normal carriers. Pulmonary disease occurs world-wide, is usually sporadic in nature, but occasionally occurs in epidemics in closed populations such as schools or mental hospitals. The estimated range of cases occurring in the US is from 150 000 to 570 000 per year. The pneumococcus is responsible for approximately 40 per cent of cases of suppurative otitis media in the US. Pneumococcal diseases are more prevalent during the late winter and spring in temperate climate zones. Humans are the only known reservoir, and the organisms are transmitted from person to person by direct contact or by droplet spread of respiratory secretions. The incubation period has been estimated to be between 1 and 3 days, but the period of communicability is unknown. Transmission probably can occur as long as the bacteria are present in the respiratory discharges. For most persons, resistance is generally high, but factors regarding suspectibility are poorly understood. Immunity following infection with a specific capsular serotype may, however, last for years.

Methods of control and prevention

In general, avoidance of crowded conditions, particularly in living quarters, may provide some method for prevention. A 23-valent pneumococcal vaccine is presently available in the US. Although currently available data show persistence of vaccine efficacy in adults for at least 5 years after immunization, vaccine efficacy and antibody titres may decline later.

Salmonellosis

Clinical picture

Salmonella causes different clinical syndromes including gastro-enteritis, enteric fever (including typhoid fever), and rarely bacteraemia and focal infections of the meninges, bones, and other organs. The most common form of salmonellosis is acute gastro-enteritis, manifested by sudden onset of diarrhoea, abdominal cramps, fever, and vomiting, occasionally severe enough to warrant hospitalization. The enteric fevers are serious infections which present non-specifically with fever, headache, myalgias, and cough.

Agent

The genus *Salmonella* includes approximately 2000 different serotypes of Gram-negative rods. The prevalence of serotypes varies by region; in the US, the most important human pathogens are *Salmonella typhimurium* and *S. enteritidis*.

Infection with *Salmonella typhi*, the primary agent of enteric fever, is rare in industrialized nations but poses unique public health problems because of its severity and the potential for development of a chronic carrier state.

Epidemiology

Salmonella infections occur world-wide and typically cause gastro-enteritis. Over two million cases of salmonellosis are estimated to occur in the US each year. The principal reservoirs of non-typhoidal salmonellae are animals, including poultry, livestock, and pets. Consequently, infection and illness generally result from eating contaminated foods of animal origin. Outbreaks have been traced to meats, poultry, inadequately pasteurized milk, and intact grade-A shell eggs. Water and vegetables can also cause infection if contaminated with excrement. Occasionally person-to-person transmission occurs. The incubation period for *Salmonella* gastro-enteritis is 6–72 hours. Communicability of non-typhoidal *Salmonella* extends throughout the course of illness.

Salmonella typhi occurs only in humans, and transmission is almost always from contamination of food or water by waste from a human carrier or an acute case. The febrile illness has a 6–30 day incubation period. Up to 5 per cent of *S. typhi* patients develop a chronic carrier state and shed infectious organisms in stool or urine for years.

Methods of control and prevention

Careful preparation, thorough cooking, and adequate storage of foods of animal origin are important preventive measures for non-typhoidal *Salmonella*. Education about handwashing, personal hygiene, and sanitary sewage disposal may also help in prevention and control. For *S. typhi*, a vaccine can enhance resistance to infection, but the quality and duration of protection is limited. Only those persons with high risk of exposure, such as household contacts of typhoid carriers, travellers to highly endemic areas, and laboratory workers who frequently work with *S. typhi*, should be considered for vaccination. Antibiotics in high dosages for prolonged periods are sometimes effective in eradicating the chronic carrier state; experimentally, quinolones show particular promise. However, often cholecystectomy is required, and even this is not always successful.

Staphylococcal infections

Clinical picture

The staphylococci can produce a wide variety of syndromes that range from the simplest of superficial skin infections to staphylococcal septicaemia and pneumonia. Staphylococcal enteritis and the most recently discovered clinical disease associated with this agent, toxic-shock syndrome, represent totally different clinical manifestations of disease. The skin lesions, characterized as a single pustule or by impetigo, are usually self-limiting infections. However, they may progress into carbuncles, furuncles, or even cellulitis where systemic signs of fever and general malaise may be prominent. As referred to in the section on nosocomial infections, staphylococcal wound infections are an important cause of morbidity within the hospital setting. Staphylococcal pneumonia has been frequently recognized as a secondary pulmonary invader following influenza infection. The most common agent in foodborne epidemics is the staphylococcus which, after incubating in certain foods, produces an enterotoxin that causes characteristic explosive nausea, vomiting, and diarrhoea. The source of the foodborne organism is usually a food-handler carrier who contaminates the food during handling and storage procedures.

Toxic-shock syndrome, a relatively new disease, is caused by one or more toxins elaborated by certain strains of staphylococcus. The syndrome is characterized by the sudden onset of high fever, vomiting, diarrhoea, and muscle aches. This is often followed by a profound drop in blood pressure and is accompanied by an erythematous 'sunburn-like' rash often with desquamation, particularly of the palms and the soles.

Agent

Staphylococcus aureus is a Gram-positive coccus.

Epidemiology

The staphylococcus is a ubiquitous bacterium normally present on the skin and in the gastro-intestinal tract. It is particularly prevalent where the use of soap and water is suboptimal and where people live in crowded accommodations. Infections usually occur sporadically but can occasionally occur in small epidemics and, of a most serious nature, in nosocomial epidemics.

Epidemics of foodborne staphylococcal gastro-enteritis occur world-wide and are usually focal.

Toxic-shock syndrome, a very rare disease originally seen primarily in children, became epidemic in scope in the US in 1980–1, appearing most frequently in menstruating women who were using tampons during menstruation.

Humans are the only known reservoir of staphylococci, and the organisms are transmitted primarily from person to person by direct contact. Incubation periods are variable but range between 4 and 10 days. Persons remain communicable as long as their lesions continue to contain the bacteria. Carriage, particularly in the anterior nares or in the anogenital region, is common and is not associated with clinical signs or symptoms. Carriers can be important common sources of infection, especially in hospitals. Susceptibility is universal, and no immunity is conferred following specific strain infection.

Methods of prevention and control

Personal hygiene and frequent handwashing are the most effective ways of preventing spread of staphylococcal disease. Proper preparation and storage of food is essential in prevention of foodborne staphylococcal outbreaks. Within the hospital setting, careful handwashing, cohorting of ill patients, and isolation may be necessary when highly resistant staphylococci cause nosocomial outbreaks.

Streptococcal infections
Clinical picture

Group A streptococci cause a variety of acute clinical diseases, the most important of which are acute tonsillitis/pharyngitis and skin infections such as impetigo and pyoderma. Other clinical infections include scarlet fever, which is simply a form of streptococcal pharyngitis with a characteristic skin rash; and erysipelas, a superficial cellulitis associated with tender, red skin lesions and fever.

The most important aspect of streptococcal upper repiratory tract infection relates to its non-suppurative sequelae, rheumatic fever and acute glomerulonephritis, both of which occur uncommonly after acute infection but may cause permanent damage to the heart, the heart valves, and the kidneys.

Group B streptococci also produce human disease. In newborn children group B organisms are an important cause of septicaemia, pulmonary infection, and meningitis. During the perinatal period, group B streptococci are also an important cause of endometritis, amnionitis, and urinary tract infection in women.

Agent

The group A beta-haemolytic streptococcus, *Streptococcus pyogenes*, comprises more than 70 serologically distinct types. Most of these strains can produce upper respiratory tract infection and subsequent rheumatic fever. However, only about a dozen serotypes have been associated with acute glomerulonephritis.

Group B streptococci comprise five serotypes, all of which can produce infections in new-borns and adults.

Epidemiology

Group A beta-haemolytic pharyngitis is world-wide in occurrence and appears sporadically, particularly in the late winter and early spring in temperate climates. Explosive outbreaks may occur, usually associated with food (especially milk and milk products) contaminated with the bacteria. Humans are the primary reservoir of beta-haemolytic streptococci, although the agent may be isolated occasionally from domestic animals. The disease is transmitted by close, intimate contact with infected persons. Rarely, streptococci are transmitted via the airborne route from carriers to susceptibles. Pyoderma and impetigo are transmitted by close physical contact.

The incubation period of streptococcal tonsillitis/pharyngitis is between 2 and 5 days, and the period of communicability may last for weeks to months in chronic carriers of the disease. Communicability of the organism is greatest during acute infection, however. Asymptomatic carriage of streptococci occurs in schoolchildren and its incidence may be as high as 20 per cent or more. Susceptibility to the streptococci is virtually universal although immunity may last for many years after specific group A streptococcal infections.

Group B streptococcal infection occurs world-wide. The

organism resides commonly in the gastro-intestinal tract and the female genital tract. New-borns acquire infection during passage through the birth canal or *in utero*; those who become ill within the next several months probably acquire infection from the environment. The period of communicability is not known but is presumed to last throughout colonization. The incubation period of early neonatal disease is less than 3 days. The incubation period of perinatal disease in the first few months of life is not known.

Methods of control and prevention

For group A beta-haemolytic streptococci, proper preparation and storage of food, particularly milk and eggs, is necessary to prevent foodborne outbreaks. Food handlers with skin lesions or respiratory illness should be excluded from work. For persons with recurrent streptococcal infection, long-term monthly injections of long-acting penicillin have proved very successful.

There are no routinely accepted measures for prevention of group B streptococcal infections.

Syphilis

Clinical picture

The venereal form of syphilis is characterized by acute, subacute, and chronic clinical manifestations. Clinical presentation may vary widely according to the predominant site of infection and the stage of disease. Characteristically, however, non-congenital syphilis is acquired by sexual contact; within about 3 weeks, a primary skin lesion appears as a papule, usually turning into a superficial chancre most commonly located on the genitalia. After 4–6 weeks, the initial lesion will disappear and often a secondary generalized skin rash will appear, sometimes with mild systemic symptoms. These skin lesion may vanish spontaneously within weeks or last for months when the latency period of infection begins. Latency may last for years before signs and symptoms of central nervous system disease or cardiovascular disease occur. The organism may also invade the bone, liver, and almost any other organ late in the disease.

Syphilis can also be acquired congenitally from an infected mother through placental infection. Congenital syphilis may result acutely in liver disease or pneumonia, or disease may appear much later with bone, cartilage, and central nervous system involvement.

Agent

Treponema pallidum, a spirochaete, is the bacterium responsible for venereal syphilis.

Epidemiology

Syphilis is a disease of world-wide occurrence with a higher frequency in urban and highly populated areas. Although over 40 000 cases were reported in the US in 1988, this represents approximately 50 per cent of the true incidence of disease; cases have increased over the past 5 years. The reservoir is exclusively humans, and disease is transmitted by direct contact with infectious exudates of primary or secondary lesions. Infectious contact takes place most frequently during intercourse, but rarely may result from kissing or other kinds of close exposure. Congenital syphilis is acquired by placental transfer of the bacterium. The incubation period varies from 10 to 90 days after exposure, but is usually 3 weeks. Communicability may also vary considerably, but is greatest during the primary and secondary stages when there are open, highly infected lesions. Communicability may last intermittently for 2–4 years. Although all ages are susceptible, disease occurs most frequently in the sexually active years. No immunity develops following infection.

Methods of control and prevention

Primary methods for prevention and control centre around health and sex education, provision of facilities for early diagnosis and treatment, and rapid identification of contacts of known patients.

Tuberculosis

Clinical picture

Infection may manifest itself in a variety of ways, but usually tuberculosis is a disease of the lungs. It very often goes unnoticed and heals spontaneously. When pulmonary disease does appear clinically, however, it usually starts with the gradual onset of low-grade fever, fatigue, weight loss, and persistent cough, often with haemoptysis. Other forms of tuberculous infection may affect organs other than the lungs, such as the meninges, lymph nodes, knees, bones, joints, larynx, and skin.

Agent

The most important species of mycobacterium causing human disease is *Mycobacterium tuberculosis*. It is a rod-shaped bacillus that can often be identified in the sputum of infected individuals using special stains (the so-called 'acid fast' quality of these bacteria).

Epidemiology

Tuberculosis is world-wide in distribution, but has been rapidly declining in the western world. The reservoir of infection is primarily humans. Tuberculosis is acquired by inhalation of the bacilli which are suspended as droplet nuclei in respiratory secretions of infected persons. The incubation period varies considerably but usually ranges from between 4 and 12 weeks. Disease remains communicable as long as individuals are excreting infectious bacilli in their respiratory secretions. People of all ages are susceptible to tuberculosis; those at greatest risk of developing disease, however, are children under 3 years of age. The most common period for development of clinical diseases is the first 6–12 months after infection. Latent infections may persist for many years. Although the US has reported progressive declines in tuberculosis rates through 1985, the incidence has slightly increased or stayed nearly stable through 1988. Thus, tuberculosis remains an important bacterial disease. Diagnosis, treatment, and follow-up are still costly and time consuming. Refugee populations with high incidence rates of antimicrobially resistant have magnified the public health hazard.

Methods of control and prevention

Patients actively excreting bacilli in the sputum can be rendered non-infectious by rapid, appropriate chemotherapy, usually within several weeks. Treatment with isoniazid prevents progression of latent infection into active clinical disease in many individuals. Bacillus Calmette–Geurin (BCG) vaccine may be an important method of tuberculosis control in certain situations, but the degree of protection has been variable in different studies and locations.

Further reading

American Academy of Pediatrics (1982). *Report of the Committee on Infectious Diseases*. American Academy of Pediatrics, Evanston, Illinois.

Benenson, A.S. (ed.) (1985). *Control of communicable diseases in man* (14th edn). The American Public Health Association, Washington, DC.

Bennett, J.V. and Brachman, P.S. (ed.) (1979). *Hospital infections*. Little, Brown and Company, Boston, Massachusetts.

Christie, A.B. (1980). *Infectious diseases: epidemiology and clinical practice* (3rd edn). Churchill Livingstone, New York.

Evans, A.S. and Feldman, H.A. (ed.) (1982). *Bacterial infections of humans. Epidemiology and control*. Plenum, New York.

Mandell, G.L., Douglas, R.G., Jr, and Bennett, J.E. (1979). *Principles and practice of infectious diseases* (2 vols). Wiley, New York.

Sanford, J.P. and Luby, J.P. (ed.) (1981). *The science and practice of clinical medicine*, Vol. 8. Grune and Stratton, New York.

References

Berkelman, R.L., Martin, D., Graham, D.R., *et al.* (1982). Streptococcal wound infections caused by a vaginal carrier. *Journal of the American Medical Association* **247**, 2680.

Fraser, D.W., Tsai, T.R., Orenstein, W., *et al.* (1977). Legionnaires' disease. Description of an epidemic of pneumonia. *New England Journal of Medicine* **297**, 1189.

Glick, T.H., Gregg, M.B., Berman, B., *et al.* (1978). Pontiac fever. An epidemic of unknown etiology in a health department: I. Clinical and epidemiological aspects. *American Journal of Epidemiology* **107**, 149.

Rosenberg, M.L., Hazlet, K.K., Schaefer, J., *et al.* (1976). Shigellosis from swimming. *Journal of the American Medical Association* **236**, 1849.

Schlech, W.F. III, Lavigne, P.M., Bortobussi, R.A., *et al.* (1983). Epidemic listeriosis—evidence for transmission by food. *New England Journal of Medicine* **308**, 203.

24

Transmissible agents

NORMAN D. NOAH

Introduction

The epidemiological methods used for the investigation of infectious disease are not substantially different from those used for any epidemiological investigation. Nevertheless, there are some investigative techniques which are particularly applicable to infectious diseases. They include the techniques of investigation of the host, agent, and environment, which are described elsewhere in this book, and techniques of surveillance and vaccine trials which are described in this chapter.

Population surveillance

The definition of population surveillance is:

ongoing scrutiny, generally using methods distinguished by their practicability, uniformity and, frequently, their rapidity, rather than complete accuracy (Last 1988).

It is thus a type of observational study (Thacker *et al.* 1983). There are several key words in the definition as provided by Last which make his the preferred definition. *Ongoing* distinguishes surveillance from a survey. *Practicability* is the essence of any workable surveillance system. *Uniformity* ensures that the data can be interpreted sensibly, especially as surveillance is all about trends. *Rapidity* is important for the system to be useful. And the words *rather than complete accuracy* sum up the rough and ready philosophy behind most surveillance systems which exist primarily for following trends in disease patterns. Many of these points, however, need to be qualified and are further discussed later in this section.

The word 'monitoring' should not be used interchangeably with surveillance: it is the ongoing evaluation of a control or management process (Eylenbosch and Noah 1988). The techniques of surveillance are usually necessary for efficient monitoring. Thus monitoring the success or failure of a vaccine policy for a disease will involve surveillance of vaccine use and surveillance of the disease, although the overall process is one of monitoring. Monitoring can become a finely-tuned measure in which outcome can be constantly evaluated to adjust the process.

Types of surveillance

In passive surveillance the recipient waits for the provider to report. In active surveillance routine checks of the provider are regularly made to ensure uniform and complete reporting. The smallpox eradication programme used active surveillance; each local health unit was 'coerced, persuaded, and cajoled' to report cases of smallpox each week, intensive further case finding was undertaken when a case was notified, and sources of information other than medical—teachers, schoolchildren, civil, and so on—were used (Henderson 1976). The reporting of a negative return is an important but not an essential part of a surveillance system. For surveillance of very rare diseases, however, completeness becomes more important, and 'negative reporting' becomes essential. Negative reporting has been used most effectively in the surveillance system for rare paediatric diseases, such as Reye's syndrome and Kawasaki disease, run by the British Paediatric Surveillance Unit in Britain (Hall and Glickman 1988).

Essentials of a surveillance system

The essential steps in any surveillance system are:

(1) the collection of data;
(2) their analysis;
(3) their interpretation;
(4) feedback of information.

These steps are similar to those taken in any scientific process. The collection of data is clearly the basic element of the system, but a failure of any of the other three steps in the process could also lead to failure of the system.

Collection of data

The collection of data has to be systematic, regular, and uniform, and in infectious diseases particularly, topical and relevant. Suppliers of information should understand clearly what needs to be reported, leaving little scope for value judgements in deciding what to report, such as including only interesting or rare cases. For clinical reporting case definitions are helpful, especially for new or rare diseases, and even for common and easily recognizable diseases when they become rare. An early case definition for AIDS was essential in initial surveillance of this infection. Measles is easily recognizable clinically, but a case definition became essential when the vaccine campaign in the USA became so successful that the infection became rare.

For laboratory reporting, likewise, the definitions of what constitutes an acceptable report need to be understood clearly by the laboratory; for example, isolation of the organism from particular sites, or a fourfold or greater rise in antibody titre, or a single antibody titre above a certain level associated with a clinical feature characteristic of that infection, may all be acceptable. Nevertheless, in infectious disease there are often difficulties in associating a correctly identified organism with a particular illness; for example, isolation of an echovirus from a patient with gastroenteritis. In these instances all laboratories should understand clearly whether every such isolation need be reported, or only one where the isolate is considered to contribute to the symptoms (which may involve a value judgement). In the laboratory reporting system run by the Communicable Disease Surveillance Centre in England and Wales, all viral isolates are reported, although there is space on the form for the laboratory to indicate if the isolate is thought not to be relevant to the condition of the patient.

If feedback is to be regular, reporting must also be regular. The Communicable Disease Report for England and Wales is produced weekly, and the laboratories report weekly. Regular reporting also helps to maintain discipline and routine, essential ingredients of any reporting system. The information required from providers must be relevant and topical, otherwise the interest of the participants is rapidly lost and their response will diminish. With infectious disease surveillance, serious infections, rare infections, or those against which a control measure is available or is being planned, tend to be most worthwhile for surveillance.

Analysis of data

The analysis of data for infectious diseases is similar to that for any other type of epidemiological data. The basic principles of analysis by time, place, and person, are fundamental.

Time

Analysis by time is necessary if trends are to be discovered from surveillance data. For infectious diseases there may be a significant delay of days or weeks between the time of acute illness and the date the infection is diagnosed and then reported. This is generally unavoidable. First, the time of onset of symptoms may be days or weeks before the illness is investigated. There will then be further delays before the disease is diagnosed, and again a delay before it is reported to the surveillance unit. With serological diagnosis, because a rise in antibody occurs only during convalescence, the infection is only confirmed when the patient is better. Even with a weekly reporting system there will be further delay before the infection is finally recorded. The burden of reporting will be increased considerably by asking reporting laboratories to record the date of onset of illness, but the date the first specimen was received in the laboratory is usually readily available and often approximates to the time of acute illness. In the analysis of such data, the interval between this date and the date of reporting may need to be taken into account. Analysing laboratory data by date of first specimen may be less helpful for immediate detection of changes in incidence. The techniques of analysis by time, and tests for seasonality or periodicity, are outside the scope of this chapter.

Place

The site of the reporting laboratory is normally taken to be the geographical location of the ill person. Except in the best organized of health-care systems, however, this is not always true. In England and Wales, for example, the statutory notifications generally relate to the area of residence of the patient, but the Local Authority to which the infection is notified is not necessarily co-terminous with the Health Authority. With laboratory data there are rarely rigid boundaries or catchment areas for hospital or public health laboratories. It may be difficult or impossible to allow for these problems in the analysis of the data, although it may only rarely be necessary to do so.

Person

Age and sex of cases reported are also important components of a surveillance report. It is possible to conduct a surveillance system based solely on numbers and geographical location, but the sensitivity of the system will be considerably reduced. In an outbreak of hepatitis B caused by a tattooist in one district of England (Limentani et al. 1979), simple analysis of notifications of 'infective jaundice' by time or place would not have uncovered the outbreak. Analysis of notifications by age and sex, however, would have revealed an increase in notified cases in males aged 15–29 years. In the British laboratory surveillance system, some important and common organisms such as salmonellas are reported without the age and sex of the patient, as it would increase greatly the burden of reporting, collection, and analysis of such data to unmanageable levels. Reliance is placed on the ability to detect an increase in number in a rare salmonella serotype, or phage type of a more common serotype. When this occurs, the age and sex of the cases are readily available from the reference laboratory and may provide some aetiological clues, as when an increase in *Salmonella ealing* infections proved to be mainly in infants and was subsequently shown to be associated with powdered milk (Rowe *et al.* 1987).

Clearly, denominator data are desirable, but they are not essential to an infectious disease surveillance system. In the

UK only the Royal College of General Practitioners Research Unit has an inbuilt rate provided with its data output. In this surveillance system, about 40 physicians in various parts of the kingdom, covering about 200 000 patients, report new episodes of illness each week to the Central Research Unit (Fleming and Crombie 1985). As each general practitioner (GP) has an age and sex distribution available of his practice population, consultation rates for the different illnesses can be calculated.

Interpretation

In the interpretation of surveillance data lies the skill of the epidemiologist. With William Farr (Langmuir 1976):

His weekly return was no archive for stale data but with his facile pen became a literate weapon for effecting change. He presented his analysis with objectivity but then stated his own interpretations forcefully . . . .

In surveillance data a detailed understanding of the reporting system is necessary before a meaningful interpretation of the statistics can be made. Some knowledge of the size and demographic characteristics of the population covered by the surveillance system is also necessary. Every data source carries its own strengths and biases (Moro and McCormick 1988). With each data source, timeliness, completeness, representativeness, and accuracy (Thacker *et al.* 1983) should be considered. To these four qualities should be added that of significance (in its sense of 'importance'). *Timeliness* may be particularly important with laboratory statistics, where there may sometimes be a considerable delay between disease onset and laboratory diagnosis. Organisms need time to grow in culture, and acute and convalescent samples of sera are needed to demonstrate the fourfold or greater rise in antibody necessary to substantiate a serological diagnosis. Rapid diagnostic methods for identification of an organism and the increasing use of IgM tests, however, have hastened considerably the diagnostic process for some infections. Notifications, although usually made on the basis of a clinical diagnosis, may not always be as prompt as one would expect, and sentinel General Practitioner clinical reporting systems tend to be more timely (Tillett and Spencer 1982).

The need for *completeness* of reporting, particularly of common infections, is often exaggerated. Detection of disease trends by time, place, and person, sufficient for meaningful epidemiological interpretation, is possible with incomplete data. Striving for completeness may waste resources. For rare diseases, however, or diseases that have become rare following a control programme, completeness grows in importance. Passive surveillance systems rarely achieve completeness; active systems are generally necessary for this.

The *representativeness* of the data collected needs thought and planning, and the advantages of collecting data from more than one source may provide ways of validating one data source against another. In the surveillance system for infectious diseases in England and Wales conducted by the Communicable Disease Surveillance Centre, for example, the three main sources of data on meningococcal meningitis—hospital, notifications, and laboratory—show similar trends over time (Fig. 24.1). Source data should be representative for time, place, and person.

The *significance* of surveillance data should also be evaluated carefully in its interpretation, and this can perhaps best be illustrated diagrammatically, using laboratory reporting of influenza as an example. If the population of a community affected by an outbreak of influenza is represented by the rectangular outline in Figure 24.2, the number actually infected will be a proportion of this, represented by circle (A). However, not all of these will have symptoms; those that do being within circle (B). Only a proportion of these will visit a doctor (C) (whether in hospital or general practice), and progressively smaller proportions will have a specimen sent for examination in a laboratory (D), specimens positive (E), and positive specimens reported to the surveillance unit (F). The biases and variables that occur during these steps need also to be considered. Those with symptomatic infections may be those never previously exposed to the particular influenza variant or subtype, or the very young and the very old, or those with a chronic disability, such as a respiratory condition. Similarly, those who visit a doctor and those whom the doctor investigates may be influenced by several factors, including social class, age, and severity of disease, and proximity to a laboratory. Laboratory success in its turn will depend on availability and cost of reagents, the interest and expertise of the laboratory, and the age and severity of disease in the patient. Finally, the accuracy, completeness, and timeliness of reporting by the laboratory will be influenced by its motivation, organization, and efficiency, as well as by the usefulness of the surveillance system and the quality and value of its feedback. A similar progression can be worked out for notification, GP, hospital, and death certification data sources (Fig. 24.3).

The *accuracy* of the data provided by the laboratory will depend not only on its interest, expertise, and motivation, but also on the clarity of the instructions for reporting provided by the surveillance unit. These include acceptable diagnostic criteria for laboratory data (especially for antibody titre measurements, the levels acceptable to the collection unit should be clarified) and clinical definition for notification and GP data. A Quality Control scheme for participating laboratories is useful, as in the system run by the Public Health Laboratory Service in England and Wales.

Dissemination of information and target groups

The logical end, and indeed the purpose of any surveillance system, is the output, and its quality and relevance are critical: not only must it be meaningful and intelligible, but it must also be directed at the appropriate targets, whether they

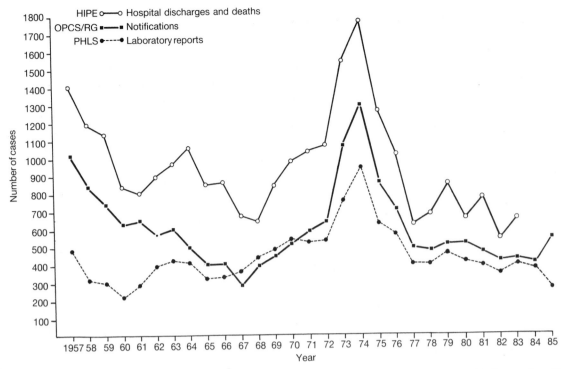

Fig. 24.1. Acute meningococcal meningitis—England and Wales 1957–85. [Source: Laboratory Reports to Communicable Disease Surveillance Centre (England, Wales and Ireland).]

be decision-makers or research workers. Moreover, feedback to those who provide the reports has an important motivating role in any surveillance system. Some surveillance and monitoring systems may even improve performance, as in the COVER programme in England and Wales. In this surveillance and monitoring system the measurement and publication of performance measures for vaccine coverage in different health districts almost certainly stimulates poorly performing districts to improve (Begg *et al.* 1989).

For infectious disease surveillance systems, weekly or monthly reports are the most appropriate. Flexibility to provide urgent information is useful, as for outbreaks, and as methods of communication continue to improve this should make urgent dissemination easier. Methods used in recent years by the Communicable Disease Surveillance Centre of England and Wales include special mailings, telex, and fax, and the use of TV/Prestel is probably not many years away. The ability to adapt to changing patterns of infection—incorporating new diseases of importance, discarding outdated or useless data—needs to be inbuilt.

Content and presentation

The summary report should not contain indigestible lists and tables, but easily understood analysis with appropriate evaluation of their significance. Especially for infectious disease, the reports need to be topical and relevant. Thus short summaries of recent trends and changes, together with more detailed reviews of subjects of interest, are important ingredients of a surveillance report, and these can be sup-

plemented by reports of outbreaks and other items of general interest by reporter participants or other contributors.

A successful report will educate and provide current scientific information for planning, prevention or change (Eylenbosch and Noah 1988).

Another function necessary for a successful surveillance system is the provision of an information service for individual inquiry. The organization of the surveillance unit and of the data collection system to provide information (as opposed to raw data) is as important a part of any surveillance feedback service as the regular report. For the sporadic inquiry of this type, appropriate interpretation of the information provided is also necessary. Encouraging regular dialogue between providers, other interested parties, and the central surveillance unit is helpful to fostering a healthy relationship between them and the long-term usefulness of the surveillance system; it can also be regarded as a form of monitoring of the surveillance system.

Ideally, the content and presentation of the output of the surveillance system needs to be adapted so as to be made intelligible to each type of target group; lay politicians and decision-makers, for example, might receive a different type of feedback from research workers, and again from the public and media. This is rarely possible with the regular surveillance report, but an information section within the central unit could tailor the response appropriately to the *ad hoc* inquiry. The increasing interest in health by the lay public, media, and politicians makes it essential to provide accurate, relevant, and topical information with skill and flair.

Content of surveillance: sources of data

When discussing the epidemiology of infection, or any other type of disease, it is useful to consider the different stages in the natural history of the disease process (Fig. 24.3). The figure is similar to that used for laboratory data (Fig. 24.2). The population again can be represented by the rectangular outline. Within this population will be a subpopulation who will be infected asymptomatically, or will be immune or carriers. A smaller proportion will develop symptomatic infection, and progressively smaller proportions will have symptomatic unreported infections, visit a doctor, be admitted to hospital, or die. To have a true measure of the total impact of a disease on a population, information at all these levels is needed. In practice it is of course rarely possible, or perhaps necessary, to be so systematic, although information at most levels can often be obtained. Serological surveys will give information on asymptomatic infection or immunity levels, and the taking of appropriate specimens or swabs from persons may detect asymptomatic carriers. Surveillance of predisposition to disease other than that determined by the absence of antibody is more difficult in the field of infectious disease, but surveillance of general functions such as growth, development, and nutritional status of children (Irwig 1976; Morley 1975; Carne 1984) may fall into this category, as will surveillance for infection in certain groups, such as tuberculosis in certain ethnic patients, HIV infection in haemophiliacs and homosexuals, cytomegalovirus infections in the immunosuppressed, or in those subject to certain procedures such as urinary catheterization. Unreported morbidity (Fig. 24.3) is not usually possible to place under passive surveillance. In a series of surveys conducted by the Office of Population Cen-

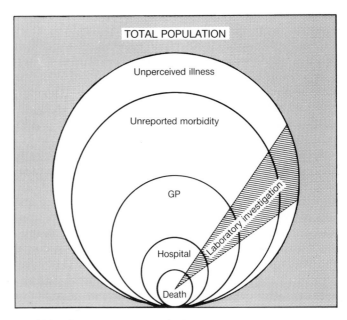

Fig. 24.3. Stages in progression of a disease.

suses and Surveys (OPCS) of England and Wales, the General Household Survey, information on unreported morbidity is obtained (Haskey and Birch 1985). Although each survey is finite, the collective information over many surveys constitutes a data base suitable for surveillance.

A General Practice surveillance system based on a sentinel reporting network was first successfully organized in the UK by the Royal College of General Practitioners in 1966 (Fleming and Crombie 1985), and since then in The Netherlands (Collete 1982) and Belgium (Thiers *et al.* 1979). General practice morbidity data are generally useful for providing information one tier in severity below that of hospital morbidity (Fig. 24.3). More specifically, in practice they produce information on clinical conditions with very low mortality not covered by notification, such as the common cold, chickenpox, or otitis media. In England and Wales, before mumps and rubella became notifiable in 1988, the Royal College of General Practitioners (RCGP) clinical reporting system was an important source of information on these common infections; it remains an important source of information on clinical influenza, which is not notifiable. Sentinel reporting systems are generally inappropriate for rare diseases. When, as in the UK, the reporting GPs keep an age–sex register, a more accurate denominator is available than that used for notification, and indeed the RCGP data are published as rates. Most countries have a notification system which usually provides and essential source of information on the important communicable diseases; the characteristics of these systems are well known and will not be discussed in detail here. Notifications cover both general practice and hospital morbidity. It is often more useful to know what the notification rate is than to strive for completeness, which is generally only

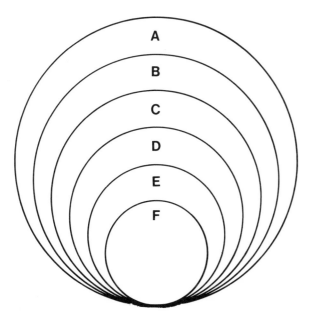

Fig. 24.2. Stages in the importing of laboratory infection.

really desirable for very rare or very serious infections. Diseases that are perceived by the notifying doctors to be either a serious public health problem and communicable are often better notified that those that are not. Thus tetanus is often poorly notified in countries such as England and Wales. It must be remembered that the primary objective of a notification system is not for surveillance but to provide an opportunity for local control, for which legal powers are usually available if necessary.

Surveillance systems using hospital admission data clearly only cover a limited, although important, phase in the natural history of an infection. Infections for which hospital data are ideal are those for which patients are usually admitted and admitted once only, and in which the diagnosis can be confirmed. If the hospital reporting system is based on a sample, the disease should not be excessively rare. Meningococcal and other forms of bacterial meningitis are examples of infections which tend to be well documented in hospital data systems. Hospital data, on the other hand, may be available late, months or even years after the event.

Death certification is virtually a universal requirement in most countries. Death certificates are clearly important to any surveillance system, but for infectious diseases their value may be somewhat limited since infections are now rare as a primary cause of death. As infections remain an important cause of morbidity, death certifications need to be supplemented by one or more of the surveillance systems detailed above.

Other sources of data

Laboratory

Laboratories provide important, perhaps essential, information for surveillance of infection. Surveillance, supported by laboratory confirmation of clinical cases reported, showed that the malaria epidemics said to be occurring in the malarious southern States of the USA did not exist, and the few cases confirmed microbiologically were imported or relapses (Langmuir 1963). In addition to this *confirmatory* role in surveillance that laboratories provide, they have an additional *qualitative* feature. Thus laboratory data can not only confirm the presence or absence of influenza, but the data can also show whether the virus is type A or type B, what the subtypes or variants are, and whether these have changed since the previous epidemic (Fig. 24.4). Similarly, atypical pneumonia may be caused by *Mycoplasma pneumoniae*, psittacosis, Q fever, several viruses, legionnella, and many other agents, and only a laboratory can distinguish these successfully. In human infections, laboratories can provide data at all levels of the disease process shown in Figure 24.3. Laboratory data often provide information on infections which may not be available by other means, for example, brucellosis, as well as important additional information on infection in vectors, animal, or other hosts, or the environment, for example, salmonellosis.

Outbreaks

Surveillance of laboratory-diagnosed infections can in itself lead to early recognition of outbreaks. Several examples of this have been reported, especially with salmonellas (Gill *et al.* 1983; Rowe *et al.* 1987; Cowden *et al.* 1989), and also with legionellas. Surveillance of outbreaks can also provide useful information, especially in assessing disease trends, and may be particularly worthwhile in countries where more sophisticated sources of data, such as laboratory data on individual infections, may not be available. Outbreak surveillance can be cheap and effective, although essentially an insensitive measure of disease trends.

Vaccine utilization

Surveillance of vaccine utilization is an important component of the process of monitoring the effect of vaccine strategy on an infection. In England and Wales the COVER programme fulfils this function (Begg *et al.* 1989) and may have an additional effect in stimulating poorly performing districts to improve coverage. Serological surveillance can provide an indirect measure of the effectiveness of a vaccine strategy (Noah and Fowle 1988).

Sickness absence

Sickness absence records may provide indications of major outbreaks in working populations. Influenza in particular may produce measurable changes in sickness absence. Sickness absence can also be monitored in special groups, such as boarding schools, Post Office workers, and so on.

Disease determinants

Biological changes in agent, vector, and the reservoirs of infection can be placed under surveillance. Surveillance of changes in an agent, such as new subtypes or variants of influenza, antibiotic resistance in bacteria or protozoa, such as plasmodium, are regularly performed. Surveillance of biological vectors include ticks and mosquitoes, and of animal reservoirs infections such as brucellosis and rabies, and is an essential component of disease control in many countries.

Susceptibility to infection can be measured by skin testing or serological surveillance. In England and Wales, antibody profiles to current circulating influenza variants and to new variants are regularly performed in small samples of the population to assess the degree of susceptibility to a new variant. The degree of immunity to vaccinatable diseases is also regularly assessed (Morgan-Capner *et al.* 1988). The use of serum banks and immunological surveys in surveillance was persuasively put forward by Raska (1971).

Objectives of surveillance

Many of the objectives of surveillance of infectious diseases have already been alluded to in the text and will be summarized here only. The main objectives of infectious disease surveillance is to monitor disease trends in time, place, and

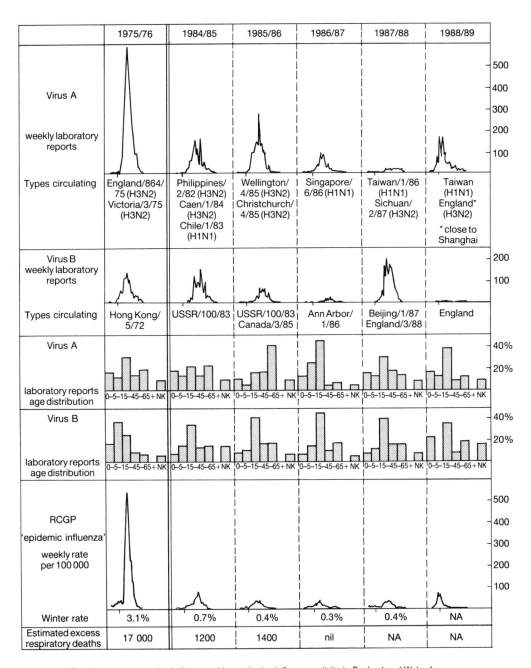

Fig. 24.4. Influenza surveillance—England and Wales. [Source: Laboratory Reports to Communicable Disease Surveillance Centre (England, Wales and Ireland).]

person. All the examples that follow are really illustrations of this.

Anticipation of changes in incidence

Many infectious diseases follow regular patterns, both seasonal and secular (Noah 1989). The RS virus follows a distinct seasonal pattern causing epidemics every year with the peak incidence, in the Northern Hemisphere, almost invari- ably shortly after the new year (Fig. 24.5). Minor variations in this pattern occur (Noah 1989). With some viruses, for example, echovirus, a failure to return to the base-line by the end of its yearly cycle signifies that a resurgence will occur the following year (Fig. 24.6; Epidemiology Research Labora- tory 1975). Some organisms, for example, *Mycoplasma pneu- moniae*, have long cycles extending over four years, but are particularly important as *M. pneumoniae* infection is

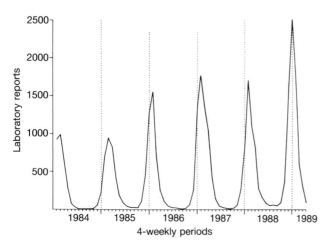

Fig. 24.5. Respiratory syncytial virus—England and Wales. [Source: Laboratory Reports to Communicable Disease Surveillance Centre (England, Wales and Ireland).]

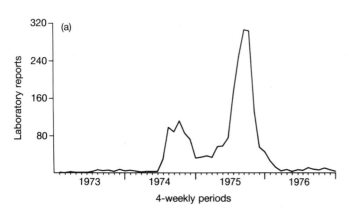

Fig. 24.6(a). Echovirus 19—England and Wales. [Source: Laboratory Reports to Communicable Disease Surveillance Centre (England, Wales and Ireland).]

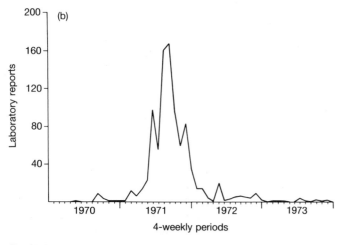

Fig. 24.6(b). Echovirus 4—England and Wales. [Source: Laboratory Reports to Communicable Disease Surveillance Centre (England, Wales and Ireland).]

treatable and early warning of an epidemic is of practical value (Fig. 24.7) (Noah 1974; Monto 1974).

Early detection of outbreaks

Outbreaks of food poisoning which have been detected only because surveillance revealed an unexpected increase in a salmonella have already been described, and reports of some of these have been published (Gill *et al.* 1983; Rowe *et al.* 1987; Cowden *et al.* 1989). Occasionally, early detection of a new strain of an organism may lead to premature action which with hindsight, turns out to have been inappropriate, as with the swine influenza episode (Silverstein 1981). Surveillance may lead to the detection of new infections, for example, Lyme disease. In an outbreak of hepatitis caused by a tattooist in England in 1978, surveillance by time alone would not have brought the outbreak to light, but only the characteristic age and sex distribution (young adult males) would have shown a change (Limentani *et al.* 1979). Analysis of surveillance data by person, time, and place is more likely to reveal changes in pattern than analysis by one of these parameters alone.

Evaluation of effectiveness of disease intervention

Surveillance techniques are used to monitor the effects of a mass vaccination programme. Not only can changes in incidence be measured by time (Hinman *et al.* 1980), but also by person; for example, age changes in measles or mumps (Cochi *et al.* 1988; Hinman *et al.* 1980) have been recorded as a result of mass immunization. There have also been examples of monitoring changes in incidence by place (district) and correlating these changes with vaccination uptake rates (Pollard 1980). Effectiveness of a vaccination programme (Noah and Fowle 1988) can also be monitored by serological surveillance. The effect of withdrawing a source of infection, such as a contaminated food, from circulation, can be monitored by surveillance, as with the *S. napoli* outbreak caused by contaminated imported chocolate in England (Gill *et al.* 1983; Fig. 24.8).

Identification of vulnerable groups

Surveillance can expose vulnerable groups; for example, by revealing ethnic differences in tuberculosis incidence. Appropriate action, such as BCG vaccination of neonates in such groups, can be taken. Serological surveillance can also identify susceptibility in particular groups of persons for selective vaccination.

Setting priorities for allocation of resources

From the examples above, it is clear that surveillance programmes can be used to provide information for setting priorities for resource allocation.

Aetiological clues

Infectious disease patterns may help to generate hypotheses for aetiology of chronic diseases. Secular variation compat-

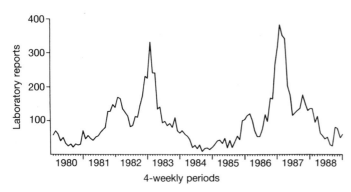

Fig. 24.7. Mycoplasma pneumoniae—England and Wales. [Source: Laboratory Reports to Communicable Disease Surveillance Centre (England, Wales and Ireland).]

ible with certain infections have been noted with sudden infant death syndrome (Helweg-Larsen *et al.* 1985), insulin independent diabetes (Gleason *et al.* 1982), deaths from asthma (Khot and Burn 1984), and anencephalius and spina bifida (Maclean and Macleod 1984).

Field investigations of vaccines

In field investigations of transmissible disease the epidemiological study of vaccines plays a major part. Epidemiological studies may be associated with vaccines in several different ways. First, the assessment of the need for a vaccine by undertaking surveillance or surveys will be necessary. Second, antibody and field trials to assess the efficacy and safety of the vaccine need to be conducted. Then follows the selection of an appropriate strategy and the implementation of the vaccine programme. Finally, the use of the vaccine and its effect on the population need to be closely monitored. These logical steps have not always been followed.

Assessment of the need for a vaccine
Morbidity and mortality of disease
The overall morbidity of the infection, and its severity, will usually be estimated from a surveillance programme. With measles the need for a preventive programme in the UK was first shown in 1963 (McCarthy and Taylor Robinson 1963). D.L. Miller (1964), by studying notifications of measles, was able to estimate the hospital admission rate (1 per cent), the respiratory and otitis media complication rate (6–9 per cent), and the neurological complication rate (0.7 per cent). C.L. Miller (1978) showed that these rates had changed little with time. Complication rates like these for an extremely common infection signified a serious health problem. Rey (1985) estimated that in France the total annual number of deaths from measles was 30, admissions to hospital was 6000, cases of encephalitis was 100–200, and SSPE was 10–20. World-wide estimates tend to be cruder but none the less possible; using World Health statistics Cutting (1983) claimed that measles

caused 1–1.5 million deaths in children in 1981. General Practice Surveillance systems may be useful in providing estimates of disease burdens, as for mumps in the UK (Research Unit 1974). Serological studies are also used to assess the overall impact of an infection on a population, and of the vulnerability of a target population. Studies on rubella in 1969 (Cockburn 1969) showed that by adulthood more than 80 per cent of a population had been infected, so that 5–15 per cent of pregnant women were still susceptible and hence vulnerable. In the USA the incidence of clinical rubella in pregnant women was found to vary from $4–8/10^4$ in endemic periods to $20/10^4$ in epidemic periods, and subclinical infection could increase by up to three times (White *et al.* 1969; Sever *et al.* 1969). Sometimes, as in polio, crude notification data on the most serious outcomes of poliovirus infection (paralysis and death), can be sufficient to point to a need for a vaccine (Fig. 24.9). For severe and very rare diseases only a selective vaccine policy can be considered, and these considerations do not apply.

Field studies

Field studies of vaccine efficacy (VE) and safety are clearly necessary before a vaccine is licensed for use. Post-licensing studies are also important, but these require a different approach and will be considered separately.

Prelicensing vaccine trials
The earliest trial (phase 1) of a new vaccine usually involves a very small number (usually 20–50) of adult volunteers with a low level of risk of acquiring the disease. Efficacy is usually tested by measurement of antibody and adverse effects are closely monitored. This tests the immunogenicity of the vaccine, following which the dose may be adjusted. Although only limited, preliminary information on safety is provided by a phase 1 trial; this is an important aspect of such a trial.

In the phase 2 trial a larger number of persons, between 100 and 200, constituting a target population is vaccinated. Their risk of acquiring the disease, again, should be low. Efficacy is again generally estimated by serological testing, and more precise information on dose–response and safety may be obtained, with some estimates of the rates of the more common side effects. Information on contra-indications may be available. If the phase 2 trial is satisfactory, a large field trial, the phase 3 trial, of more that 500 high-risk subjects can begin. Vaccine protection is tested against disease acquisition. Some of the questions that will need to be answered in a trial like this will include the efficacy of the vaccine, both in degree and duration, the rate of side-effects, both rare and common, the optimum age for giving vaccine, the dose, including how many and at what intervals, and the need for any adjuvants. Further contra-indications to the vaccine may become apparent.

Vaccine efficacy can be assessed by measuring disease incidence, by serological test, or both. An important consideration in trials based on disease incidence is how much the two

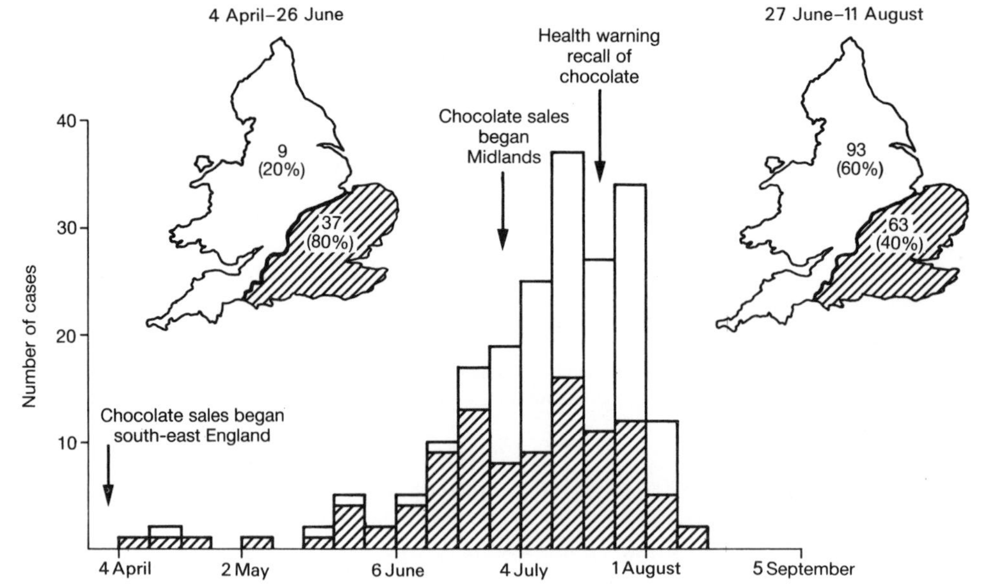

Fig. 24.8. Salmonella napoli outbreak 1982—England and Wales. Distribution of 202 primary household cases. [Source: Laboratory Reports to Communicable Disease Surveillance Centre (England, Wales and Ireland).]

groups (case and control) vary in exposure to the infection and ascertainment of infection. If, as generally believed, vaccinated groups tend to consist of those who are of higher social class, and thus groups which make better use of health services, diseases such as TB or whooping cough may be less common in them, or less severe, and hence less likely to be ascertained. On the other hand, any such disease is more likely to be reported, but these confounding factors may not cancel out each other. Allocation of persons to vaccine or control groups must be random and double-blind to minimize the effect of such biases. This is generally known as random individual allocation. Group allocation—class, village, factory—can also be used in vaccine trials, but although the control group must also be matched carefully, it is not usually possible to allocate randomly or for the trial to be double-blind. Detailed descriptions of how to conduct trials by individual or group allocation are given by Pollock (1966).

The disadvantages of a randomized controlled clinical trial are those of any longitudinal or cohort study: the necessity for large numbers of patients, the expense, the high drop-out rates that occur with long follow-up, and the fact that the vaccine is used under ideal but artificial conditions, the epidemiological equivalent of *in vitro*.

This is one of the factors which makes post-licensing vaccine trials important. Continued assessment, or monitoring, of vaccine use after licensing (*in vivo*) is mandatory. Vaccine potency may change, either because of some change at the manufacturing level, or along the chain which gets it to the user, and in storage by the user, so that testing vaccines at all levels of the chain is necessary. Rare side-effects may become apparent. In post-licensing trials the populations immunized may be more or less responsive than in the pre-licensing trials on account of a difference in age, social class, or other factors. For these reasons, complacency is unwise after a successful large scale pre-licensing trial.

Post-licensing monitoring

The success of any vaccine programme will depend on the overall population immunity level achieved. This is a function of *efficacy × uptake* of vaccine (Noah 1983). Thus post-licensing studies include not only continuous monitoring of vaccine efficacy, but also of uptake and implementation of the vaccine. Evaluation of outcome, and of side-effects, can also be studied.

Vaccine efficacy

Assessment by incidence

The incidence of the disease in a population after vaccination is compared with that before vaccination. This is a simple, fairly universally undertaken and useful method of assessing the impact of a mass vaccine programme on a population. Vaccine efficacy cannot usually be calculated from this method, only the effectiveness of the vaccine programme in broad terms. Moreover, demonstrations of a reduction in incidence following vaccination shows an association which may not be causal, and the usual rules of Bradford-Hill (1971) for assessing causality from association apply. The change in incidence may have been caused by changing social circumstances, changes in the population, or a natural variation in the incidence of the disease. In England and Wales the use, first, of Salk polio vaccine in 1957, followed by Sabin vaccine in 1963, reduced the incidence of poliomyelitis to negligible numbers (Fig. 24.9.), but the introduction of BCG in 1950 had a less than dramatic effect on an already declining disease (Fig. 24.10). This does not necessarily mean that the BCG vaccine was useless: a reduction in incidence could have been masked by more complete ascertainment or better diagnosis of cases, and accompanying increase in transmission of or susceptibility to tuberculosis, or a selective decrease in morbidity from a severe but rare form of tuberculosis, such as miliary tuberculosis or TB meningitis, too small to be detected by crude methods. Detailed surveillance would be necessary to detect such changes.

Assessment by immunological testing

Immunological testing includes skin tests, such as tuberculin or Schick, and serological testing. Tuberculin testing, generally using the Heaf or Mantoux tests, is performed to assess a person's immunity before vaccination. It has also been used to assess the quality of different strains of BCG. Schick tests for diphtheria are now performed rarely. The local reaction to smallpox vaccine was used to assess whether or not the vaccine had taken.

Serological testing can be performed using antitoxin (diphtheria and tetanus) or antibody (measles, rubella, and so on) levels. Serological tests first have to be shown to correlate with immunity to disease by disease incidence studies. Studies using such tests can be of two types, *seroconversion* or *seroprevalence* (Orenstein *et al*. 1985).

Seroconversion

Seroconversion studies are generally used in phases 1 and 2, and sometimes also in phase 3 of pre-licensing trials. They are particularly useful for assessing vaccine response at different ages, as with measles vaccine, and in different groups of persons; for example, healthy persons and those with immunosuppressive conditions respond differently to pneumococcal vaccine. Seroconversion studies are also useful for assessing vaccine efficacy for rare conditions in which antibody levels have already been shown to correlate with protection, as with tetanus, diphtheria, and polio. Moreover, seroconversion studies have the advantage that any change in antibody status can be attributed to the vaccine. On the other hand, seroconversion studies require two blood samples and laboratory back-up, but these disadvantages can be weighed against the smaller number of subjects that need to be studied. A more serious disadvantage is the absence of a reliable serological test of immunity for some infections, such as whooping cough and meningococcal infections. Seroconversion studies and incidence studies suffer from the same problems of having a case definition that is too specific (too high an antibody level) thus underestimating efficacy, or one that is too sensitive (too low an antibody level) which will overestimate efficacy.

Seroprevalence

Seroprevalence studies have the advantage of requiring only one blood test but the corresponding disadvantage is relating a 'positive' result to disease or vaccine. Seroprevalence studies can be used to monitor success of the vaccination programme, which is a function of both efficacy and uptake. Thus the prevalence of poliovirus antibody in a highly immunized population with no wild virus disease will reflect the efficacy and uptake of the vaccine, whilst the prevalence of rubella antibody in antenatal women or women of childbearing age in a country with a selective rubella vaccine policy (so

that the natural epidemics of rubella are virtually unaffected) will measure the overall immunity status of the target group, but will not easily distinguish how much of this is attributable to natural infection and how much to the success of the vaccine programme (Noah and Fowle 1988). If the expected seroprevalence of rubella antibody in this target group under conditions of natural rubella endemicity is known, then the gain in immunity attributable to vaccination of the target group can be calculated.

With seroprevalence studies the duration of protection can be evaluated also, but Marks *et al*. (1982) pointed out the importance of the timing of immunity measurements since vaccination, and also the age of vaccination. In case control studies using seroprevalence, these two parameters must be similar in cases and controls; that is, the time-lapse between vaccination and testing immunity and the age of vaccination are crucial for matching.

In seroprevalence studies the same caveats apply about sensitivity and specificity of the antitoxin or antibody levels chosen to indicate immunity. In seroprevalence studies the choice of which class of antibody to measure can also be critical. Absence of antibody may not always correlate with susceptibility.

Seroprevalence studies related to previous vaccination require accurate records of previous vaccination (including timing and number of previous doses of vaccine) for meaningful interpretation (Orenstein *et al*. 1988).

Assessment by epidemiological studies
General comments

For epidemiological studies of vaccine efficacy (VE) after licensing, a number of general comments need to be made. As with sero-prevalence studies, the sensitivity and specificity of the case definition is important, and has always to be a compromise. A highly sensitive case definition will include many spurious cases, leading to a falsely low VE, whereas too specific a case definition will lead to a falsely high VE. Case definition in both control and vaccinated groups should be undertaken with equal vigour, with due awareness of the tendency for case definition to be more specific in the vaccinated group. Exposure in the two groups should be similar: even if vaccinated and control groups have been closely matched by age, sex, social class, and geographical distribution, variations in exposure may still occur. The diagnosis should if possible be made 'blind' (Marks *et al*. 1982). In retrospective studies, ascertainment of vaccination history in the two groups must also be pursued with equal vigour, and again should preferably be done 'blind'. Those with a previous history of the infection should not be included in either group of subjects.

Cohort studies
Outbreak investigations

An outbreak often affords an opportunity to assess VE in a field (*in vivo*) setting. The basic steps are to define the cohort under investigation, ascertain all cases according to an

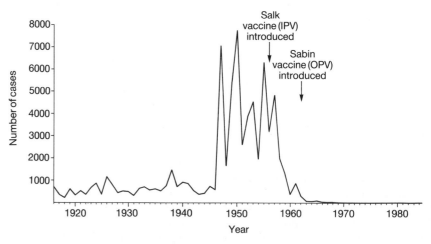

Fig. 24.9. Poliomyelitis—Notifications in England and Wales 1916–1988. [Source: Laboratory Reports to Communicable Disease Surveillance Centre (England, Wales and Ireland).]

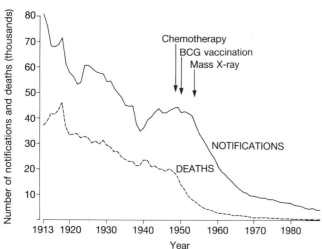

Fig. 24.10. Respiratory tuberculosis—England and Wales 1913–1988. [Source: Laboratory Reports to Communicable Disease Surveillance Centre (England, Wales and Ireland).]

acceptable case definition, ascertain vaccination histories in the entire cohort, and calculate attack rates in vaccinated and unvaccinated persons, excluding those with a previous history of disease. For very large cohorts sampling, or cluster sampling (Henderson and Sundaresan 1982), can be used. In infections in which most cases occur in fairly well demarcated age groups, the cohort under study will usually exclude those outside these age groups, as well as those too young to have been vaccinated. The optimum age of vaccination can also be ascertained from outbreak studies (Judelsohn *et al.* 1980).

Secondary attack rates in households
Measuring the attack rates in members of a household (secondary cases) following the introduction of an infection by one member of the family (primary case) affords an attractive approach to VE studies because exposures in vaccinated and unvaccinated persons are similar, thus eliminating an important bias in these studies; and also because the denominators (those exposed) can be fairly accurately deter-

mined. This type of study, however, needs more careful planning than most other post-licensing studies, and it is usually better designed as a prospective study. They also need to be planned to take place during periods of high epidemicity of the infection. Studies on pertussis vaccine in England and Wales which used secondary attack rates in households suggested that the vaccines were ineffective (PHLS Whooping Cough Committee and Working Party 1973; PHLS Epidemiological Research Laboratory 1982). This was difficult to believe because the incidence of whooping cough in the country at the time of the first study had been decreasing steadily, and other studies (Noah 1976; Pollard 1980; PHLS Epidemiological Research Laboratory 1982) suggested that the vaccine was effective. The reasons for the failure of the household-exposure studies to confirm efficacy were examined by Fine *et al.* (1988); they attributed the findings that the vaccine was ineffective primarily to the inclusion of retrospectively ascertained cases and of households in which the primary cases constituted a vaccine failure.

Cluster sampling (Henderson and Sundaresan 1982) is a modified form of this method of estimating VE. It has the advantage of being cheaper and easier to conduct than a household study, and hence particularly suitable for developing countries, but it is also less rigorous (Orenstein *et al.* 1985).

Retrospective population cohort studies
This type of study depends on the availability of fairly sophisticated disease reporting and vaccination recording systems. For each case of disease ascertained, the vaccination history is verified, and from the number and percentage of those vaccinated in the base population, the attack rates in those vaccinated and unvaccinated can be calculated and VE derived (Noah 1976).

The effectiveness of vaccination, but not its efficacy, can sometimes be checked crudely by observing the correlation between the incidence of disease in separate districts with the vaccination rates in those districts (Sutherland and Fayers 1971; Pollard 1980; Noah and Fowle 1988). Sometimes chance plays a part in affording an opportunity to estimate VE, as in 1970–71 in Texarkana, a city straddling the Texas/Arkansas state line. The Texas side differed from the Arkansas side only in not having a measles vaccine policy, so that during an outbreak of measles, the attack rate in the Texas side of the city was 48.2 per cent compared with 4.2 per cent in the Arkansas side. VE, based on ARs of 105.9 per thousand and 4.3 per thousand in the unimmunized and in the immunized respectively, was 95.9 per cent (Landrigan 1972).

Case–control studies

Case–control studies in vaccination issues are not common. The design of a case–control vaccine study is similar to that of any other case–control study. Each case of the disease is compared with one or more matched controls without the disease for a history of vaccination against the disease. The advantages of a case–control study are that it can be done more cheaply, quickly, and with fewer cases that for a cohort study. Moreover, the efficacy of vaccine given many years before can be assessed (Smith 1988). On the other hand, bias can occur as the vaccines will not have been administered at random (as, for example, if those of higher SE class have higher vaccination rates but lower expectancy of disease), and the vaccine histories in cases and controls may not be accurate (Smith 1988). The problems that need to be given attention in the design of case–control studies are similar to those in cohort vaccine studies. The case–control method, nevertheless, lends itself well to BCG efficacy studies (Smith 1988; Miceli *et al.* 1988). First, when the overall incidence of the disease is low, the case–control approach can be used on existing cases of disease to evaluate a vaccine given many years earlier, making it cheaper, quicker, and easier to do than a cohort study. Second, because of the expense of large cohort studies and the time needed to assess the efficacy of BCG, the case–control approach is a more practical one. Third, the difficulty of a 'double-blind' approach to a vaccine such as BCG, and fourth, the large variation (0–80 per cent) in estimates of efficacy of BCG with RCTs, further show the disadvantages of the cohort approach in trials of BCG.

Calculation of vaccine efficacy

Vaccine efficacy is the ratio of the observed diminution in attack rates (AR), that is, the difference in ARs between vaccinated and unvaccinated groups, to the expected attack rate, that is, that in the unvaccinated group. Thus

$$VE\% = \frac{AR\ unvaccinated - AR\ vaccinated}{AR\ unvaccinated} \times 100$$

or

$$VE\% = (1 - RR) \times 100$$

where RR is the ratio of AR vaccinated to AR unvaccinated, or relative risk.

In a case–control study, using unmatched controls and, Odds Ratios (OR), results can be arranged in a 2 × 2 table as follows:

	Cases	Controls
Vaccinated	a	b
Unvaccinated	c	d

VE is $(1 - OR) \times 100$ where OR is ad/bc

Refinements of these basic formulae can be found in Orenstein *et al.* (1985) and the impact of various types of bias on VE results in Orenstein *et al.* (1988).

Monitoring of side-effects of vaccines

Common side effects

Monitoring side-effects of vaccines begins at the very first phase of vaccine trials, running in parallel with estimations of efficacy. With randomized control trials, evaluation of side-effects after vaccination is fairly straightforward, it being usually sufficient to compare the incidence of any reactions in cases with that in controls. Where it is not practical, possible, or ethical to conduct a placebo-controlled study, some care is necessary in the interpretation of information on side-effects. In uncontrolled cohort studies, local side-effects (such as a stiff or painful arm) can still be evaluated successfully, but other symptoms, such as fever or convulsions which are non-specific to vaccines and may commonly occur from other causes in healthy children, are more difficult to evaluate without a control group. The time between the vaccination and the appearance of a side-effect also needs to be evaluated carefully. In the earliest trials of MMR vaccine meningitis caused by the mumps component was not documented until the subjects were followed up for at least 28 days.

In the early measles vaccine trials in England (Measles Vaccine Committee of the Medical Research Council 1966) 9577 children aged 10 months to 2 years were immunized. Eighteen children developed convulsions, 11 of them between post-vaccination days 6 and 9. Only 5 convulsions were reported in the control group of 16 327 children, none of them between days 6 and 9. This study established not only that convulsions were a real side-effect of measles vaccine, but also that they characteristically occurred during a fixed interval after the vaccine. The need to compare this with the incidence of convulsions after natural disease was also apparent, and they were shown to be at least 10 times commoner after the disease than after the vaccine (Miller 1978).

Unlike drug trials, the side-effects of a vaccine can sometimes be evaluated not by using a placebo in the control group but by giving a combination of vaccines omitting the vaccine under trial. The side-effects of pertussis vaccine have been investigated by comparing symptoms in the case group after diphtheria-tetanus-pertussis vaccine (DTP) with those after DT vaccine in the controls (Pollock *et al.* 1984). Even in this controlled study it was found that adverse publicity to pertussis vaccine led to a reporting bias in the cases given DTP.

Rare side-effects

Various ingenious studies have been set up to investigate the incidence of rare side-effects after vaccination. In Denmark,

Melchior (1977) compared the age-specific incidence of infantile spasms during the period when whooping cough vaccine was given at 5 months, 6 months, and 15 months with that during the period when it was given at 5 weeks, 9 weeks, and 10 months, and found no difference. In England the alleged extremely rare side-effects of serious and permanent (but non-specific) brain damage after whooping cough vaccine (with quoted rates of 1:50 000 to 1 million) were difficult to refute and the publicity engendered by this led to a dramatic fall in the pertussis vaccine uptake rate from about 79 per cent in the early 1970s to 31 per cent in 1975. A carefully controlled case–control study, the National Childhood Encephalopathy Study, was conducted. In this study the presence or absence of a history of recent vaccination was obtained in all cases of 'encephalopathy' reported to the study team and compared with controls. An association was shown between DTP given 3–7 days earlier and encephalopathy, but not between DT given at the same time and encephalopathy. It was possible to estimate from this study that the risk of persistent neurological damage in previously healthy children after pertussis vaccine was 1:310 000 immunization; but the 95 per cent confidence limits were very wide, from 1:54 000 to 1:5 310 000. The reader is referred to the original study (Department of Health and Social Security 1981) for further details of methodology, and to a balanced review for details of the whooping cough vaccine controversy (Miller 1982). Other forms of side-effect evaluation include post-marketing surveillance which may be active or passive (Stetler et al. 1987). An active surveillance system is less prone to bias but is likely to be too expensive and unwieldy; the passive system is more prone to bias.

Uptake and implementation of vaccines

One has to be a stranger in Jerusalem not to realise that public acceptance of an immunisation procedure determines its success or failure (Cohen 1978).

The need for field studies in vaccines does not cease at efficacy studies. The implementation of the vaccine is as important as its production and an effective safe vaccine with a poor uptake is of hardly greater benefit than no vaccine at all. In recent years this has been recognized and field studies to investigate the factors that influence uptake of vaccines are now common.

Studies on uptake may measure factors associated with social or service conditions. In one study (Jarman et al. 1988), social conditions associated with being underprivileged (such as overcrowding, unmarried, or single parents), living in high population density areas, being unskilled, and belonging to certain ethnic groups were found to be linked to low immunization uptake. Reasons given by family (Clarke 1980; Blair et al. 1985; Lakhani et al. 1987) or by clinic staff (Adjaye 1981; Lakhani et al. 1987) for refusal can also be investigated. Single-handed general practitioners, those aged over 65 years, or those with large list sizes and less than average expenditure on community health services were also associ-

ated with low uptake rates (Jarman et al. 1988). The efficiency of organization, quality of premises, and adequacy of staffing of three different types of immunization clinic—GP, health centre, and child health clinic—were found to affect compliance (Alberman et al. 1986).

In another study the provision of individual performance indicators was enough to stimulate GPs to improve rates (Newlands and Davies 1988). An interesting phenomenon with an annual influenza vaccination programme in industry was noted by Smith et al. (1976): in successive years the uptake of vaccine fell considerably. The reasons for this were fairly complex and were probably associated with the need for giving vaccine yearly, the low incidence of influenza during the periods over which the vaccine was given, and the general perception by the target population that the vaccine was not very effective. The message that emanates from many of these studies, however, is that it is the administrative efficiency of the immunization programme and the motivation of the professionals that are the critical features in improving uptake (Lakhani et al. 1987; Noah 1982). Communication strategies can be designed to persuade people to attend immunization campaigns (Hingson 1974).

Evaluation of factors affecting vaccine programmes

The evaluation of the outcome of a vaccine programme can be conducted in several ways. Except with the most successful programmes reasons for non-acceptance of vaccine can be investigated. Reasons for failure/poor uptake may be because of social and professional attitudes to the vaccine, including the effect of the media, or the administrative competence/managerial abilities of those responsible for the programme.

Social and professional attitudes to a vaccine or a vaccine programme may 'make or break' the programme. In the UK, reports by Kulenkampff in 1974 of alleged brain damage following whooping cough vaccine, fuelled by a television programme in the same year, and by 'the writings in the lay press and medical journals and media interviews' of a professor who was a 'leading critic of the vaccination policy' (Cherry 1986) led to a catastrophic fall in pertussis vaccine uptake rates from about 79 per cent between 1967 and 1974 to a low of 30 per cent in 1978. Television programmes were found to be especially influential (McKinnon 1979). The controversy had an effect on other routine vaccinations (DT and polio) which fortunately was transient and relatively slight. A survey of professional attitudes during the episode (Wilkinson et al. 1979) suggested that GPs, clinic doctors, or health visitors were ambivalent towards pertussis vaccine. Most had noticed increase in parental concern about the vaccine which was attributed to 'irresponsible', 'ill-controlled', or 'biased' publicity. Adjaye (1981), however, found that parental attitudes towards immunization were greatly influenced by medical and non-medical members of the health professions, while

Berkeley (1983) in a study of attitudes to measles vaccine found that the professionals themselves were unsure about the contra-indications to vaccine and to its value. In another study (Guest *et al.* 1986) many professionals offered invalid reasons for failure to give immunization.

Campbell (1983), in analysing reasons for poor uptake of measles vaccine in Britain, suggested that attitudes and ignorance, a cumbersome policy-making bureaucracy, the absence of legislation (which helped the US programme) and of a standard record card for each child, and the initiative and motivation required by mothers towards a vaccine given some months after the primary course, were factors that accounted for the unpopularity of measles vaccine in Britain. Encouraging the health professionals to take greater initiative was found to benefit uptake (Carter and Jones 1985). These and other studies (Pugh and Hawker 1986; Lakhani *et al.* 1987; Bussey and Harris 1979) suggested that education of health professionals and an efficient administration with computerized recall and records were important factors in improving uptake rates. In the US, school immunization laws were also a proven method of improving immunization uptake (Robbins *et al.* 1981).

Evaluation of outcome of immunization programmes

The case for long-term serological surveillance of immunization programmes has been forcefully advocated by Evans (1980) and Raska (1971). Evans argued that a surveillance system based on reporting of disease alone was insufficiently sensitive or specific to monitor a vaccine programme. An individual's history of immunization or disease correlated well with antibody presence in measles and mumps, but less well in rubella and poliomyelitis; moreover, for tetanus natural infection has virtually no bearing on antibody levels. The reliability of a negative history of disease or immunization was, however, poor for measles, mumps, and rubella. The serological demonstration of satisfactory levels of durable antibody in immunized persons was a more reliable measure of vaccine effectiveness than monitoring the incidence of the disease itself, and absence of antibody in particular communities or specific age groups could identify those who may need protection. Long-term surveillance of immunity after measles vaccination has also been advocated (*Lancet* 1976). Serological studies have shown that vaccine failure and not waning vaccine-induced antibody accounted for most of the low or undetectable antibody titres in immunized children, especially those immunized before 13 months of age (Yeager *et al.* 1977).

A study of measles vaccine on the survival pattern of 7–35-month-old children in Kasongo, Zaïre (Kasongo Project Team 1981) used life-table analysis to evaluate outcome of the vaccination programme in a community with a high measles incidence and measles case-fatality rate. They found that, although the measles vaccine programme undoubtedly

Table 24.1. Benefits of measles immunization over ten years (US)

Type of savings	Number
Cases averted	23 707 000
Lives saved	2 400
Cases of retardation averted	7 900
Additional years of normal and productive life by preventing premature death and retardation	709 000
School days saved	78 000 000
Physician visits saved	12 182 000
Hospital days saved	1 352 000
Net benefits	$1.3 billion

Source: Witte and Axnick (1975).

reduced the risk of measles death, overall survival was less influenced by measles vaccine after 22 months of age.

Another study found that the success of a vaccine programme against measles, mumps, and rubella using MMR vaccine was reflected in the virtual disappearance of a common complication of all three infections (encephalitis) in their population of children (Koskiniemi and Vaheri 1989). Improved intellectual performance was an outcome noted in a cohort study in children who had been immunized against pertussis compared with those who had been hospitalized for the disease, even after allowing for social differences in the two groups (Butler *et al.* 1982).

Costing studies of vaccines

Costing studies are important both in assessing the need for a vaccine and in evaluating its efficiency. The detailed methodology of costing studies for infectious disease is considered elsewhere in this book. Costs include costs of the vaccine and its administration. Costs of the vaccine may be influenced by factors as diverse as costs of development and market competition: the price of human-derived hepatitis B vaccine halved when the yeast-derived vaccine began to be marketed. Administration costs will include cost of the syringe, personnel, accommodation as well as costs of initiating and administering a record and follow-up programme. Benefits include savings on treatment costs, mortality, morbidity, the avoidance of intangibles such as pain and grief, and external benefits (Creese and Henderson 1980). It is debatable whether social benefits of successful intervention should be costed, because the value of such benefits may not be convincing to pragmatic health care managers. Patrick and Woolley (1981) have pointed out that health managers may fail to be impressed by cost–benefit studies because the costs, which are health costs, have to be weighed against benefits which are usually social. Nevertheless, the benefits of providing supportive cost calculations are an important addition to the evaluation of a vaccination programme. A list of the benefits of immunization, simply and clearly stated, often cannot fail to convince as in one of the earliest such papers (Witte and Axnick 1975) describing the results of 10 years of measles immunization in the USA in terms of lower morbidity and mortality, and improved quality of life.

Other studies (Creese and Henderson 1980; Koplan 1985) have amply shown the value for money of measles vaccine in terms of benefit–cost ratios for measles vaccine of 10–15:1. Cost studies can be successfully done in general practice (Binnie 1984). In this general practice, over a 20-year period, immunization not only brought about a small financial reward, but also reduced the number of consultations by 40 per cent, even though a home visit was often necessary to immunize a child. Examples of cost–benefit analysis of various immunizations which have demonstrated the benefit of vaccines, and to which the reader is referred for details of methodology, include: measles vaccine (White and Axnick 1975; Albritton 1978) pertussis vaccine (Koplan et al. 1979; Hinman and Koplan 1984; Hinman and Koplan 1985), mumps vaccine (White et al. 1985; Koplan and Preblud 1982), polio (Fudenberg 1973), pneumococcal vaccine (Patrick and Woolley 1981), and hepatitis B in homosexuals (Adler et al. 1983). General articles on cost–benefits of immunization programmes include Creese and Henderson (1980), and Koplan (1985). A formula has been developed for calculating vaccine profitability, defined as the economic yield, positive or negative, obtained per monetary unit invested in a campaign (Carvasco and Lardinois 1987).

References

Adjaye, N. (1981). Measles immunization. Some factors affecting non-acceptance of vaccine. *Public Health, London*, **95**, 185.

Adler, M.W., Belsey, E.M., McCutchan, J.A., *et al.* (1983). Should homosexuals be vaccinated against hepatitis B? Cost and benefit assessment. *British Medical Journal*, **286**, 1621.

Alberman, E., Watson, E., Mitchell, P., *et al.* (1986). The development of performance and cost indicators for preschool immunization. *Archives of Diseases in Childhood*, **61**, 251.

Albritton, R.B. (1978). Cost-benefits of measles eradication: effects of a federal intervention. *Policy Analysis*, **4**, 1.

Begg, N.T., Gill, O.N., and White, J. (1989). COVER (cover of vaccination evaluation rapidly): Description of the England and Wales Scheme. *Public Health*, **103**, 81.

Berkeley, M.I.K. (1983). *Measles — the effect of attitudes on immunization*. Health Bulletin. Scottish Home and Health Department, Edinburgh. 41/3: 141–147

Binnie, G.A.C. (1984). Measles immunization—profit and loss in a general practice. *British Medical Journal*, **289**, 1275.

Blair, S., Shave, N., and McKay, J. (1985). Measles matters, but do parents know? *British Medical Journal*, **290**, 623.

Bradford Hill, A. (1971). *Principles of medical statistics* (9th edn), Chapter 24. The Lancet Ltd, London.

Bussey, A.L. and Harris, A.S. (1979). Computers and effectiveness of the measles vaccination campaign in England and Wales. *Community Medicine*, **1**, 29.

Butler, N.R., Golding, J., Haslum, M., *et al.* (1982). Recent findings from the 1970 child health and education study: preliminary communication. *Journal of the Royal Society of Medicine*, **75**, 781.

Campbell, A.G.M. (1983). Measles immunization: why have we failed? *Archives of Diseases in Childhood*, **58**, 3.

Carne, S. (1984). Place of development surveillance in general practice. *Journal of the Royal Society of Medicine*, **77**, 819.

Carter, H., and Jones, I.G. (1985). Measles immunization: results of a local programme to increase vaccine uptake. *British Medical Journal*, **290**, 1717.

Carvasco, J.L. and Lardinois, R. (1987). Formula for calculating vaccine profitability. *Vaccine*, **5**, 123.

Cherry, J.D. (1986). The controversy about pertussis vaccine. *Current Clinical Topics in Infectious Diseases*, **7**, 216.

Clarke, S.J. (1980). Whooping cough vaccination: some reasons for non-completion. *Journal of Advanced Nursing*, **5**, 313.

Cochi, S.L., Preblud, S.R., and Orenstein, W.A. (1988). Perspectives on the resurgence of mumps in the United States. *American Journal of Diseases in Childhood*, **142**, 499.

Cockburn, W.C. (1969). World aspects of the epidemiology of rubella. *American Journal of Diseases in Childhood*, **118**, 112.

Cohen, H. (1978). Vaccination against pertussis, yes or no? In *International symposium on pertussis* (ed. C.R. Manclark and J.C. Hill) p.249. National Institute of Health, Bethesda, Maryland.

Collete, B.J.A. (1982). The sentinel practices system in The Netherlands. In: *Environmental epidemiology* (ed. P.E. Leaverton), p.149. Praeger Publishers, New York.

Cowden, J.M., O'Mahony, M., Bartlett, C.L.R. *et al.* (1989). A national outbreak of Salmonella typhimurium DT 124 caused by contaminated salami sticks. *Epidemiology and Infection*, **103**, 219.

Creese, A.L. and Henderson, R.H. (1980). Cost-benefit analysis and immunization programmes in developing countries. *Bulletin of the World Health Organization*, **58**, 491.

Cutting, W.A.M. (1983). Measles immunization. A review. *Journal of Tropical Pediatrics*, **29**, 246.

Department of Health and Social Security (1981). *Whooping cough*. Reports from Committee on Safety of Medicines and the Joint Committee on Vaccination and Immunization. HMSO, London.

Epidemiology Research Laboratory (1975). Echovirus 19 this summer? *British Medical Journal*, **2**, 346.

Evans, A.S. (1980). The need for serologic evaluation of immunization programs. *American Journal of Epidemiology*, **112**, 725.

Eylenbosch, W.J. and Noah, N.D. (ed.) (1988). *Surveillance in health and disease*. Oxford University Press.

Fine, P.E.M., Clarkson, J.A., and Miller, E. (1988). The efficacy of pertussis vaccines under conditions of household exposure. *International Journal of Epidemiology*, **17**, 635.

Fleming, D.M. and Crombie, D.L. (1985). The incidence of common infectious diseases: the weekly returns service of the Royal College of General Practitioners. *Health Trends*, **17**, 13.

Fudenberg, H.H. (1973). Fiscal returns of biomedical research. *Journal of Investigative Dermatology*, **61**, 321.

Gill, O.N., Bartlett C.L.R., Sockett P.N., *et al.* (1983). Outbreak of Salmonella napoli infection caused by contaminated chocolate bars. *Lancet*, **i**, 574.

Gleason, R.E., Khan, C.B. Funk, I.B. *et al.* (1982). Seasonal incidence of insulin dependent diabetes (IDDM) in Massachusetts, 1964–1973. *International Journal of Epidemiology*, **11**, 39.

Guest, M., Horn, J., and Archer, L.N.J. (1986). Why some parents refuse pertussis immunisation. *Practitioner*, **230**, 210.

Hall, S.M. and Glickman, M. (1988). The British paediatric surveillance unit. *Archives of Disease in Childhood*, **63**, 344.

Haskey, J.C. and Birch, D. (1985). Statistics from general practice: morbidity and its measurement using practice statistical reports. *Health Trends*, **17**, 32.

Helweg-Larsen, K., Bay, H., and Mac, F. (1985). A statistical analysis of the seasonality in sudden infant death syndrome. *International Journal of Epidemiology*, **14**, 566.

Henderson, D.A. (1976). Surveillance of smallpox. *International Journal of Epidemiology*, **5**, 19.

Henderson, R.H. and Sundaresan, T. (1982). Cluster sampling to assess immunization coverage: a review of experience with a simplified sampling method. *Bulletin of World Health Organization*, **60**, 253.

Hingson, R. (1974). Obtaining optimal attendance at mass immunization programs. *Health Services Reports*, **89**, 53.

Hinman, A.R. and Koplan, J.P. (1984). Pertussis and pertussis vaccine. Reanalysis of benefits, risk and costs. *Journal of the American Medical Association*, **251**, 3109.

Hinman, A.R. and Koplan, J.P. (1985). Pertussis and pertussis vaccine: Further analysis of benefits, risks and costs. *Development in Biological Standardization*, **61**, 429.

Hinman, A.R., Brandling-Bennett, A.D. Bernier, R. *et al.* (1980). Current features of measles in the United States: Feasibility of measles elimination. *Epidemiologic Review*, **2**, 153.

Irwig, L.M. (1976). Surveillance in developed countries with particular reference to child growth. *International Journal of Epidemiology*, **5**, 57.

Jarman, B., Bosanquet, N., Rice, P., *et al.* (1988). Uptake of immunization in district health authorities in England. *British Medical Journal*, **296**, 1775.

Judelsohn, R.G., Pleissner, M.L., and O'Mara, D.J. (1980). School-based measles outbreaks: Correlation of age at immunization with risk of disease. *American Journal of Public Health*, **70**, 1162.

Kasongo Project Team (1981). Influence of measles vaccination on survival pattern of 7–35 month-old children in Kasongo, Zaire. *Lancet*, **i**, 764.

Khot, A. and Burn, R. (1984). Seasonal variation and time trends of deaths from asthma in England and Wales (1960–1982). *British Medical Journal*, **289**, 233.

Koplan, J.P. (1985). Benefits, risks and costs of immunization programmes. In *The value of preventive medicine*, Ciba Foundation Symposium 110, pp. 55–680. Pitman, London.

Koplan, J.P. and Preblud, S.R. (1982). A benefit–cost analysis of mumps vaccine. *American Journal of Diseases in Childhood*, **136**, 362.

Koplan, J.P., Schoenbaum, S.C., Weinstein, M.C., *et al.* (1979). Pertussis vaccine—an analysis of benefits, risks and costs. *New England Journal of Medicine*, **301**, 906.

Koskiniemi, M. and Vaheri, A. (1989). Effect of measles, mumps, rubella vaccination on pattern of encephalitis in children. *Lancet*, **i**, 31.

Kulenkampff, M., Schwartzman, J.S., and Wilson, J. (1974). Neurological complications of pertussis inoculation. *Archives of Diseases in Childhood*, **49**, 46.

Lakhani, A.D.H., Morris, R.W., Morgan, M., *et al.* (1987). Measles immunisation: feasibility of a 90 per cent target uptake. *Archives of Diseases in Childhood*, **62**, 1209.

Landrigan, P.J. (1972). Epidemic measles in a divided city. *Journal of the American Medical Association*, **221**, 567.

Langmuir, A.D. (1963). The surveillance of communicable diseases of national importance. *New England Journal of Medicine*, **268**, 182.

Langmuir, A.D. (1976). William Farr: Founder of modern concepts of surveillance. *International Journal of Epidemiology*. **5**, 13.

Last, J.M. (ed.) (1988). *A dictionary of epidemiology* (2nd edn). Oxford University Press.

Lancet, (1976). Leading article: Vaccination against measles. *Lancet*, **ii**, 132.

Limentani, A.E., Elliott, L.M., Noah, N.D., *et al.* (1979). An outbreak of hepatitis B from tattooing. *Lancet*, **ii**, 86.

McCarthy, K. and Taylor Robinson, C.H. (1963). Immunization against measles. *British Journal of Clinical Practice*, **17**, 650.

McKinnon, J.A. (1979). The impact of the media on whooping cough immunization. *Health Education Journal*, **37**, 198.

Maclean, M.H. and Macleod, A. (1984). Seasonal variation in the frequency of anencephalus and spina bifida births in the UK. *Journal of Epidemiology and Community Health*, **38**, 99.

Marks, J.S., Hayden, G.F., and Orenstein, W.A. (1982). Methodologic issues in the evaluation of vaccine effectiveness. *American Journal of Epidemiology*, **116**, 510.

Measles Vaccine Committee of the Medical Research Council (1966). Vaccination against measles. *British Medical Journal*, **i**, 441.

Melchior, J.C. (1977). Infantile spasms and early immunization against whooping cough: Danish survey from 1970 to 1975. *Archives of Diseases in Childhood*, **52**, 134.

Miceli, I., de Kantor, I., Colaiacovo, D., *et al.* (1988). Evaluation of the effectiveness of BCG vaccination using the case-control method in Buenos Aires, Argentina. *International Journal of Epidemiology*, **17**, 629.

Miller, C.L. (1978). Severity of notified measles. *British Medical Journal*, **i**, 1253.

Miller, D.L. (1964). Frequency of complications of measles, 1963. *British Medical Journal*, **2**, 75.

Miller, D.L., Alderslade, R., and Ross, E.M. (1982). Whooping cough and whooping cough vaccine: the risks and benefits debate. *Epidemiologic Reviews*, **4**, 1.

Monto, A.S. (1974). The Tecumseh study of respiratory illness. *American Journal of Epidemiology*, **100**, 458.

Moro, M.L. and McCormick, A. (1988). Surveillance for communicable disease. In *Surveillance in health and disease* (ed. W.J. Eylenbosch and N.D. Noah) p. 166. Oxford University Press.

Morgan-Capner, P., Wright, J., Miller, C.L., *et al.* (1988). Surveillance of antibody to measles, mumps and rubella by age. *British Medical Journal*, **297**, 770.

Morley, D. (1975). Nutritional surveillance of young children in developing countries. *International Journal of Epidemiology*, **5**, 51.

Newlands, M. and Davies, L. (1988). The use of performance indicators for immunization rates in General Practice. *Public Health*, **102**, 269.

Noah, N.D. (1974). Mycoplasma pneumoniae infection in the United Kingdom, 1967–1973. *British Medical Journal*, **2**, 544.

Noah, N.D. (1976). Attack rates of notified whooping cough in immunized and unimmunized children. *British Medical Journal*, **1**, 128.

Noah, N.D. (1982). Measles eradication policies. *British Medical Journal*, **284**, 997.

Noah, N.D. (1983). The strategy of immunization. *Community Medicine*, **5**, 140.

Noah, N.D. (1989). Cyclical patterns and predictability in infection. *Epidemiology and Infection*, **102**, 175.

Noah, N.D. and Fowle, S.E. (1988). Immunity to rubella in women of childbearing age in the United Kingdom. *British Medical Journal*, **297**, 1301.

Orenstein, W.A., Bernier, R.H., Dondero, T., *et al.* (1985). Field evaluation of vaccine efficacy. *Bulletin of the World Health Organization*, **63**, 1055.

Orenstein, W.A., Bernier, R.H., and Hinman, A.R. (1988). Assessing vaccine efficacy in the field. *Epidemiologic Reviews*, **10**, 212.

Patrick, K.M. and Woolley, F.R. (1981). A cost–benefit analysis of immunization for pneumococcal pneumonia. *Journal of the American Medical Association*, **245**, 473.

PHLS Epidemiological Research Laboratory and 21 Area Health Authorities (1982). Efficacy of pertussis vaccination in England. *British Medical Journal*, **285**, 357.

PHLS Epidemiological Research Laboratory (1982). Efficacy of pertussis vaccination in England. *British Medical Journal*, **285**, 357.

PHLS Whooping Cough Committee and Working Party (1973). Efficacy of whooping cough vaccines used in the United Kingdom before 1968: Final report. *British Medical Journal*, **1**, 259.

Pollard, R. (1980). Relation between vaccination and notification rates for whooping cough in England and Wales. *Lancet*, **i**, 1180.

Pollock, T.M. (1966). *Trials of prophylactic agents for the control of communicable diseases*. World Health Organization, Monograph series No.52, Geneva.

Pollock, T.M., Miller, E., Mortimer, J.Y., et al. (1984). Symptoms after first immunization with DTP and DT vaccine. *Lancet*, **ii**, 146.

Pugh, E.J. and Hawker, R. (1986). Measles immunisation: Professional knowledge and intention to vaccinate. *Community Medicine*, **8**, 340.

Raska, K. (1971). Epidemiological surveillance with particular reference to the use of immunological surveys. *Proceedings of the Royal Society of Medicine*, **64**, 681.

Research Unit of the Royal College of General Practitioners (1974). The incidence and complications of mumps. *Journal of the Royal College of General Practitioners*, **24**, 545.

Rey, M. (1985). Eradication of measles by widespread vaccination is beneficial and feasible. *Semaine des Hôpitaux de Paris*, **61**, 21.

Robbins, K.B., Brandling-Bennett, A.D., and Hinman, A.R. (1981). Low measles incidence: association with enforcement of school immunization laws. *American Journal of Public Health*, **71**, 270.

Rowe, B., Hutchinson, D.N., Gilbert, R.J., et al. (1987). Salmonella ealing infections associated with consumption of infant dried milk. *Lancet*, **ii**, 900.

Sever, J.L., Hardy, J.B., Nelson, K.B., et al. (1969). Rubella in the collaborative perinatal research study. *American Journal of Diseases in Childhood*, **118**, 123.

Silverstein, A.M. (1981). *Pure politics and impure science: The swine flu affair*. Johns Hopkins University Press, Baltimore and London.

Smith, P.G. (1988). Epidemiological methods to evaluate vaccine efficacy. *British Medical Journal*, **44**, 679.

Smith, J.W.G., Fletcher, W.B., and Wherry, P.J. (1976). Vaccination in the control of influenza. *Postgraduate Medical Journal*. **52**, 399.

Stetler, H.C., Mullen, J.R., Brennan, J.P., et al. (1987). Monitoring system for adverse events following immunization. *Vaccine*, **5**, 169.

Sutherland, I. and Fayers, P.M. (1971). Effect of measles vaccination on incidence of measles in the community. *British Medical Journal*, **1**, 698.

Thacker, S.B., Choi, K., and Brachman, P.S. (1983). The surveillance of infectious diseases. *Journal of American Medical Association*, **249**, 1181.

Thiers, G., Maes, R., van Lierde, R., et al. (1979). Surveillance van besmettelijke ziekten door een net van peilpraktijken. *Tijdschrift voor Geneeskunde*, **12**, 781.

Tillett, H.E. and Spencer, I.L. (1982). Influenza surveillance in England and Wales using routine statistics. *Journal of Hygiene (Cambridge)*, **88**, 83.

White, C.C., Koplan, J.P., and Orenstein, W.A. (1985). Benefits, risks and costs of immunization for measles, mumps and rubella. *American Journal of Public Health*, **75**, 739.

White, L.R., Sever, J.L., and Alepa, F.P. (1969). Maternal and congenital rubella before 1964: Frequency, clinical features, and search for iso-immune phenomena. *Journal of Pediatrics*, **74**, 198.

Wilkinson, P., Tylden-Pattenson, L., and Gould, J. (1979). Professional attitudes towards vaccination and immunization within the Leeds Area Health Authority. *Public Health, London*, **93**, 11.

Witte, J.J. and Axnick, N.W. (1975). The benefits from 10 years of measles immunization in the United States. *Public Health Report*, **90**, 205.

Yeager, A.S., Davis, J.H., Ross, L.A., et al. (1977). Measles immunization. Successes and failures. *Journal of the American Medical Association*, **237**, 347.

25

Field investigations of air

ROBERT E. WALLER*

Introduction

During earlier centuries the proposition that adverse effects on health could be attributed to air pollutants originating from the combustion of fuels or industrial processes was considered many times. But it was difficult to distinguish such effects from those of microbial agents or of allergens of biological origin. Evelyn (1661), for example, in his classic treatise *Fumifugium*, said that 'New Castle Coale, as an expert physician affirms, causeth consumption, phthisicks and the indisposition of the lungs . . . ' In the nineteenth century, the Registrar-General (1845) expressed the opinion that smoke was injurious to health and one of the causes of death to which the inhabitants of towns were more exposed than those of the country. At that time he could not, however, detect short-term increases in mortality associated with high concentrations of smoke, as in London fogs. Such increases were observed later in the century and public and official concern was finally aroused by the large numbers of deaths associated with some notable episodes of fog during the present century, including that in the Meuse Valley in 1930 (Firket 1936), in Donora in 1948 (Schrenk *et al.* 1949), and in London in 1948 (Logan 1949) and in 1952 (Ministry of Health 1954).

While these incidents provided the most dramatic examples of adverse effects on health, the more insidious long-term effects on the development of respiratory disease were probably of much greater importance in the field of public health as a whole, but they were very difficult to demonstrate clearly. Thus, while there has always been a relatively high death-rate from bronchitis in the highly polluted urban areas of the UK, it is only in the past few decades, during which time carefully designed prevalence surveys have been conducted, that it has become possible to determine the relative importance of air pollution along with that of other factors such as smoking and respiratory infections. The scene has, however, been changing in the 1980s, and as Holland (1989) has reported, the large reductions achieved in the UK in concentrations of pollutants such as smoke and sulphur dioxide have led to outdoor pollution becoming of little importance in relation to the development of chronic respiratory disease, smoking being left as the overwhelming factor, with indoor pollution perhaps playing some part. The situation is similar in many other countries where effective action has been taken to curb excessive pollution from industrial or domestic fuel-burning, but problems remain elsewhere, particularly in rapidly developing parts of the world.

Air pollutants and their characteristics

The prime concern of the present discussion is the health of the general public, and interest centres, therefore, on pollutants that occur in the air of towns or other communities rather than in specific occupational environments. Such pollutants are derived mainly from the combustion of fuels for domestic heating or cooking purposes, for heating or power generation in industry, or for transport, though in some localities there may be additional pollutants from industrial process emissions. In most situations, therefore there is a complex mixture of pollutants, which may also interact with one another, and in field (as opposed to laboratory) studies it is seldom possible to ascribe any adverse effects on health to one component in particular.

There are, however, two distinct types of mixture that occur widely throughout the world. One type is the traditional 'London smog' that is associated with smoke and/or sulphur dioxide from the burning of coal or heavy oil (Meetham *et al.* 1981). This type was responsible for most of the dramatic episodes cited above. The other is the 'Los Angeles smog' or photochemical oxidant type derived largely from the incomplete combustion of petrol in motor vehicles, together with emissions from refineries or associated chemical industry, interacting with other pollutants in the presence of sunlight to produce a lachrymatory haze (Organization for Economic Co-operation and Development 1975; Photochemical Oxidants Review Group 1987). The latter type creates a major nuisance for populations exposed to it, and it can damage horticultural crops, but there has not to date

* This contribution has been seen by colleagues in the Department of Health: their comments and suggestions are gratefully acknowledged but the views expressed are the author's own and not necessarily those of the Department.

been any evidence of substantial effects on mortality or morbidity of the magnitude seen in association with the traditional pollutants.

Chemically speaking, the two types have contrasting properties, the former being 'reducing' as a result of the sulphur dioxide present from the combustion of sulphur impurities in coal or heavy oil, and the latter 'oxidizing', due to the formation of ozone and organic oxidants in the photochemical process. While the two forms can coexist, in a particular region and/or season of the year one type is often dominant, being associated with the characteristics of local sources and with climatic conditions. Thus, the most notable area of the world for photochemical pollution is the Los Angeles basin, where all the factors conducive to its formation are at a maximum. These conditions include a very high motor vehicle density, uninterrupted sunshine through much of the year, and a combination of topographical and meteorological factors that lead to light winds, frequent temperature inversions, and a very poor natural ventilation rate. Similarly, the most notable place in the world for traditional smoke/sulphur dioxide pollution at one time was London, due primarily to emissions from domestic open coal fires. The former London fogs (or 'smogs') were merely accumulations of coal smoke and other pollutants from the exceptionally inefficient combustion in such fires. Temperature inversions during cold, calm, clear periods in winter played an important part in causing episodes of high pollution. But London is not especially subject to such conditions and following the gradual elimination of coal fires under provisions of the Clean Air Act of 1956, London has proved to be a place particularly free from fog, since the 'heat island' effect (Chandler 1965) reduces the chance of natural wet fogs occurring. Although the use of open fires burning soft coal has been a particularly British habit, coal is still widely used in somewhat analogous ways for domestic heating and cooking in many parts of the world, with a risk of potentially harmful concentrations of smoke and sulphur dioxide arising in large, densely populated cities.

A summary of the principal pollutants from combustion sources together with an indication of possible effects on health is shown in Table 25.1 Methods of measuring concentrations in the air are detailed in other publications (British Standards Institution 1969; Organization for Economic Co-operation and Development 1964, 1975; World Health Organization 1976) but, as a general introduction, the nature of the individual pollutants and approaches used in their determination are outlined below.

Smoke, or suspended particulate matter

Although smoke is one of the most widespread components of urban air pollution, this is the one that leads to the most confusion in studies of effects on health. Smoke is not a chemical entity and in effect it is defined by the method of measurement. Thus, the term 'smoke' is applied to the black suspended material in the air that is dominated by products of incomplete combustion. It is measured by its blackening

power, usually by drawing a known volume of air through a white filter paper and measuring the reflectance of the resulting stain. National, or international, standard curves have been developed to convert the reflectance measurements to amounts of 'equivalent standard smoke', so that the final results can be quoted in units of $\mu g/m^3$. However, this is not necessarily equivalent to the actual amount of suspended particulate matter at the time. Thus Bailey and Clayton (1982) have shown that in many urban areas of the UK, standard smoke concentrations are now only about 40 per cent of the actual gravimetric suspended particle concentrations. However, all the material that is assessed in this way is within the respirable size range, since the sampling rate generally used is too low to collect particles much beyond 5 μm effective aerodynamic diameter, and the carbonaceous smoke aggregates that influence the result to a large extent are confined to an even smaller size range, the mass median diameters being around 1 μm (Waller et al. 1963).

The alternative approach is to make direct gravimetric measurements of the suspended particulate matter. This has been the general practice in the US, where the standard procedure for many years was to make observations of 'total suspended particulates' using a high volume sampler (World Health Organization 1976). The high sampling rate introduces a risk of including particles beyond the inhalable range, should they be present (notably in dry windy climates, where soil and other non-combustion generated dust can become re-entrained). Cyclones or other elutriators have sometimes been used to remove the larger particles, but by now the basic instrumentation for health-related observations has been redefined (Environmental Protection Agency 1987). The aim is to secure samples of particles within the inhalable range, and instruments are designed for high collection efficiency at particle sizes up to a few micrometres (μm) diameter, falling through 50 per cent efficiency at 10 μm towards zero at larger sizes. Instruments meeting this specification are termed 'PM10' (particulate matter to 10 μm) samplers. In Europe, gravimetric sampling is generally done with samplers operating at a medium flow-rate, still designed to avoid the collection of particles much beyond 10 μm (Müller 1984).

None of the routine methods of measurement characterizes the chemical composition of the material, which may vary widely from time to time and place to place, and since there is little indication as to which components might be responsible for adverse effects observed, much uncertainty remains as to which type of measurement is most appropriate in relation to health studies. The important point for the present purpose is to recognize that observations of smoke concentrations made by optical methods are quite distinct from gravimetric determinations of suspended particulates.

Sulphur dioxide

This irritant gas is generally associated with smoke or suspended particulates, since it comes from the same sources—

the burning of coal or heavy oil (the lighter distillates from oil, such as petrol, contain little sulphur). The routine method of measurement in the UK is coupled with that of smoke. The gas is collected in a bubbler beyond the smoke filter, sulphur dioxide concentration is determined by acidimetric titration, and the measurements are related to 'net acid gas'. In many urban areas sulphur dioxide is the dominant influence, but in rural areas concentrations of ammonia may be of similar magnitude, and corrections need to be made for this gas. More specific methods for sulphur dioxide estimation are also quite widely used, including a pararosaniline colorimetric method (World Health Organization 1976) and a thorin spectrophotometric method (British Standards Institution 1983).

Sulphuric acid and sulphates

In some combustion processes a small proportion of sulphur in fuel is oxidized to sulphur trioxide rather than to sulphur dioxide, which is emitted as a fine sulphuric acid mist. Sulphur dioxide itself is also liable to react with other pollutants in the air to form further sulphuric acid or sulphates. Ammonia plays an important part in these reactions, the end-product being ammonium sulphate, but there are also photochemical processes involving hydrocarbons that promote the oxidation of sulphur dioxide. All the acid and sulphate is present in the particulate phase of the pollution, and is collected on filters as used for the determination of suspended particulates. The total sulphate content of samples can be determined by turbidimetric methods, but the separate assessment of sulphuric acid, which may be of greater relevance to effects on health, is a more difficult task, and one that is not done on a routine basis. A procedure for the assessment of 'net particulate acid' (Commins 1963) has however been used regularly in London as an indicator of sulphuric acid in the air. This has shown very sharp increases during episodes of high pollution associated with increased morbidity and mortality. Increased interest is now being shown in the acid aerosol and its possible role in the adverse effects on health of mixed urban air pollutants, and an improved method of assessment has been described by Koutrakis et al. (1988).

Oxides of nitrogen

Emissions of these gases to the air are derived largely from the fixation of atmospheric nitrogen in combustion processes, although in some cases there can be contributions from nitrogenous components of fuels (this is particularly true of vegetable products, such as tobacco). At the source, nitric oxide is generally the dominant oxide present, but there is some nitrogen dioxide, and in the air further nitric oxide is oxidized gradually to nitrogen dioxide, particularly in the photochemical oxidant cycle set up in the presence of volatile hydrocarbons. Only nitrogen dioxide is of possible concern in relation to health. The most satisfactory way of measuring it is with a continuous monitor based on the chemiluminescent principle that records nitric oxide and total oxides of nitrogen, displaying nitrogen dioxide as the difference between them. While large quantities of oxides of nitrogen are emitted from chimneys, emissions from motor vehicles tend to have the most influence on concentrations at ground level (World Health Organization 1987).

Hydrocarbons

In the context of precursors of photochemical oxidant pollution, it is the volatile hydrocarbons emitted from motor vehicles or refineries that are of prime concern, notably the unsaturated aliphatic compounds. Routinely the total concentration of hydrocarbons, expressed as methane (CH_4), is monitored, but gas chromatographic techniques are required for more specific determinations. In general, these measurements are valuable as indicators of photochemical pollution problems, but at normal ambient concentrations these volatile hydrocarbons do not have any adverse effects on health. Some concern has been expressed recently about possible adverse effects of aromatic components, and in particular, benzene. Trace amounts can be present in streets either from the small proportion of benzene in petrol itself, or from the further amount produced on combustion, with larger amounts in the vicinity of petrol filling stations (World Health Organization 1987).

Ozone and organic oxidants

In the photochemical process the main secondary pollutant produced is ozone which on inhalation has adverse effects due to its strong oxidizing properties. Since some hours are required for concentrations to build up to a peak (within the daylight hours) the maxima are not necessarily at the point of emission of hydrocarbon or oxides of nitrogen precursors. Concentrations of ozone and associated organic oxidants tend to be fairly uniform over whole regions, as illustrated in the occasional episodes of high oxidant pollution in the UK (Apling et al. 1977; Photochemical Oxidants Review Group 1987).

Peroxyacetylnitrate

Many complex organic compounds produced in the photochemical pollution process contribute to its irritant and lachrymatory properties and to plant damage. Peroxyacetylnitrate is one that has been identified and determined routinely in some areas. For practical purposes in studies of adverse effects of photochemical pollutants on health it is, however, generally sufficient to take ozone as an indicator of the complex as a whole. (However, episodes with high ozone concentrations that have occurred in the UK and northern Europe, as cited above, have not been accompanied by the full range of products and effects as seen in Los Angeles, possibly because refinery and industrial sources of precursors have dominated, rather than emissions from motor vehicles.)

Table 25.1. Some urban air pollutants and their effects on health

Traditional ('reducing') pollutants from coal/heavy oil combustion

SMOKE (SUSPENDED PARTICULATES) (Some contributions from diesel traffic also)	Can penetrate to lungs, some retained: possible long-term effects. May irritate bronchi also	LONDON SMOG COMPLEX
SULPHUR DIOXIDE	Readily absorbed on inhalation: irritation of bronchi, with possibility of bronchospasm	*Short-term effects*: Sudden increases in deaths, in hospital admissions, and in illness among bronchitic patients. Temporary reductions in lung function (patients and some normals)
SULPHURIC ACID (Mainly a secondary pollutant, formed from sulphur dioxide in air)	Hygroscopic: highly irritant if impacted in upper respiratory tract. Acid absorbed on other fine particles may penetrate further to produce bronchospasm	*Long-term effects*: Increased frequency of respiratory infections (children). Increased prevalence of respiratory symptoms (adults and children). Higher death-rates from bronchitis in polluted areas
POLYCYCLIC AROMATIC HYDROCARBONS (Small contributions from traffic also)	Mainly absorbed on to smoke: can penetrate with it to lungs	*Possible carcinogenic effects*: May play some part in the higher incidence of lung cancer in urban areas

Photochemical ('oxidizing') pollutants from traffic sources, or other hydrocarbon emissions

HYDROCARBONS (Volatile: petrol, etc.)	Non-toxic at moderate concentrations	LOS ANGELES SMOG COMPLEX
NITRIC OXIDE	Capable of combining with haemoglobin in blood, but no apparent effect in humans	*Short-term effects*: Primarily eye irritation. Reduced athletic performance. Possibly small changes in deaths, hospital admissions
NITROGEN DIOXIDE Mainly secondary pollutants formed in photochemical	Neither gas is very soluble: some irritation of bronchi, but can penetrate to lungs to cause oedema (at high concentrations). Urban concentrations too low for such effects, but evidence of reduced resistance to infections in animals	*Longer-term effects*: Increased onsets of respiratory illnesses (children), increased asthma attacks (adults). No clear indication of increased bronchitis
OZONE reactions		
Aldehydes, other partial oxidation products, peroxyacetylnitrate	Eye irritation, odour	

Others from traffic

CARBON MONOXIDE (Other sources contribute, and smoking an important one)	Combines with haemoglobin in blood, reducing oxygen-carrying capacity	Possible effects on CNS (reversible unless concentrations very high). Some evidence of effects on perception and performance of fine tasks at moderate concentrations. Enhances onset of exercise angina in patients. Urban concentrations too low for specific effects
LEAD (Some industrial sources contribute to air lead, and human intake often dominated by lead in food or drink)	Taken up in blood, distributed to soft tissues, and some to bone.	Possible effects on CNS (longer time scale than in case of carbon monoxide, and not necessarily reversible). Indications of neuropsychological effects on children within overall environmental exposure range, but role of traffic lead uncertain

Carbon monoxide

Emissions of this gas are common to a wide range of sources, including motor vehicles, most other combustion sources, and cigarettes. Ground level concentrations in city streets are affected most by motor vehicle emissions, and especially those from petrol engines, since the more efficient diesel engine produces little. The gas is relatively inert, and does not play any essential part in the photochemical process, but it is of interest in its own right since it is readily absorbed by blood, forming carboxyhaemoglobin (COHb). Concentrations in air are most conveniently measured by continuous instruments based on infra-red or electrochemical principles (World Health Organization 1976).

Lead

Lead compounds are dispersed into the atmosphere mainly as fine particulates. Contributions from coal or heavy oil burning are small, but in some localities there may be emissions from industrial sources such as lead smelters (either primary production from ore or secondary production from reclaimed materials) and battery works (making lead–acid batteries for motor vehicles). The most universal source in recent decades, however, has been petrol-engined vehicles, through the use of lead alkyls as antiknock agents in the fuel. The organic lead is mainly burnt with the petrol to produce lead halides (through reactions with 'scavengers' in the additives) and other inorganic compounds such as lead oxides, carbonate, or sulphate. Samples collected for the routine determination of smoke or suspended particulates can be analysed for lead, but it is more satisfactory to collect the material specially on membrane or glass-fibre filters for subsequent analysis by atomic absorption spectrophotometry or X-ray fluorescence methods. Since, in general, cars are the dominant sources, concentrations are highest close to busy roads, falling off sharply on moving away from them (Department of Health and Social Security 1980). Only a small proportion of the lead is present in the organic (vapour) form (Harrison and Perry 1977), and although the latter has more extreme toxic properties, for all practical purposes interest is confined to the inorganic particulate component.

Polycyclic aromatic hydrocarbons

These products originate from the incomplete combustion of fossil fuels, wood, tobacco, or other organic material (referred to collectively as polycyclic or polynuclear aromatic hydrocarbons—PAHs) (WHO 1987). They mainly condense on to the particulate matter produced in the combustion process and are generally sampled on filter papers in much the same way as smoke or total suspended particulates, as described above. Some of the lower molecular weight compounds, such as phenanthrene, are fairly volatile and cannot be collected completely on filters, but, from the point of view of possible effects on health, interest is directed mainly towards higher molecular weight PAHs such as the five-ringed compound benzo(a)pyrene, which has carcinogenic properties. Even this compound has an appreciable vapour pressure at ambient temperatures so that collection may not be complete on filters. However, it is not easy to include vapour-phase collection in series with filters at the fairly high sampling rates (or extended sampling periods) generally required for determinations of the very small amounts present in urban air, and, to date, many of the studies have confined their attention to the particulate component. Samples are extracted with an organic solvent and the analytical methods now favoured are glass capillary gas chromatography combined with mass-spectrometry, or high-performance liquid chromatography (Organization for Economic Co-operation and Development 1983).

Human exposure

In most epidemiological studies on the effects of air pollution on health, advantage has generally been taken of observations made at existing monitoring sites. Sometimes these have been set up with the exposure of the general population in mind, but often there are other considerations, such as assessing the influence of local sources (factories, traffic, etc.) or effects on other aspects of the environment. The general principles to be considered in setting up a monitoring network have been discussed in a publication by the World Health Organization (WHO 1977), and a further one deals with the wider question of assessing human exposures to air pollution, using observations from networks or otherwise

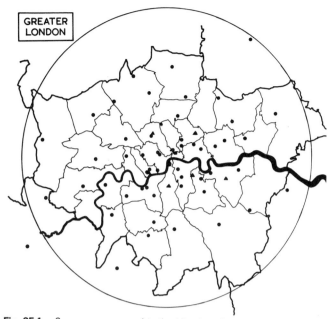

Fig. 25.1. Survey area as used in the 'diary' studies in London. ●, chest clinics and hospitals through which subjects were recruited; ▲, air pollution measuring sites.

(WHO 1982). In general, for most air pollutants it is very difficult to assess the true exposure of individuals or the average exposure of groups, and what is sought is an index that in some way reflects contrasts in exposure. The situation compares poorly even with assessments of smoking, in which a smoker can at least give a fair estimate of the number of cigarettes smoked per day, and say when he or she started.

The simplest situation to consider is where an index of day-to-day variations in traditional pollutants such as smoke and sulphur dioxide in a single locality is required. As an example, Figure 25.1 shows the set of monitoring stations in London selected for studies on short-term effects of pollution assessed by increases in daily deaths, in-hospital admissions, or exacerbations of illness among bronchitic patients. The population considered lived either within the boundary of the Greater London Council or within a slightly larger area defined by the circle. The seven sampling stations shown were originally set up by the (then) London County Council to be representative of the exposure of people living in the inner, more densely populated, part of the conurbation. There were eventually some 200 sampling stations in all in the London area, but these seven were sited close to ground level, in residential areas, and were operated in a consistent manner by a single authority from 1957 onwards, providing a virtually complete set of daily data throughout the subsequent years. They were adopted for studies in the wider area around London, on the precept that they still gave a reasonable indication of values experienced by the (adult) population as a whole in the places where they lived or worked. Whether or not the absolute values were different for people in the inner and outer areas, the important point was that day-to-day variations, being determined largely by

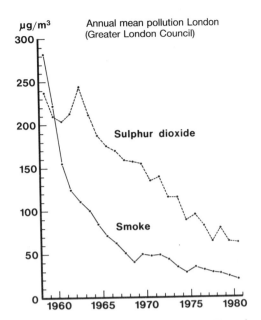

μg/m³   Annual mean pollution London
            (Greater London Council)

Fig. 25.2. Long-term trends in annual mean concentrations of smoke and sulphur dioxide in Greater London.

meteorological variables, ran parallel with one another in all areas.

For studies in which area, rather than temporal, comparisons are being made it is more important to ensure that values are representative in an absolute sense. Further, there is the added difficulty that such studies are often concerned with long-term effects, for which one might require an assessment of lifetime exposures, or possibly of exposures during a specific period earlier in life. It is then necessary to know something about the movements of the population concerned over the years and in what ways the pollution may have changed. Figure 25.2 illustrates the substantial changes in smoke and sulphur dioxide concentrations that have taken place in London over a 25-year period. This is sufficient to show at a glance that measurements made today would give little indication of possible exposure of the adult population earlier in their lives. In the case of London and some other large cities of the UK there are a few series of pollution measurements going back to the early years of the century. However, it is only in relatively recent years that extensive monitoring networks have been set up, and when earlier information is required it is sometimes necessary to resort to crude indirect indices, for example ones based on coal consumption, as used by Daly (1959) and Douglas and Waller (1966).

In some further respects the above is still a simplified discussion. It relates to pollutants such as smoke and sulphur dioxide as emitted from many thousands, if not millions, of chimneys serving solid fuel or oil-fired heating installations in homes, offices, and factories. The chimneys effect varying degrees of dispersion, depending on their height and the temperature of the emissions, but the net result in a city such as London is to create an 'area' source in which there is a reasonable degree of mixing with the surrounding air before the emissions reach the inhabitants. In such circumstances exposures are not crucially dependent on the exact location of an individual, and concentrations do not vary wildly from one moment to another within the day. The situation is different when dealing with just a few point sources, as with power plants or other large industrial emitters situated in or close to residential areas where domestic sources make little or no contribution. Ideally such sources should have stacks high enough to ensure adequate dilution of the plume before it reaches ground level, even under the most adverse meteorological conditions. But, where there is a risk of inhabited areas being 'fumigated' by poorly dispersed plumes, great variations in concentration are to be expected over short distances and over brief intervals of time, depending on wind direction and speed. Smoke is not generally a problem with such sources, since combustion is usually complete and/or arrestors may be installed to remove particulates, but there may often be sulphur dioxide. Since short-term effects of that gas seem more likely to be related to peak values than to longer-term averages (Lawther *et al.* 1975), the monitoring of 24-hour mean values at just one or two sites may provide only a poor guide to the relevant exposure of individuals. It is in fact very difficult to make satisfactory estimates of the exposure of any group under study in those circumstances.

An equally difficult situation arises with traffic sources, since emissions occur within the breathing zone of people in the streets, who may momentarily experience high concentrations from passing vehicles before dilution has proceeded to the extent normally achieved with chimney emissions. The exact place and time of exposure then becomes of crucial importance again, and at least for primary pollutants such as carbon monoxide, hydrocarbons, nitric oxide, and lead, concentrations can be far higher within the confines of a busy street than in a quiet location say 50 metres away from traffic. Fortunately, each vehicle represents a small source relative to stationary combustion sources, and most of the primary pollutants emitted are not ones for which effects are dependent on transient peak values rather than longer-term averages. For such pollutants the intermittent nature of exposure when moving around in a city does not create special problems. An exception to this statement might be made in respect of nitrogen dioxide (NO_2), for although nitric oxide dominates in the mixed oxides of nitrogen emitted from vehicles, the small proportion as nitrogen dioxide is sufficient to lead to uneven distribution, with peak values occurring close to traffic in busy streets. In general, much care is required when attempting to assess exposures to pollutants influenced by traffic sources. Kerb-side measuring sites are often set up that indicate the maximum values to which people may be exposed, supplemented by background sites away from traffic that indicate the minima. In the particular case of carbon monoxide the most satisfactory way of assessing exposure is through biological monitoring, on the population concerned. Since this gas is taken up by the blood and its effects are mediated in that way it is appropriate to measure carboxyhaemoglobin (COHb) directly on small blood samples (Commins and Lawther 1965). An alternative approach, avoiding the minor trauma of blood sampling, is to make observations on exhaled air samples (Stewart *et al.* 1976). The carbon monoxide content of alveolar air, sampled through the use of a double-bag arrangement allowing tidal air to be discarded first, is in equilibrium with that in the blood, and can be related to it via calibration experiments. A number of small portable instruments are now available commercially enabling measurements of carbon monoxide in exhaled air (or environmental concentrations) to be made directly. Whether measurements of COHb are made directly or indirectly one overwhelming problem is that the dominant contribution of carbon monoxide to the blood of people who smoke is their own smoking, so that in effect the part played by environmental sources can be assessed only in non-smokers (Lawther and Commins 1970). Exposure to traffic sources in streets cannot even be regarded as an added burden on top of smoking since the smoker who has a COHb level already above what would be in equilibrium with the concentration of carbon monoxide in the air at the time, will give out carbon monoxide to the street air rather than take more in.

A second traffic pollutant for which biological monitoring is possible is lead. Again, direct observations on blood are practicable, and now widely used, but the attribution of the lead thus found to urban air rather than to other types of source is even more difficult than in the case of carbon monoxide. Smoking is not a major problem in this case, although it does have some effect (Pocock *et al.* 1983). The principal sources of lead are the ingestion of foods (part of that lead being via contamination by airborne lead) and water (via lead pipes). Direct human exposures to the airborne component are best assessed with personal samplers, or by calculating 'weighted weekly average' exposures based on measurements in the home, in the street, or at work, and a knowledge of activity patterns (WHO 1982). While industrial sources of airborne lead are important in some localities, the major contributor in many countries has for several decades been motor vehicles, from the lead additives in petrol. There is now a general move to phase out the use of leaded petrol, and in the UK substantial reductions in lead concentrations have been observed in urban areas (McInnes 1988). There have been some heroic attempts to determine the total contribution of petrol-lead (via inhalation or ingestion) to human uptake, by changing the type of lead used for the manufacture of additives used in a whole region (Facchetti and Geiss 1983), but such procedures are beyond the scope of most researchers.

Some secondary traffic pollutants are more uniformly distributed in space than the primary pollutants, but there can still be sharp variations with time, since the intensity of sunlight is important in their production and there is thus a marked diurnal variation, with a maximum in the late afternoon (Fig. 25.3).

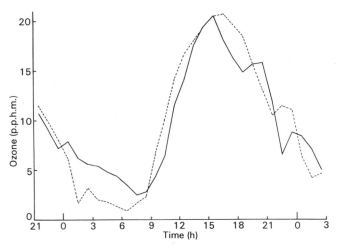

Fig. 25.3. Hourly mean concentrations of ozone in central London (solid line) and at a rural site outside London (dotted line) during an episode of photochemical pollution in the summer of 1976, showing typical diurnal pattern.

Much of the above discussions evades one particular problem—how to handle the major contrasts in exposures to pollutants indoors and outdoors. Reactive pollutants, such as sulphur dioxide and ozone, are readily lost on walls or other surfaces indoors, so that even at quite moderate ventilation rates concentrations indoors are generally less than those outside. The reverse can be true if there are sources indoors, and downdraughts in chimneys serving domestic coal fires can occasionally lead to occupants of a room being exposed briefly to extremely high concentrations of smoke and sulphur dioxide. In most epidemiological studies observations from routine (outdoor) monitoring stations have been used to provide indices of exposure of the populations concerned, even if they do not give a proper assessment of the absolute values or of the variations between individuals. This is reasonably satisfactory for studies within a region in which housing and general living conditions are similar throughout, but it would be unwise to extrapolate findings to other situations. Where there is a need to know more about the actual exposure of individuals in a study, then it may be necessary to resort to personal samplers. For most pollutants this is liable to increase costs and operating difficulties substantially, and the special problems of indoor air pollution are returned to in a later section in this chapter.

Studies of acute effects

The response to exposure to high concentrations of urban air pollution is rapid and, as seen in the major London fogs, increased deaths occur among particularly susceptible sections of the population (notably the elderly and chronic sick) within 24 hours of onset of an episode. Thus records of deaths by exact day of occurrence are required to seek evidence of possible effects. Figure 25.4 shows day-to-day variations in deaths in Greater London during the winter of

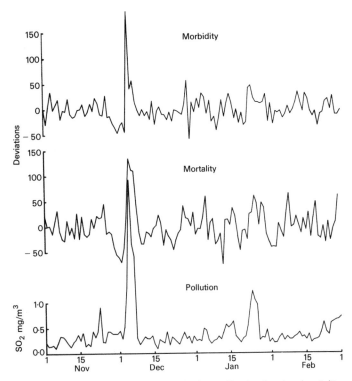

Fig. 25.4. Day-to-day variations in deaths in Greater London (mortality curve) and in emergency admissions to hospital (morbidity curve), both expressed as deviations from 15-day moving averages, and daily mean concentrations of sulphur dioxide. Winter 1962–3.

1962–3 displayed as deviations from the 15-day moving average. This technique allows sudden increases lasting for only one or two days to be seen independently of the more gradual changes associated with influenza epidemics or other seasonal factors. The mortality data clearly show the impact of a period of high pollution lasting for several days in December 1962. This was the last occasion on which any substantial increase in deaths associated with air pollution was seen in London—the Clean Air Act of 1956 had not by then had its full impact, and concentrations of smoke as well as sulphur dioxide were still very high. Examination of a long series of data from London has indicated that increases in daily deaths became detectable when 24-hour mean concentrations of smoke and of sulphur dioxide both exceeded about 750 $\mu g/m^3$ (Holland et al. 1979). However, effects were related to the complex mixture of these pollutants from coal-burning and other sources, coupled with low temperatures and high relative humidity, and the same effects are not necessarily to be expected in other circumstances. Because there is usually much 'random noise' in daily mortality records, a further limitation to the above approach is that a population base of several million is required, concentrated in a single city with reasonably uniform pollution conditions throughout.

Some authors have applied multiple regression techniques to the analysis of daily records of deaths, pollution, and weather variables, to determine which of these factors has significant effects and to develop exposure–response curves. Thus Mazumdar et al. (1982), on re-examination of the London data, found a significant effect of smoke concentrations, but uncertainty remains about the form of the relationship. In these kinds of studies much depends on the types of models developed and on the assumptions made. In general there are large seasonal variations in mortality, and epidemics of illness, notably influenza, are liable to affect the pattern from year to year. Making appropriate allowance for these longer-term fluctuations and disentangling short-term effects of air pollution and weather variables is a difficult task. Modelling techniques have been explored further, using data from Los Angeles, by Shumway et al. (1988), and Hatzakis et al. (1986) have examined mortality records in Athens on the basis of daily deviations from a sinusoidal curve.

A more sensitive measure of adverse effects can be provided by morbidity data, but these are rarely available on a routine basis. The index of morbidity displayed in Figure 25.4 is based on hospital admissions data, provided by the London Emergency Bed Service that channels all requests for admission of patients who suddenly become ill through a central office, where daily records are maintained. Such data can be handled in an analogous manner to the daily deaths, calculating deviations from 15-day moving averages, although day-of-week corrections may be required first to eliminate biases arising from the more limited medical services operating at weekends.

More generally, however, special studies are required among selected subjects in order to relate exacerbations of illness to air pollution. A technique that proved effective in London was to collect information on self-assessed changes in condition among bronchitic patients (Lawther et al. 1970). The subjects were enrolled mainly at chest clinics, selecting ones who were fit enough to be out and about, but whose condition often deteriorated from time to time during the winter months. They simply noted in small pocket diaries whether they considered their breathing to be better, the same, or worse on each day, as compared with the previous one, and the answers were scored on a numerical scale (1, 2, or 3 respectively). The mean score was then calculated on a daily basis and with groups of patients as large as 1000 the random variation in this was small, allowing occasional peaks relating to adverse environmental conditions to stand out clearly. Figure 25.5 illustrates this for a winter (1959–60) in which there were several minor episodes of high pollution. The mean score was correlated with concentrations of smoke and of sulphur dioxide more closely than with a number of weather variables that had been examined. While this approach works best with dedicated subjects living in a large city where exposures are likely to rise and fall in unison throughout the area, analogous methods have been applied in other circumstances. Cohen et al. (1972), for example, used a relatively small group of asthmatic subjects for studies related to pollution around a large point source (a power plant). This, however, is a much more difficult type of situ-

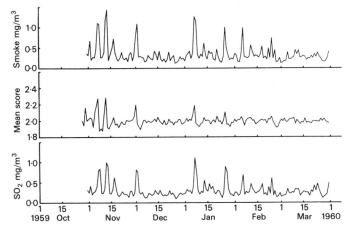

Fig. 25.5. Results from 'diary' studies in London, winter 1959–60, showing day-to-day variations in the illness score for bronchitic subjects together with mean daily concentrations of smoke and sulphur dioxide.

ation to assess, since individuals within the group are likely to be subjected to highly variable exposures, with peak values occurring at different times as the plume moves in various directions with the wind.

The data presented in Figure 25.5 represent a rather extreme example in which, with high levels of pollution at the time, associations with exacerbations of illness were evident without detailed statistical analyses. More generally there are difficulties, as with daily mortality data, in demonstrating associations between more modest variations in illness and less extreme levels of pollution, taking account of effects of weather and other factors. Lebowitz *et al.* (1987) have used time series spectral analysis to examine daily symptom diary data in the area of Tucson, Arizona, but pollution effects do not stand out clearly. In France, Charpin *et al.* (1988) have carried out a diary study among schoolchildren, and their analyses indicate an association between respiratory symptoms and sulphur dioxide levels in a coal-burning area, but not within a less polluted control area.

The inclusion of lung function measurements in studies of day-to-day variations in the health of groups of selected subjects has the potential to improve sensitivity in detecting effects of air pollution or other environmental factors, but it is costly and time consuming, requiring either the distribution of instruments to individuals or attendance at designated centres. Most studies of this type have been undertaken among children, seen conveniently in schools or at summer camps, etc. Even so, continuous daily examination is rarely practicable and attention is best focused on periods around air pollution 'alerts' as in the investigations of Dockery *et al.* (1982) in Steubenville, Ohio. In the Netherlands, Dassen *et al.* (1986) were able to detect some decrement in lung function in the course of a survey among schoolchildren that included an episode of elevated pollution levels. Once-weekly lung function measurements among schoolchildren were included as one aspect of the 'Six Cities' study in the US (Kinney *et al.* 1989), showing some link with ozone levels.

However, the significance to health of the small transient changes in lung function seen in this and most of the other studies of this type is uncertain.

Associations with chronic respiratory disease

The high death-rates from bronchitis in urban as compared with rural areas and the high overall rates in some countries (notably the UK), where there has been much air pollution, have long been taken as indicators of a role of pollution in the development of chronic respiratory disease. Studies of local or regional contrasts in death-rates can provide leads concerning the existence of a problem, but because of uncertainties about diagnostic classifications and about the effects of other factors, such as smoking and occupational exposures to dust and fumes, it is difficult to draw firm conclusions in this way.

The most important development in pursuing this matter further during recent decades has been the use of standardized questionnaires for enquiring into respiratory symptoms, so that their prevalence in defined populations can be related to a wide range of environmental and personal factors. Questionnaires of this form are discussed further in other chapters, but for studies on effects of air pollution the basic format commonly used is that of the Medical Research Council, successive versions of which were produced in 1960, 1966, and 1976 (MRC 1976). This was developed for clinical or survey use, and it sets out questions in a standardized manner on cough, phlegm production, breathlessness, wheezing, and history of respiratory illnesses. There are detailed enquiries into smoking habits, and opportunities to record occupational and residential histories, and other personal data, depending on the exact purpose of the study. Simple tests of ventilatory capacity are also commonly incorporated, such as forced vital capacity and forced expiratory volume in one second (FVC and FEV_1) or peak expiratory flow rate (PEFR).

While many of the studies based on respiratory symptoms questionnaires have been concerned with specific occupational risks, others have dealt with general population samples, so as to examine relationships between the prevalence of respiratory disease in areas of contrasting pollution levels. Van der Lende *et al.* (1973, 1986), for example, initially conducted a cross-sectional study in the Netherlands, but by repeating their surveys in the same areas every three years they were also able to look at changes in prevalence with time in relation to changes in pollution. A longitudinal study involving lung function measurements on children during three successive winters has been carried out in Italy (Arossa *et al.* 1987). Spirometric values were originally poorer in the more polluted urban area studied than in a suburban area, but with declining pollution there was finally no significant difference. To reduce the risk of differences in social class structure between areas affecting results, there are advantages in studying a single occupational group, provided it is not subject to special risks of respiratory disease itself. Thus telephone workers provided a suitable group among which to examine effects of pollution in urban and rural areas within and between the UK and the US (Holland *et al.* 1965).

Within all the studies reported to date the major factor affecting the prevalence of symptoms and lung function values has been smoking, which makes it difficult to isolate possible effects of air pollution. To circumvent this problem attention can be confined to schoolchildren although (regrettably) it is difficult to avoid the effects of smoking altogether even then (Holland *et al.* 1969*a*). Other types of studies of children have related the frequency of acute respiratory illnesses to exposures to air pollution, and in some cases examined the same group of children repeatedly as they have grown up (Douglas and Waller 1966), even continuing this into adult life (Britten *et al.* 1987; Colley *et al.* 1973).

There are some adult groups that are free from smoking, so offering the opportunity for studying possible long-term effects of exposure to air pollution. Euler *et al.* (1987) carried out a large survey of Seventh-day Adventists living in California, assessing their cumulative exposures to pollutants according to their residential histories. Associations between respiratory symptoms and sulphur dioxide/particulates exposures were shown in multiple logistic regression analyses, but other pollutants could have been involved.

The various surveys discussed above suggest that exposure to air pollution is a relatively minor factor in the development of chronic respiratory disease compared with other factors. Long term exposure stood out clearly only where there were quite large contrasts in exposure to pollution (usually expressed in terms of smoke and sulphur dioxide concentrations). Apart from smoking, one of the most important determinants of respiratory symptoms and deficits in lung function among adults is a history of acute respiratory illnesses earlier in life (Irvine *et al.* 1980). This may in turn reflect effects of urban air pollution or of parental smoking in the home (Holland *et al.* 1969*b*), but it underlines the point that the prevalence of chronic respiratory disease among adults cannot be related solely to the pollution (and other circumstances) at the time. Where pollution levels have changed substantially through the years, as they have in the UK in the wake of the Clean Air Act of 1956, it may be necessary to look back at former conditions when planning studies and interpreting findings. Because many adults have lived through periods of high pollution in the past there could still be a substantial backlog of effects of pollution detectable in surveys even where controls have now become effective. But the indications are that in populations exposed throughout their lives to only moderate levels of pollution, cross-sectional studies among subgroups living in different areas are unlikely to yield much definite information on effects of pollution.

The occurrence of chronic respiratory disease in later life may well reflect the interplay of a series of factors operating from birth onwards, involving acute respiratory infections,

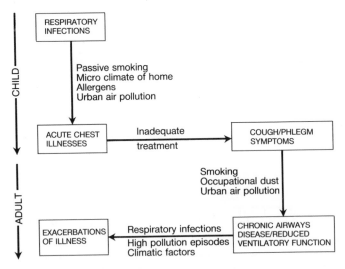

Fig. 25.6. Respiratory infections, air pollution, and other factors in the development of chronic airways disease. (Source: Waller 1989, with permission.)

exposure to indoor and outdoor air pollutants, and smoking (Fig. 25.6; Waller 1989). In most surveys only a small 'window' on these events can be examined directly, but it is clearly important to gather as complete histories as possible.

Lung cancer and air pollution

Ever since a substantial increase in lung cancer mortality was first noted, in the 1930s, it has been apparent that death-rates are generally higher in urban than in rural areas. A number of possible reasons for this have been considered through the years, including better opportunities for diagnosis in urban areas, and the presence of carcinogens such as benzo(a)-pyrene and other polycyclic aromatic hydrocarbons (PAHs) in town air, suggesting that air pollution may play some role (Lawther and Waller 1976).

Investigations into this problem must, however, take into account the effects of cigarette smoking which are the prime cause of lung cancer. Changes in smoking habits have been responsible for the long-term trends in lung cancer mortality seen in many countries, but the effect they may have had on urban/rural differentials is less clear. In the UK there is now little difference in smoking habits between urban and rural areas, but there could well have been larger contrasts in the past that would affect the overall lifetime exposure of adults in the upper age ranges. Thus great care is required in interpreting findings based purely on death certification data, without any information on smoking. The situation in England and Wales is illustrated in Table 25.2, which contrasts death-rates for one narrow age band in Greater London and the aggregate of rural districts throughout the country. While London rates were about twice those in rural districts for men born around 1891 who died in the early 1950s, this ratio is declining gradually and there may finally be little difference

Table 25.2. Lung cancer: age-specific death-rates for males aged 60–64 years in Greater London and in rural districts of England and Wales

		Deaths per 100 000	
Year of death	Year of birth	Greater London	Rural districts
1951–55	1891	350	170
1956–60	1896	420	240
1961–65	1901	450	270
1966–70	1906	405	295
1973	1911	370	290

The death-rates are approximate, being interpolated from tabulations in broader age ranges. Changes in the classification of rural districts limit this analysis beyond 1973, and data for the last quinquennium listed here are based on the single (mid) year only.

Table 25.3. Benzo(a)pyrene concentrations in the air of central London, 1949–81. All based on 24-hour samples aggregated for yearly periods

Period	Sampling site	Benzo(a)pyrene (μg/1000m^3)
1949–51	County Hall	46
1953–56	St. Bartholomew's Hospital	17
1957–64	County Hall	14
1972–73	St. Bartholomew's Medical College	4
1974–81	St. Bartholomew's Medical College	1.6

between the urban and rural rates. This could be interpreted as a direct effect of the reduction in pollution by coal smoke during the present century, which gained much momentum after the Clean Air Act of 1956, but which had been going on more slowly for a long time in London. Measurements of benzo(a)pyrene have been made during the past few decades (Table 25.3) and these show a very sharp downward trend. The problem remains, however, that little is known of changes in the pattern of smoking habits within the various urban and rural areas, and it is conceivable that cigarette smoking was at one time more popular in large urban areas such as London than in rural districts and that the situation is gradually reversing.

Only epidemiological studies which obtain information on smoking habits, residential histories, and other relevant factors (including occupation) can resolve these matters. Two types of study have proved effective, at least for demonstrating the magnitude of smoking effects and, secondarily, have shed some light on possible effects of air pollution. These are case–control studies in which enquiries are made among diagnosed cases of lung cancer and matched controls without the disease, and long-term prospective studies in which a defined population group is followed to death, gathering the required information on smoking and other factors at the outset, and at intervals of some five to ten years thereafter (Doll and Peto 1976; Hammond et al. 1977). The outcome of most studies to date has, however, been that once smoking has been taken into account, the differential between urban and rural dwellers in the incidence of lung cancer is quite small.

Clearly, to obtain any definitive results specifically on air pollution, large-scale studies encompassing areas with substantial contrasts in some suspect component (for example, coal smoke) are required. There are reports that the high lung cancer rates experienced by women in some parts of China could be related to massive exposures to coal smoke (indoors) while cooking (Chapman et al. 1988).

There has been some concern over the years that emissions from motor vehicles might have carcinogenic properties. There are traces of PAHs in the particulate components of both petrol and diesel engine exhausts, although in comparison with coal smoke the amounts are very small. Extracts of these particulates have, however, been shown to be active in short-term mutagenicity tests, and the activity does not appear to be related solely to their PAH content (currently nitro-derivatives of the PAHs are believed to contribute). Some animal inhalation studies have yielded lung tumours following lifetime exposures to massive concentrations of diesel exhaust products (Brightwell et al. 1989). While it is difficult to assess general population exposures specifically to vehicle pollutants and to identify groups with contrasting exposure, there have been some studies among occupational groups, particularly in relation to exposure to diesel fumes. One such study, among workers in bus garages subject to high concentrations of diesel smoke, has not revealed any appreciable difference in lung cancer death-rates over a 25-year period, in comparison with other categories of London Transport workers without special exposure (Waller 1981). However, investigations among railway workers in the US, with exposure to diesel locomotive fumes, have indicated an enhanced lung cancer risk (Garshick et al. 1988).

Effects of inhaled lead

A discussion on this topic is included as an example of problems that arise with a 'multi-media' pollutant. While relatively large exposures to lead, as in some occupational circumstances, can lead to a wide range of acute and chronic effects, the main concern in relation to more modest environmental exposures is with possible neuropsychological effects among young children (Department of Health and Social Security 1980). Sources of lead in air were discussed earlier. There is also lead in food, of which some results from contamination by air-lead and some natural or process sources, such as the solder in cans. In certain localities, where the drinking water supply is soft and acidic, and lead pipes are used, intake from that source becomes a major component. Children may also derive excessive amounts from a variety of adventitious sources such as old lead paint.

Thus studies on effects of lead must be related to environmental sources as a whole, and exposure is best assessed through biological measurements. Blood-lead determinations are considered to be the best indicators of recent exposure. According to experimental studies reported by Chamberlain et al. (1978) the biological half-life of lead in the blood is of the order of 18 days, so that such measurements largely reflect uptake during the preceding few weeks. Other indirect indices of lead absorption have been used. The free erythrocyte protoporphyrin (FEP) assay was recommended by the Centers for Disease Control (1978) for initial screening purposes in large populations. It is more cost-effective than blood-lead screening (Berwick and Komaroff 1982), but the relationship with the blood lead is not very close, so that the latter measurements are to be preferred from the start in epidemiological studies. Similarly, measurements of δ-amino-levulinic acid dehydratase activity have been used for some survey work (Secchi *et al.* 1973). There is an inverse relationship with blood lead, but the method does not offer advantages in terms of reliability and specificity.

A much bigger question to consider is whether any measure of recent exposure is adequate in relation to anticipated effects. The model proposed as a result of follow-up studies of clinical lead poisoning cases among children, coupled with information from biochemical and toxicological studies, is that elevated lead levels interfere with the development of the brain during infancy leading to adverse effects on cognitive function and/or behaviour that cannot in general be assessed until the primary school years are reached. There is no clear indication as to whether it is the average exposure during the early years or an acute episode within that period that is most critical, but there is some argument in favour of measurements relating to several years prior to the time when effects are assessed. It may sometimes be possible to organize a series of blood-lead measurements from birth onwards in longitudinal studies, but otherwise attention can be turned to lead in the longer-term storage compartments of the body. The major part of the body burden is relatively tightly bound in bone, and shed deciduous teeth of children provide one possibility for assessing average exposures, covering the period from the eruption of the tooth. The selection, preparation, and analysis of teeth is not easy, but it has been undertaken by research groups in several countries (Needleman *et al.* 1979; Smith *et al.* 1983; Winneke 1983). As Smith *et al.* (1983) demonstrated, between-teeth differences within any child make it necessary to standardize the position of teeth being sampled. Analysis of hair samples has been put forward as an intermediate approach, reflecting exposures over a somewhat longer period than blood lead, but it is difficult to separate external contamination from the intrinsic lead in the hair, and the method has not been widely adopted.

The assessment of effects is even more difficult than assessing exposure: carefully standardized and validated tests of the cognitive and behavioural effects considered are required and there are a great many confounding factors, related to family and social circumstances, that must be taken into account. Needleman *et al.* (1979), in their study in the US of dentine lead levels (the analysis was done on dentine rather than whole teeth), introduced teachers' assessments of behaviour in the class-room as well as standardized intelligence tests and others relating to visual motor function and motor co-ordination. The main comparisons made were between subsets at the high and low extremes of their dentine-lead distributions. For these subjects, information on a range of family factors such as mother's education, father's social class, and parental IQ were entered into the analysis of co-variance used to evaluate the results. Significant differences in IQ were seen between the high and low dentine lead groups. Blood-lead data were also available for some of the children, as measured four or five years earlier, but results were not analysed directly in relation to them. Subsequently Yule *et al.* (1981) undertook a pilot study in the UK using a range of tests analogous with those used by Needleman *et al.*, but linking findings with blood-lead measurements that had been made in a survey about a year earlier. Their results also indicated an association between attainment scores in several of the tests and the lead exposure indices (in this case, blood lead), after adjusting for social class. They did not, however, have much other information on family circumstances and in further work they planned to collect a wider range, including parental IQ tests.

As mentioned above, tooth-lead determinations have been used in studies by Winneke (1983) in Germany, and by Smith *et al.* (1983) in the UK. These show further how important it is to control adequately for non-lead variables likely to be related to performance in the intelligence tests. Thus in the German work, associations were found between IQ and tooth lead, but doubt was expressed as to whether they were truly causal, although some impairment of perceptual motor integration was considered as being caused by lead. In the British study strong associations were reported between intelligence and other psychological measures, and the social factors that had been investigated. While there were significant differences for some of these tests between groups differing in tooth-lead levels, these became small and non-significant (although still in the same direction, with high lead levels related to deficits in attainment) once the main social factors were taken into account. Continuing studies in the UK tend to support a link between blood-lead levels and ability/attainment scores in children, without any clear threshold (Fulton *et al.* 1987), while showing that other factors have a more dominant effect (Harvey *et al.* 1988; Pocock *et al.* 1987), often making it difficult to identify the role of lead.

Not all problems in the field of air pollution epidemiology are surrounded by complications of multi-media sources, but most have analogous problems of non-specificity of effects and a wide range of confounding variables. The discussion above illustrates well the complexity and magnitude of the task in trying to reach conclusions firm enough for the guidance of public health policy.

Studies on indoor pollution

Concern over possible effects of indoor air pollution has increased in recent years, as evidenced by the wide range of studies reported at successive symposia on this topic (Perry and Kirk 1988; Seifert *et al.* 1987; Spengler *et al.* 1982; World

Table 25.4. Deaths from accidental carbon monoxide poisoning in the home, England and Wales, 1968–81

Year	Number	Year	Number
1968	469	1975	95
1969	357	1976	108
1970	303	1977	101
1971	226	1978	91
1972	150	1979	145
1973	139	1980	102
1974	103	1981	85

The figures relate to cause of death codes E870, 871, 872, and 874 (8th Revision) and E867, 868.0, 868.1, and 868.3 (9th Revision)

Health Organization 1986). Reasons for this include the improved control of pollutants outdoors, bringing into closer perspective those that remain indoors, the loss (in urban areas of the UK) of the open coal fire that used to induce high ventilation rates in rooms, removing other pollutants produced within them (though at times 'fumigating' the occupants with high concentrations of smoke and sulphur dioxide), and general efforts to increase the 'tightness' of buildings in the interests of fuel economy, often reducing ventilation to well below the desirable level. In these circumstances odours and water vapour may in any case become a nuisance, but the main concern here is with some specific pollutants emitted from combustion sources or materials within the home. Sources and possible effects of such pollutants have been reviewed by Samet *et al.* (1987).

Potentially the most dangerous pollutant is carbon monoxide, released from the incomplete combustion of fuels in unflued appliances or from faulty flues in other cases. Poor combustion coupled with limited ventilation can lead to lethal concentrations in these circumstances, and as Table 25.4 shows, about 100 deaths a year have been attributed to accidental poisoning by carbon monoxide in the home in England and Wales (Office of Population Censuses and Surveys 1983). The higher figures up to the early 1970s are due to the presence of carbon monoxide in the town gas then supplied, leading to some direct toxic effects in the event of leaks, even without combustion. The natural gas now supplied contains no carbon monoxide. With a number of deaths occurring from faulty combustion sources in the most extreme conditions, it is likely that some transient morbidity may also be associated with exposure to carbon monoxide in the home, but since the symptoms (lethargy, headaches, etc.) are ill defined, this problem is difficult to investigate satisfactorily. Surveys of carbon monoxide in blood or in exhaled air can help to identify cases, but as indicated earlier there is always, for this pollutant, the complicating factor of smoking.

An associated pollutant from combustion sources that has received particular attention in relation to health is nitrogen dioxide. In streets, motor vehicles are the main sources (at least of its precursor, nitric oxide), but indoors any unflued combustion source can lead to concentrations that are often substantially higher than in streets. In countries where gas cooking is widely used and (as in the UK) hoods for removing

fumes from above the cooker are rare, that constitutes the commonest source. Concentrations are not likely to be sufficient to cause acute affects, but a number of studies have indicated possible associations between the occurrence of respiratory illnesses in children and the use of gas, as opposed to electric, cooking within the home. Sometimes it is difficult to exclude entirely social class or other factors that might also be related to risks of respiratory illness, but Melia *et al.* (1982) conducted a series of studies in an area of homogeneous housing, where gas and electric cooking was used to approximately equal extents, without any particular discrimination with respect to social class. In the homes, they measured nitrogen dioxide, using small diffusion-tube samplers that can be left in place in rooms without attention, and also air temperature and relative humidity. Despite an apparent association of respiratory illnesses in the (school-age) children with the presence of gas cookers in the home, in earlier phases of the work, no convincing evidence of a direct link with nitrogen dioxide concentrations was seen. A study done elsewhere among infants who had been followed from birth failed to reveal any association between respiratory illnesses and gas cooking (Melia *et al.* 1983). It is clear that in this field of indoor air pollution it is difficult establish cause and effect relationships, one of the basic difficulties being to make adequate estimates of exposure to the suspect agents, bearing in mind the wide range of variations from house to house, from room to room, and from time to time. Attempts have been made to investigate possible effects of transient peaks by monitoring nitrogen dioxide on an individual basis in relation to gas stove use, with simultaneous lung function measurements (Goldstein *et al.* 1988), but substantial resources would be required for any full-scale study on these lines. Portable kerosene heaters can make a large contribution to indoor nitrogen dioxide levels, and associations between respiratory illness in children and the combined use of gas cookers and kerosene heaters have been reported (Melia *et al.* 1988).

In some of the above studies, one factor consistently associated with a high prevalence of respiratory illness in young children was parental smoking, and this feature has already been noted above in relation to earlier work (Holland *et al.* 1969*b*). Certainly cigarette smoke is a widespread problem in indoor environments, leading to odour, eye and throat irritation, and to more specific effects on respiratory illnesses and lung function, at least among young children. There is also a mounting body of evidence indicating a small increase in the risk of lung cancer among non-smokers through prolonged exposures to environmental tobacco smoke (Independent Scientific Committee on Smoking and Health 1988).

A further indoor air pollutant that has gained some prominence in recent years is formaldehyde, arising in some circumstances from the urea-formaldehyde foam used for insulating purposes in cavity walls (small amounts of formaldehyde being liberated more particularly within the first few days after installation) and from binding materials and adhesives used in chipboard furniture and some other materials around the home. The most obvious effects are eye and throat irritation, and there are no indications at the present time that the concentrations liable to be encountered in the home lead to any longer-term effects.

Among other indoor pollutants of possible concern to health are various organic compounds arising from consumer products such as paints, polishes, wood preservatives and sprays, radon and its decay products that are derived from the soil and rocks beneath a house, as well as from the building materials, and asbestos that is used in some insulating and roofing materials. Some of these are considered further in other chapters of this book, and often the clearest indications of possible effects on health come first from the more extreme occupational exposures, after which studies in the community at large might be considered.

Future prospects

Enough may have already been said within this chapter to indicate that a proper sense of perspective needs to be maintained in respect of effects of air pollution on health. The problems have been highlighted in developed countries where illness associated with communicable disease has often been well controlled, leaving effects of environmental conditions, together with excesses of eating, drinking, and smoking, of more concern. At the same time, where industry has developed and towns have grown up without due regard for the efficient use of fuel and proper control of pollution, as happened in Britain during the nineteenth century, severe problems have arisen. Many of the countries that were badly polluted at one time have now taken steps to control the major pollutants, such as smoke (or suspended particulates) and sulphur dioxide, at least down to levels not expected, on the basis of previous experience, to have detectable adverse effects on health. In those circumstances there may be little more that can or need be done in the way of epidemiological studies of effects of air pollutants. As indicated in the preceding section, however, attention may still be required to the 'micro-environments' in which people live, the difficulty then being that circumstances differ so widely and change so rapidly with time that the design and conduct of epidemiological studies to investigate possible effects becomes very complex.

The micro-environment can be of great importance also in developing countries, particularly where wood, coal, or other fuels are used for cooking inside primitive dwellings, without any proper flue (Boleij et al. 1988; Sofoluwe 1968). The smoke, carbon monoxide, oxides of nitrogen, sulphur dioxide, and partial combustion products such as aldehydes can exert a variety of adverse effects on occupants, and particularly on young children, who may sometimes be carried on their mothers' backs while cooking is in progress. In these cases epidemiological studies are hardly needed to recognize the obvious dangers. As far as urban growth and industrialization is concerned in developing countries, there are lessons

from the past that might be learnt, but economic conditions and social pressures generally militate against their application, and serious problems of pollution still arise.

Vigilance is also required in relation to unforeseen problems. Photochemical pollution of the oxidant type is still relatively new and while there is no evidence that it is capable of exerting acute lethal effects on the scale of the traditional (smoke/sulphur dioxide) reducing type pollutants it may have some longer-term effects that are not fully recognized. The two types of pollution can coexist, but relatively little is known as yet about adverse effects of such mixtures, and further changes in fuel usage or industrial processes could lead to other types of pollutants or different photochemical reactions. Epidemiological studies can help to resolve problems as they arise, although pointers from toxicological work or evidence from relatively extreme occupational exposures are often required before investigations in the general population become viable.

References

Apling, A.J., Sullivan, E.J., Williams, M.L., et al. (1977). Ozone concentrations in South East England during the summer heatwave of 1976. Nature 269, 569.

Arossa, W., Spinaci, S., Bugiani, M., Natale, P., Bucca, C., and de Candussio, G. (1987). Changes in lung function of children after an air pollution decrease. Archives of Environmental Health 42, 170.

Bailey, D.L.R. and Clayton, P. (1982). The measurement of suspended particle and total carbon concentrations in the atmosphere using standard smoke shade methods. Atmospheric Environment 16, 2683.

Berwick, D.M. and Komaroff, A.L. (1982). Cost effectiveness of lead screening. New England Journal of Medicine 306, 1392.

Boleij, J.S.M. Campbell, H., Wafula, E., Thiaru, H., and de Koning, H.W. (1988). Biomass fuel combustion and indoor air quality in developing countries. In Indoor and ambient air quality (ed. R. Perry and P.W. Kirk) p. 24. Selper, London.

Brightwell, J., Fouillet, X., Cassano-Zoppi, D., et al. (1989). Tumours of the respiratory tract in rats and hamsters following chronic inhalation of engine exhaust emissions. Journal of Applied Toxicology 9, 23.

Britten, N., Davies, J.M.C., and Colley, J.R.T. (1987). Early respiratory experience and subsequent cough and peak expiratory flow rate in 36 year old men and women. British Medical Journal 294, 1317.

British Standards Institution (1969). Methods for the measurement of air pollution. BS 1747, Parts 2 and 3. BSI, London.

British Standards Institution (1983). Measurement of air pollution. BS 1747, Part 7. BSI, London.

Centers for Disease Control (1978). Preventing lead poisoning in young children: a statement by the Centers for Disease Control, Atlanta. CDC, Atlanta, Georgia.

Chamberlain, A.C., Heard, M.J., Little, P., Newton, D., Wells, A.C., and Wiffen, R.D. (1978). Investigations into lead from motor vehicles. AERE-R9198 UK Atomic Energy Authority, Harwell. HMSO, London.

Chandler, T.J. (1965). The climate of London. Hutchinson, London.

Chapman, R.S., Mumford, J.L., Harris, D.B., He, X., Jiang, W., and Yang, R. (1988). The epidemiology of lung cancer in Xuan

Wei, China: Current progress, issues and research strategies. *Archives of Environmental Health* **43**, 180.

Charpin, D., Kleisbauer, J.P., Fondarai, J., Graland, B., Viala, A., and Gouezo, F. (1988). Respiratory symptoms and air pollution changes in children: The Gardanne coal-basin study. *Archives of Environmental Health* **43**, 22.

Cohen, A.A., Bromberg, S., Buechley, R.W., Heiderscheit, L.T., and Shy, C.M. (1972). Asthma and air pollution from a coal fuelled power plant. *American Journal of Public Health* **62**, 1181.

Colley, J.R.T., Douglas, J.W.B., and Reid, D.D. (1973). Respiratory disease in young adults: Influence of early childhood lower respiratory tract illness, social class, air pollution and smoking. *British Medical Journal* **iii**, 95.

Commins, B.T. (1963). Determination of particulate acid in town air. *Analyst* **88**, 364.

Commins, B.T. and Lawther, P.J. (1965). A sensitive method for the determination of carboxyhaemoglobin in a finger-prick sample of blood. *British Journal of Industrial Medicine* **22**, 139.

Daly, C. (1959). Air pollution and causes of death. *British Journal of Preventive and Social Medicine* **13**, 14.

Dassen, W., Brunekreef, G., Hoek, G., et al. (1986). Decline in children's pulmonary function during an air pollution episode. *Journal of the Air Pollution Control Association* **36**, 1223.

Department of Health and Social Security (1980). *Lead and health.* HMSO, London.

Dockery, D.W., Ware, J.H., Ferris, B.G. Jr., Speizer, F.E., Cook, N.R., and Herman, S.M. (1982). Change in pulmonary function in children associated with air pollution episodes. *Journal of the Air Pollution Control Association* **32**, 937.

Doll, R. and Peto, R. (1976). Mortality in relation to smoking: 20 years' observations on male British doctors. *British Medical Journal* **ii**, 1525.

Douglas, J.W.B. and Waller, R.E. (1966). Air pollution and respiratory infection in children. *British Journal of Preventive and Social Medicine* **20**, 1.

Environmental Protection Agency (1987). *Ambient Air Quality Standards for Particulate Matter. Final Rules.* EPA, Federal Register **52**, (No. 126), 24634–24750. US Government Printing Office, Washington, DC.

Euler, G.L., Abbey, D.E., Magie, A.R., and Hodgkin, J.E. (1987). Chronic obstructive pulmonary disease symptom: Effects of long-term exposure to ambient levels of total suspended particulates and sulfur dioxide in California Seventh Day Adventists residents. *Archives of Environmental Health* **42**, 213.

Evelyn, J. (1661). *Fumifugium.* Reprinted 1961 by National Society for Clean Air, London.

Facchetti, S. and Geiss, F. (1983). *Isotopic lead experiment. Interim report.* Commission of the European Communities, Brussels.

Firket, J. (1936). Fog along the Meuse Valley. *Transactions of the Faraday Society* **32**, 1192.

Fulton, M., Thomson, G., Hunter, R., Raab, G., Laxen, D., and Hepburn, W. (1987). Influence of blood lead on the ability and attainment of children in Edinburgh. *Lancet* **i**, 1221.

Garshick, E., Schenker, M.B., Munoz, A., et al. (1988). A retrospective cohort study of lung cancer and diesel exhaust exposure in railroad workers. *American Review of Respiratory Disease* **137**, 820.

Goldstein, I.F., Lieber, K., Andrews, L.R., et al. (1988). Acute respiratory effects of short-term exposures to nitrogen dioxide. *Archives of Environmental Health* **43**, 138.

Hammond, E.C., Garfinkel, L., Seidman, H., and Lew, E.A. (1977). Some recent findings concerning cigarette smoking. In *Origins of human cancer. Book A: Incidence of cancer in humans* (ed. H.H. Hiatt, J.D. Watson, and J.A. Winsten) p. 101. Cold Spring Harbor Laboratory, New York.

Harrison, R.M. and Perry, R. (1977). The analysis of tetra-ethyl lead compounds and their significance as urban air pollutants. *Atmospheric Environment* **11**, 847.

Harvey, P.G., Hamlin, M.W., Kumar, R., Morgan, G., Spurgeon, A., and Delves, H.T. (1988). Relationships between blood lead, behaviour, psychometric and neuro-psychological test performance in young children. *British Journal of Development Psychology* **6**, 145.

Hatzakis, A., Katsouyanni, K., Kalandidi, A., Day, N., and Tricopoulos, D. (1986). Short-term effects of air pollution on mortality in Athens. *International Journal of Epidemiology* **15**, 73.

Holland, W.W. (1989). Chronic airways disease in the United Kingdom. *Chest* **96**, 318S.

Holland, W.W., Halil, T., Bennett, A.E., and Elliott, A. (1969a). Factors influencing the onset of chronic respiratory disease. *British Medical Journal* **ii**, 205.

Holland, W.W., Kasap, H.S., Colley, J.R.T., and Elliot, A. (1969b). Repiratory symptoms and ventilatory function: A family study. *British Journal of Preventive and Social Medicine* **23**, 77.

Holland, W.W., Reid, D.D., Seltzer, R., and Stone, R.W. (1965). Respiratory disease in England and the United States. Studies of comparative prevalence. *Archives of Environmental Health* **10**, 338.

Holland, W.W., Bennett, A.E., Cameron, I.R., et al. (1979). Health effects of particulate pollution: Re-appraising the evidence. *American Journal of Epidemiology* **110**, 533.

Independent Scientific Committee on Smoking and Health (1988). *Fourth Report of the Independent Scientific Committee on Smoking and Health.* HMSO, London.

Irvine, D., Brooks, A., and Waller, R.E. (1980). The role of air pollution, smoking and respiratory illnesses in childhood in the development of chronic bronchitis. *Chest* **77S**, 251S.

Kinney, P.L., Ware, J.H., Spengler, J.D., Dockery, D.W., Speizer, F.E., and Ferris, B.G. Jr. (1989). Short-term pulmonary function change in association with ozone levels. *American Review of Respiratory Disease* **139**, 56.

Koutrakis, P., Wolfson, J.M., and Spengler, J.D. (1988). An improved method for measuring strong acidity: Results from a nine-month study in St. Louis, Missouri and Kingston, Tennessee. *Atmospheric Environment* **22**, 157.

Lawther, P.J. and Commins, B.T. (1970). Cigarette smoking and exposure to carbon monoxide. *Annals of the New York Academy of Science* **174**, 135.

Lawther, P.J. and Waller, R.E. (1976). Coal fires, industrial pollution and smoking. *INSERM Symposium Series* **52**, 27.

Lawther, P.J., Waller, R.E., and Henderson, M. (1970). Air pollution and exacerbations of bronchitis. *Thorax* **25**, 525.

Lawther, P.J., Macfarlane, A.J., Waller, R.E., and Brooks, A.G.F. (1975). Pulmonary function and sulphur dioxide: Some preliminary findings. *Environmental Research* **10**, 355.

Lebowitz, M.D., Collins, L., and Holberg, C.J. (1987). Time series analyses of respiratory responses to indoor and outdoor environmental phenomena. *Environmental Research* **43**, 332.

Logan, W.P.D. (1949). Fog and mortality. *Lancet* **i**, 78.

McInnes, G. (1988). *Airborne lead concentrations in the United Kingdom 1984–1987.* Report LR 676 (AP). Warren Spring Laboratory, Stevenage, UK.

Mazumdar, S., Schimmel, H., and Higgins, I.T.T. (1982). Relation of daily mortality to air pollution: An analysis of 14 London winters. *Archives of Environmental Health* **37**, 213.

Medical Research Council (1976). *Questionnaire on respiratory symptoms*. MRC, London.

Meetham, A.R., Bottom, D.W., Cayton, S., Henderson-Sellers, A., and Chambers, D. (1981). *Atmospheric pollution. Its history, origins and prevention* (4th edn). Pergamon Press, Oxford.

Melia, R.J.W., Florey, C. du V., Morris, R.W., *et al.* (1982). Childhood respiratory illness and the home environment. II Associations between respiratory illness and nitrogen dioxide, temperature and relative humidity. *International Journal of Epidemiology* **11**, 164.

Melia, J., Florey, C. du V., Sittampalam, Y., and Watkins, C. (1983). *The relation between respiratory illness in infants and gas cooking in the UK: A preliminary report*. Proceedings of the 6th World Congress on Air Quality 2, 67, IUAPPA, Paris.

Melia, R.J.W., Chinn, S., and Rona, R.J. (1988). Respiratory illness and home environment of ethnic groups. *British Medical Journal* **296**, 1438.

Ministry of Health (1954). *Mortality and morbidity during the London fog of December 1952*. HMSO, London.

Müller, J. (1984). *Field measurements of suspended particulate matter*. Final report to the Commission of the European Communities, DGXI, Brussels.

Needleman, H.L., Gunnoe, C., Leviton, A., Reed, R., Peresie, H., and Maher, C. (1979). Deficits in psychologic and classroom performance of children with elevated dentine lead levels. *New England Journal of Medicine* **300**, 689.

Office of Population Censuses and Surveys (1983). *Mortality statistics*. Series DH2 (cause) and DH4 (accidents and violence) for the year 1981 and earlier annual volumes. HMSO, London.

Organization for Economic Co-operation and Development (1964). *Methods of measuring air pollution*. OECD, Paris.

Organization for Economic Co-operation and Development, (1975). *Photochemical oxidant air pollution*. OECD, Paris.

Organization for Economic Co-operation and Development (1983). *Polycyclic aromatic hydrocarbons*. Report of a Workshop held in Paris, October 1981. OECD, Paris.

Perry, R. and Kirk, P.W. (ed.) (1988). *Indoor and ambient air quality*. Selper, London.

Photochemical Oxidants Review Group (1987). *Ozone in the United Kingdom*. Department of the Environment, London.

Pocock, S.J., Ashby, D., and Smith, M.A. (1987). Lead exposure and children's intellectual performance. *International Journal of Epidemiology* **16**, 57.

Pocock, S.J., Shaper, A.G., Walker, M., *et al.* (1983). Effects of tap water lead, water hardness, alcohol and cigarettes on blood level concentrations. *Journal of Epidemiology and Community Health* **37**, 1.

Registrar-General (1845). *5th annual report, for the year 1843* (2nd edn), p. 415. HMSO, London.

Samet, J.M., Marburg, M.C., and Spengler, J.D. (1987). Health effects and sources of indoor air pollution. Part 1. *American Review of Respiratory Disease* **136**, 1486.

Schrenk, H.H., Heimann, H., Clayton, C.D., Fafafer, W.M., and Wexler, H. (1949). *Air pollution in Donora, Pennsylvania*. US Public Health Service, Washington, DC.

Secchi, G.C., Alessio, L., Cambiaghi, G., and Andreoletti, F. (1973). ALA-dehydratase activity of erythrocytes and blood lead levels in 'critical' population groups. In *Environmental health aspects of lead*, p. 595. Commission of the European Communities, Luxembourg.

Seifert, B., Esdorn, H., Fischer, M., Ruden, H., and Wegner, J. (ed.) (1987). *Indoor Air '87*. Proceedings, 4 volumes. Institute for Water, Soil and Air Hygiene, Berlin.

Shumway, R.H., Azari, A.S., and Pawitan, Y. (1988). Modeling mortality fluctuations in Los Angeles as functions of pollution and weather effects. *Environmental Research* **45**, 224.

Smith, M., Delves, T., Lansdown, R., Clayton, B., and Graham, P. (1983). The effects of lead exposure on urban children: The Institute of Child Health Southampton study. *Developmental Medicine and Child Neurology* **Supplement 47**, 1.

Sofoluwe, G.D. (1968). Smoke pollution in dwellings of infants with broncho-pneumonia. *Archives of Environmental Health* **16**, 670.

Spengler, J., Hollowell, C., Moschandreas, D., and Fanger, O. (ed.) (1982). Indoor air pollution. *Environmental International* **8**, 1.

Stewart, R.D., Scot-Stewart, R., Stamm, W., and Seelen, R.P. (1976). Rapid estimation of carboxyhaemoglobin level in firefighters. *Journal of the American Medical Association* **235**, 390.

Van der Lende, R., Schoutin, J.P., and Rijcken, B. (1986). Longitudinal epidemiological studies on effects of air pollution in the Netherlands. In *Aerosols* (ed. S.D. Lee, T. Schneider, L.D. Grant, and P.J. Verkerk) p. 731. Lewis, Chelsea, Michigan.

Van der Lende, R., Wever-Hess, J., Tammeling, G.J., de Vries, K., and Orie, N.G.M. (1973). Epidemiological investigations in the Netherlands into the influence of smoking and atmospheric pollution on respiratory symptoms and lung function. *Pneumonologie* **149**, 119.

Waller, R.E. (1981). Trends in lung cancer in London in relation to exposure to diesel fumes. *Environment International* **5**, 479.

Waller, R.E. (1989). Atmospheric pollution. *Chest* **96**, 363S.

Waller, R.E., Brooks, A.G.F., and Cartwright, J. (1963). An electron microscope study of particles in town air. *International Journal of Air and Water Pollution* **7**, 779.

Winneke, G. (1983). Neurobehavioural and neuropsychological effects of lead. In *Lead versus health* (ed. M. Rutter and R. Russell Jones) p. 249. J. Wiley, Chichester.

World Health Organization (1976). *Selected methods of measuring air pollutants*. WHO Offset Publication No. 24. WHO, Geneva.

World Health Organization (1977). *Air monitoring programme design for urban and industrial areas*. WHO Offset Publication No. 33. WHO, Geneva.

World Health Organization (1982). *Estimating human exposure to air pollutants*. WHO Offset Publication No. 69. WHO, Geneva.

World Health Organization (1986). *Indoor air quality research*. Euro Reports and Studies 103. WHO, Copenhagen.

World Health Organization (1987). *Air Quality Guidelines for Europe*. European Series No. 23. WHO, Copenhagen.

Yule, W., Lansdown, R., Millar, I.B., and Urbanowitz, M-A. (1981). The relationship between blood lead concentrations, intelligence and attainment in a school population: a pilot study. *Developmental Medicine and Child Neurology* **23**, 567.

26

Field investigations of biological and chemical hazards of food and water

FRANK A. FAIRWEATHER

Introduction

During the past decade there has been an ever increasing need to monitor food and water for potential biological and chemical hazards. The media continually review these issues and government departments and industry, sometimes separately, sometimes jointly, are endeavouring to solve the problems.

Identification of a hazard will eventually lead to review of a problem that may have been outstanding for many years. It will also highlight the need for sound scientific evidence for the continued inclusion of only microbiological or chemical contaminants of food and water that are safe for human consumption.

Field investigations play a vital role in identifying potential hazards, confirming links between outbreaks of disease and a particular element in the environment, and informing policy decisions on acceptable levels of contaminants of food and water. Field investigations of biological and chemical hazards relating to food and water call for co-operation between, among others, clinicians, epidemiologists, microbiologists, statisticians, and toxicologists, and access to toxicological services and to microbiological and analytical laboratories.

Using food as an example, it must be stressed that the industry endeavours to ensure that its products are safe by carrying out both biological and chemical tests before marketing. The legislative requirements placed on the industry are stringent and demanding, and involve acute, subacute, and chronic studies in various animals to determine the toxicological, teratological, or carcinogenic potential of the compounds included in the food. Rigid specifications are also laid down to ensure that raw material of the highest quality is used. Thus all food products are examined to ensure that they meet with toxicological, microbiological, and raw-material standards, and that the factory process is subjected to safety surveillance. This is an important contribution to the reduction of hazard.

In the following section some general issues for field investigations of food- and water-borne health hazards are discussed before attention is given to the specific clinical conditions which may occur in man.

General issues

Food

Food-borne health hazards have increased in importance following the centralization of food production, the increase in communal eating, and the development of international trade and tourism.

Food material in its natural state keeps sound and edible for only a comparatively short time. Seasonable production and poor distribution contribute to a wide disparity between production and need. As populations have become concentrated within towns and cities, the problem of the quality and supply of food has greatly intensified in some areas of the world since the consumer is remote from the field of production. The greater handling time increases the wastage of food through contamination, its destruction by pests, its inefficient utilization, and spoilage. In industrial areas, there is less home-based eating and many new preserved and processed foods are sold for convenience. Responsibility for the safety and wholesomeness of food has been moved away from the individual to industrialists of governments, creating a potential for large-scale outbreaks of food-borne diseases.

Water

Water quality influences the health and well-being of the individual, and both biological and chemical hazards may be conveyed in this medium. In considering water, both sewage and industrial effluents must also be taken into account.

Water-borne infectious diseases used to be widespread and a serious problem. With advances in medical knowledge this hazard has receded in the economically developed countries, but, with the growth of industry, chemical pollution of the aquatic environment is a growing concern for governments, scientists, and the now well-informed general public (Diamant 1979; Stevenson 1953).

Surveillance of food-borne disease

Microbiological criteria have been proposed for many varieties of food as their safety and perishability are related to microbial content. Only those for pasteurized milk have been widely adopted.

Attempts to establish microbiological food standards are frequently arbitrary and may not be related to the microbiological status of the food. However, experience in many countries has shown that the establishment of microbiological standards leads to an improvement in the microbiological status of food since food companies are encouraged to make improvements in plant hygiene and quality control. Although low microbial counts do not guarantee safety, foods that are consistently within established microbiological standards are less hazardous.

To establish a microbiological standard for a food or class or foods, the following technical and administrative aspects must be considered (WHO 1974).

1. The standard should be based on factual studies and serve one or more of the following objectives.
    (a) To determine the conditions of hygiene under which the food should be manufactured;
    (b) to minimize the hazards to public health;
    (c) to measure the perishability and storage potential of the food.
2. The standard should be attainable under practical commercial operating conditions, and should mot entail the use of excessive heat treatment or the addition of extra preservatives.
3. The standard should be determined after investigation of the processing operation.
4. The standard should be as simple and inexpensive to administer as possible—the number of tests should be kept to a minimum.
5. Details of methods to be used for sampling, examining, and reporting should accompanying all published microbiological standards.
6. In establishing tolerance levels for the permissible number of defective samples, allowance should be made for sampling and other variations due to differences in laboratory methods.

Food hygiene control

Food consumption is an essential part of life and is the route by which essential nutrients are supplied. However, because of its biological nature it also harbours micro-organisms which may contaminate it. These potential hazards have received more attention with the realization of the ubiquity of the life-threatening pathogens which are now seen in the environment. It has become clear that many are able to survive for long periods in such conditions as reduced oxygen levels and under refrigeration.

In response to the possible hazards associated with food, the industry employs a battery of processes to limit problems, and at the same time practises effective quality control and quality assurance measures for the protection of the food supply.

Processes such as pasteurization, refrigeration, dehydration, and specialized packaging all play a vital role in destroying or inactivating the bacteria or their spores. Laws and regulations concerning food hygiene aim to protect the public against injury from the foods that they consume. The sale of foods that are unwholesome, contaminated, or incorrectly processed is illegal under these laws. All workers handling food should understand and receive instruction on the basic elements of food hygiene. This not only includes personal hygiene, but also factory design and all aspects of process safety. Now, with widespread use of deep freezers and related products, it is paramount that consumers are assured that products are free from infection. Therefore, all catering staff must be aware that their hair, their clothing, as well as any open wounds or cuts can harbour bacteria that could cause disease.

The commonest clinical presentation of illness due to food-borne infection is diarrhoea, but other symptoms such as vomiting, abdominal pain, or cramps and pseudo-appendicitis may be present. However, the simple principle has always to be applied that somewhere during the cooking, handling, or storage procedure some error has taken place. In essence, food-borne diseases can be divided into two types. First, food-borne *infection* is caused by ingestion of food containing the viable organism which then multiplies and causes illness. Second, food-borne *intoxication* is caused by the consumption of food that contains a preformed bacterial toxin which results from the bacterium growing in the offending food.

Refrigeration and cooking play a part in preventing contamination of food. However, although refrigeration retards the increase in numbers of bacteria, it will not kill the bacteria. Cooking kills most pathogens, given a sufficiently long exposure time. Thus, cooking is the easiest and most convenient method for controlling the quality of the food we consume.

Over the past two decades public health has grown in importance with respect to the consumer. More people eat out either for pleasure or because their place of work is too far away to allow them to return home for the midday meal. It is essential that they are assured of quality in both food and drink. For this reason the food manufacturer, the process operator, the caterer, and the retailer have to assure themselves that they are providing food which is not infected or contaminated in any way. Public health authorities, at the same time, should frequently survey premises and individuals' work standards to ensure that the economic and social life of a country in no way suffers. It is sometimes forgotten that gastro-intestinal disorders and the like are a major cause of absence from work.

Food-borne hazards of the microbiological type are many and varied. Most are avoidable if correct procedures are adopted by the handler of the material, and it is essential that these individuals adhere to strict sanitary techniques. Microbiological contamination of food can result in clinical short-term discomfort while, at the other extreme, it can have a very severe, if not fatal, effect.

Surveillance and control of water-borne hazards and diseases

Through advances in public health, medical knowledge, and water processing, water-borne disease have largely been controlled. Windle-Taylor (1958) has led some of these measures.

1. Advances in medical care and treatment, in particular the introduction of antibiotics and eradication of carriers.
2. Separation of water sources from sewerage contamination, for example, the siting of sewage outfalls downstream of the river intakes of waterworks.
3. Protection of water sources from pollution by surveillance, legislation, and legal action.
4. Treatment of water intended for human consumption by storage, chemical clarification, filtration, and disinfection. Many modifications and combinations of these basic purification processes are in use as well as special methods for specific purposes, such as iron elimination, water softening, taste, and odour removal.
5. Prevention of external and accidental pollution of the water in the treatment plant and the distribution system to homes or factories. The latter is most important because, no matter how pure the water is when leaving the waterworks, if the distribution network of trunk main, pipes, service reservoirs, and water towers is not hygienically sound, all efforts of purification will have been in vain.
6. Advanced treatment of sewerage. The treatment of waters for subsequent use requires physical, chemical, and biological processes. These will vary from plant to plant and country to country, but will usually include distillation, gas exchange, coagulation, flocculation, sedimentation, filtration, absorption, ion exchange, and disinfection.

During this century, the incidence of infectious water-borne disease in developed countries has steadily fallen, and now usually only occurs as a result of a technical breakdown or an accident. Direct contamination of a water source by toxic waste figures in the news from time to time, and chemical pollution is now one of the most important problems to be faced.

Classical examples of chemical pollution are the organic mercurial poisoning of Minamato Bay, Japan, and the incident in which Canadian Indians suffered from tunnel vision following the consumption of fish heavily contaminated with mercury. Lead contamination of water adds to the whole-body burden of lead, and hence contributes to this potentially difficult clinical problem. Metals such as cadmium and lead have also been related to cancer, although the link is not conclusive.

Although generally the risk from water-borne infectious diseases has diminished considerably in developed countries, the large-scale use of swimming pools for recreational purposes carries specific risks of infection. These include skin, ear, eye, gastro-intestinal, and respiratory infections, and amoebic meningoencephalitis (Galbraith 1980).

Skin infections

The two most common water-borne infections are athlete's foot (a fungal infection) and warts (a viral infection). The floor surface surrounding swimming pools becomes contaminated and the disease is spread by contact. Obviously, cleanliness and disinfection of such areas are particularly important.

Ear infections

Inflammation of the outer ear, a condition known as otitis externa, is caused by *Pseudomonas aeruginosa*, which can be present in the pool water.

Eye infections

Most organisms which could cause eye infections and lead to conjunctivitis are easily controlled by adequate chlorination. However, at high concentrations chlorine can cause inflammation and soreness.

Gastro-intestinal infections

Only very rarely, if swimming pools are contaminated with sewerage, can bacteria responsible for some gastro-intestinal infections cause disease. Poliomyelitis, which in the past has been linked with the use of swimming pools, is probably by poor sanitation and overcrowding of the sanitary facilities.

Primary amoebic meningoencephalitis

This is a fatal but rare disease, first recognized in the mid-1960s. Most cases were associated with swimming in lakes or thermal springs in warm climates. An outbreak with ten fatalities, in Czechoslovakia in the early 1960s, was associated with a public swimming bath. Up to ten years later, pathogenic amoebae were isolated from the swimming pool. Amoebae probably collected in areas which drained into the pool but were not disinfected (Cerva *et al.* 1968; Kadlec *et al.* 1978). The amoeba responsible, *Naegleria fowleri*, is found in warm water. Chlorination can prevent its growth.

Specific food and water-borne hazards

Hazards associated with food and water can be classified into seven main types as shown in Table 26.1. In the remainder of this chapter we shall discuss the clinical conditions that occur from these major types.

Bacterial infections

Salmonella infections

In 1885, Dr S.E. Salmon first described a Gram-negative rod-shaped bacillus now known as *Salmonella*. This causes a paratyphoid type of infection which, in the acute form may produce a gastro-enteritis with a usual incubation period of 12–24 hours. In extreme cases this may extend to 48 hours. It is recognized that there are nearly 2000 *Salmonella* serotypes. The number of cases of *Salmonella* in humans has been increasing annually in both England and Wales and in North America. Because the disease is self-limiting, it should be appreciated that the condition is greatly under-reported.

The group of organisms has similar fermentation and cultural reactions to the paratyphoid bacillae, and specific

Table 27.1. Classification of foodborne and waterborne hazards

A. Bacterial infections
  Salmonella
  Shigella
  Streptococcus
  Campylobacter jejuni
  Listeria
  Vibrio
  Enteropathic *Escherichia coli*
  Clostridia perfringens
  Bacillus cereus
  Legionnaires' disease
B. Bacterial toxins
  Staphylococcus aureus
  Botulism
C. Viral infections
  Infectious hepatitis
  Lymphocytic choriomeningitis
  Haemorrhagic fever
D. Protozoan intestinal infections
  Amoebiasis
E. Parasite diseases
  Trichinosis (roundworm)
  Cysticerosis (tapeworm)
F. Plant and fungal hazards
  Mycetismus
  Mycotoxicosis—aflatoxins
  Enzyme inhibitors
  Phyto-haemagglutins
  Goitrogens
  Cyanogens
  Pressor amines
  Plant phenolics
  Oxalates
  Favism
G. Chemical hazards
  Aluminium
  Cadmium
  Lead
  Polychlorinated biphenyls (PCBs)
  Pesticides
  Antibiotics

identification of the organism has to be carried out using agglutination tests. The commonest organism known to human disease is *Salmonella typhimurium*, which causes typhoid fever. Other organisms seen from time to time are *S. enteriditis*, *S. heidelberg*, *S. newport*, and *S. montevideo*. Many salmonellae infect animals such as pigs, rats, and mice. The primary source of infection in most outbreaks is a farm animal, and secondary outbreaks arise from human carriers. Foodstuff such as eggs, poultry, meat, and meat products are usually involved in the spread of infection and man is always primarily responsible for the spread of salmonellosis. The onset of the clinical symptoms is usually abrupt, with headache and fever, followed by nausea, vomiting, and diarrhoea. Abdominal pain, which may be constant or colic like in nature, is experienced, and some patients may suffer great weakness and dehydration. The vomiting may continue for several days, but usually ceases after about 48 hours. The diarrhoea is often severe, and the stools are extremely offensive and may contain mucus. In approximately 10 per cent of patients septicaemia ensues, and some patients may develop complications such as meningitis, osteomyelitis, endocarditis, and pneumonia.

In view of the severity of this clinical condition, it is essential that all foodstuffs from animal sources are properly and correctly cooked, and the resulting material carefully handled. Legislation requires that all animals and meat should be inspected and checked for salmonellae, but an essential and important measure is that all people handling and involved in cooking should have a full knowledge of good hygiene practice. Infected water may be responsible for salmonellae poisoning and for this reason bacteriological checks should be made from time to time. People making home-made products, such as ice-cream, should also guard against contamination of starting materials. Eggs, for example, can be contaminated with *S. typhimurium* (Gunn and Markakis 1978). Raw milk has also been found to be a cause of human salmonellosis and *dublin* has been found in outbreaks in California (Chin 1982), and prior to 1983, reports of milk-borne infection were reported in Scotland. Pasteurization is currently the only effective method of controlling *Salmonella*.

Much has been written concerning the control of this disease, and the way in which *Salmonella* can be eliminated from food by heat processing. Education of the public with respect to the control of this condition is sadly lacking. It should be clear that during the preparation of food, whether in the home kitchen or in the large food-processing plant, there are ways of preventing the transmission of the organisms, and both the man in the street and the food service establishments should be well aware of these important and strict practices.

Shigella

Shigellosis, or bacillary dysentery as it is better known, is caused by bacteria of the genus *Shigella* (*Sh. dysenteria*, *Sh. flexneri*, *Sh. boydii*, and *Sh. zonnei*). Their normal habitat is the intestinal tract of humans and they are very easily grown from human faeces.

Although individuals may show no symptoms at all, cases may present with acute diarrhoea, vomiting, and stools streaked with blood, mucus, or pus. Because of the intense fluid loss in severe cases, there is marked dehydration and abdominal distention. The condition is most frequent in areas where malnutrition and poor sanitary conditions exist.

Salads and seafoods are the commonest foods associated with infection, and these become contaminated when handled by infected workers (Bryan 1978). Most outbreaks of food contamination in the home can be traced to carriers who lack personal hygiene. It is pleasing to note that, because of enormous progress in controlling water quality and sewerage treatment, dysentery has become very uncommon. However, elementary precautions such as keeping food covered to prevent contamination by flies is very important, particularly in developing countries (Gross *et al.* 1979).

Streptococcus infections

There are two important food-borne streptococcal infections: a haemolytic *Streptococcus faecalis* and the group A *Str. pyogenes*. *Str. faecalis* causes a fairly mild form of gastro-enteritis. The incubation period is usually 2–18 hours and the clinical picture is one of nausea, vomiting, diarrhoea, and abdominal pains. The organism has been reported as a contaminant of sausage, ham, and whipped cream.

The disease caused *Str. pyogenes* is usually confined to the respiratory system. Infection can be transmitted by the inhalation of droplets containing the bacterium or by ingestion of milk or other foods contaminated with it.

Following a 4 July celebration on an Indian reserve in Colorado, an increased incidence of sore throat and associated fever was noticed (McCormick *et al.* 1976). The Centers for Disease Control (CDC) was notified and advertisements were placed in local newspapers inviting those who had attended the picnic to fill in a questionnaire. Those who had attended the picnic and had developed a sore throat were primary cases; those who had developed a sore throat after contact with a primary case, but who had not attended the picnic, were secondary cases. Cases were compared with matched controls who had not attended the picnic and who had not had any contact with primary or secondary cases. Of those interviewed who had attended the picnic 48 per cent had a sore throat, with an incubation period ranging from two to four days. Among those who had eaten potato salad at the picnic, 59 per cent developed a sore throat. This compared with 24 per cent of those who did not eat it. Of the 139 from whom throat swabs were taken, 63 were positive for group A haemolytic *Streptococcus*, and a culture from the potato salad also yielded this organism. The 18-year-old son of one of the four who had prepared the food had complained of a sore throat the previous day. The epidemiological and microbiological evidence showed the potato salad to be the infected foodstuff, but it was never proved how it became contaminated.

Camplyobacter jejuni

This organism, previously known as *Vibrio fectus*, is becoming more recognized as a cause of human acute bacterial gastro-enteritis. It is believed that in the US campylobacteriosis is more common than *Salmonella* and *Shigella* infections combined (Foster 1986).

The disease may be seen following ingestion of small numbers of the pathogen, and the symptoms include a profuse diarrhoea, which may be blood stained, and is normally accompanied by nausea and abdominal cramps.

C. jejuni is a normal commensal of the gastro-intestinal tract of many wild and domestic animals and can be seen in poultry and cattle. For this reason raw milk and raw beef have been incriminated as vehicles in food-borne outbreaks of enteritis associated with this organism. *C. jejuni* is sensitive to drying, storage at room temperature, disinfectants, heat, and acidic conditions. It is unlikely to be a problem in pasteurized foods. The main problem is its association with transmission by raw meat or poultry.

Listeria

Up to a few years ago the disease listeriosis caused by *Listeria monocytogenes* was recognized only as the cause of abortions and encephalitis in sheep and cattle. Over the past few years, because it has become widely distributed in the environment and is known to survive for long periods under adverse conditions, it has begun to be recognized as a food-borne pathogen.

Listeria is ubiquitous in nature and water (Coyle *et al.* 1984), and for this reason is associated with fish, snails, warm-blooded animal species, and humans.

With respect to onset of disease in man, following ingestion of the bacteria, there is the onset of a mild 'influenza-like' illness, which is noted as tiredness, diarrhoea, and a low-grade fever. The enteric aspect of this phase may go unnoticed. However, during this time the bacteria invade the macrophages of the host, and any virulent strains of *Listeria* then multiply. This results in a disruption of the invaded macrophage and a consequent septicaemia in the host. The latter is a common manifestation of the disease in man, but as the organism invades various organs so the symptomatology develops.

The most common manifestation of listeriosis is meningitis; encephalitis and brain-abscess, which also involve the central nervous system, may develop. Localized listeriosis may also show itself in endocarditis, osteomyelitis, and endophthalmitis. It should be stressed that listeriosis is likely to accompany both the natural and the induced immunocompromized state.

Soft cheeses are known to be susceptible to contamination with *L. monocytogenes*, presumably because of post-pasteurization contamination. It is of great importance with respect to public health that no cross-contamination occurs between raw and finished products. Quality control must be of the highest order.

Vibrio

Three *Vibrio* species exist which are considered to be of significance in illness related to food. They are *V. cholerae, V. parahaemolyticus* and *V. vulnificus*.

V. cholerae is the organism seen in cholera outbreaks, and the zero group 01 is the one responsible for the severe watery diarrhoea seen in the epidemics. Large quantities of so-called 'rice-water' stools are produced in this condition, and this rapidly leads to dehydration, electrolyte disturbance, and death. *V. parahaemolyticus* causes attacks of acute gastroenteritis, and this is usually heralded by attacks of nausea, vomiting, abdominal cramps, low-grade fever, and watery diarrhoea. In some cases the latter may be bloody (Farmer *et al.* 1985). Raw seafood is a common carrier of this organism, and in Japan this causes 50–70 per cent of enteritis cases (Sakazaki and Barlows 1981). *V. vulnificus* is associated with wound infections and may cause life-threatening septicaemia (Blake *et al.* 1979).

These *Vibrios* occur naturally as contaminants of seafood. As all members of this genus are killed by heat, adequate cooking is essential before consumption of seafoods.

Escherichia coli

Escherichia coli is a significant cause of diarrhoea in areas of poor sanitation. Food-borne and water-borne outbreaks of *E. coli*-induced diarrhoea usually occur in areas with inadequate sanitation. There are at least four subgroups of the enteropathogenic *E. coli*. [Depending upon the type of organism, the accompanying illness may mimic cholera, shigellosis, or the typical dysentery-like illness. In some cases there may be a frank haemorrhagic colitis.]

Newborn babies, particularly premature ones, are extremely susceptible to this infection. Preventative measures include the sanitary disposal of human excreta, the boiling of milk and milk products, and adequate inspection of food manufacturing and serving areas.

Legionnaires' disease

Legionnaires' disease is a bacterial infection which, at its worst, presents as a fatal form of pneumonia. Mild forms are known to occur, for example, pontiac fever, as well as many covert subclinical infections. It can occur as an epidemic, and the name originates from an epidemic at a convention of American legionnaires in Philadelphia in 1976. There were 183 cases and 29 deaths (Lewis and Macrae 1977).

Clinically, the disease normally presents as a pneumonia, and the usual incubation period is of the order of 2–10 days. The early symptoms are those of malaise, lethargy, loss of appetite, and weakness. A majority of patients have an unproductive cough in the early stages. However, later the clinical picture changes and some patients develop a purulent sputum with dyspnoea and pleuritic pain, whereas in others haemoptysis may be seen. Nausea, vomiting, and watery diarrhoea may also be a feature of the disease.

A survey carried out in hostels and hotels showed that bacteria responsible, *Legionella pneumophilia*, is not uncommon in water supplies with buildings. The source of the infection is not the mains water supply (Tobin *et al.* 1981a,b), but the water storage tanks and plumbing systems. In the US, air-conditioning systems have been shown to harbour the organism (Broome and Fraser 1979).

Tobin *et al.* (1981b) examined an outbreak of the disease among 58 workers from a commercial firm who took part in a golfing tournament. Four golfers went down with pneumonia after the tournament, two of whom required hospitalization. Blood tests revealed an increase in the antibody titre to *L. pneumophilia*. The third man also had a raised antibody level, but this was not shown in the fourth. When compared with an average antibody titre, the two men were confirmed to have Legionnaires' disease and the third man was also assumed to have the disease. All infected men stayed at the same hotel. One employee at the hotel had a raised antibody level but had never exhibited any symptoms. The organism was later isolated from two storage tanks and shower outlets at the hotel.

Bacterial toxins

Staphylococcus aureus

This enterotoxin-producing organism is found in skin lesions, such as boils, and the nasal passages of carriers. Food handlers who are carriers or have such lesions can contaminate food. Processed meats, such as meat pies, are a common source of such infections. The toxin is produced in the food before eating, and hence the incubation period is very short, usually 1–2 hours and never more than 6 hours after consuming contaminated food. The enterotoxin produced is heat stable, and so heating contaminated foods will not prevent the ensuing food poisoning.

Nausea and vomiting of quick onset associated with upper abdominal pain are the major symptoms of this conditions. At high concentrations toxin can cause prostration and hypotension, but recovery is usually complete within 48 hours. Confirmation of the diagnosis comes from culturing large numbers of organisms from vomit and, if available, the food.

Because of the potential of foods such as sliced meats, ham, and bacon to produce this condition, it is of great importance that all such foods should receive prompt refrigeration to eliminate the reproduction of any toxin-producing staphylococci. Food handlers with pyogenic skin lesions and upper respiratory tract infections should be excluded from contact with such products.

Botulism

Botulism is a serious, but rare, clinical condition. It differs from other types of food poisoning in that the central nervous system is the prime area attacked and the gastro-intestinal tract may only exhibit minor symptoms. This fatal form of food poisoning results from an exotoxin produced by *Clostridium botulinum*, which, once absorbed by the gastro-intestinal system, acts on the central nervous system to cause widespread paralysis. *C. botulinum* is an anaerobic spore-forming Gram-positive bacillus. The disease in man is caused by the A, B, or E type, each of which has a characteristic toxin. The exotoxin is usually formed under anaerobic conditions in non-acid protein foods. The symptoms, i.e. paralysis, frequently appear up to 36 hours after ingestion of the toxin—this may be proceeded by vomiting, giddiness, and diplopia. Paralysis of cranial nerves causes difficulty in talking and swallowing. Paralysis spreads to the trunk and all limbs and is followed by death from respiratory failure.

In most cases food contaminated with *C. botulinum* and its toxin possesses a characteristic sour odour and the taste of the food is altered. These properties should warn people that the food is spoiled and therefore should not be eaten. In field investigations it is important, where possible, to salvage the contaminated food to ascertain the type of toxin that has been released, and hence to administer the correct polyvalent antitoxin.

It has always been accepted that *C. botulinum* does not grow in acidic conditions, so that, for example a product marinated in vinegar would not be suspected as the source of an outbreak. However, Rigau-Perex *et al.* (1982) recently reported an outbreak of food-borne botulism in Puerto Rico in which marinated kingfish with a pH of less than 4.6 was the vehicle. The first patient, a 46-year-old male, was admitted to hospital on 8 August 1978. He was vomiting and complained of weakness and blurred vision. No sensory abnormalities were noted. As dysphagia developed a neurologist was consulted who diagnosed botulism. The physical examination on admission revealed an alert distressed patient with sluggish responses, dry mouth, drooping eyelids, and difficulty in speaking, and inability to extrude the tongue. The illness progressed and he died on 12 August. The second patient was 24-year-old employee of the deceased. He was admitted to hospital on 10 August after four days of weakness. Examination showed similar symptoms to the first patient. By 10 August, when botulism was diagnosed, he was slightly improved and was not given antitoxin. He was discharged on 17 August, still feeling some weakness. A third patient, who was 16 years old and also an employee of the deceased, was admitted to hospital on 10 August with similar symptoms. The respiratory muscles became progressively weaker, and he was intubated and given ventilatory assistance. He was given botulinal antitoxin and discharged on 17 November, still weak but walking without assistance.

An epidemiological investigation began on 10 August, with information obtained from the first patient and his wife. Attention was quickly drawn to home-made marinated kingfish that the patient had eaten the previous day. It was then determined that patient no. 2, who had already been sick for a few days, and patient no. 3, who was absent from work that day, had also eaten this fish. The food was confiscated and other persons who might have been exposed were tracted and alerted.

The wife of patient no. 1 had prepared the fish on 22 July, frying the slices of fish and then marinating them in a mixture of oil, vinegar, onions, peppercorns, and bay leaves. She stored the fish in three large glass jars with screw caps and left them to 'cure' under a table in her husband's cafeteria pizzeria business. The fish had been prepared for personal consumption and was never sold to the public. Patients no. 2 ate some of the fish on the 2, 3 and 4 August, and began feeling ill on 5 August. Patient no. 3 ate small amounts of the fish on 8 and 9 August and began feeling ill on 9 August. The father and two other employees of patient no. 1 tasted the fish on 4 and 9 August.

The contents of the three jars of marinated fish were examined by the Anaerobe Laboratory of the Centers for Disease Control, Atlanta. Type A *C. botulinum* was isolated from enrichment cultures of fish from two of the jars.

Prompt investigation of this incident after diagnosis of the first case of botulism led to rapid detection of the causative food and other possible cases.

The reason why this organism grew in the foodstuff was investigated. There was air in the jars, but a thick layer of oil between air and solution created a virtually anaerobic environment. The measured pH in all three jars was less than 4.5, a level at which there is little risk of botulism. The vinegar may not have penetrated the fish completely or sufficient quickly, which would have left the interior pH high enough for spore germination, growth, and toxin production. This could not be proved, as too much time had elapsed between the preparation of the fish and its examination in the laboratory—equilibration would have taken place. The fact that some people ate the contaminated food and did not become ill would support the theory that the toxin was not equally spread throughout the fish.

This incident did not represent a widespread public health problem but shows how, under unusual circumstances, outbreaks of this kind can occur. It also stresses the need to monitor 'home-made products' and ensure education of the public about potential hazards. Insufficient attention has been given to this in the past.

Viral infections

Infectious hepatitis

The term hepatitis refers to an acute inflammatory disorder of the liver, and the condition may be due to a variety of pathogenic agents. Two viruses are associated with viral hepatitis. Virus A, which gives rise to infectious hepatitis, is present in both the faeces and blood of the infected person and can be transmitted orally or parenterally. Virus B causes sperm hepatitis, so called because it can only be transmitted parenterally. Virus A will be considered here as it is the form of the disease which is both food-borne and water-borne (Follett and McMichael 1981; Orenstein *et al.* 1981). The disease is spread by person-to-person contact via the oral-faecal route, and when populations are exposed to contaminated food, for example, milk or raw shellfish, an epidemic can result. It is common where conditions of hygiene are poor, and young people are particulary prone to the infection.

The disease has an incubation period of up to six weeks

and the onset is usually acute with vomiting, headache, abdominal discomfort, and tenderness over the liver. Jaundice soon follows and hepatomegaly may be found on physical examination. To prevent such cases it is essential that proper hygiene and good sanitation are practised. Shellfish should be collected from areas free of sewage disposal. It is also advisable to immunize travellers who are visiting the areas where the disease is endemic with gamma-globulin. Care must be exercised with syringes and needles that have been used on an infected patient and in the handling of food given to and excreta from infected patients.

Lymphocytic choriomeningitis

This form of meningitis is caused by an RNA virus, the natural host of which is the house mouse. Infection is transmitted via food contaminated with urine or faeces from an infected mouse. Mild infections that do not progress to meningitis may occur. Illness may commence with an influenza-like attack 8–13 days after contaminated food is eaten. The patient either recovers completely, or signs of meningoencephalitis may appear 15–21 days after the onset. Recovery usually occurs within several weeks, although severe cases occasionally prove fatal. The disease can be prevented by rodent control and good food hygiene.

Haemorrhagic fevers

The haemorrhagic fevers have been clinically diagnosed in the Far East for many years. They were first recognized during the Korean War during which seasonal outbreaks occurred in American troops. The vectors in some haemorrhagic fevers have not yet been identified. It has been suggested that in some areas field rodents act as the reservoir.

Protozoan intestinal infections

Amoebiasis

Amoebiasis, originally confined to the tropics, is seen worldwide. It occurs where conditions of habitation are cramped and unhygienic, such as prisons, institutions, and refugee camps.

Amoebiasis, or, as it is more commonly known, amoebic dysentery, is caused by a protozoan parasite invading the large intestine. Cysts of the protozoan can contaminate food and water. When consumed, protozoa emerge from these cysts and invade the intestinal wall. If the gut wall is penetrated, ulceration and abscess formation may occur. The amoebae may infect a person without giving any overt signs of infestation. Some diarrhoea may occur, but only sporadically. However, if amoebic dysentery sets in, blood-stained stools may be passed, but very few other symptoms are exhibited. The disease can come and go, and this is quite a common sign of infestation. Complications can set in, including peritonitis and, more rarely, liver abscesses.

The disease is easily spread by asymptomatic carriers, handling food or water without taking due care of personal hygiene.

Eldin (1981) reported amoebiasis as the cause of death in a 71-year-old Scottish woman taken into hospital after a 12 day history of diarrhoea. After various blood tests, including sodium, potassium, and blood urea, she was given a sigmoidoscopy and ulcerative colitis was diagnosed. Treatment was given, including tetracycline and prednisilone, but her condition deteriorated until her death. No pathogens were isolated from the stools.

At post mortem, faecal peritonitis was evident. Diagnosis of tuberculosis enteritis or amoebic colitis were considered. A frozen section of formalin-fixed tissue showed many *Enteroamoeba histolytica*, a type of amoeba widely present in healthy individuals, even in temperate climates. The case indicates that amoebiasis is always worthy of consideration when a patient presents with diarrhoea and bowel ulceration.

Food-borne and water-borne parasitic helminths

Diseases caused by helminths, once largely confined to tropical areas, have spread with increased mobility of population, both for work and tourism. They are mainly found in young children in areas where standards of public health are poor.

Again, sanitation would go a long way to eliminating these diseases, as many are spread by the faeces of the contaminated individual. There are many examples of such parasites which use man as one of their host. They include the nematodes (roundworms), the cestodes (tapeworms), and the trematodes (flukes). The sources of infection for these parasites include contaminated meat, such as pork and beef.

An outbreak of trichinosis occurred in Paris in January 1976 (Bouree *et al.* 1979). A large number of people presented with similar and severe symptoms including headache, fever, nausea, and, in some cases, periorbital oedema. The eosinophil counts were high. Very few parasites were isolated, but this was thought to be due to the fact that, within the time of infection, the larvae had not yet become encysted. A biopsy taken from one patient 18 months later revealed a living larva, showing how long-lived the parasites can be. The disease was treated with anthelmintic drugs. All the patients had consumed horse meat bought from one particular butcher. The butcher and his family were infected. It was thought that the horses must have eaten hay contaminated with rodent droppings.

Tapeworms can be found in pork or beef. Infestation with pork tapeworm, cysticercosis, can be quite a serious condition and encystment can occur in organs other than the gut (Giri 1978). The larvae can develop in the subcutaneous tissues, the heart, the eyes, or the central nervous system, where the condition often proves fatal. Again, infected faeces are the sources of infection.

Fungal hazards

Many fungi, when ingested, are toxic to man. Fungal poisoning can be divided into two categories: (a) mycetismus—the disease caused by ingestion of the fungi; (b) mycotoxicosis—ingestion of toxins produced by fungi.

Mycetismus

There are many fatalities each year associated with the consumption of fungi— only an expert can safely distinguish one type from another. The cyclopeptides found in species of *Amanita* are some of the most toxic poisons contained in fungi. Consuming a single *Amanita phalloides* fruiting body can prove fatal (Wieland 1965).

Certain mushrooms, when eaten in conjunction with other foodstuffs such as alcohol, can give a violent reaction in some people. This is because they contain alcohol synergists that, in conjunction with alcohol, block various enzyme pathways. Some mushrooms contain muscarine, a chemical which exhibits a cholinergic effect in the body. The symptoms include sweating, salivation, and lacrimation. Gastro-intestinal symptoms, headache, and visual effects may develop. Atropine may be necessary but recovery is usually spontaneous.

Ergotism is a condition caused by the ingestion of the fungus *Claviceps purpurea* which grows on wheat, rye, and other cereal grains. It has a direct effect on the artery walls causing them to contract, leading to loss of circulation to the extremities. Gangrene sets in, often resulting in the loss of fingers, toes, and even the ears and nose. In extreme cases, death occurs from neurological complications. Cereal crops are scrupulously monitored for the fungus and diseased grains are destroyed (Abdul-Haz *et al.* 1963).

Mycotoxicosis—aflatoxins

Common fungi, such as *Apergillus flavus* and *A. parasiticus*, yield a group of compounds called aflatoxins. The term is also used for metabolites of aflatoxins that are formed in the body and are further identified by a letter. For instance, an aflatoxin currently of interest as food contaminant is aflatoxin M_1, a metabolite of aflatoxin B_1 which is secreted in the milk of lactating animals. It is now possible to measure minute quantities of aflatoxin in foods. As a result there is a growing interest in their influence on public health. The B_1 aflatoxins are very toxic to animals and the manifestation is seen in the liver. Incidents of aflatoxicosis in man have been reported. In one outbreak 400 people from several villages in India were affected and over 25 per cent of these died of fatal hepatic disease (Krishnamachie 1975).

A study in Swaziland (Peers *et al.* 1976) estimated aflatoxin ingestion over a period of one year from aflatoxin levels in food samples, and related this to the incidence of primary liver cancer. It showed the importance of determining the levels of aflatoxin in the starting raw material, as ground nut is an important source of protein in Swaziland. It draws attention to the need in field operations of recording the storage and handling methods used in a particular country since this will help to judge the role of deterioration by fungi. At the same time it emphasizes the importance of having good analytical facilities for measuring the levels of a contaminant and the need to create a team spirit amongst all workers, including clinicians, toxicologists, statisticians, epidemiologists, and microbiologists.

Aflatoxins were first isolated as contaminants of ground nuts and have since been found in maize. Inadequate drying of crops before storage and poor storage conditions lead to contamination by aflatoxin-producing fungi. Ground nut is commonly used in animal feeds and hence appropriate facilities must be provided for processing such raw materials to avoid contamination of products, such as cows' milk, with aflatoxin M_1. A study by the Ministry of Agriculture, Fisheries and Food (MAFF 1980) showed that aflatoxin was present in nut products and milk, and surveillance is continuing. Governments could ban the import of materials, such as ground nuts, but this would have serious implications for other countries. Aflatoxins will remain an important public health issue for many years to come and will become much more of a problem in view of the excellent progress made by analytical chemists.

Hazards associated with plants and plant products

Many plants, vegetables, fruits, and nuts can cause illness or even death if ingested. Many are very familiar and form part of our everyday diet and environment. Some vegetables cause no problems if cooked properly, but, if inadequately cooked, they will make people very ill. Potatoes, for example, under normal conditions form a large and important part of many people's diet, but if they are eaten when they are green they contain quantities of toxic products.

Enzyme inhibitors
Trypsin inhibitors

Soya bean was the first crop to be recognized as containing a trypsin inhibitor. Since then trypsin inhibitors have also been isolated from potatoes, kidney beans, and haricot beans among others.

Amylase inhibitors

Various amylase inhibitors have been found in wheat, beans, and unripe mango.

Cholinesterase inhibitors

Solanine is a glycoalkaloid which is produced in the potato when it is exposed to light. A case of solanine poisoning among schoolboys in southeast London was described by McMillan and Thompson (1979). Seventy-eight boys became ill 7–19 hours after lunch at the beginning of the Autumn term and seventeen, of whom three were dangerously ill, required admission to hospital. Sickness began with headache, vomiting, abdominal pain, and diarrhoea. Fever was sometimes present and occasionally circulatory collapse. Prominent among the symptoms were neurological disturbances, such as apathy, restlessness, drowsiness, mental confusion, rambling, incoherence, stupor, hallucinations, dizziness, trembling, and visual alterations. Most of the boys recovered in a day or two but others took 3–6 days. A thorough investigation of the cause of the outbreak was made. The epidemiological findings were complex but their interpretation was made easier by the fact that lunch had been served to two different sets of boys, and only one set was affected. Among the potatoes served to the affected group were some that had been left over from the Summer term. When peeled, these potatoes contained 33.3 mg of solanidine alkaloids per 100 g. Potatoes left over from the meal had very high anticholinesterase activity. The toxic does of 20–25 mg could easily have been consumed in a normal helping.

Phyto-haemagglutins

These are plant proteins which clump or agglutinate red blood cells in a manner similar to antibodies. Sweet peas, lima beans, kidney beans, and soya beans, if eaten when insufficiently cooked, contain considerable amounts of the proteins.

Goitrogens

Goitrogens are natural products which cause hypothroidism with an enlargement of the thyroid. In 1958 the incidence of goitre in Nigeria and Ceylon was such that the World Health Organization was asked to make a survey of both countries (Wilson 1968). Incidence of goitre was linked to the amount of iodine in the soil, but this was only part of the story.

In 1966, a survey was carried out in Nigeria, following reports of a large number of cases in a part of the country where goitre had not been known to exist. There appeared to be no relationship between the mineral content of the soil or water and the incidence of goitre. Examination of dietary habits of the local inhabitants showed that villages with a high incidence of goitre consumed a large amount of unfermented cassava. To determine whether cassava was goitrogenic, thyroid activity was studied in the rat. Those animals fed only cassava or equal parts of stock diet and cassava showed enlarged thyroids when compared with animals fed only the stock ration. The report concluded that, for countries such as Nigeria and Ceylon, supplementation of some common food, such as salt, with an iodine compound was an urgent public health measure.

Cyanogens and cyanogenetic glycosides

Several plant foodstuffs in common use, particularly in the tropics, contain cyanide either in the form of glycosides or nitrites. Acute poisoning can occur, especially in conditions of economic stress. Chronic ill-effects from cyanide intoxication are thought to include degenerative tropical neuropathy, which has been found in African countries, especially Nigeria. The main features of the fully developed 'ataxic syndrome' include optic atrophy, nerve deafness, and sensory spinal ataxia.

Pressor amines

Certain naturally occurring phenylethylamine derivatives, such as tryptamine and dopamine, cause a marked increase in blood pressure when administered intravenously to mammals. Amines, including vasocative ones, are normally rapidly de-aminated after they enter the body, a process that is catalysed by monamine oxidase. Amines are present in numerous foods either because they are synthesized by certain plants or through microbial contamination of fermentation processes.

Bananas contain large quantities of the amines serotonin, noradrenaline, and tryptamine. Amines have been found in the following fruits and vegetables: tomato, red plum, avocado, potato, spinach, grape, orange, pineapple, lemons, and tea. Although little is known of the effects of these compounds, they are thought to be linked to migraine in man.

Plant phenolics

Some plant phenolics present in human foods are toxic. Although anthraquinones in rhubarb are mainly found in the root, high levels are also present in the leaves. Human poisoning occurs mainly from eating rhubarb leaves.

Oxalates

Oxalates occur in all forms of living matter. Certain families and species of plant contain relatively large amounts of this substance, mainly as the soluble sodium or potassium salt. Some of the common plant foods containing appreciable oxalates are spinach, rhubarb, and cocoa. Ingestion of sufficient oxalic acid can be fatal with associated corrosive gastroenteritis, shock, convulsive symptoms, and renal damage.

Favism

Fava bean, a species of broad bean, is a staple part of the diet of Mediterranean countries. When ingested by persons who have previously been sensitized, either by inhaling the pollen of this plant or by prior ingestion, severe illness may occur. The first green beans of the season are the most toxic. The toxicity is due to a nucleoside (vicine) that causes haemolysis when injected in rabbits (Noah et al. 1980).

The illness, known as favism, usually appears within 1 hour after ingestion, with the patient complaining of dizziness, vomiting, diarrhoea, and severe prostration, followed by acute fevril anaemia, jaundice, and haematuria. Treatment is symptomatic with replacement of electrolytes and blood transfusion in severe cases.

Chemical hazards

Fears are sometimes expressed that our food supply is contaminated by the addition of chemical compounds that preserve a product. Today antioxidants, preservatives, and sterile packaging have become part of our way of life, but all these compounds and processes must be tested to ensure safety before being put to general use.

Governments and expert working parties have laid down rules for the use of such materials, and the process of obtaining clearance is sometimes long and expensive because of the sophisticated toxicological work involved. The tests demanded include 90-day studies in animals to determine the 'no-effect level' to calculate the allowable daily intake for man, and to determine the metabolic pathway and the metabolites formed, and the teratogenic, mutagenic, and often carcinogenic potential of these compounds. Such tests are also required on many materials deliberately added to water and there is constant surveillance to ensure the chemical, physical, and microbiological quality of water.

Whilst legislators and toxicologists can work together to ensure the 'safety' of deliberately added chemicals, there are other uncontrolled ways in which chemicals may enter food and water. Inorganic or organic chemicals may inadvertently find their way into the food chain, for example, from soil, via both plants and animals. The microbial contamination of animal feeds was discussed above, but pesticides deserve consideration as potentially harmful chemicals that remain on foodstuffs after inadequate washing.

A number of case studies that deal with specific chemicals and also provide general lessons for the conduct of field studies of water- and food-borne contaminants are discussed below.

Aluminium

Aluminium is the third most abundant element in the earth's crust, representing about 8 per cent, with only oxygen (47 per cent) and silicon (28 per cent) being present in larger amounts. Just as silicon and oxygen combine to form silica, commonly seen as sand and as quartz in most rocks, aluminium and oxygen combine to form alumina, commonly found in soil, clays and many minerals. Bauxite, which is a combination of alumina, silica, and iron oxides, is the principal ore from which aluminium is produced.

Although aluminium is everywhere, in soil, in water, and even in the air, it is not an essential element for any form of life, and it is present only in very small amounts in living organisms. These facts, taken together, indicate that Al^{3+} possesses properties incompatible with fundamental life processes and that nature has evolved mechanisms to ensure that little, if any, aluminium is absorbed, and that any that is absorbed is either eliminated rapidly or is dealt with by some protective mechanism.

In the early 1970s, the first indication that aluminium could indeed be neurotoxic to man emerged when some patients with end-stage kidney failure, who are dependent on long-term dialysis, developed a progressive neurological disease, which became known as dialysis encephalopathy syndrome (DES) or dialysis dementia (Mahurkar 1973). An association was established between the aluminium content of the dialysis water and the incidence of DES, and reduction of the aluminium concentration in the water was followed by a marked drop in DES (Schreeder et al. 1983). Elevated concentrations of aluminium were found in the brains and other tissues of patients who had died with DES. At about the same time elevated aluminium levels were also being reported in patients with Alzheimer's disease (AD) (Crapper et al. 1975).

More recently, aluminium combined with silicon as aluminosilicates was described in cells containing neurofibrillary tangles, but not in normal cells, from both normal brains and AD brains (Perl and Brody 1980), and also in the cores of neuritic plaques (Candy et al. 1986), although both findings are disputed (Stern et al. 1986).

Excluding medical interventions which bypass the normal mechanisms and barriers of absorption, the normal intake of aluminium is via the mouth or nose, there being essentially no penetration of the skin.

Air

Other than in a few work environments, where significant aluminium oxide or aluminium silicate dust levels may exist (classed only as 'nuisance dusts' with a threshold limit value of 10 mg/m^3), normal ambient levels of aluminium in air are only a few micrograms per cubic metre or less. Intake from this source, therefore, is normally much less than 1 mg/day.

Water

Virtually all water contains some small amount of luminium, which may be somewhat higher in slightly acidic waters than others, and additionally alum (aluminium sulphate) is widely used as a flocculating agent during drinking water purification. Although aluminium levels in drinking water may reach 1 mg/l or even more in some localities, the normal range is 10–15 mg/l, and the EEC has set a standard for drinking water of 50 mg/l with a maximum acceptable level of 200 mg/l. Thus, assuming that an individual drinks 1.5 litres of water per day, the intake from this source is normally likely to be substantially less than 1 mg/day.

Food

Aluminium intake in the diet can come from three sources, namely the natural content of foodstuffs, intentional additives contain aluminium (e.g. baking powers, emulsifying agents for processed cheese, firming agents for pickled vegetables and fruits, and anticaking agents for soft and powdered products), and unintentional contamination (e.g. from packaging, cooking, or storing in metallic aluminium utensils or foil). Many surveys and estimates of aluminium intake from the diet have been made over the years in many countries and the results of many of them are not obviously comparable. The most recent surveys from three countries, however, should give some indication of average aluminium intake from food. The latest UK estimate of intake from the diet is about 6.0 mg (MAFF 1985), which includes intentional additives but not the contribution from cooking or storage in aluminium metal. In the Federal Republic of Germany (FRG), the average daily intake of aluminium has been estimated to be 11 mg/day for males and 8 mg/day for females. The use of aluminium-containing food additives is said to be limited in the FRG, but the use of aluminium metal utensils and foil, it was estimated, might result in an additional intake of about 3.5 mg/day. In the US, the most recent authoritative estimate (Greger 1985) is that Americans ingest from 3 to over 100 mg of aluminium daily, with most adults probably consuming 20–40 mg/day. The intake was estimated to be 2–10 mg/day from the natural content of food, 20–40 + mg/day from additives, and, at most, 3.5 mg/day from metallic aluminium utensils and foil.

In contrast with its abundance in the environment, normal body tissue concentrations of aluminium are low, indicating that ingested aluminium is largely excluded. Aluminium is present in all body tissues from birth, but the total body content of healthy individuals is only in the range of 20–50 mg (Ganrot 1986; Greger 1985), of which about half is in the skeleton, about a quarter in the lungs, and the remainder distributed in the other organs and tissues. Within this range there is some accumulation with age, principally in the skeleton and lungs, but also in other tissues including the brain (McDermott et al. 1977). People with impaired kidney function accumulate much more aluminium than normal individuals (Alfrey 1980).

Accurate determination of aluminium absorption and subsequent metabolism has been severely hampered by lack of a suitable radioactive isotope of aluminium and, until recently, of sufficiently sensitive analytical methodology coupled with the difficulty of avoiding contamination of samples from the ever present aluminium in the environment. However, it is clear that the gastro-intestinal tract is a very effective barrier to aluminium absorption and that the very small amount that is absorbed, estimated to be in the range 0.01–0.3 per cent (Ganrot 1986), is probably eliminated rapidly, if not entirely, from the body via the kidneys.

A substantial review of the metabolism of aluminium is provided by Ganrot (1986), who compensated for the lack of specific information on aluminium by also considering data for other chemically similar metallic ions.

The generally held scientific opinion is that aluminium is very atoxic and does not cause disease in healthy people under normal conditions (Ganrot 1986; Krueger et al. 1984). The only generally recognized health effect of high aluminium intake, e.g. from antacids, by people with normal kidney function is phosphate depletion.

The only two disease conditions which are generally accepted as being due to aluminium intoxication are DES and dialysis osteomalacia (DOM). Both of these affect patients with renal failure or severe kidney disease who are treated by dialysis with aluminium-containing dialysate solution and/or with high amounts of phosphate-binding gels. DES has been mentioned above. DOM is a disease characterized by demineralization of the bone coupled with extensive aluminium accumulation in the skeleton, and appears clinically as skeletal pain and a strong tendency to fracture (Ganrot 1986). However, there are more differences than similarities between DES and AD and, notably, there is no increase in the frequency of AD in patients with DES or DOM, or occurrence of the characteristic histological features of AD, tangles and plaques, in the brains of such patients, despite their unusually high brain levels of aluminium (Arief et al. 1979; Burks et al. 1976; Mozar et al. 1987).

Another disease complex which is characterized by PHF neurofibrillary tangles, but not plaques, is amyotrophic lateral sclerosis/Parkinsonian dementia (ALS/PD), which is observed among certain populations in the Western Pacific, namely Guam and the Kii peninsula. That the cause of the disease is environmental rather than genetic was indicated by the major drop in incidence in Guam from the 1950s onwards. The fact that this followed the drilling of deep wells to provide drinking water, replacing the use of surface water which was high in aluminium, again led to the suggestion that aluminium was the environmental toxic factor. This theory was enhanced when it was shown that aluminium was enriched in the tangle-bearing neurons in brains from ALS/PD patients (Garruto et al. 1982; Perl 1982). However, if aluminium is involved at all in ALS/PD of Guam, it is considered to be secondary to a chronic deficiency in calcium and magnesium in these localities (Ganrot 1986), and the current hypothesis is that ALS/PD was caused by the presence of an unusual exitotoxic amino acide present in cycad palms, which were used both as a source of starch for food and as a topical medicine up to and during the Second World War (e.g. Spencer et al. 1987).

A vast amount of work has been and still is being carried out to try to establish whether or not aluminium is a factor in AD. Ganrot's (1986) conclusion from his extensive and carefully argued case in favour of this hypothesis is that, 'no well documented fact associated with the disease or with Al^{3+} metabolism seems to be definitely incompatible with the hypothesis. Therefore, taking all current facts into consideration, it can be stated that, indeed, many reasons exist to couple AD with Al^{3+}'. However, other leading workers disagree and Katzman (1986), for example, concludes that 'there is no evidence that exposure to such sources of exogenous Al as Al antacids, antiperspirants or even the large amount used in renal dialysis increases the risk of AD. Thus, a direct relation between exogenous Al and Al deposits in the brains of persons with AD has not been establish'. Wisniewski et al. (1986) stated that 'Al has been implicated as a cause or an important factor in AD, ALS, ALS/PD and dialysis dementia. However, outside of the dialysis dementia syndrome, to date there is no evidence that Al has a role in the observed pathological changes, signs and symptoms in any of these diseases'. An earlier supporter of the aluminium hypothesis (Crapper McLachlan 1986) that 'evidence does not support an etiological (causative) role for Al in AD. The primary pathogenic events responsible for AD are presumed to have affected the genetically determined barriers to Al resulting in increased amounts of this toxic element to vulnerable sites'. He pointed out that there are two opposing points of view on the functional significance of aluminium in AD:

1. Aluminium merely accumulates passively in neurons compromised by the Alzheimer degenerative process and the accumulation is of no significance to the mechanisms of the disease.

2. Auminium is a plausible candidate for a neurotoxic environmental factor acting in the pathogenesis (development) of the neurodegenerative processes. However, Bartus (1986) states that 'if Al does play a role in AD, it

most likely is due to a breakdown in a protective barrier and not simply the result of over-exposure to Al-containing man-made products'.

The overall conclusion is that aluminium is not directly implicated in any disease process, other than the two dialysis diseases DES and DOM.

Cadmium

In 1979 the findings of a nationwide survey of metal concentrations in streams in the UK suggested that concentrations of cadmium in the soil were usually high in Shipham, Somerset. It was suggested that dust contamination of garden produce might put residents who consumed these products at high risk of cadmium poisoning. An enormous exercise was mounted to determine soil, dust, water, air, and food levels of cadmium, and health checks were also carried out. Only a small proportion of the residents exceeded the intake of cadmium currently considered tolerable, and no-one in the study showed any adverse health effects compared with controls (Department of the Environment 1982). This example demonstrates that studies of this nature should be carried out only if a health risk is clearly defined and that it should not figure as a political issue.

Lead

The possible health effects of lead have been debated for a long time. Food and water may become contaminated by exposure to lead solder in canning, by exposure to water used for cooking and drinking which is contaminated with lead in plumbing systems, and by exposure to lead in air either directly, e.g. crops and food, or indirectly via contamination of soil. Lead-containing pesticides are also important since they directly increase the lead level in fruit and vegetables and, if such compounds have been used for a long time, they increase the level in soil. The scientific data about this subject need to be tempered with a great deal of commonsense. There are may different sources of exposure to lead, but this problem is rarely considered in terms of the overall body burden. Contamination of vegetables due to combustion of lead-enriched petrol is one issue, but it cannot be seen in isolation.

Contaminated water supplies represent a more important problem to residents in areas where lead-lined pipes have been installed for years, because of the use of lead-contaminated water for preparation and cooking of food over a long period of time.

Mindus and Kolmodin-Hedman (1981) reported the case of a 61-year-old housewife with a history of renal disease and calculus formation for more than 10 years, accompanied by impaired renal function. She was admitted to a urological ward with a very severe intermittent pain in the right lumbar region. In the past she had been diagnosed as having a parathyroid adenoma and was treated for this. During the reported admission she was found to be confused and one morning she was no longer able to move her arms. She was in a state of general mental confusion and her reflexes were weak. The symptoms were difficult to interpret but they seemed, possibly, to be of a psychogenic nature. After various examinations it was concluded that her symptoms could be due to some toxic effect. Laboratory tests revealed that she had a very high lead blood content. It was volunteered by the patient that 10 years previously she had been advised by a urologist to drink at least 3 litres of water every day and she had followed his recommendation conscientiously. Since she did not like cold water, she filled a 1 litre jug with half warm and half cold water three times a day. The jug was analysed and excluded as the lead source. However, the source turned out to be a water boiler installed some 40 years previously and this had released very large quantities of lead.

Polychlorinated biphenyls

Polychlorinated biphenyls (PCBs) are a complex mixture of chemicals known as chlorinated hydrocarbons. They are very widespread in the environment and are used in electrical equipment such as transformers. Over the past four to five decades they have been found in rivers, fish, and birds. They cause skin lesions and liver damage in man.

In Japan, food contaminated with PCBs has produced rashes, headaches, nausea, diarrhoea, alopecia, loss of libido, and menstrual disorders. Exposure of women to PCBs during pregnancy can also harm the developing fetus. Mosher and Moyer (1981) studied 13 children born to women exposed to PCBs. Of these, one was stillborn, four were small for their gestational age, ten possessed dark skin pigmentation, pigmented gums were present in four, eight had neonatal jaundice, and nine exhibited conjunctivitis. A follow-up study on these children showed 'slight but clinically important neurological and developmental impairment'. PCBs can accumulate biologically, occurring in increasing concentrations at each step up of the food chain.

It has been noted in humans that short-term exposure to PCBs has lead to various problems involving mainly the skin, liver, and immune system. Workers exposed to such low levels as 0.07 mg/m^3 in workroom air have developed chloracne and adverse changes in the liver. Other symptoms noted included lassitude, skin, eye and upper respiratory tract irritation, nausea, anorexia, and dizziness.

Contaminated fish is another concern as it has been suggested that both hypertension and adverse reproductive effects are related to high levels of ingested PCBs. It is unfortunate that other factors such as smoking and alcohol intake were not included in these studies. Nevertheless, the need for continued monitoring should be stressed.

Pesticides

Insecticides have been employed for many years to improve the yields of agricultural crops, and their use is governed by safety criteria and testing laid down by such bodies as the Pesticides Safety Precautions Scheme in the UK. In animal tests many of these compounds are found to be both teratogenic and carcinogenic, and therefore it follows that extreme care must be exercised in the handling of these toxic chemicals. Furthermore, all such products should carry an adequate warning on the level of the materials. Contamination of food as a result of inadequate washing during preparation is a major and growing issue. People consuming such contaminated material may complain of headache, nausea, dizziness, confusion, and convulsions shortly after. In this area, adequate liaison with poison centres is essential in all field operations.

Antibiotics

Antibiotics and hormones are used in preparing certain animals for meat production. Unfortunately, antibiotics in animal feeds have contributed to a growing number of antibiotic-resistant bacteria.

Conclusion

With the ever-growing need to monitor food and water with respect to contamination, it is essential that a multidisciplinary approach is adopted to ensure that the products are wholesome. The prevention and reduction of biological and chemical contamination can only be accomplished by public health authorities ensuring integration of the field operations as outlined in this chapter.

References

Abdul-Haz, S.K., Ewald, R.A., and Kazyak, L. (1963). Fatal mushroom poisoning. Report of a case confirmed by toxicological analysis of tissue. *New England Journal of Medicine* **269**, 223.

Alfrey, A.C. (1980). *Neurotoxicology* **1**, 43.

Arief, A.L., Cooper, J.D., *et al.* (1979). *Annals of International Medicine* **90**, 741.

Blake, P.A., *et al.* (1979). Disease caused by a marine *Vibrio. New England Journal of Medicine* **300**, 1.

Bouree, P., Bouvier, J.B., Passeron, J., Galanaud, P., and Dormont, J. (1979). Outbreak of trichinosis near Paris. *British Medical Journal* **1**, 1047.

Broome, C.V. and Fraser, D.W. (1979). Epidemiologic aspects of legionellosis. *Epidemiology Reviews* **1**, 1.

Bryan, F.L. (1978). Factors that contribute to outbreaks of food-borne disease. *Journal of Food Protection* **41**, 816.

Burks, J.S., Huddlestone, J., *et al.* (1976). *Lancet* **i**, 764.

Candy, J.M., Klinowski, J., *et al.* (1986). *Lancet* **i**, 354.

Carruthers, M. and Smith, B. (1979). Evidence of cadmium toxicity in a population living in a zinc mining area. *Lancet* **i**, 845.

Cerva, L., Novak, K., and Cuthbertson, C.G. (1968). An outbreak of acute, fatal, amoebic meningoencephalitis. *American Journal of Epidemiology* **88**, 436.

Chin, J. (1982). Raw milk—an editorial. *Journal of Infectious Diseases* **146**, 440.

Coyle, H.B., *et al.* (1984). Differentiation of Bacillus anthracic and other bacillus species by lectins. *Journal of Clinical Microbiology* **19**, 48.

Crapper, D.R., Krisnan, S.S., DeBoni, U., and Tomko, G.J. (1975). *Transactions of the American Neurological Association* **100**, 154.

Crapper McLachlan, D.R. (1986). *Neurobiology of Aging* **7**, 525.

Department of the Environment Shipham Survey Committee (1982). *Final report to residents on metal contamination of Shipham, Sept 1982*. HMSO, London.

Diamant, B.Z. (1979). The role of environmental engineering in preventative control of water-borne diseases in developing countries. *Royal Society Health Journal* **3**, 120.

Eldin, G.P. (1981). Amoebiasis as a cause of death in a Scottish resident. *Scottish Medical Journal* **26**, 350.

Fallett, E.A.C. and McMichael, S. (1981). Acute hepatitis. An infection in West Scotland. *Scottish Medical Journal* **26**. 135.

Farmer, J.J., *et al.* (1985). *Vibrio.* In *Manual of clinical microbiology*, 4th edn., p. 282. American Society of Microbiology, Washington DC.

Foster, E.M. (1986). New bacteria in the news. A special symposium/*Campylobacter jejuni. Food Technology* **40**, 20.

Galbraith, N.S. (1980). Infections associated with swimming pools. *Environmental Heath* **February**, 31.

Ganrot, P.O. (1986). *Environmental Health Perspectives* **65**, 363 (and references cited therein).

Garruto, R.M., Fukatsu, R., *et al.* 1984. *Proceedings of the National Academy of Science (U.S.A.)* **81**, 1875.

Giri, I.W. (1978). Cysticercosis in Surabaya, Indonesia. *South-east Asian Journal of Tropical Medicine and Public Health* **9**, 232.

Greger, J.L. (1985). Food Technology **39**, 73.

Gross, R.J., Thomas, L.V., and Row, B. (1979). *Shigella dysentaeriae, Sh. flexneri* and *Sh. boydii* infections in England and Wales: the importance of foreign travel. *British Medical Journal* **ii**, 744.

Gunn, R.A. and Markakis, G. (1978). Salmonellosis associated with home-made ice cream. An outbreak report and summary of outbreaks in the United States in 1966 to 1976. *Journal of the American Medical Association* **240**, 1885.

Kadlec, V., Cerva, A., and Skarova, J. (1978). Virulent *Naegleria fowleri* in an indoor swimming pool. *Science* **201**, 1025.

Katzman, R. (1986). *New England Journal of Medicine* **314**, 964.

Krishnamachie, K.A.V.R. (1975). Hepatitis due to aflatoxicosis. An outbreak in Western India. *Lancet* **i**, 1061.

Krueger, G.L., Morris, T.K., *et al.* (1984). *CRC Critical Review in Toxicology* **13**, 1.

Lewis, M.J. and Macrae, A.D. (1977). Public Health Laboratory Service. Communicable Disease Report No. 45.

McCormick, J.B., Kay, D., Hayes, P., and Feldman, M.D. (1976). Epidemic streptococcal sore throat following a community picnic. *Journal of the American Medical Association* **236**, 1039.

McDermott, J.R., Smith, A.I., Iqbal, K., *et al.* (1977) *Lancet* **ii**. 710.

McMillan, M. and Thompson, J.G. (1979). An outbreak of suspected solanine poisoning in school boys. *Quarterly Journal of Medicine* **48**, 227.

MAFF (1985). *Food Surveillance paper No 15*. HMSO, London.

Mahurkar, S.D., Dhar, S.K., Salta, R., *et al.* (1973). *Lancet* **i**, 1412.

Mindus, P. and Kolmodin-Hedman, B. (1981). Told by her doctor to drink large amounts of water, suffered lead poisoning. *Acta Medica Scandinavica* **209**, 425.

Mosher, N.D. and Moyer, G. (1981). The PCB menace and breast milk. Dangerous properties of industrial materials report. Nov/Dec. In Mozar *et al.* (1987), q.v.

Mozar, H.N., Bal, D.G., and Howard, J.T. (1987), *Journal of American Medical Association* **257**, 1503.

Noah, N.D., Bender, A.E., Raeidi, G.B., and Gilbert, R.J. (1980). Food poisoning from raw kidney beans. *British Medical Journal* **ii**, 236.

Orenstein, W.A., Wu, E., Wilkins, J., *et al.* (1981). Hospital acquired hepatitis A. Report on an outbreak. *Pediatrics* **76**, 494.

Peers, F.G. and Linsell, C.A. (1973). Dietary aflatoxins and liver cancer—population study in Kenya. *British Journal of Cancer* **27**, 473.

Peers, F.G., Gilman, G.A., and Linsell, C.A. (1976). Dietary aflatoxins and human liver cancer. A study in Swaziland. *International Journal of Cancer* **17**, 167.

Pearl, A. and Brody, A. (1980). *Science* **208**, 297.

Pearl, D., Gajdusek, D.C., *et al.* (1982). *Science* **217**, 1053.

Plantanow, N.S., Funnell, H.S., Bullock, D.H., Amott, D.R., Saschenbrecker, P.W., and Grieve, D.G. (1971). Fate of polychlorinated biphenyls in dairy products processed from the milk of exposed cows. *Journal of Dairy Science* **54**, 1305.

Rigau-Perez, J.G., Hatheway, C.L., and Vatentin, V. (1982). Botulism from acidic food: first cases of botulinic paralysis in Puerto Rico. *Journal of Infectious Diseases* **145**, 783.

Riverside County Department of Health (1971). Collaborative study from Riverside Country Department of Health, Riverside, California. A water-borne epidemic of salmonellosis in Riverside California 1965. Epidemiologic aspects. *American Journal of Epidemiology* **93**, 33.

Sakazaki, R. and Barlows, A. (1981). The Geneva *Vibrio*. In *The procaryotes* (ed. M.P. Starr). Springer-Verlag, Berlin.

Schreeder, M.T., Favero, M.S., *et al.* (1983). *Journal of Chronic Diseases*. **36**, 581.

Skalsky, H.L. and Carchman, R.A. (1983). *Journal of the American College of Toxicology*. **2**, 405.

Spencer, P.S., Nunn, P.B., *et al.* (1987), *Science* **237**, 517.

Stern, A.J., Perl, D.P., *et al.* (1986). *Journal of Neuropathology and Experimental Neurology* **45**, 361.

Stevenson, A.H. (1953). Studies of bathing water quality and health. *American Journal of Public Health* **43**, 529.

Tobin, J.O'H., Swann, R.A., Bartlett, C.L.R., *et al.* (1981*a*). Isolation of *Legionella pneumophila* from water systems: methods and preliminary results. *British Medical Journal* **282**, 515.

Tobin, J.O'H., Bartlett, C.L.R., Watkins, S.A., *et al.* (1981*b*). Legionnaires' disease: further evidence to implicate water storage and distribution systems as sources. *British Medical Journal* **282**, 573.

Wieland, O. (1965). Changes in liver metabolism induced by the poisons of *Amantia phalloides*. *Clinical Chemistry* **11** (Suppl), 323.

Wilson, D.C. (1968). Goitre among the Ceylonese and Nigerians. *Nutrition Review* **26**, 77.

Windle Taylor, E. (1958). The examination of waters and water supplies (7th edn.). Churchill Livingstone, London.

Wisniewski, H.M., Moretz, R.C., and Iqbal, K. (1986). *Neurobiology of Aging* **7**, 532.

Further reading

Clean catering. A handbook on hygiene in catering establishments. HMSO, London

Drinking-water and sanitation 1981–1990. A way to health. WHO, Geneva, 1981.

Food-borne disease:methods of sampling and examination in surveillance programmes. Report of a WHO study group. WHO Technical Report Series No 543, Geneva, 1974.

Food Hygiene Codes of Practice 10: The canning of low acid foods. HMSO, London, 1981.

International standards for drinking water (3rd edn). WHO, Geneva, 1971.

Knox, E.G. (1977). Foods and diseases. *British Journal of Preventive and Social Medicine* **31**, 71.

Lead and health. Report of a DHSS Working Party on Lead in the Environment. HMSO, London, 1980.

Reports on public health and medical subjects No 71. The bacteriological examination of water supplies. HMSO, London, 1969.

Re-use of effluents: methods of waste water treatment and health safeguards. Report of a WHO meeting of experts. WHO Technical Report Series No. 517, Geneva, 1973.

Survey of Mycotoxins in the United Kingdom. Fourth report of the Steering Group on Food Surveillance. The Working Party on Mycotoxins. Food Surveillance Paper No. 4, Ministry of Agriculture, Fisheries and Food, 1980.

27

Ionizing radiation

M. C. O'RIORDAN AND A. P. BROWN

Introduction

The environment is permeated by radiation of natural origin: exposure to it is unavoidable. Human beings are also exposed to radiation of artificial origin from medical, industrial, and military procedures. Natural and artificial radiations are not different in kind or effect. In this chapter the term radiation always refers to ionizing radiations—alpha and beta particles, X-rays and gamma rays, neutrons and muons—which produce ions either directly or indirectly as they pass through matter such as tissue. As will be seen later, the dominant source of human exposure usually is radon in the home.

Radiation is inherently harmful to human beings. There are two classes of adverse effect, deterministic and stochastic. The former includes bone marrow depression and sterility for example; they will occur when a certain dose is delivered to a particular organ. The latter includes malignancies in exposed persons and hereditary defects in their descendants; these may occur with a probability related to dose.

There is much human evidence for some of these effects with complementary information from animal experiments and radiobiological investigations. The earliest observations of erythema from X-rays were made a few months after Roentgen's discovery in 1895 and were soon followed by cases of skin cancer in heavily irradiated areas (Pochin 1983). Within a few years of discovering radioactivity in 1896, Becquerel himself noticed erythema under a waistcoat pocket in which he carried a tube of radioactive material. It might be said, however, that the first observation of the harmful effects of radiation was made unwittingly by Paracelsus in his treatise 'On the miners' sickness', written in the 1530s, where he attributes the high incidence of death from lung disease among miners to a gas recognized much later as radon.

With the rapid adoption of X-rays and radium for diagnosis and therapy, there were many injuries and deaths among early patients and radiologists. National advisory committees on protection in radiology were set up in Germany, the UK and the US at the time of the First World War and were followed a decade later by international committees on radiation measurements and on protection, which exist to this day.

The next expansion of radiological protection came during the Second World War with the utilization of nuclear fission and the use of nuclear weapons at Hiroshima and Nagasaki. In the two decades after the war, there was intensive testing of weapons in the atmosphere by the US, USSR, and UK, which led to world-wide fallout of radioactive debris. France and China continued to test for another decade and more. Much of the information on risks of radiation comes from the epidemiological study of the Japanese survivors. It is probably true to say that public attitudes to radiological matters are affected to a degree by these military associations.

Meanwhile, the nuclear power industry developed strongly in many countries. Despite the arrest in the US during the 1980s, there are several hundred reactors in tens of countries producing about 20 per cent of the world's electricity (UNSCEAR 1988). In some countries such as France, Sweden, and Korea, the percentage is much higher. Workers and members of the public are exposed to radiation throughout the nuclear fuel cycle from the mining and milling of uranium, the fabrication of fuel, the operation of reactors, and the reprocessing of fuel—each of which produces radioactive waste. Epidemiological studies of uranium miners have yielded risk estimates for lung cancer from exposure to radon. There is no firm evidence of excess malignancies elsewhere in the cycle.

Radiation and radioactive materials have been widely used in general industry and continue to be so. In the UK, for example, over 20 000 persons are currently exposed at work (NRPB 1988a): in the US, over 300 000 are in this category (NCRP 1987). With one exception, there is little epidemiological evidence of late adverse effects among such workers: the exception relates to radium dial painters in the early decades of the century, who inadvertently ingested radioactive material when licking brushes to a point and later developed bone sarcomas.

Despite the initial cautionary experiences, radiological procedures in medicine have increased enormously. In countries with an advanced level of health care, there are over 500 medical and about 250 dental X-ray examinations annually per 1000 population with almost 20 nuclear medicine examinations (UNSCEAR 1988). As for therapy, there are almost three procedures or courses of treatment annually per 1000 population, mostly for malignancies, with 80 per cent or more by teletherapy and brachytherapy and the remainder by unsealed radionuclides. These procedures are of incalculable value to patients. In the past, there have been medical applications in which radiation was used for the treatment of benign conditions, such as ankylosing spondylitis and ringworm of the scalp, or for multiple fluoroscopic examinations of the chest. These episodes have also provided quantitative evidence of the adverse effects of radiation.

Most concern is felt, by professional and lay persons alike, about the induction of cancers by exposure to radiation. Much of the evidence for excess cancers comes from circumstances in which high doses were delivered quickly, the main exception being the exposure of miners to radon. For everyday circumstances of exposure to radiation at work or elsewhere, it is necessary to estimate the risk at low doses and rates. Interpolation is based on a linear relationship without threshold between dose and effect with a correction factor for dose rate in the case of X-rays and gamma rays.

Some quantities and units

Absorbed dose is the principal dosimetric quantity in radiological science: it is a measure of the energy imparted by ionizing radiation to a given mass of tissue (ICRU 1980). The special unit of absorbed dose is the gray, symbol Gy, which corresponds to 1 joule of energy per kilogram of tissue.

Equal absorbed doses do not necessarily have equal biological effects: 1 Gy from alpha radiation, for instance, is more harmful than 1 Gy from beta radiation, because alpha particles transfer energy more densely along their paths in tissue than beta particles. To put all types of radiation on a single scale of harmfulness, the radiation protection quantity dose equivalent is used: it is the absorbed dose weighted by a factor that takes account of the ionizing quality of different radiations. The special name for the unit of dose equivalent is the sievert, symbol Sv. For X-rays, gamma rays and beta particles, the quality factor is set at unity, so that absorbed dose and dose equivalent are numerically equal. For alpha particles, the quality factor is set at 20, so that an absorbed dose of 1 Gy corresponds to a dose equivalent of 20 Sv. Quality factors are being reviewed by the International Commission on Radiological Protection (ICRP) as are other quantities.

The risk of inducing malignancy per unit dose equivalent is not the same for the various tissues and organs of the body such as bone and lung, and the risk of serious hereditary damage through irradiating the testes and ovaries is different in kind and magnitude. To construct a common index of risk for non-uniform irradiation of the body, the radiation protection quantity effective dose equivalent is used: it is the sum of the dose equivalents to various organs and tissues weighted by the fractional risks to them when the whole body is uniformly irradiated. Values specified by the ICRP (ICRP 1977) are given in Table 27.1. The special name for the unit

Table 27.1. Weighting factors used in the calculation of effective dose equivalent

Tissue or organ	Factor
Gonads	0.25
Breast	0.15
Red bone marrow	0.12
Lung	0.12
Thyroid	0.03
Bone surfaces	0.03
Remainder	0.30
Whole body total	1.00

Source: ICRP (1977).

of effective dose equivalent is also the sievert, although submultiples such as mSv and μSv are more common in radiological protection, and the term is often abbreviated to dose. A change of name for this quantity is being considered by ICRP. Since the magnitudes of the risks have been reassessed recently by the United Nations Scientific Committee on the Effects of Atomic Radiation (UNSCEAR 1988) the values of the weighting factors may also be altered by ICRP.

It is often necessary to assess the risk to a group of persons or to the whole population from a particular source of radiation or from all sources. The quantity used in this context is, at present, called the collective effective dose equivalent, which is obtained by summing the effective dose equivalents to all persons in the group or population. It is expressed in units of man-sievert, symbol man Sv, and is often abbreviated to collective dose.

When radioactive material with long half-life is taken into the body by inhalation or ingestion, some may be taken up by various organs or tissues and deliver a dose to them or to adjoining parts of the body over a period of years as it physically decays and is biologically eliminated. A convention is required to take this dose into account with the dose from external sources which irradiate the body discontinuously. The quantity used is called, at present, the committed dose equivalent, which is defined as the integral of dose equivalent over a period of 50 years following a single intake of the radioactive material: it is expressed in sieverts. Fifty years represents a working lifetime, since the quantity is primarily used to assess occupational exposure (ICRP 1978a) but the concept has been extended to calculate the doses to adult persons in the general population who are exposed to radioactive materials and to infants and children by extending the integration period (NRPB 1987a). By analogy with effective dose equivalent, the quantity called, at present, committed effective dose equivalent is also used: it is the weighted sum of the committed dose equivalents to the various organs and tissues involved and is also expressed in sievert.

Activity is also an important quantity in radiological science: it is an attribute of the amount of a radionuclide, or unstable species of atom, and describes the rate at which spontaneous transformation or radioactive decay occurs. The unit of activity is the becquerel, symbol Bq, which corresponds to one transformation per second.

Sources of ionizing radiation

Natural sources of ionizing radiation include cosmic rays, terrestrial gamma rays, radioactive species such as potassium-40 in the body, and the radioactive gases radon and thoron in the air. Artificial or man-made sources include diagnostic X-rays, radioactive fall-out from weapons testing, radionuclides discharged from nuclear facilities, miscellaneous consumer products such as radioactive smoke detectors, and a wide range of sources at places of work. All these deliver doses of varying magnitude to individual persons and collective doses to communities.

International and national authorities regularly review the doses from radiation sources. Estimates of dose are based on measurements of persons, measurements in the environment, or calculations from established models of human exposure. UNSCEAR reports comprehensively on the matter (UNSCEAR 1982, 1988) as do, for example, the National Council on Radiation Protection and Measurements in the US (NCRP 1987) and the National Radiological Protection Board in the UK (NRPB 1988a).

General data for the UK, shown in Table 27.2, are typical of industrialized countries, even those without a nuclear power programme. The overall average value of the effective dose equivalent is about 2.5 mSv in a year. As Fig. 27.1(a) shows, natural sources account for 87 per cent of the dose and medical X-rays for 12 per cent: all other sources combined contribute only 1 per cent of the total. With a population of 57 million, the annual value of the collective effective dose equivalent in the UK is about 140 000 man Sv.

Averages, it is sometimes said, conceal the truth, but natural sources are still the most significant when variations in dose are considered. In employment, the variations are limited by statutory regulations, and less than 1 per cent of workers in the UK receives more that 15 mSv in a year. In medicine, doses to patients can be appreciable, with a few tens of millisieverts in some diagnostic procedures. For the other artificial sources of radiation, annual doses to members of the public are most unlikely to exceed 1 mSv. Annual values of the effective dose equivalent from natural sources can, however, exceed 100 mSv because of exposure to radon

Table 27.3. Range of annual doses in the UK from natural sources of radiation

Source	Range (mSv)
Cosmic	0.2–0.3
Gamma	0.1–1.0
Internal	0.1–1.0
Radon	0.3–100
Thoron	0.05–0.5
Overall	1–100

and its decay products in the home: the variability is shown in Table 27.3. Domestic exposure to radon has come to be regarded as a serious issue of public health.

General perception of these matters is not always related to the radiation facts (Clarke and Southwood 1989). In the UK, for example, there is much concern about the doses from nuclear discharges to the environment and in particular about liquid discharges to the Irish Sea arising from the reprocessing of reactor fuel in Cumbria. Assessment of this circumstance indicates that the overall dose to those most exposed from this practice, heavy consumers of local seafood, is about 2.9 mSv in a year, not much greater than the UK average and about the same as a frequent air traveller with increased cosmic dose, whereas the overall dose to the inhabitants of Cornwall, the most radon-prone county in the UK, is about 7.8 mSv in a year, over three times the UK average. The relative contributions of the various sources in these circumstances are shown in Figs. 27.1(b), (c) and (d). Wider appreciation of such information is beginning to affect attitudes to exposure and the allocation of resources for radiological health; this is particularly true of radon in the home.

Exposure to radon

The most important isotope of the element, radon–222, is created by the radioactive decay of the trace quantities of uranium in all rocks and soils. It is an inert gas and moves through the pores of these materials because of concentration and pressure gradients. When it emanates from the ground out of doors, it is rapidly dispersed in the atmosphere and concentrations of activity are low. In confined spaces such as mines and buildings, however, concentrations can become high because of restricted ventilation by fresh air.

Radon-222 decays in turn to solid radioactive products with short half-lives, often called radon daughters, the principal properties of which are listed in Table 27.4; the decay scheme is shown in Fig. 27.2. Most of the decay products become attached to the submicroscopic particles in air and so form a mildly radioactive aerosol. On inhalation, some of the activity is retained in the respiratory system. Of principal concern are the decay products polonium-218 and polonium-214, which emit alpha particles and irradiate the respiratory epithelium. The potential consequence is the induction of lung cancer.

Table 27.2. Average annual doses in the UK from various sources of radiation

Natural	Dose (mSv)	Artificial	Dose (mSv)
Cosmic	0.25	Medical	0.30
Gamma	0.35	Miscellaneous	0.01
Internal	0.30	Fallout	0.01
Radon	1.20	Occupational	0.005
Thoron	0.10	Discharges	< 0.001
Total	2.20	Total	0.30

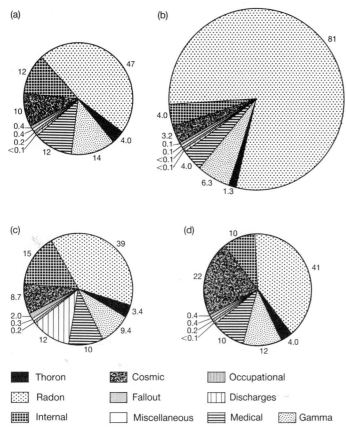

Thoron   **Cosmic**   **Occupational**

Radon   **Fallout**   **Discharges**

Internal   **Miscellaneous** **Medical**   **Gamma**

Fig. 27.1. Average annual doses to various persons from radiation sources in the UK during the late 1980s: (a) average person in the UK, 2.5 mSv overall; (b) average person in Cornwall, 7.8 mSv overall; (c) heavy consumer of Cumbrian seafood, 2.9 mSv overall; (d) frequent air traveller, 2.9 mSv overall. Pie chart areas are proportional to doses and percentage slices to sources.

Although building materials such as brick and concrete do produce radon, the main source of high indoor levels is the ground. The atmospheric pressure inside homes tends to be slightly lower than outside, and radon is drawn into the home mainly through gaps and cracks in the floor. The uranium content of the ground and its permeability to radon, the design and quality of the floor, and the living habits of the occupants combine to determine the radon levels. High values are therefore more likely to occur where geology is unfavourable, as for example around granite masses or on uraniferous shales, but levels in adjacent homes may differ markedly. The implication is that measurement is always required to determine whether a particular home is adversely affected by radon.

Many national authorities have made substantial surveys of indoor radon levels (CEC 1987; UNSCEAR 1988), and the practice is spreading. Individual authorities have adopted the principles advocated by the ICRP (ICRP 1984a) and have introduced recommendations to limit domestic exposure to radon (NRPB 1990; USEPA, 1986); a similar approach has been prescribed throughout the European Community. Limi-

tation schemes and values vary, but annual doses above 10–20 mSv are generally deemed undesirable. Remedial or preventive measures are urged for old or new homes as appropriate.

The magnitude of the radon problem in a particular country is determined by the limitation scheme in force and the geographical distribution of the population. In the UK, for example, the scheme includes an action level of 10 mSv in a year for existing homes and a requirement that the level in new homes be as low as reasonably practicable (NRPB 1990). Indoor levels are low in the conurbations and other centres of population, whereas radon-prone areas are less populous. Nevertheless, the number of homes in which remedial action is required is about one hundred thousand. Several thousand new homes built each year would have excessive radon levels unless appropriate antiradon designs were adopted.

It is a simple matter to reduce high radon levels in a home (DOE 1998a). The best approach is to prevent radon entering from the ground by drawing radon-laden air from under the floor and discharging it to the atmosphere. A small sump, duct, and fan should be installed, but gaps in the floor should be closed for best results. With new homes, suspended solid floors with good underfloor ventilation and some antiradon detailing are likely to be successful (DOE 1988b). None of these measures is expensive in comparison with the value of homes, and the cost per unit of collective dose avoided is negligible compared with the costs assigned to artificial radiation.

Biological action of ionizing radiation

The processes which ensue from the interaction of ionizing radiation with living tissue are complex, and so it is not possible to trace clearly a chain of events leading from the absorption of energy to the production of a detrimental effect. Much experimental radiobiology has dealt with the lethal action of radiation on living cells, but understanding of the more subtle and prolonged mechanisms that lead to cancer, for instance, is incomplete. Whether radiobiologists study cell killing or cell mutation by radiation, one of the principal

Table 27.4. Principal decay properties of radon-222 and its short-lived daughters

Radionuclide	Half-life	Main radiation energies and intensities					
		α		β		γ	
		MeV	%	MeV	%	MeV	%
^{222}Rn	3.824 d	5.49	100	—	—	—	—
^{218}Po	3.05 min	6.00	100	—	—	—	—
^{214}Pb	26.8 min	—	—	0.67	48	0.30	19
				0.73	42	0.35	37
^{214}Bi	19.7 min	—	—	< 1.5	32	0.61	46
				1.5–2.5	49	1.12	15
				3.27	18	1.76	16
^{214}Po	163.7 μs	7.69	100	—	—	—	—

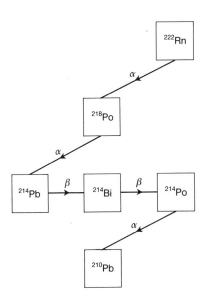

Fig. 27.2. Radon decay scheme.

objectives is to establish a quantitative relationship between the biological effect and the radiation dose administered. The quantity normally employed is absorbed dose, expressed in grays; 1 Gy is an enormous quantity of radiation as far as biological effects are concerned.

Ionization, the removal of electrons from atoms of irradiated tissue, is the major physiochemical process underlying the biological changes due to radiation (Hall 1988). Ionizations occur mainly along the tracks of individual charged particles, and the pattern of ionization depends on the physical properties—for example, charge and velocity—of the particles involved. High-speed electrons are generated when X-rays and gamma rays are absorbed in tissue, and the distribution of ionization events along the paths of these electrons is fairly sparse. In contrast, heavy charged particles, such as alpha particles, produce a dense pattern of ionization along their tracks. These differences between various forms of radiation are denoted by the term linear energy transfer (LET). X-rays and gamma rays are low-LET radiations (about 2 keV/μm), alpha particles are high-LET radiations (about 100 keV/μm), and neutrons occupy an intermediate position although they are usually classed as high-LET radiations. LET is and important factor in determining the relative biological effectiveness (RBE) of different types of ionizing radiations. High-LET radiations have a greater biological effect on higher organisms, including man, than X-rays when absorbed in equal doses: 1 Gy of neutrons is more biologically damaging than 1 Gy of X-rays because the biological outcome depends on the localized deposition of energy at crucial sites. ICRP is likely to re-emphasize the importance of RBE.

It seems probable that DNA in the cell nucleus is the critical structure that is damaged directly or indirectly by ionizing radiation. The nature of the DNA damage determines whether the cell will die prematurely when it next attempts mitotic division, or whether a non-lethal change or mutation occurs in its coded information which is then incorporated in its daughter cells and future progeny, eventually resulting in a malignant tumour. If germ cells are affected, the DNA damage could result in sterility—if sufficient cells are killed—or genetic abnormalities in any subsequent offspring.

The hypothesis that DNA mutation by radiation or other carcinogenic agents can lead to the malignant transformation of cells is supported by much clinical and experimental evidence (Hall 1988). At the microscopic rather than molecular level, chromosome aberrations caused by radiation have been studied for several decades and led to the first mathematical relationships between radiation dose and biological effect. More recently, biochemists have been able to detect radiation-induced breaks in one or both strands of a DNA molecule, the so-called single-strand or double-strand breaks: double-strand breaks may be responsible for the well-recognized chromosome aberrations.

Radiation-induced single-strand and double-strand breaks in DNA are repaired very rapidly by a variety of enzymatic processes in the cell. The DNA repair involves the recombination of genetic material within the cell nucleus, and this may give the opportunity for misrepair and mutation. The completeness of DNA repair might depend on the extent of the damage, which is, in turn, a function of the radiation dose and the rate at which it is delivered. There is abundant experimental evidence that dose and dose rate have a significant influence on the carcinogenic effect of low-LET radiation: low doses and low dose rates carry lower risks (UNSCEAR 1988). No such evidence is available for high-LET radiation.

Although the precise molecular mechanisms are unknown, it seems probable that genetic mutation due to radiation can depend on single-point changes in the DNA coding (Breimer 1988). This is similar to the alterations produced by oncogenes, and may explain why radiation does not induce any novel tumours or genetic syndromes that do not arise, so to say, spontaneously.

Carcinogenic effects of radiation

The induction of cancer by low doses of ionizing radiation is the most important public health problem in radiological protection. Although it is reasonable to suppose that DNA damage can predetermine malignant change, the development of a cancer is a multi-stage process; there are many other relevant factors, both constitutional and environmental, which may influence the outcome in any individual case.

Laboratory experiments with animals or cells in culture are valuable as a means of indicating potential hazards and in elucidating the biological processes involved; however, they cannot be expected to provide a quantitative estimate of the risks to man associated with exposure to different doses. Such quantitative evidence must be sought from epidemiological studies, where a statistically significant excess of cancers is observed in a population exposed to radiation compared

Table 27.5. Examples of human cancers induced by ionizing radiation and the populations from which the evidence was obtained

Neoplasm	Main groups studied
Leukaemia (except chronic lymphatic leukaemia)	Survivors of the atom bombs
	Patients irradiated for ankylosing spondylitis
	Patients irradiated for cancer of the cervix
Thyroid cancer (well-differentiated)	Survivors of the atom bombs
	Children irradiated for tinea capitis and supposed thymic enlargement
Breast cancer	Survivors of the atom bombs
	Women irradiated for post-partum mastitis
	Patients given fluoroscopies for tuberculosis
Lung cancer	Survivors of the atom bombs
	Miners exposed to radon decay products
Bone sarcomas	Dial painters who ingested radium
	Patients given radium injections for ankylosing spondylitis or tuberculosis
Multiple myeloma	Survivors of the atom bombs
Liver cancer	Patients given Thorotrast injections
Skin cancer	Early X-ray workers
Digestive tract	Survivors of the atom bombs
	Patients irradiated for ankylosing spondylitis
Brain, meninges	Children irradiated for tinea capitis

with a similar control population that has not been exposed. Malignant tumours that might have been induced by radiation cannot be distinguished pathologically from so-called naturally occurring cancers. In industrialized countries, some 20 per cent of all deaths are due to malignant disease; in order to detect a statistical excess above this level, one requires a large study population exposed to substantial doses. Since the population size required to detect an effect is inversely proportional to the square of the excess risk (Land 1980), the detection of adverse effects at low doses can be daunting.

Three epidemiological studies—of the survivors of the atom bombs at Hiroshima and Nagasaki in 1945, of patients with ankylosing spondylitis given X-ray treatment during the period 1935–54, and of patients irradiated for cancer of the cervix—have provided the greatest amount of information on the risk of cancer to various organs from external irradiation. The results are in general agreement as to the sites where the frequency of cancer is elevated by exposure to ionizing radiation (UNSCEAR 1988). Other smaller epidemiological studies can be used to estimate the risks of inducing cancer at various sites in the body: Table 27.5 gives examples.

The atom bomb survivors are the most important group because they have been studied prospectively since 1950, the population contains almost equal numbers of males and females—unlike patients with ankylosing spondylitis and cancer of the cervix—and they were of all ages at the time of exposure. The study population comprises about 80 000 people, and after 40 years the number of deaths from cancer was about 6000 of which 300 or so were in excess of expectation and attributable to radiation. By painstaking effort, an estimate has been made of the dose received from gamma rays and neutrons by each of the survivors in the study. The most recent reassessment of dosimetry, known as DS86, has

shown that exposure to neutrons was substantially less than had been thought previously.

Whatever the uncertainties in the dosimetry, it is clear that the dose was absorbed at a high rate and that many in the study population received relatively high doses of 1 Gy or more. Radiological protection is largely about workers, who are unlikely to accumulate doses of more than a few tenths of a gray throughout their working lives, and members of the general public, who are unlikely to accumulate more than one tenth of a gray unless they receive medical exposures or live in houses with high radon levels. As explained earlier, there is good evidence that for X-rays and gamma rays the risk per unit dose is mitigated by low doses and rates. In projecting the carcinogenic consequences to everyday circumstances of exposure to these low-LET radiations, one therefore makes the assumption that the risk per gray is the same at high and low doses but that a factor should be introduced for dose-rate effectiveness (UNSCEAR 1988; NRPB 1988b). No such factor is required for high-LET radiation.

Some general features of radiation carcinogenesis in human beings emerge from the epidemiological studies. Leukaemia is a prominent and early consequence. Its time of onset is much shorter than other radiation-induced cancers; the so-called latent period between exposure and diagnosis can be as short as a few years, whereas for other malignancies the latent period may be 20 years or more. For both the atom bomb survivors and the ankylosing spondylitis patients, the relative risk of leukaemia reached a peak within a decade and then declined.

Age is generally an important host factor; the risk of subsequent malignancy is appreciably higher when exposure occurs early in life, say under the age of 20. Cancer is relatively uncommon in the first four decades of life, and it is only recently that survivors of the atom bombs who were children or teenagers in 1945 have reached an age at which malignant diseases increase in frequency. Even four decades after the event, two-thirds of the survivors from Hiroshima and Nagasaki are still alive. Although exposure to radiation at a young age seems to carry a relatively higher risk, it will not be known until well into the next century whether this higher risk will continue to amplify the frequency of cancer as the cohort ages or whether the relative risk might diminish.

Statisticians have developed two models, the absolute and relative, to project the future risk of cancer in a population exposed to radiation. The absolute or additive risk model assumes that, after a latent period, the annual radiogenic risk becomes a constant addition to the so-called natural cancer rate until the end of life. The relative or multiplicative risk model assumes that the radiation-induced cancers will be proportional to the age-specific natural cancer rate with the excess given by a constant multiplying factor. Since the natural incidence of many cancers increases steeply until late in life, the relative risk model will yield a greater number of deaths from radiation. Whereas the absolute risk model seems to apply for leukaemia and radiation-induced bone sarcomas, for many other cancers in the Japanese and anky-

Table 27.6. Projected lifetime risks of fatal cancers from whole-body exposure to 1 Gy of low-LET radiation according to dose rate, risk model, and population

Dose rate	Model	Population	All ages (%)	Workers (%)	Reference
High	Relative	Japan	7–11	7–8	UNSCEAR (1988)
High	Absolute	Japan	4–5	4–6	UNSCEAR (1988)
High	Relative	UK	13	9.7	NRPB (1988b)
Low*	Relative	UK	4.5	3.4	NRPB (1988b)

* These values can be expressed in terms of effective dose equivalent, e.g. 4.5 per cent per Sv for the whole UK population.

Table 27.7. Risk of lung cancer for the UK and US from lifelong exposure to radon in the home

Gas (Bq/m³)	Dose (mSv/year)	Risk (%)
20	1	0.25
100	5	1.25
200	10	2.5
400	20	5.0
800	40	10

losing spondylitis studies the relative risk model seems to fit the data well and the absolute risk model does not (Muirhead and Darby 1987). UNSCEAR (1988) has adopted the relative risk model for estimating lifetime risks of most solid cancers following radiation exposure, as has the UK (NRPB 1988b). In previous UNSCEAR (1977) and ICRP (1977) reports, the additive model had been employed. The revision of the bomb dosimetry and the adoption of the relative risk model has led to an increase, by a factor of a few, in the projected radiation risks for both populations as a whole and adult workers. Table 27.6 gives values for both Japan and the UK.

The risks calculated for the UK population are higher mainly because of the disparities in cancer rates between the two populations; it is an awkward consequence of the relative risk model that the risks of radiation will vary from country to country. To obtain the risks at low dose rates for the UK, a dose rate effectiveness factor of 3 was applied to the risks for most tissues at high dose rates. The dividing line between low and high dose rates for low-LET radiation can be taken as 0.1 Gy/day. No such factor is required for high-LET radiation. For public health purposes in the UK, the most important value in Table 27.6 is 4.5 per cent/Gy, which is the lifetime risk for persons of all ages and sexes exposed to whole-body irradiation from penetrating gamma rays at low dose rates. This risk can also be expressed in terms of effective dose equivalent as 4.5 per cent/Sv. Such parameters are often called risk coefficients and are given in scientific notation e.g. 4.5×10^{-2}/Sv. The annual number of cancers from radiation is obtained by multiplying the annual collective effective dose equivalent by this number.

Although the number of fatal cancers projected by the relative risk model is higher, most of the difference between it and the absolute model will be accounted for by deaths in old age. When the relative model is applied to the risk observed after 40 years of the atom bomb study, the values for the UK in Table 27.6 are reduced by a factor of a few. It is clear that any numerical estimate of risk has a large degree of uncertainty associated with it, and this qualification must be borne in mind when interpreting published figures.

In the past few years, there has been increasing concern about the problem of radon in homes and the consequent risk of lung cancer induction. There is epidemiological evidence of excess lung cancer among uranium and other miners who were exposed to high concentrations of radon daughters at work (IARC 1988). This is supported by animal studies in which the incidence of lung tumours was found to be proportional to exposure. Epidemiological studies are being made in various countries of people exposed to appreciable radon levels in their homes, but these may well be inconclusive. Nevertheless, the radon levels in some dwellings are as high, if not higher, that in mines, so that it has been considered prudent to translate the risk estimates from mining to the domestic setting. The differences between the two circumstances of exposure and the interaction of other carcinogenic agents, especially smoking, makes the calculation of the carcinogenic risk in homes uncertain, but two independent analyses with relative risk models yield similar results for the US and UK, namely 3.5 per cent/Sv effective dose equivalent (ICRP 1987; NAS 1988). With this value, rounded estimates of lifetime risk from lifelong exposure to radon at home for these countries are given in Table 27.7.

Genetic effects of radiation

An appreciable proportion of the population suffers from inherited genetic effects or congenital anomalies which lead to clinically detectable diseases at some stage during life. Genetic defects include aberrations in chromosome structure, variations in the number of chromosomes per cell, and point mutations or changes in the genetic information of the DNA within the chromosomes. As discussed earlier, radiation can produce chromosomal aberrations and gene mutations, and so if germ cells are irradiated, there is a risk of producing hereditary effects. If the genetic diseases resulting from radiation are dominant or X-linked, they would start to appear in the first generation of children and the number of cases would be directly proportional to the rate of mutation. If the diseases were multifactorial, or recessive aetiologies were involved, the apparent excess in the first generation would be much lower and it might continue undetected for many succeeding generations. Furthermore, in these types of disease, the relationship between incidence and the underlying mutational rate would not be straightforward.

Since life has evolved against a continuous background of radiation, one would not expect exposure to further radiation, either natural or artificial, to cause any new genetic diseases, but rather to increase the rate of mutations that occur. As many genetic diseases in human beings are of the multifactorial type, it is not surprising that no epidemiological study to date has convincingly demonstrated any increase from radiation. The most recent review of data on the offspring of the atom bomb survivors in Hiroshima and Nagasaki again does not identify any statistically significant genetic effect (UNSCEAR 1988). There has been no emergence of an otherwise rare genetic disorder comparable with the emergence of leukaemia in the study of radiation-induced malignancies.

For the biological indicators studied, however, which ranged from stillbirth to rare electrophoretic variants of plasma proteins, the observed effects were compatible with the hypothesis that the survivors of the bombings sustained some genetic damage. Despite the lack of empirical evidence, there is no reason to suppose that human beings are not prone to genetic effects from radiation in the same way as other living organisms. With the exception of the Japanese indications, the genetic risk in man has been estimated only by extrapolation from experiments on other organisms.

The first experiments, in the interwar years, were conducted on the fruit fly *Drosophila*. The frequency of mutations was found to be linearly related to radiation dose, there was no threshold level below which mutations were not seen, and there appeared to be no biological repair of genetic damage. These findings when extrapolated to man led to the belief, widely held for many years, that genetic effects were the most important long-term consequence of radiation.

Experiments after the Second World War, using millions of mice studied over several generations, have radically altered these views (Hall 1988). Perhaps the most important conclusion has been that repair processes do indeed occur between the primary induction of genetic damage and its final expression. By postponing conception for a period following irradiation, therefore, the likelihood of transmitting genetic damage is lessened. There also appears to be a significant dose rate effect, so that exposure at low dose rate should reduce the genetic consequences of a given dose.

Two main methods of extrapolation have been used to scale the genetic risks of radiation from mouse to man (UNSCEAR 1988). The direct method depends on multiplying the risk of a gene mutation per unit dose in mice by the corresponding number of genes in man; this method has limited application and subsumes enormous uncertainties. The second method employs the concept of doubling dose— the dose of radiation required to double the rate of spontaneous genetic mutations in human beings. After reviewing all the available data, UNSCEAR (1988) took a doubling dose of 1 Gy and applied it to the natural frequencies of various categories of human genetic disease, excluding the multifactorial, so as to determine the excess that might be expected from radiation; the estimate for a population continuously exposed to low-LET radiation at a low dose rate of 0.01 Gy per generation is about 18 extra cases in the first generation of a million liveborn progeny. This radiation increment is related to a natural or spontaneous incidence of almost 13 000 cases of these genetic diseases per million liveborn. If time is allowed for all such genetic diseases to be expressed over future generations, so that a new elevated equilibrium is reached, the total increment is predicted to be about 120 cases per million progeny, which yields a risk estimate of 1.2 per cent/Gy for the reproductive fraction of the population. Since this fraction is about 0.4, the genetic risk for a whole population is about 0.5 per cent/Gy, which is an order of magnitude below the cancer risk for low dose rates. In the UK some allowance is retained for multi-factorial diseases and results in a higher value (NRPB 1988b). Since the risk refers to low-LET radiation, it may also be given in terms of dose equivalent to the gonads, expressed in sieverts.

It should be emphasized that genetic effects can only arise if the gonads are irradiated before or during the reproductive period, i.e. the dose must be genetically significant. When considering the genetic consequences for a particular population following exposure to radiation, one has to estimate the average child expectancy from the age and sex characteristics of the population. The parameter used to represent the risk of hereditary damage in such circumstances is the genetically significant dose (GSD); this can be defined as the nominal dose to every member of the population that would produce the same genetic harm as the actual doses received by the various persons who had been exposed. In the UK, most of the GSD to the population as a whole from artificial sources of radiation comes from diagnostic radiology. Its value is about 120 μSv in a year, on average, whereas the average gonad dose from natural radiation is about 900 μSv in a year (NRPB 1988a). The annual numbers of hereditary effects are obtained by multiplying these doses by the population size and then by 5×10^{-3} Sv^{-1}.

Just as different malignant diseases differ in their mortality rates, so genetic diseases vary in their severity, time of onset, and so on; consequently, radiation effects differ in their impact on the affected person, on his or her family, and on society at large. Practitioners of radiological protection have had to construct a scale of harm that takes into account mortality and morbidity from stochastic effects so as to develop standards of safety. They have taken the pragmatic view in the past (ICRP 1977) that severe genetic diseases in a few generations and fatal cancers are roughly equivalent in importance to affected individuals, but it is likely that non-fatal cancers (e.g. thyroid) may in future be given some weight. Such decisions are as much social as radiological.

Deterministic effects of radiation

Both cancer induction and genetic damage have been classified as stochastic effects of radiation (ICRP 1977), i.e. the probability of such an effect occurring is a function of dose, but the severity of the effect does not depend on the dose. In contrast, all other biological effects of radiation can be considered deterministic or non–stochastic, in that the severity varies with the dose; such effects depend on such large numbers of cells being killed by irradiation that there is some impairment of organ function either directly or from tissue damage by fibrosis. There is a threshold or critical dose below which no effect is seen, whereas a threshold cannot be assumed for the stochastic processes which are thought to depend on changes in a single cell. An important aim of radiological protection is to prevent non-stochastic effects by ensuring that the threshold dose for any organ is not exceeded. Non-stochastic effects are of great practical importance in radiotherapy, where the tolerance of normal tissue is often critical in determining the exact treatment to be given.

There are three well-recognized acute syndromes that occur in man following exposures to high whole-body radiation doses. Two of these, involving the central nervous system and gut, result from very high acute doses of 10 Gy or more and are uniformly fatal. The bone-marrow syndrome occurs at doses close to LD_{50}, the whole-body dose that would be fatal in 50 per cent of cases; its value for man is not accurately known, but has been estimated as 3 Gy or so by UNSCEAR (1988) for persons receiving minimal or no medical support. With conventional medical treatment such as antibiotics, experience from Chernobyl and other accidents suggests that LD_{50} might be 5 Gy or more.

Although non-stochastic effects, such as cataracts in the eye, do not occur in most organs below a threshold of a few gray, the fetal brain between the eighth and fifteenth week after conception may be a special case where the threshold is much lower, around 5 mGy (Pochin 1988). This is seen as a particularly sensitive period for the induction of serious mental retardation because of information from the follow-up on Japanese children who were irradiated *in utero* at Hiroshima and Nagasaki and survived. (See the section on irradiation during pregnancy.)

Principles of radiological protection

The central principles of radiological protection are given in the basic recommendations of the ICRP last formulated in 1977 (ICRP 1977) and likely to be reformulated in 1991. ICRP aims to maintain as much stability in the basic recommendations as new information on doses and effects allows and in particular to retain the general principles of protection. Nevertheless, there are likely to be some changes during the currency of this edition of the textbook, and readers will need to monitor developments. The ICRP also publishes subsidiary recommendations from time to time, as for example on protection of the public in radiation accidents (IRCP 1984b) which apply the principles to practical circumstances. Most industrial countries subscribe to the ICRP recommendations, and both national and supranational authorities eventually incorporate them into legislative or administrative controls with whatever variations are deemed appropriate.

The present principles of protection are three in number. Each involves social considerations and each requires the exercise of judgement: (a) no practice shall be adopted unless its introduction produces a positive net benefit; (b) all exposures shall be kept as low as reasonably achievable, with economic and social factors being taken into account; (c) the dose equivalent to individuals shall not exceed the limits recommended for the appropriate circumstances by the Commission.

These principles are often described by the terms justification, optimization, and limitation. They can only be applied in part to some sources of radiation: nothing, for example, can reasonably be done about the general level of dose from natural sources, although high doses from radon can be avoided, and it would be inappropriate to specify numerical limits on patient dose, although clinicians should have regard to the two other principles. They are applied in full, however, to the exposure of radiation workers and to the exposure of the public from man-made sources. The ability to control and the liability for control are two important considerations in any scheme of safety.

The first principle makes explicit the need to consider the potential harm before deciding whether a practice involving human exposure to radiation should be introduced and to explore the possibility of using alternative ways to achieve the same end without exposure. Costs and benefits should be analysed, including the cost of harmful radiation effects, and a judgement made about the balance between them; the procedure may be qualitative or quantitative as circumstances require or resources allow.

The second principle underlines the necessity to minimize exposures from all practices and emphasizes that mere compliance with a numerical dose limit is not sufficient: the doses actually received must be as low as social and economic circumstances allow. Costs and benefits of increasing protection and decreasing dose should be analysed so as to identify the stage at which the extra expenditure balances the drop in dose; the procedure may again be qualitative or quantative.

The third principle expresses the obligation not to expose persons or their future descendants to a degree of risk that is deemed unacceptable and thus limit any inequity that might result from unconstrained optimization. This is brought about by imposing numerical limits on the effective dose equivalent that persons may receive so as to restrict harmful stochastic effects such as cancers, or on organ dose equivalent so as to prevent harmful non-stochastic effects such as cataracts; these limits are to be observed regardless of cost. Other dose constraints related to particular sources of radiation may well be advocated by ICRP in the reformulated recommendations.

ICRP identifies various types of protection standards—primary, secondary and derived limits, and reference levels. These distinctions are reflected in rules and regulations from national authorities, which may also introduce authorized limits to reinforce controls under statutory or administrative arrangements.

The primary limits apply to the sum of the effective dose equivalent from external exposure received during one year and the committed effective dose equivalent from internal exposure arising during that year and similarly to the dose

Table 27.8. Dose limits for workers and members of the public

Quantity	Worker (mSv/year)	Public (mSv/year)
Effective dose equivalent	50	1 (5)
Dose equivalent to single organ or tissue	500	50
Dose equivalent to eye lens	150	50

Source: ICRP (1977).

equivalent plus the committed dose equivalent for organs and tissues. There are two sets of values, one for individual workers and one for individual members of the public, but the latter may sometimes be applied to the average for the most exposed or critical group of persons. Present ICRP values are given in Table 27.8.

For members of the public, the principal value of the effective dose equivalent is 1 mSv/year, but ICRP deems it permissible to use a subsidiary value of 5 mSv in a year for some years, provided that the average value over a lifetime does not exceed the principal value. For workers, the more restrictive value of the dose equivalent to the lens of the eye is to avoid the induction of cataract. (See the section regarding dose limitation during pregnancy.)

Future ICRP dose limits may be reduced to reflect the increased risk estimates (UNSCEAR 1988) and the expression of the detriment: the present subsidiary limit for the public may end, and the 50 mSv limit for workers may come down to 20 mSv or so with some transitional concessions. Some national authorities have anticipated developments of this nature; in the UK, for example, interim guidance has been issued to restrict the effective dose equivalent for workers to 15 mSv a year on the average (NRPB 1987b). Throughout the decades, there has been a steady downward trend in dose limits.

Secondary limits are given separately for external and internal irradiation; they are surrogate or intermediate quantities to facilitate compliance with the primary standards. For external exposure, they comprise a number of practical dose-equivalent quantities defined for a stylized phantom of the human body. For internal exposure, the secondary limits comprise a set of annual limits on intake, by inhalation and ingestion, for various chemical and physical forms of many radionuclides and are expressed in terms of activity (becquerels) (ICRP 1978a; NRPB,1987a).

Derived limits are designed for practical protection. As the name implies, they are obtained from the other standards and in such a cautious formalized manner that compliance is assured. It is possible to derive a limit on dose-equivalent rate in the environment (μSv/h), activity concentration of a particular radionuclide in air (Bq/m^3), or radioactive substance deposited on a given type of surface (Bq/m^2).

Reference levels can be established for a specific purpose in radiation protection; they are values of diverse quantities at which various predetermined courses of action are to be taken. For example, ICRP has recommended reference levels of dose equivalent to protect members of the public in

the event of major radiation accidents (ICRP 1984b). This concept has been adopted by national authorities which may go on to establish derived emergency reference levels for such quantities as the activity per unit mass of specific radionuclides in foodstuffs (becquerels per kilogram). (See the section on radiation accidents for further discussion.)

Practical aspects of protection

An effective system of practical protection is in place throughout the industrialized world. It derives from ICRP recommendations (ICRP 1977) and has been developed by various supranational and national authorities (CEC 1980; IRR 1985).

Exposure at work lends itself to close control. Areas within places of work are classified according to the degree of exposure that might arise: those where a certain fraction of the dose limits might be received are strictly controlled, whereas those within a lower dose band are supervised less stringently. Employees are also classified according to their working conditions: those who work in controlled areas require personal radiological surveillance, whereas general surveillance is sufficient for those in supervised areas. Adequate provision must be made for assured restriction of exposure and dose minimization.

Surveillance is in two forms, physical and medical. Regular measurements of radiation levels are required in both areas to monitor exposure potential. Workers in controlled areas also require personal monitoring to determine the dose or exposure to which they are individually subjected. Film and thermoluminescent dosimeters are widely used for external radiations; personal air samplers and track-etch devices are established for airborne activity. General principles of monitoring for the protection of workers have been elaborated by ICRP (ICRP 1982a). Medical surveillance is required for workers in controlled areas. It is based on the general principles of occupational medicine.

Many administrative requirements are also imposed on employers with regard to the commencement of operations, the supervision of sources, the reporting of untoward incidents, the maintenance of records, and so on.

Exposure of members of the public is more difficult to determine. Unlike workers, their actions cannot be controlled, and personal monitoring is not practicable. Reliance must therefore be placed on control of the source of exposure supplemented by appropriate monitoring of the environment.

Restrictions on the emission of radiation or radioactive materials to the general environment should be stringent enough to ensure that the dose limits for members of the public are not exceeded. National authorities may impose even more stringent requirements regarding the disposal of radioactive waste. All this puts a premium on good design of radiation facilities and careful pre-operational studies.

Monitoring of direct radiation from an installation is a straightforward matter: simple measurements of dose-equivalent rate with appropriate survey instruments should suffice. Monitoring of radioactive materials discharged to the environment either from an installation or a waste disposal site is more complex. It is necessary to have a clear appreciation of the mechanisms and magnitudes of the releases and to understand the movement of the materials through the environment to man. This information is best formulated in environment models, which help to identify the persons most at risk in the community, the so-called critical group. Appropriate programmes of source and environment monitoring should then be pursued.

The magnitude of monitoring programmes will be determined by the magnitude of the actual and potential releases (ICRP 1985a). For minor sources such as authorized discharges from hospitals and university laboratories, assessment is required of the activity released with occasional sampling of the receiving media and analyses for specific radionuclides. For major sources such as nuclear power stations or fuel reprocessing factories, however, continuous on-line monitoring of aerial and liquid discharges is likely to be required with continuous on-site monitoring of external radiation. It is also likely that periodic and systematic monitoring on a substantial scale will be required in the surrounding environment. National monitoring programmes by the controlling authorities may also be indicated.

An example of such a programme for a major nuclear installation is shown in Fig. 27.3 (ICRP 1985a). The physical characteristics of the environment and the habits of the populace are required to estimate doses and determine the important causes of human exposure. Sampling details are then specified and the methods of measurement selected to round out the programme. Some indicator materials which accumulate activity but are not in the human foodchain—seaweeds and mosses, for example—may usefully be monitored. It is important to maintain the quality of measurements and keep the whole programme under review so that the best estimates of dose are obtained and firm assurance can be offered about the state of the environment. If a serious accident were to occur with a large release of activity, the results from the program would be relevant to an assessment of the impact.

Some monitoring may also be required around disposal sites for solid radioactive waste (ICRP 1985b). Such disposals are subject to strict authorization which takes into account the category of waste, the management strategy, the nature of the site, the possibility of disturbance, and the potential doses. In the UK, some wastes with trivial activities may be disposed of in municipal rubbish or selected landfill sites, and occasional monitoring is carried out by the authorities.

Low-level wastes are disposed of by shallow land burial at a designated site in the UK. These contain various radionuclides with short half-lives and trace quantities of radionuclides with long half-lives in laboratory trash and other slightly contaminated materials such as building rubble; doses arising from the operation are negligible (NRPB 1988c). Systematic monitoring of the local environment is carried out by the site operator. Higher-level wastes from the

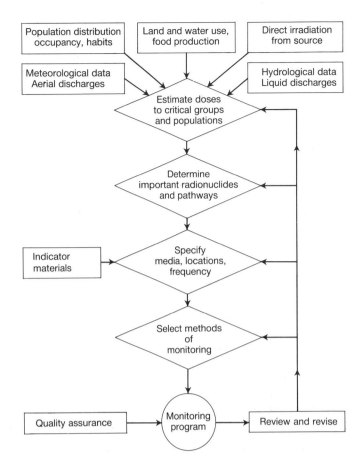

Fig. 27.3. Elements of a monitoring programme for a major source of activity.

nuclear fuel cycle are being stored in the UK until they are suitable for disposal and a disposal option, such as deep land burial, is agreed. Waste disposal has minimal implications for public health and maximal impact on public opinion.

Irradiation during pregnancy

Because of the potential risks to children irradiated in the womb, special attention must be paid to the protection of pregnant women and women of reproductive capacity. Teratological effects such as small head size and mental retardation may be caused, for which there are periods of maxium sensitivity, and cancers may also be induced. (See the section on deterministic effects.) For these reasons, ICRP recommends restrictions on the irradiation of such women at work and in diagnostic X-ray procedures (ICRP 1982b). To protect the fetus, arrangements should be made at work to ensure that gross unevenness in the rate of irradiation is avoided, and the dose in the third and fourth months of pregnancy should not exceed 1 mSv overall. It may happen that a woman is unwittingly exposed to radiation in the early stages of pregnancy—indeed before she realizes that she is pregnant. In such circumstances, it is important to estimate the dose to the embryo as accurately as possible, but it must also be remembered that the dose restriction recommended for women in general is of little import in advising an individual patient. The dose from a diagnostic procedure—the most common event—still carries a very low risk for the unborn child, and no decision should be made regarding abortion without expert counselling (NRPB 1985). ICRP is reviewing its recommendations with the broad intention of protecting the fetus as if it were a member of the public.

Radiation accidents

Fatal radiation accidents are rare. Between 1945 and 1987 some 17 events were reported with 59 fatalities, 40 among workers and 19 among members of the public (IAEA 1988a). Some 35 workers died at nuclear facilities, but virtually all the other deaths were from exposure to discrete gamma ray sources. The nuclear accident at Chernobyl in 1986 and the radiological accident at Goiânia in 1987 illustrate the potential severity of such events.

At Chernobyl in the USSR, a power reactor ruptured and released large activities of radionuclides (IAEA, 1986). Approximately 300 reactor workers and firemen were admitted to hospital, about 200 with acute radiation syndrome from gamma ray doses of 2–16 Gy to the whole body. Some men were also heavily exposed to beta rays, which caused extensive radiation burns of the skin and contributed substantially to the deaths of 29 victims; internal irradiation was not clinically significant in this group. No clinical symptoms of acute radiation exposure were seen among the 135 000 members of the public evacuated from a 30 km zone around the reactor site.

Activity released to the atmosphere was dispersed around the globe. The collective dose world-wide in all subsequent years is estimated to be 6×10^5 man Sv, of which 2×10^5 man Sv will be incurred by Soviet citizens (UNSCEAR 1988). Most of this dose will be caused by caesium-137, of which 7×10^{16} Bq or a mere 22 kg was released; iodine-131 was the second most important species. For comparison, the annual collective dose from natural radiation world-wide is over 10^7 man Sv, of which 6×10^5 man Sv arises in the USSR.

At Goiânia in Brazil, a radiotherapy machine abandoned by a private medical clinic was vandalized and the source assembly removed (IAEA 1988a). It contained 5×10^{13} Bq of caesium-137 in 90 g of caesium chloride. The source was subsequently ruptured, fragments were distributed to several families, and widespread contamination was caused. About 150 persons were eventually found to be contaminated. Twenty persons needed hospital treatment. Four died, two young men, a middle-aged woman, and a girl of six: each had received gamma ray doses of around 5 Gy and the child had ingested about 10^9 Bq. Two other persons with similar doses survived. Published guidelines on the management of exposed persons were followed, including skin decontamination, caesium decorporation, treatment of the acute radiation syndrome, treatment of local radiation injuries, general sup-

Table 27.9. Reference doses for the introduction of countermeasures to protect the public after an accident

Countermeasure	Organ	Dose equivalent (mSv)	
		Lower	Upper
Early phase projected dose			
Indoor sheltering	Whole-body	5	50
	Single	50	500
Stable iodine issue	Thyroid	50	500
Resident evacuation	Whole-body	50	500
	Single	500	5000
Intermediate phase, first year dose			
Foodstuff control	Whole-body	5	50
	Single	50	500
Resident relocation	Whole-body	50	500
Recovery phase			
No numerical criteria for return to normality			

Source: ICRP (1984b).

port, and psychotherapy. Over 40 homes required to be evacuated for decontamination and seven were demolished. Garden soil was also removed. Many vehicles and public places were also contaminated. About 3500 m³ of radioactive waste was created. Some 550 workers took part in the clean-up operation.

Supranational and national agencies have published recommendations for handling the physical and medical problems arising from nuclear accidents. In developing principles for emergency planning, ICRP recognizes three phases after a serious accident (ICRP 1984b).

1. The early phase extends from the recognition of a potential release of radioactive material to the general environment to a few hours into the release. Countermeasures such as sheltering indoors, the issuing of stable iodine tablets, and the evacuation of the local residents most likely to be at risk may be indicated.

2. The intermediate phase extends from a few hours to a few days after the onset of the accident. It is assumed that virtually all the release to the atmosphere has occurred and that significant amounts of radioactive material have been deposited on the ground. Measurements and assessments will have been made. Countermeasures such as the control of foodstuffs and the relocation of nearby residents may be indicated.

3. The recovery phase commences with the first measures to return to normal living conditions and to withdraw the early and intermediate countermeasures. A substantial decontamination programme over a prolonged period may be required.

For the early and intermediate phases, ICRP identifies reference levels of dose for introducing countermeasures: these are summarized in Table 27.9. Because of the range of circumstances which might obtain, ICRP recognizes that some discretion is required regarding intervention. This may be exercised between the lower and upper values: below the lower level, however, countermeasures are not warranted, whereas above the upper level they are almost certain to be. All values refer to the doses that would be incurred without countermeasures either during the early phase or in the first year after the intermediate phase. Numerical criteria are not set for the recovery phase: ICRP remends a considered assessment to identify the optimum dose. It recognizes, however, that the society affected by an accident may demand more stringent standards of protection on the day. National authorities have adopted and developed these recommendations.

Since the Chernobyl accident, there has been an increased emphasis on emergency preparedness. In the UK, arrangements for responding to nuclear emergencies (HSE 1990) are complemented by a national response plan for nuclear accidents overseas (DOE 1989) and there has been careful consideration of the tolerability of risk from nuclear power stations (HSE 1988). All UK nuclear installations are required by law to have comprehensive emergency plans and to rehearse them periodically. Their success depends on co-operative action between the operators' staff, local authorities, and emergency services. Arrangements also exist for appropriate central authorities to provide advice and assistance as required and to assume responsibility. A country-wide network of automatic monitoring stations has been set up to detect activity from overseas accidents, and this would be strengthened by extra monitoring should an accident occur. Elaborate arrangements have been made for data collection and analysis, for technical coordination of government and supporting agencies, and for public information.

Recommendations for the medical management of persons who may have been over-exposed to radiation or radioactive material are made by ICRP (1978b) and by the International Atomic Energy Agency (IAEA, 1988b). A range of medical assistance is required from the simple provision of first aid at the site of an accident to specialized treatment at a major medical centre. Early assessment of the severity of exposure is essential and will determine the course of action. The case of patients with contamination is more complicated than for those exposed to external radiation, and other serious injuries complicate matters further (Baranov *et al.* 1989; Gale and Reisner 1988; *Lancet* 1988). Operators of nuclear installations are required to have appropriate arrangements with the health authorities for the treatment of injured persons.

Summary

Radiological hazards are ubiquitous and radiological protection is a paradigm of protective care. Everyone is irradiated by natural sources and many also receive medical exposure. Everyone is slightly exposed to weapons fall-out and many to nuclear discharges. Some work with radiation and radioactivity. Radon in the home is the major source of human exposure.

It is necessary to protect persons from the acute and late effects of radiation. High doses delivered quickly can lead to manifest injury and death; the late effects of principal concern are the induction of malignancies and genetic defects. Quantitative estimates of risk are made for late effects under everyday circumstances of exposure and are constantly refined by international consensus. These are often challenged by antinuclear and pronuclear propagandists because the values are inferred from unusual circumstances. There is also much debate about the likelihood of nuclear accidents.

An elaborate system of radiological protection has been developed over several decades; most is in legal form. Strict control of sources and discharges and of doses to workers and members of the public is specified. Users of sources must obtain prior authorization and monitor the effectiveness of protective procedures. Regulation is facilitated by the ease with which radiation and radioactivity can be measured. Recommendations for radiological protection are constantly reviewed in a structured way at national and international level.

It would be preposterous to claim that radiological protection is perfect. Information on levels and effects is incomplete, imbalances exist in the treatment of different sources, and untoward incidents do occur. In the main, however, practitioners have achieved a high degree of protection for workers and the public.

References

Baranov, A., Gale, R. P., Guskova, A., *et al.* (1989). Bone marrow transplantation after the Chernobyl nuclear accident. *New England Journal of Medicine* **321**, 205.

Breimer, L.H. (1988). Ionizing radiation-induced mutagenesis. *British Journal of Cancer* **57**,6.

CEC (1980). Council Directive of 15 July 1980 amending the Directive laying down the basic safety standards for the health protection of the general public and workers against the dangers of ionizing radiation. 80/836/Euratom. *Official Journal of the European Communities*, L 246 Vol. 23, Luxembourg (see also L265, Vol. 27, 1984).

CEC (1987). *Exposure to natural radiation in dwellings of the European Communities.* Commission of the European Communities, Luxembourg.

Clarke, R.H. and Southwood, T.R.E. (1989). Risks from ionizing radiation. *Nature (London)* **338**, 197.

DOE (1988a). *The householders' guide to radon.* Department of the Environment, London.

DOE (1988b). *Building Regulations precautions against radon in new housing.* Department of the Environment, London.

DOE (1989). Nuclear accidents overseas. Department of the Environment, London.

Gale, R.P. and Reisner, Y. (1988). The role of bone–marrow transplants after nuclear accidents. *Lancet* **i**, 923.

Hall, E.J. (1988). *Radiobiology for the radiologist*, 3rd edn. Lippincott, Philadelphia, Pennsylvania.

HSE (1988). *The tolerability of risk from nuclear power stations.* HMSO, London.

HSE (1990). *Arrangements for responding to nuclear emergencies.* HMSO, London.

IAEA (1986). *Summary report on the post-accident review meeting on the Chernobyl accident.* SS No. 75-INSAG-1. International Atomic Energy Agency, Vienna.

IAEA (1988*a*). *The radiological accident at Goiânia*. International Atomic Energy Agency, Vienna.

IAEA (1988*b*). *Medical handling of accidentally exposed individuals*. SS No. 88. International Atomic Energy Agency, Vienna.

IARC (1988). *Evaluation of carcinogenic risks to humans. Man-made mineral fibres and radon*, Monograph Vol. 43. International Agency for Research on Cancer, Lyon.

ICRP (1977). *Recommendations of the International Commission on Radiological Protection*, Publication 26. Pergamon Press, Oxford.

ICRP (1978*a*). *Limits for intakes of radionuclides by workers*, Publication 30. Pergamon Press, Oxford.

ICRP (1978*b*). *The principles and general procedures for handling emergency and accidental exposures of workers*, Publication 28. Pergamon Press, Oxford.

ICRP (1982*a*). *General principles of monitoring for radiation protection of workers*, Publication 35. Pergamon Press, Oxford.

ICRP (1982*b*). *Protection of the patient in diagnostic radiology*, Publication 34. Pergamon Press, Oxford.

ICRP (1984*a*). *Principles for limiting exposure of the public to natural sources of radiation*, Publication 39. Pergamon Press, Oxford.

ICRP (1984*b*). *Protection of the public in the event of major radiation accidents: Principles for planning*, Publication 40. Pergamon Press, Oxford.

ICRP (1985*a*). *Principles of monitoring for the radiation protection of the population*, Publication 43. Pergamon Press, Oxford.

ICRP (1985*b*). *Radiation protection principles for the disposal of solid radioactive waste*, Publication 46. Pergamon Press, Oxford.

ICRP (1987). *Lung cancer risk from indoor exposures to radon daughters*, Publication 50. Pergamon Press, Oxford.

ICRU (1980). *Radiation quantities and units*, Report 33. International Commission on Radiation Units and Measurements, Washington DC.

IRR (1985). *The Ionizing Radiations Regulations 1985*. HMSO, London.

Lancet (1988). Brazil: radiation accident *Lancet* **i**, 463.

Land, C.E. (1980). Estimating cancer risks from low doses of ionizing radiation. *Science* **209**, 1197.

Muirhead, C.R. and Darby, S.C. (1987). Modelling the relative and absolute risks of radiation-induced cancers. *Journal of the Royal Statistical Society, Series A* **150** (2), 83.

NAS (1988). *Health risks of radon and other internally deposited alpha-emitters*, BEIR-IV. National Academy Press, Washington, DC.

NCRP (1987). *Ionizing radiation exposure of the population of the United States*, NCRP Report No. 93. National Council on Radiation Protection and Measurements, Bethesda, Maryland.

NRPB (1985). *Exposure to ionizing radiation of pregnant women: advice on the diagnostic exposure of women who are, or who may be, pregnant*, ASP 8. National Radiological Protection Board, Chilton.

NRPB (1987*a*). *Committed dose equivalent to selected organs and committed effective dose equivalent from intakes of radionuclides*, NRPB-GS7. National Radiological Protection Board, Chilton.

NRPB (1987*b*). *Interim guidance on the implications of recent revisions of risk estimates and the ICRP 1987 Como statement*, NRPB-GS9. National Radiological Protection Board, Chilton.

NRPB (1988*a*). *Radiation exposure of the UK population — 1988 review*, NRPB-R227. National Radiological Protection Board, Chilton.

NRPB (1988*b*). *Health effects models developed from the 1988 UNSCEAR report*, NRPB–R226. National Radiological Protection Board, Chilton.

NRPB (1988*c*). *Assessment of the radiological impact of disposal of solid radioactive waste at Drigg*, NRPB-M148. National Radiological Protection Board, Chilton.

NRPB (1990). *Documents of the NRPB*, Vol. 1, No. 1. National Radiological Protection Board, Chilton.

Pochin, E.E. (1983). *Nuclear radiation: risks and benefits*. Clarendon Press, Oxford.

Pochin, E.E. (1988). Radiation and mental retardation. *British Medical Journal* **297**, 153.

UNSCEAR (1977) *Sources and effects of ionizing radiation. 1977 report to the General Assembly, with annexes*. United Nations, New York.

UNSCEAR (1982). *Ionizing radiation: sources and biological effects. 1982 report to the General Assembly, with annexes*. United Nations, New York.

UNSCEAR (1988). *Sources, effects and risks of ionizing radiation. 1988 report to the General Assembly, with annexes*. United Nations, New York.

USEPA (1986). *A citizen's guide to radon*. US Environmental Protection Agency, Washington, DC.

28

Hazards of drug therapy

WILLIAM H. W. INMAN

Introduction

In 1930 the *British pharmacopoeia* listed a mere 36 synthetic drugs (Wells 1980). With very few exceptions, the cure or control of serious diseases was virtually beyond the reach of most patients, and doctors could offer only reassurance and palliatives. In the 1930s, however, the health of our society began to be transformed by the dramatic achievements of a rapidly expanding pharmaceutical industry. Sulphonamides in the 1930s, penicillin, tetracycline, and other chemotherapeutic agents in the 1940s, poliomyelitis vaccine in the 1950s, and many other similar discoveries removed much of the fear of early death from infectious disease. With the introduction of streptomycin and para-aminosalicylic acid, the mortality rate from tuberculosis fell from about 550 per million in 1946 to 12 per million in 1980. Diphtheria virtually disappeared in the 1950s, together with epidemic poliomyelitis, and pneumonia or other infections became a comparatively uncommon cause of death except in the very old or debilitated.

From the 1950s, major advances were also achieved in the treatment of chronic diseases such as hypertension or rheumatoid arthritis, and as a result of the introduction of powerful psychotropic agents there were large reductions in mental hospital admissions.

Longevity and well-being, shorter periods of illness, relief of symptoms, greater mobility, and huge savings in the cost of health care have been achieved at remarkably little cost in terms of the drugs themselves.

Survival beyond the age of 90 years is frequent and the pattern of drug use is changing, with increasing emphasis on the treatment of degenerative diseases associated with old age. In 1986, per caput expenditure on drugs prescribed by general practitioners in England and Wales averaged £34. An average of 6.5 prescriptions were issued for each person (Prescription Pricing Authority 1987).

Progress, however, cannot be achieved without risk. All effective medicines are toxic to some individuals, and some major disasters have occurred, although fortunately they have been rare.

Some of the earliest organized attempts to enquire into drug toxicity have been reviewed by Wade (1970). Frequently there was a long interval between the introduction of a drug and the recognition of its dangers. For example, sudden deaths attributed to chloroform, which was discovered in 1831, were first investigated by the British Medical Association in 1877, and this drug was not replaced by the much safer agent, halothane for more than a century. Agranulocytosis was attributed to amidopyrine in 1933, 44 years after its first use as an analgesic and antipyretic agent. A number of serious accidents have occurred with vaccines: 191 shipyard workers in Breman developed jaundice after smallpox vaccinations in 1883, 12 children died from contaminated diphtheria antitoxin in Bundaberg. Australia, in 1928, and more than 50 infants died from virulent tuberculosis after a faulty batch of BCG had been used in Lubeck in 1930. A batch of Salk polio vaccine produced by Cutter Laboratories in the USA in 1955 was responsible for paralytic poliomyelitis in several hundred vaccinated children or their contacts.

Disasters caused by the faulty manufacture of synthetic drugs have been relatively infrequent. In 1937, 107 deaths occurred when diethylene glycol was used as a solvent in a sulphanilamide mixture, in spite of the fact that its toxic properties had been recognized at least 16 years earlier (Geiling and Cannon 1938). In 1954, a pharmacist in a small town near Paris prepared an organic tin preparation for the treatment of boils, poisoning 217 people of whom 100 died.

None of these accidents, however, rival the thalidomide experience of 1961 in bringing to public attention the problem of unexpected drug toxicity, and stimulating governments to regulate drug production and to set up systems for monitoring adverse drug reactions. Approximately 6000 abnormal babies were born in Germany, 500 in the UK, and smaller numbers in other countries. In the UK, the Committee on Safety of Drugs (Dunlop Committee) was established in 1964, and a Medicines Act, to control the licensing of drugs was passed in 1968. In 1971, this Act became effective and the work of the Dunlop Committee was continued by a reconstituted Committee on Safety of Medicines (CSM). The main responsibility for monitoring the effects of marketed drugs rested on a Subcommittee on Adverse Reactions, whose activities from 1964 to 1980 have been reviewed by Inman (1980). From January 1982 this body merged with other subcommittees under the new title of Safety Efficacy and Adverse Reactions (SEAR). To a considerable extent, the identification of hazards has depended on the collection of anecdotal accounts of adverse drug reactions. Similar bodies have been established in more than 20 countries, and the World Health Organization has a central collecting office for reports derived from many national centres, currently located in Uppsala.

The arrangements that have been made to monitor drug safety after the thalidomide affair have not prevented further accidents entirely. The two most notable examples are the outbreak of subacute myelo-optic neuropathy affecting several thousand patients, mostly Japanese, who used clioquinol as an antidiarrhoeal agent (Kono 1978), and the delayed recognition of the adverse effects of practolol, used to treat heart disorders and hypertension, in many patients in the UK and other countries. This latter accident led to a general recognition of the need to develop improved methods of post-marketing surveillance, and to the establishment, in 1980, of a national scheme for post-marketing surveillance by the Drug Surveillance Research Unit at Southampton University (Inman 1981a, b).

Adverse drug reactions

Most pharmacologists recognize two kinds of adverse drug reactions, sometimes described as type A and type B reactions. Type A reactions may be explained by the pharmacological properties of the drug; they are often dose dependent, and should be identified during the course of the early laboratory and clinical studies of a new product. Type B reactions are unrelated to the drug's pharmacology and are not usually detectable during the normal pharmacological screening process. They tend to be rare and are often more serious than reactions of the first type. Type B reactions are not usually dose dependent but they may occur in certain types of individuals with acquired, or genetically determined, susceptibility; more often perhaps, they are idiosyncratic and quite unpredictable. They are occasionally caused not by the drug itself, but by excipients, such as colouring agents, included in the finished medicine.

Although the above distinction is generally satisfactory, there is almost certainly a need for a third category—type C reactions—to describe events that cannot be distinguished clinically, pathologically, or in laboratory tests from diseases or symptoms that can occur spontaneously in the absence of drug treatment. These reactions may be identified because of an increase in the incidence of a disease after the introduction of a new drug. Thrombosis in women using oral contraceptives is a classic example of a type C event.

To some extent, the method of study required to identify and measure the incidence of adverse reactions is linked to this classification. Type A reactions tend to be common and mild, and should be detected in relatively small numbers of patients during the course of clinical trials. Type B reactions

may be very serious, but because of their rarity may be brought to light only in very large studies or by spontaneous anecdotal reporting. Type C events invariably require epidemiological methods for their evaluation.

Post-marketing surveillance

In order to secure a licence for marketing, a new drug has to pass a series of rigorous toxicity tests in laboratory animals. It has to be formulated in such a way that it is suitable for use in human patients, and it has to satisfy exhaustive pharmaceutical quality assurance requirements. It is then subjected to clinical pharmacological investigations which include dose-ranging studies, screening for toxic effects, and the study of its metabolism and excretion. This work is frequently done in human volunteers. Later it is tested in clinical trials among patients suffering from the disease for which the therapeutic benefit is expected. Later still, if the results are promising, the trials are expanded, and the new drug may be compared with established products in order to determine its relative efficacy and safety and to provide a platform for subsequent commercial development. If the results of laboratory and clinical studies are deemed by the national drug regulatory agency to be satisfactory, a licence is granted, and the new drug is marketed.

The whole procedure of laboratory and clinical testing may take as long as 10 years, and only a very small proportion of the new chemical entities prepared by pharmaceutical chemists actually find their way on to the market. Laboratory animals frequently fail to predict toxic effects, which appear later in human trials, or they may identify dangers which preclude subsequent experimentation in humans. It is possible that a number of potentially useful medicines are rejected because of toxicity in animals which would not occur in humans. The clinical trails, usually in a few hundred patients, will enable the drug's efficacy to be assessed, but there are severe limitations to their ability to reveal any but the most common side-effects or adverse reactions. If, for example, the true incidence of a particular side-effect is 1 in 1000, the chance of encountering a single case in a trial of 100 patients is only 1 in 10. Even if we encountered one case during 10 trials of this size, it is more than likely that we would fail to recognize it as a drug effect. All we can reasonably assume when a new drug is first marketed is that no unacceptable hazard has been identified during its pre-marketing study. There can be no assurance that serious effects will not occur once the drug is used on a wide scale. This uncertainty arises partly because the number of patients that it is practicable to include in a clinical trial is small, partly because the trial is usually of a limited duration, and partly because the trial may have deliberately excluded certain types of patient who might be unduly sensitive to the drug—for example, pregnant women, children or the elderly.

Post-marketing surveillance is thus essential for all new drugs. A wide variety of techniques is used, ranging from spontaneous reporting of individual drug experience to highly organized intensive studies in which both the numerator—for example, adverse reaction—and the denominator—number of patients—can be estimated.

The various methods that have been developed may be listed as follows: (i) non-systematic reporting; (ii) spontaneous reporting systems; (iii) retrospective studies; and (iv) prospective studies. All these methods are non-experimental in the sense that, once a drug has been marketed, randomization of treatments is not usually possible, as it often is during the clinical trial stage.

Many classifications have been used and there is frequent confusion in terminology. It is believed, however, that all the monitoring systems that have been described can be fitted fairly easily into one or other of these four categories.

Non-systematic reporting

Many suspected adverse drug reactions have been reported for the first time in the correspondence columns of medical journals. Most frequently these take the form of abbreviated case histories. Individual case reports are often of doubtful reliability. Once published they cannot be removed, and they are often given undue weight by the popular press. Great care is needed if they are to be used as the basis for regulatory decisions or medico-legal actions. Before publication, journal editors sometime refer anecdotal reports to monitoring centres and to manufacturers for expert opinion.

Spontaneous reporting systems

A number of the industrialized nations operate spontaneous reporting systems. Doctors are invited to submit reports of suspected adverse drug reactions to a national register, usually organized by a governmental health department. Many manufacturers also keep their own registers, and in some countries special registers have been set up to deal with certain types of reaction. For example, in 1975, a National Register of Drug-Induced Ocular Side-Effects was funded by the Food and Drug Administration and established at the University of Arkansas, and a File of Adverse Reactions to the Skin was established in 1977 at the Free University of Amsterdam (Inman 1980).

In the UK, the 'yellow card' system was set up by the Committee on Safety of Drugs (later Medicines) in 1964, and its Register of Adverse Reactions now contains about 200 000 individual reports. The reports are sent to the Committee on the understanding that they will be treated with complete professional confidence and that they will never be used for disciplinary purposes or for enquiries about prescribing costs. Doctors are requested to report their suspicions rather than restrict their reports to reaction that they are reasonably certain were caused by drugs, the motto being 'when in doubt, report' (Committee on Safety of Drugs 1968).

Spontaneous reporting systems are an effective method for generating hypotheses but rarely provide material for testing them. 'Signals' are recognized either because reports of a particular reaction seem to be unexpectedly frequent or because they are of a very serious or unusual nature. The numerator (number of adverse drug reactions) is always incomplete because of under-reporting and the denominator (number of patients being treated) is usually unknown. A secondary denominator obtained from estimates of tablet sales is often available, however, and a rough approximation of the true denominator may sometimes be derived by dividing the number of tablets by the average daily dose and then by the average number of days' treatment that most patients receive. Other valuable clues may be obtained by observing the concurrence of two or more reactions—for example, rash and leucopenia—in the same patient, or the receipt of a cluster of reports of a condition that is normally thought to be very rare. Once the reports have been processed, it is also possible to use simple analytical techniques which greatly increase the sensitivity of the system. Probably the most useful of these techniques in the adverse drug reaction profile.

Adverse drug reaction profiles

Lists of reactions that have been reported to a number of therapeutically related drugs are grouped according to the systems affected, for example skin, gastro-intestinal, vascular, and so on. The sum of the reactions in each group is expressed as a percentage of the total number of reports for the drug. The group percentages are then compared, one drug with another. This procedure takes care of the fact that the actual number of adverse drug reactions reported may differ widely for different drugs, and it also enables the profiles or patterns of reactions to be compared in the absence of precise estimates of the denominator (Inman 1972).

Prepared in this way, the data may be set out most conveniently in the form of histograms, so that the profiles of therapeutically related drugs may be compared and contrasted.

Figure 28.1 shows some of the adverse reaction profiles for non-steroidal anti-inflammatory drugs. There are similarities between some chemically related drugs and some marked differences in the pattern of reaction to others. For example, it can be seen that the profiles for phenylbutazone and oxyphenbutazone are remarkably similar, with a large relative excess of blood disorders (which appear to be uncommon with the other 14 drugs). With fenbufen, benoxaprofen, alclofenac, and fenclofenac, skin reactions predominate, although with diclofenac, a close relative of the last two, the proportion of skin reactions is low. Gastro-intestinal disturbances, on the other hand appear to be prominent with nearly all these drugs, although in some cases they are overshadowed by other types of reaction, particularly skin reactions. It is important to recognize that these profiles are easily distorted by selective reporting after adverse publicity.

Use of spontaneous adverse drug reaction reports
Thrombo-embolism and the contraceptive pill

The first reports of thrombosis associated with oral contraceptives reached the Committee in May 1964 and by the end

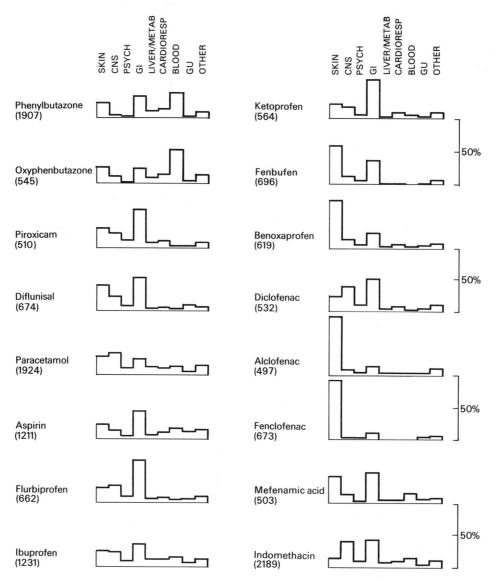

Fig. 28.1. Adverse drug reaction profiles of 16 non-steroidal anti-inflammatory drugs. The number of patients is given in parentheses. (Source: Committee on Safety of Medicines 1981, unpublished data.)

of December that year the total number of reports was 29 (Inman 1970). By 31 August 1965, 16 fatal and 95 non-fatal cases with thrombosis or embolism had been reported. At this time it was estimated that some 400 000 women, aged between 15 and 44 years, were using oral contraceptives. Table 28.1 shows how this number of reports would compare with the number expected in a population of women of this size, based on the Registrar General's report and on the data derived from the Hospital In-Patient Enquiry (Inman and Vessey 1968). There are no large differences between the number of cases reported to the Committee on Safety of Drugs and the number that would have been expected. If it could have been assumed that every case of thrombo-embolism in a contraceptive pill user had been reported, then the contraceptive pill was probably not a cause of thrombosis, but little confidence could be placed in this assumption. It

seemed probable that the reports particularly of pulmonary embolism, represented only a fraction of cases that had actually occurred.

This prompted the Adverse Reactions Sub-Committee to conduct the first study to produce statistically significant evidence that the contraceptive pill probably was a cause of thrombosis. This was a case–control study based on all deaths of women of childbearing age in England and Wales during 1966. By September, a statistically significant excess of contraceptive pill use among women dying from pulmonary embolism had been noted. In order to avoid causing public alarm, publication of this evidence was deferred for more than a year while the results were confirmed in two other studies—one small study of venous thrombosis by the Royal College of General Practitioners set up in September 1966, and a study of hospital patients by the Medical Research

Council set up in December of that year. A joint publication by the three groups appeared in 1967 (Inman and Vessey 1968; Medical Research Council 1967; Royal College of General Practitioners 1967; Vessey and Doll 1968). The Committee on Safety of Drugs was probably not given the credit it deserved for this first demonstration of a significant effect of oral contraceptives on the incidence of thrombo-embolism, but it is worth noting that the yellow card system had produced the strongest of the early signals and that the Committee on Safety of Drugs had not only set up the first of the case–control studies, but produced the earliest statistically valid evidence obtained in any study throughout the world. Moreover, it was largely because of the Committee's work that the other groups were encouraged to investigate this problem.

We have already seen how adverse reaction profiles may be constructed for the various drugs. It is also possible to construct drug profiles for the various reactions. In August 1966, one month before statistical evidence of an association between the contraceptive pill and thrombosis appeared in the report of the Committee's case–control study, an examination of the yellow cards had revealed an unexpected distribution of the reports of thrombosis associated with various brands of oral contraceptive preparations. By that time, some 940 reports of adverse reactions had been received, and of a total of 88 reports of pulmonary embolism, 63 (72 per cent) were associated with products containing mestranol, while only 25 (28 per cent) were associated with products containing ethinyloestradiol. The share of the UK market for the two types of oral contraceptive was roughly equal (Table 28.2) and, provided there were no differences between the two hormones, it would have been reasonable to expect that adverse reaction reports would be distributed in more or less the same way. Remarkably, when all other reactions were compared, their distribution was identical to that of the two hormones in the oral contraceptive market.

It was appreciated that products containing mestranol generally contained larger doses than those containing ethinyloestradiol. But, because mestranol was a weaker oestrogen, it was thought at this stage that the association with

Table 28.1. Observed and expected cases of thrombo-embolism in women using oral contraceptives (reports to the Committee on Safety of Drugs for year ending 31 August 1965)*

Diagnosis	Non-fatal		Fatal	
	Observed	Expected†	Observed	Expected‡
Pulmonary embolism	46	25	8	2
Cerebral thrombosis	32	16	2	2
Myocardial infarction	15	16	5	9
Mesenteric thrombosis	2	1	1	—
All reports	95	58	16	13

* Based on estimated population of 400 000 women aged 15–44 years using oral contraceptives.
† Derived from estimates from the Hospital In-Patient Enquiry.
‡ Derived from the Registrar General's annual report.

Table 28.2. Reports of pulmonary embolism in women using different oestrogens

	Products containing mestranol	Products containing ethinyloestradiol
Share of UK market (per cent)	52	48
Number of reports of pulmonary embolism (per cent)	63 (72)	25 (28)
Number of other suspected reactions (per cent)	441 (52)	413 (48)

Table 28.3. Relative risks of thrombo-embolism related to oestrogen dose-levels

Oestrogen dose (µg)	Mestranol				Ethinyloestradiol		Number of reports
	150	100	75–80	50	100	50	
Venous thrombo-embolism	2.4	1.6	1.2	1.5	2.5	1.0	780
Cerebral thrombosis	3.9*	1.4	0.8*	0.6*	—	1.0	79
Coronary thrombosis	3.0*	1.4	0.3*	1.3*	2.1*	1.0	61

* Values based on less than ten reports.

thrombo-embolic disease might be attributable to some subtle chemical difference between the two forms of oestrogen.

Since no clear evidence of thrombogenicity had at that time been obtained, no further action was taken until considerably more data had been accumulated, a process which took 3 years. In addition to reports accumulated by the Committee on Safety of Drugs, the Swedish and Danish Government drug monitoring centres provided data from their national registers, together with estimates of sales in those countries. In November 1969, an analysis of reports of thrombosis associated with the use of a large number of different oral contraceptive preparations revealed that there was a highly significant trend in relative risks, related, not to the two individual oestrogens, but to the total dose of oestrogen contained in each tablet. These data are shown in Table 28.3 in which the estimates of relative risk are shown for each dose level, with 50 μ taken as the unit dose for comparative purposes (Inman *et al.* 1970). Later work on cases reported to the Committee has suggested that the dose of progestogens may also be a factor contributing to the risk (Meade *et al.* 1980).

Halothane jaundice

While it had been recognized for several years that patients receiving halothane on more than one occasion might have a greater risk of developing post-operative jaundice, many anaesthetists assumed that this was simply because such a high proportion of the population had been anaesthetized with this agent. Close examination of the cases reported to the Committee confirmed beyond reasonable doubt that many were caused by halothane, although the true incidence

Table 28.4. Relation between number and timing of exposures to halothane and the onset of jaundice

Number of exposures	Days between last two exposures	Mean interval between onset of jaundice and last exposure (days)
Single exposure	—	11.4 (20)*
Multiple exposures		
Two	> 28	10.5 (19)
	< 28	5.6 (67)
Three	> 28	7.4 (11)
	< 28	4.6 (40)
Four	> 28	5.5 (8)
	< 28	3.4 (17)

* Number of cases in parentheses.

of this complication has not been measured (Inman and Mushin 1974, 1978). The majority of cases had indeed been exposed to halothane more than once and the mortality rate tended to increase with the number of exposures (Table 28.4). The most important observation, however, was that the interval between the latest exposure to halothane and the onset of jaundice was significantly shorter in those patients who had been exposed most frequently. This effect was particularly striking in the subgroup of patients in whom the two most recent exposures had been within a period of less than 28 days.

The events that followed the publication of these results were somewhat traumatic to the CSM. At least 17 anaesthetists wrote to medical journals to complain about the publication. Several doubted, even at this stage, that halothane was a significant cause of jaundice. Others attacked the Committee for publishing these results on the grounds that an anaesthetist who continued to use halothane might be held responsible if jaundice did develop. Others again suggested that, although a small but definite risk seemed to have been demonstrated, other anaesthetic agents carried a greater risk than halothane which is pleasant for the patient during recovery phase, easy to handle, and very safe even in relatively inexperienced hands.

Eventually, a group of anaesthetists, which included some of the most severe critics of the jaundice hypothesis, collected another series of 203 patients. Each history was examined by a panel of hepatologists. A reduction of the latent interval—between exposure and the onset of jaundice—was found, and the authors conceded that their results confirmed the earlier findings of Inman and Mushin (Walton et al. 1976).

Practolol and the oculomucocutaneous syndrome

The CSM's experience with the spontaneous reporting of adverse reactions to practolol contrasts strongly with the two previous examples. The yellow card system totally failed to identify the serious adverse effects of practolol. This led to an examination of the reasons why adverse events are not always reported and to a reappraisal of the need to develop new methods which did not rely on spontaneous reporting.

In June 1970, after 4 years of pre-marketing study, the cardioselective beta-blocking agent practolol—used mainly for the treatment of angina and arrhythmias—was marketed by ICI under the trade name of Eraldin. During the next few years a small number of reports of skin reactions resembling lupus erythmatosus were received by the CSM together with two reports of psoriasis and a single report of conjunctivitis. By the end of 1973, there had been a number of reports of 'exfoliative dermatitis' which would normally be regarded as a very sinister reaction carrying a high mortality. In a letter to the *British Medical Journal* in May 1974, Felix et al. drew attention to patients who had developed a rash resembling psoriasis, and they later published a full account of 21 patients (Felix et al. 1974). In six patients the rash had developed within 1 month of starting treatment with practolol, and in the remaining 15 patients it had developed during periods ranging up to 26 months. Three patients had also developed ocular symptoms, including one with bilateral corneal ulceration requiring grafting. One other patient had pleural and pericardial effusions.

In June 1974, Wright, in another letter to the *British Medical Journal*, described a number of patients who had developed ocular symptoms. He subsequently published a full report on 27 patients with eye reactions, 19 of whom had also developed skin reactions similar to those described earlier. He identified this complex of reactions to practolol as the oculomucocutaneous syndrome (Wright 1975). The eye symptoms had developed after periods ranging from 6 months to 4 years, and in a number of cases the patients had also complained of deafness and tinnitus. Characteristic of the eye symptoms were diminished tear secretion leading to conjunctival damage and, in some cases, serious fibrotic destruction of the lacrimal apparatus. Five eyes perforated but most cases showed improvement once the drug was stopped.

A third manifestation, known as sclerosing peritonitis, was reported by Brown et al. (1974). Three patients had developed peritoneal thickening and adhesions leading to abdominal symptoms and had required surgery to relieve obstruction.

Up to the time of Wright's original communication, only one eye problem was reported to the CSM. The yellow card system had thus failed to identify the oculomucocutaneous syndrome. This failure resulted not from any defect in the mechanism for reporting adverse reactions but from the fact that doctors had not used the system. This, in turn, was due to their inability to link events which, individually, were not unique, with exposure to practolol.

Once the syndrome had been described in the *British Medical Journal*, the yellow cards became extremely useful; very large numbers of reports were received and this gave a complete picture of the extent of the tragedy. About 100 000 patients had been exposed to practolol for various periods, and the Committee received nearly 2500 reports. More than 600 reports had mentioned diminished tear secretion, and more than 1000 other ocular signs, including about 400 patients with corneal ulceration, a small number of whom had lost their sight. There were more than 200 reports of sclerosing peritonitis, including 23 deaths, and 370 reports of deafness. The reporting of suspected reactions to practolol, which was almost entirely retrospective, suggested that important side-effects were occurring in at least 2 per cent of the patients. The real tragedy is that association was not identified much earlier, as soon as these events occurred.

Advantages and disadvantages of spontaneous reporting

Spontaneous reporting to a central registry is potentially the most efficient and rapid method of detecting problems with drugs and, over the years, has generated large numbers of hypotheses. It does, however, tell us little about patients except that their doctor suspects an adverse event to have been caused by a drug. The greatest defect in the system is under-reporting and this occurs for two main reasons—doctors may fail to recognize events as adverse reactions or to report those that they have recognized. The first reason is excusable in many cases because many drug effects are indistinguishable from events that occur spontaneously in the absence of treatment. This is particularly true of common events (described earlier as type C reactions) and also of reactions to drugs which have been consumed by the patient without the doctor's knowledge.

The various reasons for the failure to report recognized adverse drug reactions were described in 1976 as the 'seven deadly sins' (Inman 1978). They are: (i) *complacency*—the belief that only safe drugs are allowed on the market; (ii) *fear* of involvement of litigation; (iii) *guilt* because a patient has been harmed by a doctor's prescription; (iv) *ambition* to collect and publish a series of cases; (v) *ignorance* of the requirements for reporting; (vi) *diffidence* about reporting mere suspicions; and (vii) *lethargy*.

In the autumn of 1983, for reasons which are not clear, the manual doctor and patient indexes, which were always referred to as each report reached the CSM, were shredded. From that time, except for the very rare and serious cases which are easy to trace, the CSM has had no automatic, computer-aided check for duplication of reports when, as often happens, more than one report is sent in for the same patient. This means that statements about the number of reports may be unreliable.

In spite of these defects, voluntary reporting has proved its worth on many occasions. The action that is taken by the CSM or other similar bodies is usually undramatic, amounting to little more than the modification of manufacturers' literature or the curtailment of certain indications. Occasionally, a drug has been withdrawn from the market because of the spontaneous reports. Notable examples of this more radical action were benziodarone, a drug used for the treatment of angina, and two anti-inflammatory drugs, ibufenac and benoxaprofen, all three of which were associated with an unacceptable number of reports of jaundice. Other examples have recently included nomifensine, which caused haemolytic anaemia, and zimeldine, which was associated with transverse myelitis.

Brief reports on yellow cards can be the starting point for much more detailed investigations of an individual case. In the late 1960s, the Committee on Safety of Drugs recruited a team of about 100 medically qualified field workers who interviewed doctors and obtained much more detailed information. The same team has been used on a number of occasions to conduct a variety of epidemiological studies.

Retrospective studies

Retrospective or case–control studies normally start with the identification of patients who have developed the disease of interest matched with controls who have not. Enquiries are then made about the drugs that may have been taken and which might have been responsible for the disease. They are thus distinguished from prospective studies in which groups of exposed and unexposed patients are compared and the events that occur are recorded perhaps months or even years later. In the three examples of retrospective studies which follow, the first two were designed to test a specific hypothesis and the third to generate new hypotheses.

Thrombo-embolism and the contraceptive pill

In this study, already referred to briefly above, transcripts of 499 death certificates of women aged between 20 and 44 years, who died in England, Wales, and Northern Ireland during 1966, were used to identify cases of pulmonary, cerebral, or coronary thrombosis or embolism. Eighteen certificates indicated that the thrombosis was part of a terminal illness such as cancer. Of the remaining 481 patients, 385 were successfully investigated by medically qualified field officers. After follow-up, 51 cases were excluded for various reasons, mainly because cause of death could not be confirmed. Thirty-five medical field officers took part in the study and at each interview with a general practitioner, the records of living women in the same practice were examined to determine whether or not they had been using oral contraceptives. The very simple procedure for obtaining controls was to locate the position in the doctor's file which would have been occupied by the dead women's notes and then work alternately forwards and backwards in the file until an appropriate number of controls of the right age had been selected. Nine hundred and ninety-eight satisfactory controls were available for comparison with the 334 deaths finally available for study, a ratio of almost exactly three controls for every case.

Because the controls had been obtained in the same practices in which a death had occurred, their use of oral contraceptives up to the time of death of the index case was considered to be truly representative of the population in which the index cases had occurred. The death of the index case could not have influenced doctors' prescribing for other women in their practice before that death had occurred, but in order to exclude the possibility that more oral contraceptive users had conditions which would predispose to thrombo-embolism, cases were further subdivided into three classes. Class A patients had no identifiable predisposing

Table 28.5. Observed and expected deaths of patients with no known predisposing conditions (class A)

Diagnosis	Observed deaths	Expected deaths	p
Pulmonary embolism	16	4.2	< 0.001
Coronary thrombosis	18	11.4	0.06
Cerebral thrombosis	5	1.5	< 0.01

condition, class B had predisposing conditions such as hypertension or diabetes, and class C patients were pregnant or had been delivered during the month before the terminal episode. All the case histories were reviewed by three independent assessors who did not know which women were oral contraceptive users, and it was found that there was close agreement about the presence or absence of predisposing conditions.

The use of oral contraceptives by control patients was closely linked with parity, rising from 5 per cent in nulliparous controls, to 27 per cent in controls who had had four or more babies. Twenty-one per cent of 591 controls under the age of 34 years had been using oral contraceptives, compared with only 11.5 per cent of 407 women aged 35–44 years.

The observed and expected deaths in class A for each of the three diagnostic groups, are shown in Table 28.5. Although the number of patients without predisposing conditions is small, it can be seen that there was a highly significant excess of oral contraceptive use by women dying from pulmonary embolism, and a significant excess in the small number of women dying from cerebral thrombosis. The excess in women dying from coronary thrombosis did not quite reach statistical significance.

With the help of independent estimates of oral contraceptive use, supplied by a market research company, and with data obtained from the Registrar General on the non-pregnant female population of England, Wales, and Northern Ireland in the appropriate age group (5.7 million), it was possible first to check that the oral contraceptive use by control patients was consistent with oral contraceptive sales. Second, it was possible to calculate the relative and the attributable mortality in oral contraceptive users and non-users. In women aged between 20 and 34 years, the relative risk was 7.5 and in those aged between 35 and 44 years the relative risk was 7.8 In absolute terms, the attributable mortality was very much lower among those aged 20–34 years, the excess mortality rate being 2.2 per 100 000 users aged 20–34 years and 4.5 per 100 000 users aged 35–44 years. The statistical technique employed in this particular study has been used by Bradford Hill to illustrate a general application of the principle of standardized death-rates in epidemiological studies (Bradford Hill 1971).

Fatal subarachnoid haemorrhage and the contraceptive pill

In October 1977, the Royal College of General Practitioners published evidence from their long-term study which suggested that oral contraceptives might be strongly associated with fatal cerebrovascular disease (Royal College of General

Practitioners 1977). The trial included some 43 000 women. Ten of those who had used the contraceptive pill at any time during the 8 years of the study had died from cerebrovascular disease compared with only three control subjects. Of those ten, nine had died from subarachnoid haemorrhage compared with none in the control group. These results caused considerable alarm among oral contraceptive users. A relative risk of this size in at least three million women using oral contraceptives should have produced a considerable increase in the annual mortality rate from subarachnoid haemorrhage. During the 20 years, 1957–76, the death-rate in women of childbearing age from this cause had remained remarkably constant at about three deaths per 10 000 per annum (Office of Population Censuses and Surveys 1959–78).

In order to resolve this apparent discrepancy, a case-control study was set up along almost the same lines as the study of thrombo-embolic disease. Death entries for women aged 15–44 years who died from subarachnoid haemorrhage in England and Wales in 1976 were provided by the Office of Population Censuses and Surveys (Inman 1979). To keep the numbers within reasonable bounds, all entries for women aged 15–34 years and alternate entries for women aged 35–44 years were selected for further study. As in the previous investigation, women living in the same practice areas were selected as control subjects. Allowing for failure to make contact with general practitioners and other losses for administrative reasons, this procedure yielded a total of 134 practice-matched case-control pairs for analysis. Although the matching was based only on age, it was found that, in other respects, the case and control populations were closely similar.

In 63 per cent of the cases, the diagnosis had been confirmed at necropsy. Among the remainder, it had been confirmed by angiography or computer tomography, or was considered most likely on clinical grounds. Fifty cases had a history of hypertension with or without renal disease, or of pre-eclamptic toxaemia, in contrast with only 21 control subjects.

Although many doctors had records of prescriptions for oral contraceptives, some were uncertain whether they were in use at the time of death. Current use of oral contraceptives was known definitely in 109 of the 134 matched pairs, and the diagnosis was certain in 77 of the 109. Because of the obvious importance of hypertension in the aetiology of subarachnoid haemorrhage, the data have been arranged in Table 28.6 so that like may be compared with like. The relative risk for diagnostically proved cases was 1.22 (95 per cent confidence limits 0.46–3.33). It was concluded that, while a few patients might have developed hypertension as a result of oral contraception, subarachnoid haemorrhage was not a serious cause of concern in healthy non-hypertensive women using the contraceptive pill.

The discrepancy between the Royal College of General Practitioners' results and those from other studies suggests that the apparently strong association between oral contraceptives and subarachnoid haemorrhage was likely to have

Table 28.6. Current use of oral contraceptives by 77 women with proven subarachnoid haemorrhage and their practice-matched controls in relation to hypertension

Hypertension in	Oral contraceptives currently in use by				
	Case and control	Case only	Control only	Neither	Total number of pairs
Case and control	0	0	1	2	3
Case only	0	2	2	18	22
Control only	0	0	0	1	1
Neither member of pair	8	11	9	23	51
Total number of pairs	8	13	12	44	77

Table 28.7. Use of drugs during early pregnancy by mothers of 836 abnormal babies and 836 normal control babies

Class of drug	Case and control	Case only	Control only	Neither	Total number of pairs	Discordant pairs only
Hormonal pregnancy test	20	73	35	708	836	2.09*
Hormonal support of pregnancy	2	17	10	807	836	1.70
Oral contraceptive	0	11	10	815	836	1.10
Doxylamine†	13	63	75	685	836	0.84
Promethazine‡	4	41	37	754	836	1.11
Meclozine‡	6	28	35	767	836	0.80
Benzodiazepines	2	60	42	732	836	1.43
Antibiotics	2	60	42	732	836	1.43
Barbiturates	1	27	12	796	836	2.25**

* $p < 0.01$ $\chi^2 = 5.03$.
** $p < 0.05$ $\chi^2 = 12.68$.
† One component of Debendox.
‡ Components of other antiemetics.

arisen either from chance or from some unidentified confounding factor.

Congenital abnormalities

The two preceding accounts show how the case–control approach was used to test the hypotheses that oral contraceptives might be a cause of thrombo-embolism or subarachnoid haemorrhage. In 1969, the CSM started an investigation of the drug-histories during pregnancy of women who had given birth to children with a variety of congenital abnormalities (Greenberg *et al.* 1977). It was appreciated that, although the voluntary reporting system was working well, it would be most unlikely to identify teratogens. The long gap between exposure during early pregnancy and the delivery of the baby, and the absence of information about normal pregnancies, made it extremely doubtful that sufficient reports would be received to alert the Committee to this type of hazard. Once again the case–control approach was used, but this study was designed to generate rather than test hypotheses. The selection of children was based on the register of abnormal births maintained by the Office of Population Censuses and Surveys (OPCS). The mothers' general practitioners were identified with the help of local community physicians, and then interviewed by the Committee's medical field workers. At each interview a healthy 'control' baby, who had been born within 3 months of the index case, was selected from practice records. This ensured that, even if the abnormal birth had modified the doctor's subsequent prescribing, the change could not have affected first trimester prescriptions for the mother of the control baby.

Eight hundred and thirty-six case–control pairs which met these requirements were available for analysis. They were 'practice matched' and the matching was also very close with regard to maternal age, parity, and past history of miscarriage. Fourteen per cent of the index cases and only 3 per cent of the control group, however, had a history of congenital abnormality. Eleven of the index cases were members of twin pairs and in three the twin was also abnormal (but not included in the study). One pair of conjoined twins was treated as a single individual. Neural tube defects were found in 189 (23 per cent) of the index cases, 412 (49 per cent) had

oral clefts, 59 (7 per cent) had limb reduction deformities or other defects, and 176 (21 per cent) had other abnormalities.

The distribution of some of the drugs used by the mothers is shown in Table 28.7. A surprising finding was a significant excess of exposure to hormonal pregnancy diagnosis tests by index mothers. Although the numbers were small, there was also a significant excess of use of barbiturates, mostly as anticonculsants. When case-control pairs with a family history of abnormality were removed, however, the ratio of discordant pairs remaining was 2.0, which was not significant. In view of political pressures in the UK after a court decision in the US against the manufacturer of the antiemetic Bendectin (subsequently reversed after a retrial), the small negative relative risk for the similar product *Debendox* is of interest.

As a result of this study, the Committee recommended that, since non-invasive methods for testing were freely available, and since even a small risk was unacceptable, the hormonal test should no longer be used. This study was widely interpreted by the press and consumer activists as providing firm evidence that hormone therapy was teratogenic. More than 100 papers have appeared on the role of sex hormones in congenital abnormality, but the overall balance of the evidence appears to be strongly against this hypothesis. Test cases against the manufacturers of one of the preparations have recently been withdrawn after counsel for the plaintiffs had examined the testimony of more than 20 expert witnesses who had independently examined all these papers in preparation for the defence.

Studies of this type are liable to many forms of bias. In the CSM study, the interviews with doctors were necessarily lengthy and were unpopular with both the interviewers and the general practitioners concerned. At the start of each interview, both parties knew which baby was abnormal. Consequently, less effort may have been made to ascertain control histories. There had been earlier publicity about this problem which could have led to biased selection of mothers. There was no suggestion of any effect on a specific organ.

Cases and control subjects were not matched for social class. There was no record of the reasons why patients required a pregnancy test and no details about what attempts, if any, the women might have taken to terminate unwanted pregnancies.

Case–control studies designed to generate hypotheses may be of value, provided their limitations with regard to proving causal relationships are clearly recognized. Public action should rarely be taken on the results of a single case–control study. Experience has also shown that studies in which the total number of discordant pairs is small—that is, less than 100—or in which the relative risk is small—that is, less than 5—call for exceptional care in interpretation. Confounding factors are often responsible for statistically significant differences, especially confounding by diagnosis, selection, or observer bias.

Prospective studies

Cohorts of patients for a prospective study may be assembled in several ways. They may be groups of patients sharing certain demographic characteristics, such as married women of childbearing age, or patients suffering from certain diseases, such as diabetes or hypertension, or patients taking a particular drug.

Some epidemiologists believe that the prospective approach is superior because it may be possible to eliminate the concealed biases which often confound retrospective studies. Certainly, prospective studies have the advantage that more than one beneficial or harmful effect may be studied simultaneously. They also enable both relative and absolute risks to be determined. They also, however, suffer from certain major disadvantages. The number of patients required to enable infrequent or rare events to be detected and measured may be very large and thus the study may be very expensive. Sometimes a prospective study has to be very prolonged in order to yield answers to the questions asked. Prospective studies of carcinogenicity may take 25 years, for example, and may be an unattractive proposition for a worker who expects to retire from active research before the study is completed. Except in situations in which treatment can be randomized—that is, experimental studies—and where those who observe or record the observations are unaware of which treatment is being used, even the well-known sources of bias cannot be eliminated. In a study of 46 000 women, which was started by the Royal College of General Practitioners in 1968, for example, it seems probable that some of the differences in the incidence of diseases occurring in users of oral contraceptives and control subjects who never used them may have been due to differences which were present before entry into the trial.

In spite of these problems, non-experimental prospective observational studies can be of considerable value in post-marketing surveillance, especially as a means of generating hypotheses. In western society, nearly all patients with important signs or symptoms of disease receive some form of treatment, and untreated control subjects are hard if not impossible to find. Nearly all judgements are based on relative rather than absolute differences between groups of patients receiving various types of treatment.

Prescription–event monitoring (PEM), which was introduced in England in 1980, is the subject of the remainder of this chapter. It is the largest, by a factor of at least five, of all the national schemes anywhere in the world and it allows access to all recipients of any drug that may be selected for study within the whole population of some 47 000 000 persons in England.

PEM was devised by the director of the Drug Safety Research Unit (formerly the Post-Marketing Drug Surveillance Research Unit) which was set up within the Faculty of Medicine at the University of Southampton in 1980 and which, in 1986, became an independent charity managed by the Drug Safety Research Trust (Rawson et al. 1990). The DSRU is independent of Government and industry, though funded on a 'no strings' basis by both of them. PEM is the major activity of the DSRU and it depends on the ability of the Prescription Pricing Authority to provide copies of prescriptions for selected drugs written by general practitioners.

As soon as a medicinal product containing a new chemical entity is licensed for use in general practice, the Prescription Pricing Authority (PPA) is requested, usually by telephone, to start identifying all the patients for whom the drug has been prescribed. A 'picking list' of prescription reference numbers is produced by a computer, and clerical staff then extract the relevant prescription forms by hand and photocopy them. Each prescription identifies a doctor, a patient, and one or more drugs that have been prescribed.

PEM is restricted to England, for reasons which at first sight may appear rather curious. Many prescriptions do not include the full name of the patient or the complete address. Nevertheless, the patient's surname and initials are usually sufficient to allow a general practitioner to locate the appropriate medical records rapidly. In Wales and Scotland however, an individual practitioner often has on his or her list a large number of patients with an identical surname—for example, Jones, McDonald, and so on. This would make it difficult for the practitioner to retrieve the correct notes rapidly and, indeed, with a frequently used drug, it is probable that it will have been prescribed for a number of patients with the same surname and even the same first name.

The number of patients required in the screening programme depends on the magnitude of the risk which it is intended to detect. To be 95 per cent certain of detecting a unique event affecting 1 per cent of a population would require a minimum of 300 observations. Detection at 0.1 per cent requires a minimum of 3000 observations, and so on. Since it seemed unlikely that all doctors would respond to the questionnaires, and because some medical records would be incomplete or patients impossible to trace, it was decided that the study should comprise the first 20 000 patients who receive a new medicine.

The photocopies are posted in regional batches to the

PLEASE RETAIN THIS SECTION OF THE FORM FOR YOUR RECORDS

**DRUG SAFETY RESEARCH UNIT
PRESCRIPTION EVENT MONITORING**

CONFIDENTIAL

Professor W. H. W. Inman, FRCP, FFPHM,
Bursledon Hall,
Southampton, SO3 8BA.
Telephone: (042121) 6122/3

An **EVENT** is any new diagnosis, any reason for referral to a consultant or admission to hospital, any unexpected deterioration (or improvement) in a concurrent illness, any suspected drug reaction, or any other complaint which was considered of sufficient importance to enter in the patient's notes.

Example: A broken leg is an **EVENT**. If more fractures were associated with this drug they could have been due to hypotension, CNS effects or metabolic bone changes.

Please note that the following are essential:
1. Date of birth, indication for this drug and dates of starting and ceasing treatment with this drug.
2. Details of all events even if this drug has been discontinued.
3. Reason for stopping this drug and name of any other drug substituted.
4. Date and cause of death (if appropriate).

Ref:

The DSRU is managed by the Drug Safety Research Trust, an independent charity (No. 327206) working in co-operation with the University of Southampton. Trustees: Professor Sir Douglas Black MD FRCP, Dr. D.M. Burley FRCP, Professor D.J. Finney CBE ScD FRS FRSE, Dr. G.R. Higginson FICE FI Mech E, Professor W. H. W. Inman FRCP FFPHM.

PLEASE RETURN NO-EVENT FORMS   Ref:

SEX DATE OF BIRTH / /	WAS THIS DRUG EFFECTIVE? Yes ☐ No ☐
INDICATION FOR PRESCRIBING	PLEASE SPECIFY THE REASON FOR STOPPING THIS DRUG AND THE NAME OF ANY OTHER DRUG SUBSTITUTED
DATE PATIENT STARTED THIS DRUG / /	DATE PATIENT STOPPED THIS DRUG / /

DATE	EVENTS WHILE TAKING THIS DRUG	DATE	EVENTS AFTER STOPPING THIS DRUG

IMPORTANT: PLEASE INDICATE ANY EVENT REPORTED TO CSM OR MANUFACTURER

Fig. 28.2.   The Drug Safety Research Unit's green form for prescription–event monitoring.

DSRU. The batches are then sorted into sub-batches for individual practices. More than one doctor in the same general practice may have prescribed for the patient over a period of time. The patient's name and an abbreviated address, and the date of the prescription are entered on the computer and, as each subsequent batch of prescriptions is received, they are checked against the prescriptions previously identified in the same practice. In this way, it is possible to build up a complete month-by-month record of prescribing for individual patients.

The identifying information and certain details, such as the date of the prescription, are transferred to the DSRU's green forms (see Fig. 28.2). The doctor is asked whether the drug was effective, to provide information about the indication for treatment, whether or not treatment has been continued, and any event that may have occurred. He or she is also asked to specify the reason for stopping the drug and the name of any other drug that may have been substituted. Each form carries a definition of an event and includes a simple example, which has proved to be extremely effective. The example reads: 'A broken leg is an event. If more fractures were associated with this drug they could have been due to hypotension, CNS effects, or metabolic bone changes'.

Events include any new diagnosis, any reason for referral to a consultant or admission to hospital (for example, operation, accident, or pregnancy), any unexpected deterioration (or improvement) in current illness, any suspected drug reaction, and any other complaint which was considered of sufficient importance to enter into the patient's notes. Doctors are asked to complete the form even if they have already reported a suspected reaction to the CSM or are involved in a manufacturer's study.

Green forms may include a request to provide the patient's national health registration number so that the record may be 'flagged'. When the patient dies, the OPCS can notify the DSRU of the date and cause of death. This procedure could facilitate long-term monitoring for effects of drugs on the incidence of disease and might also be helpful in comparing the efficacy of drugs. It is possible, for example, that patients treated with one drug may survive longer than those treated with a different drug.

From the general practitioners' point of view, the procedure is very simple. Having located the notes, they need only to write the indication for the drug and then list any events that have come to their attention using 'key words' copied from their own notes. They do not need to give opinions as to whether the events are related to the use of the drug.

Event reporting requires no medical judgement, and it is perfectly feasible for a practice nurse or secretary to abstract the case notes and provide the essential information. Since the doctor need not comment on causality, there is less medico-legal risk than in the reporting of a suspected adverse reaction, which implies an admission that the drug may have been responsible for damage to the patient. If practolol had been studied by PEM it is almost certain that the syndrome it caused would have identified long before 100 000 patients had been exposed to risk. PEM on selected new drugs commenced in January 1981. Nobody can predict which new drug will present an unexpected hazard and it is considered that the most sensible policy is to study all new chemical entities routinely. The main limitation is the number of drugs for which the PPA can supply prescriptions at any one time.

PEM procedure

Figure 28.3 summarizes the various procedures that constitute PEM. The main input, amounting to up to 100 000 prescriptions per month is processed by two shifts, each comprising 12 clerks. If no immediate action is required, the prescriptions may be banked either on the computer or on microfilm so that they may be used later in retrospective enquiries. Such an enquiry might, for example, be initiated after many years because of concern about possible carcinogenicity. For the great majority of drugs, however, and for all new chemical entities, all prescriptions are immediately

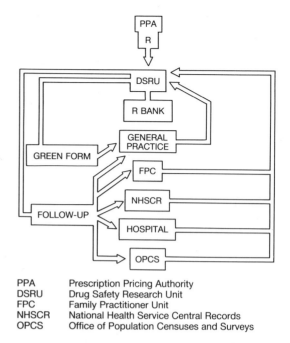

PPA	Prescription Pricing Authority
DSRU	Drug Safety Research Unit
FPC	Family Practitioner Unit
NHSCR	National Health Service Central Records
OPCS	Office of Population Censuses and Surveys

Fig. 28.3. Flow chart showing the procedures and organizations involved in prescription–event monitoring.

entered into the computer and, periodically, batches of green forms are produced and distributed to the general practitioners.

For several years the green forms were posted on the anniversary of the patient's first prescription. This had the advantage that a whole year of 'event experience' could be gathered from a single patient's records, but there was a major disadvantage. During the interval between the first prescription and the anniversary, it was possible that serious adverse reactions might occur which would not be reported to the CSM and would not appear in the PEM report for several months. For an experimental period a Red Alert Scheme was introduced in a collaborative study with the CSM. As soon as a patient was identified, a modified version of the CSM's yellow card was produced by the DSRU's computer. Each card bore a distinctive red triangle. it was personalized for the individual patient and the doctor was asked to store the card in the patient's notes, using it only if a serious adverse reaction occurred. The doctor would then send the card to the DSRU where it would be processed and immediately passed on to the CSM. The Red Alert Scheme was applied to four drugs (fluvoxamine, nabumetone, famotidine, and nizatidine). Unfortunately, many doctors misinterpreted the purpose of the scheme and used the cards to report very minor events or even to record the fact that no adverse events had occurred. In September 1988 we substituted an 'Express PEM' scheme using a conventional green form which was posted immediately after the patient had been identified, usually within 3 months of the issue of the initial prescription. The second stage of this scheme has not yet been initiated but will involve posting the same green

Table 28.8. Drugs studied by PEM (various methods from 1980 to December 1988).

Drug	Main indication	Month/year first prescriptions sent to DSRU	Drug	Main indication	Problems studied
Hypothesis-generating PEM			Hypothesis-testing PEM		
Benoxaprofen	Arthritis	4/1981	Emepronium	Incontinence	Oesophageal uleration
Fenbufen	Arthritis	4/1981	Erythromycin	Antibiotic	Jaundice
Ranitidine	Peptic ulcer	1/1982	Enalapril	Hypertension, heart faiure	Anaphylaxis
Piroxicam	Arthritis	10/1982	Mianserin	Anxiety/depression	Blood disorders, suicide risk
Flunitrazepam	Insomnia	1/1983			
Alprazolam	Anxiety	4/1983	Amitriptyline	Anxiety/depression	Blood disorders, suicide risk
Zomepirac	Pain, any cause	5/1983			
Indomethacin[1]	Arthritis	12/1983			
Diltiazem	Angina	3/1985	Prescription bank		
Betaxolol	Hypertension	4/1985	Nadolol[4]		
Enalapril	Hypertension, heart failure	6/1985	Labetalol[4]		
Acyclovir	Herpes	3/1986	Indoramin[4]		
Etodolac	Rheumatoid arthritis	3/1986	Indoprofen[5]		
Nicardipine	Angina	1/1987	Cimetidine[6]		
Gemfibrozil	Hyperlipidaemia	1/1987	Phenylbutazone[7]		
Nedocromil	Asthma	1/1987	Oxyphenbutazone[7]		
Terodiline	Incontinence	1/1987			
Fluvoxamine[2]	Anxiety	4/1987			
Nabumetone[2]	Arthritis	5/1987			
Terazosin	Hypertension	8/1987			
Nizatidine[2]	Peptic ulcer	11/1987			
Famotidine[2]	Peptic ulcer	1/1988			
Auranofin	Rheumatoid arthritis	1/1988			
Buspirone	Anxiety	3/1988			
Bisoprolol	Hypertension	5/1988			
Lisinopril[3]	Hypertension, heart failure	8/1988			
Xamoterol[3]	Heart failure	10/1988			
Fluconazole[3]	Candidiasis	11/1988			
Misoprostol[3]	NSAID-induced ulceration, prophylaxis	12/1988			
Ciprofloxacin[3]	Antibiotic	12/1988			
Tenoxicam[3]	Arthritis	12/1988			

[1] Slow-release preparation Osmosin.
[2] Included in Red Alert Scheme.
[3] Included in Express PEM scheme.
[4] A comparison of these hypotensive drugs was abandoned when one failed to 'sell'. Also retained because of remote possibility of oculomucocutaneous syndrome (cp. practolol).
[5] Withdrawn from market before PEM started.
[6] Suspicion of carcinogenicity.
[7] Retained for possible long-term study of blood dyscrasias.
PEM, prescription–event monitoring.
DSRU, Drug Safety Research Unit.
NSAID, non-steroidal anti-inflammatory drug.

form back to the doctor approximately 12 months after the issue of the first prescription requesting an update.

After processing the green forms, a number of steps may be taken to investigate 'signals' suggesting a possible hazard. Table 28.8 lists the drugs that have been studied by various methods from June 1980 to December 1988. Thirty-one drugs have been studied prospectively, 18 of the studies being active at December 1988. Four were included in the Red Alert Scheme and the six most recently introduced drugs were included in Express PEM. Five drugs have been included in hypothesis-testing studies, and for seven others the prescriptions have been 'banked' so that cohorts could be assembled for retrospective study should the need ever arise.

Direct approach to the general practitioner

This procedure is frequently followed when a group of serious or unusual events is reported which might be related in some way to the use of the particular drug. Reports of convulsions, for example, would be followed up routinely to establish whether the patient was a known epileptic or had some other predisposing condition such as a recent stroke or a brain tumour.

Family Practitioner Committee

If a patient has died, the general practitioner is asked routinely to give permission for an approach to be made to the Family Practitioner Committee, since the patient's notes will have been returned shortly after death. Almost invariably, permission is granted and the lifetime medical records of the patient are then made available to the DSRU. The information obtained in this way is proving to be of great value, not only in the investigation of drug safety problems, but also in the collection of data of general epidemiological value. It is normally difficult to obtain information about the incidence of disease in various subpopulations. It is possible, as a by-product of PEM, for example, to determine the rate of consultations for arthritic problems in patients suffering from diabetes or the incidence of heart failure in patients with arthritis or in any other subgroup characterized by a particular disease for which treatment has been prescribed. Data obtained during the course of these studies are of great value for comparative purposes.

Hospital

Where a serious event has occurred which has resulted in hospitalization, the general practitioner is often asked for the name of the hospital and the consultant so that their records may also be examined.

Office of Population Censuses and Surveys (OPCS)

As a routine, a copy of a patient's death certificate is obtained from the OPCS. Arrangements can also be made for 'tagging' populations of patients so that the cause of their ultimate death may be ascertained. Thus the effects of a drug on longevity or the incidence of cancer could be estimated.

Hypothesis testing

In addition to its use as an hypothesis-generating system, PEM can, with very slight modification, be used to test hypotheses. There have been a number of examples, notably the use of PEM to establish the role of emepronium bromide in oesophageal ulceration, the relationship between jaundice and various salts of erythromycin, and an unsuccessful attempt to measure the incidence of blood dyscrasias with mianserin and amitriptyline.

Emepronium bromide and oesophageal symptoms

In 1983, because of concern that tablets of Cetiprin (emepronium bromide) might become trapped in the lower oesophagus and cause severe irritation, the DSRU conducted a study involving more than 7000 doctors and nearly 16 000 patients. The initial green form screen suggested that 447 of these patients (2.8 per cent) had experienced discomfort after swallowing the tablets.

A special four-page questionnaire was designed and information was obtained on about 280 of the patients. Causal relationship was thought to have been probable or possible in 7 of 19 severe cases and 105 of 170 mild to moderately severe cases of oesophageal irritation. No patient had died as a result of taking Cetiprin and there was no case of oesophageal perforation. This product has since been removed from the market.

Jaundice and erythromycin

A PEM study of the relationship between jaundice and the use of erythromycin yielded results which were surprisingly different from those that had been suggested by the CSM's yellow cards (Inman and Rawson 1983). Over several years there had been a large relative excess of reports on yellow cards describing jaundice in patients who had used erythro-

mycin estolate. The excess was far greater than could possibly have been accounted for by the relatively larger sales of this formulation of the antibiotic. The risk of using the estolate appeared to be as much as 20 times as great as that associated with the use of other salts.

Twelve thousand patients were studied, with a response rate of 76 per cent. Erythromycin estolate was compared with erythromycin stearate. A total of 16 patients developed jaundice and in every case the doctor was able to provide additional information. Four of the cases were attributable to gallstones, three to cancer, six to infective hepatitis, and there were three cases in which the antibiotic might be considered as a possible cause. These three had all been treated with the stearate. Although this does not rule out the possibility that the estolate may occasionally cause jaundice, there was certainly no suggestion of a large risk associated with its use and we can be reasonably confident that the incidence of jaundice with the estolate is less that 1 per 1000. The manufacturers of the estolate (Eli Lilly) had always included warnings about jaundice in their sales literature and this might well have accounted for the huge relative excess of reports to the CSM.

Mianserin and blood disorders

Because of concern that mianserin might be associated with an unacceptably high level of risk of agranulocytosis, a large-scale comparison with amitriptyline was conducted by the DSRU. (Inman 1988, 1989a). From an initial selection of approximately 180 000 prescriptions for mianserin and amitriptyline, questionnaires were posted to 14 000 general practitioners who had prescribed mianserin for 40 000 patients and to 18 000 general practitioners who had prescribed amitriptyline for 68 000 patients. No case of agranulocytosis was reported in either group.

During the course of the study questions were also asked about attempted suicide. Among 27 000 patients in the mianserin group, 301 were reported to have attempted suicide, and among 42 000 in the amitriptyline group there were 634 attempts. Ninety-three per cent of all these reports were successfully followed up. There had been no deaths among 60 patients who had attempted suicide with mianserin alone and four deaths among 169 patients who had used amitriptyline alone. More striking, however, was the fact that there had been no admissions to hospital of patients overdosed with mianserin alone, while 56 survivors of amitriptyline overdosage had been admitted to intensive care units.

Analysis of reports to regulatory agencies suggest a probable incidence of agranulocytosis in the region of 1 in 10 000 courses of treatment. Deaths from agranulocytosis are exceedingly rare, possibly only one or two per year. Nevertheless, the CSM has attempted to restrict the use of mianserin to patients under the age of 65 years who are thought to be at lesser risk of blood disorders. Such patients are, of course, also the ones who are least likely to receive multiple drug therapy. The risk of agranulocytosis is extremely small but, when it is considered that the DSRU study was based on prescriptions written in England during only 2 weeks, these results could be extrapolated to perhaps 100 deaths and 2000 admissions to hospital each year. It, therefore, seems likely that implementation of a recommendation not to use the drug that had been suspected fo causing agranulocytosis in favour of alternatives might lead to greater mortality and morbidity. It is unfortunate that the Medicines Act 1968 does not permit the CSM to consider the relative efficacy of drugs, nor does it allow them to consider the dangers of drugs used in overdosage, whether deliberate or accidental. The toxicity of drugs in overdosage must be considered to be part of the management of patients with depressive illness and it is recommended that the Medicines Act 1968 should be reviewed and possibly changed in order to allow both types of risk to be considered in the evaluation of this type of problem.

Analysis of PEM data

PEM provides estimates of the incidence of events in the form of rates per 1000 patient-years on and off treatment, and also on a month-by-month basis. Events that are related to acute drug toxicity tend to occur in the first month, and the ratio of rates for the first month and for months 2–12 may be a useful measure for these effects. Comparison of the rates of events on and off treatment is also of interest but there is an important systematic bias in this comparison, i.e. patients visit their doctors less frequently and have fewer entries in their records when they are not taking drug treatment. Also, it is not possible to account for the effects of drugs that may have been substituted for the agent being studied. It is quite usual, therefore, for events that are almost certainly unrelated to treatment to be reported slightly more frequently whilst the patient is taking treatment. The magnitude of increase is important, however, and a truly drug-related event may be several times more common during treatment than after it.

Perhaps the most satisfactory comparisons are between drugs used for similar indications. As the number of drugs studied by PEM grows, these analyses will become more important. A comparison of the rates of various events during treatment with four cardiovascular drugs studied by PEM is shown in Figure 28.4. PEM was able to identify known drug side-effects such as cough with enalapril, flushing with nicardipine, and cold extremities with betaxolol. Although all these four drugs are hypotensive agents, postural hypotension was clearly picked out by PEM as being a problem with enalapril, rather than the other three drugs, for which the rates were similar.

The characteristics of the population being studied has an important effect on the pattern of events. The rate for cardiac surgery associated with diltiazem was high, reflecting its predominant use in ischaemic heart disease. Betaxolol, which had the lowest rate of cardiac surgery, was used almost exclusively for the treatment of hypertension. A surprising finding was that the event rate for bradycardia was lower with

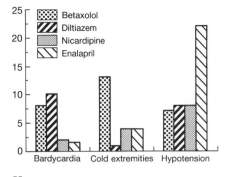

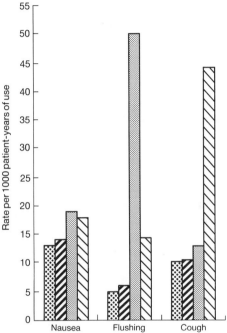

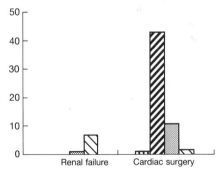

Fig.28.4. Comparison of certain events with four cardiovascular drugs. (Source: PEM studies, unpublished data.)

betaxolol than diltiazem. This beta-blocker does not have intrinsic sympathomimetic activity and would be expected to produce a much larger reduction in heart rate than diltiazem. The likely explanation for this is that, because it is an expected event, general practitioners are unlikely to record bradycardia in the notes for patients who are taking beta-blockers.

A large database such as PEM is not necessary to identify and quantitate frequent drug side-effects, since these should be identified during clinical trials. However, clinical trials have important limitations in their numbers and also tend to exclude ill patients, who may be at particular risk from adverse effects. Although it was not apparent during clinical trials or company-sponsored PMS (Cooper *et al.* 1987), PEM showed that renal failure is a rare but nevertheless important problem with enalapril (Inman *et al.* 1988). There were no cases of renal failure during treatment with the other three cardiovascular drugs, while on enalapril the rate was 7 per 1000 treatment-years. To characterize the problem, these cases were subsequently followed up in detail (Speirs *et al.* 1988).

Some general considerations

Complementary roles of spontaneous reporting and PEM

Of the many methods that have been developed, mostly since the thalidomide disaster, voluntary reporting systems and various types of intensive monitoring in hospital patients have been the most successful. Individual pharmaceutical companies have undertaken post-marketing clinical trials but, however well these may have been done, the fact that almost all of them are promotional diminishes their credibility. They are usually uncontrolled because a competitor is unlikely to agree to the use of its product for what may turn out to be an adverse comparison. Drug company post-marketing surveillance tends to be expensive, because participants are paid to collaborate.

PEM, which, except for the CSM's yellow card scheme, is the only other scheme applicable to the whole population of England (Scotland, Wales, and N. Ireland are not included in PEM), is operated by an independent, non-governmental, and non-regulatory agency, and has proved to be the most effective of all the post-marketing surveillance methods so far developed. It complements rather than replaces the voluntary reporting system, and has the advantage of being considerably less expensive than commercial schemes.

PEM, unlike voluntary reporting, cannot be applied to all drugs at the same time. Spontaneous reporting of the earliest clinical experiences is vital, but because of the publicity it attracts, it may sometimes give a distorted impression of apparent dangers. The post-marketing experience with benoxaprofen is an example of this. During its second year of use in general practice, a cluster of anecdotal reports of cholestatic jaundice appeared in the UK medical journals. Between April and June 1982, about 18 months after the drug was first marketed, 14 cases, 11 of them fatal, had been described in six letters to the *British Medical Journal* or the *Lancet* (for example see Taggart and Alderdice 1982). Remarkably, 11 of the 14 cases originated in Northern Ireland or Scotland, and only three in England. All but one of the patients were women and all were extremely elderly (average age 77 years). About half had been treated concurrently with diuretics. One doctor in Northern Ireland had reported that all of the six patients he had treated had died, an experience that was quite out of line with that of other doctors who, by that time, had probably prescribed the drug for at least half a million patients.

As a result of adverse publicity in the media, the Department of Health and Social Security decided to suspend the licence for the drug on 3 August 1982. Although intended to test the method rather than the drug, the first experience with PEM helped to put the risks of benoxaprofen into perspective. Jaundice in elderly patients taking diuretics was identified by PEM, but the signal was very weak, depending on, at most, three very doubtful histories. It was certainly not sufficient to justify special attention in a preliminary report circulated in February 1982. What the journal letters, the CSM reports, and the PEM results had in fact identified was a subgroup of patients with impaired renal function taking large doses of benoxaprofen. During the preliminary analysis of the PEM data on the drug it had been noted that about 40 per cent of the original cohort of patients treated, mostly for the first time, in January 1981, were still taking the drug in January 1982. Photosensitivity had led some to interrupt treatment temporarily, while others had continued to take the drug in spite of it because of the benefit they were experiencing. Suspension of the licence for benoxaprofen, undoubtedly an effective means of preventing future toxicity, however rare, precluded use by a small but important population of patients many of whom had found the drug to have advantages in pain relief. It is possible that benoxaprofen might even have had a disease-modifying effect, reducing the rate of progression of joint destruction.

Establishing cause and effect

While it is quite easy to generate hypotheses, either by reporting astute observations of individual patients, or during organized post-marketing surveillance (PMS). It is often quite another matter to prove a causal link between drug and event. If a patient collapses within seconds of an injection of a drug such as penicillin, to which he or she is known to have reacted on an earlier occasion, it may be reasonable to attributed the collapse to the drug. Much more frequently, however, events are clinically indistinguishable from processes that can occur spontaneously without apparent cause, such as cancer, congenital abnormalities, heart disease, or diabetes. Thus, it is almost always necessary to conduct epidemiological studies which measure the frequency of the event in treated or untreated populations, and to examine the data carefully to exclude confounding factors. Even if a statistically significant relationship between a drug and an event is demonstrated, it may still be impossible to distinguish those individuals among the drug-treated groups who suffered from taking the drug, and those who could have suffered the event without it. This difficulty is of considerable legal as well as scientific importance, since drug epidemiologists are increasingly drawn into the courts. In practice we can only be reasonably certain of causality in the individual when patients treated with the drug suffer the event very commonly, while those who are not treated develop the event very rarely. When dealing with emotive topics, such as congenital abnormality or thrombosis and the contraceptive pill, it is important to remember that both types of event are common in the general population and that it is never possible to apportion blame in an individual case.

Bias

An extensive review of the biases that may affect post-marketing epidemiological studies is beyond the scope of this chapter, but it is worth highlighting some of the biases particular likely to interfere with drug safety studies. Possibly the most important barrier to effective post-marketing surveillance is that created by premature publicity. Not only may this harm patients by causing them to lose confidence in a treatment which is generally safe, it may also make it almost impossible to set up or complete the studies required to establish safety. Occasionally, as a result of adverse publicity, the drug disappears before there is time to decide it the publicity was justified. Publicity affects patients, prescribers, and investigators, and may introduce bias at almost every phase in the development and surveillance of a drug.

Three classes of bias are particularly likely to influence investigations of drug safety. The first is *diagnostic bias* (closely related to referral bias). The question has often been asked whether or not women taking the contraceptive pill are more likely to be diagnosed as suffering from deep venous thrombosis if they are known to be using oral contraceptives, or are more likely to be admitted to hospital with somewhat less severe symptoms then would otherwise justify such a diagnosis or referral. Is a contraceptive pill user more anxious about the possible consequences and is she more likely to put pressure on her doctor to investigate her symptoms? Alternatively, a condition might be masked by treatment and therefore underdiagnosed. An example of this might be the early symptoms of gastric cancer, masked by the administration of cimetidine used to treat what was thought to be a simple peptic ulcer. Cancers could be unmasked after patients had been accepted into a trial of this drug, and it is even possible that the drug might be blamed.

The second type of bias, which is particularly difficult to allow for in case–control (retrospective) studies, is *investigator bias* brought about by previous knowledge of what treatment has been given or what abnormality has occurred. Ideally, the investigator should not be aware of whether the patient is an index case or a control, and should approach the interview or case-records of either type of patient with equal vigour. Often this is impossible because, for example, the index case may be dead and the control subject living, or it may be obvious from a coursory inspection of the notes what treatment had been taken or what abnormality occurred. Conversely, the patient's doctor may go to more trouble to provide the required information if he or she knows that a particular patient may be of special interest to the investigator.

The third, and possibly most important, bias is *recall bias*. It is especially likely to influence the results of studies which depend on patient interviews and, if the topic of interest had been the subject of publicity, it often rules out the possibility of drawing valid conclusions from the study. The mother of a baby with a congenital abnormality, for example, who has read in the papers that a possible explanation has been offered, will be more likely to recall exposure to the suspect drug than one who has borne a healthy infant.

All three sources of bias may be exaggerated by the fact that sick patients seek and receive more attention than fit ones, Sick patients will be attended by the doctor more frequently and their notes are therefore likely to be more comprehensive. They will be more knowledgeable about their diseases and therefore better able to recall important factors of past history or investigations. Both patients and doctor will be more likely to help the investigator if they know what drug or what possible complication is involved.

It is doubtful if any epidemiological study has ever been free from one or more biases of this type, but the risk of false conclusions may be substantially reduced by the application of one simple rule. Wherever possible, conclusions should be based exclusively on oberservations and that have been added to the patient's records *before* the event occurred. To this may be added 'beware of conclusions based on what the patient or the attendants think they remember after the event, especially if the event itself has been publicized'.

In the unlikely event that all these biases could be taken care of, there is still the major problem of confounding bias which may make it impossible to distinguish events caused by the disease which is being treated from those caused by the treatment itself. Congenital heart disease, for example, has been associated with the use of insulin, but is the true cause of the defect the mothers' diabetes? Women with abnormal vaginal bleeding are more likely to be admitted to hospital, more likely to be treated with sex hormones, and more likely to deliver abnormal babies. The last statement is also true when no treatment is given.

Adverse drug publicity

We have already seen how adverse drug publicity may damage or even destroy our attempts to draw scientific conclusions from epidemiological studies. Good news about a drug's efficacy rarely attracts the same attention as bad news about its alleged toxicity. The contraceptive pill, congenital abnormalitities, and fear of cancer are particularly emotive topics and there can be little doubt that much harm has been done from time to time by inaccurate and irresponsible reporting. Gillie (see Inman 1980) has admitted that while the scientist aims for 100 per cent of the truth, 'if a journalist is able to capture 75 per cent of the truth then he reckons he has done well'.

Fortunately, the effects of adverse publicity are often as ephemeral as the journalist's interest in the subject. When challenged about the accuracy of an article in a Sunday newspaper, the journalist's classic reply is 'Ah, that was yesterday's story. Now I am working on something else'. But the consequences of inaccurate publicity can be very harmful. Death or unnecessary suffering on sudden cessation of treatment after reading a news story, pregnancies after a contraceptive pill scare, and false hopes of compensation leading to prolonged misery and sometime substantial financial loss by members of patient action groups, are but a few of the results of our failure to find a satisfactory method of communicating information about drug safety and efficacy.

It is not only the journalists who get it wrong. Doctors and editors of medical journals tend to publish accounts of rare and interesting events, while undramatic accounts of efficacy or non-toxicity may not be submitted for publication or may not be published. This is termed *positive results bias* by Sackett (1979) or, if the topic is emotive, '*hot stuff bias*'. Review articles are often one-sided simply because a reviewer does not have access to the unpublished data.

During the past 5 years commercial competition has crept into post-marketing surveillance. Two separate computer systems have been advocated for use on a large scale by general practitioners, and a market research firm has started to operate in competition with PEM. A number of individual drug companies have also conducted so-called post-marketing surveillance during which doctors are offered substantial inducements to change patients' treatment.

Fragmentation of the database available for PEM by competing studies may introduce substantial delays into the detection and measurement of important hazards. Even more serious, perhaps, is the fact that serious toxicity could be experienced by patients whose treatment has been changed as a result of inducement and who are not fully informed volunteers (*PEM News* 1988).

Insurance rather than litigation would seem to be the best solution to the injured patient's problems. Whatever the cause of the injury, whether it be a spontaneous illness, a physical accident, or the result of unexpected drug toxicity, the injured person deserves the highest standard of care that society can provide. This must be instantly available and not the result of many years of struggle for compensation.

Clearly, people have a right to know about the dangers of the drugs they consume. They have an equal right, however, to expect that the information that they acquire from their doctors, the drug industry, and the media is accurate and understandable. Fear of the unknown and inability to appreciate the significance of risk-estimates are probably the most important factors. The risks of cigarette smoking, crossing the street, or defective electrical appliances are well known but frequently ignored. A sevenfold increase in risk of death from thrombo-embolism in a young woman using the contraceptive pill, where the baseline mortality is only two per million, is headline news. Many women have abandoned the use of oral contraceptives after reading such statistics. Yet the same women accept the advice of a surgeon without qualms although, as may have been explained to them, the mortality rate of an operation may be measured in whole numbers per thousand.

An essential preliminary to the release of information is peer review by experts. The role of the expert journal referee and editor is also vital in this respect. Unfortunately, there is little sense of corporate responsibility among lay journalists. Patients do suffer as a result of medical mistakes and doctors who make mistakes may get sued. But journalists, like members of parliament and some consumer activitists, enjoy the immunity provided by a free press.

Clearly, no one can deny that thalidomide was an appalling tragedy, no one suggests that the public should not be informed, no one would doubt the right of a patient, a doctor, a journalist, or a politician to question established practices, and no one would wish to delay responsible warnings about drug toxicity. In many situations, however, adverse reported drug publicity has had far more damaging consequences that the adverse drug reactions themselves. We must not forget the benefits which the development of drugs has brought in the past 50 years.

References

Bradford Hill, A. (1971). *Principles of medical statistics* (9th edn). The Lancet Limited, London.

Brown, P., Baddeley, H., Read, A.E., Davies, J.D., and McGarry, J. (1974). Sclerosing peritonitis, an unusual reaction to a β-adrenergic-blocking drug (practolol). *Lancet* ii, 1477.

Committee on Safety of Drugs (1968). *When in doubt—report. Adverse reactions Series* No. 7. HMSO, London (Unpublished document printed for the Department of Health and Social Security).

Cooper, W.D., Sheldon, D., Brown, D., Kimber, G.R., Isitt, V.L., and Currie, W.J.C. (1987). Post marketing surveillance of enalapril: experience in 11710 hypertensive patients in general practice. *Journal of the Royal College of General Practitioners* 37, 346.

Felix, R.H., Ive, F.A., and Dahl, M.G.C. (1974). Cutaneous and ocular reactions to practolol. *British Medical Journal* iv, 321.

Geiling, E.M.K. and Cannon, P.R. (1938). Pathogenic effects of elixir of sulfanilamide (diethylene glycol) poisoning. *Journal of the American Medical Association* 111, 919.

Greenberg, G., Inman, W.H.W., Weatherall, J.A.C., Adelstein, A.M., and Haskey, J.C. (1977). Maternal drug histories and congenital abnormalities. *British Medical Journal* ii, 853.

Inman, W.H.W. (1970). Role of drug-reaction monitoring in the investigation of thrombosis and 'the pill'. *British Medical Bulletin* 26, 248.

Inman, W.H.W. (1972). Monitoring by voluntary reporting at national level. In *Adverse drug reactions: Their prediction, detection and assessment* (ed. D.J. Richards and R.K. Rondel), p. 86. Churchill Livingstone, Edinburgh.

Inman, W.H.W. (1978). Detection and investigation of drug safety problems. In *Epidemiological issues in reported drug induced illness. Honolulu 1976* (ed. M. Gent and I. Shigematsu), p. 17. McMaster University Library Press, Hamilton, Ontario.

Inman, W.H.W. (1979). Oral contraceptives and fatal subarachnoid haemorrhage. *British Medical Journal* ii, 1468.

Inman, W.H.W. (1980). *Monitoring for drug safety*. MTP Press, Lancaster.

Inman, W.H.W. (1981*a*). Post-marketing surveillance of adverse drug reactions in general practice. 1: Search for new methods. *British Medical Journal* **282**, 1131.

Inman, W.H.W. (1981*b*). Post-marketing surveillance of adverse drug reactions in general practice, II: Prescription-event monitoring at the University of Southampton. *British Medical Journal* **282**, 1216.

Inman, W.H.W. (1988). Blood disorders and suicide in patients taking mianserin or amitriptyline. *Lancet* **ii**, 90.

Inman, W.H.W. (1989*a*). Mianserin. *Lancet* **i**, 382.

Inman, W.H.W. (1989*b*). PMS and Brass Tacks In *Medical risk and consent to risk* (ed. R. Mann) pp. 73–94. Parthenon Publishing, Carnforth, Lancs.

Inman, W.H.W. and Mushin, W.W. (1974). Jaundice after repeated exposure to halothane. *British Medical Journal* **i**, 5.

Inman, W.H.W and Mushin, W.W. (1978). Jaundice after halothane: A further analysis. *British Medical Journal* **ii**, 1455.

Inman, W.H.W. and Rawson, N.S.B. (1983). Erythromycin estolate and jaundice. *British Medical Journal* **286**, 154.

Inman, W.H.W. and Rawson, N.S.B. (1985) Prescription-event monitoring of five non-steroidal anti-inflammatory drugs. In *Side-effects of Anti-inflammatory drugs* (ed. K.D. Rainsford and G. Velo) Part, p. 111. MTP Press, Lancaster.

Inman, W.H.W. and Vessey, M.P. (1968). Investigation of deaths from pulmonary, coronary and cerebral thrombosis and embolism in women of childbearing age. *British Medical Journal* **ii**, 193.

Inman, W.H.W., Vessey, M.P., Westerholm, B., and Engelund, A. (1970). Thromboembolic disease and the steroidal content of oral contraceptives. *British Medical Journal* **ii**, 203.

Inman, W.H.W., Rawson, N.S.B., Wilton, L.V., Pearce, G.L., and Speirs, C.J. (1988). Post-marketing surveillance of enalapril. I: Results of prescription-event monitoring. *British Medical Journal* **297**, 826.

Kono, R. (1978). a review of SMON studies in Japan. In *Epidemiological issues in reported drug-induced illness—SMON and other examples* (ed. M. Gent and I. Shigematsu), p. 121. McMaster University Library Press, Hamilton, Ontario.

Meade, T.W., Greenberg, G., and Thompson, S.G. (1980). Progestogens and cardiovascular reactions associated with oral contraceptives and a comparison of the safety of 50– and 30–μg oestrogen preparations. *British Medical Journal* **i**, 1157.

Medical Research Council (1967). Risk of thromboembolic disease in women taking oral contraceptives. *British Medical Journal* **ii**, 355.

Office of Population Censuses and Surveys (1959–78). *Mortality statistics 1957–1976*. HMSO, London.

PEM News (1988). Issue No. 5, Drug Safety Research Unit, Southampton. Hamble Valley Press.

Prescription Pricing Authority (1987). *Annual report*. Newcastle.

Rawson, N.S.B., Pearce, G.L., and Inman, W.H.W. (1990). Prescription—Event Monitoring: Methodology and Recent Progress. *Journal of Clinical Epidemiology* **43**(5), 509–22.

Royal College of General Practitioners (1967). Oral contraception and thromboembolic disease. *Journal of the College of General Practitioners* **13**, 267.

Royal College of General Practitioners (1977). Mortality among oral contraceptives users. *Lancet* **ii**, 727.

Sackett, D.L. (1979). Bias in analytic research. *Journal of Chronic Diseases* **32**, 51.

Speirs, C.J., Dollery, C.T., Inman, W.H.W., Rawson, N.S.B., and Wilton, L.V. (1988). Post-marketing surveillance of enalapril. II. Investigation of the potential role of enalapril in deaths with renal failure. *British Medical Journal* **297**, 830.

Taggart, M.McA. and Alderdice, J.M. (1982). Fatal cholestatic jaundice in elderly patients taking benoxaprofen. *British Medical Journal* **284**, 1372.

Vessey, M.P. and Doll, R. (1968). Investigation of relation between use of oral contraceptives and thromboembolic disease. *British Medical Journal* **ii**, 199.

Wade, O.L. (1970). *Adverse reactions to drugs*. Heineman, London.

Walton, B., Simpson, B.R., Strunin, L., Doniach, D., Perrin, J., and Appleyard, A.J. (1976). Unexplained hepatitis following halothane. *British Medical Journal* **i**, 1171.

Wells, N. (1980). *Medicines: 50 years of progress, 1930–1980*. Office of Health Economics, London.

Wright, P. (1975). Untoward effects associated with practolol administration: Oculomucutaneous syndrome. *British Medical Journal* **i**, 595.

29

Traumatic injury

LEON S. ROBERTSON

Introduction

In this chapter, the contribution of traumatic injury to deficits in the public's health are reviewed. The contribution of various countermeasures to reductions in injuries is noted and a model for a public health approach to injury reduction is outlined.

From the first few months of life to mid-life, trauma is the leading cause of death in the industrialized countries. In 1986 in the US, years of potential life lost before the age of 65 years from trauma totalled 3.7 million (including 1.3 million from homicide/suicide) compared with 1.8 million from malignant neoplasms and 1.5 million from diseases of the heart (Centers for Disease Control 1988). During 1985 in the US, some 3.3 million persons were hospitalized as a result of trauma, including 1.1 million with fractures, 268 000 with intra-cranial injuries exclusive of skull fractures, and 277 000 with lacerations and open wounds (National Center for Health Statistics, unpublished data).

These are data from short-term hospital discharges and do not include multiple admissions to burn centres and other rehabilitative facilities. At the 1985 rate, about 1 in every 130 children is hospitalized for an injury annually. Assuming no repeat hospitalizations of the same child, about one of every nine children born today would be hospitalized for injury during their first 15 years should there be no reduction in the current rates.

Fortunately, fatal injury rates of small children have declined markedly in recent years and some gains have occurred among older children and adults. The 1984 death-rate of children less than 5 years old declined 19 per cent from 1980 (including a 30 per cent decline in motor vehicle related deaths). The all-age death-rate was reduced by 16 per cent during that period.

Some evidence exists on the contribution to these reductions of injury control programmes, such as restraint use laws, fencing of swimming pools, and use of smoke detectors. The evidence on successful programmes suggests that linkage of surveillance of injuries (specification of clusters and causes) with technical strategies for control can result in further acceleration of declines in severe injuries.

Motor vehicle occupants

When a moving vehicle decelerates very rapidly in a crash or sudden stop, unrestrained occupants continue to move at the predeceleration speed until they contact interior or exterior surfaces. The extent of injury is a function of the speed, the amount of energy absorbed outside the passenger compartment (e.g. in the front end of the vehicle or road crash attenuators), the energy-absorbing or energy-concentrating characteristics of the contacted surfaces, and the energy-absorbing ability of the organism.

The more prominent technologies available to reduce severity of injuries are restraint use of vehicle occupants, reduced vehicle speeds, increased vehicle stability, energy absorbing materials in the exterior perimeter and in the occupant compartments, removal of rigid roadside objects or placement of energy-absorbing materials in front of them, improved skid resistance of roads, improved road definition by striping and reflectors, improved visibility of vehicles by lighting and reflectors, improved brakes on heavier vehicles, and separation of heavier and lighter vehicles.

Since a crash on any given trip is unpredictable, restraints must be used every time a person is in a moving vehicle for the optimal benefit. For the youngest children, parents must purchase, rent, or borrow a special restraint and buckle the restraint to the vehicle seat and the child in the restraint. Older children and adults can use the seat belt that is standard equipment in most vehicles in industrialized countries. In countries that do not require belt installation, many manufacturers do not provide them.

Child restraint and belt use has proved generally resistant to voluntary change. Surveys of use of child restraints in the US during 1974 found that 93 per cent of children less than 10 years old were not restrained (Williams 1976). Adult belt use was only about 15 per cent (Roberton et al. 1972).

Several experiments involving attempts to persuade parents to use child restraints have been reported. In 1976, in an urban obstetrics hospital, mothers with new-borns were told that child restraints were available for purchase in the hospital gift shop. They were divided into three experimental groups and a control group. One group received literature on child restraint effectiveness, one group received the literature and a discussion by a person trained in persuasive techniques, and one group received literature and a free restraint. In comparison to the control group that received no information other than availability of restraints for sale, only the group that received free restraints was observed to be using them more in follow-up observations of use, about a 7–8 percentage point increase (Reisinger and Williams 1978).

A controlled study in paediatric practices found some effect of counselling by the physician. The counselling included a prescription for a child restraint and demonstration by the physician as to its proper use. In return visits, observed restraint use increased by 23 per cent in the first month, 72 per cent in the second month, but was only higher than a non-counselled control group by 9 and 12 per cent at 4 and 15 months respectively (Reisinger *et al.* 1981).

The full extent of application of these findings is unknown. Numerous hospitals and other organizations around the US have developed loaner or free restraint programmes that have increased use substantially, up to 90 per cent at hospital discharge in one state (Colleti 1984). Many such programmes have been initiated by or benefited from the 'A First Ride, A Safe Ride' promotional activities of the American Academy of Paediatrics. How many of the restraint users would have purchased restraints in the absence of the loaner programmes or used them regularly after hospital discharge is unknown, as is the extent of physician counselling, but there is undoubtedly some net increase in child restraint use from such efforts.

Older children and adults do not increase restraint use markedly in response to persuasion. For example, a controlled study of the effects of television was conducted using a dual cable television system. Advertising that would have cost US $ 7 million if done nationally in 1974 was shown for 9 months on one of the cables. Belt use, observed for 11 months before, during, and after the campaign, revealed no effect on use (Robertson *et al.* 1974).

A second approach has been the requirement by State law that restraints be used by children up to specified ages. Tennessee enacted the first child restraint use law in 1977 after effective lobbying by pediatricians and other concerned citizens. By 1985 all US States had some type of law requiring child restraint use. The Tennessee law applied to children aged less than 4 years transported by parents of legal guardians in vehicles owned by them, but exempted recreational vehicles of the truck or van type and trucks weighing more than a ton. Observed restraint use by children less than 4 years old increased from 8 to 28 per cent in the first 3 years of the law (Williams and Wells 1981).

Studies of the effects of these laws on children's injuries have indicated reductions. Fatalities to children less than 4 years old in Tennessee declined about 50 per cent from 1978 to 1983 in parallel with increases in citations by the police for non-use of child restraints (Decker *et al.* 1984). Since restraint use did not increase enough to account for that much decrease in deaths, other factors were also reducing the death-rate. A study of motor vehicle occupant injury and death to children in California before and after its 1983 law, compared with Texas before its law went in force, found about an 8 per cent reduction in injuries in the first year, but no significant reduction in deaths (Guerin and McKinnon 1985). However, the statistical tests applied were of questionable power to test for changes in deaths in 1 year. By 1986, a 19-city survey found that child restraint use had increased to about 55 per cent of children observed (National Highway Traffic Safety Administration 1986).

An analysis of the 50 State laws requiring child restraint use found that about 39 per cent of the children aged 0–5 years who were killed in the year preceding the laws were not covered by the enacted law because of age and other exemptions. The age limit is under 6 years in five States, under 5 years in 15 States, under 4 years in 25 States, under 3 years in one State, and under 2 years in four States (Teret *et al.* 1986). It has also been observed that child restraints are often not properly anchored. In 3447 observations of child restraints in vehicles in parking lots, 6 per cent were not anchored to the seat-belt and 75 per cent were improperly anchored. Of 1648 restraints with a tether attachment, 68 per cent were not attached and an additional 16 per cent were not correctly attached (Shelness and Jewett 1984).

From 1984 to the summer of 1986, required restraint use by older children and adults had been enacted into law in 27 US States, but Massachusetts and Nebraska repealed their laws by public referendum in the 1986 election. The first major jurisdiction in the world to enact belt use legislation was Victoria, Australia, following a decreased death-rate associated with required belt use on a major construction project. Deaths in Victoria declined about 10 per cent in rural areas and 20 per cent in urban areas compared with expected rates based on trends in provinces without the law (Foldvary and Lane 1974).

The results in States of the US that have had laws in force for more than a year indicate a similar experience. The first US State to enact a belt use law, New York, experienced a 16 per cent reduction in fatalities in the first year, but the reductions were less in other States—10 per cent in Michigan, 6 per cent in New Jersey, and 5 per cent in Illinois (Insurance Institute for Highway Safety 1986e). New York is the only State among these where citation for not using a seat-belt is allowed without some other violation.

Although belt use increases substantially when the laws are first enacted, it subsequently declines to less than 50 per cent without continued vigorous enforcement (Insurance Institute for Highway Safety 1986e). It is too early to estimate the long-term effect of such laws. They were not in force during the 1980–4 period when the deaths were declining. If the Massachusetts and Nebraska repeal experience spreads, belt use laws may not be sustainable politically in the US. The probability of laws in other States is also reduced.

Advertising and publicity in combination with a law does have an effect on restraint use. In Australia, child restraint use increased by about 15 per cent in association with an advertising campaign regarding the child restraint law (Freedman and Lukin 1981). In Elmira, New York, publicity regarding a crack-down by police on non-use of seat belts resulted in 77 per cent use (Insurance Institute for Highway Safety 1985). In California, where the law only allows citation for non-use of belts if some other law is violated, publicity and special enforcement in one community increased belt use from 45 to 55 per cent (Insurance Institute for Highway Safety 1986e).

Increasing restraint of occupants is expected from the installation of automatic restraints—seat-belts that automatically encircle front seat occupants or air bags that inflate when crash forces above a specified concentration are detected by sensors in the front end. After 17 years of on-again off-again attempts at regulation, and a Supreme Court decision ruling the rescision of the regulation as illegal, the US Department of Transportation's rule that automatic restraints be phased in during the 1987–90 model years is being implemented. The extent of the effectiveness of the standard will depend on the technology used. While air bags work automatically, some of the designs for 'automatic' seat-belts allow them to be detached, defeating the automatic feature. A few European manufacturers are installing driver-side air bags as standard equipment and several manufacturers offer them as an option. Other manufacturers are reported to be gearing up to offer driver-side air bags. These will not protect children, but full, front-seat protection has been promised for the future by some manufacturers (Insurance Institute for Highway Safety 1986d). Similar promises were made but unkept in the past (Kelley 1980).

The National Highway Traffic Safety Administration has crash-tested selected car models into a solid barrier at 35 miles per hour since 1979 and published the results. These tests revealed up to fivefold differences in the accelerations on the head, chests, and knees of test dummies (National Highway Traffic Safety Administration 1984). Some manufacturers have improved the energy-absorbing capability of the front ends of their vehicles and the belt and seating systems as a result of these tests.

At least part of the reductions in deaths in motor vehicles is the result of federal legislation and previous regulation. In 1966, the US Congress enacted two laws aimed at reducing motor vehicle deaths, the National Traffic and Motor Vehicle Safety Act (Public Law 89–563) and the Highway Safety Act (Public Law 89–564). The former provided authority to establish vehicle and equipment safety standards and the latter provided for federal assistance to the States.

The initial standards for 1968 and subsequent model cars included shoulder belts in outboard seating positions, energy-absorbing steering assemblies, redundant braking systems, improved door latches to reduce ejection, interior padding, seat integrity, reduced glare in drivers' eyes, and side running lights, among others. Subsequent standards included hood-latch and brake fluids in 1969, child seating systems and power-operated windows in 1971, flammability of interior materials and retread tyres in 1972, side-door strength and roof crush resistance in 1973, one-piece lap and shoulder belts in 1974, rear-end fuel system integrity and windscreen zone intrusion in 1976, and high-mounted brake-lights in 1986. In addition, there were numerous amendments to earlier standards as a result of studies of effectiveness. Lack of compliance was a problem in earlier years, but declined. Non-compliance across all standards for new cars tested during 1968–71 ranged from 19 per cent in 1969 to 11 per cent in 1971. The range for the 1974–8 models, when new, was, 1–10 per cent (National Highway Traffic Safety Administration 1969–80). Trucks were exempted from many standards until a decade later and remain exempted from some.

Estimates of the effects of these standards, based on comparison of fatalities involving vehicles in fatal crashes to which the standards did or did not apply, indicate approximately 15 000 fewer deaths per year in 1979–82 than would have occurred without the standards. Both occupant deaths and non-occupant deaths (struck by the vehicles) were reduced (Robertson 1984). While much of the death reduction due to federal safety standards was realized before 1980, some of the reductions observed in the 1980s can be attributed to the continued scrappage of vehicles that did not meet the standards.

Despite the federal standards for vehicle interiors, many, if not most, vehicles have protrusions such as knobs and tapered dashboards that concentrate energy exchanges with the faces, heads, and chests of children in crashes and sudden braking (Williams et al. 1979). One study found that 12 per cent of children's injuries in motor vehicles occurred in non-crash braking or swerving (Agran et al. 1985).

The 1986 and subsequent model cars are required by federal standard to have a new brake-light mounted above the rear truck. According to experiments in fleets where vehicles were randomly assigned to have such lights or to control groups, the extra brake-light will reduce rear-end collisions while braking by about 50 per cent (Reilly et al. 1980).

The effects on fatalities of federal funds awarded to the States for highway safety have been studied by comparing changes in fatalities among States according to their allocation of such funds. Those States that used federal funds for increased driver education in public schools had increased fatality rates while those States that allocated the funds to other projects had decreased fatality rates (Robertson 1984). Driver education in schools has been found to increase the fatal crash involvement of 16–17-year-old drivers because the trained drivers obtain licences earlier than they would have without the school programme (Robertson and Zador 1978; Shaoul 1975;). In Connecticut, when driver education was dropped from nine school districts following the elimination of State funding of the programme, licensure and crashes of 16–17-year-olds declined precipitously in those districts compared with districts that maintained the programme with local funds (Robertson 1980).

In 1973, the Highway Safety Act authorized federal assistance for improvements at railroad crossings, such as warning sounds and lights when a train was approaching. The numbers of motor vehicle fatalities at railroad crossings declined from 1128 in 1974 to 542 in 1984, a decline of 52 per cent, while potential exposures increased by 4 per cent (Dempsey 1985).

A separate federal assistance programme for highway safety is administered by the Federal Highway Administration in grants to the States for highway construction and safety improvements from the Highway Trust Fund. Although the effects of these projects in total have not been estimated, specific highway changes are associated with death reductions. Limited access roads have substantially lower death-rates per mile than so-called feeder and arterial roads. However, such roads may result in increased miles travelled so that the net reduction in deaths, if any, is less than the per mile rates suggest.

Energy-absorbing materials at sites where roadside objects concentrate energy when vehicles leave the road and hit them are known to reduce deaths (Kurucz 1984). Research comparing sites where deaths or injuries resulted from crashes with fixed objects with sites that the vehicle passed without incident has specified the characteristics of sites with high risk (Wright and Robertson 1976). Approximately 25 per cent of these deaths could be prevented by improvements on less than 8 per cent of road sections. In Georgia, selective road striping based on this research resulted in a 20 per cent reduction in deaths from that expected (Wright et al. 1982). The extent to which road improvements are based on research as to high-risk sites varies widely among States. Most of the federal funds are spent for resurfacing, which increases crashes by about 10 per cent (probably because people drive faster on smoother surfaces), rather than on the reduction of hazards (Committee for the Study of Geometric Design Standards for Highway Improvements 1987).

Attempts at speed reduction are mainly limited to law enforcement of speed limits. A National Research Council Committee estimated that the speed limit reduction to 55 miles per hour in 1973 reduced deaths by 2000–4000 from what would have been expected in 1983 (Committee for the Study of the Costs and Benefits of the 55 m.p.h. National Maximum Speed Limit 1984). Since the decline in death-rates attributable to the reduced speed limit occurred in the 1970s, the declines in the 1980s can not be attributed to the law. However, the evidence indicates that a repeal of the law on rural inter-State roads has resulted in increased deaths (Insurance Institute for Highway Safety 1987).

In the early 1970s, the National Highway Traffic Safety Administration proposed a standard that would place a limit on the top speed capability of vehicles, but the rule was not adopted. Most road vehicles are capable of speeds two to three times the maximum speed limit and manufacturers continue to advertise that capability aggressively. Radar detectors are sold to enable speeders to flaunt the law. A Maryland study found that 81 per cent of large trucks and 40 per cent of passenger cars had radar detectors (Insurance Institute for Highway Safety 1986b).

Some trucking companies install tachometers in their trucks that include records of the speed. The extent of the penalties to errant drivers and the effect of this technology on controlling speeds needs investigation.

Tractor–trailer and other truck combinations are involved in about 10 per cent of fatalities although they are only less than one per cent of registered vehicles. Truckers have been found to disconnect front tractor brakes in the mistaken belief that control of the vehicle is increased. Studies dating from 1975 have indicated that removal of front-axle brakes on tractor–trailer trucks increased stopping distances by 20–35 per cent with no improvement in vehicle control. Nearly 31 per cent of trucks inspected by the Bureau of Motor Carrier Safety in 1983 were removed from service because of safety defects; two-thirds of these had inoperative or defective brakes. Yet the Bureau reduced its inspection activities by 23 per cent from 1983 to 1984. It allocated $6.5 million to 17 States to train State employees to conduct inspections (Insurance Institute for Highway Safety 1986a, b).

The Government has allowed trucks to carry heavier cargo and has allowed two trailers per truck on federal-aid highways. Trucking firms argued that the increased loads would reduce the number of trips and have no net effect on safety but a case–control study indicated that the increased crash risk of the more unstable tandem-trailer trucks more than offset the effect of reduced trips (Stein and Jones 1988).

Technology to prevent cars from underriding trucks, shearing off the tops of the cars, and sometimes the heads of occupants, has been developed but largely unused. Current standards require an underride guard on the rear of large trucks, but these have been shown in crash tests to be ineffective, although an improved guard, not yet adopted, would reduce underride (Kelley 1978). Night-time car into truck crashes were reduced about 18 per cent in one experiment by placing reflector tape around the outline of the truck trailer (Burger et al. 1986).

Another strategy that is also largely unused would require heavy trucks to use only certain roads or restrict use on certain roads to the least hazardous times of day. The New Jersey Turnpike from New Brunswick, Jersey, to New York City has a separate channel for trucks and buses.

A strategy to increase vehicle conspicuity is to require use of lights in daytime. This requirement has been in effect for motorcycles in some States for years and new motor cycles have lights that are lit automatically when the engine is in operation. Visibility of vehicles has been found experimentally to increase when headlamps are in use. In on-road use, 2000 cars, trucks, and vans with an inexpensive relay to turn on front parking lights and rear lights at ignition were involved in 22 per cent fewer collisions than comparable vehicles without the equipment (Stein 1984).

At signalized intersections, the length of the yellow phase between green and red, relative to road design criteria, is strongly correlated with crash rates. Intersections with yellow phase lower than 10 per cent of engineering design standards had about six times the crash rates of those intersections with a yellow phase 10 per cent or more longer than design standards (Zador *et al.* 1984).

Motor cycles have remained largely immune from regulation. Motor cyclists' deaths in the US increased from 700 per year in the early 1960s to 4600 in 1986, paralleling an increase in motor cycle registrations from 0.5 to 5.6 million, mainly cheap Japanese imports. As the market for motor cycles became saturated in the 1980s, the manufacturers began marketing so-called super-bikes capable of speeds to 160 miles per hour, in the US but not in Japan. These motor cycles have twice the death-rates of ordinary motorcycles (Kraus 1988). Arguments over imbalance in international trade do not include the imbalance in death, paraplegia, quadriplegia, and brain damage. Helmet-use laws reduce motor cyclists' deaths about 30 per cent (Robertson 1976) but such laws, once in force in all but two States, have been repealed in about half the States.

A new vehicle that has increased deaths and non-fatal injuries recently is the so-called 'all-terrain vehicle'. These are motorized vehicles that, in the three-wheeled version, look like a large tricycle with balloon tyres, but are capable of speeds of 40 or more miles per hour. Usually operated off roads, these vehicles were advertised showing children as operators. Injuries associated with all-terrain vehicles, estimated from the emergency room surveillance system of the Consumer Product Safety Commission, increased from 8000 in 1982 to more than 60 000 in 1984. About 20 per cent of these injuries occurred to children less than 12 years old (Newman 1985).

This experience follows earlier increases from unlicensed motorized vehicles. Children less than 14 years old were involved in 21 per cent of 13 361 snowmobile injuries in 1977 and 51 per cent of injuries from unlicensed motor scooters and mini-bikes—small motor cycles capable of speeds of 50 miles per hour (Berger 1978). The Consumer Product Safety Commission reached an agreement with the industry that no more three-wheeled, all-terrain vehicles would be sold in the US, but those already sold will continue to injure, as will the four-wheeled version and the other-mentioned recreational vehicles that the Government has refused to regulate.

Pedestrians

Approximately half of children killed by motor vehicles were pedestrians in 1986 (957), and 5800 adult pedestrians were killed. Less research has been done on pedestrian injuries than on occupant injuries. Children as young as 3–4 years old have been observed sufficiently removed from adults and close enough to moving vehicles such that a quick dart toward the vehicles could not be prevented. No significant difference in cautionary behaviour was observed by age up to age 8 years. Children have more difficulty than adults in locating the direction of sounds and movement, as well as discerning the difference between right and left (Sandels 1970).

Attempts to educate children and drivers regarding vehicle direction, movement, and appropriate behaviour have had mixed results. Some research indicates that children acquire and retain skills in street behaviour easily (Yeaton and Bailey 1978), but other studies are more cautious. After training using model cars and roads in one study, less than half the children remembered to look behind at intersections for turning cars and half the 6-year-olds and 25 per cent of 9-year-olds did not remember to stay in the walkway marked for pedestrians. Signs showing a running child, to warn drivers in areas where there were children, were interpreted by some of the children as the place one could safely run across the street. The curb drill, 'look to the left and look to the right' is interpreted by some children as a magic incantation that will protect them (Pease and Preston 1967).

Among children aged 4–7 years, darting out into traffic is the behaviour associated with about two-thirds of pedestrian injuries. A film and television spot programme, using a character called 'Willy Whistle', has been developed concentrating on getting children to stop and look for vehicles when they come to a curb or the edge of a parked vehicle. The films and spots were distributed to schools, theatres, and television stations in three cities. A 20–30 per cent reduction in child pedestrian injuries involving dart-out was observed in the cities where the materials were distributed (Preusser 1986), yet they have not been widely used.

Identification of high-risk sites for pedestrian injury and the associated movements of the pedestrians and vehicles has also led to experimentation with site modifications to reduce risk. Depending on the actions of pedestrians and drivers at given sites, countermeasures included preventive markings, median barriers, set-back of pedestrian crossings, mid-block pedestrian crossing, barriers at parking meter posts, stop-line relocations, vendor warning signs, and bus stop relocations. With the exception of preventive markings and vendor warning signs, each of these modifications was associated with changes in driver or pedestrian actions believed to reduce risk (Berger 1975). More work of this type, including measurement of actual changes in injury rates, is needed.

Law enforcement of child behaviour has apparently not been studied, but citation of adults for illegal street-crossing behaviour has no discernible effect on the behaviour or pedestrian injuries (Singer 1969). A law allowing right turns at red lights to reduce energy consumption increased child pedestrian injuries. These laws were compared among States as they were adopted and an increase of more than 30 per cent in injuries to child pedestrians was associated with their adoption (Zador *et al.* 1982).

A detailed review of pedestrian injuries led one author to estimate that one-third of serious pedestrian injuries could be reduced by changes in vehicle design. Hard surfaces, sometimes tapered to a point like an arrow or lance on hood fronts and corners, are not uncommon on the front ends of vehicles. These designs serve no function and cost more than flat, energy-absorbing surfaces. Lower bumpers in combination with energy absorption in hood and windscreen areas, that pedestrians tend to rotate into when struck, are the most efficacious designs (Ashton 1982).

Bicycles

Bicyclists are most seriously injured in collisions with motor vehicles. About one in four injuries involving bicycle–motor vehicle collisions results in hospital admission of the bicyclist compared with 1 in 20 bicycle injuries not involving motor vehicles. Almost 40 per cent of fatalities occur when the driver runs upon a bicyclist from the rear and 27 per cent occur from bicycles darting out. Brain injury has been found to be the primary cause of death in about three-quarters of bicyclists' deaths (Friede *et al*. 1985).

Use of helmets by bicyclists is suggested as one strategy to reduce severe consequences but no study of the effects of currently manufactured helmets in use has established their adequacy. Paths separate from the road may encourage children to use their bicycles away from traffic, a strategy reported to have had some success in Scandinavia. Enforcement of traffic laws for bicyclists is probably less systematic than that for motor vehicle drivers. In several Minnesota countries, bicycle injuries were reduced in correlation with enforcement that included a request to parents that the child attend a Bike Violator's Seminar (Friede *et al* 1985). Changes in the front ends of motor vehicles from energy-concentrating points and edges to energy absorption would probably reduce severity of bicyclists' injuries in addition to those of pedestrians.

Heat and smoke

Most deaths from heat energy and accompanying smoke occur in house fires while major non-fatal injuries occur from scalds and contact with hot surfaces. The leading source of ignition in house fires (30–45 per cent) is the cigarette, usually dropped in a bed, couch, or chair and left to smoulder, often as the occupants of the household sleep (McLoughlin and Crawford 1985).

The reductions in fire-related deaths-rates are associated with great increases in the installation of smoke detectors in residences. There has been little change in the numbers of house fires (Demling 1985). According to national random sample surveys in the US, smoke detectors in households increased from 22 per cent in 1977, to 46 per cent in 1980, to 67 per cent in 1982 (US Fire Administration 1983).

Smoke detectors require much less action by the user to be effective than restraint use in cars, probably a strong factor in their more frequent use in response to persuasion. An experiment in a paediatric practice, in which an experimental group received counselling regarding the importance of installing a smoke detector, correctly installed detectors increased from 46 to 65 per cent, with no change in a control group (Miller *et al*. 1982).

The city of Baltimore, Maryland, gave away 3720 smoke detectors to those who asked for them in 1982. In a study of 231 randomly selected recipients, selected from among those that received the smoke detectors and had to instal them without assistance, 92 per cent of the detectors were found installed and 88 per cent were operating correctly 4–9 months later. Furthermore, the recipients were highly concentrated in areas of the city with the greatest fire injury rates (Gorman *et al*. 1985).

Legislation has also contributed to smoke detector use and reduced deaths where it has been attempted. In Montgomery Country, Maryland, where smoke detectors are required by law in all residences, the number of working detectors is greater and the number of residences without detectors is less than in nearby FIarfax County, Virginia, without such a law. From the 6-year period before the law to the 6-year period afterward, fire deaths in Montgomery Country declined more rapidly than in Fairfax County (McLoughlin *et al*. 1985).

Different degrees of success have been reported from attempts in increase community awareness of a variety of means to prevent fires and heat-related injury. The Missouri Division of Health identified a six-county area that had fire deaths two to five times that of other areas of the State. With funds from the US Public Health Service, a health educator and three field representatives developed contacts with a wide variety of governmental and civic groups, as well as news media. Data were gathered on the circumstances of burn injuries and the populations involved. A Burn Prevention Demonstration Unit was developed and presentations were given to community and school groups. From the 3-year period before the project to the 3-year period afterward, fire-related deaths decreased some 43 per cent (Garner *et al*. 1969). Since an area with a high rate was chosen for intervention, some of the reduction could have occurred from so-called regression to the mean, that is, areas with rates lower or higher than average during a given period tend to trend toward the average in subsequent periods. Nevertheless, it is likely that the programme contributed to at least part of the reduction.

In Massachusetts, two communities received an education programme aimed at burn prevention, delivered in news media, community groups and the schools. In comparison with control communities, self-reported knowledge of preventive actions increased, but only 13 per cent of respondents said they applied the knowledge at the moment of risk. No detectable decrease in burns could be found (McLoughlin *et al*. 1982).

Specific information on the source of burns is important in targeting control efforts. In Denmark, a surgeon identified a coffee filtering device as the source of 60 per cent of burns associated with spilled coffee. A campaign was initiated that included redesign and marketing of a new design and public information on dealing with the extant devices. The coffee scalds were reduced by two-thirds (McLoughlin and Crawford 1985).

State and federal regulation of flammability of fabrics has contributed to death reductions from that source. The Flammable Fabrics Act was enacted in 1953 and amended in 1967 to cover a wider range of fabrics. Deaths associated with clothing ignition decreased by 71–82 per cent from 1968 to 1979. Among children, the reduction was even greater, partly due to changes in clothing styles such as reductions in use of frilly dresses (S.P. Baker *et al.* 1984; M.D. Baker *et al.* 1986), and partly due to the fabric standards, particularly for children's sleepwear (Dardis *et al.* 1978).

Several potential targets for regulation to reduce burn injuries have been identified. Sprinkler systems would undoubtedly reduce deaths from fires (Bragdon 1985), but even many high-rise buildings do not have them. News accounts of a spectacular skyscraper fire in Los Angeles in 1988 noted that half such buildings in that city had no sprinkler system.

Most brands of cigarettes have design characteristics and additives that promote continued burning for up to 40 minutes when dropped. The feasibility of modifications to make cigarettes self-extinguish when dropped has been studied (Technical Study Group 1987), but no federal agency has the authority to require manufacturers to make them so. The Consumer Product Safety Commission is specifically prohibited from regulating cigarettes.

Burns from overheated tap water are a major source of children's burns. Standards for maximum temperatures for new water heaters could be set and utility companies could be required to reduce the temperature on extant water heaters during routine meter reading (Feldman 1982). Insulation of other heat sources such as heating registers, stoves, and fireplaces could be increased. Manufacture and sale of fireworks with particular heat, diffusion of heated particles, or explosive characteristics are banned in some areas.

Drowning

People drown in a variety of collections of water including bath-tubs, buckets, swimming pools, lakes, rivers, floodwaters, and, rarely, oceans. Although studies in a few States and local communities have documented the water collections involved, no national estimates are available (Pearn *et al.* 1979). There is undoubtedly local and regional variation related to access to types of water collections, and climatic conditions that influence use.

Teaching children as young as toddlers to swim has been advocated in some circles, but its effects on drowning in the aggregate are unknown. One study found that toddlers who had been trained to swim less often required retrieval from toddler pools (Spyker 1985). This does not take into consideration the extent to which swimming lessons increase the amount of swimming and potentially increase drowning in the aggregate. Research analogous to that noted for driver education is needed to resolve the question.

Drownings associated with children wandering into unsupervised swimming pools are substantially less in areas that require fences and child-proof gates around pools. The annual pool fatality rate in Honolulu, where pool fencing is required, was found to be about one-third that in Brisbane, Australia, which has a similar climate and pool-to-household ratio, but no required pool fencing (Pearn *et al.* 1980).

Poisoning

Poisoning deaths in children have become increasingly rare in the US after the adoption of containers for certain drugs and household chemicals that are resistant to being opened by children. Child poisoning deaths from aspirin declined 80 per cent from 1965 to 1975 when manufacturers voluntarily adopted container caps difficult for children to remove (Done 1978). By authority of the Poison Prevention Packaging Act of 1970, standards for specific products were promulgated from 1972 to 1980. Reported ingestion of the regulated products by children less than 5 years old, measured from the year that a given product was regulated to 1983, declined between 40 and 90 per cent (Consumer Product Safety Commission, personal communication). There were complaints initially from adults who had difficulty opening child-proof caps, but these seem to have subsided. No literature was found on the extent of this problem and the modifications, if any, that may have reduced it. Several drugs and household solvents, corrosives, and caustics continue to result in a child hospitalization rate of 5–12 per 100 000 children per year for each category of product (Trinkoff and Baker 1986).

Community poisoning prevention efforts, including community outreach seminars, school curriculum seminars, retail outreach efforts, distribution of educational materials, and publicity in mass media were associated with declines in emergency room visits for poisonings (Fisher *et al.* 1980a). Using Maternal and Child Health block grant funds as well as State and local funds, a variety of poison prevention efforts have been undertaken in the US (Fisher *et al.* 1985). Apparently no systematic evaluation of the effects of specific efforts has been done.

Other injury prevention

Efforts to prevent other types of injuries have included programmes with general themes, such as home injury prevention, and specifically targeted programmes based on surveillance of injuries and their circumstances. In general, the more specifically targeted approaches that include technology that is simple to use or protective without action of the users are the more successful.

In an attempt to reduce home injuries involving ten categories of household items, parents visiting a prepaid health plan with their children were given information on keeping the presumed hazardous items away from children. In a follow-up telephone call, parents claimed to have removed many of the items from children's access. In a home visit to these families and control families who had received no information, however, there was no difference between the two groups of families in the extent of children's access to the presumed hazardous items (Dershewitz and Williamson 1977). Falls of infants from elevated surfaces (tables, chairs, beds) were apparently reduced by a combination of messages to parents. The parents were given written material about children's falls and were counselled about them by the paediatrician. Signs reminding parents of the messages were placed above the examining table where they could be seen in subsequent visits. Falls among the infants in the group receiving the messages during the subsequent year were about 10 per cent compared with 17 per cent among a comparison group that did not receive the messages (Kravitz 1973).

Some organizations with responsibility for children's health have had significant successes. Although only 12 State health departments had any injury prevention programmes in 1981 (Fisher *et al.* 1985), the reports from some are encouraging. The New York State Health Department has been active in initiating programmes directed toward specific injuries, such as those on playgrounds where a 35–50 per cent reduction in hazards was observed (Fisher *et al.* 1989*b*). Identification of numerous hazards based on cases found by the department have been reported to the State Product Safety Commission and the federal Consumer Product Safety Commission resulting in voluntary recall of products, mandatory recall in some cases, and regulations for modifications (Fisher 1973, 1977).

Using federal Maternal and Child Health grants and State funds, the Massachusetts Department of Health has developed a multi-pronged child injury prevention effort including coding of child hospitalizations and 25 per cent of emergency visits, co-ordinated efforts among paediatricians for systematic counselling of parents about hazards, and a review of State and federal law that can be used for injury prevention (Gallagher *et al.* 1982). A pilot project that employed city health inspectors to review hazards with parents resulted in more than a 50 per cent reduction in hazards for which codes existed (Gallagher *et al* 1985). A State law enacted primarily for energy conservation and aesthetics—required deposits on beverage containers—was associated with a 60 per cent reduction in glass-related lacerations (Baker *et al.* 1986).

The Indian Health Service organized local Community Injury Control Committees in its service units, mainly on reservations, in 1982. Most of the initial efforts were directed toward education of the population in a variety of hazards. Hospitalization per population for fall injuries in 54 service units declined by 35 per cent from 1980 to 1984. The declines across service units were correlated significantly with the proportion of the population trained in general safety, recreational safety, and first aid (Robertson 1986). A more specifically targeted effort based on surveillance of severe injuries is being initiated.

Federal regulations for refrigerators, cribs, and State-required warning labels for plastic bags have been studied in relation to child suffocation and strangulation. Declines in death-rates from refrigerator entrapment and suffocation by plastic bags were found, but not in strangulation from wedging in cribs (Kraus 1985).

A local health department effort perhaps best illustrates the effect of the combination of good epidemiology, choice of a simple, effective control strategy, and employment of a mixture of persuasion and regulation in programme effort. In New York City, epidemiologists noted that children's deaths from falls were much greater than would be expected from national population data. Investigation of 201 such deaths in 1965–9 revealed that 61 per cent of the fatal falls to children less than 15 years old and 85 per cent of those to children less than 5 years old occurred to children who crawled out of windows, 96 per cent in the three of the five boroughs of the city—Bronx, Brooklyn, and Manhattan (Bergner *et al.* 1971). A barrier that could be placed over the windows, preventing children from crawling out, was the technical approach identified as most feasible. A campaign was launched in high-risk neighbourhoods to persuade parents or landlords to install the barriers (Spiegel and Lindaman 1977). Eventually, the Health Department required landlords to install the barriers when requested by tenants. In association with these efforts, child deaths in falls from windows declined from 30–60 per year in the mid-1960s to 4 in 1980 (Bergner 1982).

Subsequently, as families moved and children were born in new families, the fatal falls increased. In July, 1986, the city changed the regulation to require barriers in windows in buildings where there were children less than 11 years old, without the necessity of parental request.

The epidemiological model

Many of the successes in injury control have evolved from a conceptual shift in the scientific view of injury. Long thought to be primarily a behavioural problem, in the 1960s the concept of injury as the result of inordinate energy exchanges with human hosts was suggested (Gibson 1961).

Injuries are analogous to acute infectious diseases. Such diseases have been characterized by epidemiologists as the interaction of infectious biological 'agents' (bacteria, viruses, parasites) with 'host' individuals. Often the agent is conveyed to the host by some animate carrier, such as a mosquito or another human being, called a 'vector', or is conveyed by an inanimate carrier, such as water, called a 'vehicle'. Individuals who have not developed resistance may become infected. Successful control of many infectious diseases that historically plagued humankind has been based on agent control, vehicle or vector control, or increasing resistance of the host. If the agents, vehicles or vectors, and host factors that contribute to injuries ar sufficiently identified and modified, control of injuries can be accomplished.

The agent of injury is physical energy at concentrations outside the range of human resilience. The forms of energy are mechanical, chemical, electrical, heat, and ionizing radiation. The major vehicles of energy damage to people in the US are predominantly motor vehicles, guns, cigarettes (the major cause of house fires), water (in drownings), and gravity (in falls usually from constructions). In the case of asphyxiation, including drowning and smoke-related, it is too little energy produced by oxidation, rather than too much, that results in injury.

To prevent injury or reduce severity, the logical approach is to identify the agents and vehicles that are causing particu-

Table 29.1. Factors and phases of the injury process

Phases	Factors		
	Human	Vehicle	Environment
Pre-event	1	2	3
Event	4	5	6
Post-event	7	8	9

lar subsets of injuries and find a strategy that prevents or reduces the interaction of the agent with the host (Haddon *et al.* 1964). Traditionally, injury prevention was called 'accident prevention' and focused on the behaviour of the injured person or others in proximity who increased the risk of injury. Since many accidents are not injurious and behaviour is often more resistant to change than agents or vehicles, the accident prevention approach, focused on behaviour change, had very limited success.

Factors and phases

Modern injury epidemiology focuses on the full range of factors in terms of what happened before the injury, what happened at the time of the injury, and what happened afterwards that could have prevented the incidence or reduced the severity. Intervention to reduce injury severity many focus on any combination of the factors at any phase. This model is summarized in Table 29.1.

Consider injuries to child pedestrians. Pre-event–human factors are the behaviours and human characteristics that increase the probability of exposure to damaging energy, such as children playing in the vicinity of motor vehicle traffic. Pre-event, vehicle factors include the braking capacity and condition of brakes on motor vehicles. Pre-event–environmental factors include parked vehicles or other objects that reduce drivers' visual perception of children. Event–human factors include conditions of the children that increase the damage when impacted, such as haemophilia. Event–vehicle factors include sharp points and edges that concentrate energy at the point of impact, and raised bumpers that result in impacts to heads and chests rather than less vital legs. Event–environmental factors include hard road surfaces and other objects that a child pedestrian contacts in rebound from the impact and road and street designs that increase exposure to vehicles. Post-event–human factors include first-aid abilities of persons in the vicinity. Post-event–vehicle factors refer to property damage which is irrelevant to the child's injury. Post-event–environmental factors include the rapidity of response of the emergency medical system.

Viewed from this perspective, the behavioral factors that contributed to the injury are only a small part of the factors that could potentially be modified to reduce injury or severity. As was noted in the literature review, focus on factors other than 'accident prevention' has been remarkably successful in reducing certain types of injuries.

Surveillance

Identification of interventions to reduce injuries effectively and efficiently depends on research into the clustering of homogeneous subsets in space and time relative to specifically identified factors and phases of the injuries. This type of research is called surveillance.

Again using the example of child pedestrian injuries, surveillance research would specify the types of streets, roads, or driveways on which the injuries most frequently occur, the time of day and day of week, what the children were doing at the time, the vehicles most frequently involved, the condition of the vehicles, the surfaces on the vehicle or in the environment that concentrated the energy exchanges to the children, the actions and conditions of the drivers, the post-injury actions to provide first aid to children at the scene, and the emergency medical response, and treatment and rehabilation of those who survive.

There are very few injuries that have undergone surveillance research in that degree of detail. In some instances, enough is known to select effective and efficient countermeasures, but too often countermeasures are employed without adequate surveillance.

Much less is known about non-fatal injuries than about fatal injuries. Although hospital discharge data indicate the type of injury, the cause is often not specified. Inclusion of the E-codes, from the International Classification of Diseases indicating source of injury, in the hospital discharge data would greatly facilitate the investigation and control of non-fatal, severe injuries.

Haddon (1970) identified ten logically distinct categories of countermeasures that are applicable to hazards in general, biological or physical. These are:

1. Prevent the creation of the hazard in the first place.
2. Reduce the amount of the hazard brought into being.
3. Prevent the release of the hazard that already exists.
4. Modify the rate or spatial distribution of release of the hazard from its source.
5. Separate, in time or space, the hazard and that which is to be protected.
6. Separate the hazard and that which is to be protected by interposition of a material barrier.
7. Modify relevant basic qualities of the hazard.
8. Make that which is to be protected more resistant to damage from the hazard.
9. Begin to counter the damage already done by the hazard.
10. Stabilize, repair, and rehabilitate the object of the damage.

All of these may not be applicable to every hazard, but systematic review of each may result in identification of one or more countermeasures that is more effective or efficient than those traditionally accepted. Consider examples of those that might be applied to children's drownings, identified by Haddon's numbering system:

1. Prohibit private, unsupervised swimming pools.
2. Reduce the number or permitted depth of private, unsupervised swimming pools.
3. Teach all children to swim.
4. Place sensors in dams and levees to signal need for release of waters at a controlled rate.
5. Place playgrounds at a distance from streams or reroute streams away from playgrounds.
6. Place unscalable fences and locked gates around swimming pools.
7. Not applicable. Water is not modifiable such that it can be breathed by human beings.
8. Require children to exercise to increase lung capacity.
9. Place underwater lights in pools.
10. Provide rehabilitation services to children with brain damage from anoxia.

These are only examples. Based on this conceptualization, long lists of options for a wide variety of injuries have been noted by various authors (Dietz and Baker 1974; Feck *et al.* 1977; Haddon 1980; Robertson 1983).

This technical options analysis does not take into account need, feasibility, or relative effectiveness. Clearly, without adequate research, adoption of any one of these options might be unneeded or ineffective. If no children are drowning in streams, options relevant to such bodies of water are unneeded. If children who drown know how to swim, training in swimming would be ineffective. Furthermore, without adequate research on the effect of such training, there is no guarantee that the training will not increase the amount of swimming and associated drowning. Therefore, surveillance research is necessary to identify need in specific populations and research on the effectiveness of a given option is necessary before widespread adoption.

Technical approaches can be implemented in one or more of three ways:

1. Change the behaviour of the hosts or other persons (parents, teachers, coaches, drivers, etc.).
2. Require behavioural change of the hosts or other persons in proximity to hazards by law or administrative rule.
3. Provide automatic protection by modifying the agents and vehicles of injury (Robertson 1975).

The first two require action by millions of people while the third requires action by manufacturers and processors of hazardous energy and its carriers. In general, as well illustrated in the literature review, legal or administratively required behavioural change is usually more successful than education or persuasion aimed at behavioural change, and modification of agents and vehicles is usually the most successful strategy. That is not to say that a legal or regulatory strategy is available or can be developed for all hazards. Several factors contribute to the relative success of each of these approaches.

The effect of education or persuasion strategies is largely dependent on the frequency of the behaviour and whether or not the behavioural change increases exposure to hazards. Generally, the more frequently people must act to protect themselves or children, the greater the difficulty in educating or persuading them to do so.

The effect of laws and administrative rules is dependent to some degree on the frequency of the required behaviour. Also important is the public observability of the behaviour, which affects the probability of imposition of sanctions, and the extent of augmentation of enforcement by the community rather than sole dependence of police enforcement. Thus, laws directed at police detection of drunk driving, which is not directly observable, tend to be only temporarily successful (Ross 1982). Increased licensing age or curfews for teenaged drivers and increased drinking ages are more successful (Preusser *et al.* 1984; Williams *et al.* 1983a), apparently because they are mainly enforced by parents and alcohol sales outlets.

For automatic protection to be successful, the manufacturers and producers of potential hazards must be aware of and use technical strategies to reduce the hazardous characteristics of the agents and vehicles. This may occur based on in-house expertise or may be accomplished by governmental regulation. If governmental regulations are necessary, the regulators must have expertise in the characteristics of agents and vehicles that are hazardous and must have the political support to impose and maintain appropriate standards.

References

Agran, P.F., Dunkle, D.E., and Winn, D. (1985). Noncrash motor vehicle accidents: Injuries to children in the vehicle interior. *American Journal of Diseases of Children* **139**, 304.

Ashton, S.J. (1982). Vehicle design and pedestrian injuries. In *Pedestrian accidents* (ed. A.J. Chapman, F.M Wade, and H.C. Foot). John Wiley and Sons, London.

Baker, M.D., Moore, S.E., and Wise, P.H. (1986). The impact of 'Bottle Bill' legislation on the incidence of lacerations in childhood. *American Journal of Public Health* **76**, 1243.

Baker, S.P., O'Neill, B., and Karpf, R.S. (1984). *The injury fact book*. Lexington Books, Lexington, Massachusetts.

Berger, W.G. (1975). Urban pedestrian countermeasures experimental evaluation. US Department of Transportation, Washington, DC.

Berger, L.R. (1978). Modern motor milestones. *Pediatrics* **62**, 1037.

Bergner, L. (1982). Environmental factors in injury control: Preventing falls from heights. In *Preventing childhood injuries* (ed. A.B. Bergman), p. 65. Ross Laboratories, Columbus, Ohio.

Bergner, L., Mayer, S., and Harris, D. (1971). Falls from heights: A childhood epidemic in an urban area. *American Journal of Public Health* **61**, 90.

Bragdon, C.A., Jr. (1985). Prevention of residential fire fatalities. *Hearing before the Subcommittee on Science, Research and Technology*, US House of Representatives, Washington, DC.

Burger, W.J., Mulholland, M.U., and Smith, R.L. (1986). Improved commercial vehicle conspicuity and signalling systems—Task III field test of vehicle reflectorization effectiveness. US Department of Transportation, Washington, DC.

Centers for Disease Control (1988). Estimated years of potential life lost (YPLL) before age 65 and cause-specific mortality, by cause of death—United States, 1986. *Morbidity and Mortality Weekly Reports* **37**, 255.

Colleti, R.B. (1984). A state-wide hospital-based program to improve child passenger safety. *Health Education Quarterly* **11**, 207.

Committee for the Study of the Benefits and Costs of the 55 m.p.h. National Maximum Speed Limit (1984). 55: *A decade of experience*. Transportation Research Board, National Research Council, Washington, DC.

Committee for the Study of Geometric Design Standards for Highway Improvements (1987). *Designing safer roads: Practices for resurfacing, restoration, rehabilitation*. Transportation Research Board, National Research Council, Washington, DC.

Dardis, R., Aaronson, S., and Ying-Nan, L. (1978). Cost–benefit analysis of flammability standards. *American Journal of Agricultural Economics* **60**, 695.

Decker, M.D., Dewey, M.J., Hutcheson, R.H., and Schaffner, W. (1984). The use and efficacy of child restraint devices. *Journal of the American Medical Association* **252**, 2571.

Demling, R.H. (1985). Burns. *New England Journal of Medicine* **313**, 1389.

Dempsey, W.H. (1985). *Extension of the nation's highway, highway safety, and public transit programs*. Hearings before the *Subcommittee on Surface Transportation of the Committee on Public Works and Transportation*, US House of Representatives, Washington, DC.

Dershewitz, R.A. and Williamson, J.W. (1977). Prevention of childhood household injuries: A controlled clinical trial. *American Journal of Public Health* **67**, 1148.

Dietz, P.E. and Baker, S.P. (1974). Drowning: Epidemiology and prevention. *American Journal of Public Health* **64**, 303.

Done, A.K. (1978). Aspirin overdosage: Incidence, diagnosis and management. *Pediatrics* (Supplement), 890.

Feck, G., Baptiste, M.S., and Tate, C.L., Jr. (1977). An epidemiologic study of burn injuries as a basic for public policy. New York Department of Health, Albany, New York.

Feldman, K.W. (1982). Controlling scald burns. In *Preventing childhood injuries* (ed. A.B. Bergman). Ross Laboratories, Columbus, Ohio.

Fisher, L. (1973). New York leads injury control activities. *Journal of Environmental Health* **36**, 213.

Fisher, L. (1977). Communication and home safety. *Proceedings of the 81st Annual Conference of the Association of Food and Drug Officials*. Association of Food and Drug Officials, Portland, Oregon.

Fisher, L., Lawrence, R.A., and Sinacore, J. (1980*a*). Highlight results of the Monroe poison prevention demonstration project *Veterinary and Human Toxicology* **22**, 15.

Fisher, L., Harris, V.G., VanBuren, J., Quinn, J., and DeMaio, A. (1980*b*). Assessment of a pilot child playground injury prevention project in New York State. *American Journal of Public Health* **70**, 1000.

Fisher, L., Greensher, J., Mack, R.B., *et al.* (1985). Poison controls legislation and state government funding in the United States. *Veterinary and Human Toxicology* **27**, 120.

Foldvary, L.A. and Lane, J.C. (1974). The effectiveness of compulsory wearing of seat-belts in casualty reduction. *Accident Analysis and Prevention* **6**, 59.

Freedman, K. and Lukin, J. (1981). Increasing child restraint use in NSW, Australia: The development of a successful media campaign. *Proceedings of the American Association for Automotive Medicine*. American Association for Automative Medicine, Morton Grove, Illinois.

Friede, A.M., Azzara, C.V., Gallagher, S.S., and Guyer, B. (1985). The epidemiology of injuries to bicycle riders. *Pediatric Clinics of North America* **32**, 141.

Gallagher, S.S., Guyer, B., Kotelchuck, M., *et al.* (1982). A strategy for the reduction of childhood injuries in Massachusetts. *New England Journal of Medicine* **307**, 1015.

Gallagher, S., Hunter, P., and Guyer, B. (1985). A home injury prevention program for children. *Pediatric Clinics of North America* **32**, 95.

Garner, L.M., Love, D.M., and Jones, S.B. (1969). *A community action approach for prevention of burn injuries*. US Public Health Service, Bethesda, Maryland.

Gibson, J.J. (1961). The contribution of experimental psychology to the formulation of the problem of safety. In *Behavioral approaches to accident research*, p. 77. Association for the Aid of Crippled Children, New York.

Gorman, R.L., Charney, E., Holtzman, N.A., and Roberts, K.B. (1985). A successful city-wide smoke detector giveaway program. *Pediatrics* **75**, 14.

Guerin, D. and McKinnon, D.P. (1985). An assessment of the California child passenger restraint requirement. *American Journal of Public Health* **75**, 142.

Haddon, W., Jr. (1970). On the escape of tigers: An ecologic note. *Technology Revue* **72**, 44.

Haddon, W., Jr. (1980). Advances in the epidemiology of injury as a basis for public policy. *Public Health Reports* **95**, 411.

Haddon, W., Jr., Suchman, E.A., and Klein, D. (ed.) (1964). *Accident research: Methods and approaches*. Harper and Row, New York.

Insurance Institute for Highway Safety (1985). Elmira, NY campaign increases seat belt use to 77 per cent. *Status Report, December 7*.

Insurance Institute for Highway Safety (1986*a*). Large trucks should have front brakes. *Status Report, February 22*.

Insurance Institute for Highway Safety (1986*b*). Truck crashes for first 8 months of '85 are 6 per cent. higher than all of '84. *Status Report, April 26*.

Insurance Institute for Highway Safety (1986*c*). 285 road deaths averted in four states with seat belt laws. *Status Report, August 13*.

Insurance Institute for Highway Safety (1986*d*). Air bags in '88 models: The pace picks up. *Status Report, November 8*.

Insurance Institute for Highway Safety (1986*e*). Belt use law: Success is tied to type of enforcement permitted. *Status Report, December 13*.

Insurance Institute for Highway Safety (1987). Caution: 65 m.p.h. speed is hazardous to your life. *Status Report, December 26*.

Kelley, A.B. (1978). *Underride* (Film). Insurance Institute for Highway Safety, Washington, DC.

Kelley, A.B. (1980). GM and the air bag: A decade of delay. *Business and Society Revue* (Fall).

Kravitz, H. (1973). Prevention of accidental falls in infancy by counseling mothers. *Illinois Medical Journal* **144**, 570.

Kraus, J.F. (1985). Effectiveness of measures to prevent unintentional deaths of infants and children from suffocation and strangulation. *Public Health Reports* **100**, 231.

Kraus, J.F. (1988). Motorcycle design and crash injuries in California. *Bulletin of the New York Academy of Medicine* **64**, 788.

Kurucz, C.N. (1984). An analysis of the injury reduction capabilities of breakaway light standards and various guard rails. *Accident Analysis and Prevention* **16**, 105.

McLoughlin, E. and Crawford, J.D. (1985). Burns. *Pediatric Clinics of North America* **32**, 61.

McLoughlin, E., Vince, C.J., Lee, A.M., and Crawford, J.D. (1982). Project burn prevention: Outcome and implications. *American Journal of Public Health* **72**, 241.

McLoughlin, E., Marchone, M., Hanger, S.L., German, P.S., and Baker, S.P. (1985). Smoke detector legislation: Its effect on owner occupied homes. *American Journal of Public Health* **75**, 858.

Miller, R.E., Reisinger, K.S., Blatter, M.M., and Wucher, F. (1982). Pediatric counseling and subsequent use of smoke detectors. *American Journal of Public Health* **72**, 392.

National Highway Traffic Safety Administration (1969–80). *Standards enforcement tests reports index*. US Department of Transportation, Washington, DC.

National Highway Traffic Safety Administration (1984). *New Car assessment program, results as of May 7, 1984*. US Department of Transportation, Washington, DC.

National Highway Traffic Safety Administration (1986). *19 city safety belt and child safety seat observed usage*. US Department of Transportation, Washington, DC.

Newman, R. (1985). *Survey of all terrain vehicle related injuries*. Consumer Product Safety Commission, Washington, DC.

Pearn, J., Brown, J., and Hsia, E.Y. (1980). Swimming pool drownings and near-drownings involving children: A total population study from Hawaii. *Military Medicine* **145**, 15.

Pearn, J.H., Richard, Y.K., Wong, M.D., *et al.* (1979). Drowning and near-drowning involving children: A five-year total population study from the city and county of Honolulu. *American Journal of Public Health* **69**, 450.

Pease, K. and Preston, B. (1967). Road safety education for young children. *British Journal of Educational Psychology* **37**, 305.

Preusser, D.F. (1986). Pedestrian safety. *Proceedings of the Utah Child Injury Conference*, Utah Department of Health, Salt Lake City, Utah.

Preusser, D.F., Williams, A.F., Zador, P.L., and Blomberg, R.D. (1984). The effects of curfew laws on motor vehicle crashes. *Law and Policy* **6**, 115.

Reilly, R.E., Kurke, D.S., and Bukenmaier, C.C., Jr. (1980). *Validation of the reduction of rear-end collisions by a high mounted auxiliary stoplamp*. National Highway Traffic Safety Administration, Washington, DC.

Reisinger, K.S. and Williams, A.F. (1978). Evaluation of programs designed to increase protection of infants in cars. *Pediatrics* **62**, 280.

Reisinger, K.S., Williams, A.F., Wells, J.K., John, E., Roberts, T.L., and Podgainy, H.J. (1981). The effect of pediatricians counseling on infant restraint use. *Pediatrics* **67**, 201.

Robertson, L.S. (1975). Behavioral research and strategies in public health: A demur. *Social Science and Medicine* **9**, 165.

Robertson, L.S. (1976). An instance of effective legal regulation: Motorcycle helmet and daytime headlamp laws. *Law Society Revue* **10**, 456.

Robertson, L.S. (1980). Crash involvement of teenaged drivers when driver education is eliminated from high school. *American Journal of Public Health* **70**, 599.

Robertson, L.S. (1983). *Injuries: Causes, control strategies and public policy*. DC Health, Lexington, Massachusetts.

Robertson, L.S. (1984). Federal funds and state motor vehicle deaths. *Journal of Public Health Policy* **5**, 376.

Robertson, L.S. (1986). Community injury control programs of the Indian Health Service: An early assessment. *Public Health Reports* **101**, 632.

Robertson, L.S. and Zador, P.L. (1978). Driver education and fatal crash involvement of teenaged drivers. *American Journal of Public Health* **68**, 959.

Robertson, L.S., O'Neill,B., and Wixom, C.W. (1972). Factors associated with observed safety belt use. *Journal of Health and Human Behaviour* **13**, 18.

Robertson, L.S., Kelley, A.B., O'Neill, B., Wixom, C.W., Eiswirth, R.S., and Haddon, W., Jr. (1974). A controlled study of the effect of television messages on safety belt use. *American Journal of Public Health* **64**, 1071.

Ross, H.L. (1982). *Deterring the drinking driver*. DC Heath, Lexington, Massachusetts.

Sandels, S. (1970). Young children in traffic. *British Journal of Educational Psychology* **40**, 111.

Singer, S. (1969). *Pedestrian regulation enforcement and the incidence of pedestrian accidents*. US Department of Transportation, Washington, DC.

Shaoul, J. (1975). *The use of accidents and traffic offenses as criteria for evaluating courses in driver education*. University of Salford, Salford.

Shelness, A. and Jewett, J. (1984). Observed misuse of child restraints. *Physicians for Automotive Safety News* (winter).

Spiegel, C.N. and Lindaman, F.C. (1977). Children can't fly: A program to prevent childhood mortality and morbidity from window falls. *American Journal of Public Health* **67**, 1143.

Spyker, D.A. (1985). Submersion injury: Epidemiology, prevention and management. *Pediatric Clinics of North America* **32**, 113.

Stein, H.S. (1984). *Fleet experience with daytime running lights in the United States—preliminary results*. Insurance Institute for Highway Safety, Washington, DC.

Stein, H.S. and Jones, I.S. (1988). Crash involvement of large trucks by configuration: A case–control study. *American Journal of Public Health* **78**, 491.

Technical Study Group (1987). *Toward a less fire-prone cigarette*. Consumer Product Safety Commission, Washington, DC.

Teret, S., Jones, A.S., Williams, A.F., and Wells, J.K. (1986). Child restraint laws: Analysis of gaps in coverage. *American Journal of Public Health* **76**, 31

Trinkoff, A.M. and Baker, S.P. (1986). Poisoning hospitalizations and deaths from solids and liquids among children and teenagers. *American Journal of Public Health* **76**, 657.

US Fire Administration (1983). *Residential smoke and fire detector coverage in the United States: Findings from a 1982 survey*. Federal Emergency Management Agency, Washington, DC.

Williams, A.F. (1976). Observed child restraint use in automobiles. *American Journal of Diseases in Children* **130**, 1311.

Williams, A.F. and Wells, J.K. (1981). The Tennessee child restraint law in its first year. *American Journal of Public Health* **71**, 163.

Williams, A.F., Karpf, R.S., and Zador, P.L. (1983*a*). Variations in minimum licensing age and fatal motor vehicle crashes. *American Journal of Public Health* **73**, 1401.

Williams, A.F., Wong, J., and O'Neill, B. (1979). Occupant protection in interior impacts: An analysis of Federal Motor Vehicle Safety Standard No. 201. *Proceedings of the Twenty-third Conference of the American Association for Automotive Medicine*. American Association for Automotive Medicine, Morton Grove, Illinois.

Williams, A.F., Zador, P.L, Harris, S.S., and Karpf, R.S. (1983*b*). The effect of raising the legal minimum drinking age on fatal crash involvement. *Journal of Legal Studies* **12**, 169.

Wright, P.H. and Robertson, L.S. (1976). Studies of roadside hazards for projecting fatal crash sites. *Transportation Research Record* **609**, 1.

Wright, P.H., Zador, P.L., Park, C.Y., and Karpf, R.S. (1982).

Effect of pavement markers on night-time crashes in Georgia. Georgia Institute of Technology, Atlanta, Georgia.

Yeaton, W.H. and Bailey, J.S. (1978). Teaching pedestrian skills to young children: An analysis and one-year follow-up. *Journal of Applied Behavior Analysis* **11**, 315.

Zador, P.H., Moshman, J., and Marcus, L. (1982). Adoption of right turn on red: Effect on crashes at signalized intersections. *Accident Analysis and Prevention* **14**, 219.

Zador, P.H., Stein, H., Shapiro, S., and Tarnoff, P. (1984). *The effect of signal timing on traffic flow and crashes at signalized intersections.* Insurance Institute for Highway Safety, Washington, DC.

30

Laboratory and technical support of field investigations of chemical, physical, and biological hazards

JAMES L. PEARSON

Introduction

In April of 1987, two children aged 4 and 7 years were admitted to a rural North Dakota hospital because of the acute onset of marked somnolence, vomiting, and ataxia. After developing haematuria, the children were transferred to the paediatric intensive care unit of a Fargo hospital. They were given intravenous fluids and recovered fully within two days. Urinalysis for each child revealed calcium oxalate crystals. Consultation with the state toxicology laboratory resulted in urine being found positive for ethylene glycol.

On the day they became ill, the two children had been at a picnic attended by approximately 400 persons at a rural fire-hall. Three hundred and fifty-four (91 per cent) of the 391 attenders identified were interviewed by telephone about symptoms and the foods and beverages they consumed at the picnic. Those persons who reported marked fatigue (felt or acted more tired than usual and slept two or more hours before 10 p.m. that evening) or ataxia on the evening of the picnic and who had not drunk beer were considered as having met the case definition of acute illness.

Twenty-nine (8 per cent) of the 354 persons interviewed met the case definition. One item, a non-carbonated soft drink, was strongly associated with illness (relative risk 31.0). Among those who consumed the soft drink, 18 per cent (28/159) became ill, while among non-consumers, 0.6 per cent (1/176) became ill.

Chemists at the North Dakota State Public Health Laboratory were consulted about sampling and transport of specimens by the epidemiologists who were conducting the investigation. Subsequently, samples of the water used to prepare the powdered beverage analysed at the public health laboratory (PHL) showed an ethylene glycol concentration of 9 per cent.

The source of the water was the spigot nearest the fire-hall's heating system. The heating system used a combination of water and antifreeze. An investigation by an environmental health practitioner (EHP) showed a cross-connection from the potable water system to the heating system that was regulated by a single valve. It was uncertain whether the valve had been closed during the preparation of the beverage. Recommendations were made to permanently separate the heating system from the potable water supply (Centers for Disease Control 1987).

Successful completion of this investigation required co-operative efforts between the epidemiologists and several disciplines, including toxicology, chemistry, and environmental health.

Field investigations begin when surveillance indicates that endemic levels of a health condition have been exceeded, and preliminary investigation shows that a health risk may exist (the source of the surveillance may have initially been laboratory data). Laboratory records should be an integral part of any surveillance system. In selecting a laboratory to provide support for field investigations, several factors should be considered, including certification or licensure, availability of service, cost, technical expertise, quality of analysis, turnaround time, record of past performance, and quality control/assurance. In some situations, confirmation of results by another laboratory may be indicated, particularly if litigation is possible.

It is critical that the field investigator consult with the laboratory before starting an investigation. The laboratory can provide information on the number and type of samples to collect, background laboratory data for the disease in question, how to transport (including the type of transport container and preservatives), the time required for analysis, the form of report, and interpretation of results. The laboratory should also be appraised of investigational findings as the study proceeds. Release of preliminary findings needs to be co-ordinated with the designated spokesperson for the investigation. The laboratory does not lead the search but should be a partner in the process.

Selecting a laboratory

Laboratory support for investigations should be cultivated before there is need for an investigation. The epidemiologist should contact the laboratories in his surveillance area to assure that potential changes in endemic patterns will be regularly reported. In soliciting data for surveillance, the investigator can also assess the laboratories and plan for the future. When an outbreak occurs, the laboratory should already have been selected. Since the results of any investigation could be part of litigation, the laboratory used for studies relating to the investigation needs to be credible. One way of establishing the credentials of a laboratory is to review their certification or licensure status.

Environmental laboratories conducting work on drinking water are certified by the Environmental Protection Agency (EPA) in the United States. Some States are providing certification to other environmental areas. To achieve this status, a laboratory must utilize uniform methodologies, have an acceptable on-site inspection of facilities, have careful record keeping, quality assurance procedures, and perform acceptably on annual check samples in each category for which the laboratory is certified. In the United States, certification is conducted by States that have assumed the responsibility (primacy), or by the Federal Government where primacy has not been assumed.

Uniform methodologies are used by certified environmental laboratories to help assure that results reported by one laboratory are comparable to results another laboratory would report if testing the same sample. If the method of analysis is different, the results may vary considerably. The appropriate method for an analyte is dictated by the EPA under the certification requirements of programs such as drinking water, water pollution, or hazardous waste.

On-site inspection of a facility is an integral part of the EPA certification process. An on-site visit by a trained laboratorian can provide a valuable insight into the general operating quality of a laboratory. The size and layout of the facility are addressed, analysts are quizzed on methodologies to assess their depth of knowledge, laboratory records of tests and quality control are reviewed, and quality assurance and procedure manuals are examined.

Quality assurance is a critical part of good laboratory practice. Sample handling policies must be followed, including sample collection, field preservation if necessary, conditions of transport, and holding times.

In EPA certified environmental chemistry laboratories, at least 10 per cent of samples must be duplicates. These duplicate samples help to assure that numbering errors or other mix-ups have not occurred. They are also useful in monitoring the precision of the analysis, although duplicates are unreliable at or near the detection limits of the procedure.

Another 10 per cent of samples must be spiked. These spikes are known quantities of the analyte in question added to selected samples to monitor the accuracy of the analysis. This verifies that the sample matrix has not altered the slope of the standard curve. If spike recovery is between 85 and 115 per cent the standard curve is valid for the sample.

To achieve and maintain certification, laboratories must also perform acceptably on proficiency evaluation (PE) samples submitted by the certifying agency. These proficiencies are analysed by the appropriate methodologies and the results are compared to the results achieved by other laboratories. If a laboratory fails to identify and quantify correctly the PE samples, they may lose or have their certification status downgraded.

The Contract Laboratory Program through the EPA provides another type of certification. Contract laboratories have achieved certification, and also have demonstrated the capability to provide documentation of results that will satisfy the court without expert testimony.

When choosing a clinical laboratory to support an outbreak investigation, look for certification or licensure from agencies such as Joint Commission on Accreditation of Hospitals (JCAH), state, federal, or national governments or other approved accrediting agencies.

To provide support for field investigations, a laboratory needs to provide a full range of standard laboratory procedures. Experimental or non-standardized procedures have a place in some investigations; however, findings using experimental procedures need to be confirmed by more standard procedures whenever possible. Not all laboratories provide 24-hour or stat(express)services. Some analyses require extensive studies or long incubation periods. Consultation with the laboratory at the time sampling is first considered can help prevent misunderstandings later. Preliminary reports can often be made available during laboratory studies, and these findings may be useful in conducting the investigation. Arrangements should be made in advance with the laboratory director if preliminary results are to be released, and to whom. In all cases, the final report is definitive.

The PHL can be an excellent source of information about the types of services offered by laboratories in the jurisdiction, and often will provide technical and laboratory support for field investigations. In the United States, the PHL also has access to the services provided by the Centers of Disease Control (CDC), including identification of pathogenic microorganisms and technical advice on transport and identification. Requests for these services must come through the State Laboratory Directors as by an agreement between CDC and the Association of State and Territorial Public Health Laboratory Directors. Previous chapters in this section have mentioned the technical support available through State Epidemiologists and the Epidemiology Program Office at CDC for conduct of investigations.

Another factor to consider when selecting a laboratory is the availability of technical expertise, not just in performance of the laboratory analyses, but also in interpretation of the results. Laboratory testing is becoming increasingly complex, and the methodologies are changing rapidly. Laboratory test results may be presented in different formats by different laboratories, and in some circumstances the reports may lead to confusion.

The quality of analysis provided by a laboratory may be difficult to assess by the epidemiologist. State licensure agencies in those states who have laboratory regulatory authority can be a good source of information about laboratories within their jurisdiction. Records of performance on proficiency testing samples, or even enrolment in recognized proficiency testing services (i.e. CAP, CDC, etc.) shows commitment on the part of the laboratory administration to providing good service. Quality assurance programs should be in place in all laboratories, and should include spiked samples, duplicates, standards, or controls with each run. Certified environmental laboratories must meet quality assurance guide-lines that include approximately 20 per cent of samples as quality control. Clinical laboratories do not have to meet the same criteria; however, guide-lines are in place for proficiency testing, and in some jurisdictions, personnel standards. On-site inspections conducted by outside agencies provide information about staffing, record keeping, safety, and proficiency test results.

Co-ordination with the laboratory

When conducting investigations of possible environmental hazards, the state environmental protection agency (this may be the Department of Health, Natural Resources, Environment or some other agency) can provide a list of certified laboratories in the area. Not all laboratories have the full capabilities to provide analyses for all potential environmental hazards. Also, interpretation of contaminant levels found in samples may be difficult, since maximum contaminant levels (MCLs) have not been established for the majority of chemicals. The investigator should be prepared to discuss with the laboratory director or analyst which analytes to look for, the detection limits that can be achieved in the matrix for the chemicals in question, measures of precision, accuracy, maximum holding times, and the time required for studies. The investigator should avoid the 'shotgun' approach (whatever you can find, as low as you can go). That approach dramatically increases the cost of the analysis, and may produce spurious data that can lead to erroneous conclusions. The epidemiologic investigation should point to causation, and the laboratory data provide confirmatory support.

The laboratory should be given as much lead time as possible when sampling is contemplated, particularly when chemical contaminants are suspected. In many instances, standards are not immediately available to the laboratory, and must be obtained before samples can be processed. Since analytical methods have not been published for all chemicals, methods may need to be developed.

In some situations, particularly if product liability is possible, or if death, serious illness or injury are being investigated, confirmatory studies by another laboratory should be considered.

Reference

Centers for Disease Control (1987). Ethylene glycol intoxication due to contamination of water systems. *Morbidity and Mortality Weekly Report* **36**, 611.

Further reading

Environmental Monitoring and Support Laboratory (1976). *Quality assurance handbook for air pollution measurement systems*. US Environmental Protection Agency, Research Triangle Park, North Carolina.

Environmental Monitoring and Support Laboratory (1979). *Handbook for analytical quality control in water and wastewater laboratories*. US Environmental Protection Agency, Cincinnati, Ohio.

Office of Solid Waste and Emergency Response (1986). *Test methods for evaluating solid waste* (3rd edn). US Environmental Protection Agency, Washington, DC.

Schmidt, N.J and Emmons, R.W. (ed.) (1989). *Diagnostic procedures for viral rickettsial, and chlamydial infections* (6th edn). American Public Health Association, Washington, DC.

Wentworth, B.B. (ed.) (1987). *Diagnostic procedures for bacterial infections* (7th edn). American Public Health Association, Washington, DC.

Wentworth, B.B. (ed.) (1988). *Diagnostic procedures for mycotic and parasitic infections* (7th edn). American Public Health Association, Washington, DC.

Index

NOTE: Since the major subject of this book is public health, entries under this keyword have been kept to a minimum, and readers are advised to seek more specific index entries. Likewise, with United Kingdom and United States of America references. This is an index to all three volumes of the *Oxford Textbook of Public Health*. The first digit in any reference gives the **volume number** (1–3); the digits after the decimal point give the page number.

Abbreviations used in subentries:

CHD	Coronary heart disease	MHC	Major histocompatibility complex
EPA	Environmental Protection Agency	NHS	National Health Service
GDP	Gross domestic product	UK	United Kingdom
GNP	Gross national product	US	United States of America

Heterick Memorial Library
Ohio Northern University

DUE	RETURNED	DUE	RETURNED
1.		13.	
2.		14.	
3.		15.	
4.		16.	
5.		17.	
6.		18.	
7.		19.	
8.		20.	
9.		21.	
10.		22.	
11.		23.	
12.		24.	

HETERICK MEMORIAL LIBRARY

3 5111 00397 8081